T0205323

Proceedings in Adaptation, Learning and Optimization

Volume 1

Series editors

Yew Soon Ong, Nanyang Technological University, Singapore
e-mail: asysong@ntu.edu.sg

Meng-Hiot Lim, Nanyang Technological University, Singapore
e-mail: emhlim@ntu.edu.sg

About this Series

The role of adaptation, learning and optimization are becoming increasingly essential and intertwined. The capability of a system to adapt either through modification of its physiological structure or via some revalidation process of internal mechanisms that directly dictate the response or behavior is crucial in many real world applications. Optimization lies at the heart of most machine learning approaches while learning and optimization are two primary means to effect adaptation in various forms. They usually involve computational processes incorporated within the system that trigger parametric updating and knowledge or model enhancement, giving rise to progressive improvement. This book series serves as a channel to consolidate work related to topics linked to adaptation, learning and optimization in systems and structures. Topics covered under this series include:

- complex adaptive systems including evolutionary computation, memetic computing, swarm intelligence, neural networks, fuzzy systems, tabu search, simulated annealing, etc.
- machine learning, data mining & mathematical programming
- hybridization of techniques that span across artificial intelligence and computational intelligence for synergistic alliance of strategies for problem-solving
- aspects of adaptation in robotics
- agent-based computing
- autonomic/pervasive computing
- dynamic optimization/learning in noisy and uncertain environment
- systemic alliance of stochastic and conventional search techniques
- all aspects of adaptations in man-machine systems.

This book series bridges the dichotomy of modern and conventional mathematical and heuristic/meta-heuristics approaches to bring about effective adaptation, learning and optimization. It propels the maxim that the old and the new can come together and be combined synergistically to scale new heights in problem-solving. To reach such a level, numerous research issues will emerge and researchers will find the book series a convenient medium to track the progresses made.

More information about this series at http://www.springer.com/series/13543

Hisashi Handa · Hisao Ishibuchi
Yew-Soon Ong · Kay Chen Tan
Editors

Proceedings of the 18th Asia Pacific Symposium on Intelligent and Evolutionary Systems – Volume 1

Editors
Hisashi Handa
Department of Infomatics
Faculty of Science and Technology
Kindai University
Higashi-Osaka
Japan

Hisao Ishibuchi
Department of Computer Science
and Intelligent Systems
Osaka Prefecture University
Osaka
Japan

Yew-Soon Ong
School of Computer Engineering
Nanyang Technological University
Singapore

Kay Chen Tan
Department of Electrical and Computer
Engineering
National University of Singapore
Singapore

ISSN 2363-6084 ISSN 2363-6092 (electronic)
ISBN 978-3-319-38611-9 ISBN 978-3-319-13359-1 (eBook)
DOI 10.1007/978-3-319-13359-1

Springer Cham Heidelberg New York Dordrecht London

Softcover reprint of the hardcover 1st edition 2015

Printed on acid-free paper

Springer is part of Springer Science+Business Media (www.springer.com)

Preface

This book contains a collection of the papers accepted for presentation in the 18th Asia Pacific Symposium on Intelligent and Evolutionary Systems (IES 2014), which was held in Singapore from 10th to 12th November 2014. IES 2014 was sponsored by the Memetic Computing Society and co-sponsored by the SIMTECH-NTU Joint Lab, and the Center for Computational Intelligence at the School of Computer Engineering, Nanyang Technological University, and supported by the National University of Singapore and Nanyang Technological University. The book covers the topics in intelligent systems and evolutionary computation, and many papers have demonstrated notable systems with good analytical and/or empirical results.

The Editors
Hisashi Handa
Hisao Ishibuchi
Yew-Soon Ong
Kay Chen Tan

Contents

Impact of Stubborn Individuals on a Spread of Infectious Disease under Voluntary Vaccination Policy

Eriko Fukuda* and Jun Tanimoto

Interdisciplinary Graduate School of Engineering Sciences, Kyushu University,
Kasuga-koen, Kasuga-shi, Fukuoka 816-8580, Japan
eriko.fukuda@kyudai.jp,
tanimoto@cm.kyushu-u.ac.jp

Abstract. Achievement of the herd immunity is essential for preventing epidemics of vaccine-preventable diseases. However, an individual's decision-making whether or not to be vaccinated depends on several factors, such as perceived risks of vaccination and infection, her self-interest, and response of others to vaccination under voluntary vaccination policies. In this study, we consider the case where "stubborn individuals" are presented in lattice populations, who consistently hold the vaccination strategy (stubborn vaccinated individuals) or the no-vaccination strategy (stubborn unvaccinated individuals). We investigate individuals' decision-making process with vaccination by means of modeling the dynamics for epidemic spreading applied to evolutionary game theory. As a result, we find that the presence of stubborn ones, even if it accounts for a small fraction, significantly affect the epidemic spreading and vaccination behavior.

Keywords: Vaccination, Evolutionary game theory, Infectious diseases, Mathematical epidemiology.

1 Introduction

To immunize susceptible individuals pre-emptively against the vaccine-preventable diseases is one of the primary public health measures for the control and preventing epidemics of the infectious diseases such as the flu [1]. However, under voluntary vaccination policy, it produces a conflict between the vaccination behavior of each individual according to self-interest and achieving a social-optimum level of vaccination coverage, thus the level for herd immunity [2,3,4,5,6]. This is a well-known vaccination dilemma in epidemiology: With increasing vaccination coverage over the population, the remaining unvaccinated individuals are quite less likely to be infected because they can benefit from the herd immunity as a public goods without taking the perceived risks by vaccination and indirectly protected by the vaccinated individuals [1], [7]. Thus, they have less incentive to be vaccinated. In consequence, this

* Corresponding author.

H. Handa et al. (eds.), *Proc. of the 18th Asia Pacific Symp. on Intell. & Evol. Systems – Vol. 1*,
Proceedings in Adaptation, Learning and Optimization 1, DOI: 10.1007/978-3-319-13359-1_1

dilemma makes it difficult to eradicate the infectious disease because the critical level of vaccination coverage sufficient to prevent epidemics cannot be sustained.

From this viewpoint, in the decade, many works regarding the vaccination dilemma have applied to evolutionary game theoretic framework for the population in which each individual tries to maximize her own payoff, and many fruitful results have been reported [8,9,10,11,12,13,14,15,16,17,18]. Fu et al. modeled an imitation dynamics of vaccination behavior in various structured populations to investigate the effect of individual's adaptation behavior and population structure [12]. In Ref. [12], they found that individual's vaccination behavior sensitively depends on both the population structure and the cost of vaccination. In many evolutionary game theoretic studies regarding to the vaccination dilemma, it assumes that individuals have the same perceived risk of infection in the face of an epidemic. In reality, some individuals overestimate the risks of infection, and those oversensitive ones can consistently take vaccination. Xiao et al. investigated that how the stubborn (or committed) vaccinated individuals who always hold the vaccination strategy affects the vaccination dynamics in structured populations [16]. In Ref. [16], they found that a small fraction of the stubborn vaccinated individuals can promote the vaccination behavior in the population. However, an attitude which stubborn individuals stick to is not necessarily "taking vaccination" in the face of an infectious disease. In fact, it is reported that the fraction of the vaccinated individuals may be reduced by underestimating the risks of infection due to the lack of knowledge about the disease and/or by overestimating vaccine risks based on scientifically groundless information [19]. That is, separate and aside from the stubborn vaccinated individuals, the stubborn unvaccinated individuals who always take no-vaccination strategy can exist in a population.

Thus, in this study, we investigate the impact of the presence of the stubborn (vaccinated / unvaccinated) individuals on the vaccination behavior and epidemic spreading in the population by using the model which implements an epidemiological process into a decision-making process for taking vaccination. In our model the dynamics is modeled as a two-stage process [12], [16], [18]. The first stage corresponds to a vaccination campaign. Each individual in the population makes a decision whether or not to be vaccinated before any epidemic spreading. The second stage corresponds to an epidemic season. For describing epidemiological dynamics on a structured population, susceptible-infected-recovered (SIR) dynamics on a network is adopted. In the SIR model, a population is separated into three groups: susceptible, infected, and recovered. Additionally, recovered individuals as well as vaccinated ones have acquired immunity to the infectious disease. Now, temporal development of all those sub-populations is governed by a certain mathematical structure. Those who decide not to be vaccinated are included in the susceptible group. At the end of the epidemic season, each susceptible individual is determined to be either infected or not. According to the final epidemic state, a stipulated payoff is assigned to each individual. Subsequently, each individual reexamines her strategy on vaccination via an imitation process (except for stubborn individuals) [16]. The details of our model are described in the following section.

2 Models and Methods

2.1 Base Model

We consider a population including the stubborn individuals who consist of the stubborn vaccinated individuals (SVs) and the stubborn unvaccinated individuals (SUs). All the individuals are placed on a square lattice with von Neumann neighborhood as a simple structured population. Seasonal and periodical infectious diseases, such as flu, are assumed to spread through such a population. The protective efficacy of a flu vaccine persists for less than a year because of waning of antibodies and year-to-year changes in the circulating virus. Therefore, under a voluntary vaccination program, individuals must decide whether to be vaccinated every year. Thus, the dynamics of our model consists of two stages: the first stage is a vaccination campaign, and the second is an epidemic season [12], [16], [18].

First Stage: The Vaccination Campaign
Here, in this stage, each individual makes a decision whether to take vaccination before the beginning of the seasonal epidemic, i.e., before any individuals are exposed to the epidemic strain. Vaccination imposes a cost C_v on each individual (including SV) who decides to be vaccinated. The cost of vaccination includes the monetary cost and other perceived risks, such as adverse side effects. For simplicity, we assume that the vaccination provides perfect immunity to an individual against the disease during a season; however, an unvaccinated individual (including SU) faces the risk of being exposed to infection during a season.

Second Stage: The Epidemic Season
Here, at the beginning of this stage, the epidemic strain enters the population, and randomly selected susceptible individuals I_0 are identified as initially infected ones. Then, the epidemic spreads according to SIR dynamics.

SIR Dynamics in Structured Populations
The classic Kermack-McKendrick SIR model is given by coupled (integro-) differential equations and does not assume any spatial structure for the population. Using Kermack-McKendrick SIR model, a short-range and local epidemic outbreak of infectious diseases such as plague are modeled [20]. Here, we use an extended SIR model that involves a spatial structure for the whole population. This structure is represented by a network consisting of nodes and links [9]. The dynamics of SIR on a spatially structured population is not captured by a system of differential equations; thus, we numerically simulate an epidemic spreading on the square lattice by using the Gillespie algorithm [21] in the extended SIR model.

In the model, the whole population N is divided into three groups: susceptible (S), infected (I), and recovered (R) individuals. The disease parameters are β, which is the transmission rate per day per person, and γ, which is the recovery rate per day (i.e., the inverse of the mean number of days required to recover from the infection). In this study, we calibrate the value of β such that the final proportion of infected individuals across the square lattice will be 0.9 [12], [16], [18]. Accordingly, we set $\beta = 0.46$ day^{-1} person^{-1} and the recovery rate $\gamma = 1/3$ day^{-1}. A typical flu is assumed to determine these disease parameters.

An epidemic season lasts until no infection exists in the population. Each individual (including infected-SU) who gets infected during the epidemic season incurs the cost of infection, C_i. However, the cost paid by a "free-rider" (Including healthy-SU) is zero, who does not vaccinate and still is free from infection. For simplicity, we set $C_i = 1$, and renormalize these costs (payoffs) by defining the relative cost of vaccination $C_r = C_v/C_i$ $(0 \leq C_r \leq 1)$. Then, the payoff for every individual after the end of an epidemic season is summarized according to her state in Table 1.

Table 1. Payoff for the three individual's strategies and state in the population after the epidemic season

Strategy \ State	Healthy	Infected
Vaccination	$-C_r$	
No-vaccination	0	-1

Strategy Adaptation

After the end of above two stages, every individual reexamines her vaccination decision-making strategy at the beginning of the next season. The rule for strategy adaptation is as follows. A certain individual i randomly selects a neighboring individual j. Let π_i and π_j denote the payoffs of individuals i and j, respectively. The probability $P(s_i \leftarrow s_j)$ that the individual i (whose strategy is s_i) imitates the strategy s_j of individual j is determined by a pairwise comparison of their payoff difference according to the Fermi function [22,23],

$$P(s_i \leftarrow s_j) = \frac{1}{1 + \exp\left[\dfrac{\pi_i - \pi_j}{\kappa}\right]}, \tag{1}$$

where the term "strategy" implies an individual's decision to be vaccinated and κ is the sensitivity of individuals to the difference in the payoffs. For $\kappa \to \infty$ (weak selection pressure), an individual i is insensitive to the payoff difference $\pi_i - \pi_j$ against another individual j and the probability $P(s_i \leftarrow s_j)$ approaches 1/2 asymptotically, regardless of the payoff difference. For $\kappa \to 0$ (strong selection pressure), individuals are sensitive to the payoff difference, and they definitely copy the successful strategy that earns the higher payoff, even if the difference in the payoff is very small.

This updating rule has been widely accepted in evolutionary game theory [24,25]. In the present study, we set $\kappa = 1$, which has been used as a typical selection pressure in many previous studies. This value of κ implies that, in most situations, individuals adopt any successful strategy; however, occasionally they end up imitating a worse performer with a lower payoff. Such erratic decision making is a reflection of irrationality or mistakes made by ordinary individuals. Note that, in Eq. (1), if an individual i is a stubborn individual, then $P(s_i \leftarrow s_j) = 0$, regardless of the j's strategy or their payoff difference since she always holds her own strategy. Fig. 1 shows the flow of the model described thus far.

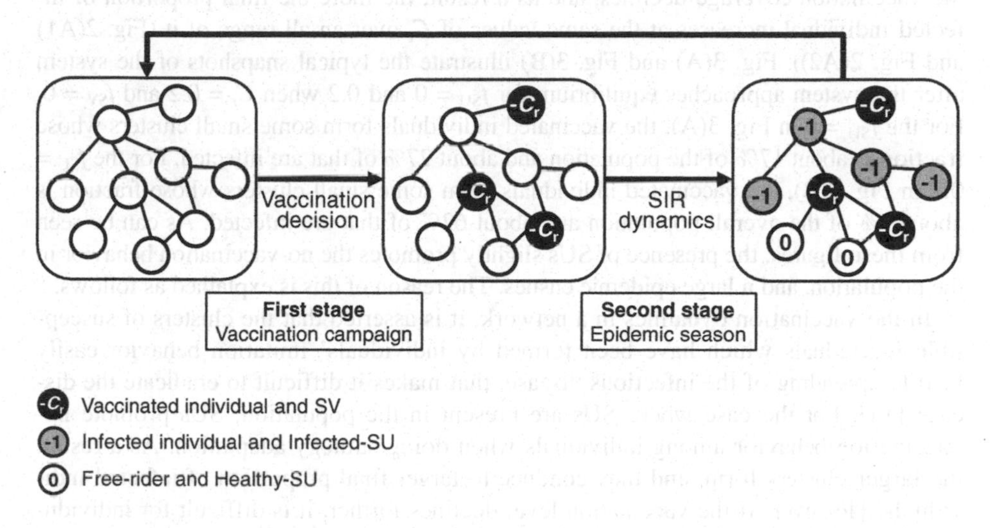

Fig. 1. Flow of our model in which the dynamics is modeled as a two-stage process

2.2 Simulation Setting

Initially, equal fractions of the vaccinated (including a certain fraction f_{SV} of SVs) and unvaccinated individuals (including a certain fraction f_{SU} of SUs) are randomly distributed over the population allocated on the square lattice, and the population size $N = 100 \times 100$. Note that, the stubborn individuals, SVs and SUs, occupy the same nodes with the same state throughout a simulation, i.e., from the 1st season to the final season, the population. After that, the epidemic strain infects the initial number $I_0 = 10$ of randomly selected susceptible individuals, and then, the epidemic spreads according to SIR dynamics. The vaccination coverage and the fraction of infected individuals are updated by iterating each two-stage process (the vaccination campaign and the epidemic season). The equilibrium (steady-state) results shown in Fig. 2 represent average fractions over the last 1000 from among 3000 iterations in 100 independent simulations.

3 Results and Discussion

3.1 $f_{SV} = 0$

Fig. 2(A1) and Fig. 2(A2) show equilibrium values for the vaccination coverage and the final proportion of infected individuals as functions of the relative cost of vaccination C_r, for different fractions of SUs f_{SU}. In this case the SVs are absent (the fraction of SVs $f_{SV} = 0$). The case where $f_{SU} = 0$, i.e., the absence of stubborn individuals in Fig. 2(A1) and Fig. 2(A2) corresponds to the results in Ref. [12]. The more f_{SU} increases, the more the vaccination coverage declines, and as a result, the more the final proportion of infected individual increases at the same values of C_r over an all range of it (Fig. 2(A1) and Fig. 2(A2)). Fig. 3(A) and Fig. 3(B) illustrate the typical snapshots of the system after the system approaches equilibrium for $f_{SU} = 0$ and 0.2 when $C_r = 0.2$ and $f_{SV} = 0$. For the $f_{SU} = 0$ in Fig. 3(A), the vaccinated individuals form some small clusters whose fraction is about 17% of the population and about 27% of that are infected. For the $f_{SU} = 0.2$ in Fig. 3(B), the vaccinated individuals form some small clusters whose fraction is about 9% of the overall population and about 62% of that are infected. As can be seen from these figures, the presence of SUs slightly promotes the no-vaccination behavior in the population, and a large epidemic ensues. The reason of this is explained as follows.

In the vaccination dynamics in a network, it is asserted that the clusters of susceptible individuals which have been formed by individuals' imitation behavior easily lead to spreading of the infectious disease, that makes it difficult to eradicate the disease [14]. For the case where SUs are present in the population, SUs promote no-vaccination behavior among individuals when doing strategy adaptation. As a result, the larger clusters form, and they conduce to larger final proportion of infected individuals. However, if the vaccination level declines further, it is difficult for individuals to take a free ride on the benefit, that is, the unvaccinated individuals are indirectly protected by the vaccinated ones without any vaccination. Therefore, the SUs cannot encourage individuals not to make voluntary vaccination greatly; hence, the large reduction of vaccination level doesn't occur.

3.2 $f_{SV}= 0.1$

Fig. 2(B1) and Fig. 2(B2) show equilibrium values for vaccination coverage and final proportion of infected individuals as functions of C_r, for different values of f_{SU} while keeping $f_{SV} = 0.1$. The results for the case where $f_{SU} = 0$, i.e., the absence of SUs in Fig. 2(B1) and Fig. 2(B2), correspond to the results in Ref. [16]. As is the same as the case where $f_{SV} = 0$, the more f_{SU} increases, the more the vaccination coverage declines over an all range of C_r (Fig. 2(B1) and Fig. 2(B2)). However, in comparison with the obtained results for the case where there is no SUs nor SVs in the population, only a small fraction of SVs greatly reduces the final proportion on infected individuals at a wider range of C_r (approximately $C_r > 0.04$), even if the SUs are present in the population. Fig. 3(C) and Fig. 3(D) illustrate the typical snapshots of the system after the system approaches equilibrium for $f_{SU} = 0$ and 0.2 when $C_r = 0.2$ and $f_{SV} = 0.1$. For the case where $f_{SU} = 0$ in Fig. 3(C), the vaccinated individuals whose fraction is about

63% of the total population form large clusters which are distributed evenly throughout the lattice. And about 1% of the total population are infected. For the case where f_{SU} = 0.2 in Fig. 3(D), the vaccinated individuals whose fraction is about 27% of the total population form small clusters which are distributed evenly throughout the lattice. And about 5% of the total population are infected. As can be seen from these figures, specifically Fig. 3 (D), the SVs who are distributed randomly throughout the lattice population promotes vaccination behavior of the total population, and helps to form the clusters of vaccinated individuals evenly in the population, although those occupy only a small fraction. And its impact inhibits SUs from seducing the whole population to avoid vaccination. As a result, epidemic is greatly prevented. The detailed reason of this is explained as follows.

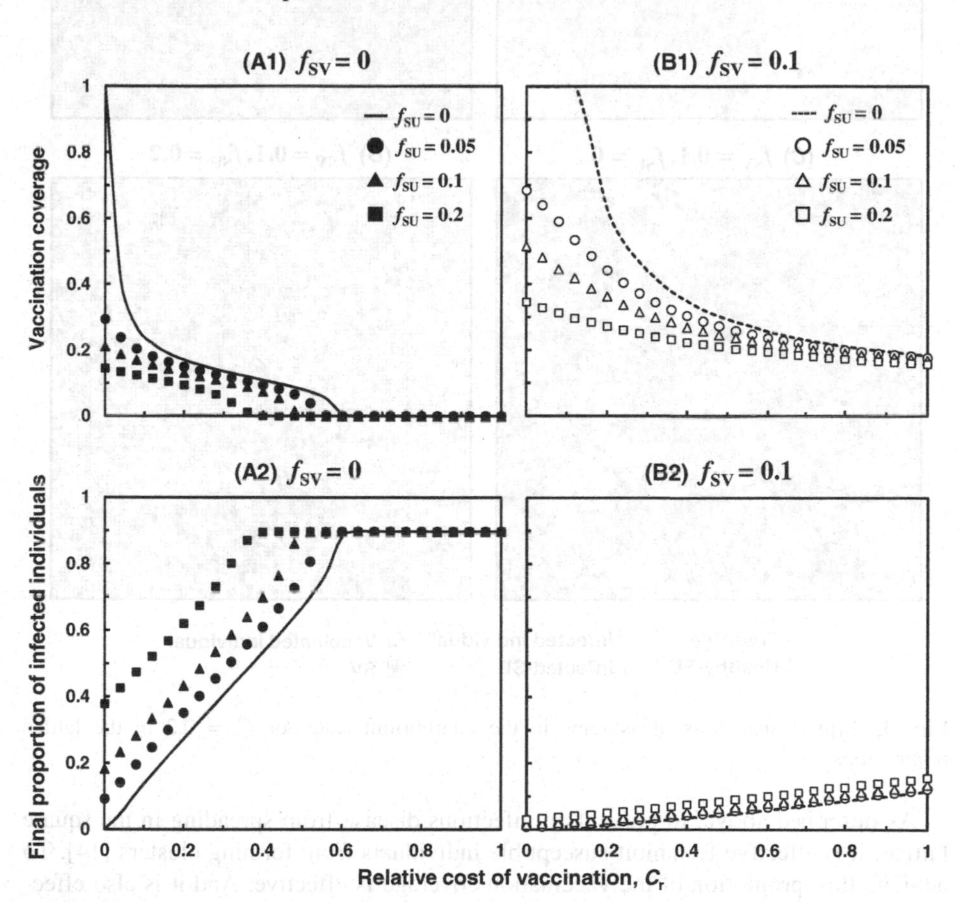

Fig. 2. Vaccination coverage (upper two panels) and final proportion of infected individuals (lower two panels) as functions of relative cost of vaccination C_r for different fraction of SVs f_{SV} and fraction of SUs f_{SU} in the lattice population

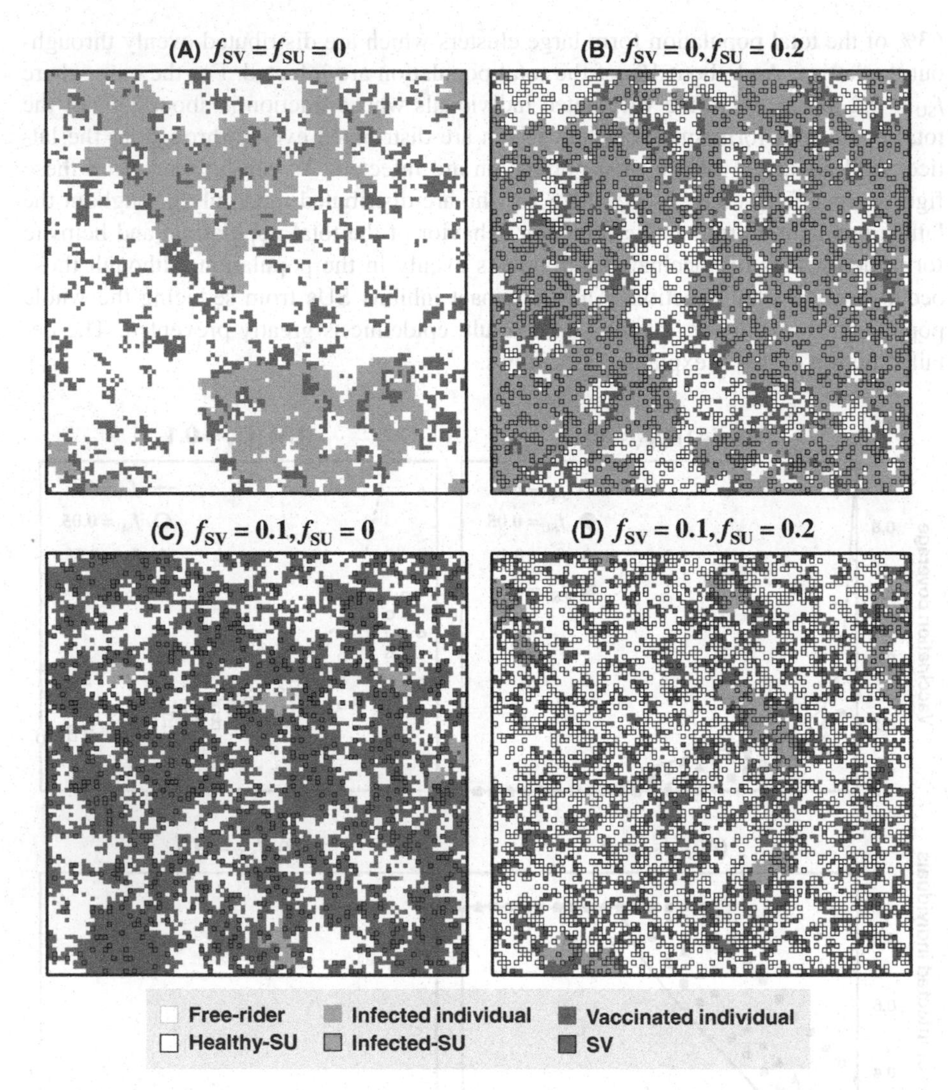

Fig. 3. Typical snapshots of systems in the equilibrium state for $C_r = 0.2$ in the lattice populations

As described above, to prevent an infectious disease from spreading in the square lattice, it is effective to inhibit susceptible individuals from forming clusters [14]. To achieve this, promotion of the vaccination coverage is effective. And it is also effective to distribute the clusters of the vaccinated individuals evenly in the network, because any of these cluster shuts out the initial infected individuals who appear everywhere in the network. For the case where $f_{SV} = f_{SU} = 0$ in Fig. 3(A), the infectious disease spreads widely in the population because the clusters of vaccinated individuals are small (thus, the vaccination coverage is low), and they distribute not evenly in the network. In contrast, for the case where $f_{SV} = 0.1, f_{SU} = 0.2$ in Fig. 3(D), not only it is promoted the imitation behavior with vaccination but also it helps to form and distributed evenly the clusters of vaccinated individuals, even if the sizes of clusters are small, avoid the clustering of susceptible individuals thanks to the presence of SVs who are randomly distributed in the population. Hence, the small fraction of SVs has a large impact on the suppression of epidemics thanks to the imitation process and the local interactions of the structured populations (For detailed discussions, see Ref [16].). On the other hand, as mentioned before, the SUs cannot seduce individuals not to make voluntary vaccination greatly. Therefore, the obtained results where the both SVs and SUs are present are the combined results of the effects of the presence of both SVs and SUs.

4 Conclusions

In this study, we investigated how the presence of stubborn individuals who always hold her own strategy affects an individual's decision to get vaccinated against a spread of infectious disease and the aggregate vaccination behavior of the population. Consequently, we found that the presence of SVs affects more than the presence of SUs in a population, i.e., the small fraction of SVs promotes individuals' vaccination behavior, and epidemic is greatly inhibited.

For simplicity, we assumed that individuals are in lattice populations in the face of the flu-like disease. However, both the network structure and the epidemic parameters such as transmission rate β have a profound effect on the appearance of epidemic spreading [26]. For example, in a heterogeneous network such as scale-free networks, an infectious disease spreads more easily than a lattice population because of the presence of hub individuals who sometimes work as super-spreaders. In future work, we must investigate the impact of the presence of stubborn individuals under various network structures and epidemic-parameter conditions.

Acknowledgements. This study was partially supported by a Grant-in-Aid for Scientific Research by JSPS, Japan, awarded to Professor Tanimoto (Grant No. 25560165), Tateishi Science & Technology Foundation. We would like to express our gratitude to these funding sources.

References

1. Anderson, R.M., May, R.M.: Infectious Diseases of Humans: Dynamics and Control. Oxford University Press, New York (1991)
2. Cullen, J., West, P.: The Economics of Health: An introduction. Martin Robertson, Oxford (1979)
3. Fine, P.E., Clarkson, J.A.: Individual versus public priorities in the determination of optimal vaccination policies. Am. J. Epidemiol. 124, 1012–1020 (1986)
4. Geoffard, P., Philipson, T.: Disease eradication: private versus public vaccination. Am. Econ. Rev. 87, 222–230 (1997)

5. Bauch, C.T., Galvani, A.P., Earn, D.J.D.: Group interest versus self interest in smallpox vaccination policy. Proc. Natl. Acad. Sci. U.S.A. 100, 10564–10567 (2003)
6. Bauch, C.T., Earn, D.J.D.: Vaccination and the theory of games. Proc. Natl. Acad. Sci. U.S.A. 101, 13391–13394 (2004)
7. Brisson, M., Edmunds, W.: Economic evaluation of vaccination programs: the impact of herd-immunity. Med. Decis. Making. 23, 76–82 (2003)
8. Bauch, C.T.: Imitation dynamics predict vaccinating behavior. Proc. R. Soc. B. 272, 1669–1675 (2005)
9. Vardavas, R., Breban, R., Blower, S.: Can influenza epidemics be prevented by voluntary vaccination? PLoS Comput. Biol. 3(5), e85 (2007)
10. Breban, R., Vardavas, R., Blower, S.: Mean-field analysis of an inductive reasoning game: Application to influenza vaccination. Phys. Rev. E. 76, 031127 (2007)
11. Breban, R.: Health Newscasts for Increasing Influenza Vaccination Coverage: An Inductive Reasoning Game Approach. PLoS One 6(12), e28300 (2011)
12. Fu, F., Rosenbloom, D.I., Wang, L., Nowak, M.A.: Imitation dynamics of vaccination behaviour on social networks. Proc. R. Soc. B. 278, 42–49 (2011)
13. Bauch, C.T., Bhattacharyya, S.: Evolutionary Game Theory and Social Learning Can Determine How Vaccine Scares Unfold. PLoS Comput. Biol. 8(4), e1002452 (2012)
14. Ndeffo Mbah, M.L., Liu, J., Bauch, C.T., Tekel, Y.I., Medlock, J., Meyers, L.A., Galvani, A.P.: The Impact of Imitation on Vaccination Behavior in Social Contact Networks. PLoS Comput. Biol. 8(4), e1002469 (2012)
15. Zhang, H., Fu, F., Zhang, W., Wang, B.: Rational behavior is a 'double-edged sword' when considering voluntary vaccination. Physica A 391, 4807–4815 (2012)
16. Liu, X.T., Wu, Z.X., Zhang, L.: Impact of committed individuals on vaccination behavior. Phys. Rev. E. 86, 051132 (2012)
17. Zhang, H.F., Wu, Z.W., Xu, X.K., Small, M., Wang, L., Wang, B.H.: Impacts of subsidy policies on vaccination decisions in contact networks. Phys. Rev. E. 88, 012813 (2013)
18. Fukuda, E., Kokubo, S., Tanimoto, J., Wang, Z., Hagishima, A., Ikegaya, N.: Risk assessment for infectious disease and its impact on voluntary vaccination behavior in social networks. Chaos, Solitons Fractals (in press, 2014), doi:http://dx.doi.org/10.1016/j.chaos.2014.07.004
19. Jansen, V.A., Stollenwerk, N., Jensen, H.J., Ramsay, M.E., Edmunds, W.J., Rhodes, C.J.: Measles Outbreaks in a Population with Declining Vaccine Uptake. Science 301, 804 (2003)
20. Kermack, W.O., McKendrick, A.G.: A contribution to the mathematical theory of epidemics. Proc. R. Soc. Lond. A. 115, 700–721 (1927)
21. Gillespie, D.T.: Exact stochastic simulation of coupled chemical reactions. J. Phys. Chem. 81, 2340–2361 (1977)
22. Szabo, G., Toke, C.: Evolutionary prisoner's dilemma game on a square lattice. Phys. Rev. E. 58, 69–73 (1998)
23. Traulsen, A., Pacheco, J.M., Nowak, M.A.: Pairwise comparison and selection temperature in evolutionary game dynamics. J. Theor. Biol. 246, 522–529 (2007)
24. Wang, Z., Perc, M.: Aspiring to the fittest and promotion of cooperation in the prisoner's dilemma game. Phys. Rev. E. 82, 021115 (2010)
25. Wang, Z., Kokubo, S., Tanimoto, J., Fukuda, E., Shigaki, K.: Insight into the so-called spatial reciprocity. Phys. Rev. E. 88, 042145 (2013)
26. Keeling, M.J., Eames, K.T.D.: Networks and epidemic models. J. R. Soc. Interface. 2, 295–307 (2005)

Cognitive Bias, ABM and Emergence of China Stock Market

Guo-cheng Wang

Institute of Quantitative & Technical Economics (IQTE),
Chinese Academy of Social Sciences (CASS),
Beijing, 100732, China
wanggc@aliyun.com

Abstract. Why, what and how real behavior(s) should be incorporated into ABM (Agent-Based Modeling), and is it appropriate and effective to use ABM with HS-CA collaboration and micro-macro link features for complex economy/finance analysis? Through deepening behavioral analysis and using computational experimental methods incorporating HS (Human Subject) into CA (Computational Agent), which is extended ABM, based on the theory of behavioral finance and complexity science as well, we constructed a micro-macro integrated model with the key behavioral characteristics of investors as an experimental platform to cognize the conduction mechanism of complex capital market and typical phenomena in this paper, and illustrated briefly applied cases including the internal relations between impulsive behavior and the fluctuation of stock's, the asymmetric cognitive bias and volatility cluster, deflective peak and fat-tail of China stock market.

Keywords: Cognitive Bias of Investors, Behavioral Macro-financial Model, ABM.

1 Introduction

The complexity and variability in modern economic activity, especially complicated investment decisions in the financial market, have attracted much attention. Agent-based modeling (ABM) has progressed strongly in the economic/financial research. There are more and more rising call and urgent demand from new analytical and cognitive perspective to explore complexity in contemporary economy and finance, for their increasingly complex practice, widening lag gap of modern financial theory, and poorer tools; on the other hand, it will be occurring and breaking out for saving potential energy that accelerates more supporting and more effective theory and tools, driven by the interests of people in economy and finance and promoted by the big-data technique and so on. Due to realistic demand, it forces us to study and reveal the mystery of actual economy/finance from the visual angle of real behaviors of investors, to use the experimental method with the combination of HS (Human Subject) and CA (Computational Agent) and advanced research tools (Arthur, 1993; Duffy, 2006), to penetrate and understand the volatility of stock market based on individual investing behaviors, to strive to find the inner relationship between the real and key characteristics of behaviors for

H. Handa et al. (eds.), *Proc. of the 18th Asia Pacific Symp. on Intell. & Evol. Systems – Vol. 1*,
Proceedings in Adaptation, Learning and Optimization 1, DOI: 10.1007/978-3-319-13359-1_2

investors at the micro-level and stylized facts or amount anomalies at the macro-level, the threshold value and sensitivity of critical point(s), and structural evolutionary process and so on.

Behavioral economics (finance), fateful arising and vigorous development, focuses on the qualitative analysis of micro individual behaviors up to now (Levin, 2012), and is based on the time series analysis and quantitative finance or financial econometrics method that is taking the zero mean, I.I.D, stationary and martingale process as the premise, logical starting point and analytic conditions, and ignoring or concealing the heterogeneous characteristics of behaviors for individual investors; whereas these neglected differences maybe are just the root causes of the complexity of financial markets. In the basis of deepening behavioral analysis, it is straightforward and driving tendency and new scientific approach to build integrated models with micro (behavior)-macro (output) link for uncovering the complexity of actual finance (economy), such as from the CAPM to (S) BCAP (Stochastic Behavioral Asset Pricing Models) and DSGE etc., that is the typical representative and foreshowing of this kind of cases (Lux, 2009). This paper aims at discussing and exploring key behavioral characteristics and conducting mechanism each other surrounding the bias focus and the center axes of the relationship between micro-investing behaviors and macro-stylized facts (anomalies) in the stock market.

This paper is organized as follows. The following section contains the extended cognitive bias model and gives a brief analysis of micro-structural behavior causes for the complexity of financial market; then the principle of computational experimental finance with the method and implemental process of incorporating HS into CA is summarized, a benchmark model of behavioral analysis with micro-macro link is built and basic testing rules are discussed in Section3; some relative work and applied cases in recent years as well as preliminary experience are given in Section 4. The final section contains concluding remarks.

2 Cognitive Biases and Micro-behavior Causes of Financial Market Complexity

Anormalies frequently occurring in stock market and financial activities, such as the volatility cluster, deflective peak and fat-tail, and drastic fluctuation, are boosting up risks and crisis that already have existed. A measure that could foresee and avoid risks and similar situations could be found only if the root cause of such complexity is perceived.

2.1 Enriching Cognitive Biases

Cognitive bias, we think, is materially to reflect the uneven and asymmetric of investor's preferences and behavioral representations on the investment products, so it can be understood as cognitive bias for a lot of bounded rationality or "irrational" real investment behaviors. Such as policy-sensitive, preferred stocks from emotion and psychology, herding effect, calendric effects examples in behavioral finance. Therefore, it is a good starting point to understand and to reveal the complexity of financial markets by exploring the cognitive bias of investors.

2.1.1 CH Model

Hierarchy is an internal intrinsic property in the development of nature and human society that conforms to the law of cognitive development. Hierarchy represents dynamic changes in human behavior; thus, its existence is very rational and necessary. The complexity of human behavior can be determined by applying hierarchy to social science.

The CH（Cognitive Hierarchy）model begins with 0-step players whose strategies are randomized equally. k-step players ($k \geq 1$) believe that all other players use only 0 to k-1 steps. Assuming that k-type beliefs $g_k(h)$ regarding the proportions of lower-step h types denote a normalized true distribution,

$$g_k(h) = f(h) \Big/ \sum_{l=0}^{k-1} f(l)$$

$$g_k(h) = 0, \text{for } h \geq k$$

Therefore, both "partially rational" expectations and $g_k(h)$ approaches $f(h)$ are obtained as k increases. Expected payoffs are computed, and the best responses are selected. Camerer et al.(2004) state that $f(h)$ follows the Poisson distribution; Shao (2010) has developed an investment game model based on CH under the framework of bounded rationality from an investor group perspective. The CH model has been widely applied in economic studies and it is a feasible approach to study cognitive bias with CH model.

2.1.2 CH Behavioral Analysis

Different properties are reflected in the CH model in consideration of the various cognitive abilities possessed by individuals. Our model focuses on two influential factors: first, stock price is a very important factor that influences investor decisions; second, policy adjustments in China's stock market are of great concern. Both factors differ based on investors; however, determining the factor that is more significant to Chinese investors is an interesting topic. The heterogeneity of individual investors generally results in stock market diversity. Heterogeneous behavior causes the so-called leverage effect; that is, changes in stock prices or adjustments in the stock market are negatively correlated with volatility. Stock market heterogeneity often results in asymmetric, non-stationary, and nonlinear reactions. Stock forecasting is known as an uncertain, nonlinear, and non-stationary time-series problem; thus, accurately predicting the market via traditional methods is difficult. ABM can therefore be used to address arcane problems. Generalizing real stock market behavior is necessary to characterize heterogeneity.

2.1.3 Extended CH Model

The financial market is often considered the self-organization of groups of interactive, learning, and bounded-rational agents in ABM. Investor behavior parameters for empirical data can be determined, and heterogeneous behavior can be calibrated using the given data. Hommes (2006) examines important stylized facts in financial time series by using simple economic and financial heterogeneous agent models.

Real macroeconomic phenomena can generally be described by Bayesian inferences based on linear and nonlinear behavior models. Integrated models are constructed based on human cognition, decision, and execution, and the transition from concrete individuals to the gross population is analyzed. We focus on endogenous weighted analysis and determine various behavioral attribute parameters in our model. The behavioral parameter λ_i could be considered as the cognizing level in the benchmark model and can thus be used to describe stock market complexity as a whole. The cognitive parameter λ_i can be influenced as follows: first, the differences among the natural instincts of individuals are significant; second, the responsiveness to external factors varies from individual to individual; third, reactions to interactions among individuals differ. Individual to economic CH is strongly related to previous macroeconomic phenomena.

2.2 Internal Relation between Complex Financial Market and Individual Behavior

Any phenomenon (result) occurred in the financial market is the interactions made by different individual investors and affected by many factors. The decision made by the individual investor changes as macro-policies, market signals and external impacts change. The individual cognition, belief and strategy constantly adjust and change during the process of such cycle; meanwhile, the proportion, structure, ways, track and process of evolution are synergistically changed as well. The behavior property, assets amount, response model to the external condition changed and differences in its intensity of an individual investor shall not be considered at many cases, nor these things are simplified to a random error or disturbance with mean value of zero. The investment behavior in Chinese capital market shall be considered more cautiously; otherwise it just copies other's behavior or falls into penny-wise misconceptions in thought.

The complexity in financial activities caused by individual subjective initiative (humanistic complexity) has some same parts with that of substances change in natural world but they have essential difference. Thus we need to further discuss the internal relation (a corresponding relation form could be built) between micro behavior characteristics and stylized facts. The individual behavior property, assets amount, opportunities and subjective wishes in humanity complexity are of significant difference and the causal chain among them is two-way, interdependent and interweaved with many factors and situations, which couldn't be speculated or reproduced according to some natural rules. However, some basic facts and objective attribute don't vary from person to person or from event to event (meet the individual irrelevance). For example, the whole economy/society is constituted by numerous individuals (dissoluble). The whole phenomenon is the individual behavioral outcome under the influence of individual behavior property and external conditions (constant relation). The behavioral process (rules), related behavior model, parameter, threshold and relational structure followed and presented by individual decisions could be perceived, simulated and reproduced.

This could be represented geometrically: the complicated financial market in reality is a compounded polyhedron. The theoretical mapping at any viewing angle and level is complicated and hard to define. The modern finance is a mapping system on a rational behavior section of the polyhedron built by an efficient market mechanism (see Figure 1); it also could spread to other behavior section (Bounded or non-rationality) with more feasible conduction and operation mechanism. Thus, a financial market of multiple dimensions could be obtained so as to find the internal relation between micro behavior and macro phenomena.

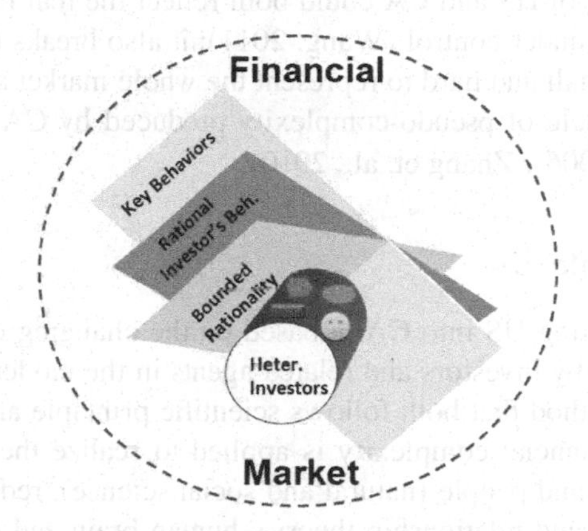

Fig. 1. M-dimension behavioral sections for complex financial market

The relation between financial market outcome and investor behavior manifests such feature. Different individuals' investment behaviors are diverse as situation changes (depending on situation) as well as the response model and intensity. The path and influence of relations in economic network is asymmetric (Schweitzer, 2009). The conduction mechanism and evolution path are complicated and various. However, there is recognizable internal relation and needs material promotion of theory and method.

More focusing on the issue studied and promoting by scientific-technical progress, a synthetically analytical method that is people-oriented and covering all existed theories and methods is required and feasible. The influence of loss aversion, herd behavior, over-confidence, stock premium of outstanding bounded rationality (non-rationality) on stock market is hard to explain due to mixed affections of many factors and kinds of subjects by the traditional quantitative finance of modern financial theory based on the empirical approach. The behavioral finance couldn't be quantified with high cost and small coverage; the agent-based computational economics/finance (ACE/ACF) is in its initial stages (Farmer & Foley, 2009), as well as some superior techniques like noise trading and wavelet analysis. Based on the above methods, all these methods lay a good foundation for independent and partial analysis but their limitation is been revealed day by day. It is better to set the investor's behavior as the logical starting point to analyze the theoretic root of financial complexity in reality. The conduction mechanism and evolution path from individual to group behavior are not single and changeless (Miller & Page, 2007) so that a deepening, integrated model and coordinating analytical method are required.

3 Agent-Based Computational Finance with Incorporating HS into CA

If we want to find the root cause and to understand furthermore the stock market, it is necessary to focus on the investors' behaviors and mutual relations (including response to rules, institutions and information). It's difficult to rationally portray and accurately quantify characteristics of real investing behavior. Comparatively speaking, the combination of HS and CA could both reflect the real investing behavior and be tested repeatedly under control (Wang, 2011); it also breaks the limitation that the HS sample size is small and hard to represent the whole market as well as overcoming the substantial obstacle of pseudo-complexity[1]produced by CA-based computational finance (LeBaron, 2006 ; Zhang et. al., 2010).

3.1 Basic Principle

ACF with incorporating HS into CA is based on the changing character and rules of real behaviors made by investors and related agents in the modern financial activities. An experimental method that both follows scientific principle and reflects humanistic spirit and related financial complexity is applied to realize the integration of micro and macro, material and people (natural and social science), reductionism and holism (individual standard and relationship theory), human brain and computer (reality and virtual world, human wisdom and mechanical efficiency) as well as several model techniques. The evolutionary process could be reflected gradually; it also promotes the close relation between theory and reality so that they can verify each other.

3.2 General Method and Steps

Aimed at breaking the financial complexity, ACF with HS and CA could be divided into three steps according to modern finance, behavioral finance, complexity science and related knowledge:

(1) Analysis of micro body behavior. The human behaviors in social activities can't be divided without linear transformation and similarity on the change of property and quantitative relation, and sometimes they are irreversible. What's more, the relation between individual and group rationality is complicated and its inconsistency may be the root of various social paradoxes. The driving forces, mode of action and conduction mechanism from individual to social action are complex. Increasing from individual to group amount is not only about quantity but also about the change of nature and direction (Wang, 2012; 2013).

Seeing from the historical evolution of basic behavioral assumptions in economics, people have constantly been deepening and enriching their cognition on behavior property. The following figure shows the achievement and logical network of each stage, so that the developing path of investor's behavior could be compared.

---

[1] The computational complexity under given structural relationship is not real humanistic complexity; it is similar to the pseudo-random number generated by computer.

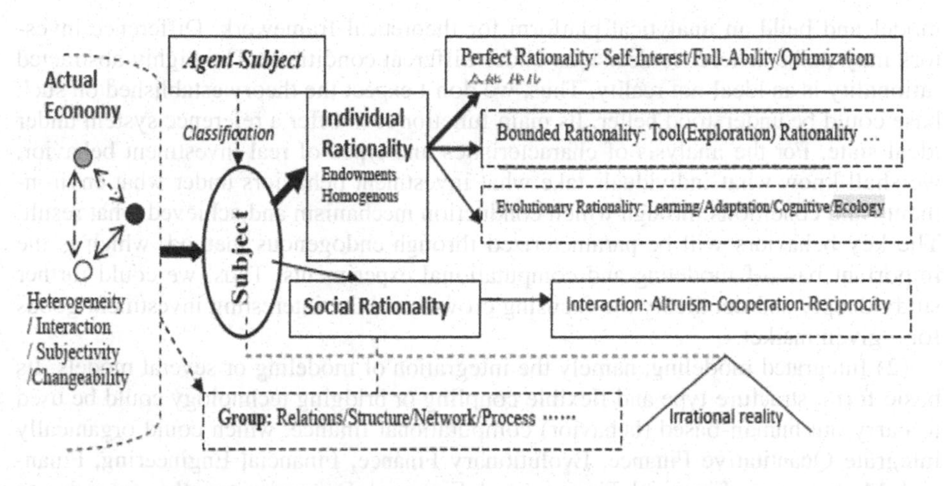

Fig. 2. The spectrum of economic behavioral characteristics

By virtue of formalized language, the cognition depth of behavior could be classified and defined longitudinally to highlight the nature of behavior. For a given capital market E (including information and institutions), stock or investment goods set X, investor's set N, way and process of behavior F, investment behavior results set Y.

a) Deterministic type;
b) Stochastic type;
c) Complex type.

The first two types of behaviors select representative bodies, implying a homogeneity assumption, namely all bodies only consider a behavior property and they follow the same decision-making strategy. Only one mechanism model from individual to group behavior is set. For the third type of complexity behavior, considering the difference and interaction among individuals, subjective consciousness and dependency on situation, the condition and interaction that permit multi-attribute and selectivity behavior could be described (Wang, 2012).

The assumptions of rational investors in modern finance theory are revealed the limitation during the process of practice and challenge, so that the description on real investing behavior changes which needs observation, experiment, record, summarization, abstraction and classification. The essential attribute of human behavior could be perceived through behavioral expression. The micro behavior could be analyzed based on the experiment and two aspects of functions could be realized. One is to observe real response of human in various situations so as to further understand and describe behavior property; another is to find the way that how individual behavior comes to group behavior and how the overall characteristics of stoke market and economic activity affect. Without the support of modern computing technology, we only can rationally simplify the investment behavior and set an analytical model to find an optional solution.

The basic investment decision is nothing more than decision selection (buying, holding and selling) and the corresponding state could be recorded as {-1, 0, 1}. The totality and individual may form an alternative relation set so as to identify the basic model and build an analytical platform for theoretical framework. Difference investors may make difference decisions under different conditions. The highly abstracted rationality is an ideal not reality. Thus, we don't expect the theory established on such base could be understood better. Its main function is to offer a reference system under ideal state. For the analysis of characteristics and types of real investment behavior, we shall know what individuals take what investment behaviors under what environments and conditions through which conduction mechanism and achieved what result. The key behaviors will be parameterized through endogenous method, which is the important base of modeling and computational experiments. Thus, we could further study deeply and detailedly the focusing crowds and the interesting investment goods for a given market.

(2) Integrated modeling, namely the integration of modeling or several models. Its basic form, structure type and flexible coupling or bridging technology could be used to carry out human-based (behavior) computational finance, which could organically integrate Quantitative Finance, Evolutionary Finance, Financial Engineering, Financial Mathematics, Financial Physics, and Financial Dynamics together in order to reveal the mystery of financial complexity.

The integrated model is the real response model and behavioral pattern of a vivid body in the form of scientific manner, establishing behavior rules, equation of relation and behavior (Dynamics). Based on the objective and overall analysis of subject behavior and centralized on the process of decision-making behavior, different models are established according to its constituent parts, stages and factors. Then each part and type of models is integrated through flexible coupling or bridging according to the natural process of realistic problems. Thus, such model is quite inclusive and flexible which could contain and connect existing models as per the need of problems and feasible conditions. It could selectively focuses on any special locality problem and integrates the distributed simulation and overall analytic method.

The meaning of soft-linkage or flexible coupling is the simulated mapping and links between parts of human and the nature and other complex relationships on the decision node. It is the join and verification between different models, namely bridging or coupling. For example, the output of previous step is translated into the input of next step; or changing name of variable, establishing similar instrumental variable, piecewise function of dummy variable in different forms; serving as the external environment parameter of subject behavior. Such connected models will stay the same as things change on the whole. By virtue of integrated model, the internal relation between micro and macro is revealed through construction of social function and related parameters adjusted in the steady-state process. The integrated modeling on the basis of subject behavior is a carrier or platform that calculates the social economy. The effectiveness, accuracy and sensitivity of the modeling could be evaluated by the actual degree of coincidence.

(3) Computational experiments. The extreme assumption and implicated premise in the modern finance's empirical measurement couldn't summarize humanity features, such as human's experiences, intuition, comprehensive analysis, judgment and response. When the difference in risk preference, gaming instinct, price sensibility, interacting decision-making behavior, division or cluster of different investors group can't be ignored, data of actual investor's behavior type shall be gathered and screened, setting the original value of controlled variable, distribution pattern and

range of variation.[2] The data are input into the computer virtual subject as well as integrated model and computational experiment platform. By virtue of software such as MatLab, NetLogo, and genetic or bionic algorithm and comparing different scenarios, conditions and experiments, the micro behavior that causes market anomalies or stylized facts could be speculated by the output results. Various behavior pattern occurred in reality and conduction mechanism are sorted to calibrate and determine the key behavior parameters, set models, foresee potential market phenomena as the basis of investment decision.

The practicing experimental process is as follow: selecting all elements and possible corresponding combination (theoretically practical) among market scenario, behavior property, alternative mechanism, external condition and stylized facts setting. The conduction chain related to the internal link is calibrated and sorted according to generalized Bayesian Decision Theory so as to obtain a proper, real and scientific conclusion of persuasion and explanation power as well as brief and formal computational experiment result. Thus, the internal relation between individual investment behavior and market stylized facts or anomalies in the real capital market and the function relationship corresponding to the mechanism of evolving from individual to total amount are found. For the design and implementation of the experiment, results analysis, improvement and other necessary routine work, refer to related content of experimental economics, finance and ACF.

4 Benchmark Model and Behavioral Test

Cognition also includes the agent's perception and judgment on the environmental conditions and gross change, it is natural and better to need the support of the micro-macro integrated model. This part is applied in the reference model of capital market with micro-macro link. It improves the integration of former research results (DeGrauwe, 2010; Scheffknecht and Geiger, 2011; Lengnick and Wohltmann, 2013) and basic representation as well as micro behavior, and then discusses how to test and to analyze deeply the classical behavioral hypothesis.

4.1 Benchmark Model

Just considering a giver stock market in China, let y_t express net the output gap in period t, r_t be the nominal interest rate, and π_t be the rate of inflation, then we can obtain a series of behavioral equations of depicting macro-states with variables and relationship of between factors as follows:

Aggregate demand equation

$$y_t = a_1 \tilde{E}_t^0[y_{t+1}] + (1-a_1)y_{t-1} - a_2(r_t - \tilde{E}_t^0[\pi_{t+1}] + \varsigma_t) + u_t \tag{1}$$

[2] Practicing process is similar to the field-experiment; how to select samples, in-time observing and tracking in investors, and how to depict and grasp main characteristics during applications, will be seen in the further paper.

Aggregate supply equation

$$\pi_t = b_1 \tilde{E}_t^0[\pi_{t+1}] + (1-b_1)\pi_{t-1} + b_2 y_t + v_t \tag{2}$$

and the market behavior following the amending Taylor's rules

$$r_t = c_1 r_{t-1} + (1-c_1)[c_2(\pi_t - \pi_t^*) + c_3 y_t + c_4^{\mathrm{T}} \chi_t] + w_t \tag{3}$$

Where : $\tilde{E}_t^0$ represents the expectation of the overall market (pseudo-player) at t time, $a., b. and$ c. are factors or parameters to be estimated ; π^* is the expected control targets of the increase ; χ_t is a vector, which include all other factors impacting the yield, ζ is the risk and risk-free real interest rate spread ; u_t, v_t and w_t are (random) disturbance/error term or white noise disturbance term.

The parameters could be obtained from the market empirical date according to traditional measuring empirical approach (related to the two fundamental assumptions, rational investor and EMH in modern financial theory), or be inferred from the investor's actual behavior in the market when there is Bias. Observing the actual investment behavior occurred in the stock market, an integrated model will be formed through construction of related behavior equation and connection with macro state equation, and then the computational experiment will be conducted to find out the potential correspondence between micro behavior and stylized facts (anomalies) in the stock market (see figure 5). For the perfectly competitive market of general goods, the general equilibrium is the exception of game equilibrium under the condition of materialization. At this point, neutral system, complete information, individual rationality are satisfied with agreement of individual and collective rationality; if the assumption of perfectly competitive market is significantly deviated, a similar method could be used to find the internal relation between actual micro behavior and abnormal complex phenomena in the market. Thus, the complex economic and financial problems now could be better explained to find a way of quantifying evidences and realized the technology. The description and extension of micro investment behavior are as follows:

$$I_B(i) \rightarrow \begin{cases} S_b & \text{at Prob. } p_1 \\ S_h & \text{at Prob. } 1-p_1-p_2 \\ S_s & \text{at Prob. } p_2 \end{cases} \quad \Longrightarrow \quad I_B(i) \rightarrow \begin{cases} S_b & \text{at Thres. Value } TB_b \\ S_h & \text{at Thres. Value } TB_h \\ S_s & \text{at Thres. Value } TB_s \end{cases}$$

The behavior property, critical change and threshold difference reflect individual diversity (heterogeneity). The determination of threshold connects the combined action of numerous agents' behavior and external conditions change, including the influence among agents, namely endogenesis and interaction; the threshold is determined according to the total available market received by the individual, other investors' strategies and individual characteristic parameters. It could also be used in various key behavior characteristics for the agents in general market.

4.2 Hypothesis Testing of Basic Behavior

The behavior, structure and output (variable, equation and parameters etc.) are integrated in the above equation and description. The change of macro aggregate in the specific application is expressed as actual situation and its key point is to determine the basic attribute and type. For the estimation of parameters, macro equation is still a common method for econometrics and its main change is that the data sources are based on the macro abnormal situations and related data and integration method. The micro behavior parameters are obtained from the personalized features from agent's behavior through endogenalization and game experimental method. The hypothesis of basic behavior could be gradually released and expanded to make up the logical losses by key features among complete rationality, limited rationality and actual behaviors. The inspection of basic behavior hypothesis is divided into two directions and ways: Prior Testing and Posterior Testing(Wang, 2014).

5 Applications

Combining with the research on Chinese stock market quotation and investors' behavior, we want to find what kinds of behaviors correspond to the stylized facts and anomalies, taking Shanghai Stock Exchange's actual data as example and using current financial database and professional software to carry out tentative application. We choose parts of cases for positive supporting.[3]

5.1 Investment Impulsion and Share Price Change

Based on the actual investment behavior of Chinese investors and taking 254 actual historical data from Jan.4, 2007 to Jan.18, 2008 as example. The types of behaviors are analyzed as the original value, using Matlab software to simulate. Take 1 business day (or unity of time) as step length and 500 business days as the total length of the simulation. Firstly, the static simulation on the basic model constructed by comparing investors' normal behaviors is studied to obtain the result of basic model analysis. Then, considering the dynamic evolution of various groups, we focus on the investment impulsion behaviors. Paying attention to the learning adaptability, difference and interaction of investor's investment decision-making and supposing other factors that influence behaviors unchanged, simplification or discard will be made (according to requirements and increasing or changing behaviors characteristics and influencing factors).

Specific method: the reaction speed (change the time of decision rule lasting) and strength (the proportion of individual trading volume in its total assets) are used to describe the individual investment impulsion behavior; when the external information or market trend changed, the trade that decision rule changed is called as first-class impulsion behavior; the investment behavior that is changed two days or more later is

[3] The author gratefully acknowledge Dr. Fei Liu and Dr. Yuntao Long for conducting examples, seeing their dissertation (Liu, 2009; Long, 2012) respectively.

called as second-class impulsion behavior. These two kinds of behaviors affect and transmute into each other. Taking the max daily fluctuations of stock prices 8% as critical value, the fluctuating value (mental threshold) mentally bore by the individual is recorded as w, deciding w and different impulsion behavior and taking 5% as the contrast threshold; when first-class (second-class) behavior dominates, the stock market will boom (slump). The time of duration for these two behaviors before the stock market changes is quite different (see figure 4); the length of variable section near the threshold is subdivided and narrowed to observe the sensibility and critical effect of investor's behaviors. If the investor could make decision before the critical point of market fluctuation, the earnings are considerable. The new findings in this researching application are as follows:

(1) The influence and relation between the investment behavior characteristics and price fluctuation in the stock market are analyzed from the micro-view; comparatively speaking, individual investment impulsion behavior is internally related to the stock market booming (slumping); the first-class behavior may cause booming and the second-class behavior may cause slumping;

(2) Trying to explore the multi-dimension description and classified quantitative method of micro motivation and to use investors' actual behavior to study the investment impulsion behavior;

(3) The computational experiment is used to explore the micro-cause of stock market complexity in China, trying to reveal the mechanism of overall complex phenomena caused by the potential individual behavior model. The main performance of investor's mental fragility is sensibility around the critical point(s).

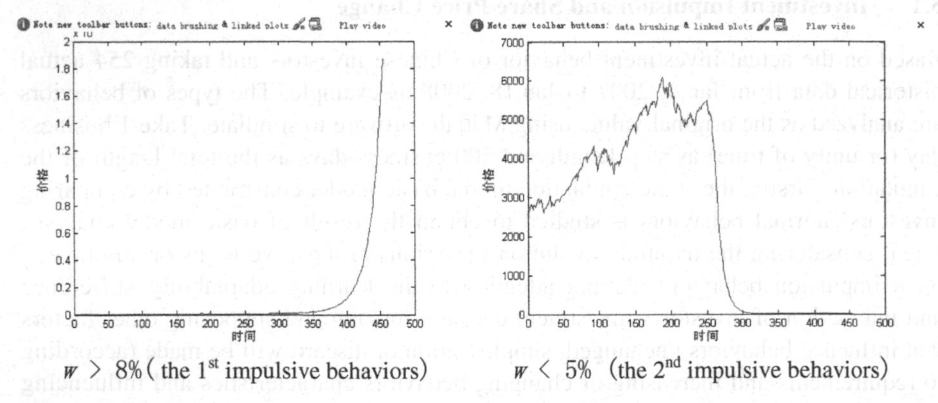

$w > 8\%$ (the 1st impulsive behaviors) $w < 5\%$ (the 2nd impulsive behaviors)

Fig. 3. The relation simulation between different types of impulsive behaviors and sharp stock-price

(Note: Vertical axis is price; Horizontal axis is time)

What's more, from the point of methodology, acceptance or adjustment of rational behavior assumption, establishment of causality or theoretical logic model and utilization of empirical data could not fit the need of actual complexity. However, micro-behavior-based computational experiment analysis focuses on the actual investment behavior of different agents and development of group structure, simulating different results of convergence, violent shock, booming or slumping. It shows the necessity to further study different agents' inter-behaviors characteristics. This method could be one of the best ways to discover and perceive the complex essence of abnormal economy and financial activities.

5.2 Type of Information Preference and Market Sensitiveness

The developing stock market in China prefers what kinds of information. General quantitative finance and statistic testing may result in large errors due to lack of inspection of behavior basic hypothesis. We use computational experimental finance to combine HS and CA based on the investor's actual behavior and real data of the stock market. By comparing different responses by different investors to various policies and market signals, policy sensitivity parameters and tendency parameters influenced by the vicinity are designed, taking proportions of different investors as controlled variable in order to observe the different preference type and market effect identified and responded by the investor's behavior biases and market structure change to the policy and market signals; these were further parameterized and changed according to the potential situations to know whether the numerous investors' behaviors coincide with the stylized facts occurred in reality, focusing on internal relation between cognitive behavior and gathered fluctuation. The specific simulation has been launched in NetLogo 5.0.3 version (Wang & Long, 2014).

5.3 The Essence of Behavior and Integrative Effect of the Market

The asymmetry on benefits and losses in people's behavior is inherent. Several behavior-bias and their combination of key behavioral characteristics illustrated in the behavioral finance could be the main reason that induces some stylized facts or anomalies in the market (see figure 5). The problem is that what reason could convincingly prove it.

Asymmetric investment behavior (key behavior characteristic) has a complicate and internal relation with market anomalies (stylized facts), such as impulsive action vs. share price fluctuation, asymmetric cognition vs. fat-tail and drastic moves and stock market differentiation, herd behavior, psychological complex vs. blue-chip gathering, non-rational, irregular investment, overconfidence vs. stock market bubble, information preference and market awareness, overall features, individual response, behavior pattern vs. trend and development of the stock market. Observing and judging the situation, we could take measures before group behavior and market generate results or other investors find this, so as to decide our investment behavior. For the actual behavior of Chinese investors, they have to face the market complexity and stylized facts and integrate theories by changing ideas, which is well-suited and effective and also fit the essence of market volatility.

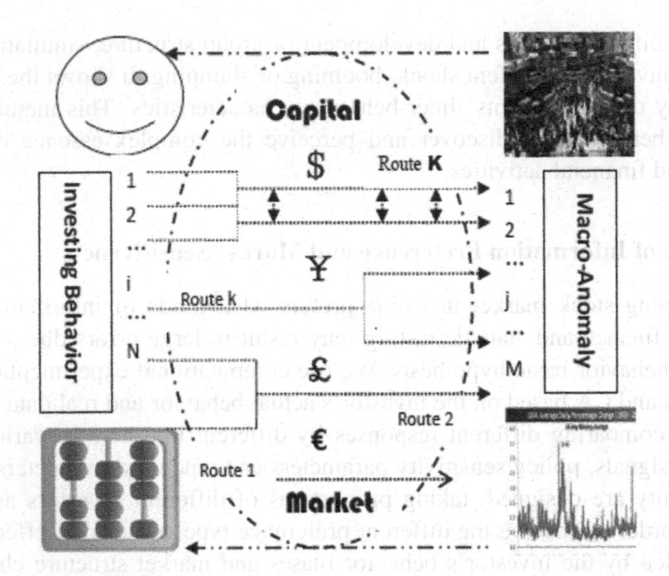

Fig. 4. Mapping from Actual Investing Behavior to Macro-Anomaly in Stock Market

6 Conclusions

The modern finance undoubtedly is more and more complicated but it could be recognizable. Computational experimental finance with agent-based and integrated modeling is scientific, acceptable and effective. Viewing from investor's behavior, Chinese financial (stock) market has significant behavioral biases, presenting as policy sensibility (depending on conditions and structures) but irregularly responds to the market signal. Thus, the theoretical method based on rational investment and efficient markets hypothesis is prudently used to study Chinese problems. It's necessary to consider the adaptation of method aiming at solving problems.

The market effect of potential actual investment behaviors shall be further studied. Different investors shall determine their behavioral characteristics by themselves and choose a proper way of investing. The theoretical method in this paper aims at the stylized facts of capital market, realistic investor's behavior and economic environment in China. It's worth deepening and promoting as it is constantly proof-testing.

Acknowledgment. The author is grateful to the National Basic Research Program of China (973 Program) (Serial No. 2012CB955802) and NSFC(Grant No. 71471177)for funding that supported this research.

References

1. Arthur, B.: Economic Agents that Behave like Human Agents. Journal of Evolutionary Economics 2000(3), 1–22 (1993); Reprinted in The Legacy of Joseph A. Schumpeter, H. Hanusch (ed.) Edward Elgar Publishers (2000)

2. Camerer, C.F., Ho, T., Chong, J.K.: A cognitive hierarchy model of games. Quarterly Journal of Economics 119(3), 861–898 (2004)
3. Duffy, J.: Agent-based models and human subject experiments. In: Tesfatsion, L., Judd, K.L. (eds.) Handbook of Computational Economics, vol. 2, pp. 949–1011. Elsevier, North-Holland (2006)
4. Farmer, J.D., Foley, D.: The economy needs agent-based modeling. Nature 460, 685–686 (2009)
5. De Grauwe, P.: Top-Down versus Bottom-Up Macroeconomics. CESifo Economic Studies 56, 465–497 (2010)
6. Hommes, C.H.: Heterogeneous agent models in economics and finance. In: Tesfatsion, L., Judd, K.L. (eds.) Handbook of Computational Economics. Agent-based computational economics, vol. 2. Holland/Elsevier, Amsterdam (2006)
7. LeBaron, B.: Agent-based computational finance. In: Tesfatsion, L., Judd, K.L. (eds.) Handbook of Computational Economics, pp. 1187–1233. Elsevier (2006)
8. LeBaron, B.: Heterogeneous gain learning and the dynamics of asset prices. Journal of Economic Behavior and Organization 83, 424–445 (2012)
9. Lengnick, M., Wohltmann, H.W.: Agent-based financial markets and New Keynesian macroeconomics: a synthesis. Journal of Economic Interaction and Coordination 8(1), 1–32 (2013)
10. Levin, D.: Is Behavioral Economics Doomed? The Ordinary versus the Extraordinary. Openbook Publishers, UK (2012)
11. Lux, T.: Stochastic Behavioral Asset-Pricing Models and the Stylized Facts. In: Hens, T., Schenk-Hoppé, K.R. (eds.) Handbook of Financial Markets: Dynamics and Evolution, ch. 3, pp. 161–215 (2009); De Grauwe, P.: Top-Down versus Bottom-Up Macroeconomics. CESifo Economic Studies 56, 465–497 (2010)
12. Miller John, H., Scott, E., Page: Complex Adaptive Systems: An introduction to computational models of social life. Princeton University Press, NJ (2007)
13. Scheffknecht, L., Geiger, F.: A behavioral macroeconomic model with endogenous boom-bust cycles and leverage dynamcis. FZID Discussion Papers 37-2011, University of Hohenheim, Center for Research on Innovation and Services (2011)
14. Schweitzer, F., et al.: Economic Networks: The New Challenges. Science 325(5939), 422–425 (2009)
15. Shao, P.: Bounded rationality, cognitive hierarchy and investment game. The Journal of Quantitative & Technical Economics 10, 145–155 (2010)
16. Wang, G.: Exploring complex economy to develop quantitative economics from Micro-behavior perspective. Journal of Quantitative Economics 2(1), 102–120 (2011)
17. Wang, G.: The evolvement and beyond of rationalism in modern economics. Social Sciences in China (7), 66–82 (2012)
18. Wang, G.: Deepening micro-behavioral analysis and exploring the complexity of macro-economy. Jiangsu Social Sciences (3), 20–28 (2013)
19. Wang, G.: Behavioral Macro-Financial Modeling from Investor's Bias with Applications — Based on the Experiment of Incorporating HS and CA. Journal of Management Science & Statistical Decision 11(1), 24–40 (2014)
20. Wang, G., Long, Y.: Study on Emergence of Capital Market with Cognitive Hierarchy and Extensive Agent-Based Modeling— An Application of e-Science in Social Sciences. E-Science Technology & Application 5(1), 83–92 (2014)
21. Zhang, W., Zhang, Y.J., Xiong, X.: Agent-based Computational Finance. Science Press, Beijing (2010)

Collective Behavior in Cascade Model Depend on Turn of Choice

Saori Iwanaga and Akira Namatame

Department of Maritime Safety Technology, Japan Coast Guard Academy,
5-1 Wakaba, Kure, Hiroshima, 737-8512, Japan
s-iwanaga@jcga.ac.jp

Abstract. There are growing interests for studying collective behavior including the dynamics of markets, the emergence of social norms and conventions, and collective phenomena in daily life such as traffic congestion.

We showed that collective behavior is affected in the structure of the social net-work and theta, and the collective behavior was stochastic in previous work. Moreover, collective behavior is almost same as Schelling model, though the decision is not interactive and simultaneously. Then, we found that the collective behavior in Schelling model is similar to cascade model. That is, our results with heterogeneous rules or heterogeneous networks are possible to apply for cascade model. In this paper, we analyzed that the collective behavior of population is stochastic, although decisions of agents are deterministic. We focused on the effect of network degree. And we found that turn of decision is effect on collective behavior not the first choices.

Keywords: collective behavior, agent, social network, degree, threshold.

1 Introduction

There are growing interests for studying collective behavior including the dynamics of markets, the emergence of social norms and conventions, and collective phenomena in daily life such as traffic congestion. Many researchers have pointed out that an equilibrium analysis does not resolve the question of how individuals behave in a particular interdependent decision situation. It is often argued "it is hard to see what can advance the discussion short of assembling a collection of individual, putting them in the situation of interest, and observing what they do"[1]

In examining collective behavior, we shall draw heavily on the interactions of individuals. We also need to work on two different levels: the microscopic level, where the decisions of the individual agents occur, and the macroscopic level where collective behavior can be observed [2]. The greatest promise lies in analysis of linking microscopic behavior to macroscopic behavior [3]. What makes collective behavior interesting and difficult is that the entire aggregate outcome is what has to be evaluated, not merely how each person does within the constraints of her own environment. The performance of the collective system depends crucially on the type of interaction as well as the heterogeneity in preference of agents [4].

H. Handa et al. (eds.), *Proc. of the 18th Asia Pacific Symp. on Intell. & Evol. Systems – Vol. 1*,
Proceedings in Adaptation, Learning and Optimization 1, DOI: 10.1007/978-3-319-13359-1_3

Feng et al. [5] brings together agent-based models and stochastic models of complex systems in financial markets and show how individual decisions give rise to macroscopic actions. Additionally, the heterogeneity in agents' investment horizons gives rise to long-term memory in volatility. Using market data, Kenett et al. [6] provides new information about the uniformity preset in the world's economies. From their analysis, it becomes evident that this uniformity does not only stem from an increase of correlation between markets, but that there has also been an ongoing simultaneous shift towards uniformity in each single market.

There are many situations where interacting agents can benefit from coordinating their behavior. Coordination usually implies that increased effort by some agents leads the remaining agents to follow suit, which gives rise to multiplier effects. Examples where coordination is important include trade alliance, the choice of compatible technologies or conventions such as the choice of a software or language. These situations can be modeled as coordination games in which agents are expected to select the strategy the majority do [7]. The traditional game theory, however, is silent on how agents know which equilibrium should be realized if a coordination game has multiple equally plausible equilibria, where these can be Pareto ranked [8]. This silence is all the more surprising in games with common interest since one expects that agents will coordinate on the Pareto dominant equilibrium [9]. The game theory has been also unsuccessful in explaining how agents should behave in order to improve an equilibrium situation [10].

Often an individual's decision depends on the decisions of others because they have limited information about the problem or limited ability to process the information [7, 11]. An individual's payoff is a function of the actions of others [1, 10]. In particular, in the diffusion of a new technology [12], early adopters impose externalities on later ones by rationally choosing technologies to suit only themselves. Then, individual has an incentive to pay attention to the decisions of others. This is known as binary decisions with externalities [7].

Schelling model or threshold model [7] has been postulated as one explanation for the contagion. Schelling shows by example of attendance the optional Saturday morning review session. For people, attendance depends on the percentage of attendance. The critical point which the benefit exceeds the cost of attendance is threshold. Schelling assumes that all people know the others' decision and deals with stability of equilibrium of several threshold distributions.

Contagion is said to occur if one behavior can spread from a finite set of agents to the whole population. When can behavior that is initially adopted by only an infinite set of agents spread to the whole population? Morris [13] shows that maximal contagion occurs when local interaction is sufficiently uniform and there is low neighbor growth, i.e., the number of agents who can be reached in k steps does not grow exponentially in k. Lopez-Pintado [14] showed that there exists a threshold for the degree of risk dominance of an action such that below the threshold, contagion of the action occurs. He also showed that networks with intermediate variance (where the connectivity of the lowest connectivity nodes are not so low) are best for diffusion purposes. Meanwhile, Watts [15] showed that when the network of interpersonal influences is sufficiently sparse, the propagation of cascades is limited by the global connectivity of the network; and when it is sufficiently dense, cascade propagation is limited by the stability of the individual nodes. Therefore, the rate at which a social innovation spread depends on three factors: the topology of network, the payoff gain of the innovation and the amount of noise in the best response processes [16]. Montanari [17] shows that innovation spreads much more slowly on well-connected network structure dominated by long-range links than in low-dimensional ones dominated and Komatsu [18] obtains the optimal network for good cascade using genetic algorithm and they show the network have a sufficient number of vulnerable nodes and hub node of medium size.

To illustrate how important spatial structure is to the emergence of cooperation in society, Nowak [19] and Axelrod [20] have investigated lattice models of agents confronted with a social dilemma. At the other extreme, most of human social networks were regarded as random networks whose nodes are connected randomly because of its large scale and complexity. In reality, Barabási et al. found that many complex networks have a scale-free structure [21]. Moreover, another kind of network structure small-world has been defined and researched [22]. Of course, the number of individual is large and the relationship is assumed to be complex. However, the world is much smaller than we think. We deal with population with scale-free network.

On the other hand, Hasan et al. [23] deal with a threshold model of social contagion originally proposed in network science literature. And they show that faster propagation of warning is observed in community networks with greater inter-community connections.

In our previous work [24], we showed that collective behavior is affected in the structure of the social network, the initial collective behavior and diversity of theta. In this paper, we focus on scale-free network and investigate the effect of number of interaction on collective behavior. And we compared Schelling model with Cascade model [25]. And we showed that collective behavior is affected in the structure of the social network and theta and the collective behavior was stochastic. Moreover, collective behavior is almost same as Schelling model, though the decision is not interactive and simultaneously. Then, we showed that the collective behavior in Schelling model is similar to cascade model.

In this paper, we analyze that the collective behavior of population is stochastic, although decisions of agents are deterministic. We focus on the effect of network degree. And we find that turn of decision is effect on collective behavior not the first choices.

2 Model

We consider the following dynamics to describe the evolution of agents' choices through time. At time t, each agent plays a 2×2 game with each neighbor and chooses an action from the space S ={S1, S2}[14,15]. The assumption that an agent cannot make his action contingent on his neighbor's action is natural in this context. Otherwise, the behavior of an agent would be independent of the network structure. Payoffs from each interaction in each period are given by a function(s, s') , where $s, s' \in S$

and they are summarized in the following symmetric matrix as shown in Table 1, where $0 \leq \theta_i \leq 1$. In this matrix, if agent Ai chooses same strategy as the other agent, it can get the positive payoff $1-\theta_i$ or θ_i, otherwise it receives nothing. It means coordination game. This is also called as Conformity Model. This payoff matrix can be translated from Stag Hunt game. Nash equilibria are (S1, S1) and (S2, S2). Agent Ai's payoff from playing $s_i \in \{S1, S2\}$ when the strategy profile of the remaining agents s_{-i} is given by $\Pi(s_i, s_{-i}) = \sum_{j \in N_i} \pi(s_i, s_j)$. Thus, an agent's payoff is simply the sum of the payoffs obtained across all the bilateral games in which he is involved. Agents select the action that maximizes his benefits given the action of others in the previous period (a myopic best response).

Table 1. Payoff matrix of agent Ai ($0 \leq \theta_i \leq 1$).

Choice of agent Ai	Choice of other agents S1	S2
S1	$1-\theta_i$	0
S2	0	θ_i

Here we define $p_i(t)$ as the proportion of agent Ai's neighbors who choose S1 at time *t*. Fig.1 explain the social network and the proportion of agent's neighbors who choose S1 at time *t*. Node means agent and link means neighboring social network. Each agent has theta θ_i and can decide the choice S1 or S2 at given time step according to the proportion of neighbors who choose S1, $p_i(t)$. Because each agent has idiosyncratic theta θ_i and the proportion of neighbors who choose S1, that is $p_i(t)$, is different each other, the decision at next time step is deferent each other.

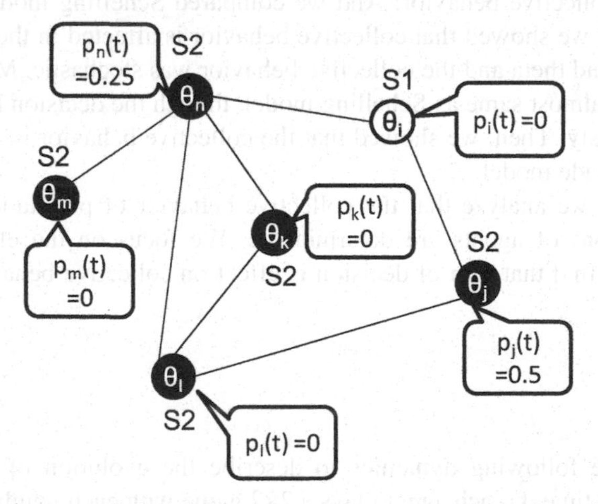

Fig. 1. Social network and the proportion of agent's neighbors who choose S1 at time *t*

If the proportion of neighbors $p_i(t)$ is higher than or equal to theta θ_i, then agent Ai's best response is to choose S1. Otherwise, Ai chooses S2. The value of theta θ_i, namely the degree of risk dominance of action S1, specifies a lower bound for the fraction of individuals that must be choosing S1 in order to make action S1 preferred to action S2. If $\theta_i < 1/2$, action S1 is risk dominant. Also, the more risk dominant action S1 is the lower the value of θ_i. This rule is given by these functions.

$$p_i(t) \geq \theta_i \text{ : Agent Ai chooses S1} \quad (1)$$

$$p_i(t) < \theta_i \text{ : Agent Ai chooses S2} \quad (2)$$

The model differs from cascade models which Watts or Lopez-Pintado deals with in some respects. All these features; simultaneity, heterogeneous rule, interactive interaction and network heterogeneity are essential to collective behavior.

1. Interactive interaction: Each agent can revise his behavior both of two alternatives, that is the decision is two ways. But, in cascade model, once an agent has switched on one alternative S1, it remains on S1 for the duration of the dynamics.
2. Simultaneity: Each agent decides his behavior depend on the neighbors' behavior in previous time step for each time step. While in cascade model, a certain probability agents are chosen each period to revise their strategy.
3. Heterogeneous networks: typically modeled on regular lattices, here we are concerned with heterogeneous networks; networks in which individuals have different numbers of neighbors.
4. Heterogeneous rules: Each agent has an idiosyncratic threshold. According to the threshold, agent's behavior is different from each other. That is, the decision rule is homogeneous in cascade model.

3 Former Results

3.1 Settings

In previous work [25], we dealt with these features except heterogeneous rule.

At first, we defused the interactive interaction and focus on the effect of two way interaction. That is, agent can only change the behavior for S1 and the decision rule is given by this function.

$$p_i(t) \geq \theta_i \text{ : Agent Ai chooses S1} \quad (3)$$

Next, we defused the simultaneity and focus on the effect of simultaneity. That is, all agents cannot decide at the same time, randomly chosen agent can only decide. This is cascade model.

At last, we dealt with a heterogeneous network. We make a social network for population of 1000 agents as scale-free network by arranging the regular network with degree of 10 using Kawachi algorism [26]. Then, their average degrees are 10. Kawachi proposed generation algorithm from regular network to various networks by each agent's with a link of the same number changing a link. That is, a node whose number of links is large must be much larger and a node whose number of links is small must be much smaller. When all links of each node have been considered once, the procedure is repeated several times. For scale-free network, we set probability as 1.0 and times as 20. The scale-free network is organized as shown in Fig. 2. And it is shown by log-log graph.

We set that all agents have same payoff matrix, that is, population is homogeneous. And thetas of all agents are set as 0.1. Then, considering pair of agents, Nash equilibrium is for both of them to choose S1 or for both of them to choose S2. So, for population, Nash equilibrium is for all agents to choose S1 or to choose S2.

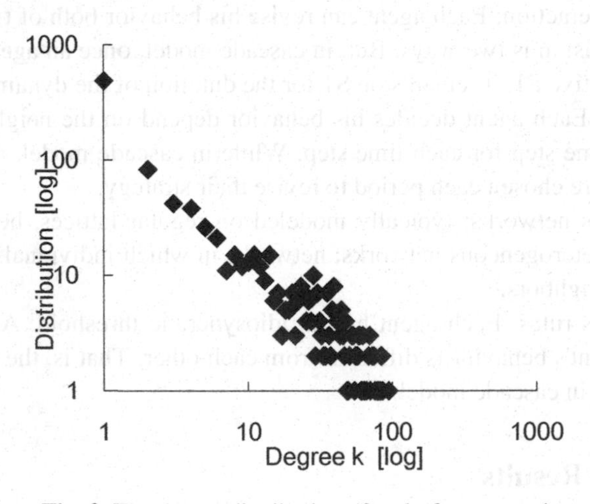

Fig. 2. The degree distribution of scale-free network

3.2 Results

We denote the collective behavior p(t) that the proportion of agents having chosen S1 in whole population at time *t*. Here, we set the initial collective behavior as 0.01, which means only small agents choose S1 at first time step. Watts [22] sets that a single node chooses S1 at first time step in 10000 nodes and simulates. If there are not node chooses S1 at all at first, choice of S1 doesn't spread according to function (3). A single node or small nodes need for cascade or contagion. And we assume that all agents choose at random at first time step. Then, each agent makes decision depend on rule of cascade model given by function (3) each time step, and then collective behavior turns.

Fig. 3 shows the simulation results in scale-free network, where theta is 0.1. When initial collective behavior is 0.01, which means low proportion, final collective behaviors depends on trials. And final collective behaviors sometimes converges to 0.01 and a few agents choose S1 at last, but other times converges to1.0 and all agents choose S1 at last. In latter case, there occur contagion and triggered by small proportion of S1 spread to the whole population, which is rare case.

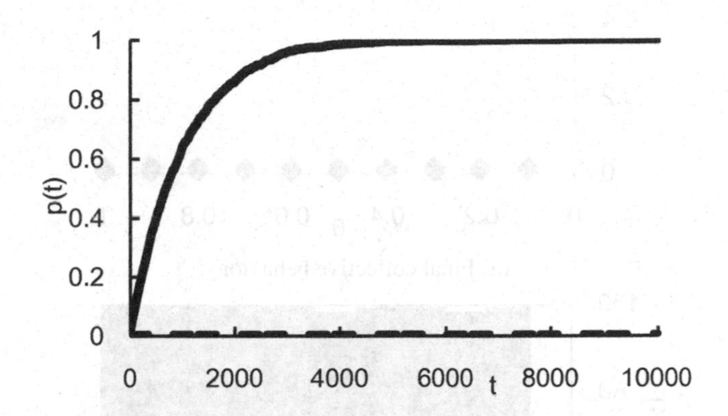

Fig. 3. The transition of collective behavior in scale-free network

We set theta θ at intervals of 0.1 from 0.0 to 1.0. We simulate until 10000 time step, we call this as a trial. Then, we simulate 100 trials per theta θ and investigate the final collective behavior. Fig. 4(a) shows the simulation results in scale-free network. The *x*-axis represents theta θ and the *y*-axis represents the final proportion of agents who choose S1, which we define final collective behavior as p ∗. And 100 trials are plotted for each theta θ. And the collective behavior depends on theta θ and becomes 0.01 or 1.0, stochastically. Because agent cannot change for S2, there few agents who choose S1 at first time step remain at last time step.

If theta is greater or than equal to 0.6, collective behavior converges to 0.01 and almost agents choose S2 at last in any trials. Otherwise theta is 0.0 collective behavior converges to 1.0 and all agents choose S1 at last. But, if theta is between 0.1 and 0.5, collective behaviors become 0.01 or 1.0 depending on the trials.

Other viewpoints of results are shown in Fig. 4(b). In this figure, the *x*-axis represents theta and the *y*-axis represents the histogram of the final proportion collective behavior p ∗. And white area means final collective behavior p ∗ become 1.0 and black area means final collective behavior p ∗ become 0.01. We found that when theta between 0.1 and 0.5; there are chances that all agents come to choose S1, though, it is rare case.

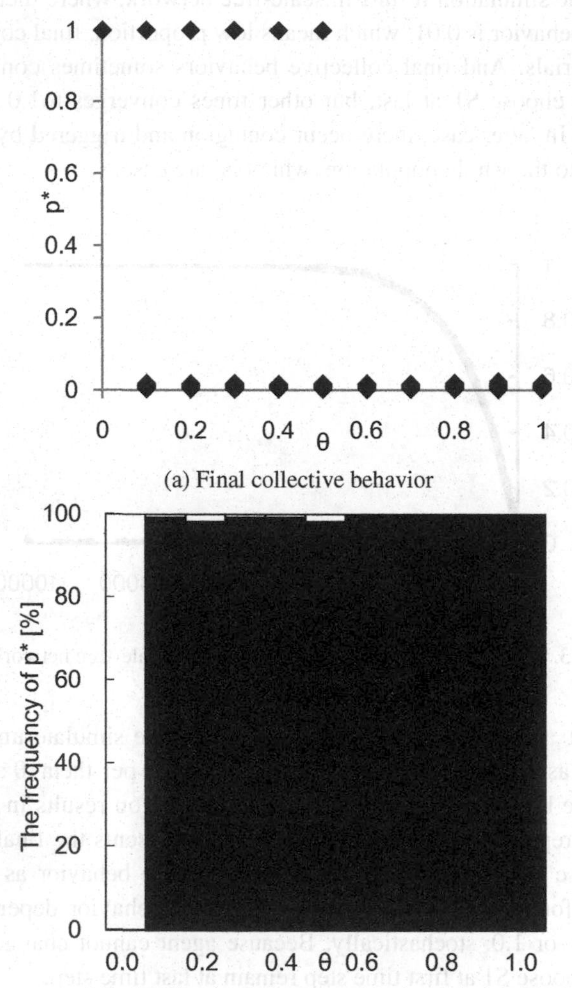

(a) Final collective behavior

(b) The histogram of final collective behavior: withe bar: p*=1.0, black bar: p*=0.01

Fig. 4. Collective behavior in scale-free network

On the other hands in regular network, the simulation result is similar to scale-free network as shown in Fig. 5. But, the range that collective behavior becomes 0.01 or 1.0 depending on the trials is narrower than scale-free network and the range is between 0.1 and 0.2.

We showed that collective behavior is affected in the structure of the social network and theta. We found that the collective behavior in Schelling model is similar to cascade model. In population, each agent has an idiosyncratic threshold or networks in which individuals have different numbers of neighbors.

Morris [13] deals with m-dimension lattice and shows that contagion of the action occurs below contagion threshold. And in a homogeneous network, where all nodes have the same connectivity, the contagion threshold equals the inverse of the connectivity k. And it is very approximate for our simulation results of regular network.

Moreover, López-Pintado by mean-field approach showed that contagion threshold of homogeneous network is low than that of scale-free network. They are very approximate for our simulation results.

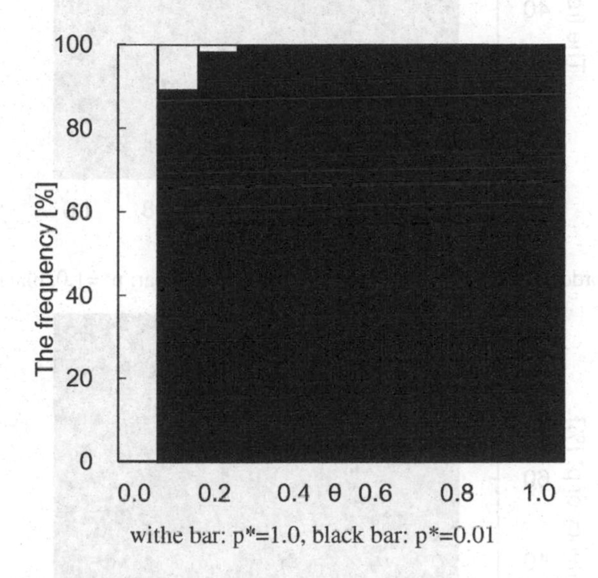

Fig. 5. The histogram of final collective behavior in regular network

4 Where Is Stochastic From

4.1 Initial Decision Depend on Network Degree

The collective behavior of population is stochastic, although decisions of agents are deterministic. We analyze this point. In previous simulation, agents decide at random at initial time step and each time step. Because there are agents with each degree in scale-free network, we focus on the effect of network degree. We set agents' initial behavior as ascending order or descending order with depending on degree. That is, in ascending order, agent with low degree chooses S1 preferentially at initial step. In descending order, agent with high degree chooses S1 preferentially at initial step.

We show the simulation results in Fig. 6. The characteristic of collective behavior is similar to the former results. We found that first decision is not effect on collective behavior.

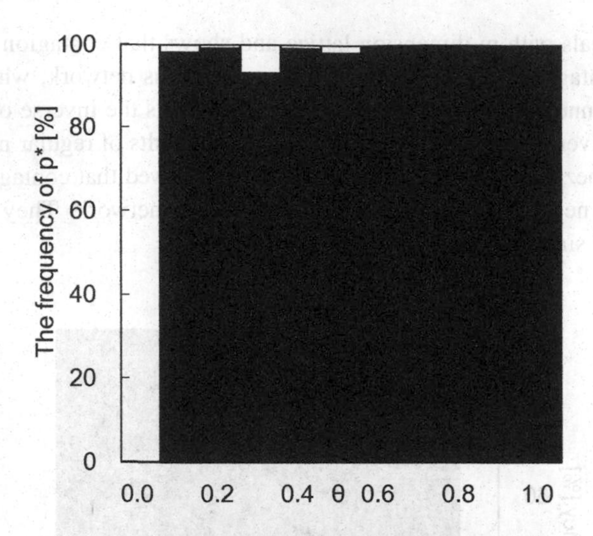

(a) Ascending order of network degree at initial step: withe bar: p*=1.0, black bar: p*=0.01

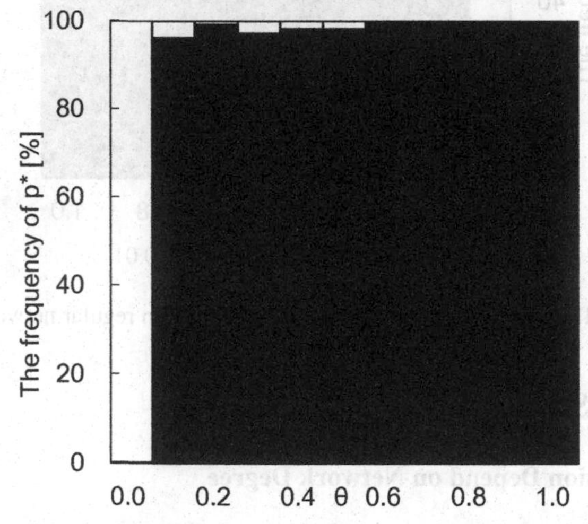

(b) Descending order of network degree at initial step: withe bar: p*=1.0, black bar: p*=0.01

Fig. 6. Collective behavior about initial decision in scale-free network

4.2 Turns Depend on Network Degree

We set agents' turn of choice as ascending order or descending order with depending on degree. That is, in ascending order, agent with low degree decides S1 preferentially each step. In descending order, agent with high degree decides S1 preferentially each step.

We show the simulation results in Fig. 7. The characteristic of collective behavior is different from the former results. We found that turn of decision is effect on collective behavior.

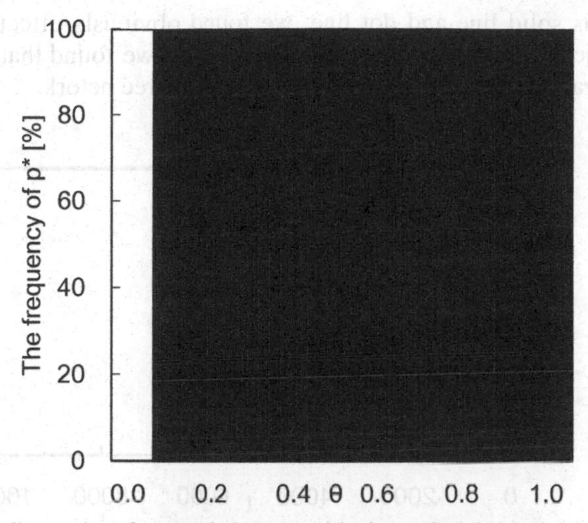

(a) Ascending order of network degree: withe bar: p*=1.0, black bar: p*=0.01

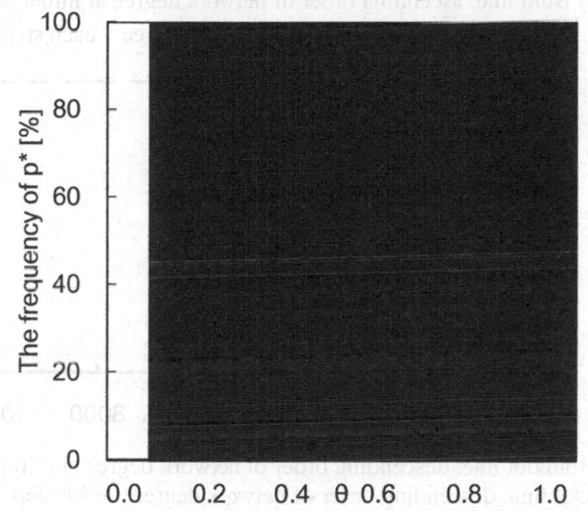

(b) Descending order of network degree: withe bar: p*=1.0, black bar: p*=0.01

Fig. 7. Collective behavior about turns of choice in scale-free network

Hasan et al. investigate effect of the initial seed on cascade propagation for uniform degree distribution. And they show that the local neighborhood will be relatively smaller than the affected cluster size which makes the cascade to propagate fast.

We also investigated the speed of contagion and show the results in Fig. 8. There are four case of contagion, ascending order of network degree at initial step and network degree each step, and descending order of network degree at initial step and network degree each step. In the first and the third case, that is bold line and bold dot line, the transitions are almost same as Fig. 3. And the speed of contagion is not effected the initial seed in scale-free network. On the other hand, in the second and the fourth case, that is solid line and dot line, we found obviously effect of the turns of choice on contagion speed in scale-free network. Then, we found that turns of choice effect on the congation speed than initial seed in scale –free netork.

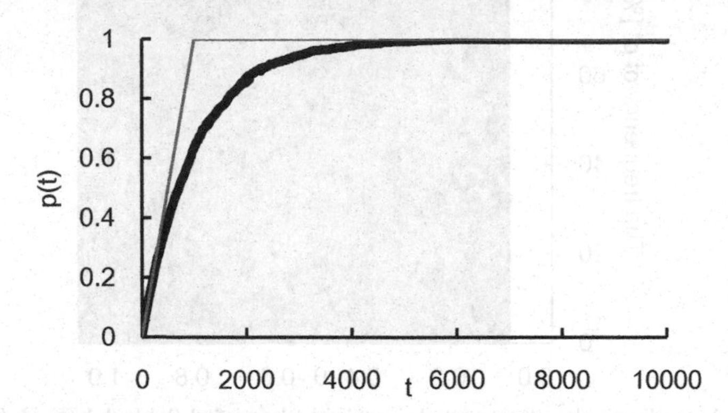

(a) Bold line: ascending order of network degree at initial step,
Slid line: ascending order of network degree each step

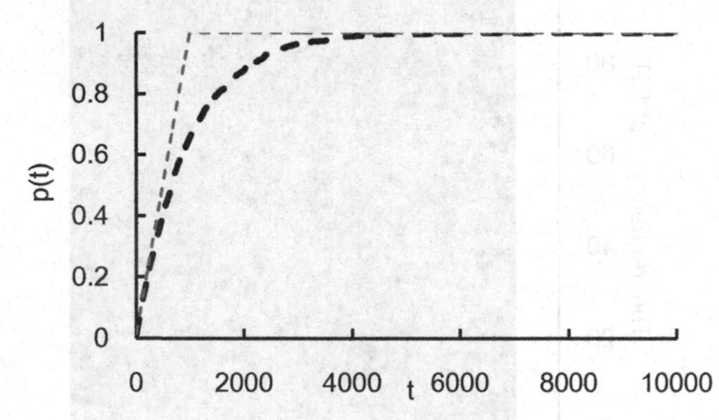

(b) Bold dot line: descending order of network degree at initial step,
Dot line: descending order of network degree each step

Fig. 8. Contagion speed of four cases

5 Conclusion

We showed that collective behavior is affected in the structure of the social network and theta, and the collective behavior was stochastic in previous work. Moreover, collective behavior is almost same as Schelling model, though the decision is not interactive and simultaneously. Then, we found that the collective behavior in Schelling model is similar to cascade model. That is, our results with heterogeneous rules or heterogeneous networks are possible to apply for cascade model.

In this paper, we analyzed that the collective behavior of population is stochastic, although decisions of agents are deterministic. We focused on the effect of network degree. And we found that turn of decision is effect on collective behavior not the first choices.

References

1. Huberman, B.A., Glance, N.S.: Diversity and Collective Action. Interdisciplinary Approaches to Nonlinear Complex Systems. Springer Series in Synergetics, vol. 62, pp. 44–64 (1993)
2. Sipper, M.: Evolution of Parallel Cellular Machines: The Cellular Programming Approach. Springer, Heidelberg (1997)
3. Schweitzer, F.: Brownian agent models for swarm and chemotactic interaction. In: Fifth German Workshop on Artificial Life. Abstracting and Synthesizing the Principles of Living Systems, pp. 181–190 (2002)
4. Kirman, A.P.: The Economy as an Interactive System. In: Arthur, W.B., Durlauf, S.N., Lane, D. (eds.) The Economy as an Evolving Complex System II, Perseus (1997)
5. Feng, L., Li, B., Podobnik, B., Preis, T., Stanley, H.E.: Linking agent-based models and stochastic models of financial markets. National Academy of Sciences of the United States of America (PNAS) 109(2), 8388–8393 (2012)
6. Kenett, D.Y., Raddant, M., Lux, T., Ben-Jacob, E.: Evolvement of Uniformity and Volatility in the Stressed Global Financial Village. PLoS One 7(2), e31144 (2012)
7. Schelling, T.: Micromotives and Macrobehavior, Norton (1978)
8. Arthur, W.B.: Inductive reasoning and bounded rationality. American Economic Review 84, 406–411 (1994)
9. Hansarnyi, J., Selten, R.: A Game Theory of Equilibrium Selection in Games. MIT Press (1988)
10. Fudenberg, D., Levine, D.: The Theory of Learning in Games. The MIT Press (1998)
11. Rubinstein, A.: Modeling Bounded Rationality. The MIT Press (1998)
12. Arthur, W.B.: Competing Technologies, Increasing Returns, and Lock-In by Historical Events. The Economic Journal 99(394), 116–131 (1989)
13. Morris, S.: Contagion. Review of Economic Studies 67(1), 57–78 (2000)
14. López-Pintado, D.: Contagion and coordination in random networks. International Journal of Game Theory 34(3), 371–381 (2006)
15. Watts, D.J.: A simple model of global cascades on random networks. National Academy of Sciences of the United States of America (PNAS) 99(9), 5766–5771 (2002)
16. Young, H.P.: The dynamics of social innovation. National Academy of Sciences of the United States of America (PNAS) 108(4), 21285–21291 (2011)
17. Montanari, A., Saberi, A.: The spread of innovation in social networks. National Academy of Sciences of the United States of America (PNAS) 107(47), 20196–20201 (2010)
18. Komatsu, T., Namatame, A.: Evolutionary optimized network topology for maximizing cascade. In: The Special Interest Group Technical Reports of Processing Society of Japan (IPSJ SIG Technical Report), 2012OICS-166(3), 1–6 (2012)
19. Nowak, M.A.: Evolutionary Dynamics: Exploring the Equations of Life. Belknap Press of Harvard University Press (2006)
20. Axelrod, R.: The Evolution of Cooperation. Basic Books (1985)
21. Albert, R., Barabási, A.L.: Topology of evolving networks: Local events and universality. Physical Review Letters 85(24), 5234–5237 (2000)
22. Watts, D.J.: Small-Worlds: The Dynamics of Networks between Order and Randomness. Princeton University Press (1999)
23. Hasan, S., Ukkusuri, S.V.: A threshold model of social contagion process for evacuation decision making. Transportation Research Part B: Methodological 45(10), 1590–1605 (2011)

24. Iwanaga, S., Namatame, A.: Collective behavior and diverse social network. International Journal of Advancements in Computing Technology 4(22), 320–321 (2012)
25. Iwanaga, S., Namatame, A.: Collective Behavior in Cascade and Schelling Model. In: 17th Asia Pacific Symposium on Intelligent and Evolutionary Systems (IES 2013), vol. 24, pp. 217–226 (2013); Procedia Computer Science
26. Kawachi, Y., Murata, K., Yoshii, S., Kakazu, Y.: The structural phase transition among fixed cardinal networks. In: the 7th Asia-Pacific Conference on Complex Systems, pp. 247–255 (2004)

Improved Performance of a Cooperative Genetic Algorithm When Solutions Were Presented as Cartoon Faces

Sean R. Green and Joshua S. Redford

Department of Psychology, University at Buffalo-SIM Programme, Singapore
{srgreen,redford}@buffalo.edu

Abstract. Genetic algorithms are often able to identify solutions within a multidimensional problem space that elude human detection. Humans sometimes have difficulty retrospectively identifying the features or feature combinations that lead to the success of an individual solution. However, humans are adept at identifying configural patterns within faces. We hypothesized that mapping the features of a problem space onto a cartoon face space would help humans visualize patterns and contribute to the operation of a genetic algorithm.

In this study, human participants viewed foragers in a virtual environment. The genetic code of each forager was presented either as a list of features or as a cartoon face. Participants selected either one or both parents for mating at each generation. Foraging improved significantly when forager attributes were presented as a face, compared to trials in which attributes were presented as a list. This improvement was found for both human-directed and cooperative genetic algorithms.

Keywords: evolutionary, human-computer interaction, perception, decision-making.

1 Introduction

A genetic algorithm is a problem-solving technique in which solutions to a problem are represented as sequences of genes, and in which the genes are subject to an evolutionary selection process to arrive at a solution [see 1]. They have proven a successful tool for identifying novel solutions to problems that humans may overlook, particularly when the solution to the problem relies upon a large number of features within a multidimensional problem space. However, systems that arise through evolutionary processes can be complex, and the evolutionary process is often a challenge to describe and interpret. Human efforts to retrospectively solve and learn from these evolutionary adaptations are often time-consuming and cognitively challenging, though computer simulations can make these problems more tractable (e.g., [2-3]).

H. Handa et al. (eds.), *Proc. of the 18th Asia Pacific Symp. on Intell. & Evol. Systems – Vol. 1*,
Proceedings in Adaptation, Learning and Optimization 1, DOI: 10.1007/978-3-319-13359-1_4

The complexity of solutions offered by genetic algorithms can interfere with human-computer interaction. As Kowaliw et al. [4] have noted, the number of variables to consider may overwhelm the human ability to keep track of them. This may not be a concern if a genetic algorithm's solution can be applied to a problem with minimal human involvement, but many problems require the user to interact with or retrospectively interpret genetic algorithm outputs. This could occur because some aspects of the problem (e.g. the aesthetic desirability of a product design) cannot be evaluated automatically, because the user needs to draw general conclusions from the solutions, or simply because the number of solutions is large and the user needs to narrow the set of possibilities ([5]; see also [6-7]). In many situations, users benefit not only by determining which solutions are best but also by attaining some theoretical or intuitive sense of why the solutions work best.

One method for facilitating the interaction between human and computer is by mapping potentially unwieldy data onto a recognizable, tractable pattern [e.g., 8]. Facial configurations are ideal for this purpose as humans are adept at interpreting features on a human or cartoon face [9].

The study was organized in the following way. A foraging task was devised in which a simple object (a grey circle) moved around in an environment containing food objects (green squares). The direction and speed of the forager was determined by a 31-bit binary code. The code for each forager was converted into a list of features and, separately, a cartoon face. Each feature from the list was mapped onto a facial feature, so that both the list and the face represented the same information. Participants in the study observed foragers and, after watching 25 candidate foragers for each population, selected either both parents (the Human condition) or one parent (the Hybrid condition). In the Hybrid condition, the forager that quantitatively found the most food was selected as the second parent. The Human condition was included to evaluate the effectiveness of the face in helping humans evaluate the foragers, while the Hybrid condition was included to evaluate the effect of human-computer interactions on performance.

The paper was organized as follows: Section 2 reviews relevant literature and presents a context for the present study; Section 3 describes the forager, foraging environment, and genetic algorithm. Section 4 describes task performed by human participants and the procedure for human-computer interaction; Section 5 describes the statistical analysis, and Section 6 discusses the results; Section 7 considers the implications of the current findings and the potential for future research.

2 Related Work

Over the past two decades, many systems have allowed humans to contribute to a genetic algorithm or similar evolutionary process. Some allow the user to add candidates to the population in place of automated selection [10], while others have the user shape the parameters of the selection process or both [11]. Takagi [12] used human evaluation within psychological space as a surrogate for automated fitness evaluation within an evolutionary algorithm. Subsequent studies [13-14] used 2D visualizations of a genetic algorithm to aid the human decision maker. Other alternative approaches to allowing human decisions to shape the evolutionary process include crowdsourcing the evaluations [15] or creating a model of human decision making [16].

Several researchers have applied genetic algorithms to the search of face spaces. Caldwell and Johnston [17], considered witness identification as a search through a face space. Secretan et al. [18] used an evolutionary breeding process to allow users to develop pictures, some of which contained faces and face-like elements. To our knowledge, the current study is the first to employ a face space to represent attributes of a problem space that are not intrinsic properties of faces in the context of a genetic algorithm.

3 Methods

3.1 The Foraging Task and Environment

Foraging took place within a 300 x 300 pixel white square. The forager was depicted as a grey circle with a 25 pixel radius that started in the centre of the foraging space. Twenty five food items (represented by green squares 25 pixels on a side) were placed at random locations within the foraging area at the start of each trial. During each time step in the simulation the forager moves in one of eight directions (north, south, east, west and appropriate combinations thereof). Any food items within a given radius of the forager at the endpoint of motion are considered consumed. Onscreen, they flashed red and disappeared. Food items were replaced in a random location within the foraging area before the next time step. After 25 time steps, the trial ended. Total food consumed in the 25 time steps was recorded as an indicator of fitness.

3.2 The Forager

Foraging behaviour was defined by a 31-bit genetic code that determined the following forager attributes (see Table 1 for a summary and Figure 1 for a screenshot):

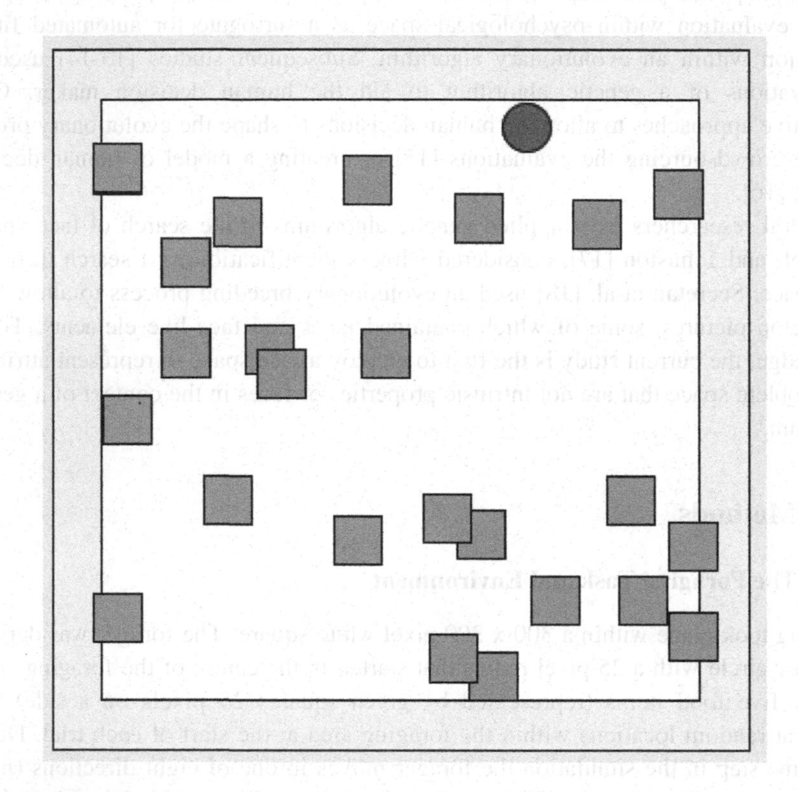

Fig. 1. The foraging screen, showing the forager (circle, grey) in the upper right and food items (square, green)

Visual Field Width. Each eye detects food within a region extending outward from the centre of the forager and reaching to the edge of the foraging area. The width of this region in degrees ranged from 10-40 degrees, with the value equal to 10 plus twice the value of the binary digit defined by genes 1-4. For example, if genes 1-4 were (0110), the Visual Field Width would be 10+8+4 or 22.

Antennae. Each forager possessed antennae which detected food omnidirectionally within a defined radius from the forager. Genes 5-8 code for the antennae. The antenna detection radius was 8 times the value of the four-bit binary (ranging from 0 to 120 pixels).

Reach. (Mouth Size) Reach determines how close a forager has to be to a food item in order for it to be eaten. The forager's range is coded by genes 9-12, and ranged from 10-40 in the same manner as visual field width.

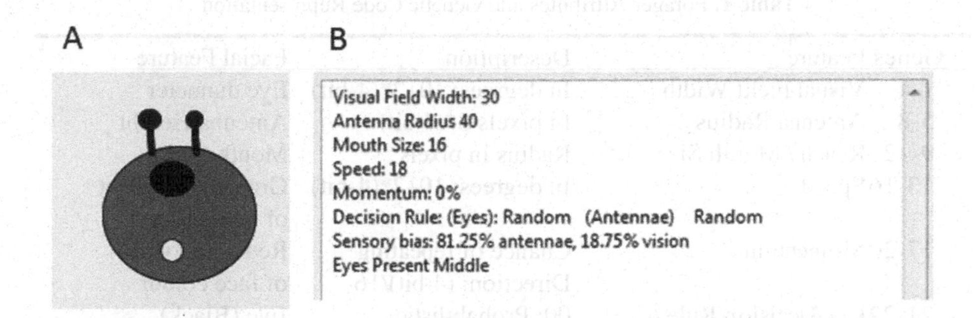

A
B
Visual Field Width: 30
Antenna Radius 40
Mouth Size: 16
Speed: 18
Momentum: 0%
Decision Rule: (Eyes): Random (Antennae) Random
Sensory bias: 81.25% antennae, 18.75% vision
Eyes Present Middle

Fig. 2. Examples of forager attribute presentation in the (a) face and (b) list conditions. Note that the colour of the face in this example was blue.

Speed. Speed indicates the distance that the forager travels per cycle. Speed ranged from 10 to 40 pixels, according to genes 13-16, analogous to visual field width. For the sake of simplicity and avoiding rounding errors, diagonal movement consisted of movement in the horizontal axis followed by movement in the vertical rather than Euclidean diagonal motion. This means that diagonal motion covered a greater distance by a factor of a square root of two.

Momentum. Momentum determined the likelihood that the forager would go in a particular direction once it had gone in that direction in the previous cycle. Genes 17 through 20 coded for momentum, which ranged from 0 to 15. Each cycle, if a random integer between 1 and 16 was less than momentum, inputs from the eyes and antennae were ignored, and the forager continued moving in the same direction.

Eye Decision Rule. Probabilistic: If genes 21 and 22 were both 0, the eye direction was chosen by selecting one unit of activity at random, so that direction was weighted in favour of the directions in which more food items were detected. For example, if the forager has two eyes, and 4 food items are within view of one eye and one food item is within view of the second, then the decision rule would have an 80% chance of indicating motion along the line of sight of the first eye and a 20% chance of motion along the light of sight of the second eye. If the forager had no eyes or could not detect food along the line of sight of any of its eyes, a random direction of motion was chosen.

Summation: If genes 21 and 22 were either 10 or 01, the direction with the highest activity is chosen. In the example above, the forager would move along the line of sight of the first eye.

Opponent: If genes 21 and 22 were both 1, then each eye activated two opposing directions as described above. If no eyes were present, a random direction was selected. If at least one pair of directions was active, a decision rule was applied to determine the direction that the eyes would lead the forager to move toward.

Table 1. Forager Attributes and Genetic Code Representation

Genes	Feature	Description[a]	Facial Feature
1-4	Visual Field Width	In degrees: 10+2*(4-bit)	Eye diameter
5-8	Antenna Radius	In pixels 8*(4-bit)	Antenna Height
9-12	Reach / Mouth Size	Radius in pixels	Mouth Size
13-16	Speed	In degrees: 10+2*(4-bit)	Green component of face colour
17-20	Momentum	Chance of repeating Direction: (4-bit)/16	Red component of face colour
21-22	Eye Decision Rule	00: Probabilistic 01/10: Winner-take-all 11 – Opponent Process	Eye (Black) Eye (Grey)
23-24	Antenna Decision Rule	00: Probabilistic 01/10: Winner-take-all 11 – Opponent Process	Tip (Black) Tip (Grey) Tip (White)
25-28	Sensory Bias	Chance of using antenna input (4-bit)/16	Red component of face colour
29-31	Eye Presence	Left / Middle / Right	Eye visible

[a]bit value in parenthesis indicates a binary value derived from genes (e.g.: 0110 = 6).

Antennae Decision Rule. Genes 23 and 24 coded for the antenna decision rule. The computation for the Probabilistic and Winner-take-all decision rules were the same as the eye decision rule, but with the antenna inputs. Note that the antennae could detect motion in all eight directions rather than three. This was done by counting the number of food items on one side of each of the forager's axes of symmetry. For the Opponent condition, the direction in which the number of food items detected exceeded the average of its neighbours (e.g. north vs northeast and northwest) was chosen.

Sensory Bias. In cases where the eye and antenna directions conflict, one is chosen over the other. Genes 25 through 28 code for sensory bias in the same manner as for momentum. As with momentum, sensory bias ranges from 0 (in which the vision is always favoured), to 15, in which there is a 15 in 16 chance of favouring the antennae. The range of bias values slightly favoured vision. However, blind foragers are represented, as 12.5% of the initial population were expected to lack eyes (see below).

Eyes Present. Genes 29-31 represented the left, middle and right eyes respectively. The middle of the line of sight for the eyes were 45° counter-clockwise, 0°, and 45° clockwise relative to the direction of the forager's last motion (i.e., the direction it was facing). The eye was present when the gene was 1 and absent when it was 0.

Note that the mappings between the face stimulus and the foraging attribute that it represents may be more intuitive in some cases (e.g. mouth size) and less intuitive in others (e.g., face colour). The individual contribution of specific face attributes to performance was not assessed in this study but remains an avenue for future research.

3.3 Genetic Algorithm

Selection was done in one of the following two ways depending on the experimental condition:

In the human selection condition, the participants selected the two foragers that they believed to be most effective based on observation of their behaviours and observation of a representation of each forager's genetic code. These foragers became parents to the next generation. In the hybrid condition, the participants selected one forager and that selection was paired with the forager with the highest fitness, where fitness was defined as the number of food items that it consumed during the 25 turns.

A cut point was randomly selected from 0 to 31 for each child forager. The genes from one of the parents (randomly determined) were transmitted to the child up to the cut point. Genes past the cut point were transmitted from the other parent. Two random genes were flipped from 0 to 1 or from 1 to 0.

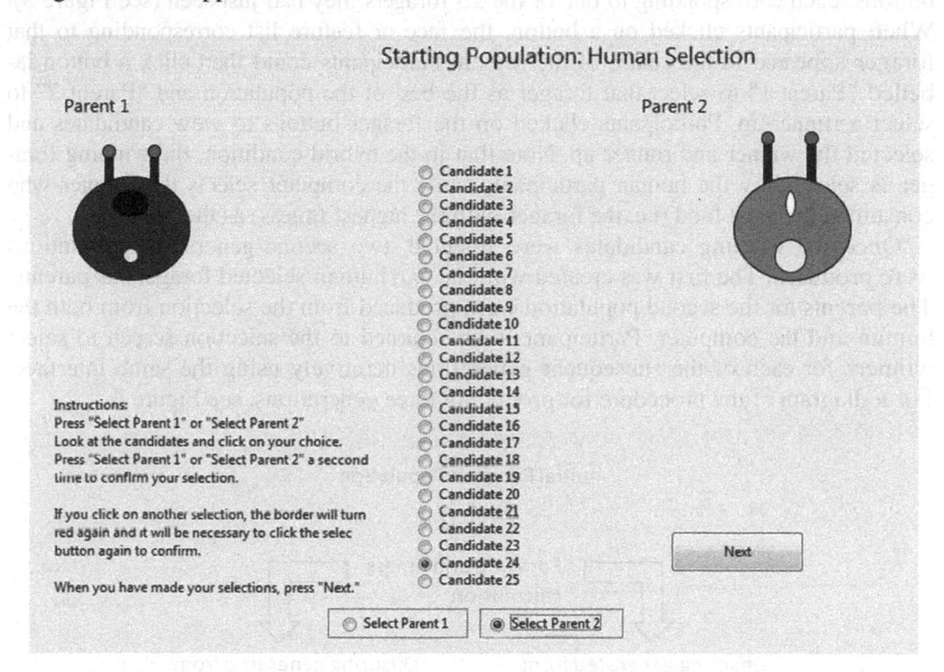

Fig. 3. The selection screen

4 Experimental Procedure

The participants in this study were 46 undergraduate University at Buffalo students in Singapore who participated in exchange for credit toward a psychology course. Participants were seated a comfortable distance from a computer. They watched an animation showing the foraging screen and foraging process and subsequently explained each of the forager attributes. Next, participants watched five practice trials in which a random forager searches for 25 time steps. In the "face" condition, the face was shown for each, and in the list condition, the list of attributes was shown for each. Participants were instructed to identify good foragers.

The process of the genetic algorithm was explained to the participants, so they were aware that the foragers they chose would be the starting point for creating a new generation of foragers.

Participants were shown five practice trials in which a forager moved for 25 time steps within the foraging screen (See Figure 1). In the "face" condition, participants were told that the features would be illustrated using a cartoon face, while in the "list" condition, participants would be shown a list of features for each forager.

Participants then observed 25 foragers, sequentially, as each searched for 25 time steps. As per the instructions, the face and list were present alongside the foraging screen in their respective conditions. All participants participated in each trial. Twenty-two participants encountered the list trials first, while twenty four encountered the face trials first.

Following the foraging sequence, participants were shown a selection screen with 25 buttons, each corresponding to one of the 25 foragers they had just seen (see Figure 3). When participants clicked on a button, the face or feature list corresponding to that forager appeared in the centre of the screen. Participants could then click a button labelled "Parent 1" to select that forager as the best of the population and "Parent 2" to select a runner-up. Participants clicked on the forager buttons to view candidates and selected the winner and runner up. Note that in the hybrid condition, the winning forager is selected by the human participant, while the computer selects the forager who consumed the most food (i.e. the forager with the highest fitness) as the runner-up.

Once the winning candidates were selected, two second-generation populations were produced. The first was created with the two human-selected foragers as parents. The parents for the second population were produced from the selection from both the human and the computer. Participants then returned to the selection screen to select winners for each of the subsequent generations iteratively using the same interface. For a diagram of the procedure for producing three generations, see Figure 4.

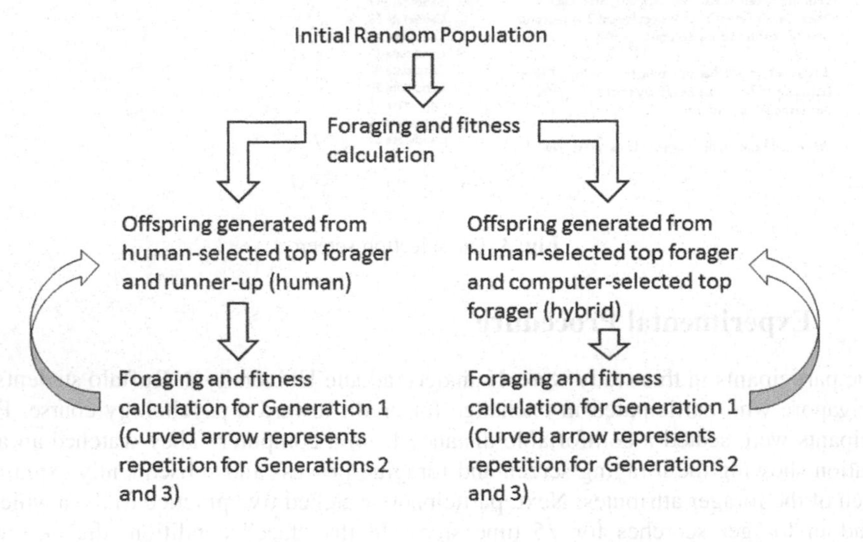

Fig. 4. Trial sequence for genetic algorithm procedure

5 Analysis

Success of the algorithm for each new generation was measured as the mean fitness for all 25 members of the generation. As the parent generation was random, each participant's list and face condition began from a different starting population. To account for this, a difference score was computed by subtracting the mean fitness of the parent (Generation 0) population from the mean fitness of each child population. Comparisons were made between the three conditions at each generation and also between corresponding conditions from the "face" and "list" trial blocks.

A three-way (Presentation, Selection, and Generation) repeated measures analysis of variance was conducted in SPSS. The factor of Presentation refers to how the forager features were presented (i.e., via a list or a face) and Selection refers to whether parents were selected by the human participant or a combination of the two.

Two hypotheses are considered:

1) The hybrid condition produced more effective foragers than the human condition illustrating the potential importance and benefit of human-computer cooperation.

2) Presenting features as a face produced more effective foragers than features presented as a list in both the human and hybrid selection condition.

6 Results

A three-way repeated-measures ANOVA of the difference scores revealed significant effects of Presentation ($F(1, 45) = 8.01$, $MSE = 46.28$ $p < .01$, partial $\eta^2 = 0.15$), Generation ($F(1, 45) = 49.71$, $MSE = 2.87$, $p < .001$, partial $\eta^2 = 0.53$) and Selection ($F(1, 45) = 49.45$, $MSE = 19.35$, $p < .001$, partial $\eta^2 = 0.52$). A significant interaction was observed between Selection and Generation (F(2, 90) = 10.20, $MSE = 3.97$ $p < .001$, partial $\eta^2 = 0.185$). No significant Presentation x Selection ($F(1,45) < 1$), or Presentation x Generation ($F(2,90) = 1.1$) interaction, nor was any three-way interaction found ($F(2,90) < 1$).. As shown in Table 1, performance significantly declined in the human selection condition when forager features were presented as a list, but significantly improved when features were presented as a face. In the hybrid condition, a significant increase in fitness was found for both face and list presentations. Figure 5 shows difference scores between each of the three child generations and the parent generation (Generation 0) for the hybrid trials, while Figure 6 shows (on the same scale) difference scores for the human trials. The significant interaction between Selection and Generation was explored with post-hoc t-tests (shown in Table 2). When fitness scores were collapsed across face and list conditions, the human trials showed no significant average improvement (as the gains in the face condition were presumably cancelled out the decrement in the list condition), while the hybrid condition showed fitness gains from the first child generation.

The magnitude of the improvement was similar across conditions, suggesting that the problem in the list condition may have been an initial difficulty in selecting appropriate foragers in the initial parent generation. Alternatively, the recovery in the list condition could reflect regression to the mean rather than genuine improvement. Selections made in generations two and three may have, by chance, mitigated maladaptive choices in the first generation. No such decrement was found in the face condition, suggesting that participants were (a) able to identify good foragers (or at least avoid selecting bad ones) early on, and (b) were able to build upon these initial selections in generations two and three to improve fitness.

Table 2. Planned *t* tests comparing the fitness of child generations to the parent generation, collapsed across three generations

Presentation	Selection	Mean	SD	*t*(137)	*P*
Face	Generation 0	7.52	1.07	-	-
	Human	8.42	3.45	3.00	0.003
	Hybrid	10.92	4.05	10.06	<0.0001
List	Generation 0	8.02	1.28	-	
	Human	7.16	3.74	-2.61	0.01
	Hybrid	9.92	3.87	5.92	<0.0001

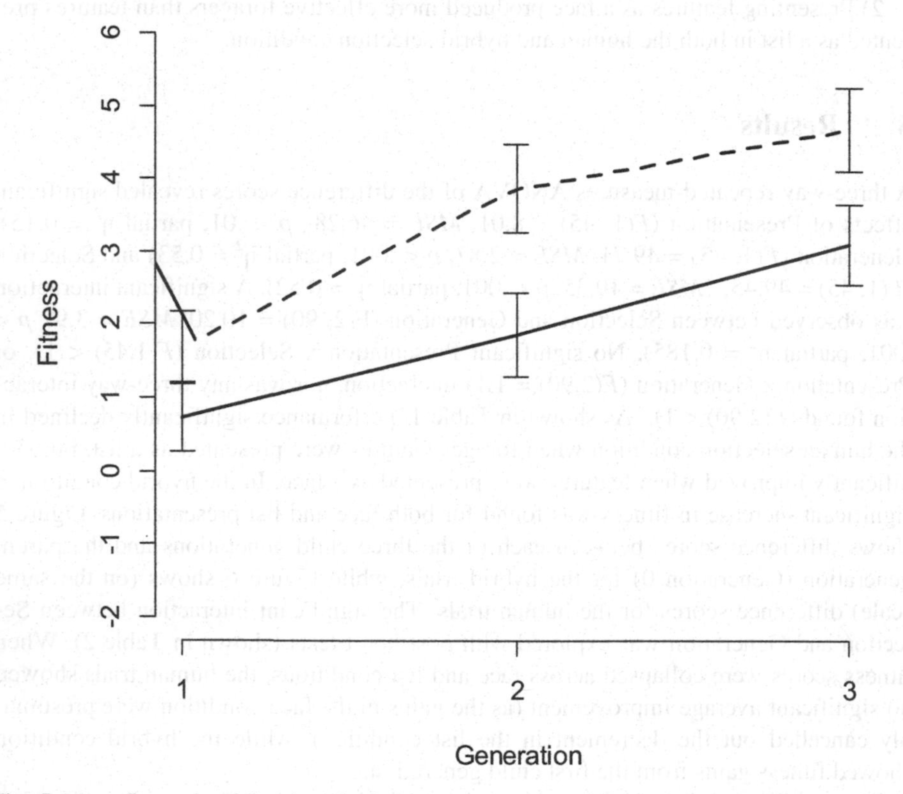

Fig. 5. Mean fitness in trials where both parents were selected through human-computer cooperation (hybrid condition). Error bars indicate standard error.

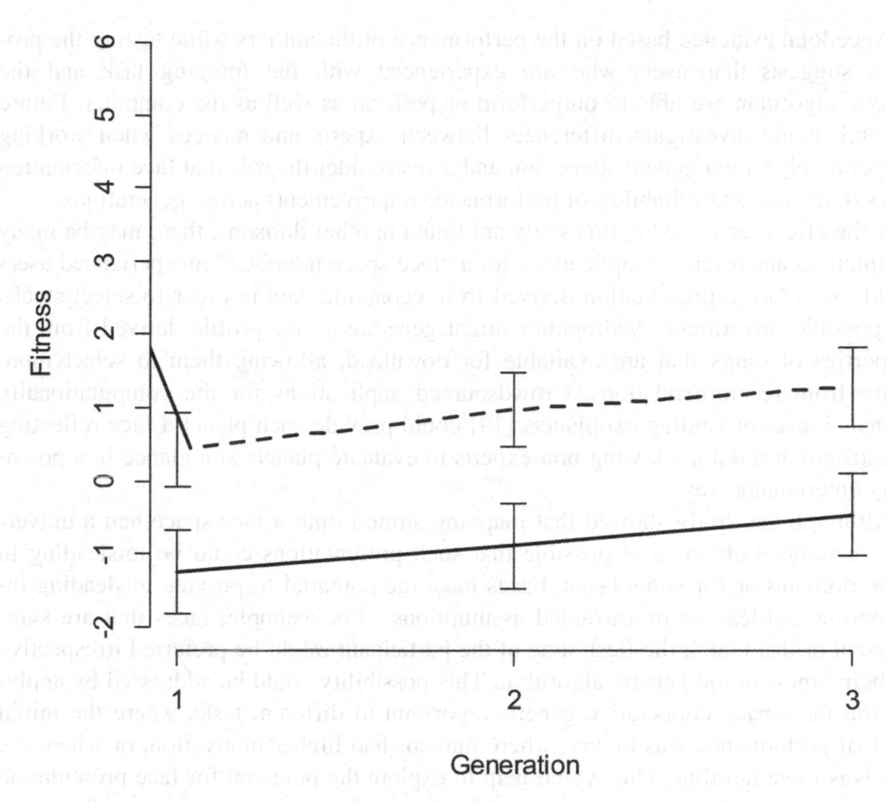

Fig. 6. Mean fitness for each generation in trials where both parents were selected by humans (human only condition). Error bars indicate standard error.

7 Conclusion

Presentation of genetic information as faces led to increased fitness in conditions that featured both human-only and cooperative genetic algorithms. Face recognition depends on the holistic interpretation of numerous features that change depending on viewpoint, distance, occlusion and other factors [9]. Presenting data as a face may have allowed participants to perceive or at least respond to combinations of features in parallel, improving genetic algorithm speed and effectiveness.

These results suggested that the computer's performance overshadowed human performance overall, so improved performance in the cooperative condition could simply reflect the effectiveness of the computer's selection criteria rather than a synergistic effect of human-computer interaction. However, the advantage of the face condition versus list carried over into the cooperative conditions, suggesting that participant selections based on the facial features are not wholly incongruent with the computer selections.

Anecdotal evidence based on the performance of the authors while testing the program suggests that users who are experienced with the foraging task and the genetic algorithm are able to outperform or perform as well as the computer. Future research could investigate differences between experts and novices when working cooperatively with a genetic algorithm and also consider the role that face information plays in the rate and reliability of performance improvements across generations.

If the effects observed in this study are found in other domains, there may be many commercial and research applications for a "face-space interface" Inexperienced users could use a face representation derived from economic data in order to select stocks for possible investment. Audiophiles might generate a face profile derived from the properties of songs that are available for download, allowing them to select more easily from recommendations. Crowdsourced applications for the computationally intensive tasks of finding exoplanets [19] could provide each planet a face reflecting its astronomical data, allowing non-experts to evaluate planets at a glance in a potentially entertaining way.

Although this study showed that mapping stimuli onto a face space had a universally beneficial effect, it is possible that such presentations could be misleading in some domains or for some tasks. Faces have the potential to provide misleading information and lead to unwarranted assumptions. For example, faces that are symmetrical or that match the flesh tone of the participant might be preferred irrespective of their fitness in the genetic algorithm. This possibility could be addressed by applying the face-space cooperative genetic algorithm to different tasks where the initial level of performance was higher, where humans had higher motivation, or where the task was more familiar. This would help to explore the potential for face presentation to synergistically improve human-computer interaction and in the process allow subtle biases associated with face processing to potentially impact performance.

Future research might address whether an intuitive mapping between the face attributes and the attributes of a genetic algorithm contributes to the benefits of face-space mapping. In this study, some mappings were presumed to be intuitive (e.g., a large mouth signifies an increased ability to consume food), while others were less intuitive or not intuitive (e.g., green signifying speed). Human sensory and perceptual processing tends to be most effective when it is applied to a domain that it evolved to deal with, so it is reasonable to expect that we might be biologically prepared to make use of some mappings but not others.

However, numerous lines of research, ranging from sensory substitution [20] to the recovery of visual segmentation [21] and stereopsis [22] show that human cognition is plastic. With the advent of human-computer interaction, augmented reality and brain computer interfaces (BCI), humans may exploit this plasticity to improve decision-making and executive functions by offloading some of the cognitive load to brain areas that have not previously had a domain-general cognitive role. As computers become faster and more complex, representing information so that it suits human perceptual capabilities is one step toward facilitating more efficient human-computer interaction.

References

1. Holland, J.H.: Adaptation in Natural and Artificial Systems. University of Michigan Press (1975)
2. De Jong, H.: Modeling and simulation of genetic regulatory systems: a literature review. Journal of Computational Biology 9(1), 67–103 (2000)
3. Dill, K.A., Ozkan, S.B., Weikl, T.R., Chodera, J.D., Voelz, V.A.: The protein folding problem: when will it be solved? Current Opinion in Structural Biology 17, 342–346 (2007)
4. Kowaliw, T., Dorin, A., McCormack, J.: Promoting creative design in interactive evolutionary computation. IEEE Transactions on Evolutionary Computation 16(4), 523 (2012)
5. Battiti, R., Passerini, A.: Brain-computer evolutionary multi-objective optimization (PC-EMO): a genetic algorithm adapting to the decision making. IEEE Transactions of Evolutionary Computation 14(5), 671–687 (2010)
6. Noda, E., Freitas, A.A., Lopes, H.S.: Discovering interesting prediction rules with a genetic algorithm. In: Proceedings of the 1998 Congress on Evolutionary Computation, vol. 2 (1999)
7. Dehuri, S., Mall, R.: Predictive and comprehensible rule discovery using a multi-objective genetic algorithm. Knowledge Based Systems 19, 413–421 (2006)
8. Domas, C.: The 1s and 0s behind cyber warfare (2013), `http://www.ted.com/talks/chris_domas_the_1s_and_0s_behind_cyber_warfare`
9. Farah, M.J., Wilson, K.D., Drain, N., Tanaka, J.N.: What is "special" about face perception. Psychological Review 105(3), 482–498 (1998)
10. Qian, Z.Q., Teng, H.F., Xiong, D.L., Sun, Z.G.: Human-computer cooperation genetic algorithm and its application to layout design. In: Proceeding of the 4th Asia-Pacific Conference on Simulated Evolution and Learning, Singapore, pp. 299–302 (2002)
11. Kosorukoff, A.: Human based genetic algorithm. In: IEEE International Conference on Systems, Man and Cybernetics, vol. 5, pp. 3464–3469 (2001)
12. Takagi, H.: Interactive evolutionary computation: system optimization based on human subjective evaluation. In: IEEE International Conference on Intelligent Engineering Systems (INES 1998), pp. 17–19 (1998)
13. Boschetti, F., Takagi, H.: Visualization of EC landscape to accelerate EC conversion and evaluation of its effect. In: Proceedings of the 2001 Congress on Evolutionary Computation, vol. 2, pp. 880–886 (2001)
14. Hayashida, N., Takagi, H.: Acceleration of EC convergence with landscape visualization and human intervention. Applied Soft Computing 1(4), 242–256 (2002)
15. Bao, J., Sakamoto, G., Nickerson, J.V.: Evaluating design solutions using crowds. In: Proceedings of the Seventeenth Americas Conference on Information Systems, Detroit, Michigan (2011)
16. Guo, Y.-.N., Cheng, J.: Adaptive evaluation strategy based on surrogate models. In: Maurtua, I. (ed.) Human Computer Interaction, pp. 259–277. InTech (2007)
17. Caldwell, C., Johnston, V.S.: Tracking a criminal suspect through face space with a genetic algorithm. In: Proceedings of the Fourth International Conference on Genetic Algorithm, pp. 416–421 (1991)
18. Secretan, J., Beato, N., D'Ambrosio, D.B., Rodriguez, A., Campbell, A., Stanley, K.O.: Picbreeder: evolving pictures collaboratively online. In: Proceedings of the Computer Human Interaction Conference, pp. 1759–1768 (2008)
19. Stump, C.: Project PANOPTES: crowdsourcing the search for exoplanets. In: American Astronomical Society Meeting Abstracts 22,

20. Bach-y-Rita, P., Kercel, S.W.: Sensory substitution and the human-machine interface. Trends in Cognitive Sciences 7(12), 541–546 (2003)
21. Ostrovsky, Y., Meyers, E., Ganesh, S., Sinha, P.: Visual parsing after recovery from blindness. Psychological Science 20, 1484–1491 (2009)
22. Barry, S.: Fixing my Gaze, p. 272. Basic Books (2009)

Parallel Particle Swarm Optimization Using Message Passing Interface*

Guang-Wei Zhang[1,2,3], Zhi-Hui Zhan[1,2,3,**], Ke-Jing Du[4], Ying Lin[2,3,5], Wei-Neng Chen[2,3,4], Jing-Jing Li[6], and Jun Zhang[1,2,3,4]

[1] Department of Computer Science, Sun Yat-sen University, China, 510275
[2] Key Lab. Machine Intelligence and Advanced Computing, Ministry of Education, China
[3] Engineering Research Center of Supercomputing Engineering Software, MOE, China
[4] School of Advanced Computing, Sun Yat-sen University, China, 510275
[5] Department of Psychology, Sun Yat-sen University, China, 510275
[6] School of Computer Science, South China Normal University, China
zhanzhh@mail.sysu.edu.cn

Abstract. Parallel computation is an efficient way to combine the advantages of different computation paradigms to obtain promising solution. In order to analyze the performance of parallel computation techniques to the particle swarm optimization (PSO) algorithm, a parallel particle swarm optimization (PPSO) is proposed in this paper. Since the theorem of "no free lunch" exists, there is not an optimization algorithm that can perfectly tackle all problems. The PPSO provides a paradigm to combine different variants of PSO algorithms by using the Message Passing Interface (MPI) so that the advantages of diverse PSO algorithms can be utilized. The PPSO divides the whole evolution process into several stages. At the interval between two successive stages, each PSO algorithm exchanges the achievement of their evolution and then continues with the next stage of evolution. By merging the global model PSO (GPSO), the local model PSO (LPSO), the bare bone PSO (BPSO), and the comprehensive learning PSO (CLPSO), the PPSO achieves higher solution quality than the serial version of these four PSO algorithms, according to the simulation results on benchmark functions.

Keywords: Parallel particle swarm optimization (PPSO), evolutionary algorithm, evolution stage, Message Passing Interface (MPI).

1 Introduction

Particle swarm optimization (PSO) [1], introduced by Kennedy and Eberhart, was inspired by the bird flocking and fish schooling. The PSO system is made up of a group of individuals (particles). These particles fly through the search domain while pursuing optimal solution to the problem. During the flying, each particle remembers

* This work was supported in part by the National High-Technology Research and Development Program (863 Program) of China No.2013AA01A212, in part by the National Natural Science Fundation of China (NSFC) with No. 61402545, 61332002, and 61300044, and in part by the NSFC for Distinguished Young Scholars with No. 61125205.
** Corresponding author.

H. Handa et al. (eds.), *Proc. of the 18th Asia Pacific Symp. on Intell. & Evol. Systems – Vol. 1*,
Proceedings in Adaptation, Learning and Optimization 1, DOI: 10.1007/978-3-319-13359-1_5

the best solution it has found. Moreover, the particle shares these promising good solutions among the group. In this way, each particle takes advantage of the best solution it has found as well the best solution among the neighbors for seeking for the new position. Known for its simplicity and efficiency, PSO has been applied widely to many domains of real-world optimization problems [2]-[7].

However, as PSO is a population-based heuristic stochastic optimization algorithm, it may be time-consuming for obtaining satisfactory results [8]. Moreover, particles in the group are often attracted to the local optima. Tackling the above problems have always been the two ever-lasting goals with the development of the PSO [9]. Many researchers have proposed a variety of PSO algorithms for addressing the two goals. Nevertheless, according to the theorem of "no free lunch" [10], one algorithm cannot fit every kind of problems perfectly. One algorithm may outperform the other one on some functions, but may be beaten on other functions.

To combine several algorithms and make full use of their respective advantages, a parallel particle swarm optimization (PPSO) algorithm is proposed with the help of message passing interface (MPI) in this paper. Four well-known PSO algorithms are adopted in the simulation, i.e., the global PSO (GPSO) [11], the local PSO (LPSO) [12], the bare bone PSO (BPSO) [14], and the comprehensive learning PSO (CLPSO) [22]. The reason for choosing these four algorithms is because they have quite different searching behaviors. This paper aims to analyze whether a proper parallel computation paradigm can achieve good performance by combining the advantages of different PSO variants. The comparisons of these PSOs' performance on the serial architecture and the parallel architecture that combines them are presented.

The rest of the paper is structured as follows. Section 2 introduces the original PSO briefly and shows some well-known variants of PSO. The proposed PPSO algorithm is described in Section 3. Section 4 presents the test results and discussions. Conclusions are given in Section 5.

2 PSO

In the PSO, a swarm of particles are maintained to represent the potential solutions of the problem to be solved. Each particle in the swarm preserves its current position and the best solution it has found so far. Two vectors are employed to characterize the *i*th particle, i.e., the velocity vector $\mathbf{v}_i = [v_i^1, v_i^2, \cdots, v_i^D]$ and the position vector $\mathbf{x}_i = [x_i^1, x_i^2, \cdots x_i^D]$, where *D* is the dimensions of the solution space. While moving, the velocity v_i^d and the position x_i^d of the *d*th dimension of the *i*th particle are updated as follows:

$$v_i^d(g+1) = wv_i^d(g) + c_1 rand_1^d(pBest_i^d(g) - x_i^d(g)) + c_2 rand_2^d(gBest^d(t) - x_i^d(t)) \tag{1}$$

$$x_i^d(g+1) = x_i^d(g) + v_i^d(g+1) \tag{2}$$

$$w = w_{\max} - (w_{\max} - w_{\min})\frac{g}{G} \tag{3}$$

where *g* is the generation index, w is the inertia weight, c_1 and c_2 are the acceleration coefficients, and $rand_1^d$ and $rand_2^d$ are the two independent random numbers uniformly distributed in the range of [0,1] for the *d*th dimension. The $pBest_i$ is the best position the *i*th particle has found so far and the $gBest$ is the best previous position among a specified topology structure of the neighborhood which is kept in the repository. The *w* was first introduced by Shi and Eberhart [11], and the w_{max} and w_{min} were proposed to be 0.9 and 0.4 respectively in [8]. The *g* in (3) represents the current index of the generation and the *G* is the maximum number of generations.

Generally, there are two major variants of PSO algorithms depending on the topology they choose [15]. The *gbest* model (GPSO) shares information among the whole swarm and the $gBest$ is the best position among all the particles. The *lbest* model (LPSO), whose neighborhood is constructed with a small group of particles updates the velocity with the following equation:

$$v_i^d(g+1) = wv_i^d(g) + c_1 rand_1^d(pBest_i^d(g) - x_i^d(g)) + c_2 rand_2^d(lBest^d(g) - x_i^d(g)) \tag{4}$$

where $lBest$ is the best among the neighborhood in the topology. The ring topology [12] is as the local topology for LPSO [13]. In the ring topology, each particle connects with only two of the other particles in the ring and only makes use of the information from the two particles while updating its own position.

Kennedy [14] introduced a simpler version of PSO, named bare bone PSO (BPSO). In BPSO, the velocity vector $V_i = [v_i^1, v_i^2, ..., v_i^D]$ is totally removed and the position vector is updated with the following equations.

$$x_i^d(g+1) = N(\mu, \sigma) \tag{5}$$

$$\mu = 0.5 \times (pBest_i^d(g) + gBest^d(g)) \tag{6}$$

$$\sigma = |pBest_i^d(g) - gBest^d(g)| \tag{7}$$

whcre $N(\mu, \sigma)$ is a Gaussian random number with the mean of μ and the deviation of σ. In this way, BPSO is famous for its simplicity and does not need the three parameters commonly used in PSO.

Because of its simplicity and efficiency, PSO has attracted wide interest since its first introduction by Kennedy and Eberhart [1], and has derived many interesting and efficient variants. Along with the development, three of the most promising approaches have been far and wide discussed, i.e., how to control the parameters of the algorithm, how to make combination with the other evolutionary algorithms, and how to develop efficient topologies [16][17].

Since the inertia weight w and the acceleration coefficients c_1, c_2 are the three main parameters in the original PSO, many researchers have tried to control these three parameters to gain a better variant. In [18] Shi and Eberhart proposed a fuzzy method and changed the inertia weight nonlinearly. In [9], a novel parameter automation strategy was presented and changed the three parameters adaptively with the process of evolution. To combine the PSO with other evolutionary algorithms, like GA, evolutionary operations have been inserted into the original PSO. These operators can help the PSO to converge quickly, to maintain the diversity of the group or keep the swarm from being trapped into local optimal when solving multimodal functions [19][20]. Except for the simple star and ring topologies, some researchers have come up with some efficient topologies. Suganthan in [21] suggested that the dynamically changing topologies could enhance the performance of PSO. In [22], a comprehensive learning PSO (CLPSO) was proposed, where the particles choose different *pBest* values from the swarm to update the position as:

$$v_i^d(g+1) = w \times v_i^d(g) + c \times rand^d(pBest_{f(i)}^d(g) - x_i^d(g)) \tag{8}$$

where *f*(*i*) is the particle index that used to guide the flying of the *d*th dimemsion.

3 PPSO

Although there are many variants for the PSO, there is not a perfect one. One variant may be good at dealing with unimodal function but perform badly on multimodal functions. Some PSO algorithms may converge quickly but can be easily trapped to local optima. In the GPSO, all particles in the swarm learn from the information of the whole group and make use of *gBest* and *pBest* while updating to the next position. This results in a fast convergence speed but also suffers from local optimal. This is especially evident when the dimension of the function is high and the landscape has multiple local optima. The LPSO, on the contrary, converges slower but as the particles in the swarm are not attracted to the same seemingly best position, it can avoid being attracted to the local trap on some level [15]. The BPSO adopts the Gaussian distribution of the *gBest* and *pBest* information and can be implemented in a very simple way, but it also suffers from its premature in the multimodal function [23]. The CLPSO has promising performance on multimodal functions but converges slow on unimodal functions [22]. In order to take advantages of these four aforementioned PSOs and avoid the drawbacks of them, a PPSO is proposed in this paper to run these PSO variants distributed and in parallel.

3.1 Parallel Strategy

To combine the above four algorithms together, a parallel strategy needs to be chosen. The message passing interface (MPI) is adopted herein. In this strategy, five virtual processors are used. The processor, whose role is to control the whole evolution process, is called MASTER. The other four processors, responsible for each of the four aforementioned PSO algorithms respectively, are acted as SLAVEs.

If only the SLAVEs evolve and they never exchange information with each other, the parallel strategy will be ineffective because they perform as independent serial processes. Therefore, information exchange is important in parallel computation, and the SLAVEs need to interchange information with each other at some periods.

However, how to communicate and what information is needed to be exchanged are the two core issues in the parallel algorithm design [24]. In [25], the SLAVE processors are only responsible for each PSO algorithm, and after only one stage of the evolution, they send the result of the evolution to the MASTER. The MASTER processor, after receiving the evolution result from the SLAVEs, carries on with the evolution. In this way, the advantages of each PSO algorithms are not fully exploited. Moreover, the diversity of the swarm cannot be guaranteed, because the SLAVEs send only the best or worst particles to the MASTER.

In the PPSO algorithm we propose in this paper, the evolution process is divided into several stages. At the end of each stage, every SLAVE processor sends the best solution information it has found in this stage to the MASTER. The MASTER selects the best solution from the received solutions and sends this best solution to all the SLAVE processors. For each SLAVE processor, once receiving the best solution from the MASTER, it replaces a random selected particle by this received particle and continues the evolution with the best solution information. This procedure goes on until the end of the final stage.

In this way, the advantages of each PSO algorithms can be better taken, because the SLAVEs are more involved into the evolution. The MASTER acts as a controller and helps every SLAVE to find a better "exemplar" to guide the search in the next stage of evolution. Each SLAVE, at the same time, can learn from other SLAVE processors and inspire the particles to converge to a better position.

3.2 The Implementation of PPSO

Based on the analysis above, the key pseudo-code of PPSO is presented in Fig. 1. The *STAGE* in the pseudo-code is the number of the stages for the whole procedure.

```
For stage=1 to STAGE
    Switch (processor id):
      SLAVE 1 :
        GPSO( );
        Send the best solution to the MASTER;
        Receive the best solution from the MASTER;
      End SLAVE 1
      SLAVE 2 :
        LPSO( );
        Send the best solution to the MASTER;
        Receive the best solution from the MASTER;
      End SLAVE 2
      SLAVE 3 :
        BPSO( );
        Send the best solution to the MASTER;
        Receive the solution from the MASTER;
      End SLAVE 3
      SLAVE 4 :
        CLPSO( );
        Send the best solution to the MASTER;
        Receive the solution from the MASTER;
      End SLAVE 4
      MASER :
        Receive the best solution from each SLAVE;
        Select the optimal one from these bests;
        Send the optimal one to every SLAVE;
      End MASTER
    End Switch
End For
```

Fig. 1. Key Pseudo-code of PPSO

4 Experiments and Analysis

4.1 Benchmark Functions

Since the proposed PPSO consists of four different kinds of PSO algorithms and each of them is expert in some functions, we need to select different benchmark functions to test the PPSO. From the standard set of benchmark functions available in the literature, 13 different functions are chosen to test the performance of the PPSO. The benchmark functions are listed in Table I. Among these functions, f_1~f_7 in Table I are unimodal problems and f_8~f_{13} are multimodal problems.

Table 1. The Benchmark Functions

Test function	Search Space	f_{min}
$f_1=\sum_{i=1}^{D} x_i^2$	$[-100,100]^D$	0
$f_2=\sum_{i=1}^{D}\lvert x_i\rvert+\prod_{i=1}^{D}\lvert x_i\rvert$	$[-10,10]^D$	0
$f_3=\sum_{i=1}^{D}(\sum_{j=1}^{i} x_j)^2$	$[-100,100]^D$	0.
$f_4=\min\lvert x_i\rvert$	$[-100,100]^D$	0
$f_5=\sum_{i=1}^{D-1}\left[100(x_{i+1}-x_i^2)^2+(x_i-1)^2\right]$	$[-30,30]^D$	0
$f_6=\sum_{i=1}^{D}(\lfloor x_i+0.5\rfloor)^2$	$[-100,100]^D$	0
$f_7=\sum_{i=1}^{D} ix_i^4+random[0,1)$	$[-1.28,1.28]^D$	0
$f_8=\sum_{i=1}^{D}-x_i\sin(\sqrt{x_i})$	$[-500,500]^D$	-12569.5
$f_9=\sum_{i=1}^{D}[x_i^2-10\cos(2\pi x_i+10)]$	$[-5.12,5.12]^D$	0
$f_{10}=-20\exp(-0.2\sqrt{1/D\sum_{i=1}^{D}x_i^2})-\exp(1/D\sum_{i=1}^{D}\cos 2\pi xi)+20+e$	$[-32,32]^D$	0
$f_{11}=1/4000\sum_{i=1}^{D}x_i^2-\prod_{i=1}^{D}\cos(xi/\sqrt{i})+1$	$[-600,600]^D$	0
$f_{12}=\frac{\pi}{D}\{10\sin^2(\pi y_1)+\sum_{i=1}^{D-1}(y_i-1)^2[1+10\sin^2(\pi y_{i+1})]+(y_D-1)^2\}+\sum_{i=1}^{d}u(x_i,10,100,4)$ where $y_i=1+\frac{1}{4}(x_i+1)$, $u(x_i,a,k,m)=\begin{cases}k(x_i-a)^m, & x_i>a\\ 0, & -a\le x_i\le a\\ k(-x_i-a)^m, & x_i<-a\end{cases}$	$[-50,50]^D$	0
$f_{13}(x)=0.1\{\sin^2(3\pi x_1)+\sum_{i=1}^{D-1}(x_i-1)^2[1+\sin^2(3\pi x_{i+1})]+(x_D-1)[1+\sin^2(2\pi x_D)]\}+\sum_{i=1}^{D}u(x_i,5,100,4)$	$[-50,50]^D$	0

4.2 Experimental Settings

In order to validate the efficiency of PPSO in combining the advantages of different PSO variants, we compare the performance of PPSO and the respective serial PSO variants (i.e., GPSO, LPSO, BPSO, and CLPSO) in this paper. In the experiments, all these four PSO variants use 20 particles in the population and the number of total function evaluations (FEs) is assigned to be 3×10^5. The other parameters are set according to their respective references. That is, the inertia weight w is initialized to 0.9 at the beginning and linearly decreases to 0.4 at the end according to (3) for all these four PSOs. The accelerations coefficients c_1 and c_2 are set as 2.0 for GPSO and LPSO, while set as 1.49445 for CLPSO. For the PPSO, we use four SLAVE processors to run GPSO, LPSO, BPSO, and CLPSO respectively, and use one MASTER processor to control the whole evolution. All the parameters of these four PSO algorithms in the four SLAVEs are set the same as the ones when they are running in serial environment. The number of stage *STAGE* is chosen to be 4. Therefore, the number of FEs for each stage is 7.5×10^4.

It should be noticed that two successive stages in PPSO are related independent. Therefore, the inertia weight w is initialized to 0.9 at the beginning of every stage and linearly decreases to 0.4 at the end of the stage according to (3). The GPSO updates positions as (1) and (2). The LPSO updates position as (4) and (2). The BPSO updates by (5), (6), and (7). The CLPSO updates by (8) and (2).

All the experiments are carried out in computers with the same configurations. That is, with Intel Pentium (R) Dual 2.40GHz CPU, 3.00GB memory, and a Win7 x32 operation system. For PPSO, three computers are enough because each computer has two CPUs. To make the results more convincing, the algorithm is run for 25 independent times and the results of every run are recorded.

4.3 Results of Comparisons

The results of the proposed PPSO with the parameters set above are listed in Table II. The results of GPSO, LPSO, BPSO, and CLPSO are compared with PPSO in this table in terms of the 'Best' solution, 'Mean' solution of the 25 runs, and the 'Std. Dev' of the 25 runs. The best results of the five algorithms are marked with **boldface**.

It can be clearly seen that the PPSO algorithm that runs distributed and in parallel by the four PSO algorithms outperform the serial PSO variants, no both the umimodal functions and the multimodal functions. The interesting observation is that BPSO is promising on unimodal functions while CLPSO performs better than GPSO, LPSO, and BPSO on multimodal functions. However, no a single PSO variant does best on both unimodal and multimodal functions. For our proposed PPSO, the results show that it can combine the advantages of these four PSO variants. Therefore, for the unimodal functions, PPSO can perform similar or even better than BPSO, while for the multimodal functions, PPSO can obtain the global optima on all the tested functions, and is even better than CLPSO on some functions.

Table 2. Results Comparisions Between the PPSO and the Serial PSOs

F	Property	GPSO	LPSO	BPSO	CLPSO	PPSO
f_1	Best	$1.96*10^{-84}$	$1.24*10^{-20}$	$3.43*10^{-225}$	$1.23*10^{-77}$	**$4.09*10^{-226}$**
	Mean	$3.76*10^{-77}$	$3.09*10^{-18}$	**$3.35*10^{-218}$**	$1.19*10^{-74}$	$3.75*10^{-216}$
	Std. Dev	$1.18*10^{-76}$	$4.47*10^{-18}$	**0**	$2.61*10^{-74}$	**0**
f_2	Best	$1.24*10^{-49}$	$1.14*10^{-14}$	$5.37*10^{-156}$	$5.68*10^{-51}$	**$3.11*10^{-158}$**
	Mean	$2.00*10^{-41}$	$2.36*10^{-13}$	**$2.39*10^{-152}$**	$4.16*10^{-47}$	$9.33*10^{-152}$
	Std. Dev	$6.07*10^{-41}$	$3.05*10^{-13}$	**$4.09*10^{-152}$**	$7.84*10^{-47}$	$2.69*10^{-151}$
f_3	Best	$3.29*10^{-4}$	24.36	**$1.55*10^{-14}$**	$2.47*10^{-2}$	$2.85*10^{-14}$
	Mean	$1.66*10^{-3}$	499.85	**$4.83*10^{-13}$**	$8.14*10^{-2}$	$1.37*10^{-12}$
	Std. Dev	$4.67*10^{-4}$	255.425	**$5.54*10^{-13}$**	$4.23*10^{-2}$	$1.87*10^{-12}$
f_4	Best	$1.08*10^{-1}$	4.33	$3.18*10^{-06}$	0.19	**$1.38*10^{-06}$**
	Mean	0.22	9.39	$7.41*10^{-05}$	0.27	**$1.48*10^{-05}$**
	Std. Dev	0.12	3.40	$8.51*10^{-05}$	$9.30*10^{-3}$	**$2.26*10^{-05}$**
f_5	Best	$2.54*10^{-2}$	0.22	**$2.22*10^{-3}$**	21.54	$8.23*10^{-3}$
	Mean	44.60	37.02	10.08	32.72	**2.85**
	Std. Dev	51.13	34.36	20.75	24.17	**3.24**
f_6	Best	**0**	**0**	**0**	**0**	**0**
	Mean	**0**	**0**	3.1	**0**	**0**
	Std. Dev	**0**	**0**	4.06	**0**	**0**
f_7	Best	$3.97*10^{-3}$	$1.30*10^{-2}$	$9.23*10^{-04}$	$6.25*10^{-4}$	**$4.00*10^{-04}$**
	Mean	$8.31*10^{-3}$	$3.61*10^{-2}$	$4.36*10^{-03}$	**$7.50*10^{-4}$**	$8.30*10^{-04}$
	Std. Dev	$3.01*10^{-3}$	$1.85*10^{-2}$	$2.67*10^{-3}$	**$1.35*10^{-4}$**	$3.11*10^{-4}$
f_8	Best	2033.22	2078.67	671.17	**$1.34*10^{-2}$**	**$1.34*10^{-2}$**
	Mean	3089.33	4242.06	1342.33	43.44	**$1.34*10^{-2}$**
	Std. Dev	745.64	1193.72	536.02	97.10	**0**
f_9	Best	17.90	22.06	43.78	**3.98**	4.97
	Mean	26.47	37.63	75.92	7.96	**6.87**
	Std. Dev	6.89	19.88	18.87	2.72	**1.65**
f_{10}	Best	$7.69*10^{-15}$	$3.43*10^{-11}$	$4.14*10^{-15}$	$4.14*10^{-15}$	**$4.00*10^{-15}$**
	Mean	$9.11*10^{-15}$	$2.20*10^{-09}$	0.93	$6.98*10^{-15}$	**$5.77*10^{-15}$**
	Std. Dev	$2.48*10^{-15}$	$3.03*10^{-09}$	1.36	$2.97*10^{-15}$	**$1.87*10^{-15}$**
f_{11}	Best	**0**	**0**	**0**	**0**	**0**
	Mean	$2.80*10^{-2}$	$2.84*10^{-2}$	$1.77*10^{-2}$	$1.48*10^{-3}$	**$1.23*10^{-3}$**
	Std. Dev	$2.06*10^{-2}$	$3.10*10^{-2}$	$2.75*10^{-2}$	**$3.33*10^{-3}$**	$3.90*10^{-3}$
f_{12}	Best	$1.57*10^{-32}$	$3.65*10^{-20}$	$2.21*10^{-32}$	**$1.57*10^{-32}$**	**$1.57*10^{-32}$**
	Mean	$1.04*10^{-2}$	$4.15*10^{-2}$	0.19	$1.67*10^{-32}$	**$1.63*10^{-32}$**
	Std. Dev	$3.28*10^{-2}$	$7.25*10^{-2}$	0.47	$2.31*10^{-33}$	**$1.64*10^{-33}$**
f_{13}	Best	$1.59*10^{-32}$	$4.40*10^{-19}$	$1.80*10^{-31}$	$1.35*10^{-32}$	**$1.34*10^{-32}$**
	Mean	$3.29*10^{-3}$	$1.10*10^{-3}$	$7.69*10^{-3}$	$1.50*10^{-32}$	**$1.44*10^{-32}$**
	Std. Dev	$5.30*10^{-3}$	$3.47*10^{-3}$	$1.37*10^{-2}$	$2.03*10^{-33}$	**$1.54*10^{-33}$**

5 Conclusions

In this paper, several PSO algorithms in the literature are run distributed and in parallel with the help of MPI. By taking advantages of GPSO, LPSO, BPSO, and CLPSO, the parallel versions of the PSO algorithms show significant promising performance as shown in the benchmark tests. Although promising results are obtained, there are still many works to do in the future work. Firstly, as the total FEs are not the same in PPSO and the serial PSOs, the running time of PPSO and the serial PSOs need to be investigated although PPSO runs on five processors while serial PSO runs on only one processor. Secondly, the number of stages should be investigated because it may influence the performance of PPSO. Moreover, the investigation of PPSO on efficient cloud computing environment [26] is a promising research topic.

References

[1] Kennedy, J., Eberhart, R.C.: Particle swarm optimization. In: Proc. IEEE Int. Conf. Neural Networks, pp. 1942–1948 (1995)

[2] Krohling, R.A., dos Santos Coelho, L.: Coevolutionary particle swarm optimization using Gaussian distribution for solving constrained optimization problems. IEEE Trans. Syst., Man, Cybern. B, Cybern. 36(6), 1407–1416 (2006)

[3] Zhan, Z.-h., Zhang, J., Fan, Z.: Solving the optimal coverage problem in wireless sensor networks using evolutionary computation algorithms. In: Deb, K., Bhattacharya, A., Chakraborti, N., Chakroborty, P., Das, S., Dutta, J., Gupta, S.K., Jain, A., Aggarwal, V., Branke, J., Louis, S.J., Tan, K.C. (eds.) SEAL 2010. LNCS, vol. 6457, pp. 166–176. Springer, Heidelberg (2010)

[4] Zhan, Z.H., Li, J., Cao, J., Zhang, J., Chung, H., Shi, Y.H.: Multiple populations for multiple objectives: A coevolutionary technique for solving multiobjective optimization problems. IEEE Trans. Cybern. 43(2), 445–463 (2013)

[5] Zhan, Z.H., Li, J.J., Zhang, J.: Adaptive particle swarm optimization with variable relocation for dynamic optimization problems. In: Proc. IEEE Congr. Evol. Comput., pp. 1–7 (2014)

[6] Shen, M., Zhan, Z.H., Chen, W.N., Gong, Y.J., Zhang, J., Li, Y.: Bi-velocity discrete particle swarm optimization and its application to multicast routing problem in communication networks. IEEE Trans. Ind. Electron 61(12), 7141–7151 (2014)

[7] Ji, C., Liu, F., Zhang, X.: Particle swarm optimization based on catfish effect for flood optima operation of reservoir. In: Proc. IEEE Int. Conf. Neutral Netw., pp. 1192–1201 (2011)

[8] Zhang, J., Zhan, Z.H., Lin, Y., Chen, N., Gong, Y.J., Zhong, J.H., Chung, H., Li, Y., Shi, Y.H.: Evolutionary computation meets machine learning: A survey. IEEE Comput. Intell. Mag. 6(4), 68–75 (2011)

[9] Zhan, Z.H., Zhang, J., Li, Y., Chung, H.: Adaptive Particle swarm optimization. IEEE Trans. Syst., Man, Cybern. B 39(6), 1362–1381 (2009)

[10] Wolpert, D.H., Macready, W.G.: No free lunch theorems for optimization. IEEE Trans. Evol. Comput. 1(1), 67–82 (1997)

[11] Shi, Y., Eberhart, R.C.: A modified particle swarm optimizer. In: Proc. IEEE World Congr. Comput. Intell., pp. 69–73 (1998)

[12] Kennedy, J., Mendes, R.: Population structure and particle swarm performance. In: Proc. IEEE Congr. Evol. Comput., pp. 1671–1676 (2002)
[13] Zhan, Z.H., Zhang, J., Li, Y., Shi, Y.H.: Orthogonal learning particle swarm optimization. IEEE Trans. Evol. Comput. 15(6), 832–847 (2011)
[14] Kennedy, J.: Bare bone particle swarms. In: Proc. IEEE Swarm Intelligence Symposium, pp. 80–87 (2003)
[15] Bratton, D., Kennedy, J.: Defining a standard for Particle Swarm Optimization. In: Proc. IEEE Swarm Intelligence Symposium, pp. 120–127 (2007)
[16] Eberhart, R.C., Shi, Y.: Guest editorial—Special Issue Particle Swarm Optimization. IEEE Trans. Evol. Comput. 8(3), 201–203 (2004)
[17] Li, Y.H., Zhan, Z.H., Lin, S., Wang, R.M., Luo, X.N.: Competitive and cooperative particle swarm optimization with information sharing mechanism for global optimization problems. Information Sciences (accepted, 2014)
[18] Shi, Y., Eberhart, R.C.: Fuzzy adaptive particle swarm optimization. In: Proc. IEEE Congr. Evol. Comput., pp. 101–106 (2001)
[19] Angeline, P.J.: Using selection to improve particle swarm optimiza-tion. In: Proc. IEEE Congr. Evol. Comput., pp. 84–89 (1998)
[20] Lovbjerg, M., Rasmussen, T.K., Krink, T.: Hybrid particle swarmoptimizer with breeding and subpopulations. In: Proc. Genetic Evol. Comput. Conf., pp. 469–476 (2001)
[21] Suganthan, P.N.: Particle swarm optimizer with neighborhood operator. In: Proc. Congr. Evol. Comput., pp. 1958–1962 (1999)
[22] Liang, J.J., Qin, A.K., Suganthan, P.N., Baskar, S.: Comprehensive learning particle swarm optimizer for global optimization of multimodal functions. IEEE Trans. Evol. Comput. 10(3), 281–295 (2006)
[23] Krohling, R.A., Mendel, E.: Bare bone particle swarm optimization with Gaussian or Cauchy jumps. In: Proc. Congr. Evol. Comput., pp. 3285–3291 (2009)
[24] Vanneschi, L., Codecasa, D., Mauri, G.: An empirical study of parallel and distributed particle swarm optimization. In: Fernandez de Vega, F., Hidalgo Pérez, J.I., Lanchares, J. (eds.) Parallel Architectures & Bioinspired Algorithms. SCI, vol. 415, pp. 125–150. Springer, Heidelberg (2012)
[25] Deep, K., Sharma, S., Pant, M.: Modified parallel particle swarm optimization for global optimization using Message Passing Interface. In: Proc. Bio-Inspired Computing: Theories and Application, pp. 1451–1458 (2010)
[26] Liu, X.F., Zhan, Z.H., Du, K.J., Chen, W.N.: Energy aware virtual machine placement scheduling in cloud computing based on ant colony optimization approach. In: Proc. Genetic Evol. Comput. Conf., pp. 41–47 (2014)

An Automatic Clustering Algorithm Based on a Competition Model of Probabilistic PCA*

Yunxia Li[1], Jian Cheng Lv[2], and Xiaojie Li[2]

[1] School of Automation, University of Electronic Science and Technology of China, Chengdu 610065, P.R. China
yunxiali@uestc.edu.cn

[2] The Machine Intelligence Laboratory, College of Computer Science, Sichuan University, Chengdu 610065, P.R. China
lvjiancheng@scu.edu.cn

Abstract. A number of mixture models of local Principal Component Analysis (PCA) have been developed to analyze data distributed in space. Most of these models require the users to determine the number of the local PCA models, i.e., the number of clusters for clustering analysis. This is not a reasonable requirement in practical applications. This paper proposes an automatic clustering algorithm to analyze data based on a competition model of probabilistic PCA. Without identifying the number of clusters in advance, the algorithm automatically evolves to partition a given data set into some small clusters in terms of the *empirical rule* of Gaussian distribution. It is shown the algorithm will not only group data but also can explore the hierarchical structure of a given data.

Keywords: Automatic clustering, probabilistic principal component analysis, Gaussian empirical rules, probability competition.

1 Introduction

Principal component analysis (PCA) is a useful tool for dimensionality reduction. It has been widely used for data compression, feature extraction, visualization, pattern recognition, regression and time series prediction [1–6]. It is well known that for Gaussian data, PCA eigenvectors represent the axes of the underlying distribution, whereas the eigenvalues represent the variance along these axes. In high-dimensional data space, the Gaussian distribution is usually used to describe the local structure of subset in an embedded low-dimensional subspace with hyper-ellipsoidal shapes [7]. Therefore, the local PCA can be used to present the subset with fewer dimensions. The mixture of local PCA can partition a data set into several small subsets and find linear expressions of the data subset in a low-dimensional space [8–12]. The mixture models of PCA have been used for clustering analysis to capture the local structure of a given data set. Meanwhile,

* This work was supported by Specialized Research Fund for the Doctoral Program of Higher Education under Grant 2010081110053.

H. Handa et al. (eds.), *Proc. of the 18th Asia Pacific Symp. on Intell. & Evol. Systems – Vol. 1*, Proceedings in Adaptation, Learning and Optimization 1, DOI: 10.1007/978-3-319-13359-1_6

an important factor to guarantee validity of clustering is the number of clusters [5]. In the mixture models of PCA, the number of local PCA, i.e., the number of clusters for clustering analysis, must be determined in advance. This is not a reasonable requirement in practical applications.

In this paper, based on a competition model of probabilistic PCA, an automatic clustering algorithm is proposed to group data automatically. Without identifying the number of clusters, the proposed algorithm can automatically evolve to partition a given data into some small clusters by means of three basic processes: probability competition, generating a cluster and merging two clusters. The probability density is used as the proximity measure of competition and the *empirical rule* of Gaussian distribution is used as the criterion of generating and merging clusters. The proposed algorithm will not only partition a given data, but also it can explore the hierarchical structure of the data by using the different empirical values.

The rest of this paper is organized as follows. Section 2 describes a competition model of probabilistic PCA. In Section 3, an automatic clustering algorithm is presented. Simulation results and discussions are presented in Section 4. Finally, conclusions are drawn in Section 5.

2 Probability Competition Model of PPCA

Consider an observed data set $D = \{\mathbf{y}(k)|\mathbf{y}(k) \in R^n(k = 1, 2, \cdots, K)\}$. Let $\mu = \frac{1}{K}\sum_{k=1}^{K}\mathbf{y}(k)$ be the sample mean vector. The sample covariance matrix is presented as: $C = E\{(\mathbf{y}(k) - \mu)(\mathbf{y}(k) - \mu)^T\} = \frac{1}{K}\sum_{k=1}^{K}(\mathbf{y}(k) - \mu)(\mathbf{y}(k) - \mu)^T$. Clearly, the covariance matrix C is a symmetric nonnegative definite matrix. Let $\lambda_1, \lambda_2, \cdots, \lambda_n$ be all the eigenvalues of the matrix C ordered by $\lambda_1 \geq \lambda_2 \cdots \geq \lambda_n \geq 0$. The unit eigenvectors of C associated with the corresponding eigenvalues λ_i construct an orthonormal base in $\mathbb{R}^n$, denoted by $\{\mathbf{v}_i|i = 1, \cdots, n\}$. Let $U = [\mathbf{v}_1, \mathbf{v}_2, \cdots, \mathbf{v}_n]$ and $\Lambda = diag\{\lambda_1, \lambda_2, \cdots, \lambda_n\}$. It follows that $C = U\Lambda U^T$.

2.1 Review of PPCA

In [11, 12], the probabilistic PCA is presented. The latent variables are defined as: $\mathbf{x} \sim N(0, I_{q\times q})(q < n)$. The observation vector $\mathbf{y}$ is obtained: $\mathbf{y} = W\mathbf{x} + \mu + \epsilon$, where W is a $n \times q$ matrix as a linear operator and $\epsilon \sim N(0, \sigma^2 I_{n\times n})$. Then, the marginal distribution of $\mathbf{y}$ is given by

$$p(\mathbf{y}) = \int p(\mathbf{y}|\mathbf{x})p(\mathbf{x})d\mathbf{x} = \frac{1}{(\sqrt{2\pi})^n\sqrt{|S|}}exp\left\{-\frac{1}{2}(\mathbf{y} - \mu)^T S^{-1}(\mathbf{y} - \mu)\right\}, \quad (1)$$

where $S = \sigma^2 I_{n\times n} + WW^T$. The maximum-likelihood estimators of the model parameters have been obtained as:

$$\mu_{ML} = \frac{1}{K}\sum_{k=1}^{K}\mathbf{y}(k), \sigma^2_{ML} = \frac{1}{n-q}\sum_{j=q+1}^{n}\lambda_i, W_{ML} = U_{n\times q}\sqrt{(\Lambda_{q\times q} - \sigma^2 I_{q\times q})}R_{q\times q},$$

where $U_{n\times q} = [\mathbf{v}_1, \mathbf{v}_2, \cdots, \mathbf{v}_q]$, $\Lambda_{q\times q} = diag\{\lambda_1, \lambda_2, \cdots, \lambda_q\}$, and $R_{q\times q}$ is an arbitrary orthogonal matrix.

2.2 Probability Competition Model

In the competition model of PPCA, the number of local PCA varies as the algorithm evolves. Suppose at time k, the number of local PCA, i.e., the number of clusters for clustering analysis, is $M(k)$. $\mu_i(k_i)$, $C_i(k_i)(i = 1, \cdots, M(k))$ are the mean and the covariance matrix of the cluster i, respectively, where k_i denotes $k_i th$ observed data for cluster i. For the input $\mathbf{y}(k)$, the probability density associated with cluster i will be calculated by

$$p(\mathbf{y}(k)|i) = \frac{exp\left\{-\frac{1}{2}(\sum_{j=1}^{q}\frac{z_{ij}^2(k_i)}{\lambda_{ij}} + \frac{1}{\sigma_i^2(k_i)}\sum_{j=q+1}^{n} z_{ij}^2(k_i))\right\}}{(\sqrt{2\pi})^n\sqrt{\prod_{j=1}^{q}\lambda_{ij}(k_i) + (\sigma_i^2(k_i))^{n-q}}}, \tag{2}$$

where $\lambda_{ij}(k_i)$ is eigenvalues of the covariance matrix $C_i(k_i)$, $z_{ij} = \mathbf{v}_{ij}(k_i)(\mathbf{y}(k) - \mu_i(k))$. $\mathbf{v}_{ij}(k_i)$ is the eigenvector associated with the $\lambda_{ij}(k_i)$. If $\lambda_{ij}(k_i) < \sigma_i(k_i)$, let $\lambda_{ij}(k) = \sigma_i(k_i)$. In the appendix, the deduction of (2) is given.

In the competition model, the probability density is used as the proximity measure of competition. The probability density of the input $\mathbf{y}(k)$ given the clusters $i(i = 1, \cdots, M(k))$ will be calculated by (2). The competition rule is given by

$$j = arg\max_i\{p(\mathbf{y}(k)|i), i = 1, \cdots, M(k)\}. \tag{3}$$

The updated rules are given as follows:

$$\begin{cases} \mu_j(k_j+1) = \mu_j(k_j) - \frac{1}{k_j+1}(\mu_j(k_j) - \mathbf{y}(k)), \\ \tilde{C}_j(k_j) = (\mathbf{y}(k) - \mu_j(k_j+1))(\mathbf{y}(k) - \mu_j(k_j+1))^T, \\ C_j(k_j+1) = C_j(k_j) - \frac{1}{k_j+1}(C_j(k_j) - \tilde{C}_j(k_j)), \\ k_j = k_j + 1. \end{cases} \tag{4}$$

For every $C_i(k_i)$, there is a big condition number in the starting phase of its evolution. In this phase, a very small $\sigma_i^2(k_i)$ may make $p(\mathbf{y}(k)|i)$ very small and there exists a big error between $p(\mathbf{y}(k)|i)$ and the final $p(\mathbf{y}(k)|i)$ as $k_i \to +\infty$. Therefore, to amend the effect of the ill-condition, σ_i^2 is designed as:

$$\sigma_i^2(k_i) = \tau + \frac{1}{tan(f(k_i))}, \tag{5}$$

where τ is a small constant, $f(k_i)$ is a monotonically increasing function and $\frac{\pi}{4} < f(k_i) \leq \frac{\pi}{2}$. Clearly, $\sigma_i^2(k_i) = \tau$ as $k_i \to +\infty$, as shown in Figure 1.

3 An Automatic Clustering Algorithm

An automatic clustering algorithm is proposed and presented in Figure 2. Without identifying the number of clusters, the proposed algorithm can evolve to partition data into some small clusters by means of three basic processes: probability competition, generating a new cluster and merging clusters.

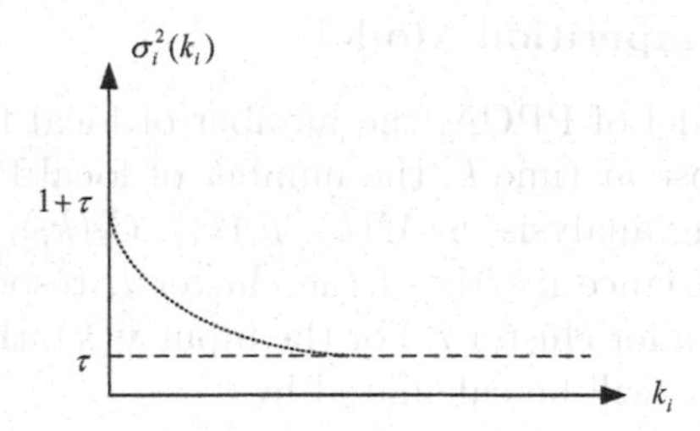

Fig. 1. $\sigma_i^2(k_i)$ approximates the τ as k_i increases

In the proposed algorithm, the *empirical rule* of Gaussian distribution is used as the measure of generating a cluster and merging clusters. The *empirical rule* states that for a Gaussian distribution, almost all values lie within three standard deviations of the means. Let θ present how many standard deviations from the means. It will be shown that an appropriate clustering result will be obtained if $\theta = 3$ or 4 for the data with Gaussian distribution.

For the local PCA model i (cluster i for clustering analysis), $\theta \cdot \lambda_{ij}(j = 1, \cdots, q)$ or $\theta \cdot \sigma_{ij}^2(j = q+1, \cdots, n)$ denote the distance from the mean. Thus, from (2), the probability density of the θ standard deviations for the cluster i at time k can be calculated as

$$p(\theta\sigma|i, k_i) = \frac{exp\{-\frac{1}{2}(n \cdot \theta)\}}{(\sqrt{2\pi})^n \sqrt{\prod_{j=1}^{q} \lambda_{ij}(k_i) + (\sigma_i^2(k_i))^{n-q}}}. \tag{6}$$

Figure 2 depicts the procedure of the clustering algorithm with the following basic steps.

Initializing the parameters. We initialize the first cluster $C_1(0) = \beta \cdot I$, where β is a small constant. Let $\mu_1(0) = \mathbf{y}(0)$, $\mathbf{y}(0)$ is selected arbitrarily from the given data set. The parameter τ is a small constant. Usually, let $\theta = 3$ or 4 for Gaussian data according to the *empirical rule.*

Competing with probability. The probability density is used as the proximity measure of clustering analysis. The cluster j with maximum probability density will win at time k by (3). If $\mathbf{y}(k)$ is inside of the θ standard deviations from the mean $\mu_j(k_j)$, i.e, $p(\mathbf{y}(k)|j, k_j) > p(\theta\sigma|j, k_j)$, the observed data $\mathbf{y}(k)$ is partitioned into the cluster j and the cluster j will be updated by (4).

Generating a new cluster. If $\mathbf{y}(k)$ is outside of the θ standard deviations of the mean $\mu_j(k_j)$, i.e, $p(\mathbf{y}(k)|j) < p(\theta\sigma|j, k_j)$, then a new cluster will be generated by

$$\begin{cases} M(k) = M(k) + 1, k_{M(k)} = 1, \\ C_{M(k)}(0) = \beta \cdot I, \mu_{M(k)}(0) = \mathbf{y}(k), \\ \sigma_{M(k)}^2(k_{M(k)}) = \tau + \dfrac{1}{tan(f(k_{M(k)}))}. \end{cases} \tag{7}$$

· Initialization:
1) Set a small τ and empirical value θ. Let $k_1 = 0$.
2) Let $C_1(k_1) = \beta I$, $\mu_1(k_1) = \mathbf{y}(k_1)$
and $\sigma_1^2(k_1) = \tau + \frac{1}{tanh(f(k_1))}$.
3) Set the number of cluster $M(1) = 1$.
· For $k = 1, \cdots, K$:
1) Calculate the $p(\mathbf{y}(k)|i)$ by (2),
for all $i(i = 1, \cdots, M(k))$
2) Compete with probability density by (3).
The cluster j with the maximum probability density wins.
Then, the $p(\theta\sigma|j, k_j)$ is calculated by (6).
3) If $p(\mathbf{y}(k)|j) - p(\theta\sigma|j, k_j) > 0$
Update the cluster j by (4) and (5).
else
Generate a new cluster by (7).
end
4) Merge clusters by (8) and update $M(k)$.
· Final results

Fig. 2. An automatic clustering algorithm

Merging the two clusters. For two cluster i, j, if $p(\mu_i(k_i)|j, k_j) - p(\theta\sigma|j, k_j) > 0$ or $p(\mu_j(k_j)|i, k_i) - p(\theta\sigma|i, k_i) > 0$, the two cluster will be merged into the cluster i by

$$\begin{cases} C_i(k_i) = \dfrac{k_i \cdot C_i(k_i) + k_j \cdot C_j(k_j)}{(k_i + k_j)}, \\ \mu_i(k_i) = \dfrac{k_i \cdot \mu_i(k_i) + k_j \cdot \mu_j(k_j)}{(k_i + k_j)}, \\ k_i = k_i + k_j. \end{cases} \tag{8}$$

Then, delete the cluster j, and $M(k) = M(k) - 1$.

4 Simulations and Discussions

4.1 Example 1

A data set with Gaussian distribution is generated artificially as shown in Figure 3 (top left). Let $\beta = 0.1$ and $\tau = 0.000001$. The Figure 3 shows the results of clustering analysis. Clearly, with different values of θ, a hierarchical structure of these data is obtained. It is also shown that an appropriate result of clustering analysis is obtained as $\theta = 4$.

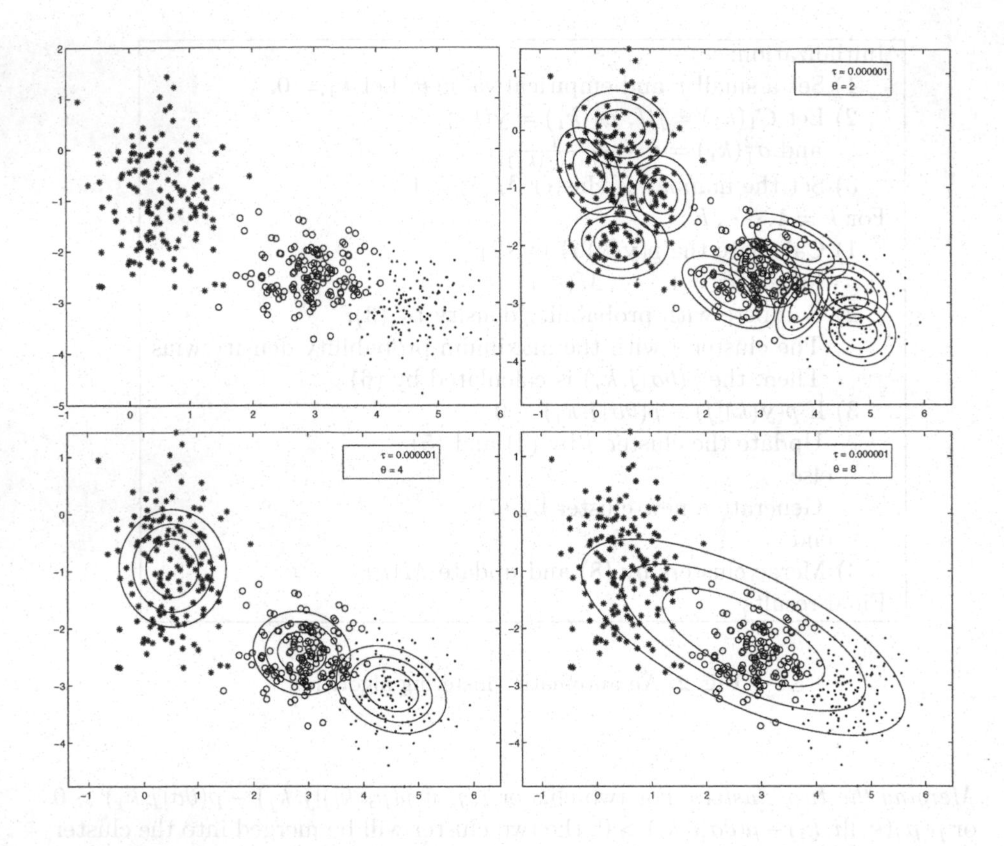

Fig. 3. The result of clustering analysis for a two-dimensional data set

4.2 Example 2

The iris data set (*http://www.ics.uci.edu/ mlearn/MLRepository.html*) is used to illustrate the performance of the proposed algorithm. The data set includes three iris categories: Iris Setosa, Iris Versicolour, Iris Virginica. Each has 50 patterns with four features. In this simulation, the first three features are used and shown in Figure 4 (top left). The other pictures of Figure 4 shows the result of clustering analysis. Clearly, with the different empirical values θ, a hierarchical structure of the data set is obtained. It is also seen that when $\theta = 3$, Figure 4 (middle right) shows an appropriate result of clustering.

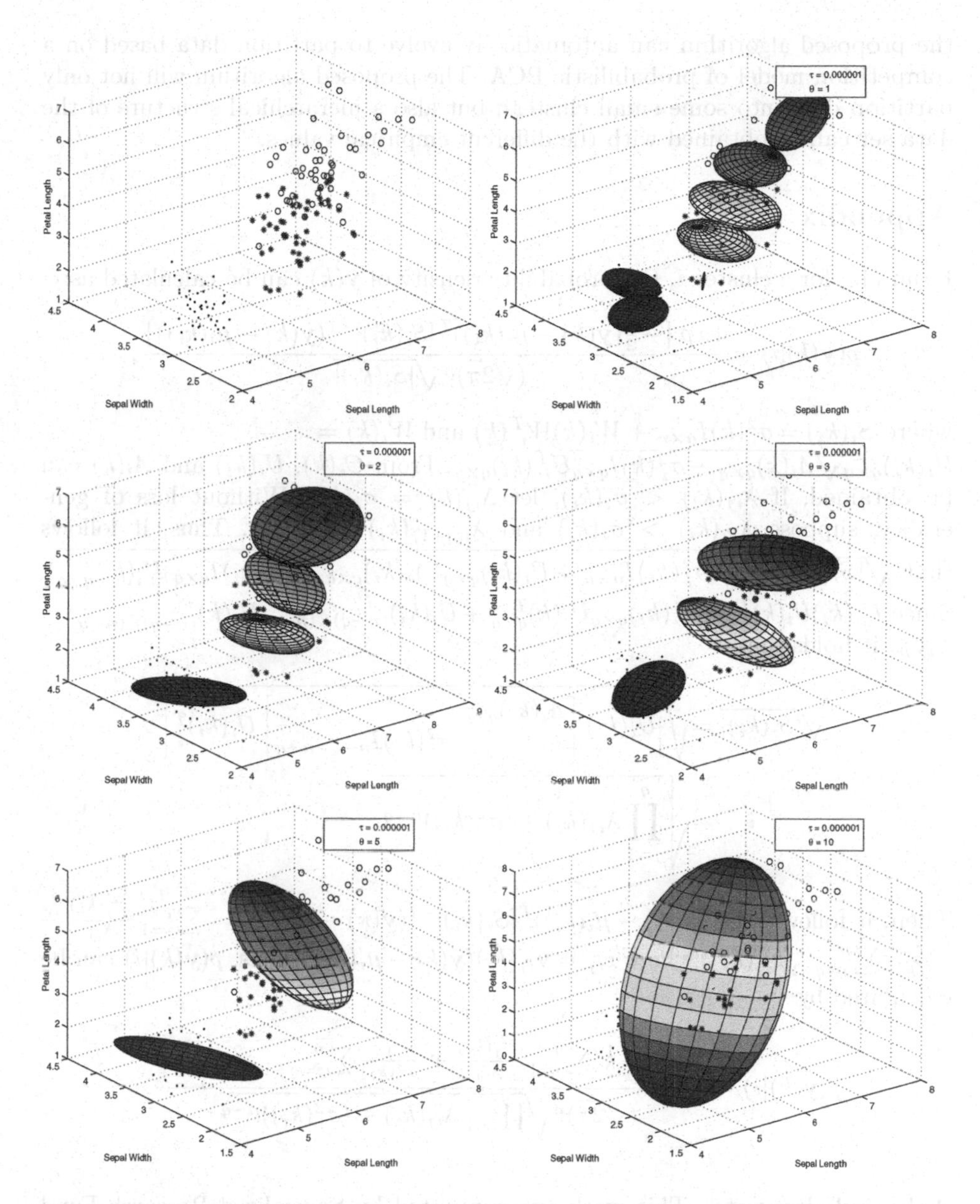

Fig. 4. The results of clustering analysis for iris data set; the original data of three iris categories(top left); The results of clustering analysis with $\theta = 1$ (top right), $\theta = 2$ (middle left), $\theta = 3$(middle right), $\theta = 5$ (bottom left) and $\theta = 10$ (bottom right)

5 Conclusions

An automatic clustering algorithm is proposed to explore the distribution structure of a given data set. Without identifying the number of clusters in advance, the proposed algorithm can automatically evolve to partition data based on a competition model of probabilistic PCA. The proposed algorithm will not only partition data into some small clusters, but also a hierarchical structure of the data set can be obtained with the different empirical values.

Appendix

From (1), for a cluster i, the probability density of $\mathbf{y}(k)$ can be calculated as:

$$p(\mathbf{y}(k)|i) = \frac{exp\left\{-\frac{1}{2}(\mathbf{y}(k) - \mu_i(k_i))^T[S_i(k_i)]^{-1}(\mathbf{y}(k) - \mu_i(k_i))\right\}}{(\sqrt{2\pi})^n\sqrt{|S_i(k_i)|}},$$

where $S_i(k_i) = \sigma_i^2(k)I_{n\times n} + W_i(k)W_i^T(k)$ and $W_i(k) = U_i(k_i)_{q\times q}\sqrt{\Lambda_i(k)_{q\times q} - \sigma_i^2(k)I_{q\times q}}U_i^T(k_i)_{q\times q}$. From $C_i(k)$, $U_i(k_i)$ and $\Lambda_i(k)$ can be obtained. If $\lambda_{ij}(k_i) < \sigma_i(k_i)$, let $\lambda_{ij}(k) = \sigma_i(k_i)$. Without loss of generality, suppose $\lambda_{iq}(k_i) > \sigma_i(k_i)$ and $\lambda_{i(q+1)}(k_i) < \sigma_i(k_i)$. Thus, it follows that $\sqrt{|S_i(k_i)|} = \sqrt{|\sigma_i^2(k_i)I_{n\times n} + U_i(k_i)_{q\times q}(\Lambda_i(k_i)_{q\times q} - \sigma_i^2(k_i)I_{q\times q})U_i(k_i)_{q\times q}^T|}$. Since $U_i(k_i)U_i(k)^T = U_i(k_i)_{q\times q}U_i(k)_{q\times q}^T + U_i(k_i)_{(n-q)\times(n-q)}U_i(k)_{(n-q)\times(n-q)}^T = I_{n\times n}$, it holds that

$$\begin{aligned}\sqrt{|S_i(k_i)|} &= \sqrt{\left|U_i(k_i)\begin{bmatrix}\Lambda_i(k_i)_{q\times q} & \\ & \sigma_i^2(k_i)I_{n-q\times n-q}\end{bmatrix}U_i(k_i)^T\right|}\\ &= \sqrt{\prod_{j=1}^{q}\lambda_{ij}(k_i) + (\sigma_i^2(k_i))^{n-q}}.\end{aligned}$$

Then, it follows that $(\mathbf{y}(k) - \mu_i(k_i))^T[S_i(k_i)]^{-1}(\mathbf{y}(k) - \mu_i(k_i)) = \sum_{j=1}^{q}\frac{z_{ij}^2(k_i)}{\lambda_{ij}} + \frac{1}{\sigma_i^2(k_i)}\sum_{j=q+1}^{n}z_{ij}^2(k_i)$, where $z_{ij} = \mathbf{v}_{ij}(k_i)(\mathbf{y}(k) - \mu_i(k))$. Thus, $p(\mathbf{y}(k)|i)$ can be calculated by

$$p(\mathbf{y}(k)|i) = \frac{exp\left\{-\frac{1}{2}(\sum_{j=1}^{q}\frac{z_{ij}^2(k_i)}{\lambda_{ij}} + \frac{1}{\sigma_i^2(k_i)}\sum_{j=q+1}^{n}z_{ij}^2(k_i))\right\}}{(\sqrt{2\pi})^n\sqrt{\prod_{j=1}^{q}\lambda_{ij}(k_i) + (\sigma_i^2(k_i))^{n-q}}}.$$

Acknowledgments. This work was supported by Specialized Research Fund for the Doctoral Program of Higher Education under Grant 2010081110053.

References

1. Honda, K., Ichihashi, H.: Regularized Linear fuzzy clustering and probabilistic PCA mixture models. IEEE Transactions on Fuzzy Systems 13(4), 508–516 (2005)
2. Lv, J.C., Yi, Z., Zhou, J.: Subspace Learning of Neural Networks. CRC Press, Taylor & Francis Group (2011)

3. Lv, J.C., Tan, K.K., Yi, Z., Huang, S.: A family of fuzzy learning algorithms for robust principal component analysis neural networks. IEEE Transactions on Fuzzy Systems 18(1), 217–226 (2010)
4. Lv, J.C., Tan, K.K., Yi, Z., Huang, S.: Convergence Analysis of Hyvärinen and Oja's ICA Learning Algorithms with Constant Learning Rates. IEEE Transactions on Signal Processing 57(5), 1811–1824 (2009)
5. Xu, R., Wunsch II, D.: Survery of clustering algorithms. IEEE Transactions on Neural Networks 16(3), 645–678 (2005)
6. Zhao, J., Jian, Q.: Probabilistic PCA for t distribution. Neurocomputing 69, 2217–2226 (2006)
7. Bruneau, P., Gelgon, M., Picarougne, F.: Parsimonious reduction of gaussian mixture models with a variational-bayes approach. Pattern Recognition 43(3), 850–858 (2010)
8. Archambeau, C., Delannay, N., Verleysen, M.: Mixtures of robust probabilistic principal component analysizers. Neurocomputing 71, 1274–1282 (2008)
9. Kambhatla, N., Leen, T.K.: Dimension reduction by local principal component analysis. Neural Computation 9(7), 1493–1516 (1997)
10. Möller, R., Hoffmann, H.: An extension of neural gas to local PCA. Neurocomputing 62, 305–326 (2004)
11. Tipping, M.E., Bishop, C.M.: Probabilistic principal component analyzers. J. Roy. Statist. Soc. Ser. B. 63, 611–622 (1999)
12. Tipping, M.E., Bishop, C.M.: Mixtures of probabilistic principal component analyzers. Neural Computation 11, 443–482 (1999)

Bayesian Inference to Sustain Evolvability in Genetic Programming

Ahmed Kattan[1] and Yew-Soon Ong[2]

[1] AI Real-World Application Lab, UQU, Saudi Arabia
[2] School of Computer Engineering, Nanyang Technological University, Singapore
ajkatta@uqu.edu.sa, asysong@ntu.edu.sg

Abstract. This paper proposes a new framework, referred to as Recurrent Bayesian Genetic Programming (rbGP), to sustain steady convergence in Genetic Programming (GP) (i.e., to prevent premature convergence) and effectively improves its ability to find superior solutions that generalise well. The term 'Recurrent' is borrowed from the taxonomy of Neural Networks (NN), in which a Recurrent NN (RNN) is a special type of network that uses a feedback loop, usually to account for temporal information embedded in the sequence of data points presented to the network. Unlike RNN, our algorithm's temporal dimension pertains to the sequential nature of the evolutionary process itself, and not to the data sampled from the problem solution space. rbGP introduces an intermediate generation between each subsequent generation in order to collect information about the offspring's fitness distribution of each parent. Placing the collected information into a Bayesian model, rbGP predicts the probability of any individual to produce offspring fitter than its parent. This predicted probability (calculated by the Bayesian model) is used by the tournament selection instead of the original fitness value. Empirical evidence, from 13 problems, against canonical GP, demonstrates that rbGP preserves generalisation in most cases.

1 Introduction

In our previous work [5], we introduced a new framework for Genetic Algorithm (GA) referred to as *Recurrent Genetic Algorithms* (RGA). RGA guided the evolutionary process of GA using a reverse form of fitness inheritance [3]. Smith et al. [13] first introduced the technique of fitness inheritance identifying two types of inheritance: the first which takes the average of the fitness values of the two parents and the second takes the weighted average according to the similarity between offspring and their parents. While the standard notion of fitness inheritance presented as a reward for offspring based on their parents' performance (assuming a level of smoothness in the search space), RGA uses a reversed concept of the standard fitness inheritance in which the parents' fitness values are readjusted based on the fitness of their offspring, thus presenting an indication of individuals' level of evolvability. To this end, RGA uses an intermediate population (called $\hat{P}$) between each subsequent generation. This

H. Handa et al. (eds.), *Proc. of the 18th Asia Pacific Symp. on Intell. & Evol. Systems – Vol. 1*,
Proceedings in Adaptation, Learning and Optimization 1, DOI: 10.1007/978-3-319-13359-1_7

intermediate population is used as a feedback loop that recurrently adjusts the fitness values of individuals in population P, at the ith generation, based on the fitness of their offspring in population $\hat{P}_i$ (i.e., the intermediate population). Empirical evidence illustrated that this recurrent process of fitness adjustment reinforces the evolvability of subsequent generations by ensuring that parents at P_i are rewarded for producing fit offspring and then given a second chance to reproduce.

In this work, we extend the RGA framework presented in [5] to Genetic Programming (GP) [12] and present a new framework referred to as *Recurrent Bayesian Genetic Programming* (rbGP). rbGP, also, introduces an intermediate population between each subsequent generation. However, away from the reversed fitness inheritance concept adopted by RGA, rbGP uses a Bayesian model [4] to readjust individuals' fitness values based on their probability to produce fitter offspring. To this end, rbGP forces each selected individual in population P_i, where i is the number of generation, to produce k number of offspring to generate population $\hat{P}_i$ (i.e., the intermediate population). Hence, rbGP collects information about the offspring fitness distribution of each selected parent and utilise this information to build a Bayesian model. rbGP employs the Bayesian model as method of inference to readjust the fitness values of individuals in population P_i. Thus, rbGP uses each individual's probability of producing fitter offspring (as measured by the k offspring when generating the intermediate population), and the likelihood of the population to produce fitter offspring, in the Bayesian model to rank individuals. To this end, individuals that may lead the search to premature convergence (i.e., their immediate fitness gain may not lead to long-term improvement in the search) receive lower rankings in order to prevent sudden premature convergence. Details of this process are provided in Section 3. In this paper, the term 'successful parent' will refer to parents that can produce fitter offspring.

This paper is organised into six sections. Section 2 reviews some related works. Section 3 explains rbGP in detail. Sections 4 and 5 discuss the experimental setup and the results, respectively. Finally, some conclusive remarks and future directions for research are presented in Section 6.

2 Related-Work

As mentioned earlier, rbGP uses Bayesian model as a method of inference to readjust the fitness values of individuals in order to prevent premature convergence. The whole process adopted by rbGP sustains evolvability. Therefore, the literature review focuses on previous works related to works that define the concept of evolvability and Bayesian models in GP.

2.1 Evolvability

The notion of "evolvability" is defined as *"the ability of a population to produce variants fitter than any yet existing"* [1]. Hence, generally, the choice of selection, search operator and representation is vital to the performance of GP because they control the creation of new individuals throughout the evolutionary process. One aim of researchers in the Evolutionary Computation (EC) field is to discover new methods for increasing evolvability of evolutionary systems. The term evolvability does not only refer to how often offspring are fitter than their parents but also to the entire distribution of fitness values among offspring produced by a group of parents [1].

The concept of evolvability has been an active research area in both evolutionary biology and computer science for the past several decades. Hu and Banzhaf in [9] have argued that adopting new knowledge about natural evolution generated in areas such as molecular genetics, cell biology, developmental biology, and evolutionary biology would benefit the field of evolutionary computation. The authors discussed evolvability and methods for accelerating artificial evolution by introducing notions from biology and their potential in designing new algorithms in EC.

It has been recorded that the evolvability property has good effect on the search process. For example, in [2], the authors suggested that evolvability can effectively reduce the bloat in evolutionary algorithms that use variable length representations. In their work, the authors noted the similarity of bloat causes and evolvability theory, thus, they argue that reproductive operators with high evolvability will be less likely to cause bloat.

With the importance of evolvability as a research topic, several measurements have been proposed to quantify it. Wang and Wineberg [14], suggested two measures of evolvability one based on fitness improvement and the other based on the amount of genotypic change. The authors divided the population into three sub-populations, where the size of each sub-population is determined dynamically. The first sub-population uses selection based on fitness directly; the second sub-population is based on the fitness-improvement-ratio; finally, selection for the third sub-population is based on genotypic change. Each sub-population is filled by selecting chromosomes from the parent's generation under its own selection functions. Thereafter, the three sub-populations are merged, and the standard GA search operators are applied to form the next generation. Experiments with several continuous optimisation functions showed that the proposed approach has higher evolvability (and consequentially achieves better solutions) than standard GA.

Hu in [8], proposed a new measurement for evolvability called "rate of genetic substitutions". This measurement method was used to investigate the effects of four major configuration parameters in EC (namely, mutation rate, crossover rate, tournament selection size, and population size) to show the effectiveness of these parameters with respect to evolution acceleration. In his work, Hu has developed a new indicator based on this proposed measurement for adjusting population size dynamically during evolution.

2.2 Bayesian Models for GP

Bayesian probability model is an interpretation of the concept of probability and belongs to the category of evidential probabilities. To evaluate a hypothesis' probability, the Bayesian probability model needs to specify a prior distribution of probabilities (i.e., training data), which can then be updated in the light of new relevant data. Relatively few works in the literature have used Bayesian probability model to enhance GP process.

Zhang [16,17] proposed a Bayesian framework for GP based on the Bayesian approach in which, under GP, individuals are viewed as models of the fitness data. Bayes theorem is used to estimate the posterior probabilities of programs based on their prior probabilities and likelihood of fitness in observed cases. Offspring programs are then generated by sampling from the posterior distribution by using genetic operators. This work presented two methods for Bayesian GP: 1) GP with the adaptive Occam's razor designed to evolve parsimonious program, and 2) GP with incremental data inheritance designed to accelerate evolution by active selection of training cases. All these methods are implemented as adaptive fitness functions that take into account the dynamics of evolutionary processes.

Yanai and Iba [15] proposed Estimation of Distribution Programming (EDP) based on GP extension. In their work, a probability distribution expression using a Bayesian network was used to generate individuals instead of standard search operators and the Bayesian network described the dependency relationship of probabilistic nodes. Truncation selection selects the individuals with the top fitness, analyses their structure, and estimates the probability distribution of these superior individuals is estimated. Later, Hasegawa and Iba [7] introduced a new tree-like program evolution algorithm employing a Bayesian network for generating new individuals. It employs a special chromosome called the expanded parse tree, which significantly reduces the size of the conditional probability table.

As can be seen, most previous works used a Bayesian model or Bayesian network as a method to create individuals, in a similar manner to Estimated Distribution Algorithms (EDA). In this paper, we utilise the Bayesian model to enhance the GP evolutionary process in a novel way that, to the best of our knowledge, has never been proposed before. To this end, Bayesian rule is used to prevent premature convergence that could occur in the canonical GP iteration process. Further details of this process are provided in Section 3.

3 Recurrent Bayesian Genetic Programming

Generally, premature convergence occur because selection pressure may encourages dense congregations of homogeneous solutions, a key characteristic of premature convergence [10]. It is reasonable to hypothesise that mature convergence is inhibited by the loss of potentially useful genetic material due to the replacement strategies undertaken by search operators wherein worst individuals are replaced by new offspring. This dynamic may allow some individuals, that seems potential, to exist for multiple evolutionary cycles within the population and may hinder the exploration of superior areas in the search space. rbGP techniques aim to reduce premature convergence by ranking individuals in the population based on their level of evolvability relative to the performance of the whole population.

The process adopted by rbGP is broadly outlined in figure 1. Similar to standard GP, rbGP starts by randomly initialising a population P_0, where the number of generations is $i = \{0, ..., n\}$, and calculates their fitness values using the given fitness measure. However, unlike standard GP, instead of driving the population to generate the next generation, rbGP generates an intermediate population $\hat{P}_i$ to collect observations about the population's performance. To this end, rbGP applies standard tournament selection to identify potential individuals where it forces each individual to generate k number of offspring to constitute $\hat{P}_i$ (i.e., the intermediate population). Hence, the size of the $\hat{P}_i = k \times size(P_i)$. rbGP uses standard tournament selection to select the parents of the individuals in $\hat{P}_i$. Naturally, some individuals in P_i might never be selected while other individuals might be selected more than once. During the creation process of $\hat{P}_i$, rbGP notes the number of successful and unsuccessful offspring generated by each selected individual participated in $\hat{P}_i$. By 'successful' offspring we mean the ones that their fitness values are better than their parents while 'unsuccessful' means the opposite. Using the collected observations, rbGP builds two sets for the population P_i. First, *numbers of successful offspring* and second, *numbers of failure offspring*. Let the set of successful offspring for P_i be represented as $D_s(P_i) = \{|o_0^s|, |o_1^s|, ..., |o_m^s|\}$ where $|o_j^s|$ is the number of successful offspring generated by the jth individual participated in generating $\hat{P}_i$ and m is the population size. In addition, let the set of failure offspring of P_i be denoted as $D_f(P_i) = \{|o_0^f|, |o_1^f|, ..., |o_m^f|\}$ where $|o_j^f|$ is the number of failure offspring generated by the jth selected individual from P_i. Note that both sets $D_s(P_i)$ and $D_f(P_i)$ are built using information from the selected individuals from P_i and those individuals that never selected will be ignored from both sets. The sets $D_s(P_i)$ and $D_f(P_i)$ can represent the convergence state of the population P_i. Hence, naturally, a high mean of $D_s(P_i)$ and a low mean of $D_f(P_i)$ may indicate that the individuals in P_i are scattered in the search space and that the population remains far from the global optimum. A low mean of $D_s(P_i)$ and a high mean of $D_f(P_i)$ though, might indicate the opposite, that P_i individuals are already approaching optimum solutions (perhaps a local optimum), and thus, it is difficult to find further superior offspring.

Using a Bayesian model, rbGP analyses the *probability of evolvability* (i.e., probability of success) for each individual participated in constitution of the intermediate population based on their number of successful offspring versus their number of failure offspring relative to the whole population and ranks them accordingly. This ranking process is calculated as following:

$$P(I_j^f|P_{if}) = \frac{P(P_{if}|I_j^f)P(I_j^f)}{P(P_{if}|I_j^f)P(I_j^f) + P(P_{is}|I_j^s)P(I_j^s)} \tag{1}$$

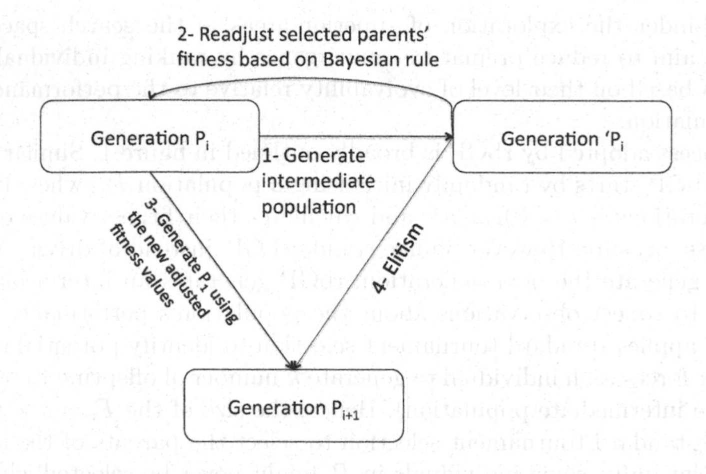

Fig. 1. rbGP process outline

where $P(I_j^f)$ is the probability of the jth individual to produce inferior offspring. This variable is calculated as $\frac{|o_j^f|}{|o_j^f|+|o_j^s|}$. The term $P(P_{if}|I_j^f)$ refers to the likelihood of individual I_j produce inferior offspring given P_{if} (i.e., P_{if} refer to the probability of the whole population P_i to produce inferior offspring). The term $P(I_j^s)$ refer to the probability of the individual I_j producing better offspring. The $P(I_j^s)$ variable can be calculated as $\frac{|o_j^s|}{|o_j^f|+|o_j^s|}$. Finally, $P(P_{is}|I_j^s)$ is the likelihood of individual I_j produce better offspring given P_{is} (i.e., P_{is} refer to the probability of population P_i producing better offspring).

We assumed that the set $D_f(P_i)$ have a Gaussian distribution and that the likelihood is calculated as follows:

$$P(P_{if}|I_j^f) = \frac{1}{\sqrt{2\pi\sigma_f}} e^{\frac{(|o_j^f|-\mu_f)^2}{2\sigma_f}} \tag{2}$$

where μ_f and σ_f denote the mean and variance of $D_f(P_i)$, respectively. Also, the likelihood of $P(P_{if}|I_j^f)$ in $D_s(P_i)$ is calculated in similar manner.

$$P(P_{is}|I_j^s) = \frac{1}{\sqrt{2\pi\sigma_s}} e^{\frac{(|o_j^s|-\mu_s)^2}{2\sigma_s}} \tag{3}$$

where μ_s and σ_s denote the mean and variance of $D_s(P_i)$, respectively. Now, according to Equation 1, individuals are ranked based on their probability of evolvability in relation to the whole population. Individuals that have a higher potential evolvability level than the whole population receive lower ranks. To this end, rbGP readjusts individuals' fitness values according to their probability of evolvability relative to the whole population and their fitness values as:

$$Rank(I_j) = P(I_j^f|P_{if}) \times fitness(I_j) \tag{4}$$

Individuals that have been ignored by the selection process during generating the intermediate population will automatically receive rank of 0. This is because the system has no information about their level of evolvability.

The main disadvantage of rbGP is that it requires to produce several offspring for each selected parent when generating the intermediate generation (in our case it is 100 offspring) in order to constitute meaningful distributions for $D_f(P_i)$ and $D_s(P_i)$ which may be computationally expensive. However, as we will see in the experiments in Section 5, rbGP managed to evolve good solutions in small number of generations.

3.1 Elitism

The ranking process, described in Equation 4, could underestimate some potential solutions, that appear in early stages of the search, and assign them lower ranks, which will reduce their chances of participating in P_{i+1}, thus hindering progress of the search. Therefore, rbGP copies the best individual from P_i to P_{i+1} to preserve potentially useful genetic material from being lost.

In addition, as illustrated in figure 1, rbGP copies the best individuals from $\hat{P}_i$ to P_{i+1}. The logic for this is that rbGP has already devoted considerable computational efforts to generate the intermediate population ($\hat{P}_i$) and it is reasonable to utilise this efforts in the search process.

4 Experimental Setup

Experiments have been devised to compare the proposed rbGP model against standard GP. The main aim of the experiments is to evaluate the performance of the rbGP and to assess the algorithms behaviour under a variety of circumstances. Our experimental study included 15 problems, 12 symbolic regression problems and 3 time-series problems. The symbolic regression covered a variety of functions: polynomial, trigonometric, logarithmic and square-root and complex functions. These functions selected because they represent different landscapes. Thus we stress rbGP under different search conditions. For each symbolic regression problem, we uniformly sampled 200 data points from the interval $[-5, 5]$. These points were divided into 50 for training, 50 for validation, and 100 for testing. Table 1 shows the problems included in our experimental study. GP evolved solutions to minimise the average absolute error on the training set. The best individual in each generation is further tested on the validation set and the best individual across the whole run (i.e., the one with best performance on the validation set) is tested with the testing set.

For the real-world time-series problem, we used data from *Google Trends* service [6], a free service offering data about the search terms that people enter into Google's search engine. The service provides free downloadable historical time-series data about any keyword. It, also, offers the flexibility to restrict the search by country. One use of Google Trends is for E-Marketing managers to monitor how often people type certain keywords related to their products at different times of the year. Using this information, E-Marketing managers can determine the best time to release their marketing campaigns so their advertisements coincide with peoples searches and eventually achieve higher hit rates. For the purpose of our experiments, we imported time-series data about searches for the following keywords; *Jobs, Holidays,* and *Cinema* and we restricted the search to get data from *USA, USA* and *UK*, respectively. All the imported data from Google Trends represent the weekly frequencies of these keywords between January 2004 and May 2013. The data yielded 490 data points. We used a sliding window of size 5 to capture the average values of 5 consecutive weeks, which was input to GP in order to predict the value of the next week. GP received inputs of averages for weeks $w_i : w_{i+5}$ where $i = \{1...490\}$ and the expected output is the value of point in location w_{i+6}. Data were divided into 50% Training, 20% Validation, and 30% Testing sets. Here, the aim is to predict the frequency of keywords searches (treating the testing set as unseen weekly observations in the future) so as to help employers to select the best time to advertise their new jobs, travel agencies to predict the best time to release holidays packages, and media companies to preview new shows at the most appropriate time.

We compared rbGP against standard GP (SGP). Both systems received exactly the same settings and the same number of evaluations, to ensure fair comparison, as illustrated in table 2. In our experiments, we considered the number of consumed evaluations in the intermediate generations in rbGP and allocated exactly the same evaluation budget to SGP. To this end, rbGP was set to search

Table 1. Test problems included in the experimental study

Problem	*Notation*	*Type*	*Variables*
F0	$f(x) = 5x^3 + 2x^2 + x + 5$	Polynomial	1
F1	$f(x) = 5x^2 + 2x^2 + x$	Polynomial	1
F2	$f(x) = tan(x) + sin(x)$	Trigonometric	1
F3	$f(x) = 5\sqrt{(\lvert x\rvert)}$	Square root	1
F4	$f(x) = 1 - log(x^2 + x + 1)$	Logarithmic	1
F5	$f(x) = \frac{1}{100+log(x^2)+\sqrt{\lvert x\rvert}}$	Logarithmic	1
F6	$f(x) = log(x^3)$	Logarithmic	1
F7	$f(x, y) = sin(atan(y, x)\sqrt{x^2 + y^2} \times 6\pi)$	Complex	2
F8	$f(x) = 5x^4 + 5x^3 + 2x^2 + x + 5$	Polynomial	1
F9	$f(x, y) = (1.5 - x + xy)^2 + (2, 25 - x + xy^2)^2 + (2.625 - x + xy^3)^2$	Complex	2
F10	$f(x_n) = 10 \times \sum_n^{i=1} x_i{}^2 - 10 \times cos(2\pi x_i)$	Complex	5
F11	$f(x_n) = 10 \times \sum_n^{i=1} x_i{}^2 - 10 \times cos(2\pi x_i)$	Complex	10
F12	Time-Series, Google Trends (Jobs, USA)	Prediction	490
F13	Time-Series, Google Trends (Holidays,USA)	Prediction	490
F14	Time-Series, Google Trends (Cinema, UK)	Prediction	490

Table 2. Parametric settings of the algorithms considered in the experiments

Parameter	*Standard GP Setting*
Sub-tree Mutation	70%
Sub-tree Crossover	30%
Tournament size	2
Population Size	5050
Generations	20
Elitism	2%
Parameter	*rbGP Settings*
Sub-tree Mutation	70%
Sub-tree Crossover	30%
Tournament size	2
Population Size	50
Intermediate Population Size	5000 (100 offspring for each parent)
Generations	20
Elitism ($\hat{P}_i$)	1%
Elitism (P_i)	1%

the search space using 50 individuals. For each selected individual, rbGP generates 100 different offspring to constitute the intermediate generation. SGP was set to search the search space using 5050 individuals. Thus, received exactly the same search budget as rbGP. For every problem, we tested each system through 50 independent runs.

5 Results

Tables 3, 4 and 5 summarise 1500 independent runs. As stated, for each problem, we tested and compared each system in 50 independent runs and report the *mean* and *median* of the best evolved solutions by each system. In addition, we report the *best* solution found by each system across the whole 50 runs. As can be seen, for the symbolic regression problems, rbGP achieved the best mean and median in 7 problems and the best solution in 8 problems. Both rbGP and SGP almost have similar performance in problem $F7$. We observed that in the cases that rbGP outperforms its competitor, it finds solutions better than SGP at margins varying from 0.02% to 54%. Note that results are obtained from the performance of the best tree on an unseen testing set to reflect the generalisation ability of each system. To further verify the statistical significance of our results, a Kolmogorov-Smirnov two-sample test [11] has been performed. Table 6 reports the *P-value* for the tests. In all cases where rbGP achieved better results, P-value is statistically significantly superior to SGP at the standard 95% significance level. Interestingly, the P-value also show a statistical significance in the three problems where rbGP was outperformed by SGP (namely, $F2$, $F10$, and $F11$).

Table 3. Summary results of 600 independent runs (for problems F0 - F5). Results are sampled from 50 independent runs form each system in each problem.

	Mean	Median	Best
F0			
rbGP	**12.558**	**5.252**	**0.469**
GP	22.628	16.985	3.533
F1			
rbGP	**1.254**	**3.342E-01**	**4.586E-05**
GP	1.882	1.900E+00	0.318
F2			
rbGP	2.532	6.693E-01	0.001
GP	**1.748**	**6.374E-01**	0.001
F3			
rbGP	**0.009**	**2.457E-06**	**8.839E-07**
GP	0.020	1.159E-02	6.018E-05
F4			
rbGP	**0.232**	**0.174**	**0.052**
GP	0.350	0.329	0.159
F5			
rbGP	**0.329**	**0.309**	**0.060**
GP	0.493	0.482	0.286

***Bold** numbers are the lowest.

Table 4. Summary results of 600 independent runs (for problems F6 - F11). Results are sampled from 50 independent runs form each system in each problem.

	Mean	Median	Best
F6			
rbGP	**0.292**	**0.289**	**0.038**
GP	0.296	0.292	0.170
F7			
rbGP	**0.866**	0.867	0.762
GP	0.868	0.867	**0.737**
F8			
rbGP	163.436	**48.896**	**3.898**
GP	**115.343**	111.453	13.628
F9			
rbGP	53430.451	13927.500	**848.112**
GP	**13706.556**	**12422.200**	4300.640
F10			
rbGP	30.875	24.376	15.949
GP	**21.714**	**18.614**	**15.789**
F11			
rbGP	37.730	36.177	26.759
GP	**33.159**	**33.377**	**25.031**

***Bold** numbers are the lowest.

The only two cases that P-Value shows statistical significance below 95% is in problems $F6$ and $F7$.

For the time-series prediction problems, rbGP served best as measured by mean and median in 2 out of 3 problems and achieved the best solution in only one problem. rbGP's improvement margins varying from 0.02% to 6% and the loss margins from 0.04% to 0.007%. Generally, both SGP and rbGP achieved almost equal performance on the three time-series prediction problems.

5.1 Discussion

The results are encouraging in the sense that rbGP is already doing well compared to SGP and it is not too far behind in the cases that it loses the compression. We believe rbGP still has room for further improvements. Apart from

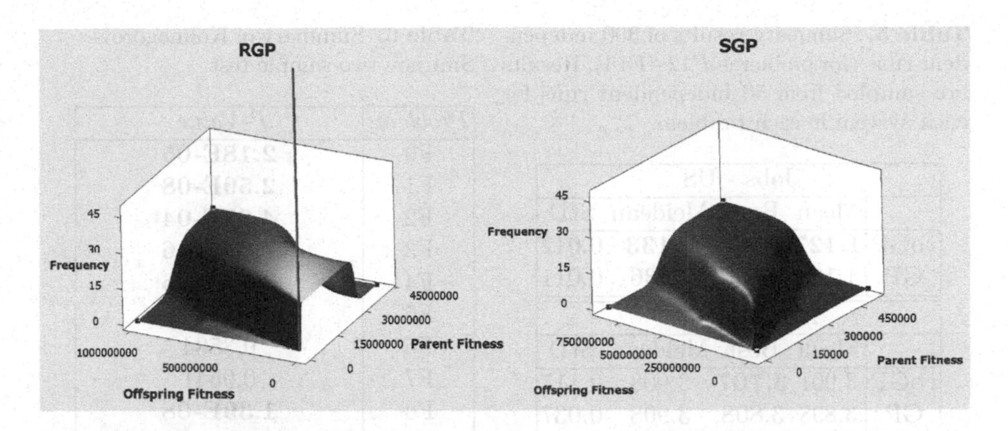

Fig. 2. Fitness distribution for problem $F0$ accumulated from 50 independent runs. The graph show the fitness values of selected parent against the fitness values of their offspring and the frequency of their occurrence in the search process.

rbGP's good generalisation ability, we noted that rbGP is outperforming SGP in all the experiments on the training cases. Naturally, the performance on unseen testing data is much more important than the performance on any given training set. However, it should be noted that despite being competitive on unseen testing sets, rbGP shows remarkable resistance to the over-fitting problem.

Surprisingly, we observed that rbGP produces larger trees, in some cases, than SGP. This is interesting because rbGP effectively explores the search space using 50 individuals while using the intermediate generation as indicator re-rank individuals based on their level of evolvaiblity (as described in Section 3) while SGP explores the search space using 5050 individuals (the same exploration budged allocated to rbGP). Therefore, it is natural to assume that SGP's population will bloat faster. However, as the demonstrated by the experiments, this is not the case. We believe that rbGP bloat at faster rate than its competitor for two reasons. The first reason, as stated in Section 3.1, is because rbGP copies the best individuals from the intermediate generation $\hat{P}_i$ to P_{i+1}. This can accelerate the bloat. The second reason is that the whole process undertaken by rbGP ensures to enhance the evolvability. The fitness improvement, generally, occur across the whole population. This is further confirmed in figure 2 where the fitness distribution of both rbGP and SGP for problem $F0$ is visualised. The figure shows the fitness values of the selected parents against the fitness values of their offspring and the frequency of their occurrence in the search process, accumulated from 50 runs. For SGP (on the left side of the figure), the figure shows that SGP's search, generally, dominated by poor parents that produce poor offspring and then some parents produce good offspring which make the search converges toward an optimum. In SGP most of the search budget is wasted in areas where poor parents produce poor offspring. However, in rbGP (on the right side of the figure) parents that produce good offspring are more and then the whole population become dominated by good individuals (shown by the large peak on

Table 5. Summary results of 300 independent runs (for problems $F12$ - $F14$). Results are sampled from 50 independent runs for each system in each problem.

Jobs - US				
	Mean	Best	Meidean	StD
rbGP	**1.127**	1.102	**1.133**	0.012
GP	1.133	1.100	1.136	0.011
Holidays - US				
	Mean	Best	Meidean	StD
rbGP	3.901	**3.707**	3.908	0.117
GP	3.898	3.808	3.908	0.037
Cinema - UK				
	Mean	Best	Meidean	StD
rbGP	**9.051**	**8.449**	**9.039**	0.334
GP	9.096	8.632	9.050	0.226

**Bold* numbers are the lowest.

Table 6. Summary of Kolmogorov-Smirnov two-sample test.

Problem	*P-Value*
F0	**2.18E-05**
F1	**2.59E-08**
F2	**4.23E-04**
F3	**7.23E-16**
F4	**3.63E-06**
F5	**1.08E-08**
F6	0.3584
F7	0.9541
F8	**1.39E-08**
F9	**6.43E-01**
F10	**1.26E-07**
F11	**4.89E-05**
F12	0.8253
F13	**0.0440**
F14	**0.0440**

***Bold** numbers are less than 5%.

the figure's corner). This indicates that most of the allocated search budget was well utilised and rbGP directed the search effectively. Thanks to the re-ranking process of parents' fitness values.

6 Conclusions and Future Work

In this paper, introduced a new framework, referred to as Recurrent Bayesian Genetic Programming (rbGP), to sustain evovability. rbGP generates an intermediate population to collect statistical observations about the populations performance and incorporate this information into a Bayesian model. The Bayesian model is trained with the collected observations and used as a method of inference to readjust fitness values of individuals based on their performance and likelihoods of driving the population into premature convergence.

To verify the usefulness of rbGP, we conducted an experimental study that included 15 non-trivial problems. rbGP has been compared against standard GP. The results were promising in the sense that rbGP outperformed its competitor in most cases and when it lost the comparison it is not too far behind. Moreover, we believe that rbGP performance can still be improved in future work. Furthermore, results indicate the Bayesian ranking process make rbGP remarkably resistance to the over-fitting.

For the future work, one direction is to explore the idea of using a metric that establishes the best GP trees to breed from in one stage method rather than this two-stage process. Another direction to explore is eliminating the intermediate population and allowing the Bayesian model to continuously learn population distributions as new evidences emerges.

References

1. Altenberg, L.: The evolution of evolvability in genetic programming. In: Kinnear Jr., K.E. (ed.) Advances in Genetic Programming, ch. 3, pp. 47–74. MIT Press (1994)
2. Bassett, J.K., Coletti, M., De Jong, K.A.: The relationship between evolvability and bloat. In: Proceedings of the 11th Annual Conference on Genetic and Evolutionary Computation, GECCO 2009, pp. 1899–1900. ACM, New York (2009)
3. Ducheyne, E., De Baets, B., De Wulf, R.: Is fitness inheritance useful for real-world applications? In: Fonseca, C.M., Fleming, P.J., Zitzler, E., Deb, K., Thiele, L. (eds.) EMO 2003. LNCS, vol. 2632, pp. 31–42. Springer, Heidelberg (2003)
4. Ellison, A.M.: Bayesian inference in ecology. Ecology Letters 7(6), 509–520 (2004)
5. Fakeih, A., Kattan, A.: Recurrent genetic algorithms: Sustaining evolvability. In: Hao, J.-K., Middendorf, M. (eds.) EvoCOP 2012. LNCS, vol. 7245, pp. 230–242. Springer, Heidelberg (2012)
6. Google. Google insights (June 2013), `http://www.google.com/trends/`
7. Hasegawa, Y., Iba, H.: A Bayesian network approach to program generation. IEEE Transactions on Evolutionary Computation 12(6), 750–764 (2008)
8. Hu, T.: Evolvability and Rate of Evolution in Evolutionary Computation. PhD thesis, Department of Computer Science, Memorial University of Newfoundland, ST. John's, Newfoundland, Canada (May 2010)
9. Hu, T., Banzhaf, W.: Evolvability and speed of evolutionary algorithms in light of recent developments in biology. J. Artif. Evol. App. 2010, 1:1–1:28 (2010)
10. Murphy, G.P.: Manipulating Convergenc. In: Evolutionary Systems. PhD thesis, University of Limerick, Ireland (May 19, 2009)
11. Peacock, J.A.: Two-dimensional goodness-of-fit testing in astronomy. Royal Astronomical Society, Monthly Notices 202, 615–627 (1983)
12. Poli, R., Langdon, W.W.B., McPhee, N.F.: Field Guide to Genetic Programming. Lulu Enterprises Uk Limited (2008)
13. Smith, R.E., Dike, B.A., Stegmann, S.A.: Fitness inheritance in genetic algorithms. In: Proceedings of the 1995 ACM Symposium on Applied Computing, SAC 1995, pp. 345–350. ACM, New York (1995)
14. Wang, Y., Wineberg, M.: The estimation of evolvability genetic algorithm. In: The 2005 IEEE Congress on Evolutionary Computation, vol. 3, pp. 2302–2309 (September 2005)
15. Yanai, K., Iba, H.: Estimation of distribution programming based on bayesian network. In: The 2003 Congress on Evolutionary Computation, CEC 2003, vol. 3, pp. 1618–1625 (2003)
16. Zhang, B.-T.: Bayesian genetic programming. In: Haynes, T., Langdon, W.B., O'Reilly, U.-M., Poli, R., Rosca, J. (eds.) Foundations of Genetic Programming, Orlando, Florida, USA, pp. 68–70 (July 13, 1999)
17. Zhang, B.-T.: Bayesian methods for efficient genetic programming. Genetic Programming and Evolvable Machines 1(3), 217–242 (2000)

Synthesis of Clock Signal from Genetic Oscillator

Chia-Hua Chuang and Chun-Liang Lin

Department of Electrical Engineering, National Chung Hsing University,
Taichung 402, Taiwan
chunlin@dragon.nchu.edu.tw

Abstract. This paper attempts to design a genetic frequency synthesizer circuit with counter to synthesize a clock signal whose frequency is a multiple of that of an existing synthetic genetic oscillator. A genetic waveform-shaping circuit constructed by Buffers in series is used to reshape a genetic oscillation signal into a pulse-width-modulated (PWM) signal with different duty cycle. Design of the Buffers and accompanied genetic logic gates is based on the use of the real genetic structural genetic algorithm. By assembling different PWM signals, a series of clock pulses is synthesized as the rising and falling edges of the desired clock signals triggering the counter and the clock signal with the integer multiple of base frequency is generated from the same oscillation signal. Simulation results show that the proposed genetic frequency synthesizer circuit is effective to realize a variety of genetic clocks.

Keywords: circuit synthesis, biology, repressilator, logic gate, synthesizer.

1 Introduction

Synthetic biology is developed to construct an artificial genetic circuit using the approaches of mathematics and engineering [1-3]. Several synthetic genetic circuits have successfully been built to achieve the basic functions, e.g. genetic oscillator generates a periodic oscillation signal and genetic logic circuit performs biological logical computation which is an important device for constructing the more complicated bio-computers [4-7]. Based on a bottom-up approach, more complicated bio-computing processes can be expected to perform specific functions, like very-large-scale integration (VLSI) circuits in electronic systems.

There are many engineering methods proposed to convert a synthetic design problem into a tracking optimization problem. Based on those methods, a class of synthetic genetic circuits is implemented for new tasks. In [8], a robust design approach based on H_∞ optimization theory is proposed to construct a robust synthetic genetic oscillator with the desired sustained periodic oscillation behavior under the stochastic perturbational environment. For embedding the synthetic genetic circuit into the host cell, a combined parameter and structure optimization formulation has been attempted. A real structural genetic algorithm (RSGA) has been applied to synthesize a class of genetic logic circuits with the minimal number of genes while ensuring acceptable performance [9, 10]. A variety of clock signals is synthesized

H. Handa et al. (eds.), *Proc. of the 18th Asia Pacific Symp. on Intell. & Evol. Systems – Vol. 1*,
Proceedings in Adaptation, Learning and Optimization 1, DOI: 10.1007/978-3-319-13359-1_8

from the base-frequency oscillation signal using genetic frequency synthesizer circuits [11, 12]. One may refer, for example, to [13] for other applications oriented from the RSGA. In which, comparison of a class of GA-based algorithms can also be found.

This paper presents a genetic frequency synthesizer circuit with counter to generate a genetic clock signal based on the existing synthetic genetic oscillator. Frequency of the clock signal is multiple to that of the genetic oscillator. A genetic waveform-shaping circuit is firstly constructed by Buffers in series to reshape genetic oscillation into an approximate clock signal with explicit discrimination between low and high logic levels. To regulate the different threshold levels of a "Buffer", a pulse-width-modulated (PWM) signal with different duty cycles in a sinusoidal cycle can be generated on the same oscillation signal. Based on the specific feature, the rising and falling edges of an ideal clock signal can be determined and a series of clock pulse signals is obtainable by assembling the different PWM signals. Appling the generated clock pulse signal to trigger a 1-bit genetic counter constructed by JK flip-flop, the desired genetic clock with the integer number of frequency of genetic oscillator can be realized.

Different from [11], an edge-triggered genetic frequency synthesizer circuit is designed to construct the clock signal with a multiple of frequency of the genetic oscillator. Simulation results *in silico* show performance of the synthetic genetic clock with base frequencies while operating at a genetic oscillator.

2 Model Description

In biological systems, the synthetic genetic network with L genes is described by the following nonlinear Hill differential equation [6, 11, 12]

$$\dot{p}_i = \alpha_i f_i(u) - \beta_i p_i + \alpha_{0,i},\ i = 1,\ldots,L \tag{1}$$

where p_i is the concentration of protein for gene i, α_i, β_i and $\alpha_{0,i}$ are, respectively, the synthesis, degradation and basal rates, $f_i(\cdot)$ denotes the promoter activity function used to describe the nonlinear transcriptional logic reactions, and u is the concentration of transcription factor (TF) from other gene's produces or inducers to control the gene expression.

For a gene with an operator site, the promoter activity functions for the genetic logic NOT and the Buffer are described, respectively, as

$$f_{\mathrm{NOT}}(u) = \frac{1}{1 + \left(\frac{u}{K}\right)^n} \tag{2}$$

and

$$f_{\mathrm{Buffer}}(u)=\frac{\left(\frac{u}{K}\right)^{n}}{1+\left(\frac{u}{K}\right)^{n}} \tag{3}$$

where f_{NOT} and f_{Buffer} are the promoter activity functions for logic NOT and Buffer, respectively, u is the concentration of a repressor or activator TF, n is the Hill coefficient, and K is the Hill constant.

For a gene with two operator sites, the promoter activity functions for the genetic logic AND, OR, NAND, NOR and XOR gates are described as

$$f_{\mathrm{AND}}(u_1,u_2)=\frac{\left(\frac{u_1}{K_1}\right)^{n_1}\left(\frac{u_2}{K_2}\right)^{n_2}}{1+\left(\frac{u_1}{K_1}\right)^{n_1}+\left(\frac{u_2}{K_2}\right)^{n_2}+\left(\frac{u_1}{K_1}\right)^{n_1}\left(\frac{u_2}{K_2}\right)^{n_2}} \tag{4}$$

$$f_{\mathrm{OR}}(u_1,u_2)=\frac{\left(\frac{u_1}{K_1}\right)^{n_1}+\left(\frac{u_2}{K_2}\right)^{n_2}+\left(\frac{u_1}{K_1}\right)^{n_1}\left(\frac{u_2}{K_2}\right)^{n_2}}{1+\left(\frac{u_1}{K_1}\right)^{n_1}+\left(\frac{u_2}{K_2}\right)^{n_2}+\left(\frac{u_1}{K_1}\right)^{n_1}\left(\frac{u_2}{K_2}\right)^{n_2}} \tag{5}$$

$$f_{\mathrm{NAND}}(u_1,u_2)=\frac{1+\left(\frac{u_1}{K_1}\right)^{n_1}+\left(\frac{u_2}{K_2}\right)^{n_2}}{1+\left(\frac{u_1}{K_1}\right)^{n_1}+\left(\frac{u_2}{K_2}\right)^{n_2}+\left(\frac{u_1}{K_1}\right)^{n_1}\left(\frac{u_2}{K_2}\right)^{n_2}} \tag{6}$$

$$f_{\mathrm{NOR}}(u_1,u_2)=\frac{1}{1+\left(\frac{u_1}{K_1}\right)^{n_1}+\left(\frac{u_2}{K_2}\right)^{n_2}+\left(\frac{u_1}{K_1}\right)^{n_1}\left(\frac{u_2}{K_2}\right)^{n_2}} \tag{7}$$

and

$$f_{\mathrm{XOR}}(u_1,u_2)=\frac{\left(\frac{u_1}{K_1}\right)^{n_1}+\left(\frac{u_2}{K_2}\right)^{n_2}}{1+\left(\frac{u_1}{K_1}\right)^{n_1}+\left(\frac{u_2}{K_2}\right)^{n_2}+\left(\frac{u_1}{K_1}\right)^{n_1}\left(\frac{u_2}{K_2}\right)^{n_2}} \tag{8}$$

where f_{AND}, f_{OR}, f_{NAND}, f_{NOR} and f_{XOR} are, respectively, the promoter activity functions for logic AND, OR, NAND, NOR and XOR reactions, u_1 and u_2 are the concentrations of two repressor or activator TFs, K_1 and K_2 are Hill constants for

u_1 and u_2, respectively, and n_1 and n_2 are the corresponding Hill coefficients. Determination of the key parameters in (4) to (8) was obtained using the RSGA [10].

3 Synthesis of Clock Signal

3.1 Realization of Genetic Waveform-Shaping Circuit

A waveform-shaping circuit in electronics is constructed to convert the oscillation signal to the clock signal with the explicit logic edge. As the red line shown in Fig. 1, the input signal is mapped to the low level when it is less than a threshold level y_T while it is mapped to the high level when it exceeds the threshold level. In reality, this curve in biological systems doesn't exist. A more reasonable input and output (I/O) characteristic curve of a sigmoid function (blue line in Fig. 1) is thus considered. There are two operational regions: saturation region and transition region. The input signal in the saturation region is mapped to the high level or the low level. In the transition region, the gain in the operation point y_T must be larger than (normalized) 1 to ensure that the input will be amplified when it is larger than the threshold level and decayed when it is less than the threshold level. By connecting a series of sigmoid functions, the input signal will gradually reach the saturation region and stay at the high or low level.

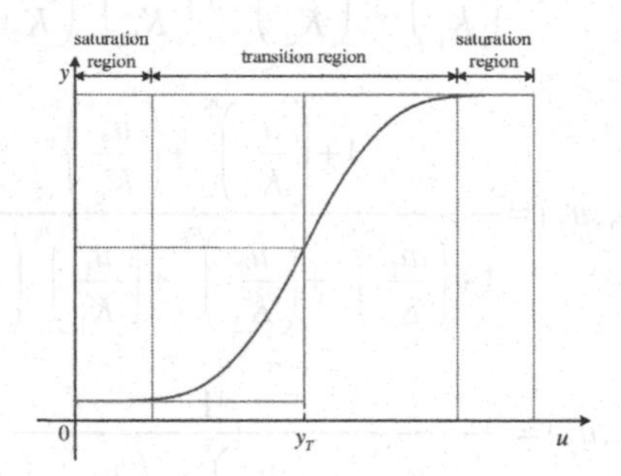

Fig. 1. I/O characteristic curve

According to this idea, we use several cascaded Buffers to realize the genetic waveform-shaping circuit described by [11, 12]

$$\dot{p}_{B_k} = \alpha_{B_k} f_{\text{Buffer},k}\left(u_k, K_k, n_k\right) - \beta_{B_k} p_{B_k} + \alpha_{B_{0,k}}, \; k = 1, \ldots, M \tag{9}$$

and the steady-state solution is easily obtained as

$$p_{B_{k,ss}} = \frac{\alpha_{B_k}}{\beta_{B_k}} f_{\text{Buffer},k}\left(u_k, K_k, n_k\right) + \frac{\alpha_{B_{0,k}}}{\beta_{B_k}} \tag{10}$$

where p_{B_k} is the output concentration of the kth Buffer, $p_{B_{k,ss}}$ denotes its steady-state concentration, u_k, K_k and n_k are respectively the input concentration, Hill constant, and Hill coefficient of the kth Buffer, and α_{B_k}, β_{B_k} and $\alpha_{B_{0,k}}$ are, respectively, synthesis, decay and basal rates. The second term in the right side of (10) is minimal level and α_{B_k}/β_{B_k} is the difference between the minimal level and the maximal level. The output concentration of the Buffer is the half maximal output concentration when the input concentration equals K_k and K_k refers to the threshold level y_T.

The gain in the operation point K_k is obtained by

$$A_{B_k} = \left.\frac{\partial p_{B_{k,ss}}}{\partial u_k}\right|_{u_k = K_k} = \frac{\alpha_{B_k} n_k}{4\beta_{B_k} K_k} \tag{11}$$

where A_{B_k} is the gain of the kth Buffer. It is observed that the gain is proportional to the Hill coefficient n_k and is inversely proportional to the Hill constant K_k at the operation point $u_k = K_k$. To achieve the genetic waveform-shaping circuit design, the necessary condition of the gain at the operation point K_k should be more than 1.

In each stage, the corresponding input signals and threshold levels are given by

$$u_k = \begin{cases} A\sin(\omega_0 t + \varphi) + y_{d,0}, & k = 1 \\ p_{B_{k-1}}, & k > 1 \end{cases} \tag{12}$$

and

$$K_k = \begin{cases} y_T, & k - 1 \\ \dfrac{\alpha_{B_{k-1}} + \alpha_{B_{0,k-1}}}{2\beta_{B_{k-1}}}, & k > 1 \end{cases} \tag{13}$$

where A, ω_0, φ and $y_{d,0}$ are, respectively, amplitude, basal frequency, phase, and base level of the desired oscillation signal. In the first stage, the input signal is oscillation signal produced by the genetic oscillator and described by a sinusoidal function. In the next stage, the input signal and the threshold level are respectively the output and the half maximal output level in the previous stage. Figure 2 shows topology of the designed genetic waveform-shaping circuit. The produced protein of the first gene activates the transcription of the second gene whose production activates the next gene.

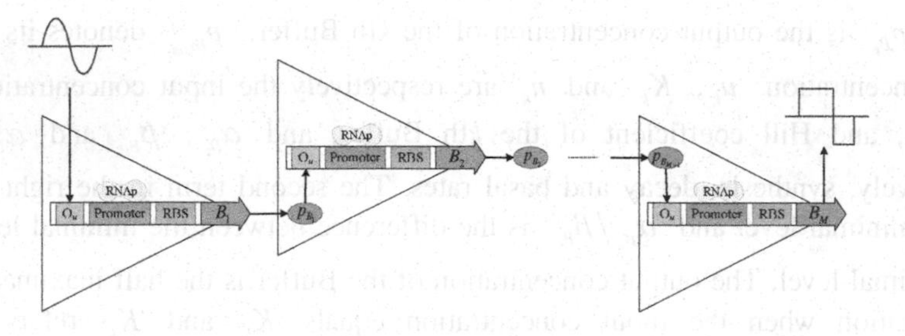

Fig. 2. Topology of the genetic waveform-shaping circuit

3.2 Regulation of Threshold Level

To regulate different threshold levels in (13), one can generate the PWM signal with different duty cycle defined by

$$D = \frac{T_{\text{on}}}{T_0} \times 100\% \tag{14}$$

where D is the duty cycle, T_0 is the basal period of oscillation signal with $2\pi/\omega_0$ and T_{on} is the period of logic high in a basal period. For the different duty cycle, the corresponding threshold value is selected by

$$y_T = A\sin\left(\omega_0 t + \varphi\right) + y_{d,0},\ t = t_h \pm \frac{T_{\text{on}}}{2} \tag{15}$$

and $t_h = \frac{1}{\omega_0}\sin^{-1}(1) - \frac{\varphi}{\omega_0}$, $t_h \in [0\ \ T_0]$. Clearly, the PWM signal has 50% duty cycle when $y_T = y_{d,0}$.

3.3 Design of Genetic Frequency Synthesizer Circuit

Frequency synthesizer is designed to generate an output signal whose frequency is a multiple of that of input signal in electronics. Based on the feature, one can generate the clock pulses with frequency multiple to that of the genetic oscillator by using the genetic waveform-shaping circuit mentioned above. To construct a clock signal with *N*-fold frequency of genetic oscillator, a series of clock pulses should be stimulated by the threshold levels:

$$y_{T_\varepsilon} = A\sin\left(\omega_0 t_\varepsilon + \varphi\right) + y_{d,0},\ \varepsilon = 1, \ldots, N+1 \tag{16}$$

with $t_\varepsilon = t_h + \frac{T_0}{2N}(\varepsilon - 1)$, $T_0 = \frac{2\pi}{\omega_0}$, where y_{T_ε} is the value of threshold level for the synthesis of frequency synthesizer and t_h in (15). For example, the clock pulses become activated in the threshold levels $y_{d,0}+A$ and $y_{d,0}-A$ for the clock signal with 50% duty cycle shown in Figs. 3(a) and 3(b). For the clock signal with double basal frequency, the clock pulses become simulated in the threshold levels $y_{d,0}+A$, $y_{d,0}-A$, and $y_{d,0}$. To generate the clock pulse in the threshold level $y_T = y_{d,0}$, a genetic logic XOR gate is used to combine two PWM signals with threshold levels $y_{d,0}+\Delta y_T$ and $y_{d,0}-\Delta y_T$ where Δy_T is a small variation.

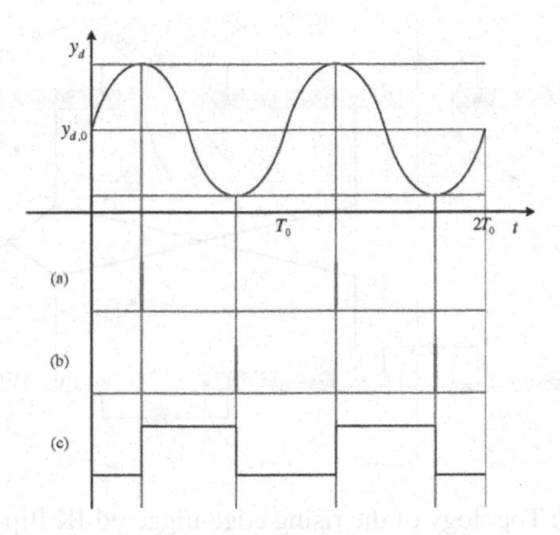

Fig. 3. Ideal signals for the design of clock signal with 50% duty cycle. (a) clock pulse in $y_T = y_{d,0}+A$; (b) clock pulse in $y_T = y_{d,0}-A$; and (c) clock signal.

After generating the clock pulses via the proposed genetic waveform-shaping circuit, the 1-bit genetic counter shown in Fig. 4 is triggered by the clock pulse to synthesize the genetic clock with the multiple frequencies to the genetic oscillator. Figure 5 displays the topology of 1-bit genetic counter with the model constructed as

$$\begin{aligned}
\dot{p}_W &= \alpha_W f_{\text{AND}}\left(p_K, p_{CLK}\right) - \beta_W p_W, \\
\dot{p}_V &= \alpha_V f_{\text{AND}}\left(p_J, p_{CLK}\right) - \beta_V p_V, \\
\dot{p}_R &= \alpha_R f_{\text{AND}}\left(p_W, p_Q\right) - \beta_R p_R, \\
\dot{p}_S &= \alpha_S f_{\text{AND}}\left(p_V, p_{\bar{Q}}\right) - \beta_S p_S, \\
\dot{p}_Q &= \alpha_Q f_{\text{NOR}}\left(p_R, p_{\bar{Q}}\right) - \beta_Q p_Q, \\
\dot{p}_{\bar{Q}} &= \alpha_{\bar{Q}} f_{\text{NOR}}\left(p_S, p_Q\right) - \beta_{\bar{Q}} p_{\bar{Q}}
\end{aligned} \tag{17}$$

where p_{CLK} is the clock pulse signal, $p_K = p_J = 1$, p_W, p_V, p_R, p_S, p_Q and $p_{\bar{Q}}$ are respectively the concentrations of productions of the genes, α_W, α_V, α_R, α_S, α_Q and $\alpha_{\bar{Q}}$ are synthesis rates, and β_W, β_V, β_R, β_S, β_Q and $\beta_{\bar{Q}}$ are decay rates.

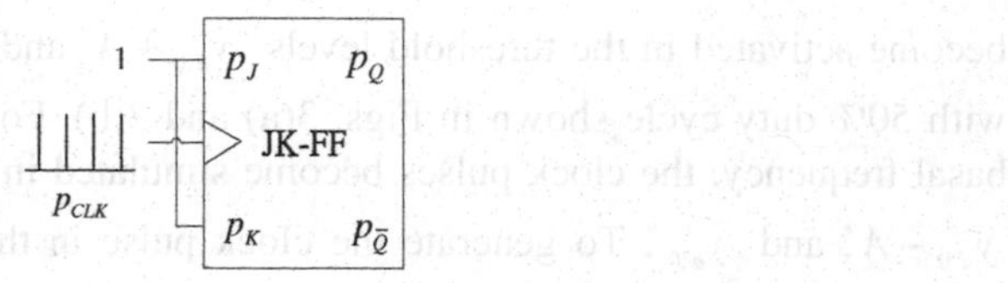

Fig. 4. A 1-bit genetic counter circuit with the rising edge-triggered JK flip-flop

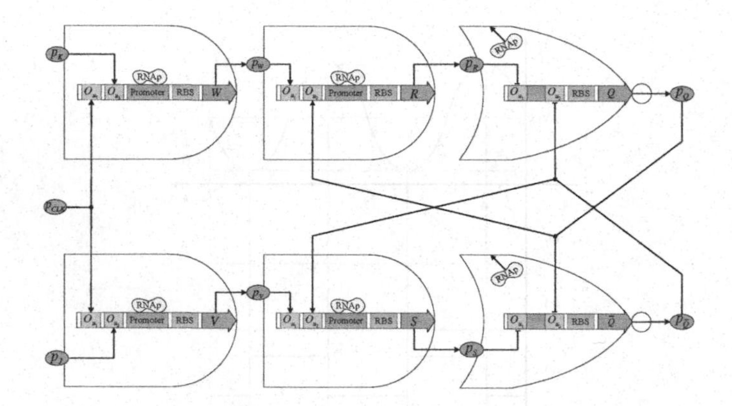

Fig. 5. Topology of the rising edge-triggered JK flip-flop

4 Simulation Results

An example of the existing synthetic genetic oscillator, known as a repressilator, is illustrated to confirm performance of the designed genetic logic circuit. The dynamic model of the repressilator [4] is given by

$$\begin{aligned}
\dot{p}_{lacI} &= 0.2877\frac{1}{1+p_{cI}^4} - 0.0974\,p_{lacI},\quad p_{lacI}(0) = 0.7, \\
\dot{p}_{tetR} &= 0.2877\frac{1}{1+p_{lacI}^4} - 0.0974\,p_{tetR},\ p_{tetR}(0) = 1.2, \\
\dot{p}_{cI} &= 0.2877\frac{1}{1+p_{tetR}^4} - 0.0974\,p_{cI},\quad p_{cI}(0) = 1.7
\end{aligned} \tag{18}$$

where p_{lacI}, p_{tetR} and p_{cI} are, respectively, the produced proteins of the repressor genes $lacI$, $tetR$ and cI. This oscillation signal has the basal period $T_0 = 39$ sec, the amplitude $A = 0.63$, and the base level $y_{d,0} = 1.1731$. Assume that p_{cI} is the oscillation input with concentration response shown in Fig. 6.

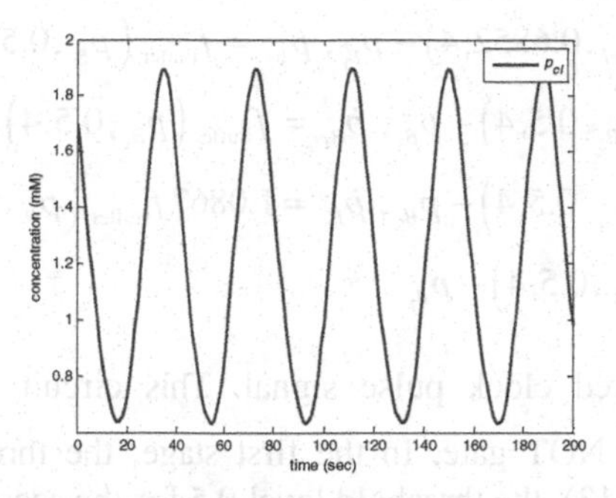

Fig. 6. Response of concentration of p_{cl}

To synthesize the clock pulse in the threshold level $y_{d,0}+A$, the designed genetic waveform-shaping circuit is obtained as

$$\begin{aligned}
&\dot{p}_{B_1}=f_{\text{Buffer}}\left(p_{cl},1.7462,4\right)-p_{B_1},\ \dot{p}_{B_2}=f_{\text{Buffer}}\left(p_{B_1},0.5,4\right)-p_{B_2},\\
&\dot{p}_{B_3}=f_{\text{Buffer}}\left(p_{B_2},0.5,4\right)-p_{B_3},\ \dot{p}_{B_4}=f_{\text{Buffer}}\left(p_{B_3},0.5,4\right)-p_{B_4},\\
&\dot{p}_{B_5}=f_{\text{Buffer}}\left(p_{B_4},0.5,4\right)-p_{B_5},\ \dot{p}_{B_6}=1.1073f_{\text{Buffer}}\left(p_{B_5},0.5,4\right)-p_{B_6}
\end{aligned} \tag{19}$$

where p_{B_6} is the clock pulse signal. This circuit consists of six cascaded Buffers. In the first stage, the threshold level 1.7462 is designed. According to (13), the threshold level 0.5 for the second to the sixth Buffers is chosen. To compensate the output of maximal level, the appropriate rate constants for the last Buffer is selected by (10). The concentration response of the designed clock pulse signal is shown in Fig. 7.

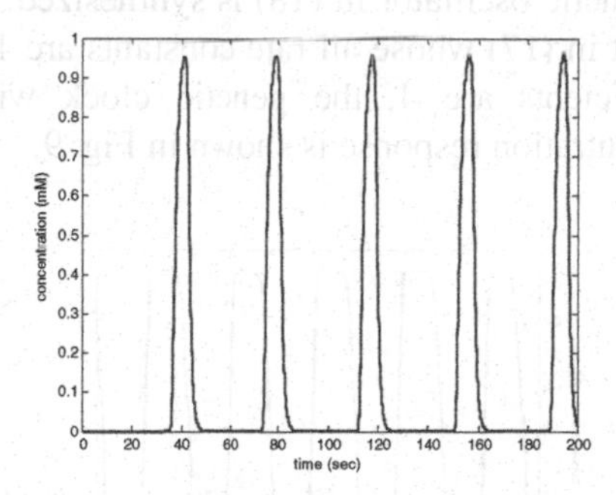

Fig. 7. Response of the designed clock pulse signal in the threshold level $y_{d,0}+A$

Similarly, the genetic waveform-shaping circuit for design of the clock pulse in the threshold level $y_{d,0}-A$ is obtained as

$$\begin{aligned}
&\dot{p}_{B_1} = f_{\text{Buffer}}\left(p_{cl}, 0.6852, 4\right) - p_{B_1},\ \dot{p}_{B_2} = f_{\text{Buffer}}\left(p_{B_1}, 0.5, 4\right) - p_{B_2}, \\
&\dot{p}_{B_3} = f_{\text{Buffer}}\left(p_{B_2}, 0.5, 4\right) - p_{B_3},\ \dot{p}_{B_4} = f_{\text{Buffer}}\left(p_{B_3}, 0.5, 4\right) - p_{B_4}, \\
&\dot{p}_{B_5} = f_{\text{Buffer}}\left(p_{B_4}, 0.5, 4\right) - p_{B_5},\ \dot{p}_{B_6} = 1.0867 f_{\text{Buffer}}\left(p_{B_5}, 0.5, 4\right) - p_{B_6}, \\
&\dot{p}_{B_7} = f_{\text{NOT}}\left(p_{B_6}, 0.5, 4\right) - p_{B_7}
\end{aligned} \tag{20}$$

where p_{B_7} is the desired clock pulse signal. This circuit is constructed by six cascaded Buffers and a NOT gate. In the first stage, the threshold level 0.6852 is designed. According to (13), the threshold level 0.5 for the second to the sixth Buffers is chosen. Figure 8 shows the concentration response of the designed clock pulse.

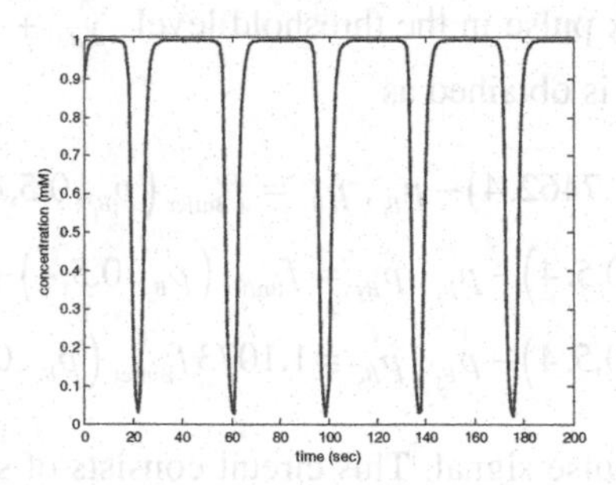

Fig. 8. Response of the designed clock pulse signal in the threshold level $y_{d,0} - A$

Using the logic AND gate in (4) with $\alpha = \beta = 1$, $n_1 = n_2 = 4$, and $K_1 = K_2 = 0.5$ to integrate the designed clock pulses in (19) and (20), the clock pulse with double base-frequency of the genetic oscillator in (18) is synthesized. By the clock signal to trigger the counter circuit in (17) whose all rate constants are 1, all Hill constants are 0.5, and all Hill coefficients are 4, the genetic clock with base frequency is synthesized and its concentration response is shown in Fig. 9.

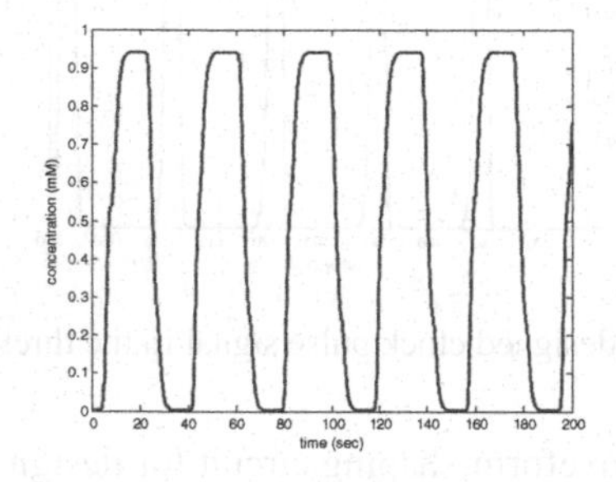

Fig. 9. Response of the designed genetic clock with base frequency

5 Conclusion

A synthetic genetic frequency synthesizer circuit with counter has been proposed to synthesize the genetic clock with the multiple to frequency of the genetic oscillator. Through applying a waveform-shaping technique, genetic oscillation is shaped to becoming an ideal logic signal. By regulating the different threshold levels, the PWM signals with different duty cycles are obtained. A genetic counter is triggered by the clock pulse generated by the proposed genetic waveform-shaping circuit to synthesize the desired genetic clock, whose frequency is a multiple to that of the genetic oscillator, is realized.

Experimental realization of the proposed network is a potential issue worthy of further investigation. In addition, assembling the clock signal with a variety of sequential genetic logic circuits to realize artificial logic functions is recommended for future research.

Acknowledgment. This research was sponsored in part by National Science Council, Taiwan, ROC under the Grants NSC 102 -2218-E- 005- 012 and NSC-101-2221-E-005-015-MY3.

References

1. Andrianantoandro, E., Basu, S., Karig, D.K., Weiss, R.: Synthetic biology: new engineering rules for an emerging discipline. Molecular Systems Biology 2, 1–14 (2006)
2. Lu, T.K., Khalil, A.S., Collins, J.J.: Next-generation synthetic gene networks. Nature Biotechnology 27, 1139–1150 (2009)
3. Khalil, A.S., Collins, J.J.: Synthetic biology: applications come of age. Nature Reviews Genetics 11, 367–379 (2010)
4. Elowitz, M.B., Leibler, S.: A synthetic oscillatory network of transcriptional regulators. Nature 403, 335–338 (2000)
5. Buchler, N.E., Gerland, U., Hwa, T.: On schemes of combinatorial transcription logic. PNAS 100(9), 5136–5141 (2003)
6. Zabet, N.R., Hone, A.N.W., Chu, D.F.: Design principles of transcriptional logic circuits. In: Proceedings of the Alife XII Conference, Odense, Denmark (2010)
7. Wang, B., Kitney, R.I., Joly, N., Buck, M.: Engineering modular and orthogonal genetic logic gates for robust digital-like synthetic biology. Nature Communications 2, 1–9 (2011)
8. Chen, B.S., Chang, C.H., Wang, Y.C., Wu, C.H., Lee, H.C.: Robust model matching design methodology for a stochastic synthetic gene network. Mathematical Biosciences 230, 23–36 (2011)
9. Chang, Y.C., Lin, C.L., Jennawasin, T.: Design of synthetic genetic oscillators using evolutionary optimization. Evolutionary Bioinformatics 9, 137–150 (2013)
10. Chuang, C.H., Lin, C.L., Chang, Y.C., Jennawasin, T., Chen, P.K.: Design of synthetic biological logic circuits based on evolutionary algorithm. IET Systems Biology 7(4), 89–105 (2013)

11. Chuang, C.H., Lin, C.L.: A novel synthesizing genetic logic circuit: frequency multiplier. IEEE/ACM Transactions on Computational Biology and Bioinformatics 11(4), 1–12 (2014)
12. Chuang, C.H., Lin, C.L.: Synthesizing genetic sequential logic circuit with clock pulse generator. BMC Systems Biology 8(63), 1–15 (2014)
13. Lin, C.L., Huang, C.H., Tsai, C.W.: Structure-Specified Real Coded Genetic Algorithms with Applications. In: Advanced Knowledge Based Systems: Model. Applications & Research, Technomathematics Research Foundation, vol. 1, pp. 160–187 (2010)

A Multi-agent System for Smartphone Intrusion Detection Framework

Abdullah J. Alzahrani and Ali A. Ghorbani

University of New Brunswick, Faculty of Computer Science,
Fredericton NB, Canada
{a.alzahrani,ghorbani}@unb.ca

Abstract. There has been significant growth in the number of malware using the Android platform, especially malware that target Short Message Services (SMS). The mobile botnet has been using SMS as a channel to distribute spam, send unauthorized SMS messages without user knowledge, use command and control (C&C) channel, and attach malicious URLs. With the limitation of Android smartphone resources, a multi-agent technology can make our framework to be more robust and efficient. In this paper, we propose a multi-agent system that is currently being developed using JADE platform for observing Android Smartphone features and monitoring SMS services, as well as creating Android profiles. Our framework applies hybrid detection approaches in order to counteract botnet attacks, by investigating damaging SMS botnet activities through the examination of Smartphone behaviour. These approaches utilize multi-agent technology to recognize malicious SMS and prevent users from opening these messages, by applying behavioural analysis to find the correlation between suspicious SMS messages and the profiles reported by the agents.

Keywords: Multiagent Systems, Agent, SMS, Botnet Detection, Smartphone.

1 Introduction

Mobile botnets present some of the most serious security threats to mobile devices. This is a serious issue due to the increasing worldwide trend in the use of handheld mobile communication phones. The mobile botnet is a threat that contains a network of compromised Smartphones universally controlled by an attacker, also known as a 'bot master'. The bot master uses a command and control (C&C) channel with the intention of performing malicious attacks on the mobile users [14]. Given the complex and distressing nature of the botnets, past research has already placed a great emphasis on elucidating different concepts of botnets' functionality, including its threats, models, control designs, strategies, and botnet detection [13,22,27]. Given the fact that mobile phones have advanced at a fast pace over the past number of years, society is striving to get the most out of the capabilities and functionalities that the new phones provide. As greater

H. Handa et al. (eds.), *Proc. of the 18th Asia Pacific Symp. on Intell. & Evol. Systems – Vol. 1*,
Proceedings in Adaptation, Learning and Optimization 1, DOI: 10.1007/978-3-319-13359-1_9

utilization comes into these platforms, it is expected that malicious behaviour will follow at almost the same pace [25]. Mobile botnets are said to be difficult to detect [14]. This difficulty occurs because mobile botnets are progressing quickly and becoming more and more refined. Moreover, features of today's Smartphones contain opportunities for attackers to capitalize on mobile botnets in order to perform illegal actions. These actions will impact negatively the victims themselves. One of the main components of botnet is the C&C channel, which is used by attackers to carry out command and control messages. With the availability of short message services (SMS) on Smartphones, SMS messages are used to transfer C&C commands, send SMS spam, send premium-rate SMS messages without user knowledge [18] and distribute the malware as propagation vectors. These actions have a negative impact on the victims themselves.

With the limitation of the Smartphone resources and changes in the characteristics of connectivity changes, there is need to use intelligent approaches to elaborate SMS botnet and Android Smartphone behaviours. The development of a multi-agent system will meet this need.The JADE Platform has the ability to be adapted to the characteristics of a deployment environment [6]. Past research has proven that a multi-agent system is successful for intrusion detection system [20]. The main advantages of using a multi-agent system are pro-activity, autonomy, and self-awareness. These advantages have been extensively discussed by by Carabelea et al. [8].

Our work focuses on detection of SMS botnets specifically, we have proposed a framework with two components, an intrusion detection system and a multi-agent system. In our work, we focus on multi-agent system design and the generally usage of our framework. The main component of our research involves developing a multi-agent system that is composed of multiple interacting agents. To provide robustness against failure of our framework, a multi-agent system must be embedded to share the responsibilities among the different agents [23]. Agents have the ability to adjust their behavior in order to be alerted to resource constraints, such as network bandwidth and battery capacity. Botnet detection is most effective when it utilizes multi-agent systems that allow for more logical botnet diagnosis through agents' communication within the network. Our framework uses multi-agent system technology to achieve accurate detection without exhausting Smartphone resources such as battery and memory.

The second main component is an intrusion detection system with signature-based and anomaly-based detection modules. Signature-based detection is also called misuse detection or knowledge-based detection. Traditionally, the signature-based approach extracts the features from traffic and detects malicious activities by comparing incoming traffic to the signatures of the attacks. Signatures are patterns or sets of rules that can uniquely identify an attack. The anomaly-based detection approach can also be referred to as behavior-based detection. This approach builds models of normal data and then attempts to detect the deviation from the norm in observed data; this deviation is considered an anomaly [21]. The key advantage of this approach is that it can detect new types of threats when there are deviations from normal data. To improve the capability of intrusion detection, researchers have proposed hybrid detection, which is a combination of signature-based detection and anomaly-based detection. We have adapted signature-based detection in the Smartphone and anomaly-based detection and behaviour analysis in the central server. This hybrid approach detects known attacks with high accuracy using signature-based detection, while also being able to detect unknown attacks through the use of anomaly-based detection. We extensively explored this part in detail [4].

In this paper, we propose a multi-agent system that is developed using the JADE platform for observing Android Smartphone features and monitoring SMS services, as well as creating Android profiles. Our framework applies hybrid detection approaches in order to counteract botnet attacks, by investigating damaging SMS botnet activities through the examination of Smartphone behaviour. These approaches utilize multi-agent technology to recognize malicious SMS and prevent users from opening these messages.

The rest of this paper is organized as follows: Section 2 presents some related work; Section 3 presents the proposed framework focusing on multi-agent system design; and Section 4 explains how the proposed framework can be evaluated. Finally, the conclusion and future work are summarized in Section 5.

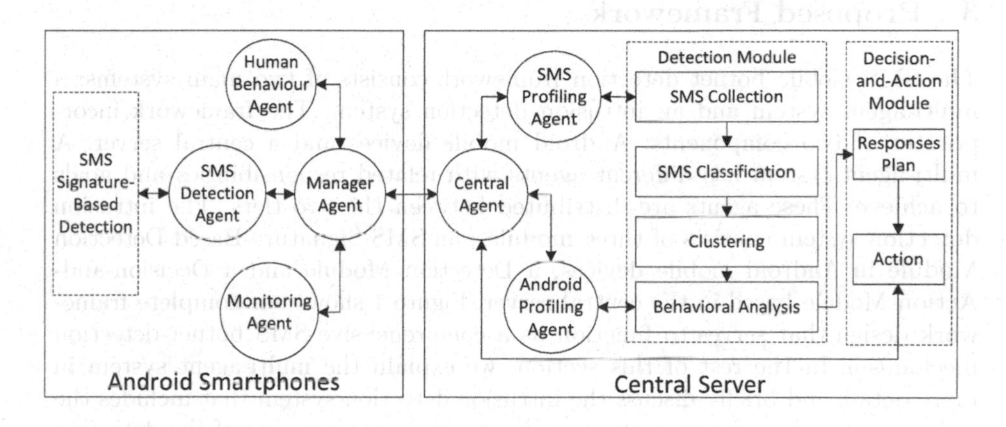

Fig. 1. SMS Mobile Botnet Detection Framework using a Multi-Agent System

2 Related Work

The concept of adapting a multi-agent approach to Smartphones is a very new area of research that focuses on the Android platform. Currently, several multi-agent platforms have been developed for Smartphones, and research is still ongoing. Multi-agent systems for botnet detection require a generation of agents set with some structure and functionality. Based on the work by Savenko et al. [20], agents are utilized when their results are transmitted to the effectors as a means of influence on the computer system. If malware is detected, an agent, through effectors, blocks the processes that are responsible for performance of some malware and then notifies the user about the infection.

Carabelea et al. [8] present an overview of several multi-agent platforms that have been created for use in small devices, and identify characteristics of these platforms. Frantz et al. [12] propose a micro-agent framework for the Android platform that has the ability to interface with Android platform facilities, in order to allow agent-based applications to access Android functionality. Agüero et al. [2] present the Agent Platform Independent Model (APIM) for mobile devices. Their proposed approach has been implemented and tested on the Android platform with some limitations. JaCa-Android [19] presents the use of Agent-Oriented Programming (AOP) technologies to develop a Smartphone application for the Android platform. However, the approach has some weaknesses that need to be addressed, such as devising a notion of type for agents and artifacts, improving modularity in agent definition and improving the integration with the OO layer. Cheng [9] proposes a multi-agent security system for the Android platform that uses agents to collect data from Android devices, and then send it to agent service providers to make a decision. The study shows multi-agent systems can be adapted to the Android platform with its inherent resource limitations.

3 Proposed Framework

The SMS mobile botnet detection framework consists of two main systems: a multi-agent system and an intrusion detection system. The framework incorporates of two components: Android mobile devices and a central server. A multi-agent system has different agents with related responsibilities and goals to achieve. These agents are distributed between the two tiers. The intrusion detection system consists of three modules: an SMS Signature-Based Detection Module in Android mobile devices, a Detection Module and a Decision-and-Action Module based in the central server. Figure 1 shows our complete framework design that serves to function as a comprehensive SMS botnet-detection mechanism. In the rest of this section, we explain the multi-agent system in more detail, and briefly discuss the intrusion detection system that includes the SMS signature-based approach, describe the main components of the detection module, and finally, illustrate the functions of the decision-and-action module.

3.1 Multi-agent System Design

A multi-agent is a system composed of multiple interacting agents [28]. Our proposed framework requires a multi-agent system with extensive knowledge about distributed systems and required agent interactions in order to observe, monitor, and handle the data exchange. There are several multi-agent system frameworks in existence. Among the platforms in use, such as AgentBuilder [1], JACK [7], and Cougaar [29], one of the most well-known is the JADE platform [24] as describe below. The Foundation for Intelligent, Physical Agents (FIPA) has defined a list of specifications for agents [11]. Agent management and agent communication are at the core of this list, and need to be included in any multi-agent system framework in order to develop an agent platform [5].

JADE is a software development platform that complies with FIPA specifications. JADE platform is an open source software and object-oriented language, Java, which provides basic middleware-layer functionalities and libraries that fully support the Android platform [16]. The advantage of the JADE platform is the provision of ready-to-use and easy-to-customize core functionalities.

Given the features of JADE, we selected this platform to implement our multi-agent system using the methodology that was developed for JADE [17]. Our proposed system involves multi-agents distributed on two levels: four agents in each Android mobile device and three agents in the central server, as shown in Figure 1. The basic idea of how a multi-agent system works using JADE is a framework in which the main container is the central server and the agent migrates to Android devices. Our JADE-based multi-agent system has the ability to perform tasks.

The SMS mobile botnet detection framework allows an Android device user to install our application (app) in order to obtain all services. The agents monitor and observe Android device activities, and then capture and report suspicious behaviours to the detection part of the central server. We define each agent's tasks and responsibilities in more detail as outlined below. Figure 2 shows how our framework use case achieves a specific goal, by interacting with the end user and other agents or system, utilizing the most popular specification, Unified Modeling Language (UML) [17].

1. Android Smartphone Agents and Their Responsibilities
An Android mobile user must subscribe to the central agent providers in order to obtain all the defined services and to maintain the interaction between local agents. The agents monitor incoming and outgoing SMS, and observe Smartphone behaviour and resources. There are four major agents within Android mobile devices, including a Manager Agent, an SMS Detection Agent, a Monitoring agent, and a Human-Behavior Agent, as outlined below. Table 1 illustrates all agent types with their relevant functionalities.

a) **Manager Agent.** This is the agent that establishes a connection with the central agent in the central server by subscription. The Manager Agent plays a critical role in the system as it creates a channel that allows communication from the actual phone to the central server. The Manager Agent obtains the agent identification from central service provider and establishes the interaction. It reads the status of the Smartphone, including SMS delivery and battery power. Basically, as the name states, the Manager Agent supervises the interactions between the local agents for Smartphone Android devices and is in charge of responding to any request received from central server. In addition, the Agent can exchange the data with the Android Profiling Agent. The Android Mobile user can unsubscribe from the central service provider at any time. The Manage Agent notifies the user when a new threat is detected.
b) **SMS Detection Agent.** The main task of this Agent is to monitor incoming and outgoing SMS, then send requests to the SMS signature-based

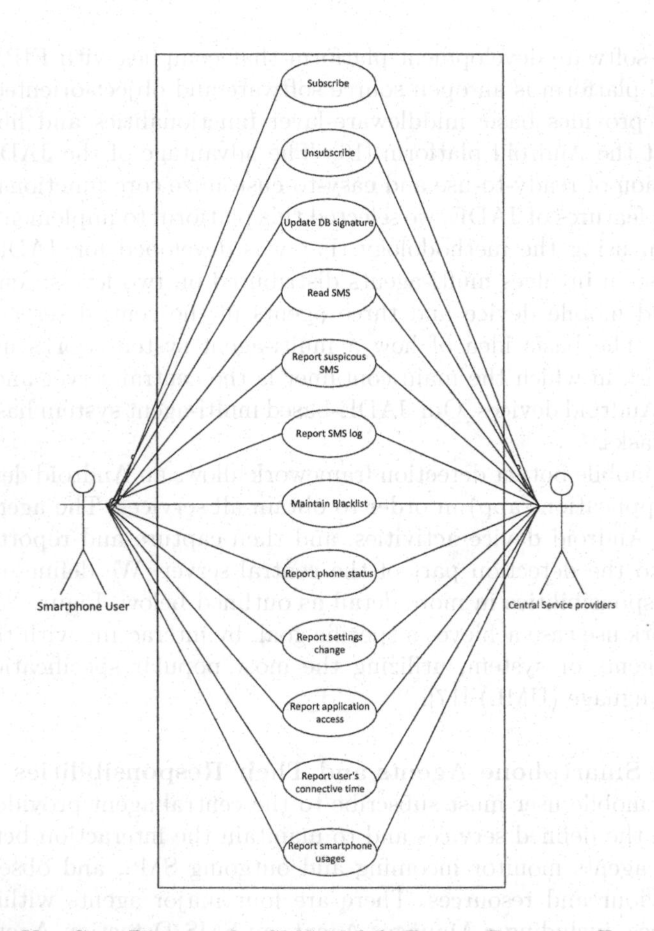

Fig. 2. Use Case Diagram for SMS Mobile Botnet Detection Framework

detection module that reports suspicious SMS to the SMS Detection Agent. The system first takes action by registering itself on the SMS profiling service through the Central Agent, and then obtaining updates on the SMS signature database. Additionally, this agent is responsible for monitoring SMS logs and then reporting to the SMS Profiling Agent. This Agent gets the results from the SMS signature-based detection module and perform one of the following actions: deliver, delete, or send the suspicious SMS to the SMS Profiling Agent.

c) **Monitoring Agent.** The Monitoring Agent registers with the Android profiling service in the central server. This Agent is responsible for observing phone settings and phone activities. The Monitoring Agent also plays a role in reporting accessibility in the browser and other installed applications. In addition, it checks the Internet connection since connectivity is an important factor that likely contributes to the realization of SMS botnet attacks. Reports regarding all the agent activities are then passed through to the Manager Agent.

d) **Human-Behaviour Agent.** Malicious activities usually wait until the smart- phone is in an ideal mode or after reboot. The Human-Behaviour Agent registers with the Android profiling service in the central server. This Agent is in charge of monitoring user connectivity time, maintaining the whitelist and blacklist, reporting daily usage of the mobile phone and responding to the Manager Agent as required.

Table 1. Android Smartphone Agents and their Responsibilities

Agent Type	Responsibilities
Manager Agent	1. Smartphone user is required to subscribe to the central service provider. 2. Read Smartphone status. 3. Obtain the Agent identification from central service provider to establish the interaction. 4. Respond to requests from Central Agent. 5. Manage the interaction communication between local agents. 6. Send data to Android Profiling Agent. 7. Unsubscribe from central service provider. 8. Notify the user when new threat is detected.
SMS Detection Agent	1. Register with SMS profiling service in central server. 2. Obtain SMS signature update. 3. Read incoming and outgoing SMS. 4. Get the result from SMS Signature-Based Detection. 5. Monitor SMS logs. 6. If SMS is normal, deliver it to SMS application. 7. If SMS is malicious, delete SMS and notify the user. 8. If SMS is suspicious, send a copy of suspicious SMS to SMS Profiling Agent.
Monitoring Agent	1. Register with Android profiling service in central server. 2. Report any access to browser or other apps when SMS application tries to access. 3. Check WiFi status and Internet access. 4. Monitor Smartphone status including battery usage, apps that are running, memory usage, etc. 5. Spot any setting changes.
Human-Behavior Agent	1. Register with Android profiling service in central server. 2. Observe user connectivity time. 3. Maintain the whitelist and blacklist. 4. Report daily usage of Android mobile.

2. Central Service Agents and Their Responsibilities

As shown in Figure 1, the central server has three agents and two modules that are used to process the data in order to detect suspicious SMS, make intelligent decisions, and perform actions. In the central server, there are three major agents that perform the majority of the activities of the detection system: the Central Agent, the Android Profiling Agent and the SMS Profiling Agent, as outlined below. Based on the results of behaviour analysis, these agents provide service and offer further analysis to achieve a high detection rate and make intelligent decisions, in order to detect SMS botnet activities. The descriptions and roles of the Central Service Agents are outlined below. Table 2 illustrates all agent types with their relevant functionalities.

a) **Central Agent.** The Central Agent is in charge of handling and responding to Smartphone devices and adding them to the subscriber list. The Central

Agent also performs activities that are relevant to Android mobile device agents, such as managing, updating, blocking, deleting and controlling. The Central Agent sends the updated signature database to the Manager Agents, and then sends commands to the Manager Agents, entailing the decision that has been obtained and established in the decision-and-action module. The Central Agent manages all the local agents situated in the central server. The Central Agent primarily obtains profile updates and then forwards them on to the Android profiling service provider. It also has to obtain copies of SMS and SMS logs, and then send them to the SMS profiling service provider, which manages all suspicious SMS and log profiling.

b) **SMS Profiling Agent.** The SMS Profiling Agent handles all the incoming and outgoing SMS that are considered to be suspicious, and maintains the SMS logs. The SMS Profiling Agent receives reports on suspicious SMS from the SMS Detection Agent. These received suspicious SMS data and logs are then forwarded to the detection module to verify whether they are deemed to be botnets.

c) **Android Profiling Agent.** Once the profile updates are received by other agents, this particular Agent will then maintain and update the profile for all subscribing Smartphones. Additionally, this agent updates the received changes from the detection module and the other agents. It responds to detection module requests, which are findings and actions that need to be acted upon. Finally, this Agent can request more information from Monitoring and Human Behavior Agents if required.

Table 2. Central Service Agents and their Responsibilities

Agent Type	Responsibilities
Central Agent	1. Respond to Smartphone devices and add them to the subscriber list. 2. Update, block, and delete Smartphone agents as appropriate. 3. Manage the interaction communication between local agents. 4. Update the signatures database. 5. Send commands to start new agents or preform an action on Android devices if certain conditions are met. 6. Forward the Android profile, suspicious SMS, and SMS logs to Android profiling provider and SMS profiling provider.
SMS Profiling Agent	1. Handle the received suspicious SMS and then send it to Detection Module. 2. Maintain an updated signature for each SMS detection agent. 3. Handle SMS logs and request an update within specific time. 4. Interact with Detection Module.
Android Profiling Agent	1. Maintain the profile database for all subscribing Smartphones. 2. Update the profile changes when message received from other agents. 3. Respond to Detection Module requests. 4. Request more information from monitoring and human behavior agents if needed.

Agent-Resource Interactions. For Smartphones, agents need to interact with other agents or with Smartphone resources. There are four agents as shown in Figure 3: the Smartphone Agents, required to interact with remote server and with Smartphone resources, such as battery and phone status; the SMS Detection Agents, needed to communicate with SMS application and SMS signature-based

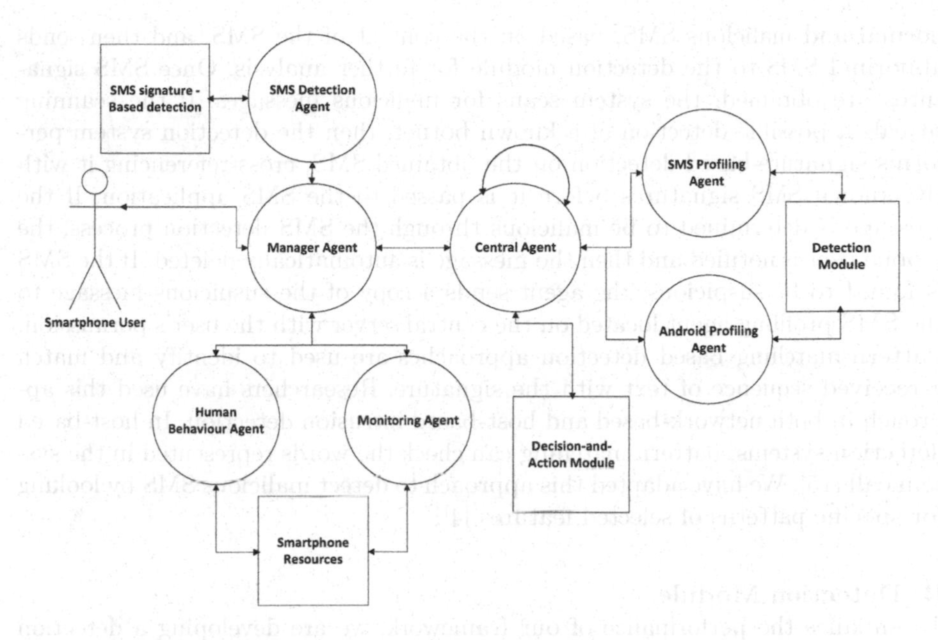

Fig. 3. Agent Diagram for SMS Mobile Botnet Detection Framework

detection module; the Monitoring Agents, to interact with other applications, internet access and the Smartphone Agent; and the Human-Behaviour Agents, to interact with Smartphone resources, identifying Smartphone status, time-phone wake-up, for example, and observes user interaction with the Smartphone.

3.2 Intrusion Detection System

The second part of our SMS botnet detection framework includes three components: an SMS Signature-Based Detection Module in Android mobile devices, a Detection Module and a Decision-and-Action Module found in the central server, as shown in Figure 1. These components employ a hybrid detection approach using different algorithms and methods in order to detect SMS botnets. Intentionally, the involvement of agents ensures that our framework will have a high detection rate and detect unknown malicious activities on the Android platform by observing Smartphone behaviour. More detail for each component are outlined below.

A. SMS Signature-Based Detection Module in Android Device

Focusing on incoming and outgoing SMS messages, our proposed design for Android mobile devices uses a signature-based detection algorithm to identify SMS botnets. SMS signatures are obtained and copied from the SMS signature database where signatures of known botnets and malware are stored. Our system uses real-time content-based signature detection to differentiate between normal and malicious SMS, based on the content of the SMS, and then sends abnormal SMS to the detection module for further analysis. Once SMS signatures are obtained, the system scans for malicious messages. If the scanning reveals a possible detection of a known botnet, then the detection system performs signature-based detection on the obtained SMS, cross-referencing it with the known SMS signatures before it is passed to the SMS application. If the message is determined to be malicious through the SMS detection process, the mobile user is notified and then the message is automatically deleted. If the SMS is found to be suspicious, the agent sends a copy of the suspicious message to the SMS profiling agent located on the central server with the user's permission. Pattern-matching based detection approaches are used to identify and match a received sequence of text with the signature. Researchers have used this approach in both network-based and host-based intrusion detection. In host-based detection systems, pattern matching can check the words represented in the system call [15]. We have adapted this approach to detect malicious SMS by looking for specific patterns of selected features [4].

B. Detection Module

To enhance the performance of our framework, we are developing a detection module that consists of four elements. The first element is an SMS collector, which is responsible for gathering, combining, storing and retrieving the data from the agents. The second is an SMS classifier that categorizes messages into normal, suspicious, and malicious SMS. The third involves clustering algorithms that group SMS messages based on two main factors, the content itself, and the phone number. The final element addresses behavioural analysis. Using information from the Android Profiling Agent, the system performs behaviour-profiling analysis on the output of the clusters to locate suspicious behaviour. It identifies if there are correlations between alerts from the clusters and any abnormal activities in the Android devices. The results from the detection module are then forwarded to the Decision-and-Action Module.

In summary, the detection module is where SMS collection, SMS classification, anomaly-based detection and behaviour-profiling analysis are conducted. Once the Detection Agent receives suspicious SMS from the SMS Profiling Agent, it performs the anomaly-based detection, through specifically-created and manipulated algorithms. Once SMS are deemed to be malicious, content analysis is performed in order to check the URLs. All of the Android mobile profiles that contain the same SMS are grouped together.

C. Decision-and-Action Module

The Decision-and-Action Module uses the output received from the Detection Module to make logical decisions based on the set of rules that have been established by a Human-Network Manager. There are two main components in this module, response plan and action. The primary component, response planning, involves the actions taken when all the necessary criteria are satisfied. There are two main approaches for action to respond to an attack [26]. The first approach is to identify the malicious correspondent's phone number and block these numbers. The second approach is to identify the similar characteristics of malicious SMS and group them by their common features. The common features include, for example, FromPhone#, ToPhone#, URLs, commands, and phone# in SMS, size of SMS, and time. We will use both approaches, identifying phone numbers and examining/monitoring content.

4 Evaluation

To evaluate the efficiency of our proposed system, selecting relevant datasets is critical. We aim to use as many malicious SMS messages as possible. Because of privacy concerns, it is hard to find standard SMS datasets with all the required features. Fortunately, there are two well-known public datasets that we considered in our experimental study. SMS Spam Collection dataset [3] contains labeled spam SMS and normal SMS, which proved to be the most useful for our purposes. MIT Reality dataset [10] contains data that we could use to assess our system, although these datasets do not contain all SMS threats and required features. A variety of SMS malware, used to send premium rate SMS without user knowledge, is being analyzed. These SMS are being extracted, along with ToPhone#, and added to our dataset. Finally, we are building behaviour profiling for each Android device and employing light-weight agents to collect certain data. As a result, we are creating some new benchmark datasets to evaluate the performance of the proposed system and are considering the following SMS threats: new SMS spam, SMS containing malicious URLs, SMS Command and Control (C&C), and malware sending premium-rate SMS messages without user knowledge.

5 Conclusions and Future Work

In this paper, we have devised a multi-agent system that is integrated with an intrusion detection system, to detect malicious SMS botnets and identify ways to block the attacks in order to prevent damage caused by these attacks. To identify SMS botnets in Android mobile devices, we have defined and designed a multi-agent system that has the ability to monitor and observe Android device activities, and then capture and report suspicious behaviour to a central server. The agents are capable of interacting with each other and other systems to achieve specified goals. Our framework provides a model that applies signature-based detection to Smartphone SMS messages, and behaviour detection on collected data at the central server. Profiling behavioural analysis is conducted in the central server in order to spot unknown SMS botnet using a multi-detection system. This is done in order to collect data from mobile devices by the agents and send the collected data to the central server. This leads to a response from the Decision-and-Action Module, which finally sends an action to be performed on the Smartphone.

This is work in progress. The next steps involve further development of the framework to ensure its viability. We will test the efficiency, scalability and accuracy of our framework using a large dataset. The main challenges we face include feature selection and production of a suitable dataset. Feature selection within the Detection Module is an important key to achieve significant and accurate classification results. We will carry out several experiments based on behavioural profiling to identify the method for feature extraction for SMS botnets. To produce a suitable dataset, we will define the method and build an SMS botnet dataset, taking into consideration user privacy.

Acknowledgment. The first author graciously acknowledges the funding from University of Hail in Saudi Arabia.

References

1. Acronymics. An integrated toolkit for constructing intelligent software agents,agentbuilder user's guide, http://www.agentbuilder.com/
2. Agüero, J., Rebollo, M., Carrascosa, C., Julián, V.: Developing intelligent agents on the android platform. Universidad Politecnica de Valencia, Spain (2010)
3. Almeida, T., Hidalgo, J.M.G., Silva, T.P.: Towards sms spam filtering: Results under a new dataset. International Journal of Information Security Science 2(1) (2013)
4. Alzahrani, A.J., Ghorbani, A.A.: Sms mobile botnet detection using a multi-agent system: Research in progress. In: Proceedings of the 1st International Workshop on Agents and CyberSecurity, ACySE 2014, pp. 2:1–2:8. ACM, New York (2014)
5. Bellifemine, F., Caire, G., Greenwood, D.: Developming multi-agent systems with jade (2007)
6. Bellifemine, F., Caire, G., Poggi, A., Rimassa, G.: Jade: A software framework for developing multi-agent applications. lessons learned. Information and Software Technology 50(1), 10–21 (2008)
7. Busetta, P., Rönnquist, R., Hodgson, A., Lucas, A.: Jack intelligent agents-components for intelligent agents in java. AgentLink News Letter 2(1), 2–5 (1999)
8. Carabelea, C., Boissier, O., et al.: Multi-agent platforms on smart devices: Dream or reality. In: Proceedings of the Smart Objects Conference (SOC 2003), Grenoble, France, pp. 126–129. Citeseer (2003)
9. Cheng, Z.: A multi-agent security system for android platform (2012)
10. Eagle, N., Pentland, A.: Reality mining: sensing complex social systems. Personal and Ubiquitous Computing 10(4), 255–268 (2006)
11. FIPA. Fipa agent management specification (2004), http://www.fipa.org/specs/fipa00023/SC00023K.html
12. Frantz, C., Nowostawski, M., Purvis, M.K.: Micro-agents on android: interfacing agents with mobile applications. In: Dechesne, F., Hattori, H., ter Mors, A., Such, J.M., Weyns, D., Dignum, F. (eds.) AAMAS 2011 Workshops. LNCS (LNAI), vol. 7068, pp. 488–502. Springer, Heidelberg (2012)
13. Geer, D.: Malicious bots threaten network security. Computer 38(1), 18–20 (2005)
14. Geng, G., Xu, G., Zhang, M., Guo, Y., Yang, G., Wei, C.: The design of sms based heterogeneous mobile botnet. Journal of Computers 7(1), 235–243 (2012)
15. Ghorbani, A., Lu, W., Tavallaee, M.: Network intrusion detection and prevention: concepts and techniques, vol. 47. Springer (2010)

16. Giovanni, C., Giovanni, I., Michele, I., Kevin, H.: Jade tutorial: Jade programming for android (2012), http://jade.tilab.com/doc/tutorials/JadeAndroid-Programming-Tutorial.pdf
17. Nikraz, M., Caire, G., Bahri, P.A.: A methodology for the analysis and design of multi-agent systems using jade (2006)
18. Rosenberg, D.: Carrieriq: The real story (2011), http://vulnfactory.org/blog/2011/12/05/carrieriq-the-real-story/
19. Santi, A., Guidi, M., Ricci, A.: Jaca-android: an agent-based platform for building smart mobile applications. In: Dastani, M., El Fallah Seghrouchni, A., Hübner, J., Leite, J. (eds.) LADS 2010. LNCS (LNAI), vol. 6822, pp. 95–114. Springer, Heidelberg (2011)
20. Savenko, O., Lysenko, S., Kryschuk, A.: Multi-agent based approach of botnet detection in computer systems. In: Kwiecień, A., Gaj, P., Stera, P. (eds.) CN 2012. CCIS, vol. 291, pp. 171–180. Springer, Heidelberg (2012)
21. Shyu, M.-L., Chen, S.-C., Sarinnapakorn, K., Chang, L.: A novel anomaly detection scheme based on principal component classifier. Technical report (2003)
22. Song, L.-P., Jin, Z., Sun, G.-Q.: Modeling and analyzing of botnet interactions. Physica A: Statistical Mechanics and its Applications 390(2), 347–358 (2011)
23. Stone, P., Veloso, M.: Multiagent systems: A survey from a machine learning perspective. Autonomous Robots 8(3), 345–383 (2000)
24. Tilab. Jade - java agent development framework (2011), http://jade.tilab.com/
25. Traynor, P., Lin, M., Ongtang, M., Rao, V., Jaeger, T., McDaniel, P., La Porta, T.: On cellular botnets: measuring the impact of malicious devices on a cellular network core. In: Proceedings of the 16th ACM Conference on Computer and Communications Security, pp. 223–234. ACM (2009)
26. Weaver, N., Staniford, S., Paxson, V.: Very fast containment of scanning worms. In: Proceedings of the 13th Conference on USENIX Security Symposium
27. Dagon, D., Lee, W., Wang, C.: Botnet Detection: Countering the Largest Security Threat. Springer US, New York (2008)
28. Wood, M.F., DeLoach, S.A.: An overview of the multiagent systems engineering methodology. In: Ciancarini, P., Wooldridge, M.J. (eds.) AOSE 2000. LNCS, vol. 1957, pp. 207–221. Springer, Heidelberg (2001)
29. Zinky, J.: Cougaar overview (2009), http://cougaar.org/wp/

A Risk-Averse Inventory Cost Model Using CVaR

Jasmine Jiamin Lim[1], Allan N. Zhang[1], Yew Soon Ong[2], and Puay Siew Tan[1]

[1] Singapore Institute of Manufacturing Technology, Singapore
{jasmine-lim,nzhang,pstan}@SIMTech.a-star.edu.sg
[2] Nanyang Technological University, Singapore
asysong@ntu.edu.sg

Abstract. This paper studies the situation where a risk-averse manager, whose company produces goods catering to uncertain demand, has to decide on the quantity to be produced with the objective of minimizing his cost. Demand uncertainty is a major issue faced by many inventory managers, as overstocking or understocking, will result in the company incurring extra unwanted cost. To tackle this problem, an inventory cost model is proposed in this paper. The model is an extension of the newsvendor framework and is formulated as a minimization problem via conditional Value at Risk (CVaR). The solution to the model is presented and evaluated via a case study. Through this model, the optimal inventory level can be determined and this will serve to be useful for the risk-averse inventory manager.

Keywords: Newsvendor, Risk-averse Supplier, conditional Value at Risk (CVaR), Demand Uncertainty, Inventory Cost Model.

1 Introduction

We consider an inventory cost model where a risk-averse manager, whose company produces goods catering to random demand, decides on the quantity to be produced with the objective of minimizing Conditional Value-at-Risk (CVaR). Value-at-Risk (VaR) and CVaR are two of the many types of risk measurements used in financial risk management (Chen et al., 2009). These two measures have emerged and been used broadly in current few years (Wu et al., 2013; Qiu et al., 2014). Given a confidence level β, the VaR of a portofolio is defined to be the lowest amount α such that the loss will not exceed α with the specified probability β, while the CVaR is defined as the conditional expectation of losses above α. CVaR is seen as a better tool for financial risk as compared to VaR (Yao et al., 2013). The former serves as a better risk measure than the latter, as it is more coherent (Artzner et al., 1999), has better computational characteristics (Chen et al., 2009) and is consistent with higher order stochastic dominance (Ogryczak and Ruszczynski, 2002). As such, the risk measurement tool chosen in this paper will be CVaR.

A supply chain is a complex network consisting of independent and economically rational members (Ryu and Yucesan, 2010). Each member in a supply chain has a role, be it a supplier, a retailer or a customer and so on. The aim of a supplier, or rather a manufacturer, is to maximize his own profits through the sales of his products while

H. Handa et al. (eds.), *Proc. of the 18th Asia Pacific Symp. on Intell. & Evol. Systems – Vol. 1*,
Proceedings in Adaptation, Learning and Optimization 1, DOI: 10.1007/978-3-319-13359-1_10

minimizing his endogenous costs, such as the production cost (Ryu and Yucesan, 2010). The sales are dependent on the demand on the products, which is uncertain and that imposes a risk on the manufacturer as he could incur additional unnecessary costs due to overproduction or underproduction. The newsvendor problem is a good example of an inventory model which is dependent on uncertainty in demand. Overproduction or underproduction of goods can lead to overstocking or understocking. That can give rise to two types of inventory risk: the risk of excessive inventory and the risk of insufficient supply (Cachon, 2004). Such inventory risks are common in the supply chains, as most supply chains are not capable of matching supply and demand perfectly (Cachon, 2004).

Besides, as supply chains are getting more and more complex, they are also more vulnerable to disruptions, which can result in risks that bring along negative impacts (Roh *et al.*, 2013; Pettit *et al.*, 2013). These risks may result in greater demand uncertainties and thus become more problematic to the manufacturer, as demand uncertainties pose as threats to him imposing more unnecessary costs which can be translated to be seen as a form of loss. Hence, there exists a motivation to seek an optimal inventory level based on demand uncertainties and this will in turn minimize the cost and at the same time also help to reduce the inventory risks mentioned earlier. This is done so through the formulation of a cost-based inventory model. The goal of the model is to minimize inventory cost via CVaR minimization. The model is formulated through the extension of the related newsvendor problem.

2 Related Work

The papers reviewed in this section all use the newsvendor models, either basic or various newsvendor-type versions, for their optimization problems. Most papers employ the profit performance measurement and only a handful touches on the cost performance measurement as the objective function. Zhou *et al.*, (2007) propose and solve a profit-based model for multi-product with CVaR constraints. Ryu and Yucesan (2010) introduce fuzzy parameters to their model and solve the fuzzy profit maximization problem based on three coordination policies (quantity discounts, profit sharing and buyback) between manufacturer and retailer. Ozen *et al.* (2011) apply game theory to newsvendors forming a coalition, model the profit of the coalition and study on the convexity of such newsvendor games. Chen (2011) examines the profit functions of the manufacturer and retailer under a wholesale-price-discount contract. Wang *et al.* (2012) study on the effectiveness of several contract formats between manufacturer and retailer by evaluating profit functions. Jeong and Leon (2012) considers a supply chain with multiple members are serially connected and evaluates the profit functions of the members based on complete and partial information of safety stocks. Xanthopoulos *et al.* (2012) propose optimal policies for dual-sourcing supply chains based on a risk management framework. Using the profit model for newsvendor problem, Wen and Qin (2013) study the solution of the newsvendor problem under VaR. Wu *et al.* (2013) study the profit maximization problem with random shortage cost under VaR and CVaR criteria. Jammernegg and Kischka (2013) use the newsvendor framework under two conflicting constraints – the service and loss constraints and present solution based on the profit function. Wu (2013) studies the coordination of competing supply chains and the relevant profit functions, and shows that the buyback contract strategy between the manufacturer and retailer can lead to higher profits. Qiu *et al.* (2014) propose three models with incomplete demand information: expected profit maximization, CVaR-based profit maximization and a combination of the first two. The proposed solution approaches for the three models are validated and said to be robust.

Gotoh and Takano (2007) are the few which introduce a cost-based model. The cost-based model compliments the profit-based model used in the paper for optimization of product quantity. The paper uses CVaR and mean-CVaR to optimize both the cost-based and profit-based models and presents the proposed solutions.

Literature on newsvendor models using cost performance measurement as objective function is not common and the profit performance measurement is often used instead. In this paper, we will look at the optimization problem from a different perspective – to use a cost-based approach to formulate a cost minimization problem, instead of the usual profit-based approach where the focus is on the maximization of profit. Also, we are not only using the traditional newsvendor setting; we further extend the newsvendor model to fit that of a manufacturing company's situation. We then aim to minimize the cost using CVaR. This distinguishes the model to be different from the papers discussed above.

The rest of the paper is structured as follows: we introduce the classic newsvendor model and our extended model in Section 3, formulate the extended model with CVaR and propose an optimization solution in Section 4, evaluate the proposed solution in Section 5 and lastly give a summary and discuss future research directions in Section 6.

3 Model Description

3.1 The Classic Newsvendor Model

The newsvendor problem is a tricky situation faced by a newsvendor who tries to decide on the number of newspapers to stock on a newsstand before the start of the day's sales. The newsvendor is at a risk of overstocking if he orders too much or understocking if he orders too little to cater to the market demand for the day. When the newsvendor overstocks, he sells the remainder at a salvage price; similarly, when he understocks, he incurs a backorder cost due to stockout. So, the newsvendor profit function, which is commonly mentioned in most papers, is formulated as follows:

$$\pi(Q,X) = p\min(Q,X) + s\max(Q-X,0) - b\max(X-Q,0) - cQ \qquad (1)$$

where

Q is the daily order quantity for newspaper,
X is the random demand for newspaper during the selling season,
p is the selling price per unit for the newspaper,
s is the salvage price per unit for the newspaper,
b is the backorder cost per unit for the newspaper and
c is the cost price per unit for the newspaper.

Assume that $p > c > s$ and $b \geq 0$. Let F be the continuous distribution function of random demand for the newspaper. The expected value of $\pi(Q, X)$ is:

$$E[\pi(Q, X)] = \int_0^\infty \pi(Q, x)\, dF(x) \tag{2}$$

The optimal order quantity Q^* can be obtained by maximizing the expected profit shown in equation (2). Assuming that F has an inverse, solving $\frac{\partial E[\pi(Q,X)]}{\partial Q} = 0$ will get Q^* as:

$$Q^* = F^{-1}\left(\frac{p+b-c}{p-s+b}\right) \tag{3}$$

Notice that the product newspaper can be replaced by any other product. Thus, the newsvendor problem may be viewed as an inventory problem for any single product.

3.2 The Inventory Cost Model

The newsvendor problem can be extended to one whereby the newsvendor now produces newspapers to sell to the market. The newsvendor may in fact be viewed as a company which serves as a supplier that produces goods (aka a manufacturer) in a supply chain. Consider that the company produces single product instead of multiple products for simplicity. Instead of using the profit function, we look at the cost function for the company. The cost function of the company is formulated as follows:

$$\omega(q, x) = c_f + c_p(q - y) + b \max(x - q, 0) + h \max(q - x, 0) \tag{4}$$

where

q is the product quantity in inventory and it also includes initial inventory,
x is the demand for the product and it is a continuous non-negative random variable,
c_f is the fixed cost for each production,
c_p is the production cost per unit for the product,
y is the initial inventory level,
b is the backorder cost per unit for the product and
h is the holding cost per unit for the product.

Assume that c_f, c_p, y, b, $h \geq 0$ and $b \geq c_p$. Let G be the cumulative distribution function of the demand for the product. The expected value of $\omega(q, x)$ is similar to equation (2) and is expanded as follows:

$$\begin{aligned} &E[\omega(q, x)] \\ &= c_f + c_p(q - y) + b\, \mathrm{E}[\max(x - q, 0)] + h\, \mathrm{E}[\max(q - x, 0)] \\ &= c_f + c_p(q - y) + b \int_q^\infty (x - q) g(x) dx + h \int_0^q (q - x) g(x) dx \end{aligned} \tag{5}$$

The optimal product quantity q^* can be obtained in similar fashion as section 3.1 by minimizing the expected cost. Assuming that G has an inverse, solving $\frac{\partial E[\omega(q,x)]}{\partial Q} = 0$ will yield:

$$q^* = G^{-1}(\frac{b-c_p}{b+h}) \quad (6)$$

The inventory cost model formulated above is our model of interest and will be used throughout the whole discussion of this paper.

4 CVaR in the Inventory Cost Model

4.1 Definition of CVaR

The results derived from newsvendor model problems, which mainly focus on either maximizing expected profit or minimizing expected cost, have been shown to be unsatisfactory (Qiu *et al.*, 2014). Newsvendor model problems that incorporate several risk preferences are preferred (Qiu *et al.*, 2014). Thus, in this section, we consider introducing the risk measure of CVaR to the inventory cost model. The objective of the risk-averse inventory manager of the company is to minimize the CVaR of the cost.

Let $\varphi(y,z)$ be a loss function associated with random variable z for fixed y. The distribution function of $\varphi(y,z)$ is denoted by $\Phi(y,\eta) := P\{z|\varphi(y,z) \leq \eta\}$. Let β be the confidence level where $\beta \in [0,1)$. The β-VaR, denoted by $\alpha_\beta(y)$, is defined as follows (Rockafellar and Uryasev, 2000):

$$\alpha_\beta(y) = \min\{\alpha \in \mathbb{R}: \Phi(y,\alpha) \geq \beta\} \quad (7)$$

This means that the loss φ is expected to exceed α_β only in probability of $(1-\beta)$ (Gotoh and Takano, 2007).

The β-CVaR with confidence level β and denoted by $\gamma_\beta(y)$, is defined as the mean of the β-tail distribution of $\varphi(y,z)$ (Rockafellar and Uryasev, 2002):

$$\gamma_\beta(y) = \int_{-\infty}^{\infty} \alpha \, d\Phi_\beta(y,\alpha) \quad (8)$$

where

$$\Phi_\beta(y,\alpha) = \begin{cases} 0 & for\ \alpha < \alpha_\beta(y), \\ \dfrac{\Phi(y,\alpha)-\beta}{1-\beta} & for\ \alpha \geq \alpha_\beta(y) \end{cases}$$

In order to minimize the CVaR, (Rockafellar and Uryasev, 2002) proposed an auxiliary function defined by:

$$F_\beta(y,\alpha) = \alpha + \frac{1}{1-\beta} E\{[\varphi(y,z)-\alpha]^+\} \quad (9)$$

where

$$[s]^{+} = \max\{0, s\}$$

Also, they provided an optimization shortcut for minimizing CVaR as follows (Rockafellar and Uryasev, 2002):

$$\min_{y \in Y} \gamma_{\beta}(y) = \min_{(y,\alpha) \in Y \times \mathbb{R}} F_{\beta}(y, \alpha) \tag{10}$$

The optimal solution α^{*} in this minimization problem is equal or almost equal to $\alpha_{\beta}(y^{*})$ (Rockafellar and Uryasev, 2002). Function (9) is convex if the loss function $\varphi(y, z)$is convex, as mentioned by (Gotoh and Takano, 2007).

4.2 Optimization of CVaR of Inventory Cost Model

Using the cost function shown in equation (4) as the loss function, we formulate the optimization problem for the model based on equations (9) and (10) as follows:

$$\min_{q} CVaR_{\beta}(q) = \min_{\alpha,q}\{\alpha + \frac{1}{1-\beta}\int_{0}^{\infty}[\omega(q,x) - \alpha]^{+} g(x)dx\} \tag{11}$$

Proposition 1 The cost function $\omega(q, x)$ is convex with respect to q. Also, the minimization problem stated in (11) is a convex problem.

Proof

We refer to the following theorem from Rockafellar (1997): Let *f* be a real continuous function. Let it be twice differentiable on an open interval (δ, ε). Then *f* is convex if and only if its second derivative *f''* is non-negative throughout the open interval (δ, ε).

We shall apply this theorem. By taking the second derivative of the expected value of the cost function $\omega(q, x)$ stated in equation (5), one can easily get the following result:

$$\frac{\delta^{2} E[\omega(q,x)]}{\delta q^{2}} = (b + h)g(q) \geq 0 \; for \; all \; q \geq 0 \tag{12}$$

Using our assumptions stated in section 3.2, *(b+h)* is nonnegative and thus equation (12) is nonnegative. Since the second derivative is nonnegative, the expected value of the cost function is said to be convex for fixed *x*. Thus, this makes the cost function $\omega(q, x)$ convex with respect to q too.

We use the following proposition from Rockafellar and Uryasev (2002): If the loss function $\varphi(y, z)$ is convex with respect to *y*, then the function $\gamma_{\beta}(y)$ in equation (10) is also convex with respect to *y*. Furthermore, the function $F_{\beta}(y, \alpha)$ in equation (9) is jointly convex in (y, α).

Based on this proposition, we can hence say that equation (11) is a convex problem since the cost function $\omega(q, x)$ is shown to be convex with respect to q. ∎

We assume that the inverse for the distribution function of the product demand exists. Let $H_\beta(q,\alpha)$ be the terms on the right hand side of equation (11):

$$H_\beta(q,\alpha) = \alpha + \frac{1}{1-\beta}\int_0^\infty [\omega(q,x) - \alpha]^+ g(x)dx \tag{13}$$

For $x \in [0,q]$, the cost function $\omega(q,x)$ becomes $c_f + c_p(q-y) + h(q-x)$; similarly for $x \in [q,\infty)$, the cost function $\omega(q,x)$ becomes $c_f + c_p(q-y) + b(x-q)$. Thus we can expand the integral part of equation (13) to be:

$$\int_0^q [c_f + c_p(q-y) + h(q-x) - \alpha]^+ g(x)dx + \int_q^\infty [c_f + c_p(q-y) + b(x-q) - \alpha]^+ g(x)dx \tag{14}$$

Proposition 2 The optimal solution (q^*, α^*) for the minimization of CVaR of inventory cost model is as follows:

$$\begin{cases} q^* = A + B \\ \alpha^* = c_f - c_p y + (c_p + h)A + (c_p - b)B \end{cases} \tag{15}$$

where

$$A = \frac{b}{h+b} G^{-1}\left(\frac{\beta(c_p+h)+b-c_p}{b+h}\right),$$
$$B = \frac{h}{h+b} G^{-1}\left(\frac{(1-\beta)(b-c_p)}{b+h}\right)$$

In particular, when $\beta = 0$, any α^* that satisfies the inequality $\alpha^* \le c_f + c_p(q^* - y)$ also satisfies optimality.

Proof

We solve the problem in same manner as (Gotoh and Takano, 2007) did. The integrals in (14) can be evaluated based on 3 cases for α as shown in Figure 1. The three cases are:

1. $\alpha < c_f + c_p(q-y)$
2. $c_f + c_p(q-y) \le \alpha \le c_f + c_p(q-y) + hq$
3. $\alpha > c_f + c_p(q-y) + hq$

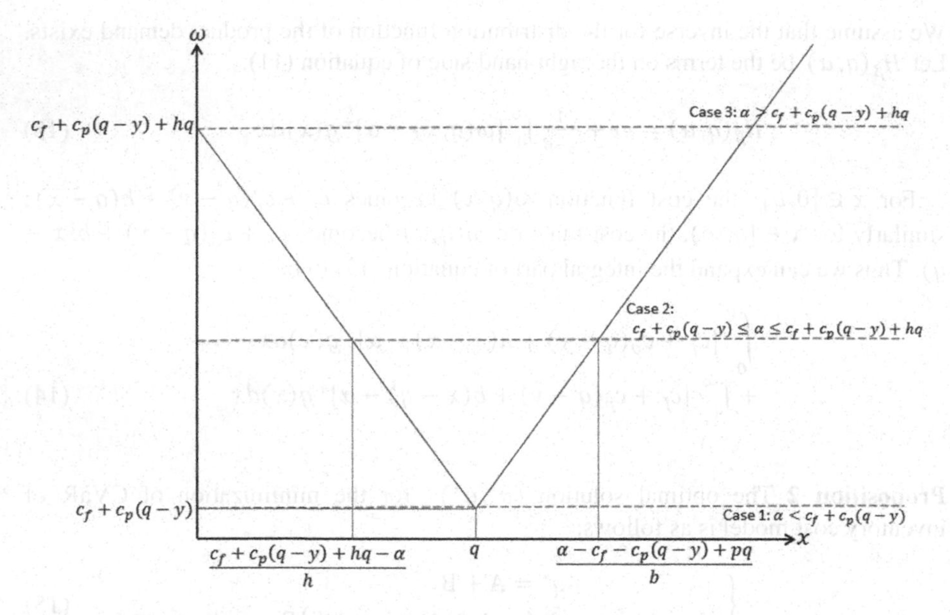

Fig. 1. Three cases in minimization of CVaR of the inventory cost model

From Figure 1, the graph is drawn as: $\omega(q,x) = c_f + c_p(q-y) + h(q-x)$ for $x \in [0,q]$ and $\omega(q,x) = c_f + c_p(q-y) + b(x-q)$ for $x \in [q,\infty)$. Case 1 refers to the range of α below the cost function ω, case 2 refers to the range of α between $c_f + c_p(q-y)$ and $c_f + c_p(q-y) + hq$, and lastly case 3 refers to the range of α above $c_f + c_p(q-y) + hq$.

Case 1: $\alpha < c_f + c_p(q-y)$

Since the cost ω is greater than α for any demand x, equation (13) becomes:

$$\alpha + \frac{1}{1-\beta}\left[\int_0^q (c_f + c_p(q-y) + h(q-x) - \alpha)g(x)dx + \int_q^\infty (c_f + c_p(q-y) + b(x-q) - \alpha)g(x)dx\right] \tag{16}$$

Taking derivatives of equation (16) with respect to q and α respectively and setting them to zeroes, i.e. $\frac{\delta H_\beta(q,\alpha)}{\delta q} = 0$ and $\frac{\delta H_\beta(q,\alpha)}{\delta \alpha} = 0$, we obtain the solution (q^*, α^*) to be $q^* = G^{-1}(\frac{b-c_p}{b+h})$ and $\alpha^* < c_f + c_p(q^* - y)$ only when $\beta = 0$. Notice that the q^* obtained here is exactly the same as the optimal quantity obtained in equation (6).

Case 2: $c_f + c_p(q-y) \leq \alpha \leq c_f + c_p(q-y) + hq$

We set the cost ω when there is no demand (*x=0*) to be the upper limit and the cost ω when product quantity meet exactly the demand (*x=q*) to be the lower limit for the range of α. Also, by equating the cost ω with α, we set ω=α and get

$x = \frac{c_f + c_p(q-y) + hq - \alpha}{h}$ for $x \in [0, q]$ and $x = \frac{\alpha - c_f - c_p(q-y) + bq}{b}$ for $x \in [q, \infty)$. In this case, equation (13) will then become:

$$\alpha + \frac{1}{1-\beta}\Big[\int_0^{\frac{c_f + c_p(q-y) + hq - \alpha}{h}} \big(c_f + c_p(q-y) + h(q-x) - \alpha\big) g(x) dx + \int_{\frac{\alpha - c_f - c_p(q-y) + bq}{b}}^{\infty} \big(c_f + c_p(q-y) + b(x-q) - \alpha\big) g(x) dx\Big] \quad (17)$$

Taking derivatives of equation (17) with respect to q and α respectively and setting them to zeroes, i.e. $\frac{\delta H_\beta(q,\alpha)}{\delta q} = 0$ and $\frac{\delta H_\beta(q,\alpha)}{\delta \alpha} = 0$, we obtain the solution (q^*, α^*) to be:

$$\begin{cases} q^* = A + B \\ \alpha^* = c_f - c_p y + (c_p + h)A + (c_p - b)B \end{cases} \quad (18)$$

where

$$A = \frac{b}{h+b} G^{-1}\left(\frac{\beta(c_p+h) + b - c_p}{b+h}\right),$$
$$B = \frac{h}{h+b} G^{-1}\left(\frac{(1-\beta)(b-c_p)}{b+h}\right)$$

Notice that setting $\beta = 0$ to q^* in (18) will also yield the same solution as q^* found in case 1.

Case 3: $\alpha > c_f + c_p(q-y) + hq$

In this last case we consider having α to be greater than the cost ω at the point of no demand (*x=0*). The left integral disappears in this case and equation (13) then becomes:

$$\alpha + \frac{1}{1-\beta}\Big[\int_{\frac{\alpha - c_f - c_p(q-y) + bq}{b}}^{\infty} \big(c_f + c_p(q-y) + b(x-q) - \alpha\big) g(x) dx\Big] \quad (19)$$

There is no optimal solution derived from the first-order condition of equation (18) with respect to q and α respectively.

Therefore from the results of these three cases, we conclude the following optimal solution (q^*, α^*) for the minimization of CVaR of inventory cost model to be as stated in equation (15). ■

5 Case Study

The company in this case study is a prominent chemical provider in the paint industry. It distributes raw materials for paints to its customers. As the company is a retailer rather than a manufacturer in the supply chain, we define both parameters c_f and c_p from our inventory cost model to be the ordering cost and cost price (per unit product) respectively. In this new definition, the inventory cost model will still work as the company can still act as a supplier who will incur both holding and backorder costs depending on its customers' demand. In this case study, we will evaluate our model based on two types of demand distributions – Normal and Gamma. The aim of the case study is not only to help the company to determine the optimal values q^* and α^* but to also test on the feasibility of the model and its proposed solution.

Due to the confidentiality of the said company, data used in this case study will be fictitious but representative so that they will still resemble the real data given by the company. We set the parameters in the model to be: $c_f = 5.3, c_p = 0.15, b = 0.83, h = 0.35, y = 100$. The demand data are shown in Table 1, sorted in ascending order.

Table 1. Data Points Used for the Demand Scenarios

Demand	27	34	38	45	144	149	153	157	161

We consider two types of demand scenarios for our model. As the normal distribution is a commonly used distribution, we first assume that the demand is nominal. This assumption is feasible as inventory models can be implemented based on nominally distributed demands (Johansen and Hill, 2000; Guijarro *et al.*, 2012; Schrijver *et al.*2013). Thus, this becomes our first scenario for the demand.

By inspection, the demand seems to be irregular as the demand can go as high as 157 or 161 and as low as 27 or 34. To model such irregular demand, it is acceptable that we choose the gamma distribution (Nenes *et al.*, 2010; Khaniyev *et al.*, 2013). This shall be our second scenario.

Calculations done will be based on 3 confidence levels: 90%, 95% and 99%. These three levels are the most commonly used confidence levels (Rockafellar and Uryasev, 2000). Based on the data used in Table 1, the results for the model after computation are summarized (rounded off to 2 decimal places) in Table 2.

Table 2. Summarized Results after Computation

Normal Distribution for Demand				Gamma Distribution for Demand			
Confidence Level	q^*	α^*	VaR based on q^*	Confidence Level	q^*	α^*	VaR based on q^*
0.9	144.39	59.36	53.03	0.9	163.07	61.66	63.92
0.95	151.36	69.45	63.31	0.95	182.78	74.02	84.28
0.99	165.23	89.21	85.13	0.99	228.14	100.21	125.23

Looking at the results from Table 2, we can see that the values of α^* and VaR calculated based on q^* are quite close. This is in line with what (Rockafellar and Uryasev, 2002) mentioned about the optimal solution α^* being almost equal to $\alpha_\beta(q^*)$ (which is VaR calculated based on q^*). This could imply that α^* can be used as a rough estimate for the actual VaR based on q^*.

We also compare the optimal solutions q^* from both scenarios with optimal quantity q^* obtained from equation (6), which is the conventional way of solving newsvendor problems. The optimal quantity q^* from equation (6) is derived using the same distribution functions from both scenarios. We use cost as the basis for our comparison and this is done by taking the average of costs calculated based on the parameters defined above, demand data from table 1 and q^* from the 2 scenarios and equation (6). The results are summarized in Tables 3 and 4.

Table 3. Summary of Results for q^* and Average Costs based on Normal Distribution

Type	Normal Distribution for Demand (Scenario 1)			Equation (6) with Nominal Distribution
Confidence Level	0.9	0.95	0.99	N.A
q^*	144.39	151.36	165.23	112.12
Average cost	32.75	32.89	37.61	37.72

Table 4. Summary of Results for q^* and Average Costs based on Gamma Distribution

Type	Gamma Distribution for Demand (Scenario 2)			Equation (6) with Gamma Distribution
Confidence Level	0.9	0.95	0.99	N.A
q^*	163.07	182.78	228.14	100.78
Average cost	35.31	44.09	64.38	39.48

From Tables 3 and 4, with the exception of scenario 2 at confidence level of 95% and 99%, we can see that the average costs for the q^* obtained from our model are lower than the average costs for q^* obtained from equation (6). This seems to suggest that our model is capable of getting better optimal quantities q^* as compared to the conventional solution from equation (6). Also, by looking at all the average costs obtained from the 2 scenarios, it is easy to see that the costs are smaller than their respective α^* on various confidence levels. This is in line with the definition of VaR where the cost is not expected to exceed the VaR at the probability level β.

6 Discussion and Conclusion

6.1 Discussion

The results based on the case study in section 5 suggest that the model is feasible and can be applied on real world situations. However, this is just a standalone case and more case studies will need to be conducted with the proposed model to confirm its applicability. Also, the sensitivities of the parameters of the model are not yet known.

More research, such as sensitivity analysis, will need to be done so that necessary modifications or constraints can be imposed on the parameters. This would also further validate our model. Last but not least, more demand types and their distribution types should be considered for the model. For instance, an irregular demand may also be seen as being intermittent, sporadic, lumpy and so on. This might require other distributions to model the demand even more accurately. Also, considering other distribution types can help to investigate which type of distributions may or may not suit the model.

Notice that the demand distribution function of the model requires past demand data and this would, however, also highlight another downside of this model, which is that past demand data are a pre-requisite for the model to work. To tackle this, one could either use Monte Carlo simulation to generate such data or to think of another solution to deal with incomplete data. The latter will be touched on as a future research direction in Section 6.2.

One limitation of this proposed model is that it is an unconstrained problem. In other words, the solution is only limited to the objective function and may not apply when there are constraints, like having restrictions on the variables, being added to the problem. When there are more constraints being added, the problem may become quite different from the original problem and it will require other algorithms to solve for the optimal values.

6.2 Conclusion and Future Research

This paper presents a risk-averse inventory model extended from the newsvendor model with the objective of minimizing the cost function via CVaR. It helps a risk-averse manager, whose company produces goods catering to uncertain demands, make a better decision on the inventory level in order to minimize inventory risk. A case study discussed in this paper shows that it enables the company to reduce its inventory cost by as high as 13.17% in an average.

The proposed model might be applied during disruptions too, where the company can use past demand data based on past disruptions to deduce a demand distribution or predict a demand distribution type which might occur during disruptions. Thus, the model can also be seen as a risk management tool. However, this point will need to be further explored in future works to test on its feasibility.

We may consider incorporating the concept of incomplete demand information into the model, like what (Qiu *et al.*, 2014) did. In the absence of demand data, this concept will come in handy. This will greater enhance our model especially since the model requires complete demand information or the demand distributions to be known and such information may not be available at times. Also, it will be beneficial to look at how supply chain disruptions may impact other operations of the company, such as restricting the amount of goods to be kept as inventory or affecting prices due to reduction in raw materials for production etc., and to examine the relationship between disruptions and the parameters of our model. This will hence allow us to redefine our model and to better determine our optimal quantities based on disruptions.

Acknowledgement. This work is partially supported under the A*STAR TSRP funding, the Singapore Institute of Manufacturing Technology and the Centre for Computational Intelligence (C2I) at Nanyang Technological University.

References

1. Xanthopoulos, A., Vlachos, D., Iakovou, E.: Optimal Newsvendor Policies for Dual-sourcing Supply Chains: A Disruption Risk Management Framework. Computers and Operations Research 39, 350–357 (2012)
2. Artzner, P., Delbaen, F., Eber, J.M., Heath, D.: Coherent Measures of Risk. Mathematical Finance 9, 203–208 (1999)
3. Wu, D.: Coordination of Competing Supply Chains with News-vendor and Buyback Contract. International Journal of Production Economics 144, 1–13 (2013)
4. Guijarro, E., Cardos, M., Babiloni, E.: On the Exact Calculation of the Fill Rate in a Periodic Review Inventory Policy under Discrete Demand Patterns. European Journal of Operational Research 218, 442–447 (2012)
5. Yao, F., Wen, H., Luan, J.: CVaR Measurement and Operational Risk Management in Commercial Banks According to the Peak Value Method of Extreme Value Theory. Mathermatical and Computer Modelling 58, 15–27 (2013)
6. Cachon, G.P.: The Allocation of Inventory Risk in a Supply Chain: Push, Pull, and Advance-Purchase Discount Contracts. Management Science 50, 222–238 (2004)
7. Nenes, G., Panagiotidou, S., Tagaras, G.: Inventory Managment of Multiple Items with Irregular Demand: A Case Study. European Journal of Operational Research 205, 313–324 (2010)
8. Jeong, I.-J., Leon, V.J.: A Serial Supply Chain of Newsvendor Problem with Safety Stocks Under Complete and Partial Information Sharing. International Journal of Production Economics 135, 412–419 (2012)
9. Roh, J., Hong, P., Hokey, M.: Implementation of a Responsive Supply Chain Strategy in Global Complexity: The Case of Manufacturing Firms. Internationl Journal of Production Economics 147, 198–210 (2013)
10. Wang, J.-C., Lau, A.H.-L., Lau, H.-S.: Practical and Effective Contracts for the Dominant Retailer of a Newsvendor Product with Price-Sensitive Demand. International Journal of Production Economics 138, 46–54 (2012)
11. Chen, J.: Returns with Wholesale-price-discount Contract in a Newsvendor Problem. Internation Journal of Production Economics 130, 104–111 (2011)
12. Gotoh, J.-Y., Takano, Y.: Newsvendor Solutions via Condtional Value-at-Risk Minimization. European Journal of Operational Research 179, 80–96 (2007)
13. Ryu, K., Yucesan, E.: A Fuzzy Newsvendor Approach to Supply Chain Coordination. European Journal of Operational Research 200, 421–438 (2010)
14. Wu, M., Zhu, S.X., Teunter, R.H.: Newsvendor Problem with Random Shortage Cost under a Risk Criterion. International Journal of Production Economics 145, 790–798 (2013)
15. Wen, P., Qin, L.: The Solution of Newsvendor Problem based on Value-at-Risk. In: 25th Chinese Control and Decision Conference, pp. 1029–1033 (2013)
16. Tyrrel Rockafellar, R.: Convex Analysis, Part I Section 4, Reprint Revised edn., p. 26. Princeton University Press, NJ (1997)
17. Tyrrell Rockafellar, R., Uryasev, S.: Optimization of Conditional Value-at-Risk. Journal of Risk 2(3) (2000)

18. Tyrrel Rockafellar, R., Uryasev, S.: Conditional Value-at-Risk for General Loss Distributions. Journal of Banking and Finance 26, 1443–1471 (2002)
19. Qiu, R., Shang, J., Huang, X.: Robust Inventory Decision under Distribution Uncertainty: A CVaR-based Optimization Approach. International Journal of Production Economics 153, 13–23 (2014)
20. Johansen, S.G., Hill, R.M.: The (r, Q) Control of a Periodic-Review Inventory System with Continuous Demand and Lost Sales. International Journal of Production Economics 68, 279–286 (2000)
21. De Schrijver, S.K., Aghezzaf, E.-H., Vanmaele, H.: Aggregate Constrained Inventory Systems with Independent Multi-product Demand: Control Practices and Theoretical Limitations. International Journal of Production Economics 143, 416–423 (2013)
22. Khaniyev, T., Turksen, I.B., Gokpinar, F., Gever, B.: Ergodic Distribution for a Fuzzy Inventory Model of Type (s, S) with Gamma Distributed Demands. Expert Systems with Applications 40, 958–963 (2013)
23. Pettit, T.J., Croxton, K.L., Fisksel, J.: Ensuring Supply Chain Resilience: Development and Implementation of an Assessment Tool. Journal of Business Logistics 34, 46–76 (2013)
24. Ozen, U., Norde, H., Slikker, M.: On the Convexity of Newsvendor Games. International Journal of Production Economics 133, 35–42 (2011)
25. Jammernegg, W., Kischka, P.: Risk Perferences of a Newsvendor with Service and Loss Constraints. International Journal of Production Economics 143, 410–415 (2013)
26. Ogryczak, W., Ruszczynski, A.: Dual Stochastic Dominance and Related Mean-Risk Models. Society for Industrial and Applied Mathematics Journal on Optimization 13, 60–78 (2002)
27. Zhou, Y.-J., Chen, X.-H., Wang, Z.-R.: Optimal Ordering Quantities for Multi-products with Stochastic Demand: Return-CVaR Model. International Journal of Production Economics 112, 782–795 (2007)
28. Chen, Y.(F.), Xu, M., Zhang, Z.G.: A Risk-Averse Newsevendor Model Under the CVaR Criterion. Operations Research 57, 1040–1044 (2009)

Multi-objective Heterogeneous Capacitated Vehicle Routing Problem with Time Windows and Simultaneous Pickup and Delivery for Urban Last Mile Logistics

Chen Kim Heng[1], Allan N. Zhang[1], Puay Siew Tan[1], and Yew-Soon Ong[2]

[1] Singapore Institute of Manufacturing Technology, A*STAR, Singapore
{hengck,nzhang,pstan}@simtech.a-star.edu.sg
[2] Nanyang Technological University, Singapore
asysong@ntu.edu.sg

Abstract. The Urban Last Mile Logistics (LML) is known to be the most expensive, least efficient and most polluting section of the supply chain. To that extent, a multi-objective heterogeneous capacitated vehicle routing problem with time windows and simultaneous pickup and delivery (MoHCVRP-TWSPD) is formulated and solved to cater to this section of the supply chain. The proposed model is solved through two proposed methods that are based on exact methods. A small benchmark was adopted from the current literature to test the proposed methods and computational results are reported. Based on the computational results, a number of insights into the MoHCVRP-TWSPD problem are provided.

Keywords: last mile, vehicle routing, time window, simultaneous pickup and delivery, capacitated, heterogeneous, logistics.

1 Introduction

The Vehicle Routing Problem (VRP) was first introduced by Dantzig and Ramser [1]. Throughout the years, various variations of VRP have been studied, each variation having different or multiple attributes. An attribute is defined as the feature considered in the problem such as heterogeneous fleet or customers' time windows.

The Urban LML; the last leg of the supply chain of much concern to logistics service providers (LSPs), is known to be the most expensive, least efficient and most polluting section of the supply chain. The LSPs typically have a mixed fleet of vehicles with differing capacities to handle day to day operations that include criteria such as customer requirements for pickup and delivery and customers' preferred times. Despite the maturity in the study of VRP, according to our literature review, there has yet to be work that directly addresses all of the aforementioned concerns. In order to cater to the multi-objective nature of the Urban LML problem faced by LSPs, in this paper, we present a VRP variant referred to as Multi-objective Heterogeneous Capacitated Vehicle Routing Problem with Time Windows and Simultaneous Pickup and

H. Handa et al. (eds.), *Proc. of the 18th Asia Pacific Symp. on Intell. & Evol. Systems – Vol. 1*,
Proceedings in Adaptation, Learning and Optimization 1, DOI: 10.1007/978-3-319-13359-1_11

Delivery (MoHCVRP-TWSPD). To our best knowledge, there has yet to be work in this multi-objective multi-attribute VRP problem.

The MoHCVRP-TWSPD is a problem of finding optimal heterogeneous fleet routes to serve pickup and delivery orders of customers according to the customers' preferred time windows, departing from the depot and returning to the depot once serving all of their assigned customers. During the course of execution, each of the vehicle's capacity cannot be exceeded and the time window of the depot cannot be violated as well. Optimality of vehicle routes is defined as per the objective function values. In our study, we attempt to optimise the total travelling distance, total travel time and the total emission caused by the execution of the fleet routes.

Our contributions in this paper are three folds. First, we present a mixed integer formulation of MoHCVRP-TWSPD. Second, we present two solving methods based on exact methodologies; a One-step method that provides a more realistic solution through the discretisation of vehicle speed and a Two-step method that provides a glimpse of the consequences of the current industry practice of prioritizing minimization of distance over other objectives such as emission. Finally, we provide a few insights gained from the results of our computations. Section 2 will provide the reader with an overview of related works in this variant of VRP. Section 3 will present the reader with our formulation of the problem and the proposed solving methods. Section 4 houses the computational results obtained by us and insights into the meaning of the results. Finally, Section 5 is where we will conclude the paper and their findings.

2 Related Works

In line with the aim to better fit research work to real world problems, there has been a surge in VRPs with multiple attributes such as Capacitated VRP with Time Window and Simultaneous Pickup and Delivery (CVRPTWSPD)[2-6] and Capacitated VRP with Time Window and Pickup and Delivery(CVRPTWPD)[7].

As mentioned in the introduction, to our best knowledge, there hasn't been any works in MoHCVRP-TWSPD. The closest variant that has been discovered in the literature review is CVRPTWSPD which only considers a single objective as opposed to the problem being studied in this paper. We first present the reader with a brief literature review of CVRPTWSPD. The reader will then be provided with an overview of works in Green VRP, Multi-Objective VRP and truckload factor considerations in VRP.

CVRPTWSPD is one of the multi-attribute problems that has been increasingly studied over the years. In 2002, Angelelli and Marsini[2] presented a set covering formulation for CVRPTWSPD and solved the problem using branch-and-price and branch-and-bound algorithms. The benchmark from Solomon[8] was modified to test their algorithms. Chang et al.[3] looked into real time CVRPTWSPD and solved it by iteratively solving a mixed integer programming model on a rolling time horizon.

The methodology employed by them consisted of route construction and improvement heuristics and TS. They tested their algorithm by adopting a benchmark from Gelinas et al.[9]. More recently, Gutiérrez-Jarpa et al.[4] presented a branch-and-price algorithm and Mingyong and Erbao[5] studied a minimization of total travel distance and employed differential evolution algorithm, both to solve the CVRPTWSPD problem. In 2012, Wang and Chen [6] designed a coevolution genetic algorithm to solve the problem and adopted the Solomon[8] benchmark to test their algorithm.

Given recent initiatives taken by governments all over the world to improve the environment, number of studies in Green VRP has been picking up over the years. Bektaş and Laporte [10] looked into the Pollution Routing Problem (PRP) in 2011, an extension of VRP that considers other factors such as greenhouse emissions. They considered fuel consumption and greenhouse gas emission as their eco-indicators. In order to solve for the solution, they employed CPLEX 12.1 with its default settings and tested their model with three classes of problems with varying number of cities as nodes. Interested readers may refer to a survey done by Lin et al.[11].

Jozefowiez et al.[12] presented a review on Multi-objective VRP (Mo-VRP). According to the review, some of the most studied objectives include minimization of travel distance[13], travel time[14], number of vehicles[15], waiting times of vehicles[15], makespan of routing solution[16], deviation from or number of violations of constraints such as time window constraints[17] and risk in transporting of hazardous materials[18]. There have also been studies on optimizing the balance of travel time between vehicles[13], load balance between the various vehicles in the fleet[19] and balance of tour lengths between the various vehicles[20].

According to our literature review, research on active optimization of truckload factor in VRP is scarce. The search terms used by us in Google Scholar are inclusive of 'truckload factor vehicle routing problem', 'capacity utilization vehicle routing problem', 'balance capacity vehicle routing problem', 'balance load vehicle routing problem', 'balance load factor vehicle routing problem', 'resource utilization vehicle routing problem', 'resource maximization vehicle routing problem', 'deadhead vehicle routing' and 'volume utilization vehicle routing problem'. Among all the results obtained by the author, only a few works were found to have focused on maximizing truckload factor of vehicles. Tavakkoli-Moghaddam[21] considered maximizing vehicle utilization in a variant of Capacitated Vehicle Routing Problem (CVRP). In the field of Mo-VRP, Sutcliffe and Board[13] considered maximization of vehicle utilization in their paper besides minimization of travel distance while Moura[22] considered the objective in the VRPTW with Loading Problem.

From the literature review performed by us, it can be concluded that the literature on MoHCVRP-TWSPD is scarce. Therefore, this paper attempts to bridge the gap between the current literature and the needs of the industry through the analysis of the MoHCVRP-TWSPD problem. For an example, one of the major gaps concerns with the balance between objectives such as total travelled distance and total emission which is looked into in this paper.

3 Problem Definition and Formulation

In order to tackle the MoHCVRP-TWSPD problem, we modified and extended the formulation presented by Cordeau et al.[23] and produced the following mixed integer programming formulation as follows:

We assume that there is only one depot and a set of customers where the number of customers and number of vehicles are assumed to be at least two. The capacities of the vehicles are assumed to be non-homogeneous, that is there exists at least two vehicles where their capacities are different from each other. The vehicles are assumed to be of heavy duty with weights in between 3500 kg and 7500 kg. It is also assumed that the vehicles depart from the depot at the start of their trip and return to the depot upon completion of their trip. Finally, each customer is assumed to be visited only once by one vehicle and has a time period of which they can only be served within this time period. In this paper, we are going to focus on several objectives inclusive of minimization of travel distance, travel time and emission while maintaining a certain level of vehicle utilization.

Sets

- $V = \{0, 1, 2, ..., n+1\}$
  - Set of nodes of G (0: index of node representing departure of depot, 1 ~ n: indices of nodes representing customers, $n+1$: index of node representing destination of depot)
- $N = V \backslash \{0, n+1\}$
  - Set of customers
- $K = \{1, ..., k\}$
  - Set of vehicles
- $A = \{(i,j) | \, i, \, j \in V, \, i \neq j\}$
  - Set of arcs of G

Parameters

- $c_{i,j}$: Travel cost or distance from node i to node j
- d_i: Delivery demand of customer i
- p_i: Pickup demand of customer i
- $[a_i, b_i]$: Time window for node $i \in N$
- $[a_0, b_0]$: Time window for the depot
- s_i: Service time for node i
- Ca^k: Capacity of vehicle k
- V_{ij}^{max} : Maximum speed limit travelling from node i to node j
- V_{ij}^{min} : Minimum speed limit travelling from node i to node j
- TF: Minimum desired truckload factor
- M: A large constant

Variables

- $x_{i,j}^k$: Binary variable, 1 if arc $(i,j) \in A$ belongs to the optimal routes by vehicle k, 0 otherwise
- $l_{i,j}^k$: Load of vehicle k travelling from node i to node j
- $v_{i,j}^k$: Vehicle k's speed travelling from node i to node j
- w_i^k: Start time of vehicle k at node i

$$\text{Min} \quad \sum_{k \in K} \sum_{(i,j) \in A} c_{i,j} x_{i,j}^k \tag{1a}$$

$$\text{Min} \sum_{k \in K} \sum_{(i,j) \in A} \left(\frac{-8.8125 \times 10^{-7}}{{v_{ij}^k}^2} \left(v_{ij}^k - 539.376\right)\left(v_{ij}^k - 1.04928\right)\left({v_{ij}^k}^2 - 100.34 v_{ij}^k + 6495.77\right)\left({v_{ij}^k}^2 + 100.34 v_{ij}^k + 3572.38\right) \right) c_{i,j}\, x_{i,j}^k \tag{1b}$$

$$s.t. \quad \sum_{k \in K} \sum_{j \in V} x_{i,j}^k = 1, \forall i \in N, \tag{2}$$

$$\sum_{j \in V} x_{0,j}^k = 1, \forall k \in K, \tag{3}$$

$$\sum_{i \in V} x_{i,n+1}^k = 1, \forall k \in K, \tag{4}$$

$$\sum_{i \in V} x_{i,h}^k - \sum_{j \in V} x_{h,j}^k = 0, \forall h \in N, \forall k \in K, \tag{5}$$

$$\sum_{i \in V} l_{0,i}^k = \sum_{i \in N} d_i \sum_{j \in V} x_{i,j}^k, \forall k \in K, \tag{6}$$

$$\sum_{i \in V} l_{i,n+1}^k = \sum_{i \in N} p_i \sum_{j \in V} x_{i,j}^k, \forall k \in K \tag{7}$$

$$l_{j,i}^k + p_i - d_i - Ca^k\left(1 - x_{i,j}^k\right) \le l_{i,m}^k, \forall i \in N, \forall j, m \in V, \forall k \in K, \tag{8}$$

$$w_i^k + s_i + \frac{c_{ij}}{v_{ij}^k} - w_j^k \le \left(1 - x_{i,j}^k\right) M, \forall (i,j) \in A, \forall k \in K, \tag{9}$$

$$a_i \le w_i^k \le b_i, \forall i \in V, \forall k \in K, \tag{10}$$

$$l_{i,j}^k \le Ca^k x_{i,j}^k, \forall i,j \in V, \forall k \in K \tag{11}$$

$$x_{i,j}^k \in \{0,1\}, \forall i,j \in V, \forall k \in K \tag{12}$$

$$l_{i,j}^k \geq 0, \forall i,j \in V, \forall k \in K \tag{13}$$

$$V_{ij}^{\min} \leq v_{ij}^k \leq V_{ij}^{\max} \tag{14}$$

$$\frac{\sum_{(i,j)\in A} c_{i,j} \frac{l_{i,j}^k}{Ca^k} x_{i,j}^k}{\sum_{(i,j)\in A} c_{i,j} x_{i,j}^k} \geq TF \tag{15}$$

The objective functions of the formulation above are represented by 1a and 1b. 1a represents the total travel distance of the fleet. 1b represents the total emission produced by the fleet. Our emission formula is adopted from the formula presented by Hickman[24]. It is of utmost importance to note that travel time is decided by the speed of the vehicle and distance covered by the vehicle. Through the relationship between the 3 variables, it follows that travel time and travel distance are positively proportional to each other i.e. optimization of travel distance implies optimization of travel time and vice versa. Constraint 2 specifies that each customer is served by exactly one vehicle. Constraint 3 necessitates that each vehicle will leave node 0 for a total of one time while constraint 4 ensures that each vehicle returns to node n+1. Constraint 5 specifies that each vehicle that enters a customer node has to leave the customer node. Constraint 6 specifies that each vehicle leaves the depot with the total delivery load assigned to it. Constraint 7 ensures that each vehicle returns to the depot with the total pickup load assigned to it. Constraint 8 is the load balance constraint. Constraint 9 ensures that the start time of the service at the next node is not earlier than the earliest time possible given the start time of the current node. Constraint 10 is the time window constraint. Constraint 11 is the vehicle capacity constraint. Constraint 12 is the binary variable constraint while constraint 13 is a non-negativity constraint. Constraint 14 is the speed limit constraint on each arc. Constraint 15 specifies that the weighted average of the utilization of the vehicle is at least of the desired minimum truckload factor. Note that Constraint 15 is an ε-constraint method to optimise the utilization of each vehicle in the fleet.

Note that the formulation can be easily altered to only consider a subset of the attributes considered in this formulation. The reader is reminded to note that the formulation provided above is non-linear.

4 Solution of MoHCVRP-TWSPD

In order to solve the aforementioned problem in CPLEX, the formulation above is solved using two methods which are explained below.

4.1 One-Step Optimisation

Assume that the speed variables of the vehicles on each arc are discrete variables. Using linearization methods provided by Bisschop [25], objective 1b, constraint 9 and constraint 15 are linearised. The outcome of the linearization is as provided below. Unless otherwise stated, the variables, parameters and sets in the following expressions, equations and inequalities are defined as before.

Sets

- $speed = \{10,20,\ldots,80\}$: Set of discrete speeds for each vehicle on each arc where $speed_u$ refers to element u of the set $speed$.
- $SL = \{1,2,\ldots,8\}$: Set of indices of the set $speed$

Parameters

- M': A large constant

Variables

- $v_{i,j}^{u}$: Binary indicator specifying whether the vehicle transversing arc (i,j) is travelling at $speed_u$
- $y_{i,j}^{k}$: $x_{i,j}^{k} \times l_{i,j}^{k}$

$$\text{Min } \sum_{u \in SL} \sum_{(i,j) \in A} \left(\frac{-8.8125 \times 10^{-7}}{{speed_u}^2} (speed_u - 539.376)(speed_u - 1.04928)\left({speed_u}^2 - 100.34 speed_u + 6495.77\right)\left({speed_u}^2 + 100.34 speed_u + 3572.38\right) \right) c_{i,j}\, v_{i,j}^{u} \quad (1c)$$

$$w_i^k + s_i + \frac{c_{ij}}{speed_u} - w_j^k - \left(1 - x_{i,j}^k\right)M \le \sum_{z \in SL, z \neq u} M' v_{i,j}^{z},\ \forall (i,j) \in A, \forall k \in K, \forall u \in SL \quad (16)$$

$$\sum_{u \in SL} v_{i,j}^{u} - \sum_{k \in K} x_{i,j}^{k} = 0,\ \forall (i,j) \in A,\ \forall k \in K,\ \forall u \in SL \quad (17)$$

$$\frac{\sum_{(i,j) \in A} c_{i,j} \frac{y_{i,j}^k}{Ca^k}}{\sum_{(i,j) \in A} c_{i,j} x_{i,j}^k} \ge TF \quad (18)$$

$$y_{i,j}^k \le \min\{Ca^k x_{i,j}^k, l_{i,j}^k\},\ \forall (i,j) \in A,\ \forall k \in K, \quad (19)$$

$$y_{i,j}^k \ge \max\{0, l_{i,j}^k - Ca^k\left(1 - x_{i,j}^k\right)\},\ \forall (i,j) \in A,\ \forall k \in K, \quad (20)$$

Objective 1b is linearised into objective 1c while Constraint 9 is linearised into constraint 16 and constraint 17. Finally, constraint 9 is linearised into constraint 18-20. A new objective 1d is then formed through the summation of objective 1a and 1c.

$$\text{Min } \sum_{k\in K}\sum_{(i,j)\in A} c_{i,j}x_{i,j}^k + \sum_{u\in SL}\sum_{(i,j)\in A}\left(\frac{-8.8125\times10^{-7}}{speed_u{}^2}(speed_u - 539.376)(speed_u - 1.04928)(speed_u{}^2 - 100.34speed_u + 6495.77)(speed_u{}^2 + 100.34speed_u + 3572.38)\right)c_{i,j}\, v_{i,j}^u \tag{1d}$$

The resultant formulation is solved using CPLEX.

4.2 Two-Step Optimisation

Using linearization methods provided by Bisschop [25], constraint 15 is linearised into constraint 18-20. The first step involves solving a HCVRPTWSPD formulation with minimization of total travel distance as the objective. A new travel time matrix is formed through the normalization of the distance matrix. The formulation is then simplified to a HCVRPTWSPD formulation through the replacement of the speed variable by the corresponding travel time parameter and removal of constraint 14. This results in constraint 9 being formulated into constraint 21 as given below. Unless otherwise stated, the variables, parameters and sets in the following expressions, equations and inequalities are defined as before.

Parameters:

- $t_{i,j}$: Travel time from node i to node j

$$w_i^k + s_i + t_{ij} - w_j^k \le (1 - x_{i,j}^k)M, \forall (i,j) \in A, \forall k \in K, \tag{21}$$

The resultant simplified problem is then solved using CPLEX to obtain an initial solution.

The second step involves the optimisation of emission of the initial solution through the optimisation of travel speed on each arc. Note that the solution produced by step 2 will have total distance value corresponding to the one provided by step 1. Given that the emission function (objective 1b) is a convex function from 10km/h to 100km/h, it can be obtained that the optimal speed for minimal emission without any constraints is approximately 55.1771km/h. Through the employment of goal programming, a new formulation is formed as provided below. Unless otherwise stated, the variables, parameters and sets in the following expressions, equations and inequalities are defined as before.

$$\text{Min} \quad \sum_{k\in K}\sum_{(i,j)\in A} c_{i,j} \times \max(55.1771 \times t_{ij} - c_{i,j}, c_{i,j} - 55.1771 \times t_{ij}) \tag{1e}$$

$$s.t. \quad w_i^k + s_i + t_{ij} - w_j^k \le (1 - x_{i,j}^k)M, \forall (i,j) \in A, \forall k \in K, \tag{21}$$

$$a_i \leq w_i^k \leq b_i, \forall i \in V, \forall k \in K, \tag{10}$$

$$V_{ij}^{\min} \times t_{ij} \leq c_{i,j} \leq V_{ij}^{\max} \times t_{ij} \tag{22}$$

The objective is to minimize the deviation of optimal feasible speed on each arc from the aforementioned value while ensuring that constraint 9, 10 and 14 are satisfied. Note that constraint 9 and 14 is represented by constraint 21 and constraint 22 respectively as time based constraints. The resultant formulation is solved through CPLEX again.

5 Computational Results and Discussions

In order to test the methods proposed, we modified the benchmark provided by Mingyong and Erbao [5]. CPLEX was run on a 64-bit Windows 7 computer with a 2.7Ghz quad core processor and 4Gb of RAM. We tested both methods at TF = 0, TF = 0.3 and TF = 0.8 respectively to represent negligence, low and high requirement in truckload factor. Note that we assume the objectives considered in this paper are unitless.

Table 1. Objective Values and Computation Time

Settings	Objective Values & Computation Time	Single Step Optimisation	Two-Step Optimisation
TF = 0.8	Distance	795	795
	Emission	≈295686.65	≈293468.685
	Computation Time(seconds)	6.46	1.21(Step 1)
			0.50(Step 2)
TF = 0.3	Distance	795	790
	Emission	≈295686.65	≈313441.06
	Computation Time(seconds)	8.66	1.23(Step 1)
			0.51(Step 2)
TF = 0	Distance	795	790
	Emission	≈295686.65	≈313441.06
	Computation Time(seconds)	10.49	1.27(Step 1)
			0.63(Step 2)

Table 1 shows the computational results obtained by us. A notable observation from the results would be the drastic increase in the number of units of emission as a tradeoff with a small decrease in unit distance in the case of using Two-step optimisation method to solve the problem with the setting TF = 0.3 and TF = 0. This implies that a prioritization of minimisation of distance in solving the MoHCVRP-TWSPD might lead to a drastic increase in emission. The observation also suggests that a low truckload factor requirement might lead to a solution with high emission. Given the results above, we have summarised the advantages and disadvantages of the suggested methods as provided in Table 2.

Table 2. Advantages and disadvantages of proposed methods

	One-Step Optimisation	Two-step Optimisation
Advantages	• Optimise all the 3 objectives on the same level of priority • A more granular discretisation leads to more optimal solution	• Fast computation
Disadvantages	• Slow computation • Discretisation of velocity decision variable generates more constraints	• Possibly less optimal than single step optimisation • Emission as secondary objective

Table 2 shows a comparison of both methods presented in this paper in terms of the advantages and disadvantages. From our analysis, even though the Single step optimisation method optimizes all the considered objectives with equal priority, it should be noted that a more optimal solution in objective 1b would necessitate more constraints which would lead to slower computation time. On the other hand, Two-step optimisation shows a relatively shorter computation time compared to Single step optimisation at the possible expense of a less optimal objective value for objective 1b.

6 Conclusion and Further Research

In this paper, we have presented a formulation for MoHCVRP-TWSPD to address complexity of urban last mile logistics and presented 2 approaches based on exact methods to show that the problem is solvable. The One-step method provides a more realistic solution through the discretisation of vehicle speed. The Two-step method on the other hand provides a glimpse of the consequences of the current industry practice of prioritizing minimization of distance over other objectives such as emission. The results indicate that a prioritisation of minimization of travel distance might lead to a solution with substantially higher emission and a possible inverse relationship between truckload factor and emission level.

For further research, a number of promising directions can be pursued. In order to better serve the logistics industry, the inclusion of total waiting time into the total travel time of the objective will serve to make the problem more realistic. Inclusion of other objectives such as balance of load or travel time or maximization of customer satisfaction will bolster the usefulness of the study of this problem to the real world. Finally, research on more efficient multi-objective evolutionary algorithms is our next research effort.

Acknowledgement. This work is partially supported under the A*STAR TSRP funding, the Singapore Institute of Manufacturing Technology and the Centre for Computational Intelligence (C2I) at Nanyang Technological University.

References

1. Dantzig, G.B., Ramser, J.H.: The truck dispatching problem. Management Science 6(1), 80–91 (1959)
2. Angelelli, E., Mansini, R.: The vehicle routing problem with time windows and simultaneous pick-up and delivery. In: Quantitative Approaches to Distribution Logistics and Supply Chain Management, pp. 249–267. Springer (2002)
3. Chang, M.-S., Chen, S., Hsueh, C.-F.: Real-time vehicle routing problem with time windows and simultaneous delivery/pickup demands. Journal of the Eastern Asia Society for Transportation Studies 5, 2273–2286 (2003)
4. Gutiérrez-Jarpa, G., et al.: A branch-and-price algorithm for the vehicle routing prob-lem with deliveries, selective pickups and time windows. European Journal of Operational Research 206(2), 341–349 (2010)
5. Mingyong, L., Erbao, C.: An improved differential evolution algorithm for vehicle routing problem with simultaneous pickups and deliveries and time windows. Engineering Applications of Artificial Intelligence 23(2), 188–195 (2010)
6. Wang, H.-F., Chen, Y.-Y.: A genetic algorithm for the simultaneous delivery and pickup problems with time window. Computers & Industrial Engineering 62(1), 84–95 (2012)
7. Desrosiers, J., et al.: Time constrained routing and scheduling. In: Handbooks in Operations Research and Management Science, vol. 8, pp. 35–139 (1995)
8. Solomon, M.M.: Algorithms for the vehicle routing and scheduling problems with time window constraints. Operations Research 35(2), 254–265 (1987)
9. Gélinas, S., et al.: A new branching strategy for time constrained routing problems with application to backhauling. Annals of Operations Research 61(1), 91–109 (1995)
10. Bektaş, T., Laporte, G.: The Pollution-Routing Problem. Transportation Research Part B: Methodological 45(8), 1232–1250 (2011)
11. Lin, C., et al.: Survey of Green Vehicle Routing Problem: Past and future trends. Expert Systems with Applications 41(4), 1118–1138 (2014)
12. Jozefowiez, N., Semet, F., Talbi, E.-G.: Multi-objective vehicle routing problems. European Journal of Operational Research 189(2), 293–309 (2008)
13. Sutcliffe, C., Boardman, J.: Optimal solution of a vehicle-routeing problem: transporting mentally handicapped adults to an adult training centre. Journal of the Operational Research Society, 61–67 (1990)
14. Hong, S.-C., Park, Y.-B.: A heuristic for bi-objective vehicle routing with time window constraints. International Journal of Production Economics 62(3), 249–258 (1999)
15. Sessomboon, W., et al.: A study on multi-objective vehicle routing problem considering customer satisfaction with due-time (the creation of Pareto optimal solutions by hybrid genetic algorithm). Transaction of the Japan Society of Mechanical Engineers (1998)

16. Murata, T., Itai, R.: Local search in two-fold EMO algorithm to enhance solution similarity for multi-objective vehicle routing problems. In: Obayashi, S., Deb, K., Poloni, C., Hiroyasu, T., Murata, T. (eds.) EMO 2007. LNCS, vol. 4403, pp. 201–215. Springer, Heidelberg (2007)
17. Geiger, M.J.: Genetic algorithms for multiple objective vehicle routing. arXiv preprint arXiv:0809.0416 (2008)
18. Giannikos, I.: A multiobjective programming model for locating treatment sites and routing hazardous wastes. European Journal of Operational Research 104(2), 333–342 (1998)
19. Bowerman, R., Hall, B., Calamai, P.: A multi-objective optimization approach to urban school bus routing: Formulation and solution method. Transportation Research Part A: Policy and Practice 29(2), 107–123 (1995)
20. El-Sherbeny, N.: Resolution of a vehicle routing problem with multi-objective simulated annealing method. Faculté Polytechnique de Mons (2001)
21. Tavakkoli-Moghaddam, R., Safaei, N., Gholipour, Y.: A hybrid simulated annealing for capacitated vehicle routing problems with the independent route length. Applied Mathematics and Computation 176(2), 445–454 (2006)
22. Moura, A.: A multi-objective genetic algorithm for the vehicle routing with time windows and loading problem. In: Intelligent Decision Support, pp. 187–201. Springer (2008)
23. Cordeau, J.-F., et al.: VRP with time windows. The Vehicle Routing Problem 9, 157–193 (2002)
24. Hickman, J., et al.: Methodology for calculating transport emissions and energy consumption (1999)
25. Bisschop, J.: AIMMS-optimization modeling. Lulu. com (2006)

Orthogonal Predictive Differential Evolution*

Yue-Jiao Gong[1,2], Qi Zhou[1], Ying Lin[2,3], and Jun Zhang[2,4,**]

[1] Department of Computer Science, Sun Yat-sen University
[2] Key Laboratory of Machine Intelligence and Advanced Computing, Ministry of Education
[3] Department of Psychology, Sun Yat-sen University
[4] School of Advanced Computing, Sun Yat-sen University
Guangzhou 510006, China
junzhang@ieee.org

Abstract. In traditional differential evolution (DE) algorithms, the perturbation direction of mutation is not sophisticatedly designed, which performs ineffectively or inefficiently for optimizing some complex and large-scale problems. This paper designs an orthogonal predictive mutation scheme to solve this problem. The mutation investigates the landscape near the individuals by using orthogonal experimental design, and then applies factor analysis to predict a promising direction for the individuals to evolve. With a clear sense of search direction, the efficiency of DE is improved. Moreover, the step length of the proposed mutation is adaptively adjusted according to the effect of the prediction, which helps to balance the exploration and exploitation abilities of DE. By employing such a mutation scheme, a novel DE algorithm termed orthogonal predictive DE (OPDE) is proposed in this paper. As OPDE can adopt different kinds of classical mutation schemes for choosing the base vector and calculating the differential vector, we further develop an OPDE family including various OPDE variants. Experimental results demonstrate the effectiveness and high efficiency of the proposed algorithm.

Keywords: Differential evolution, evolutionary computation, global optimization, orthogonal experiment design, factor analysis.

1 Introduction

Differential Evolution (DE) is a very competitive population-based stochastic optimization algorithm proposed by Storn and Price in 1997 [10]. Similar to the other Evolutionary Algorithms (EAs), DE performs mutation, crossover, and selection on the population at each generation to search for the global optimum. The major difference of DE with the other EAs (such as Genetic Algorithm,

* This work was supported in part by the National High-Technology Research and Development Program (863 Program) of China No.2013AA01A212, in part by the NSFC for Distinguished Young Scholars 61125205, in part by the NSFC No. 61332002 and No.61300044.

** Corresponding author.

H. Handa et al. (eds.), *Proc. of the 18th Asia Pacific Symp. on Intell. & Evol. Systems – Vol. 1*,
Proceedings in Adaptation, Learning and Optimization 1, DOI: 10.1007/978-3-319-13359-1_12

Evolutionary Programming, Evolutionary Strategy, etc.) lies in its mutation operation, which utilizes the difference of individuals to explore the problem landscape.

A series of studies have been conducted and reported in the literature on designing effective and efficient mutation operation for DE in order to improve its performance. In addition to the original "DE/rand/1" mutation scheme, Storn and Price [9] later suggested four new forms of DE mutation termed "DE/rand/2", "DE/best/1", "DE/best/2", and "DE/target-to-best/1". In [3], a trigonometric mutation scheme was proposed to increase the convergence speed of DE. By modifying the topology of DE, a neighborhood-based mutation was presented in [1] to balance the exploration and exploitation ability. Recently, Epitropakis *et al.* [2] proposed a proximity-based mutation also incorporating the information of the neighboring individuals to improve classical DEs.

Although DE is a very competitive algorithm among EAs, it still has some drawbacks. On the one hand, for multimodal problems, DE is probable to converge into local optima. This is because, in the mutation of DE, the perturbation direction is based on the difference of the parameter vectors (chromosomes), which does not cover all the direction in the problem space. Hence, the individual may not be able to reach the global optimum via mutation. Moreover, as the scalar factor F that determines the proportion of the perturbation length to the differential length is traditionally set smaller than 1, once all the individuals converge into a same local optimum, the population can no longer jump out of the valley. On the other hand, as traditional DE mutation simply adds the scaled differential vector to the base vector, it is hardly to know that the result is evolution or degradation. In the theoretical analysis part of this paper, we will find that the probability of the direction used in traditional DE mutation being the optimal direction is only $1/3^D$, where D is the dimensionality of the problem space. Therefore, DE may still suffer from low efficiency with the increase of the problem scale. Most existing works of DE focus on enhancing the performance of DE on normal-scale problems whose dimensionality is not larger than 100. However, due to the curse of dimensionality, the problems which are easy to solve in low dimensions may become quite difficult to tackle in high-dimensional space.

In this paper, we propose an orthogonal predictive mutation scheme for DE to solve the above described problems. By using orthogonal prediction method (OPM), the mutation samples a set of representative points uniformly distributed in the environment and predicts a promising direction for the perturbation. Similar to traditional DE, the perturbation length is based on the difference of individuals in the population, which is large at the beginning and gradually decreases with the convergence of the population. In addition, the scalar factor F_{op} for the orthogonal predictive mutation is adaptively adjusted according to the prediction effect, which can be larger than 1 if needed. By adopting such a mutation scheme, a novel orthogonal predictive DE algorithm termed OPDE is proposed in this paper. The advantages of OPDE are three-fold. First, the mutation with OPM costs polynomial times of evaluations (which will be proved later) to enable the individual search the problem space with a promising direction. The search efficiency of DE is improved. Second, because the search direction is no longer strictly restricted by the current parameter vectors, and the scalar factor is adaptively adjusted which could be larger than the traditional setting of F, the exploration ability of OPDE is greatly enhanced. Third, the OPM investigates a number of representative vectors around the parameter vectors, which can be considered as a local search process. This mechanism helps to enhance the exploitation ability of OPDE.

The rest of the paper is organized as follows. Section 2 presents the backgrounds and related works. In Section 3, first our methodology is introduced and then the implementations of the proposed mutation and OPDE are detailed. Experiments are conducted and results are thoroughly analyzed in Section 4, followed by a conclusion in Section 5.

2 Backgrounds

2.1 Orthogonal Prediction Method

Orthogonal design method (ODM) is a very popular experimental design method that has been widely applied in various applications for many years. In addition, orthogonal prediction method (OPM) consists of an ODM part and a prediction process using factor analysis based on the ODM part.

Orthogonal Design Method: Suppose that an experiment consists of k factors and each factor has s levels. Then a complete experiment contains s^k combinations. When the evaluation cost of each combination is heavy and the value of k or s is large, it becomes impossible to conduct such a complete experiment. In contrast, by using the ODM, only a small subset of representative combinations need to be tested in order to obtain a high quality combination.

Let $L(N, k, s)$ denote an orthogonal array (OA) with N rows and k columns. Each column of the array stands for a factor in the experiment, while the entities in the column shows the level of the factor (each factor has s levels). Therefore, N is the number of representative combinations to be tested in the experiment. The selected combinations are uniformly distributed over the space of all possible combinations.

Let us take a biological experiment as example to make the ODM easier to understand. Shown in Table I, the experiment is to find the best level combination of the three factors, namely, temperature (Temp.), CO2 concentration (CO2 Conc.), and illumination (Illu.), while each factor contains three levels. For this instance, a complete experiment requires $3^3 = 27$ tests. On the other hand, if ODM is applied, we can use the OA $L(9, 4, 3)$ shown in Table II, where the forth column is a redundant column that can be ignored (an OA without some columns is still an OA). By using the ODM, only 9 combinations need to be tested.

Table 1. A Biological Experiment with Multiple Factors

Level \ Factor	Temp.(°C)	CO2 Conc.(%)	Illu. (lx)
1	30	5%	200
2	40	10%	300
3	50	15%	400

Table 2. An Orthogonal Array $L(9, 4, 3)$

Combination \ Factor	1	2	3	4
C_1	1	1	1	1
C_2	1	2	2	2
C_3	1	3	3	3
C_4	2	1	2	3
C_5	2	2	3	1
C_6	2	3	1	2
C_7	3	1	3	2
C_8	3	2	1	3
C_9	3	3	2	1

Prediction Based on Factor Analysis: After evaluating the N combinations, factor analysis is applied to generate a new combination which may be very promising for the experiment. The process is very simple. For each factor, the average evaluation values to each of its s levels are calculated, and then the level with the smallest average is selected to form a new combination.

In the biological experiment described above, the experimental results of the nine combinations based on $L(9, 4, 3)$ is shown in Table III. Then, for the first factor, as the average evaluation values to levels 1, 2, and 3 are 37, 38, and 40 respectively, level 1 is selected. In a same way, level 3 is selected for factor 2, and level 1 is selected for level 3. Finally, a new combination $P = (1, 3, 1)$ is predicted. The evaluation value of P is compared with the above nine orthogonal combinations, and the best one is considered as the final result of OPM. This combination is always a solution of high-quality, even though it may not be the real optimal combination in the complete experiment.

2.2 Related Works

In the literature, ODM has been widely used to improve the performance of EAs such as genetic algorithm (GA) [17], particle swarm optimization (PSO) [15,7], ant colony optimization (ACO) [8,5,16], etc. Considering DE, Gong *et al.* [4] and Wang *et al.* [12] applied ODM to the crossover operator of DE to make DE faster and more robust. Zhao [18] introduced ODM into the mutation operator of DE and hybridized it with the Guo Tao algorithm.

Table 3. Illustration of the Orthogonal Prediction Method

Level \ Factor	Temp.(°C)	CO2 Conc.(%)	Illu. (lx)	Death Rate (%)
C_1	(1) 30	(1) 5	(1) 200	$r_1 = 30$
C_2	(1) 30	(2) 10	(2) 300	$r_2 = 45$
C_3	(1) 30	(3) 15	(3) 400	$r_3 = 36$
C_4	(2) 40	(1) 5	(2) 300	$r_4 = 39$
C_5	(2) 40	(2) 10	(3) 400	$r_5 = 42$
C_6	(2) 40	(3) 15	(1) 200	$r_6 = 33$
C_7	(3) 50	(1) 5	(3) 400	$r_7 = 51$
C_8	(3) 50	(2) 10	(1) 200	$r_8 = 39$
C_9	(3) 50	(3) 15	(2) 300	$r_9 = 30$
Level	Factor Analysis			
L_1	$(r_1 + r_2 + r_3)/3 = 37$	$(r_1 + r_4 + r_7)/3 = 40$	$(r_1 + r_6 + r_8)/3 = 34$	
L_2	$(r_4 + r_5 + r_6)/3 = 38$	$(r_2 + r_5 + r_8)/3 = 42$	$(r_2 + r_4 + r_9)/3 = 38$	
L_3	$(r_7 + r_8 + r_9)/3 = 40$	$(r_3 + r_6 + r_9)/3 = 33$	$(r_3 + r_5 + r_7)/3 = 43$	
P	(1) 30	(3) 15	(1) 200	$r(P) = 24$

Compared with these literature works embedding ODM into DE [4,12,18], there are four major differences in our work. First and the most important, existing works use ODM only to sample neighbors, whereas the proposed algorithm contains a predicting procedure based on the sampled orthogonal neighbors. Experimental study in Section 4.4 will validate the effectiveness of the predicting operation, i.e., even though the sampled orthogonal neighbors are all of low quality, the predicted points can still be very promising. Second, the proposed algorithm adaptively adjusts the orthogonal step length based on the current condition of the search process so as to improve the performance of OPM. Third, different from some studies that let all individuals conduct the ODM, the proposed mutation is only performed on the best individual to guide the search. In this way, the rest individuals in the population adopt traditional mutation to improve diversity and to save the computational cost of OPM. Last, the proposed mutation is a general framework that can adopt various existing mutation schemes to enhance its performance.

3 Orthogonal Predictive Differential Evolution

3.1 Methodology

The mutation operation of DE can be generally characterized by a form of

$$V_i = Xbase_i + F \cdot Diff_i \tag{1}$$

For example, considering the "DE/rand/1" mutation scheme, the base vector $Xbase_i$ is $X_{r_{i,1}}$, while the differential vector is calculated as $Diff_i = (X_{r_{i,2}} -$

$X_{r_i,3}$). For another example, in the "DE/best/2" scheme, it can be found that $Xbase_i$ is the best individual in the population while $Diff_i$ equals to $(X_{r_i,1} - X_{r_i,2}) + (X_{r_i,3} - X_{r_i,4})$. Traditional DE mutation simply adds the $Xbase_i$ and $Diff_i$ together to generate the mutant vector, which actually does not fully consider the direction of the perturbation. Here without loss of generality, we use the "DE/rand/1" mutation for illustration. Consider a case that $X_{r_i,2}$ and $X_{r_i,3}$ locate on opposites of $Xbase_i$. Suppose $X_{r_i,2}$ is worse than $Xbase_i$ whereas $X_{r_i,3}$ is better than $Xbase_i$, then the result of the mutation is deterioration rather than evolution.

In fact, to fully explore the neighborhood of the base vector, when a differential vector $Diff_i$ is generated, for each dimension j of the individual, the component after perturbation could be $Xbase_i - F \cdot Diff_i$, $Xbase_i$, and $Xbase_i + F \cdot Diff_i$. The corresponding super-cuboid in the D-dimensional space contains 3^D possible combinations, which represent 3^D different direction for the base vector to vary. Assume that the probability of reaching any point in the super-cuboid is the same, if there is only one optimal combination (direction), the probability of the direction used in traditional DE mutation being the optimal direction is only $1/3^D$. When the problem dimensions are 30 and 1,000 respectively, the expected numbers of mutation conducted to achieve the best direction are up to 1,014 and 10,477, respectively. The OPM described in Section 2.1 provides an efficient way to predict a high-quality direction by sampling only a few representative points in the super-cuboid. In fact, for a problem with D dimensions, the sampling times N (the number of rows in the OA) in OPM equals to $3^{\lceil log_3(2D+1) \rceil}$ [6]. As there exists a

$$2D + 1 \leq (N = 3^{\lceil log_3(2D+1) \rceil}) < 3^{log_3(2D+1)+1} = 6D + 3 \tag{2}$$

N is a polynomial of the problem dimensions. For the 30D and 1,000D problems, N can be calculated as 81 and 2,143, respectively.

The traditional DE mutation does not take the direction of perturbation into consideration, which may costs exponential times of mutation to achieve a good direction. When the problem dimensions increases to a very large value such as 1,000, the efficiency of DE decreases dramatically. By using OPM for prediction, we can develop a novel mutation framework for DE, which searches the problem space with a clear sense of direction. The mutation costs only polynomial times of evaluations to predict a good direction and is capable of improving the search efficiency of DE. Noticed that the OPM process is independent with how the base and differential vectors are generated, the proposed mutation framework can employ all the traditional mutation schemes, such as "DE/rand/1", "DE/rand/2", "DE/best/1", "DE/best/2", etc., to generate $Xbase_i$ and $Diff_i$. The DE mutation scheme adopting this framework with OPM is named orthogonal predictive mutation. A DE algorithm employing an orthogonal predictive mutation is then termed orthogonal predictive DE (OPDE). Detailed implementations of the proposed mutation and the OPDE algorithm will be described in the following subsections.

3.2 Implementations of Orthogonal Predictive Mutation

For a D-dimensional problem, an OA $L(N, k, 3)$ is constructed, where N is the number of points to be tested, k is the number of columns satisfying $k \geq D$, 3 stands for each factor has three levels (-1, 0, and $+1$). As discussed in Section 3.1, N is a polynomial of the problem dimensions. For an individual i in the population ($i = 1, 2, \ldots, NP$), the process of the proposed orthogonal predictive mutation performed on the individual is described as follows.

Step 1) Randomly select D different columns from the OA $L(N, k, 3)$ and record them as $c_1, c_2, \ldots, c_D$.

Step 2) Generate N orthogonal points $O_1, O_2, \ldots, O_N$ in the neighborhood of the base vector $Xbase_i$. For each $O_k = [o_{k,1}, o_{k,2}, \ldots, o_{k,D}]$ $(k = 1, 2, \ldots, N)$, $o_{k,j}$ $(j = 1, 2, \ldots, D)$ is calculated as

$$o_{k,j} = Xbase_{i,j} + L_{k,c_j} \cdot F_{op} \cdot Diff_{i,j} \tag{3}$$

where L_{k,c_j} is the level value in the kth row and c_jth column of the OA, F_{op} is the scalar factor for the orthogonal predictive mutation.

Step 3) Select $N - 1$ neighboring points randomly and evaluate their cost values. Notice that we do not evaluate all the N neighboring points. On the one hand, this mechanism introduces noise into the mutation so as to discourage some extreme cases, e.g., if the problem has a uniform landscape, the effects of different levels will be offset. On the other hand, as only one point is neglected, the orthogonal property of the sampling points is maximally maintained, which will be further analyzed in the experimental part of this paper.

Step 4) Calculate the average effect of each level on each factor based on the $N - 1$ neighboring points and combine the levels with the smallest averages on each factor to form a predicted combination $P = [p_1, p_2, \ldots, p_D], p_j \in \{-1, 0, 1\}$. Therefore, $P \cdot Diff_i$ indicates a promising direction for individual i to mutate.

Step 5) An orthogonal predicted point $OP = [op_1, op_2, \ldots, op_D]$ in the predicted direction is calculated according to (4). Then, the cost of OP is evaluated.

$$op_j = Xbase_{i,j} + p_j \cdot F_{op} \cdot Diff_{i,j} \tag{4}$$

Step 6) In this step, the scalar factor F_{op} is adjusted according to the effect of the orthogonal prediction. Select the best point among OP and the $N - 1$ neighboring points and record the point as BP. If BP is not worse than the base vector $Xbase_i$, the orthogonal scale factor F_{op} is expanded, otherwise it is contracted.

$$F_{op} = \begin{cases} v_e \cdot F_{op}, & \text{if } f(BP) \leq f(Xbase_i) \\ v_c \cdot F_{op}, & \text{otherwise} \end{cases} \tag{5}$$

where v_e is the expansion rate and v_c is the contraction rate empirically setting as $v_e = 1.3$ and $v_c = 0.7$, respectively.

Step 7) Moreover, in order to further investigate the predicted direction, a random point $RP = [rp_1, rp_2, \ldots, rp_D]$ located between OP and the base point $Xbase_i$ is generated according to

$$rp_j = Xbase_{i,j} + p_j \cdot F_{op} \cdot rnd_j \cdot Diff_{i,j} \tag{6}$$

where rnd_j is a random number in (0, 1). Evaluate the cost of RP, and compare it with that of BP. Set the mutant vector V_i to the winner between BP and RP.

The test of the random point RP on the predicted direction further adds flexibility to the prediction process, which helps to improve the effectiveness and efficiency of the proposed method.

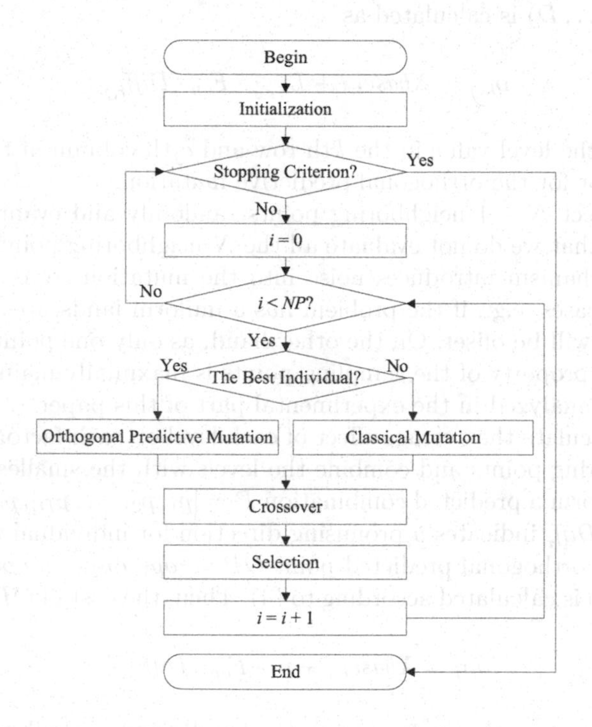

Fig. 1. The flowchart of OPDE

3.3 Orthogonal Predictive Differential Evolution (OPDE)

A general flowchart of the proposed OPDE algorithm is shown in Fig. 1. It can be seen that the algorithm follows the basic structure of the classical DE. The main difference is the mutation process. In OPDE, the best individual in each generation undergo the proposed orthogonal predictive mutation whereas the other individuals undergo classical DE mutation. After generating the mutant vector, the crossover operation is performed. The mutant and target vectors

Table 4. Benchmark Functions

No.	Function Name	Search Space
f_1	Sphere	$[-100, 100]^D$
f_2	Schewefel 2.22	$[-10, 10]^D$ $(D = 30)$ $[-5, 5]^D$ $(D = 1,000)$
f_3	Shifted Schwefel 2.21	$[-100, 100]^D$
f_4	Step	$[-100, 100]^D$
f_5	Noise	$[-1.28, 1.28]^D$
f_6	Shifted Rosenbrock	$[-100, 100]^D$
f_7	Shifted Rastrigin	$[-5, 5]^D$
f_8	Shifted Ackley	$[-32, 32]^D$
f_9	Shifted Griewank	$[-600, 600]^D$
f_{10}	Schewefel 2.26	$[-500, 500]^D$
f_{11}	Penalized 1	$[-50, 50]^D$
f_{12}	Penalized 2	$[-50, 50]^D$

exchange some of their components to form a trial vector. Then, the cost value of the trial vector is evaluated. In the selection, the target vector and the trial vector competes to survive in the next generation. For the best individual in the population, the mutant vector, whose cost has been evaluated in the orthogonal mutation, also participates in the competition. Hence, for the best individual, the cost of three vectors, namely, the mutant vector, the trial vector, and the target vector are compared, and the best one survives in the next generation.

As described in Section 3.1, the proposed mutation method can adopts all kinds of classical DE mutation for the generation of base and differential vectors. Following the traditional naming convention of DE algorithms, the OPDEs that embed OPM into the "DE/rand/1", "DE/rand/2", "DE/best/1", "DE/best/2", and "DE/target-to-best/1" are denoted as "OPDE/rand/1", "OPDE/rand/2", "OPDE/best/1", "OPDE/best/2", and "OPDE/ target-to-best/1", respectively.

4 Experimental Verification and Comparisons

4.1 Experimental Setup

In the experiment undertaken, 12 benchmark functions widely used in the literature [14,11], with dimensions set to 30 and 1,000 respectively, are tested. Listed in Table 1, functions f_1 to f_6 are unimodal functions, whereas functions f_7 to f_{12} are multimodal functions. As as pointed out in [13] that the simulation of function f_2 may lead to storage overflow, the variable range of f_2 is set to [-10, 10] for $D = 30$ and [-5, 5] for $D = 1,000$.

We compare the OPDEs with the corresponding classical DEs to minimize the above described objective (cost) functions. All the compared algorithms set F to 0.5 as suggested in most literatures, and test different settings of CR including $CR = 0.1$, 0.5, and 0.9. For the 30D functions, the population size NP is set

Table 5. Comparisons Between OPDEs and the Corresponding DEs on 30D Functions

	BEST/1/bin						RAND/1/bin					
	DE CR=0.1 average best	OPDE CR =0.1 average best	DE CR=0.5 average best	OPDE CR=0.5 average best	DE CR=0.9 average best	OPDE CR=0.9 average best	DE CR=0.1 average best	OPDE CR =0.1 average best	DE CR=0.5 average best	OPDE CR=0.5 average best	DE CR=0.9 average best	OPDE CR=0.9 average best
f_1	5.31E-35	**2.62E-91**	4.49E-94	**1.06E-125**	2.38E-195	**3.09E-206**	1.25E-18	**5.22E-58**	1.13E-15	**3.00E-84**	3.89E-09	**1.13E-97**
	1.83E-35	**9.28E-97**	4.59E-96	**3.84E-129**	1.01E-198	**9.25E-211**	8.20E-19	**3.17E-62**	6.76E-16	**5.22E-92**	9.26E-10	**6.10E-111**
f_2	1.18E-19	**8.36E-24**	1.63E-48	**1.82E-67**	6.91E-83	**1.70E-116**	1.69E-11	**3.87E-28**	1.83E-09	**3.82E-26**	3.68E-04	**3.69E-33**
	6.70E-20	**1.28E-24**	2.16E-49	**1.37E-68**	2.52E-93	**3.44E-118**	1.26E-11	**3.67E-32**	1.50E-09	**5.70E-27**	1.39E-04	**3.89E-34**
f_3	2.25E-01	**6.75E-08**	1.33E+00	**6.48E-08**	**1.07E-13**	**1.00E-13**	1.15E-01	**1.47E-08**	8.89E-02	**2.77E-09**	7.04E-02	**2.26E-09**
	2.92E-04	**1.13E-08**	3.55E-02	**2.84E-13**	**5.68E-14**	**5.68E-14**	9.38E-02	**2.21E-09**	5.73E-02	**1.52E-11**	4.27E-02	**1.63E-10**
f_4	**0.00E+00**	**0.00E+00**	**0.00E+00**	**0.00E+00**	**0.00E+00**	**0.00E+00**	**0.00E+00**	**0.00E+00**	**0.00E+00**	**0.00E+00**	**0.00E+00**	**0.00E+00**
	0.00E+00	**0.00E+00**	**0.00E+00**	**0.00E+00**	**0.00E+00**	**0.00E+00**	**0.00E+00**	**0.00E+00**	**0.00E+00**	**0.00E+00**	**0.00E+00**	**0.00E+00**
f_5	**7.17E-03**	9.00E-03	**1.94E-03**	2.90E-03	**3.26E-03**	4.40E-03	1.53E-02	**9.28E-03**	1.40E-02	**3.99E-03**	1.33E-02	**3.98E-03**
	3.88E-03	5.62E-03	**1.03E-03**	1.35E-03	**9.05E-04**	1.99E-03	7.78E-03	**5.36E-03**	8.14E-03	**1.32E-03**	5.94E-03	**1.39E-03**
f_6	**5.31E+01**	7.55E+01	3.00E+01	**7.59E+00**	1.59E+00	**1.12E+00**	4.23E+01	**2.44E+01**	2.35E+01	**1.59E+00**	1.89E+01	**3.06E+00**
	1.94E+01	**5.90E-02**	2.03E-01	**5.68E-14**	**0.00E+00**	**0.00E+00**	2.89E+01	**1.04E-04**	2.32E+01	**0.00E+00**	1.80E+01	**0.00E+00**
f_7	5.62E-01	**2.79E-01**	2.40E+01	**6.25E+00**	8.61E+01	**1.67E+01**	1.44E-08	**0.00E+00**	1.02E+02	**1.74E+01**	1.65E+02	**2.52E+00**
	0.00E+00	**0.00E+00**	1.39E+01	**1.99E+00**	4.52E+01	**4.97E+00**	3.56E-09	**0.00E+00**	8.98E+01	**2.55E-07**	1.25E+02	**0.00E+00**
f_8	3.30E-14	**2.84E-14**	3.73E-02	**2.84E-14**	3.91E+00	**1.46E-01**	3.80E-10	**2.84E-14**	7.09E-09	**2.84E-14**	1.27E-05	**2.84E-14**
	2.84E-14	**2.84E-14**	**2.84E-14**	**2.84E-14**	**2.84E-14**	**2.84E-14**	2.85E-10	**2.84E-14**	5.88E-09	**2.84E-14**	7.19E-06	**2.84E-14**
f_9	**2.96E-04**	**2.96E-04**	6.70E-03	**1.68E-03**	1.49E-01	**4.23E-03**	9.09E-15	**0.00E+00**	2.50E-14	**0.00E+00**	**5.37E-09**	1.67E-03
	0.00E+00	**0.00E+00**	**0.00E+00**	**0.00E+00**	**0.00E+00**	**0.00E+00**	**0.00E+00**	**0.00E+00**	**0.00E+00**	**0.00E+00**	1.80E-09	**0.00E+00**
f_{10}	1.57E+02	**1.42E+01**	1.98E+03	**6.82E+02**	3.85E+03	**1.22E+03**	1.67E-12	**-3.64E-12**	5.19E+03	**2.37E+01**	6.74E+03	**1.14E+02**
	-3.64E-12	**-3.64E-12**	9.48E+02	**3.55E+02**	2.43E+03	**7.11E+02**	-1.82E-12	**-3.64E-12**	4.90E+03	**-3.64E-12**	6.33E+03	**-3.64E-12**
f_{11}	**1.57E-32**	**1.57E-32**	7.05E-02	**4.98E-02**	4.50E-01	**1.54E-01**	3.71E-20	**1.57E-32**	2.05E-15	**1.57E-32**	**7.31E-10**	1.66E-02
	1.57E-32	**1.57E-32**	**1.57E-32**	**1.57E-32**	**1.57E-32**	**1.57E-32**	2.11E-20	**1.57E-32**	1.06E-15	**1.57E-32**	1.47E-10	**1.57E-32**
f_{12}	**1.35E-32**	**1.35E-32**	1.32E-03	**4.39E-04**	2.27E-01	**1.76E-03**	2.63E-19	**1.35E-32**	6.06E-15	**1.35E-32**	4.17E-09	**1.35E-32**
	1.35E-32	**1.35E-32**	**1.35E-32**	**1.35E-32**	**1.35E-32**	**1.35E-32**	1.39E-19	**1.35E-32**	3.25E-15	**1.35E-32**	1.30E-09	**1.35E-32**

to 200, while the maximum number of evaluations (*MaxFEs*) is set to 300,000. For the 1,000D problems, *NP* and *MaxFEs* are set to 1,000 and 5,000,000. (Note that the additional evaluations consumed by the OPM of OPDE are included in the *MaxFEs*.) Each algorithm runs 25 times independently and the average error values as well as the best error values are recorded. For clarity, the results of the best algorithms are marked in boldface.

4.2 Comparisons on Normal-Scale (30D) Functions

Table 2 tabulates the average and best results of DE algorithms compared with the corresponding OPDEs on 30D functions. For the six unimodal functions, OPDEs greatly improve the solution accuracy than the corresponding classical DEs for most functions. This is because the proposed mutation scheme makes an orthogonal learning of the environment around the base point and predict a promising direction for the individual to evolve. With this favorable guidance, the search efficiency of the individuals is greatly improved and so that the algorithm is capable of achieving much better results. However, when optimizing the noisy function f_5, the results of the OPDE/best/1 may be a little bit worse than those of the DE/best/1. This is because the noisy environment may mislead the orthogonal learning process, which makes it difficult for the OPM to give a correct prediction.

When optimizing the multimodal functions, it can be observed in Table 2 that OPDEs also outperform the corresponding DEs on almost all benchmarks. There are two reasons for this improvement. First, as we know, the orthogonal mutation samples a set of representative points which uniformly distributed in the neighborhood of an individual in order to fully investigate the environment. This mechanism results in better exploration ability of OPDEs. Second, the improvement of OPDEs also owes to the adaptive adjustment of the orthogonal step length F_{op}. In the orthogonal predictive mutation, F_{op} could be larger than the

Table 6. Comparisons Between OPDEs and the Corresponding DEs on 1,000D Functions

	BEST/1/bin						RAND/1/bin					
	DE CR=0.1 average best	OPDE CR =0.1 average best	DE CR=0.5 average best	OPDE CR=0.5 average best	DE CR=0.9 average best	OPDE CR=0.9 average best	DE CR=0.1 average best	OPDE CR =0.1 average best	DE CR=0.5 average best	OPDE CR=0.5 average best	DE CR=0.9 average best	OPDE CR=0.9 average best
f1	4.81E+03	**1.64E-68**	8.06E+02	**1.07E-163**	1.13E+01	**1.50E-287**	8.51E+05	**1.85E-31**	2.31E+06	**2.41E-71**	8.61E+04	**1.29E-78**
	3.06E+03	**1.08E-73**	8.81E-02	**6.74E-168**	6.06E-03	**0.00E+00**	8.30E+05	**1.03E-33**	2.19E+06	**1.16E-75**	6.82E+04	**5.26E-87**
f2	5.42E+01	**9.88E-14**	1.34E+01	**6.38E-85**	9.68E+00	**3.13E-160**	1.07E+03	**2.28E-13**	1.98E+03	**3.21E-13**	3.70E+02	**2.70E-21**
	4.31E+01	**1.13E-15**	1.83E-01	**1.84E-86**	4.61E+00	**5.62E-178**	1.05E+03	**2.61E-15**	1.92E+03	**1.86E-13**	3.51E+02	**1.55E-21**
f3	1.41E+02	**1.30E+02**	1.41E+02	**1.39E+02**	**1.19E+02**	1.25E+02	1.17E+02	**9.00E+01**	1.57E+02	**9.32E+01**	1.23E+02	**9.50E+01**
	1.35E+02	**1.24E+02**	1.36E+02	**1.34E+02**	**1.12E+02**	1.18E+02	1.15E+02	**8.65E+01**	1.56E+02	**8.93E+01**	1.18E+02	**9.16E+01**
f4	6.03E+03	**0.00E+00**	3.90E+04	**0.00E+00**	1.33E+05	**0.00E+00**	8.51E+05	**0.00E+00**	2.30E+06	**0.00E+00**	8.56E+04	**0.00E+00**
	4.72E+03	**0.00E+00**	2.08E+04	**0.00E+00**	9.82E+04	**0.00E+00**	8.27E+05	**0.00E+00**	2.20E+06	**0.00E+00**	7.56E+04	**0.00E+00**
f5	1.99E+02	**4.07E-01**	2.12E+01	**4.41E-01**	7.44E+01	**5.14E-01**	4.36E+04	**4.06E-01**	2.08E+05	**3.65E-01**	2.23E+05	**2.90E-01**
	8.15E+01	**3.03E-01**	4.52E+00	**3.10E-01**	3.19E+01	**3.95E-01**	4.11E+04	**3.38E-01**	1.94E+05	**3.07E-01**	1.59E+05	**2.09E-01**
f6	2.80E+08	**3.79E+02**	1.01E+09	**8.39E+01**	5.18E+08	**2.07E+00**	4.93E+10	**1.39E+02**	7.04E+11	**2.09E+02**	4.00E+10	**1.86E+01**
	1.28E+08	**1.80E+01**	1.86E+07	**1.76E-12**	2.25E+07	**1.14E-13**	4.79E+10	**5.85E+00**	6.34E+11	**2.87E+00**	3.15E+10	**2.08E-11**
f7	9.08E+03	**2.82E+00**	6.17E+03	**2.88E+02**	9.77E+03	**8.07E+02**	1.01E+04	**1.99E-01**	1.41E+04	**2.35E+00**	1.20E+04	**8.12E+00**
	8.69E+03	**0.00E+00**	5.17E+03	**2.47E+02**	9.04E+03	**6.88E+02**	9.93E+03	**0.00E+00**	1.39E+04	**0.00E+00**	1.18E+04	**1.98E+00**
f8	4.14E+00	**1.46E-13**	2.07E+01	**1.38E-13**	1.96E+01	**1.76E+00**	1.34E+01	**8.98E-14**	2.00E+01	**5.91E-14**	1.47E+01	**6.71E-14**
	3.78E+00	**1.14E-13**	1.91E+01	**1.14E-13**	1.95E+01	**1.45E+00**	1.33E+01	**5.68E-14**	1.99E+01	**5.68E-14**	1.42E+01	**5.68E-14**
f9	1.27E+01	**3.41E-14**	4.24E+01	**2.84E-14**	1.83E+01	**2.84E-14**	1.31E+03	**2.50E-14**	1.19E+04	**7.96E-15**	2.27E+03	**1.14E-15**
	9.38E+00	**2.84E-14**	6.30E+00	**2.84E-14**	1.93E+00	**2.84E-14**	1.27E+03	**0.00E+00**	1.12E+04	**0.00E+00**	1.98E+03	**0.00E+00**
f10	3.07E+05	**3.32E+03**	1.31E+05	**3.47E+04**	1.68E+05	**5.29E+04**	3.27E+05	**4.74E+00**	3.69E+05	**2.51E+02**	3.81E+05	**3.23E+03**
	2.80E+05	**2.25E+03**	1.24E+05	**3.04E+04**	1.51E+05	**4.46E+04**	3.24E+05	**1.57E-09**	3.67E+05	**1.57E-09**	3.77E+05	**1.30E+03**
f11	7.29E+07	**1.24E-03**	3.13E+02	**3.73E-03**	5.51E+02	**3.73E-03**	1.09E+10	**1.74E-03**	3.45E+10	**1.49E-03**	3.45E+10	**8.71E-04**
	3.93E+07	**4.71E-34**	2.06E+01	**4.71E-34**	6.67E+00	**5.10E-34**	1.03E+10	**4.71E-34**	3.29E+10	**4.71E-34**	3.26E+10	**4.71E-34**
f12	1.01E+08	**8.79E-04**	4.35E+03	**3.08E-03**	5.81E+05	**2.27E-01**	1.86E+10	**4.39E-04**	6.33E+10	**2.64E-03**	6.36E+10	**1.32E-03**
	5.10E+07	**1.35E-32**	1.71E+03	**1.35E-32**	5.17E+02	**2.85E-31**	1.80E+10	**1.35E-32**	6.15E+10	**1.35E-32**	6.09E+10	**1.35E-32**

general setting $F = 0.5$. As the differential vector used in the proposed mutation is the same as the corresponding classical mutation, a large step length means that the algorithm can make a broader search during the mutation process.

4.3 Comparisons on Large-Scale (1,000D) Functions

Considering the classical DEs, when the function dimensions increases to 1,000, even the DE/best/1 with a fast converging speed exhibits a poor performance on the simplest unimodal function f_1. This is because the search space increases exponentially with the function dimensions, it becomes much more difficult for the original DE to search efficiently. On the contrary, it can be observed in Table 3 that, for unimodal functions f_1, f_2 and f_4, the proposed OPDE/best/1 with $CR = 0.9$ can achieve solution accuracy up to $10^{-287}, 10^{-160}$, and 0 respectively, whereas the average results found by DE/best/1 with $CR = 0.9$ cannot even reach accuracy 1. In addition, the best result of OPDE/rand/1 can reach the global optima of all the multimodal functions whereas DE/rand/1 is far from the convergence stage. Generally speaking, the performance of the classical DEs on all the large-scale functions is far from satisfactory. In contrast, OPDEs with different parameter settings significantly outperform the corresponding DE variants on almost all the functions.

4.4 Analysis on the Effect of the Orthogonal Predictive Mutation

In this section, we investigate whether the orthogonal neighbors or the predicted vector can bring improvement on the base point. Denote $cost_{base}$ as the cost of the base vector, and $cost_{nei}$ and $cost_{pre}$ as the cost values of the best neighbor based on the OA and the best predict vector respectively. A log-ratio r to describe

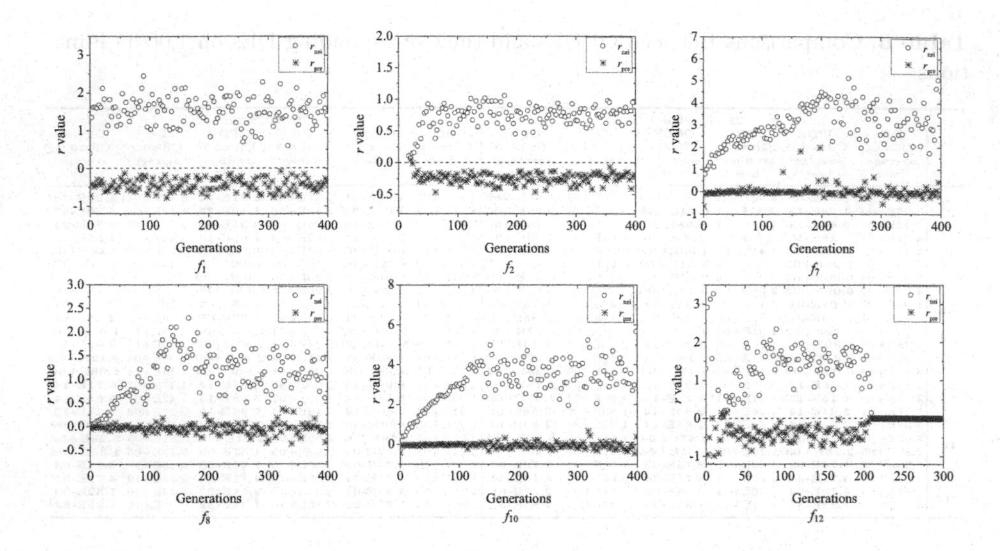

Fig. 2. The values of r_{nei} and r_{pre} during one run of OPDE in optimizing functions f_1, f_2, and $f_9 - f_{12}$.

the relationship between a cost value *cost* ($cost_{nei}$ or $cost_{pre}$) and the $cost_{base}$ is defined as

$$r = ln(\frac{cost}{cost_{base}}) \tag{7}$$

In this way, a negative value of r means that *cost* is smaller than $cost_{base}$, whereas a positive value of r means *cost* is larger than $cost_{base}$. Moreover, the difference between *cost* and $cost_{base}$ increases with $|r|$. The r_{nei} of $cost_{nei}$ to $cost_{base}$ and the r_{pre} of $cost_{pre}$ to $cost_{base}$ are calculated, which reflect the improvement brings by the best orthogonal neighbor and by the best predicted vector, respectively. The distributions of r_{nei} and r_{pre} on functions $f_1, f_2, f_7, f_8, f_{10}$, and f_{12} in one run are illustrated in Fig. 2.

It can be seen in Fig. 2 that both on unimodal functions f_1 and f_2 and on multimodal functions f_8, f_9, f_{11}, and f_{12}, r_{nei} is much larger than 0 in nearly every generation. This indicates that the best neighbor is far worse than the the base vector in almost all the generations. The severe inferiority of the best neighbor to the base vector indicates that, without a clear sense of direction, the individuals undergo classical DE mutation are more likely to degenerate rather than evolve. Therefore, it costs a high price to make an improvement on the current solution by employing the classical DE mutation. When the function dimensions increases, this problem becomes even worse.

On the contrary, r_{pre} is always smaller than 0 for both unimodal and multimodal functions, which demonstrates that the predicted vector outperforms the base vector in most cases. Therefore, the prediction process in the proposed mutation scheme is very effective, which can provide a promising direction for individuals to evolve. Moreover, as described in Section 3.1, in the proposed mutation scheme, the number of sampling points is a polynomial of D. The prediction process of the mutation is not only effective but also very efficient.

5 Conclusion

In this paper, an orthogonal predictive differential evolution termed OPDE has been developed. The OPDE employs an orthogonal predictive mutation scheme that makes an orthogonal learning of the environment around the base vector and predicts a potentially promising direction for DE to evolve. The learning and prediction process of OPDE is very efficient, of which the time complexity is proved polynomial. Moreover, by adopting different methods to calculate the base and differential vectors, an OPDE family consisting of different OPDE variants is defined. The effectiveness of the orthogonal predictive mutation scheme that it can improve both the exploration and exploitation abilities of DE has been validated by comparing OPDEs with the corresponding classical DEs. Moreover, it can be found that the enhancement brought by the proposed mutation becomes more significant with the increase of the function dimensions.

References

1. Das, S., Abraham, A., Chakraborty, U.K., Konar, A.: Differential evolution using a neighborhood-based mutation operator. IEEE Transactions on Evolutionary Computation 13(3), 526–553 (2009)
2. Epitropakis, M.G., Tasoulis, D.K., Pavlidis, N.G., Plagianakos, V.P., Vrahatis, M.N.: Enhancing differential evolution utilizing proximity-based mutation operators. IEEE Transactions on Evolutionary Computation 15(1), 99–119 (2011)
3. Fan, H.Y., Lampinen, J.: A trigonometric mutation operation to differential evolution. Journal of Global Optimization 27(1), 105–129 (2003)
4. Gong, W., Cai, Z., Jiang, L.: Enhancing the performance of differential evolution using orthogonal design method. Applied Mathematics and Computation 206(1), 56–69 (2008)
5. Gong, Y.J., Xu, R.T., Zhang, J., Liu, O.: A clustering-based adaptive parameter control method for continuous ant colony optimization. In: IEEE International Conference on Systems, Man and Cybernetics, pp. 1827–1832. IEEE (2009)
6. Hedayat, A.S., Sloane, N.J.A., Stufken, J.: Orthogonal arrays: theory and applications. Springer (1999)
7. Ho, S.Y., Lin, H.S., Liauh, W.H., Ho, S.J.: Opso: Orthogonal particle swarm optimization and its application to task assignment problems. IEEE Transactions on Systems, Man and Cybernetics, Part A: Systems and Humans 38(2), 288–298 (2008)
8. Hu, X.M., Zhang, J., Chung, H.H., Li, Y., Liu, O.: Samaco: variable sampling ant colony optimization algorithm for continuous optimization. IEEE Transactions on Systems, Man, and Cybernetics, Part B: Cybernetics 40(6), 1555–1566 (2010)
9. Price, K., Storn, R.M., Lampinen, J.A.: Differential Evolution: A Practical Approach to Global Optimization. Springer (2006)
10. Storn, R., Price, K.: Differential evolution–a simple and efficient heuristic for global optimization over continuous spaces. Journal of Global Optimization 11(4), 341–359 (1997)

11. Tang, K., Yao, X., Suganthan, P.N., MacNish, C., Chen, Y.P., Chen, C.M., Yang, Z.: Benchmark functions for the cec2008 special session and competition on large scale global optimization. Nature Inspired Computation and Applications Laboratory, USTC, China (2007)
12. Wang, Y., Cai, Z., Zhang, Q.: Enhancing the search ability of differential evolution through orthogonal crossover. Information Sciences 185(1), 153–177 (2012)
13. Yang, Z., Tang, K., Yao, X.: Large scale evolutionary optimization using cooperative coevolution. Information Sciences 178(15), 2985–2999 (2008)
14. Yao, X., Liu, Y., Lin, G.: Evolutionary programming made faster. IEEE Transactions on Evolutionary Computation 3(2), 82–102 (1999)
15. Zhan, Z.H., Zhang, J., Li, Y., Shi, Y.H.: Orthogonal learning particle swarm optimization. IEEE Transactions on Evolutionary Computation 15(6), 832–847 (2011)
16. Zhang, J., Chung, H.H., Lo, A.L., Huang, T.: Extended ant colony optimization algorithm for power electronic circuit design. IEEE Transactions on Power Electronics 24(1), 147–162 (2009)
17. Zhang, Q., Leung, Y.W.: An orthogonal genetic algorithm for multimedia multicast routing. IEEE Transactions on Evolutionary Computation 3(1), 53–62 (1999)
18. Zhao, Z.-F., Liu, K.-Q., Li, X., Zhang, Y.-H., Wang, S.-L.: Research on hybrid evolutionary algorithms with differential evolution and guo tao algorithm based on orthogonal design. In: Huang, D.-S., Zhao, Z., Bevilacqua, V., Figueroa, J.C. (eds.) ICIC 2010. LNCS, vol. 6215, pp. 78–85. Springer, Heidelberg (2010)

Content-Adaptive Analysis and Filtering of Microblogs Traffic for Event-Monitoring Applications

Claudia Meda[1], Federica Bisio[1], Paolo Gastaldo[1], Rodolfo Zunino[1], Roberto Surlinelli[2], Eugenio Scillia[2], and Augusto Vincenzo Ottaviano[2]

[1] Dept. of Electric, Electronic and Telecommunications Engineering and Naval Architecture (DITEN), University of Genoa, Genoa, Italy
{federica.bisio,claudia.meda}@edu.unige.it
{paolo.gastaldo,rodolfo.zunino}@unige.it
[2] Compartimento Polizia Postale e delle Comunicazioni per la Liguria, Ministry of the Interior - Department of Public Security, Italian National Police, Genoa, Italy
roberto.surlinelli@poliziadistato.it

Abstract. The paper presents an integrated methodology for the continuous monitoring of traffic from Twitter, to support analysts in the early detection and categorization of critical situations. The main purpose of the developed technology is to monitor the information flow and track sensible situations or look for unexpected events. The main architecture relies on specific technologies covering language detection, text-mining, clustering and semantic labeling. The concept of 'mission profile' allows the analyst to specify target concepts and features related to the scenario of interest. Upon clustering and semantic labeling of running traffic, a novel refinement algorithm retrieves and aggregates relevant contents that would otherwise escape plain semantic analysis. Experimental results in different scenarios prove the method effectiveness in enhancing content retrieval and in greatly enhancing efficiency in contents presentation to the analyst.

Keywords: Text Mining, Social Network Analysis, Semantic labeling, Clustering.

1 Introduction

Social Networks are a popular mean to share data and ideas, and witness an ever-increasing diffusion. The amount of data generated in 30 seconds on the Internet [1] is about 600 GB of traffic: this information confirms that social networks have become a source of big data. Accordingly, the increasing of number of individuals in the network contributes to an increasing amount of data in the web.

In the specific case of the Twitter community, every minute more than 320 new accounts are created and more than 98,000 tweets are posted. This makes the analysis of Twitter micro-blogging a topmost and significant domain for business intelligence and cyber security. A multiplicity of users populate this network, sharing different types of information: the average age of users on Twitter ranges from fourteen to sixty years, equally distributed among individuals of both sexes.

H. Handa et al. (eds.), *Proc. of the 18th Asia Pacific Symp. on Intell. & Evol. Systems – Vol. 1*,
Proceedings in Adaptation, Learning and Optimization 1, DOI: 10.1007/978-3-319-13359-1_13

Among the multitude of tweets, an analyst may want to retrieve information associated with specific relevant topics. Such a challenging environment calls for new methods to analyze and extract information from Twitter traffic. The related technologies can be an important source of information for both commercial applications and Open Source Intelligence; in both cases, methods to analyze and classify large volumes of data in real-time are required. The literature offers a variety of approaches to analyze Twitter traffic for topical clustering and event monitoring [2].

This paper presents a methodology for the continuous monitoring of social-network traffic, and the extraction of significant information about high-impact situations in real-time, covering natural events, such as earthquakes, floods, tsunamis, as well as crowd behavior in critical situation. Target applications are ensuring the safety of citizens in natural events and detecting security risks in critical crowded scenarios. The method relies on a Text Miner environment, the SLAIR suite, [3] which features clustering and classification methods for real-time processing of large corpora of documents. Recent advances endowed the basic framework with Web-exploration capabilities for focused crawling and sentiment analysis [4, 5].

The basic approach consists in defining a "profile" for the target scenario, which allows the analyst to specify relevant information about the expected event features and possible topics of interest. A semantic knowledge base augments the basic profile description and enriches the profile contents by inferring the analyst's expectations. This process is mostly general and can include any semantic descriptor that characterizes flowing information. The present approach features topic identification but sentiment analysis [6] is applicable and particularly appropriate in this context. At run time during traffic monitoring, a clustering algorithm preliminarily yields content grouping. Clustering results are then analyzed and labeled semantically according to the relevance to the nature of the monitored event. Finally, the augmented description of groups drives the selection of affine traffic data and the rearrangement of groups into more compact clusters that best adhere to the expected relevant contents.

The novelty of the proposed approach consists, first, in the application of the semantic domain knowledge to the refinement of the scenario profile, which greatly improves accuracy in selecting target contents; secondly, in the development of a novel algorithm for group refinement, which leads to compact clusters and facilitates immediate and intuitive inspection.

Experimental results prove that the integrated profiling/semantic refinement approach can notably enhance effectiveness in retrieving relevant contents which would otherwise be dispersed in the vastness of real-time traffic. At the same time, yielding compact groups (i.e., groups holding a considerable set of homogeneous contents that are considered relevant to the specific interests) reduces the complexity for the analyst to get a clear picture of ongoing events.

Section 2 overviews the specific literature on the subject of interest; Section 3 presents the general model and operation of the Twitter language processor. Section 4 describes the profile-based approach, analyzes the specific mining process and highlights the novel algorithms adopted. Section 5 presents the experimental sessions and discusses the empirical results obtained. Concluding remarks are drawn in Section 6, also exploring options for future extensions of the presented research.

2 Specific Background Literature

The study of Twitter traffic is a relatively recent research topic, and the analysis of tweets for security and the control of high impact events roughly starts from 2011. The research presented in [7] treated traffic downloaded from Twitter Streaming API [8] to collect information about two natural disasters occurred in the U.S.A., and specifically exploited geo-localization information drawn from users' profiles. The work [9] focused on a Twitter-like Chinese media (Sine-Weibo) to extract messages related to an earthquake in China, filtering tweets that contained specific words about the disaster sites. The application of traffic analysis for earthquake monitoring [10] involved users' real-time interactions on the social network. After defining an event as an arbitrary classification of a space-time interval, each event included several properties, such as the scale extension or its influence on people's daily life; events were categorized accordingly. The graph-based retrieval algorithm proposed in [2] identified messages concerning popular subjects from among the set of tweets returned by the Twitter research engine in response to a keyword-based query. Topical clustering of tweets supported the actual retrieval method. The approach described in [11] evaluated the reliability of information carried by tweets related to fourteen high-impact news events in 2011 around the world. As a significant result, on average, a fraction of 30% of the overall traffic posted about an event actually contained situational information, whereas a significant portion (14%) was spam. A supervised machine learning approach supported the prediction of a tweet reliability; it used regression analysis and identified the important content and source-based features. The analyzed Twitter traffic as a source of big data [12] allowed users to get access to different type of information, such as mild flu epidemic, most frequent topics and the reaction to tweets. In this scenario, a Twitter traffic analysis can also detect event related to malicious users and network. The approach shown in [13] applied a novel, weakly supervised method for mining cybercriminal network and facilitating cybercrime forensics. That research relied on a probabilistic generative model enhanced by a novel context-sensitive Gibbs sampling algorithm. At last the impact of sentiment analysis, especially when applied to social and crowd events [6], is clearly useful and promising.

3 Basic Text-Preprocessing and Mining Framework

In the present approach, a 'document' is defined as an atomic data containing some information. This means that it is not restricted to textual format but may also include images, videos, audio information and raw numerical data. The SeaLab Advanced Information Retrieval (SLAIR) is a software framework intended to the management of large masses of documents for clustering and classification purposes [3]. The SLAIR suite adopts specific enhancements of adaptive machine-learning algorithms, such as Kernel K-Means and Support Vector Machines, for both clustering and classifying documents. From a semantic perspective, it relies on the lexical multilingual database EuroWordNet [14, 15].

When applying the above interpretation of document to the analysis of Twitter traffic, a tweet therefore supports a document made up of (at most) 140 characters in a natural language. The SLAIR framework treats short messages such as tweets by a text preprocessing procedure, and a sequence of proper text-mining technologies, including a series of steps characterized by increasing complexity in representation, as shown in Figure 1.

- Step 0 - *Language Identification*: the first step involves the detection of the original language in which the document is expressed; the Infinity Gram algorithm [18] supports this operation, which is crucial to the effectiveness of the subsequent steps.
- Step 1 – *Stemming*: all tweets undergo a stemming process, in which SLAIR uses language-specific stemming algorithm (at this time, Italian and English languages benefit from dedicated, accurate stemmer modules), in order to have a consistent and compact representation of lemmata. For other languages the popular Porter stemming algorithm [16] is applied.
- Step 2 – *Stopword removal*: in this step the application removes Stop and Common Words, taking out low informative lemmas and returning a better vector space for any processed document of the corpus.
- Step 3/4 – *MultiWord feature, Indexing and Quering*: multiword features are identified, reducing several lemmata to a single term; upon construction of the related dictionary, the Text Miner embeds additional features, such as classical reverse-indexing functions, which support conventional full-text search and retrieval of documents.

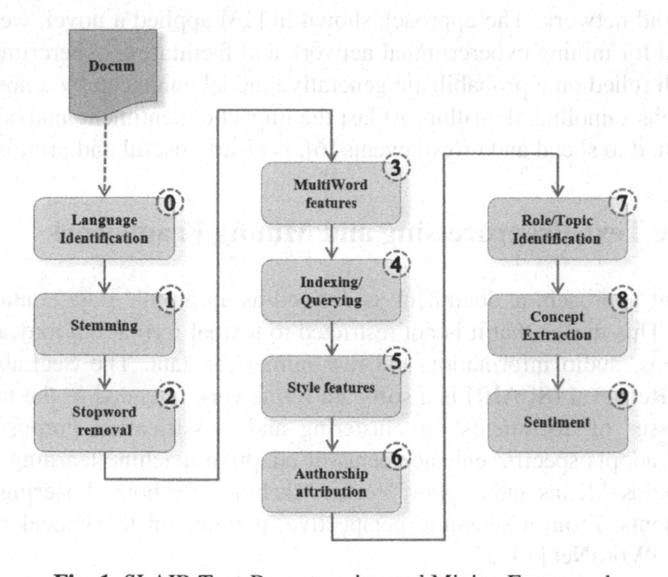

Fig. 1. SLAIR Text-Preprocessing and Mining Framework

The following steps are more properly related to text-mining activities, and are performed once text characterization and structuring has been completed.

- Step 5/6 – *Style feature and Authorship attribution*: the characterization of style has been proved to enhance significantly both accuracy in representation and discriminating power for document categorization and ultimately leading to authorship attribution, allowing to ascribe a document to an author or a group based on the order and signify of words.
- Step 7/8 – *Role or Topic Identification and Concept Extraction*: the semantic-related steps characterizing each document consist in the identification of the main topic, the extraction of relevant concepts, the geo-referentiation of contents whenever applicable. For the semantic analysis, SLAIR relies on the lexical multilingual database Euro WordNet [17] for text processing.
- Step 9 – *Sentiment Analysis*: the application adopts Euro WordNet-Affect database [14] for sentiment analysis and, in addition, a specific semantic-orientation network based on adjectives [19], categorizes a tweet mood, defined as positive, negative or neutral.

In the last three critical steps, specific and comprehensive databases such as SenticNet [20] can be used as well; in the present research, without loss of generality, the EuroWordNet repository was tested because of the built-in multilingual capabilities of the database itself, allowing easy extension to a wide selection of European languages.

Topic tracking [21] represents an important process in traffic analysis, as monitoring approaches use topic to extract the most common and current tweets and relevant issues. This is mostly useful in trend analysis and commercial applications. Topic tracking by itself, however, is not sufficient for the identification and run-time monitoring of high impact events, because of the specificity and possibly limited amount of relevant messages. This technology plays a complementary role in the specific application that is being pursued in this research.

The resulting corpus of documents, including a lexical and semantic characterization according to the above multi-dimensional analysis, supports both clustering and classification functionalities. The former typically aims to provide the analyst with a real-time, hierarchical presentation of the contents embedded in a large and unstructured corpus of documents; applications range from security applications to computer forensic support. Classification tools discriminate the possible categories and messages, and reply on supervised training models; applications mainly involve span detection and mail leak prevention.

4 Microblog Analysis for Event Monitoring

The use of text-mining and semantic analysis tools for tracking critical events and situations in microblogs (e.g. Twitter) poses two main requirements. First, the method must be able to handle large masses of documents in real-time (that is, within a time frame that allows the control operator to react promptly). Secondly, the analysis process must exhibit a considerable discrimination ability, to pick out significant contents from among the vast spectrum of observed traffic.

4.1 Assembling Scenario Parameters into a Profile

In principle, one might envision novelty-detection methods, for example based on "term acceleration", to pinpoint emerging contents within a certain time slot by observing anomalies in the relative frequencies of terms or even concepts. Such an approach does prove effective in limited-scope domains such as newsgroups or blogs, where the subjects and the community of contributors are, to a certain extent, more focused. When dealing with unbounded traffic monitoring, however, general novelty-detection methods suffer from the vastness of the involved sources and the related interests, hence even proper emerging topics may be irrelevant to the target at hand and therefore should be treated as background noise.

In view of these issues, it seems that an effective strategy requires the analyst to interact deeply with the monitoring tool in order to drive focusing and presentation. More in detail, the analyst should be able to specify the relevant, target concepts and situations; at the same time, one cannot expect that such a crucial configuration task might be supported by some semi-automated process.

The SLAIR platform assembles all the analysis-related goals and settings within a "Profile" structure, which embeds lexical information (relevant terms and expressions), semantic information (concepts and geo-referentiation data), and metadata about the individual messages (e.g. hashtags).

The basic attention-focusing mechanism that drives content selection operates on a watch-list of relevant terms. It is worth noting that the watch-list-based extractor does not operate at the mere lexical level by plain keyword matching. Upon definition and refinement of the basic watch-list by the analyst, the SLAIR engine exploits the semantic knowledge available to the text miner to infer the user's expectations, and fill the watch-list accordingly. This process first retrieves both synonyms and hypernyms of basic terms from the semantic net; then inflates the watch-list itself by adding those terms that best correlate with the entire scenario (set of terms+concepts) provided by the analyst. As a result, the operational watch-list is an augmented set of terms and expressions which does not suffer from disambiguation issues and actually refines the expected contents. This integrated lexical/semantic mechanism, driven by limited user intervention and supported by a semantic engine is a novelty point of the method proposed in this paper. In fact it resembles the approach developed in [4] for focused crawling, where the active guidance by the analyst lead to faster and more accurate search process in web-space exploration.

Another relevant role of semantics consists in the determination of geo-referencing information about the involved locations. Message geo-referencing is a critical task because only a limited fraction of Twitter traffic (about 5%) exhibits a valid geo-referenced field by the provider. In addition, there is no assurance about the actual reliability of geo-referencing (for example, original geo-referencing often stems from the user's profile settings, which may be misplaced to the event or the actual user location). The geo-referencing approach adopted in the present research follows a hybrid two-fold strategy: first, the system checks whether the tweet metadata contains valid geographical fields about the originating device: in this case, this information is privileged. Otherwise, the text miner tries to extract geographical information from the tweet contents: unlike [7], the analysis mechanism relies on a geographical database and matches geographical denominations with the tweet contents.

In summary, the analyst can adjust all settings by a user interface, which allows to manage a "scenario profile" embedding edit five categories of parameters:

- The *black-list* option allows to skip any tweet containing at least one term specified in the black list.
- In the *watch list*, the analyst can draw the text-miner's attention to messages containing specific terms as explained previously.
- The *common-words* set holds a list of high-frequency terms that the analyst marks as irrelevant in order to remove noise (this differs from blacklist filtering because messages are retained after common-word removal).
- *Hashtags*: this option gives the possibility to keep or drop tweets depending on specific hashstags. It should be noted that this feature, although popular, may prove of limited use in event monitoring applications, especially when hashtagging is missing or unreliable.
- Finally, the analyst can limit traffic inspection to specific geographic areas based on the original or inferred geo-referenced *locations*.

The concept definition of "scenario profile" provides a suitable basis for massive traffic inspection. In fact, one can define several, independent profiles and apply them in parallel to the run-time traffic segment that is being observed. This makes it possible to set up a "Repository of Profiles" and support a corresponding monitoring panel, where the relevance results for each profile can be displayed and presented simultaneously.

4.2 Inductive Methods for Profile-Based Event Monitoring

The traffic analysis process relies on the user-defined profiles to identify relevant contents. The overall principle of operation involves three main steps which are iterated during every time span of observed traffic. First, traffic raw data are arranged and structured by a clustering algorithm, whose progression is driven by a user-supplied metric criterion; then documents and clusters are labeled according to their adherence with respect to a relevance criterion (in the present application, this implies a watch-list-based criterion); finally, a novel re-grouping algorithm specifically conceived to maximize density of relevance in group presentation, rearranges clustering results and creates novel groups featuring a significantly higher density (density is intended as the ratio of relevant documents in a group to the overall group cardinality). Figure 2 illustrates the cyclic process of continuous traffic analysis.

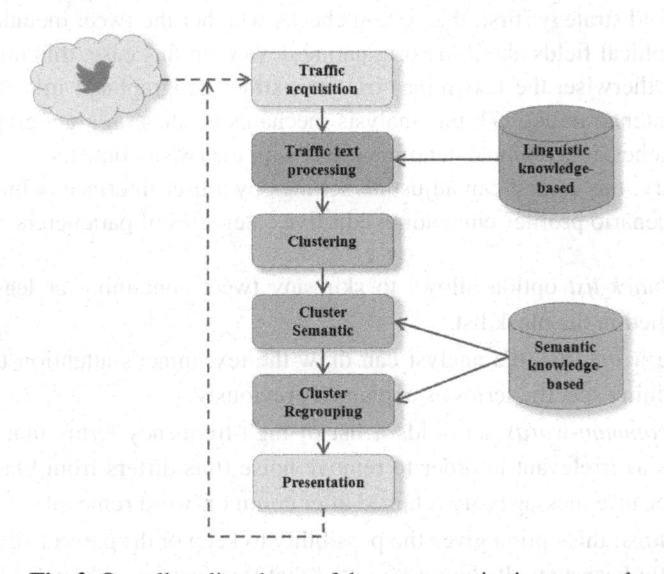

Fig. 2. Overall cyclic scheme of the event-monitoring approach

Clustering, in its hierarchical variant, supports the preliminary operation that is carried out on incoming traffic for relevant content detection. Thanks to the Kernel-based approach to clustering in SLAIR [19], the overall process is entirely application-dependent, since the analyst can set the grouping criterion by defining a mission-specific metric. A metric measures the affinity between a pair of documents and drives the grouping process in the overall clustering algorithm. Specific metrics have been defined for text processing [3, 19] and be applied when setting up a scenario profile together with the other user-supplied parameters. In conventional applications, text clustering typically aims to categorize documents according to the implicit distribution of contents, and/or to provide the analyst with a structured presentation of the underlying traffic [4]. This implies a top-down inspection process on the hierarchy of groups, in which the analyst can focus his/her attention from macro-groups to smaller clusters at increasing levels of detail. In the Twitter analysis domain, however, this approach proves ineffective because messages carrying relevant contents represent a marginal fraction of the overall information flow. A bottom-up inspection strategy, which focuses on the hierarchy 'leaves' appears appropriate, under the assumption that interesting messages joined a restrained set of small groups. The considerable rate of incoming information makes an exhaustive inspection process practically unfeasible for a human operator. A mission-oriented focusing mechanism is thus required to sift relevant contents and, in fact, to rank the hierarchy leaves. A semantic labeling of clusters seems a viable and consistent approach toward that end.

The second step of the automated traffic analysis therefore exploits the target information held in scenario profile(s): a relevance score marks every message, and leaf clusters are labeled accordingly. The rank of each leaf cluster just results from accumulating the relevance scores of its members. In principle, any semantic-based labeling mechanism applies in this operation. For real-time computational reasons and without loss of generality, the research presented in this paper adopts the (semantic) watch-list-based strategy described in the previous Section.

4.3 Mission-Oriented Algorithm for Refining Clusters

The labeling and ranking approach may suffer from two main limitations. First, in the presence of large, noisy corpus of documents, the clustering process may spread relevant contents over a number of leaf clusters, thus bringing about a kind of 'dispersion' effect which hinders the analyst's monitoring task. Secondly, by applying a plain watch-list-based labeling strategy, even in its semantically augmented version, one runs the risk of over-fitting the inspection criterion. In other words, the analyst is left the burden either to specify a detailed watch-list or to rely on an over-detailed (often missing) ontology. To overcome those potential issues, the approach proposed in this paper applies a novel re-grouping algorithm. The algorithm input is the set of leaf groups that result from the clustering+labeling process described in the previous Section. The main goal is, on one hand, the aggregation of the results of the detection process into fewer and denser groups. On the other hand, the algorithm can 'attract' to these groups those promising messages which are affine to relevant ones but would escape a straightforward watch-list-based selection.

The principle of operation of the algorithm assumes that a relevant message, after being labeled by the watch-list-based ranking, might still convey useful information in the part of its contents that does *not* match the watch-list-based criterion. Typically, one expects that such a portion includes relevant terms and concepts that are not covered by the augmented watch-list, but yet adhere to the analyst target contents. The 'non-watch-listed' portion of contents supports a search basis over the entire corpus of current traffic, to identify and 'attract' messages that match relevant documents from a complementary perspective. This process is performed for every ranked message, and yields a list of possibly attracted documents; thanks to the semantic homogeneity of contents, the same message is often 'attracted' by several scored documents, thus giving a measure of its associated adherence to the target scenario. The resulting candidates are selected according to a matching threshold and eventually grouped together with the original labeled documents. The regrouping algorithm is outlined as follows:

Cluster Regrouping algorithm

Input: the set, C, of labeled leaf clusters;
the corpus, K, of labeled documents with associated watch-list scores
a relevance threshold, τ;
the augmented watch-list of concepts, W

/// Each Cluster has a relevance score, summing the watch-list-based scores of the members

For each cluster, $\chi \in$ C

/// Only relevant clusters are considered

If (score(χ) <= 0)

skip χ;

/// Prepares the set, A, of attracted documents and the regrouped Cluster, Z

Set A := $\varnothing$; Z := $\varnothing$;

/// Only relevant documents (having score $s_j > 0$) are considered for attracting other docs

Select R = { $D_j \in \chi : s_j > 0; j=1,\ldots,N_R$ };
where s_j is the score associated with document D_j

/// Relevant documents are attracted straight away

Set Z := Z $\cup$ R;

*/// *** The actual attraction-based mechanism ****

For each document D $\in$ R

/// Extract the lexical set, L, of terms t that <u>mismatch</u> the watch-list contents

Extract L = {$t \in$ D : $t \notin$ W}; (*)

/// ReverseIndexing retrieves all documents covering L in the current corpus K

Fetch RI(L) = { X$\in$K : $\exists\, t : t \in$ X $\cap$ L };

/// Accumulates frequency scores

Join A := A $\cup$ RI(L); (frequencies of duplicates are accumulated)

/// Scans the list of candidate attracted documents

For each document Q $\in$ A

/// Ignore documents with low attraction scores

If (frequency(Q) < τ)

skip Q;

/// Otherwise attract the document into the newly forming cluster

Set Z := Z $\cup$ {Q};

/// Remove the attracted document from it's original cluster

Set $\lambda := \lambda \setminus$ {Q}; (where λ is the cluster Q belongs to)

/// Uses document group fo form new Cluster replacing χ

Create new cluster θ := Z , replacing χ.

The crucial step in the algorithm itself is the application of reverse-indexing to those terms that do not hit the expected relevance criterion. This makes it possible to recover documents that *per se* would not exhibit inherent significance, but are cross-referenced by a meaningful subset of relevant documents. The above algorithm was implemented according to a lexical-based matching strategy between terms and concepts at step (*), but any semantic procedure (such as Latent Semantic Analysis) capable to provide a matching measure can apply toward that end.

This sort of indirect voting mechanism yields the main effect to greatly enhance the resulting cluster density, which proves much higher than the density of the original clusters. This can be verified when considering that all regrouped clusters necessarily contain all the original relevant documents, and the newly added members are a (quite limited due to the cross-referencing constraint) subset of the original corpus of documents.

5 Experimental Results

This section reports on the experimental procedure adopted to verify the effectiveness of the framework for Twitter traffic analysis. Traffic acquisition was supported by two services providers, namely, Datasift [22], and Twitter Streaming API [8]. Both sources were chosen so as to collect and process a large amount of tweets in real-time. The experiments involved three main scenarios to test the approach effectiveness under different conditions; for each scenario, several experiments were performed, and this Section presents a sample result for the sake of simplicity and without loss of generality.

- *Crowded event scenario*: this experiment tested the approach effectiveness at continuous monitoring a crowded event, for the early detection of critical situations. A public demonstration scheduled in Rome on October 31st, 2013 provided the test case. The collected traffic (Datasift) originated from the entire Italy area, and a time span for one cycle of traffic analysis (as shown in Figure 2). The sample case shown in this Section included a total of 21603 tweets, acquired during one of the most intense spans of the event.
- *Specific topic tracking*: this test aimed to verify the method's behavior when observing contents and discussions about a highlighted relevant topic. In this case, the attention focused on the war episodes taking place in the Gaza strip during July, 2014. The presented test case relied on a traffic database (Twitter Streaming API) collected on July 23rd, 2014, and covered a time span of 105 minutes yielding 11096 tweets.
- *Popular trend monitoring*: this scenario simulated the monitoring of relevant contents and trends in mass opinions. In this case, the database contained 10452 messages, downloaded on 23 July 2014, and related to popular trending habits of Italians during summer holidays.

The experimental procedure was designed to reflect the procedure suggested (and followed in practice) by professional analysts. Basic operational constraints, in terms of both real-time performance and focusing capability, limit the analyst's attention to a subset of traffic, and a support system should facilitate the operator's inspection process by highlighting most relevant results first.

Therefore, the interface providing the operator with the results of clustering, semantic labeling and further refinement, limited to the 10 top-relevant clusters, which were presented in a decreasing order of significance (relevance was measured in terms of watch-list scores). To attain a consistent and objective comparison, the same traffic segments underwent two alternative processing schemes: the former scheme just applied semantic labeling, the second also included the cluster-regrouping process detailed in Section 4.3.

Tables 1, 2 and 3 present the obtained results for the test scenarios. For each cluster presented to the analyst, three quantities were measured: 1) the cluster cardinality (CC), 2) the Relevant Scored (RS) count, that is, the number of tweets that were highlighted as a result of the plain semantic-based labeling, and 3) the Relevant Attracted (RA) count, which summed up the tweets that were indeed consistent with the overall analysis goal, but had not been covered by the semantic labeling mechanism in the first place. The latter quantity was measured by direct observation.

Three main performance parameters were worked out to assess the relative contribution of the regrouping algorithm. The crucial aspect that was being observed was the capability of the overall process to present the analyst a 'dense' set of significant traffic elements, in an integrated, compact fashion; an important contribution consisted, in particular, in the aggregation of those messages that would escape a semantic-only labeling selection. Presentation Effectiveness (*PE*) measured the ratio of the overall relevant contents (RA+RS) to the total traffic considered; this parameter clearly related to the practical value of presentation results at supporting the analyst's operation. Attraction Effectiveness (*AE*) measured the ratio of RA to the cluster cardinality; this quantity yielded an overall score to represent the effectiveness of the clustering mechanism at prompting contents that would have been lost if a semantic-labeling mechanism had only been applied. Grouping Effectiveness (*GE*) was computed as the ratio of RA to the number relevant scored tweets, RS; this performance parameter gave a direct measure of the relative impact, in the subspace of relevant-only traffic, of the grouping algorithm with respect to the semantic-labeling mechanism.

The set of experimental measurements and the associated performance parameters were worked out individually on each presented cluster, and on the overall presentation contents. Experimental evidence points out the beneficial effects of the grouping algorithm from two different viewpoints: first, the number of attracted contents is significantly higher, thus providing a richer and more complete collection of significant results. At the same time, the distribution of contents within the set of presented clusters changes significantly, since most relevant contents are aggregated and presented in the topmost clusters; this greatly facilitating the analyst's inspection and prevents a dispersion of contents throughout the clusters.

Figure 3 provides a visual presentation of the aggregate numerical outcomes, as a comparison between semantic-labeling only and grouping-assisted performances.

Table 1. Semantic-labeling and cluster-regrouping results for the crowded event scenario

SEMANTIC LABELING ONLY							SEMANTIC LABELING WITH CLUSTER REGROUPING						
Cl.	CC	RS	RA	*PE*	*AE*	*GE*	Cl.	CC	RS	RA	*PE*	*AE*	*GE*
1	15	3	0	*0.2*	*0*	*0*	1	40	21	18	*0.98*	*0.45*	*0.86*
2	20	4	0	*0.2*	*0*	*0*	2	79	17	62	*1*	*0.78*	*3.65*
3	9	2	0	*0.22*	*0*	*0*	3	33	15	17	*0.97*	*0.52*	*1.13*
4	16	4	1	*0.31*	*0.06*	*0.25*	4	8	1	2	*0.38*	*0.25*	*2*
5	19	4	0	*0.21*	*0*	*0*	5	6	0	0	*0*	*0*	*0*
6	18	2	1	*0.17*	*0.06*	*0.5*	6	4	1	2	*0.75*	*0.5*	*2*
7	18	3	2	*0.28*	*0.1*	*0.67*	7	1	1	0	*1*	*0*	*0*
8	8	1	1	*0.25*	*0.13*	*1*	8	3	0	0	*0*	*0*	*0*
9	7	1	0	*0.14*	*0*	*0*	9	1	1	0	*1*	*0*	*0*
10	5	1	1	*0.4*	*0.2*	*1*	10	1	1	0	*1*	*0*	*0*
Total	135	25	6	*0.23*	*0.04*	*0.24*	Total	176	58	101	*0.90*	*0.57*	*1.74*

Table 2. Semantic-labeling and cluster-regrouping results for the topic-tracking scenario

SEMANTIC LABELING ONLY							SEMANTIC LABELING WITH CLUSTER REGROUPING						
Cl.	CC	RS	RA	*PE*	*AE*	*GE*	Cl.	CC	RS	RA	*PE*	*AE*	*GE*
1	15	1	0	*0.07*	*0*	*0*	1	8	5	3	*1*	*0.38*	*0.60*
2	18	1	0	*0.06*	*0*	*0*	2	6	4	0	*0.67*	*0*	*0*
3	16	2	0	*0.13*	*0*	*0*	3	2	2	0	*1*	*0*	*0*
4	14	1	0	*0.07*	*0*	*0*	4	9	2	1	*0.33*	*0.11*	*0.5*
5	8	1	0	*0.13*	*0*	*0*	5	3	3	0	*1*	*0*	*0*
6	10	1	0	*0.10*	*0*	*0*	6	1	1	0	*1*	*0*	*0*
7	11	1	0	*0.10*	*0*	*0*	7	1	1	0	*1*	*0*	*0*
8	15	1	0	*0.07*	*0*	*0*	8	2	2	0	*1*	*0*	*0*
9	16	1	0	*0.06*	*0*	*0*	9	2	2	0	*1*	*0*	*0*
10	18	1	0	*0.06*	*0*	*0*	10	2	2	0	*1*	*0*	*0*
Total	141	11	0	*0.08*	*0*	*0*	Total	36	24	4	*0.78*	*0.11*	*0.17*

Table 3. Semantic-labeling and cluster-regrouping results for the trend-monitoring scenario

Semantic Labeling Only							Semantic Labeling with Cluster Regrouping						
Cl.	**CC**	**RS**	**RA**	*PE*	*AE*	*GE*	**Cl.**	**CC**	**RS**	**RA**	*PE*	*AE*	*GE*
1	11	2	1	*0.27*	*0.09*	*0.5*	1	65	6	41	*0.72*	*0.63*	*6.83*
2	18	1	0	*0.06*	*0*	*0*	2	2	1	0	*0.5*	*0*	*0*
3	17	2	1	*0.18*	*0.06*	*0.5*	3	10	2	3	*0.5*	*0.3*	*1.5*
4	10	0	0	*0*	*0*	*0*	4	20	0	13	*0.65*	*0.65*	*0*
5	13	0	0	*0*	*0*	*0*	5	2	1	1	*1*	*0.5*	*1*
6	14	0	0	*0*	*0*	*0*	6	1	1	0	*1*	*0*	*0*
7	14	1	0	*0.07*	*0*	*0*	7	1	1	0	*1*	*0*	*0*
8	12	0	0	*0*	*0*	*0*	8	1	0	0	*0*	*0*	*0*
9	13	0	0	*0*	*0*	*0*	9	3	1	0	*0.33*	*0*	*0*
10	15	0	0	*0*	*0*	*0*	10	1	1	0	*1*	*0*	*0*
Total	137	6	2	*0.06*	*0.01*	*0.33*	Total	106	14	58	*0.68*	*0.55*	*4.14*

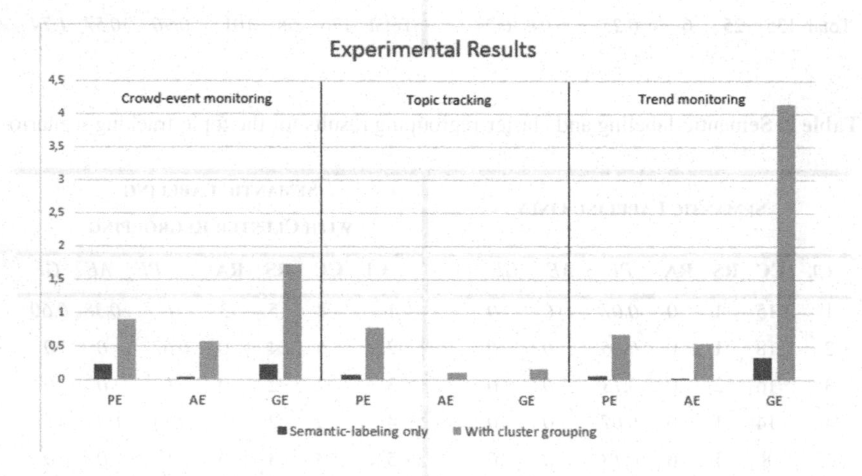

Fig. 3. Comparison of overall performance indicators in the three test scenarios

It is worth stressing that, when the cluster regrouping algorithm is not applied, both PE and GE values decrease. This means that the percentage of (relevant) attracted tweets is a significant fraction of the total number of tweets; the consequent increase in PE conveys a positive effect on the cluster density. At the same time, the increased weight of attracted tweets with respect to semantic-scored ones witnesses the beneficial effect of the grouping mechanism.

An intriguing property of the analysis process was that the clustering regrouping algorithm could aggregate tweets belonging to the same conversation among groups of users, when no information about the user ID was ever involved throughout the process. This might eventually support a community analysis methodology. This functionality was found out experimentally even though it was not a core target of the developed methodology.

6 Conclusions

The main challenge during the analysis of social network traffic, and in particular microblog, lies in the filtering of (limited) relevant messages from the vastness of 'noisy' traffic. The approach presented in this paper attains this goal, first, by setting up a targeted-focusing approach, in which the analyst defines a profile of interest and applies semantic information to label traffic elements accordingly. Then a novel grouping mechanism uses the overall contents of messages (not only semantic labeled ones) to retrieve and aggregate portion of traffic that would otherwise escape the semantic filter. Experimental evidence proved that combination of these mechanisms can boost the effectiveness at recovering relevant information, and therefore greatly improve the overall method impact when presenting results to the analyst for real-time event monitoring. Besides, the overall methodology is quite general hence any semantic knowledge base can be applied to label and categorize traffic.

References

[1] The Internet in Real Time, `http://pennystocks.la/internet-in-real-time/`

[2] Khan, M., Bollegala, D., Liu, G., Sezaki, K.: Multi-Tweet Summarization of Real-Time Events. In: SocialCom (2013)

[3] Sangiacomo, F., Leoncini, A., Decherchi, S., Gastaldo, P., Zunino, R.: SeaLab Advanced Information Retrieval. In: IEEE Int. Conf. Semantic Computing, ICSC 2010, Pittsburgh, USA, pp. 444–445 (September 2010)

[4] Zunino, R., Bisio, F., Peretti, C., Sangiacomo, F., Surlinelli, R., Scillia, E., Ottaviano, A.V.: An Analyst-Adaptive Approach to Focused Crawlers. In: Int. Symp. Foundations of Open Source Intelligence and Security Informatics, FOSINT-SI 2013, Niagara Falls, Canada, pp. 1073–1077 (August 2013)

[5] Bisio, F., Cambria, E., Gastaldo, P., Peretti, C., Zunino, R.: Data Intensive Review Mining for Sentiment Classification across Heterogeneous Domains. In: Int. Symp. Foundations of Open Source Intelligence and Security Informatics, FOSINT-SI 2013, Niagara Falls, Canada, pp. 1061–1067 (August 2013)

[6] Elsevier KBS Special issue on Big Social Data Analysis. Knowledge –Based Systems, `http://dx.doi.org/10.1016/j.knosys.2014.07.002`

[7] Vieweg, S., Hughes, A.L., Starbird, K., Palen, L.: Microblogging During Two Natural Hazards Events: What Twitter May Contribute to Situational Awareness. In: CHI 2010 (2010)

[8] Twitter Streaming API, `https://dev.twitter.com/docs/api/streaming`

[9] Qu, Y., Huang, C., Zhang, P.: Microblogging after a Major Disaster in China: A Case Study of the 2010 Yushu Earthquake. In: CSCW 2011 (2011)

[10] Sakaki, T., Okazaki, M., Matsuo, Y.: Earthquake Shakes Twitter Users: Real-Time Event Detection by Social Sensors. In: WWW (2010)

[11] Gupta, A., Kumaraguru, P.: Credibility Ranking of Tweets during High Impact Events. In: PSOSM 2012 (2012)
[12] Sang, E., Bosch, A.: Dealing with big data: The case of Twitter. Computational Linguistics in the Netherlands Journal (2013)
[13] Lau, R.Y.K., Xia, Y., Ye, Y.: A Probabilistic Generative Model for Mining Cybercriminal Networks from Online Social Media. IEEE Computational Intelligence Magazine (2014)
[14] WordNet-Affect, `http://wndomains.fbk.eu/wnaffect.html`
[15] Miller, G.A.: Wordnet: A lexical database for english. Communications of the ACM (1995)
[16] Porter, M.F.: An algorithm for suffix stripping. Program 14, 130–137 (1980)
[17] EuroWordNet, `http://www.illc.uva.nl/EuroWordNet/`
[18] Short Text Language Detection with Infinity Gram. NARA Institute of Science and Technology
[19] Leoncini, A., Sangiacomo, F., Decherchi, S., Gastaldo, P., Zunino, R.: Semantic Oriented Clustering of Documents. In: Liu, D., Zhang, H., Polycarpou, M., Alippi, C., He, H. (eds.) ISNN 2011, Part III. LNCS, vol. 6677, pp. 523–529. Springer, Heidelberg (2011)
[20] SenticNet, `http://sentic.net/`
[21] Yang, S., Kolcz, A., Schlaikjer, A., Gupta, P.: Large-Scale High-Precision Topic Modeling on Twitter. In: KDD 2014, NY, USA (2014)
[22] Datasift, `http://datasift.com/`

A New Supervised Learning Algorithm for Spiking Neurons*

Malu Zhang, Hong Qu, Jianping Li, and Xiurui Xie

School of Computer Science and Engineering,
University of Electronic Science and Technology of China,
Chengdu 610054, P.R. China

Abstract. Training spiking neurons to output desired spike train is a fundamental research in spiking neural networks. The current article proposes a novel and efficient supervised learning algorithm for spiking neurons. We divide the running time of spiking neurons into two classes: desired output time and not desired output time. Our learning method makes the membrane potential equal to threshold at desired output time, and makes the membrane potential lower than threshold at not desired output time. For efficiency, at not desired output time, we just calculate the membrane potential at some special time points where the spiking neuron is most likely to output a wrong spike. The experimental results show that the learning performance of the proposed method is better than the existing methods in accuracy and efficiency.

Keywords: Spiking neurons, Membrane potential, Supervised learning.

1 Introduction

The view that information in the brain is represented by explicit timing of spikes rather than mean firing rates has received increasing attention [1]-[4]. Just using the firing rates concept may be too simplistic to express the brain information and will result in loss of information in the form of precise timing of spikes [5]. Therefore spiking neurons whose operating model is much closer to biological neurons and encode information by precise timing of spikes have been proposed and expected to have a better performance than traditional neural networks [6]-[8]. There are a lot of applications based on spiking neurons [9]-[12], but powerful computing capability and broad application prospects are not fully exploited. One of the reasons is that there is no effective learning algorithm for spiking neural networks(SNNs). Although the exact mechanism of supervised learning in biological neurons remains unclear [13], in order to explore a suitable learning algorithm for SNNs, many researchers have done a lot of work and achieved certain results. The existing supervised learning algorithm for spiking neurons

* This work was supported by National Science Foundation of China under Grant 61273308, 61370073 and the Fundamental Research Funds for Central Universities under Grant ZYGX2012J068.

H. Handa et al. (eds.), *Proc. of the 18th Asia Pacific Symp. on Intell. & Evol. Systems – Vol. 1*,
Proceedings in Adaptation, Learning and Optimization 1, DOI: 10.1007/978-3-319-13359-1_14

can be classified into two types: single-spike learning and multi-spike learning. In the following paragraphs some typical learning methods of these two types will be introduced briefly.

SpikeProp [14] is one of the most typical methods of singe-spike learning, it extends the traditional BP algorithm to SNNs and can be applied to classification problems. However, SpikeProp has obvious drawbacks: it only applies to single-spike. Though many researchers improved SpikeProp, the drawbacks are still not resolved well. In singe-spike learning field, some other researchers have also done a lot of contributions. Although singe-spike learning has good application capability, networks with only singe-spike output will have limitations in the capacity and diversity of information that they transmit [5].

Multi-spike learning methods can control multiple spikes. ReSuMe [13] (Remote Supervised Method) is derived from Widrow-Hoff rule, and the practical learning algorithm composed of two weight update processes: strengthening synaptic weights by STDP based on input spike trains and desired output spike train, weakening the synaptic weights by anti-STDP based on input spike trains and actual spike trains [5]. PSD [15] is also derived from Window-Hoff rule, the biggest difference between ReSuMe and PSD is that they apply different learning windows. The Chronotron E-learning rule [16] and the SPAN rule [17] are both based on an error function of the difference between the actual output spike train and desired spike train. PBSNLR [18] first transforms the supervised learning into a classification problem and solves the problem by using the perceptron learning rule. The biggest drawback of PBSNLR is that it needs many learning epochs to achieve a good learning result when time step is precise.

One common disadvantage of these methods is that the learning efficiency and accuracy are relatively low. The reason of low efficiency is that all of these methods need to calculate the membrane potential continuously. However, when membrane potential does not reach threshold at not desired output time, the spiking neurons will not output a wrong spike, so the calculations of membrane potential are unnecessary and will cause a waste of time. At not desired output time, our method does not calculate the membrane potential continuously, but just calculates the membrane potential at some special time points where the spiking neuron is most likely to output a spike. In this way, the learning efficiency will increase much. Most of the supervised learning methods of spiking neurons adjust the synaptic weights according to the difference between the desired output time and actual output time. However, membrane potential can be regarded as a parameter of firing function and synaptic weights are parameters of membrane potential. Adjusting synaptic weights directly by the difference between desired output time and actual output time will cause the adjustment of synaptic weights indirectly. In this paper, we also put forward a new synaptic weights update rule which adjusts the synaptic weights by comparing membrane potential with threshold at different time. Experimental results show that the proposed method has higher learning accuracy and efficiency over the existing learning methods.

The rest of this paper is organised as follows. In section 2, the spiking neuron model and spiking neural networks used in this paper is formally defined. Our learning method is shown in section 3. In section 4, some comparison experiments are given to investigate the learning performance of the proposed learning method. Conclusion is presented in section 5.

2 Spiking Neuron and the SNN Model

2.1 Spiking Neuron Model

There are many spiking neuron models like LIF, HH, and SRM [6]-[8] which aim to explain the running mechanism of a biological neuron. The internal state of the SRM model can be expressed intuitively, as a result, the SRM model is easy for depicting our learning method and our method is valid for SRM model.

The SRM model used in this paper is much like the one used in [5]. The membrane potential of neuron is represented by a variable u. When there is no spike transmits from the presynaptic neurons, the variable u is at its resting value, $u_{rest} = 0$. When each spike arrives, a postsynaptic potential(PSP) will be induced in the neuron. The function ε describes the time course of the response to an incoming spike. After the summation of the effects of several incoming spikes, if μ reaches the threshold ϑ an output spike is triggered. If a spiking neuron has N input synapses and the ith synapse transmits G_i spikes whose arrival times at the neuron are denoted as $g_i = (t_i^{(1)}, t_i^{(2)}, ..., t_i^{(G_i)})$. The time of the most recent output spike is t^{f_r}. The membrane potential of the spiking neuron is expressed as

$$u(t) = \sum_{i=1}^{N} \sum_{\substack{t_i^{(g)} \in g_i \\ t_i^{(g)} > t^{(f_r)} + R_a}} \omega_i \varepsilon(t - t_i^{(g)}) + \eta(t - t^{f_r}), \tag{1}$$

where ω_i is the synaptic weight of the ith synapse. The membrane potential of the neuron must be calculated using input spikes with arrival time after $t^{f_r} + R_a$, where R_a is the length of the absolute refractory period. The PSP induced by one spike is determined by the spike response function $\varepsilon(t)$

$$\varepsilon(t) = \begin{cases} \frac{t}{\tau} e^{1-\frac{t}{\tau}} & if \quad t > 0 \\ 0 & if \quad t \leq 0, \end{cases} \tag{2}$$

When neuron fires an output spike at t^{f_r}, at the same time, the membrane potential immediately starts dropping to its resting value, $u_{rest} = 0$. Then two processes will affect membrane potential: absolute refractory period and relative refractory period. The refractoriness function $\eta(t - t^{f_r})$ depicts the relative refractory period. The function $\eta(t)$ is expressed as

$$\eta(t) = \begin{cases} -2\vartheta e^{\frac{t}{\tau_R}} & if \quad t > 0; \\ 0 & if \quad t \leq 0. \end{cases} \tag{3}$$

During the absolute refractory the neuron can not output a spike no matter how many PSPs arrive, and during the relative refractory the neuron is hard to output a spike. In this paper, we simplify the SRM model by just considering about the absolute refractory. During the absolute refractory, the membrane potential is set to 0.

2.2 Spiking Neural Networks Model

The network architecture used in this paper is shown in Fig. 1, which consists of three components, encoding part, supervised learning part, and output part. This network of spiking neurons has been proved can do perfect work. Between the encoding part and learning part, a reservoir network or Liquid State Machine(LSM) could be added. In this case, we can increase the learning capability of a neural system significantly.

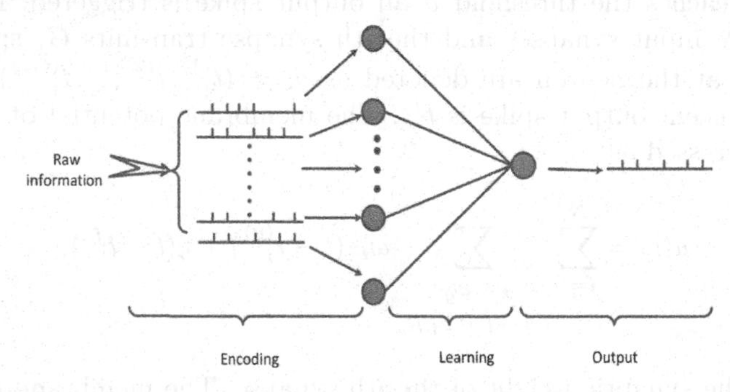

Fig. 1. Network architecture used in this paper. It contains three functional parts: encoding, learning, and readout. A stimulus is converted into spatiotemporal spikes by the encoding neurons. The input spiking train is passed to the next layer for learning. The final decision is represented by the readout layer.

In this network architecture, the learning part consists of two layers spiking neurons, so local supervised learning method like [13] [18] can be applied here, and the method proposed in this paper is also a local supervised learning algorithm.

3 Our Learning Method

In the first part of this section, a new weight update rule is proposed which pays attention to membrane potential instead of firing time. The spiking neurons are most likely to output a spike at some special time points, and these special time points include stationary points and discontinuous points (Here after we call stationary points and discontinuous points listen points(LPs)). In the second part of this section, the processes of solving LPs are shown. In the last part of this section, we give a figure to illustrate our learning method.

3.1 Weight Update Rule

Spiking neuron does not emit a spike until the membrane potential reaches threshold. Our update rule adjusts synaptic weights to make membrane potential equal to threshold at desired output time and lower than threshold at not desired output time.

At Desired Output Time. At desired output time, membrane potential is required to be equal to threshold by constructing error function below

$$E_{td} = \frac{1}{2}[\mu(t) - \vartheta]^2 \qquad if \quad t = t_d, \tag{4}$$

where $\mu(t)$ is the membrane potential of spiking neuron, and ϑ is the threshold. What calls for special attention is that when we calculate $\mu(t)$, the t^{f_r} (in Eq. (1)) is not the most recent actual fire time but the most recent target fire time. If t^{f_r} is the most recent actual fire time, it will increase the difficulty to train and make the training process more instable. An incorrect actual output time leads absolute refractory to happen at a wrong time, which will affect the subsequent membrane potential and make the adjustment after actual output time invalid.

The weight update rule is expressed as

$$\begin{aligned}\Delta\omega_i &= -\beta_1 \frac{\partial E_{td}}{\partial \omega_i} \\ &= -\beta_1[u(t) - \vartheta] \sum_{\substack{t_i^{(g)} \in g_i \\ t_i^{(g)} > t^{(f_r)} + R_a}} \varepsilon(t - t_i^{(g)}),\end{aligned} \tag{5}$$

where β_1 is the learning rate, and ω_i is the synaptic weight of synapse i.

At Not Desired Output Time. At not desired output time, membrane potential is required to be lower than threshold. For the sake of efficiency, we do not compute the membrane potential at not desired output time continuously, but just at LPs. The weight update rule is the same as PBSNLR when the membrane potential of LPs is not less than threshold

$$\omega_i = \omega_i - \beta_2 \sum_{\substack{t_i^{(g)} \in g_i \\ t_i^{(g)} > t^{(f_r)} + R_a}} \varepsilon(t - t_i^{(g)}) \qquad if \quad t = LPs, \quad t \neq t_d \quad and \quad \mu \geq \vartheta, \tag{6}$$

where ω_i is the synaptic weight of synapse i, β_2 is the learning rate, and LPs are the stationary points or discontinuous points.

3.2 Processes of Solving LPs

In the SRM model which is introduced in section 2, the membrane potential must be calculated using input spikes with arrival time after $t^{f_r} + R_a$, and t^{f_r} is the most recent desired output time in our weight update rule. We need to calculate the membrane potential at LPs which are between two sequential t^{f_r} (Here after we called the two sequential firing time t^{f_r-1} and t^{f_r}). During the sequential t^{f_r} period, the function of membrane potential is not continuous, so LPs should be discussed in different intervals.

Assuming that there are m input spikes($t^{f_r-1}+R_a < t_1 < t_2 < ..., t_m < t^{f_r}$) between t^{f_r-1} and t^{f_r} period, and ω_{t_i} means the weight of synapse which transmits the t_i spike. In the following, the processes of solving LPs in different intervals are shown.

LPs in $[t_1, t_2)$. The membrane potential $\mu(t)$ between t_1 and t_2 can be calculated as

$$\mu(t) = \omega_{t_1}\varepsilon(t - t_1) = \omega_{t_1}\frac{t - t_1}{\tau}e^{1-\frac{t-t_1}{\tau}} \qquad t_1 \le t < t_2 \tag{7}$$

Taking the derivative of $\mu(t)$ with respect to t.

$$\mu'(t) = \frac{\omega_{t_1}}{\tau}e^{1-\frac{t-t_1}{\tau}}(1 - \frac{t - t_1}{\tau}) \tag{8}$$

when $\mu'(t) = 0$, we can get that $t = t_1 + \tau$. $t_1 + \tau$ is a LP (stationary point) if $(t_1 + \tau) < t_2$, and t_1 is a LP, because t_1 is a discontinuity point of membrane potential.

LPs in $[t_{n-1}, t_n)$. The membrane potential $\mu(t)$ between t_{n-1} and t_n can be calculated as

$$\mu(t) = \sum_{N=1}^{n-1} \omega_{t_N}\varepsilon(t - t_N) \tag{9}$$

Taking the derivative of $\mu(t)$ with respect to t and defining $f(x) = e^{1-\frac{t-x}{\tau}}(1 - \frac{t-x}{\tau})$

$$\mu'(t) = \sum_{N=1}^{n-1} \frac{\omega_{t_N}}{\tau} f(t_N) \qquad t_{n-1} \le t < t_n \tag{10}$$

When $\mu'(t) = 0$, it is easy to get Eq. (11) from Eq. (10).

$$\begin{cases} \dfrac{\sum_{N=1}^{n-2} \frac{\omega_{t_N}}{\tau} f(t_N)}{-\frac{\omega_{t_{n-1}}}{\tau} f(t_{n-1})} = 1 \quad \omega_{t_{n-1}} \neq 0 & \text{(11a)} \\ \dfrac{\sum_{N=1}^{n-3} \frac{\omega_{t_N}}{\tau} f(t_N)}{-\frac{\omega_{t_{n-2}}}{\tau} f(t_{n-2})} = 1 \quad \omega_{t_{n-1}} = 0 \;\; and \;\; \omega_{t_{n-2}} \neq 0 & \text{(11b)} \end{cases}$$

Defining $g(x) = \frac{f(x)}{f(t_{n-1})} = e^{\frac{x-t_{n-1}}{\tau}} \frac{\tau - t + x}{\tau + t_{n-1} - t}$. Taking Eq. (11a) as an example, Eq. (11a) can be expressed as

$$\sum_{N=1}^{n-2} \frac{\omega_{t_N}}{\omega_{t_{n-1}}} g(t_N) = -1. \tag{12}$$

After some basic transforms, Eq. (12) can be written as

$$\sum_{N=1}^{n-2} \omega_{t_N} g(t_N) = -\omega_{t_{n-1}} \frac{\tau + t_{n-1} - t}{\tau + t_{n-1} - t}. \tag{13}$$

Combining the defined function $g(x)$ and Eq .(13), we can get

$$t = \frac{\sum_i^{n-2} \frac{\omega_{t_i}}{\omega_{t_{n-1}}} e^{\frac{t_i - t_{n-1}}{\tau}} (t_i + \tau) + \tau + t_{n-1}}{\sum_i^{n-2} \frac{\omega_{t_i}}{\omega_{t_{n-1}}} e^{\frac{t_i - t_{n-1}}{\tau}} + 1}. \tag{14}$$

t is a LP (stationary point) if $t < t_n$. t_{n-1} is a discontinuity point of membrane potential, so it is a LP.

LPs in $[t_m, t^{f_r})$. The calculations of LPs at $[t_m, t^{f_r})$ is similar to $[t_{n-1}\ , t_n)$. Here we just give the results

$$t = \frac{\sum_i^{m-1} \frac{\omega_{t_i}}{\omega_{t_m}} e^{\frac{t_i - t_m}{\tau}} (t_i + \tau) + \tau + t_m}{\sum_i^{m-1} \frac{\omega_{t_i}}{\omega_{t_m}} e^{\frac{t_i - t_m}{\tau}} + 1}. \tag{15}$$

t is a LP (stationary point) if $t^{f_r} > t \geq t_m$. t_m is a discontinuity point of membrane potential, so it is a LP.

3.3 An Illustration of Our Learning Method

Fig. 2 gives an illustration of our learning algorithm. At desired output time, membrane potential is required equal to threshold, and the synaptic weight update rule is abide by Eq. (3) and Eq. (4). The membrane potential of t_{d1} is lower than threshold, so the synaptic weight should be increased. At t_{d2}, the membrane potential is above threshold, synaptic weight is reduced to make the membrane potential equal to threshold. At LPs, membrane potential is required lower than threshold, so if the membrane potential is not below threshold, the synaptic weights should be reduced by Eq. (5). The membrane potentials of the first and forth LP are below threshold, then the synaptic weight is no need to update. However, at second, third and fifth LP, the membrane potential is not less than threshold, so the synaptic weight should be reduced as shown in Fig. 2.

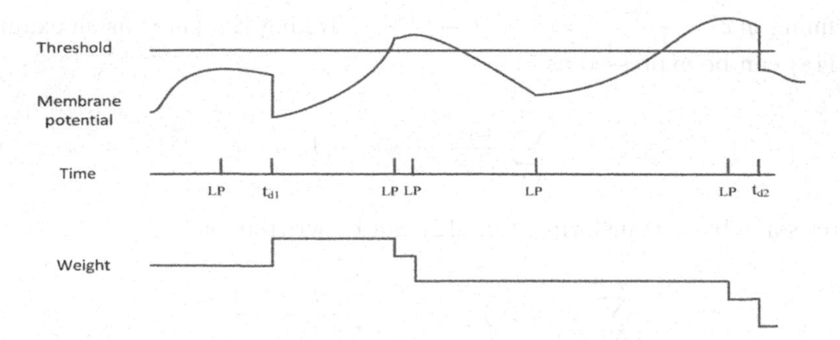

Fig. 2. A simple example to illustrate the learning mechanism of our method

4 Experiment Results

To quantitatively evaluate the learning performance, we introduce a correlation-based measure proposed in [19] which is a method to evaluate the similar degree between the desired spike trains and actual spike trains. The metric is calculated after each learning epoch according to

$$C = \frac{v_d \cdot v_o}{|v_d||v_o|}, \tag{16}$$

where v_d and v_o are vectors representing a convolution(in discrete time) of desired and actual output spike trains with a Gaussian low-pass filter. The measure C equals one for the identical spike trains and decreases towards zero for loosely corrected trains.

4.1 Learning Results

In the first example, a SRM neuron with 200 synaptic inputs is trained to emit a desired spike train with the length 500 ms. Every input spike train and desired output spike train are generated randomly according to the homogeneous poisson process with rates $r = 10$ Hz and 100 Hz, respectively. The desired output spike trains are required to be learnable in the experiments by setting the time interval of two adjacent spikes 3 ms because the length of the absolute refractory period is 3 ms. The learning performance is shown in Fig. 3. (See the appendix for details on the experimental strategy and parameter settings.)

Fig. 3(a) and 3(b) show the initial synaptic weights and synaptic weights at the end of the learning, respectively. Fig. 3(c) illustrates the learning process during the consecutive learning epochs, which includes the desired output spike train denoted by ∘ and the actual output spike trains after each learning epoch denoted by •. At first the actual output spike train is very different from the desired output spike train. As the training progresses, the gap gets smaller and smaller. At about 55 epochs the actual output spike train is the same as the desired one. The measure C plotted as a function of the learning epoch is shown

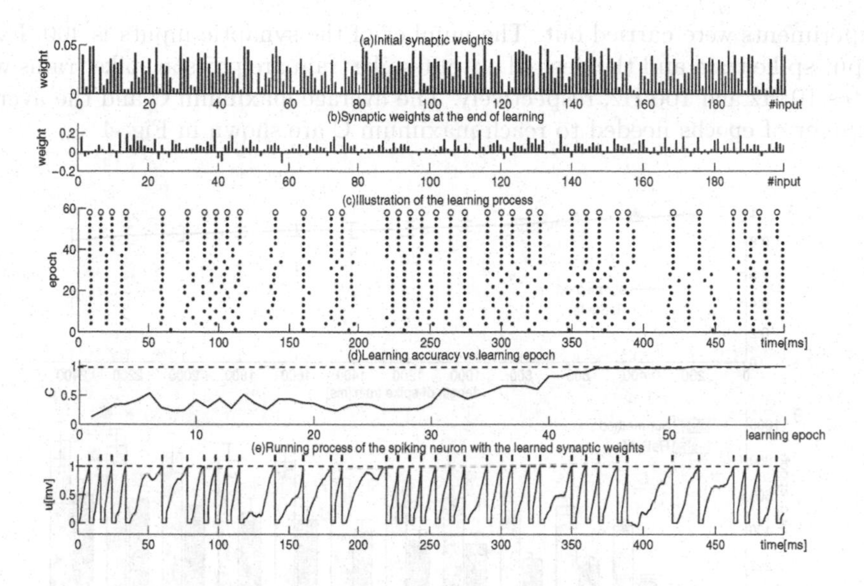

Fig. 3. The learning performance of our method. A spiking neuron with 200 synaptic inputs is trained on a 100 Hz poisson target spike train of length 500 ms. The actual output spikes after each learning epoch are shown by •. ○ denotes the desired output spikes. The trained spiking neuron can emit the desired spike train after about 55 learning epochs.

in Fig. 3(d). At the beginning of the learning, the value of C is close to 0.18, after about 55 learning epochs $C = 1$. Fig. 3(e) shows the running process of the trained spiking neuron with the learned synaptic weights. We can see that the firing time of the spiking neuron is the same as the desired output time which is represented by the short bar.

4.2 Comparison Results

ReSuMe has good performance and is widely used as a supervised learning method for spiking neurons. The learning performance of our method can be measured intuitively by comparison with ReSuMe. In this subsection, three groups of comparison experiments were carried out between ReSuMe and our method. In the first group, the length of the desired output spike trains was varied from short to long. In the second group, the firing rate of desired output spike trains was varied from low to high. In the third group, the number of synaptic inputs was varied from less to more. Longer spike train, higher firing rate, and less synaptic inputs are used to test the merits of supervised learning algorithm of spiking neurons widely [5] [18]. In the following experiments, the time step which simulates continuous time is set as 0.01 ms.

In the first group of experiments, the length of the desired output spike trains was varied from 200 ms to 2400 ms with an interval of 200 ms. In each case, 50 experiments were carried out. The number of the synaptic inputs is 400. Every input spike train and the desired output spike train are poisson spike trains with rates 10 Hz and 100 Hz, respectively. The average maximum C and the average number of epochs needed to reach maximum C are shown in Fig. 4

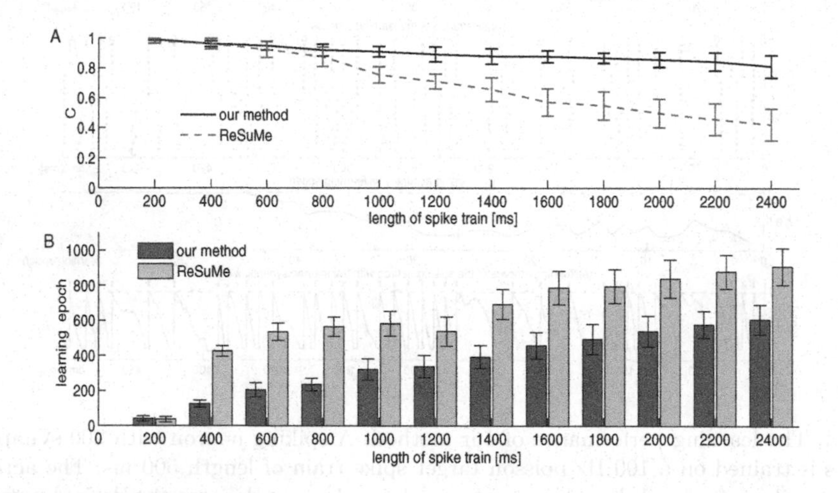

Fig. 4. The comparison of the learning performance between our method and ReSuMe when the length of desired spike trains increases gradually. (A) The learning accuracy of our method and ReSuMe. (B) The learning epoch of our method and ReSuMe.

Fig. 4(A) investigates the learning accuracy. The accuracy of our method and ReSuMe is very high when the length of desired spike trains is not very long. The accuracy of both our method and ReSuMe decreases when the length of the desired spike train increases. However, the accuracy curve of ReSuMe decreases earlier than our method and the decrease extent of our method is less than ReSuMe. Overall, the accuracy curve of our method is higher than ReSuMe when the length of desired spike train varies from 200-2400 ms.

Fig. 4(B) investigates the learning efficiency. The number of learning epoch increases as the length of desired output spike train increases. However, the number of learning epoch of our method is fewer than ReSuMe significantly. More importantly, the time cost of one learning epoch of our method is much less than ReSuMe. In Fig. 4, we do not mark the time cost of one learning epoch, as different experiments hardware conditions and different codes will affect the time cost. We give a theoretical analysis for time complexity. Computing the membrane potential spends the majority of experimental time. ReSuMe needs to calculate the membrane potential continuously, while our method just need to calculate the membrane potential at LPs, so the time cost of one learning epoch of our method is much less than ReSuMe. For efficiency, our method need less learning epochs to reach the maximum C, and the time cost of one learning epoch of our method is less than ReSuMe, so the convergence speed and learning efficiency is much better than ReSuMe.

In the second group of experiments, the firing rate of the desired output spike trains was varied from 80 Hz to 150 Hz with an interval of 10 Hz. In each case, 50 experiments were carried out. The synaptic inputs number is 400. Every input spike train is poisson spike trains with rate 10 Hz. The length of the desired output spike train is 1000 ms. The experimental results are shown in Fig. 5.

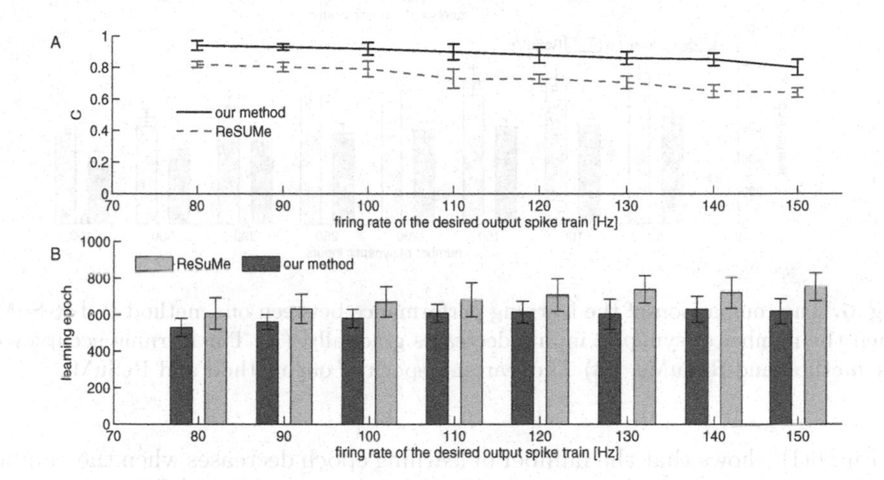

Fig. 5. The comparison of the learning performance between our method and ReSuMe when the firing rate of desired spike trains increases gradually. (A) The learning accuracy of our method and ReSuMe. (B) The learning epoch of our method and ReSuMe.

First, we investigate the learning accuracy. Fig. 5(A) shows that the learning accuracy of our method is higher than ReSuMe when the firing rate varies from low to high(80Hz-150Hz).

We next investigate the learning efficiency. From Fig. 5(B), we can see that the learning epoch of our method is fewer than ReSuMe obviously. As we have analysised before, the time cost of one learning epoch of our method is much less than ReSuMe. When the firing rate varies from low to high, the learning performance is better than ReSuMe.

In the third group of experiments, the number of synaptic inputs was varied from 50 to 4000 with an interval of 50. In each case, 50 experiments were carried out. Every input spike train and the desired output spike train are poisson spike trains with rates 10 Hz and 100 Hz, respectively. The length of the desired output spike train is 1000 ms. The experimental results are shown in Fig. 6.

In Fig. 6(A), the learning accuracies of the two methods are very low when the number of the synaptic inputs is small and the accuracy curve of ReSuMe is above our method. Accuracy increases when the number of synaptic inputs increases gradually, the accuracy curve of our method increases relatively early and steeply. Finally, the two curves intersect. The learning accuracy of our method is significantly higher than the learning accuracy of ReSuMe with a range of the number of synaptic inputs (200-400).

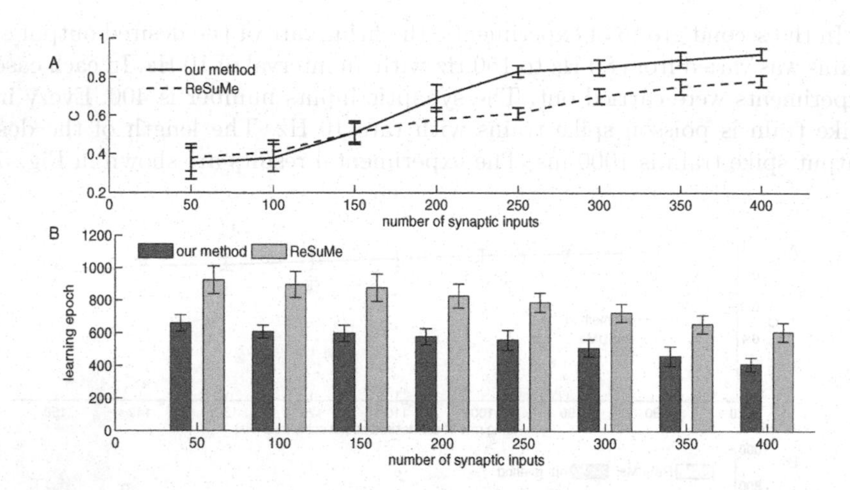

Fig. 6. The comparison of the learning performance between our method and ReSuMe when the number of synaptic inputs decreases gradually. (A) The learning accuracy of our method and ReSuMe. (B) The learning epoch of our method and ReSuMe.

Fig. 6(B) shows that the number of learning epoch decreases when the number of synaptic inputs is gradually increases, our method needs fewer learning epoch than ReSuMe, so the convergence speed of our method is better than ReSuMe.

5 Conclusion

This paper presented an efficient and novel supervised learning algorithm for spiking neurons. In the experiments, the learning accuracy and efficiency of our method are much better than ReSuMe. Learning accuracy and efficiency are two extremely important performance indexes of the learning methods and play a key role in the practical applications. Therefore, our method is significant to the practical applications and theoretical research of spiking neurons. ReSuMe has wide applicability for different spiking neuron models, such as LIF and HH. However, our method is limited to SRM model. How to extend our method to different spiking neuron models is our future work. Our method is a local multi-spike learning method. If we can improve it to multi-layer, the learning and memory capacity of spiking neural networks will enhance a lot and will reduce the size of the networks, this is future work.

Appendix: Details of Simulations

In this experiment of Fig. 3, the parameter values of the neurons and learning rates are listed in Table 1, and the values of τ, ϑ and R_a keep the same in other simulations.

The learning rates of our method in Fig. 4 to Fig. 6 are shown in Table 2.

Table 1. Parameters of the SRM neuron and learning rates

Parameter	τ	β_1	β_2	ϑ	R_a
Value	7ms	0.05	0.005	1mv	3ms

Table 2. Learning Rates of the Experiments

Learning rates	β_1	β_2
Figure 4	0.02(200ms)-0.003(2400ms)	0.002(200ms)-0.0005(2400ms)
Figure 5	0.01(80Hz)-0.007(150Hz)	0.001(80Hz)-0.0006(150Hz)
Figure 6	0.01(400)-0.008(50)	0.001(400)-0.0008(50)

References

1. Reinagel, P., Reid, R.C.: Temporal coding of visual information in the thalamus. The Journal of Neuroscience 20(14), 5392–5400 (2000)
2. Wyeth, B., Koch, C.: Temporal precision of spike trains in extra striate cortex of the behaving macaque monkey. Neural Computation 8(6), 1185–1202 (1996)
3. Mainen, Z.F., Sejnowski, T.J.: Reliability of spike timing in neocortical neurons. Science 268, 1503–1506 (1995)
4. Wolfgang, M.: Noisy spiking neurons with temporal coding have more computational power than sigmoidal neurons
5. Xu, Y., Zeng, X., Han, L., Yang, J.: A supervised multi-spike learning algorithm based on gradient descent for spiking neural networks. Neural Networks 43, 99–113 (2013)
6. Wulfram, G., Kistler, W.M.: Spiking neuron models: Single neurons, populations, plasticity. Cambridge university press (2002)
7. Samanwoy, G.-D., Adeli, H.: Spiking neural network. International Journal of Neural Systems 19(04), 295–308 (2009)
8. Wolfgang, M.: Networks of spiking neurons: the third generation of neural network models. Neural Networks 10(9), 1659–1671 (1997)
9. Hu, J., et al.: A spike-timing-based integrated model for pattern recognition. Neural computation 25(2), 450–472 (2013)
10. Buhmann, J.M., Lange, T., Ramacher, U.: Image segmentation by networks of spiking neurons. Neural Computation 17(5), 1010–1031 (2005)
11. Wade, J.J., McDaid, L.J., Santos, J.A., et al.: SWAT: a spiking neural network training algorithm for classification problems. IEEE Transactions on Neural Networks 21(11), 1817–1830 (2010)
12. Kasabov, N., Feigin, V., Hou, Z.G., et al.: Evolving spiking neural networks for personalised modelling, classification and prediction of spatio-temporal patterns with a case study on stroke. Neurocomputing 134, 269–279 (2014)
13. Ponulak, F., Kasinski, A.: Supervised learning in spiking neural networks with ReSuMe: sequence learning, classification, and spike shifting. Neural Computation 22(2), 467–510 (2010)
14. Bohte, S.M., Kok, J.N., La, P.H.: Error-backpropagation in temporally encoded networks of spiking neurons. Neurocomputing 48(1), 17–37 (2002)

15. Yu, Q., Tang, H., Tan, K.C., Li, H.: Precise-Spike-Driven Synaptic Plasticity: Learning Hetero-Association of Spatiotemporal Spike Patterns. PloS One 8(11), e78318 (2013)
16. Florian, R.V.: The Chronotron: a neuron that learns to fire temporally precise spike patterns. PloS One 40233, e40233 (2012)
17. Mohemmed, A., Schliebs, S., Matsuda, S., Kasabov, N.: Span: Spike pattern association neuron for learning spatio-temporal spike patterns. International Journal of Neural Systems 22(04) (2012)
18. Xu, Y., Zeng, X., Zhong, S.: A new supervised learning algorithm for spiking neurons. Neural Computation 25(6), 1472–1511 (2013)
19. Schreiber, S., Fellous, J.M., Whitmer, D., Tiesinga, P., Sejnowski, T.J.: A new correlation-based measure of spike timing reliability. Neurocomputing 52, 925–931 (2003)

An Optimization Method of SNNs for Shortest Path Problem*

Hong Qu, Zhi Zeng, Changle Chen, Dongdong Wang, and Nan Yao

School of Computer Science and Engineering,
University of Electronic Science and Technology of China,
Chengdu 610054, P.R. China
hongqu@uestc.edu.cn

Abstract. In this paper we propose a new optimization method of integrate-and-fire neural model for shortest path problem and compare it with Dijkstra algorithm. The proposed algorithm improves the speed of path planning by parallel property of spike spreading in the network and makes efficiency independent from the number of connections of the network which is only related to the length of the shortest path in the network with determined number of edges. Mathematical analysis and simulations demonstrate that planing shortest path by utilizing connections between transmission time in integrate-and-fire model and edge weights is feasible. The comparisons with Dijkstra algorithm manifest that the new algorithm does have superiority in some application scenarios.

Keywords: Shortest path problem, Spike spreading, Integrate-and-fire model.

1 Introduction

Shortest path (SP) problem, the fundamental network optimization problem, is often raised as a sub-problem in both practical applications and research problems [1] such as routing strategy [2], geographic information system [3], mobile navigation [4], wireless network routing [5]. Researches on SP Algorithm, considered as SP Problem's kernels in computer science, are mainly focused on dynamic performance, efficiency optimization, model innovation [6], etc. Typically, Dijkstra algorithm [7] provides a universal solution for SP problem. It is fairly simple when implemented on normal scale networks. But for large scale networks, high costs on CPU time has limited its application [8].

Spiking neural networks (SNNs), emerged as the new generation of pulsed neuron networks, take the benefits of computational properties of biological neurons [9]. They focus on model's temporal structure, using spikes to carry information. [10] Classical spiking neuron models such as spike response model (SRM),

* This work was supported by National Science Foundation of China under Grant 61273308 and the Fundamental Research Funds for Central Universities under Grant ZYGX2012J068.

H. Handa et al. (eds.), *Proc. of the 18th Asia Pacific Symp. on Intell. & Evol. Systems – Vol. 1*,
Proceedings in Adaptation, Learning and Optimization 1, DOI: 10.1007/978-3-319-13359-1_15

integrate-and-fire model (IF), Hodgkin-Huxley model (HH) process information in a novel way which are logically and biologically feasible. There are a lot of applications based on spiking neural networks [11] [12]. Our work on this topic has presented a novel shortest-path algorithm based on IF model.

Section 2 introduces the model used in our algorithm and does the theoretical analysis to match the model with the SP problem. Algorithms are described in Section 3. Simulations and comparisons are conducted out in the following section. Finally, conclusions are drawn in the last section.

2 The Proposed Spiking Neuron Model

Biologically, the communication between two cells is usually in the form of spiking. When one cell fires, the spiking transmits to another cell through synapse which will fire after a certain time. Ignore transmission time through axon and only consider the time that a cell takes to fire which is just the time that the spiking takes to transmit through synapse. In short, the time of spiking transmission from source to destination is that all the cells on the transmission path take to fire.

Since the axon can stretch over several millimeters, it is common that one cell connects to several cells. However, the transmission through one specific synapse is unidirectional which makes it necessary to build two synapses while the transmission between two nodes is bidirectional.

In our model, we use the time of transmission from source to destination to weight the length of paths, comparing to Dijkstra. To solve the SP problem successfully, it is necessary to prove that the edge weight w has proportional relations with transmission time. How to make connections between edge weights w and parameters in spiking neural model becomes an important question. In this section we will introduce the model used in our algorithm first and then analyze the relations between weights and time.

2.1 Integrate-and-Fire Model

As one of the most well-known and widely-used spiking neuron model, the integrate-and-fire model, which is also known as the linear integrate-and-fire model or leaky-integrate-and-fire model referring to its basic model, assumes the membrane to be leaky, and simulates the ion transfer in the ion channels of the membrane. The ion transfer from all presynaptic neurons nearby is represented as the temporal summation of all contributions to the membrane potential $u(t)$. The postsynaptic neuron will fire while a threshold θ is reached by this contribution. Such that the membrane potential repeatedly transfers from the threshold potential to the reset potential. By the RC circuit, a typical neuron of IF model is composed as below.

The main compartment of the model is the RC circuit in the circle of the right-hand side shown in Fig. 1. With the circuit input current I, when the voltage U across the capacitor C reaches the threshold θ, a pulse will be transmitted to

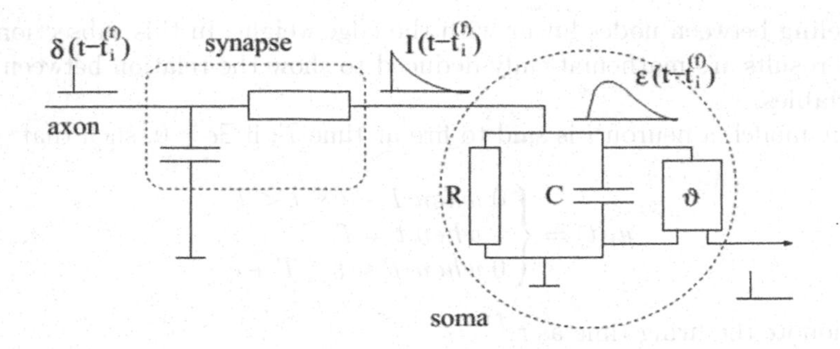

Fig. 1. Integrate-and-fire neuron model [13]

other neurons as an output of the circuit shown in the lower right of Fig. 1. The pulse sent out by the presynaptic neuron will travel through the axon of another neuron shown in the left of Fig. 1. The pulse transmitted will be filtered in the low-pass circuit shown as the synapse in the middle of Fig. 1. Then it will be transmitted as the driving current $I(t - t_i^{(f)})$ into the integrate-and-fire circuit. The driving current will be split into two components, the capacitor and the resistor, yielding the Eq. (1).

$$I(t) = \frac{u(t)}{R} + C\frac{d(u)}{dt} \tag{1}$$

Where the voltage $u(t)$ across the capacitor C is the membrane potential. Making the time constant $\tau_m = RC$, which is called the membrane time constant, and rewrite Eq. (1) in the form of Eq. (2).

$$\tau_m \frac{d(u)}{dt} = -u(t) + RI(t) \tag{2}$$

The Eq. (2) describes the first-order linear differential characteristics of the model. To present its full neuronal spiking behavior, a threshold condition need to be added as follow.

A pulse is emitted at time $t = t^{(f)}$ when the membrane potential reaches the threshold θ with $u(t^{(f)} = \theta)$. The potential will be reset to u_{reset} defined by Eq. (3) immediately after the emitting and integrate again from this point.

$$\lim_{t \to t^{(f)};t>t^{(f)}} u(t) = u_{reset} \tag{3}$$

The proposed model used in our algorithm is mainly ruled by Eq. (2) and (3).

2.2 Theoretical Analysis

To apply neural spiking model, the weight of topological graph need to be transfered to the spiking time. Thus we have to make the time which spiking spends on traveling between nodes linear with the edge weight. In this subsection, theoretical results are mathematically deduced to show the relation between these two variables.

In our model, a neuron i is said to fire at time T, if $\exists \epsilon \geq 0$, such that

$$u_i(t) = \begin{cases} 0 \; when \; T - \epsilon \leq t < T \\ 1 \; when \; t = T \\ 0 \; when \; T < t \leq T + \epsilon \end{cases} \tag{4}$$

We denote the firing time as $t_i^{(f)}$.

For any neuron i in IF model, whose parent neuron p fires at time $t_p^{(f)}$ while neuron i itself has not fired, if it fires at time $t_i^{(f)}$ with $t_i^{(f)} > t_p^{(f)}$, then the firing time t_1 and t_2 of neuron i in two individual experiments is in a linear relation with their respective weights w_1 and w_2 in the topological graph, presenting as the Eq. (5).

$$w_2 - w_1 = \frac{t_2 - t_1}{\tau_m} \tag{5}$$

Here is the proof of Eq. (5).

In the IF model mentioned in subsection 2.1, with a constant input I_0 and a reset potential equaling to zero, that $I(t) = I_0$ and $u_{reset} = u(t_p^{(f)}) = 0$, the relation between the membrane potential and the spiking time can be presented as Eq. (6) by integrating Eq. (2).

$$u(t) = RI_0[1 - e^{(-\frac{t - t_p^{(f)}}{\tau_m})}] \tag{6}$$

Since neuron i fires at time t_1 and t_2 when membrane potential reaches the threshold θ_1 and θ_2 respectively, the Eq. (7) can be concluded as Eq. (6).

$$\begin{cases} \theta_1 = RI_0[1 - e^{(-\frac{t_1 - t_p^{(f)}}{\tau_m})}] \\ \theta_2 = RI_0[1 - e^{(-\frac{t_2 - t_p^{(f)}}{\tau_m})}] \end{cases} \tag{7}$$

Make w a function of θ as Eq. (8).

$$w = -ln(RI_0 - \theta) \tag{8}$$

Eq. (7) can be presented as

$$\begin{cases} e^{w_1} = e^{-\frac{t_1 - t_p^{(f)}}{\tau_m}} \\ e^{w_2} = e^{-\frac{t_2 - t_p^{(f)}}{\tau_m}} \end{cases} \tag{9}$$

Divide these two equations, getting Eq. (10).

$$e^{-(w_2 - w_1)} = e^{-(\frac{t_2 - t_1}{\tau_m})} \tag{10}$$

Thus

$$w_2 - w_1 = \frac{t_2 - t_1}{\tau_m} \tag{11}$$

This completes the proof. From the deduction, the connection between topological weights and parameters in spiking neural model has been made as Eq. (8). Also, from the Eq. (11), we can conclude that Δw is in proportion to Δt which means the applying of shortest path problem is practicable by setting the neural threshold θ according to Eq. (12).

$$\theta = RI_0 - e^{-w} \tag{12}$$

3 The Novel Shortest-Path Algorithm

The temporal encoding type of spiking neural model makes spiking neural network an applicable network to analysis the time-structured data. Consider a random graph, we take every neuron of the neural network mapping as a node in the graph respectively. If there is an edge between two nodes, it indicates that one neuron cell's axon connects to another cell's dendrites through synapses bidirectionally. Specially when there is no edge between two nodes, the weight is set to be infinite. Initially fire the neuron corresponding to the starting node, triggering the pulse propagation of the neural network following the shortest path. The shortest path tree will result when all the neurons have fired successfully.

Here we use a real graph illustrated in Fig. 2 to present the mapping relation.

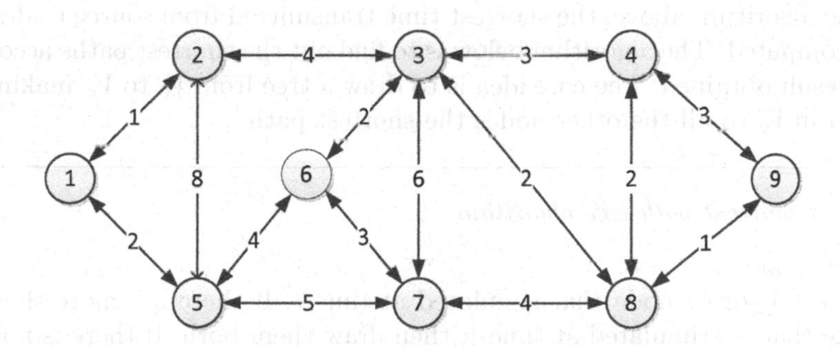

Fig. 2. A random graph for demonstration

All the neurons stay in refractory period at beginning. When neuron 1 corresponding to node 1 is activated by current I_0 initially, neuron 2 corresponding to node 2 starts getting the pulse from neuron 1 with threshold setting as $RI_0 - e^{-1}$. After time τ_m, neuron 2 starts emitting pulses to the inactivated neighbor neurons with input current I_0 since the threshold is reached. Step by step, once neuron A is activated, neighbor neuron B's threshold will be set as $RI_0 - e^{-w_{AB}}$ according to the edge weight between node A and node B respectively. After time $w_{AB}\tau_m$, neuron B will be activated. Neuron A sends out the pulse to all the neighbor neurons simultaneously, and will be reset to refractory period after all the neighbor neurons being activated. Infinite weight will be set if neuron D is not activated by neuron C. Shortest path tree will then be formed by the strengthen networks.

Using this mapping method, topological information is successfully transferred into spiking neuron networks. Conjecture can be presented that an area processing the routing process exists in biological neuron networks. A routing task will result in fast neural mapping and processing. The mapping relation will be strengthened with the better familiarity and judgment of the geographical environment. The specific description of the algorithm is given as below, with an example based on Fig. 2.

Firstly we set $V_n = V_0, t = 0$. Then compute the shortest path by the following two algorithm.

Searching shortest time algorithm

1. Stimulate V_n at time t.
2. $t = t + 1$, $w = w - 1$. (One node is stimulated but another is not.)
3. If there exists $w_m = 0$, then stimulate another node V_k(connect to V_n). If there are more than one w that equals 0, then stimulate all nodes connecting to them. If there is no w that equals 0, then turn to step 2.
4. Record V_k at time t. If there are more than one node that is excited at time t, then record them all. Then turn to step 2.

In the algorithm above, the shortest time transmitted from source to destination is computed. The algorithm below is to find out the shortest paths according to the result obtained. The core idea is to draw a tree from V_0 to V_n making the paths form V_0 to all the other nodes the shortest path.

Searching shortest path tree algorithm

1. Draw node V_0, $t = 0$.
2. $t = t + 1$, draw node that simulated at time t. If there are more than one node that is stimulated at time t, then draw them both. If there is no node stimulated at time t, turn to step 2.
3. If w between the node stimulated at time t and another node stimulated before time t equals 0, then point the node stimulated before time t to the node stimulated at time t. Turn to step 2.

To clarify our algorithms more clearly, the shortest path tree of Fig. 2 is computed step by step as an example. Fig. 2 illustrates a real network, where node 1 is the starting node. All neurons have not fired yet at the initial time, and there is no descending of any edge weight. The computation process is shown below where $w_{i,j}$ indicates the weight of edge between node i and node j.

- At time 0, neuron 1 fires, thus $w_{1,2}$ and $w_{1,5}$ start to diminish, $w_{1,2} = 1$ and $w_{1,5} = 2$.
- At time 1, $w_{1,2}$ reaches 0, hence neuron 2 is stimulated by node 1. Then $w_{2,5}$ and $w_{2,3}$ start to diminish, $w_{2,5} = 8$ $w_{2,3} = 4$ and $w_{1,5} = 1$.
- At time 2, $w_{1,5}$ reaches 0, hence neuron 5 is stimulated by node 1. Then $w_{5,6}$ and $w_{5,7}$ start to diminish, $w_{2,5} = 7$ $w_{2,3} = 3$ $w_{5,7} = 5$ and $w_{5,6} = 4$.
- Then there is no neuron fire until time 5. At time 5, $w_{2,3}$ reaches 0, hence neuron 3 is stimulated by neuron 2. Then $w_{3,6}$ $w_{3,7}$ $w_{3,8}$ and $w_{3,4}$ start to diminish, and $w_{2,5} = 4$ $w_{5,7} = 2$ $w_{5,6} = 1$ $w_{3,6} = 2$ $w_{3,7} = 6$ $w_{3,8} = 2$ and $w_{3,4} = 3$.
- At time 6, $w_{5,6}$ reaches 0, hence neuron 6 is stimulated by neuron 5. Then $w_{6,7}$ start to diminish, $w_{2,5} = 3$ $w_{5,7} = 1$ $w_{3,6} = 1$ $w_{3,7} = 5$ $w_{3,8} = 1$ $w_{3,4} = 2$ and $w_{6,7} = 3$.
- At time 7, $w_{5,7}$ and $w_{3,8}$ reach 0. As neuron 6 has fired, only neuron 7 and 8 fire, which are respectively stimulated by neuron 5 and 3. Then $w_{7,8}$ $w_{8,9}$ and $w_{4,8}$ start to diminish, $w_{2,5} = 2$ $w_{8,9} = 1$ $w_{7,8} = 4$ $w_{3,7} = 4$ $w_{3,8} = 1$ $w_{4,8} = 2$ and $w_{6,7} = 2$.
- At time 8, $w_{3,4}$ and $w_{8,9}$ reach 0, hence neuron 4 and 9 are respectively stimulated by neuron 3 and 8. So far, all neuron have fired, and the pulse propagation stops. Then we can get a shortest-path tree as illustrated in Fig. 3.

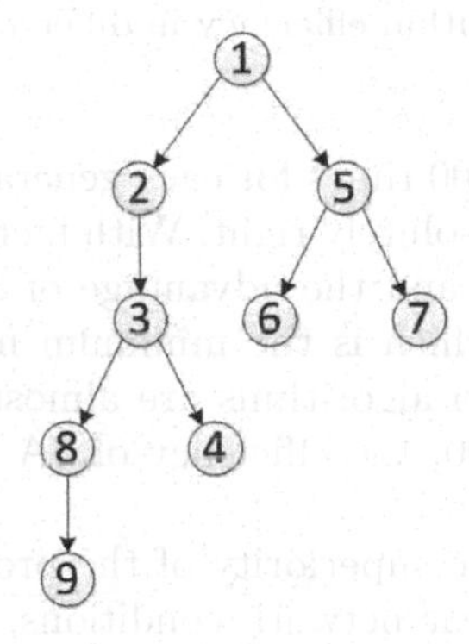

Fig. 3. The shortest-path tree obtained by

4 Simulations and Comparisons

In this section, simulations will be given to testify the feasibility of the Spiking algorithm (SA) and to compare its efficiency with the Dijkstra algorithm (DA). The graph generation method is introduced as a prerequisite. Main interest focuses on scale and connectivity of different networks. These programs are coded in MATLAB 7.14, and run in a compatible PC with Core 2 Duo CPU T6670@2.20GHz and 2GB RAM.

Firstly, generate a proper network: N nodes are generated within a 100*100 square randomly and connect the two nodes when the Euclidean distance of the nodes is shorter than 30. Suppose that the weigh is equal to the distance and N is range from 80 to 1000. The simulation selects a node randomly as the starting node and calculates the path to all other nodes afterwards. The result of the comparison of SA and DA is indicated in Fig. 4(a).

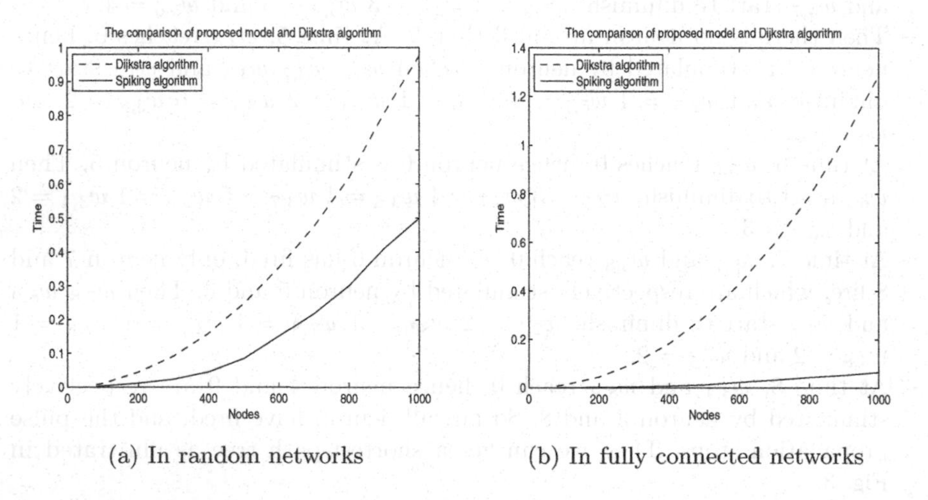

(a) In random networks (b) In fully connected networks

Fig. 4. Algorithm efficiency in different networks

We ran SA and DA both 100 times for each generated network. The comparison confirm that the SA is absolutely right. With increasing network complexity, the time used will be longer and the advantage of SA is more obvious. When the number of nodes is 80 which is the minimum nodes ensuring a connected network, the efficiency of two algorithms are almost same. However when the number of nodes reaches 1000, the efficiency of SA is almost twice as good as DA, regarding the CPU time.

To further demonstrate the superiority of the proposed algorithm, more experiments are done in different network conditions. In the Fig. 4(b), the fully connected network is generated for efficiency comparison and the result indicated that SA is so much better than DA. Seemingly that he performance is proportional to intensity of connections of the network. We designed an experiment resulting as Fig. 5.

The simulation is performed 100 times utilizing networks with 500 nodes and same average weights. The intensity of the network is described as a variety r which we called as connection probability. Supposing that two nodes connect only if their Euclidean distance less than $r*100$ (r is ranging from 0.3 to 1.4). The larger the connection probability, the better the network connectivity, resulting the higher efficiency of SA, while the efficiency of DA stays almost the same.

To further analysis the algorithm, we structured some networks with 500 nodes and changing weights, making the weights randomly spread in $[w, w+10]$ with w ranging from 0 to 200. Those network was used to do the simulation

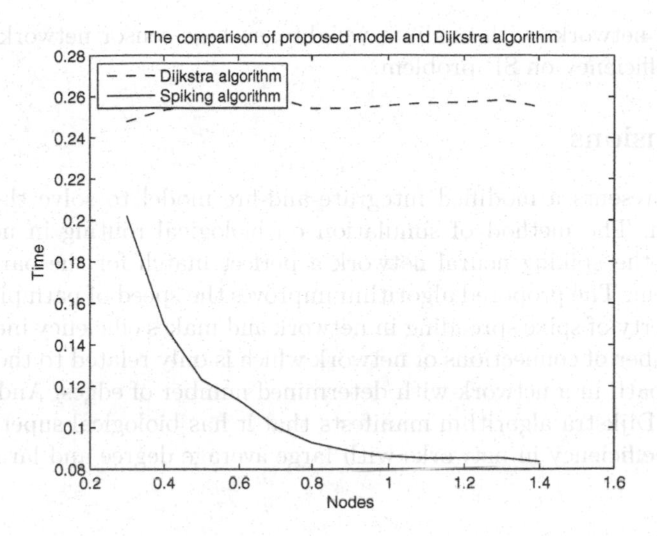

Fig. 5. Relation between algorithm efficiency and connection probability

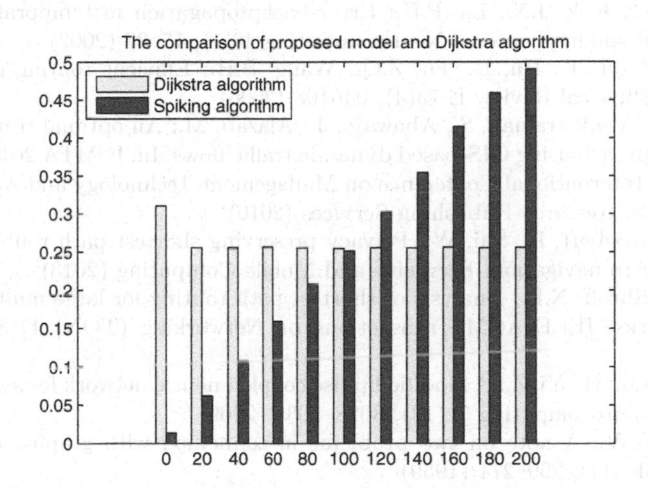

Fig. 6. Relation between algorithm efficiency and average weights

100 times also. According to the results we obtained a bar graph as Fig. 6. It shows that when $w = 100$, the CPU time of SA and DA stays almost the same. Superiority of SA is more obvious on relatively smaller weights networks.

Simulation results show that all the shortest path trees computed by SA are verified to be right. In the computational neuron model, every neuron is taken as a sensor to transfer the firing event to neighbor neurons without comparing. Comparison experiment shows that SA is better than DA when the network is with more nodes and more complex connections. However, SA needs more time than DA when the network's average weight becomes larger. In the condition of large scale networks with medium weights, such as sensor networks, SA will have better efficiency on SP problem.

5 Conclusions

This paper presents a modified integrate-and-fire model to solve the shortest path problem. The method of simulation on biological routing in neural networks makes the spiking neural network a perfect match for the parallel computing problem. The proposed algorithm improves the speed of path planning by parallel property of spike spreading in network and makes efficiency independent from the number of connections of network which is only related to the length of the shortest path in a network with determined number of edges. And the comparison with Dijkstra algorithm manifests that it has biological superiority and computation efficiency in networks with large average degree and large amount of nodes.

References

1. Bohte, S.M., Kok, J.N., La, P.H.: Error-backpropagation in temporally encoded networks of spiking neurons. Neurocomputing 48(1), 17–37 (2002)
2. Yan, G., Zhou, T., Hu, B., Fu, Z.Q., Wang, B.H.: Efficient routing on complex networks. Physical Review E 73(4), 046108 (2006)
3. Alazab, A., Venkatraman, S., Abawajy, J., Alazab, M.: An optimal transportation routing approach using GIS-based dynamic traffic flows. In: ICMTA 2010: Proceedings of the International Conference on Management Technology and Applications, pp. 172–178. Research Publishing Services (2010)
4. Xi, Y., Schwiebert, L., Shi, W.: Privacy preserving shortest path routing with an application to navigation. Pervasive and Mobile Computing (2013)
5. Kwon, S., Shroff, N.B.: Analysis of shortest path routing for large multi-hop wireless networks. IEEE/ACM Transactions on Networking (TON) 17(3), 857–869 (2009)
6. Wang, X., Qu, H., Yi, Z.: A modified pulse coupled neural network for shortest-path problem. Neurocomputing 72(13), 3028–3033 (2009)
7. Dijkstra, E.W.: A note on two problems in connexion with graphs. Numerische mathematik 1(1), 269–271 (1959)
8. Fuhao, Z., Jiping, L.: An algorithm of shortest path based on Dijkstra for huge data. In: Sixth International Conference on Fuzzy Systems and Knowledge Discovery, FSKD 2009, vol. 4, pp. 244–247. IEEE (August 2009)
9. Maass, W., Zador, A.M.: Dynamic stochastic synapses as computational units. Neural Computation 11(4), 903–917 (1999)
10. Kasabov, N.: To spike or not to spike: A probabilistic spiking neuron model. Neural Networks 23(1), 16–19 (2010)
11. Hu, J., et al.: A spike-timing-based integrated model for pattern recognition. Neural Computation 25(2), 450–472 (2013)
12. Kasabov, N., Feigin, V., Hou, Z.G., et al.: Evolving spiking neural networks for personalised modelling, classification and prediction of spatio-temporal patterns with a case study on stroke. Neurocomputing 134, 269–279 (2014)
13. GMaass, W., Bishop, C.M. (eds.): Pulsed neural networks. MIT Press (2001)

C^2: Adaptive Load Balancing for Metadata Server Cluster in Cloud-Scale Storage Systems

Quanqing Xu[1], Rajesh Vellore Arumugam[1], Khai Leong Yong[1], Yonggang Wen[2], and Yew-Soon Ong[2]

[1] Data Storage Institute, A*STAR, Singapore
{Xu_Quanqing,Rajesh_VA,YONG_Khai_Leong}@dsi.a-star.edu.sg
[2] Nanyang Technological University, Singapore
{ygwen,asysong}@ntu.edu.sg

Abstract. Big data is an emerging term in the storage industry, and it is data analytics on big storage, i.e., Cloud-scale storage. In Cloud-scale storage systems, load balancing in request workloads across a metadata server cluster is critical for avoiding performance bottlenecks and improving quality of services. Many good approaches have been proposed for load balancing in distributed storage systems. Some of them pay attention to global namespace balancing, making metadata distribution across metadata servers as uniform as possible. However, they do not work well in skew request distributions, which impair load balancing but simultaneously increase the effectiveness of caching and replication. In this paper, we propose *Cloud Cache* (C^2), an adaptive load balancing scheme for metadata server cluster in Cloud-scale storage systems. It combines adaptive cache diffusion and replication scheme to cope with the request load balancing problem, and it can be integrated into existing distributed metadata management approaches to efficiently improve their load balancing performance. By conducting a performance evaluation in trace-driven simulations, experimental results demonstrate the efficiency and scalability of C^2.

1 Introduction

Big data depends on a key foundation, which is how to make much data durable and dependable, and access to as much data as possible[1]. It brings a challenging problem, which is how to efficiently store all that data on a Cloud scale. Modern Cloud-scale storage systems that store EB-scale (10^{18} or 2^{60} bytes) data [1, 2], separate file data access and metadata transactions to achieve high performance and scalability. EB-scale data is managed with a distributed storage system in a data center to support many computations, e.g., the large synoptic survey telescope[2], in which there are more than 10^{18} files. Data is stored on a storage cluster including numerous servers directly accessed by clients via the network, while metadata is managed separately by a metadata server (MDS) cluster consisting of a few dedicated servers. The dedicated MDS cluster manages the global namespace and the directory hierarchy of file system, the mapping from

[1] http://storiant.com/resources/Storiant-CIO-Survey-Report.pdf
[2] http://www.lsst.org/lsst

H. Handa et al. (eds.), *Proc. of the 18th Asia Pacific Symp. on Intell. & Evol. Systems – Vol. 1*,
Proceedings in Adaptation, Learning and Optimization 1, DOI: 10.1007/978-3-319-13359-1_16

files to objects, and the permissions of files and directories. The MDS cluster just allows for concurrent data transfers between large numbers of clients and storage servers, and it provides efficient metadata service performance with specific workloads, e.g., thousands of clients updating to the same directory or accessing the same file.

Compared to the overall data space, the size of metadata is relatively small, and it is typically 0.1% to 1% of data space[3], but it is relatively large in EB-scale storage systems, e.g., 1PB to 10PB for 1EB data. Besides, 50% to 80% of all file system accesses are to metadata [3]. Therefore, in order to achieve high performance and scalability, a careful MDS cluster architecture must be designed and implemented to avoid potential bottlenecks caused by metadata requests. To efficiently handle the workload generated by a large number of clients, metadata should be properly partitioned so as to evenly distribute metadata traffic by leveraging the MDS cluster efficiently. At the same time, to deal with the changing workload, a scalable metadata management mechanism [4] is necessary to provide highly efficient metadata performance for mixed workloads generated by tens of thousands of concurrent clients. The concurrent accesses from a large number of clients to Cloud-scale distributed storage will cause request load imbalance among metadata servers and inefficient use of metadata cache. Distributed caching is a widely deployed technique to handle request load imbalance and reduce request latency, and it is both orthogonal and complementary to the load balancing technique proposed in [4]. Meanwhile, distributed replication is also able to decrease the retrieve latency of metadata items. There are two insights based on our experience: 1) replicas on cached metadata items can balance request workload, and 2) increasing the number of replicas does help handle bursts of workloads.

In this paper, we propose an adaptive load balancing approach named C^2 to solve the above problems. We consider how to find an efficient caching and replication scheme, which automatically adapts to changing workload in EB-scale storage systems. By analyzing a running workload of requests to metadata items, it calculates a new load-balancing plan and then migrates them when their request rates are more than the request capacity of the node that maintains them. The input to our migration plan consists of an initial state of metadata items in virtual nodes and a given requirement of load balancing, and their request capacities. Our goal is to find a migration plan that moves the metadata from the initial state to the final state of load balancing with the minimum rounds. Moreover, the overlay network topology and metadata access information are utilized for metadata replication decisions.

The rest of the paper is organized as follows. Section 2 describes the problem definition. The adaptive cache diffusion mechanism is presented in Section 3. Section 4 introduces the adaptive replication scheme in C^2. In Section 5 we present performance evaluation results of C^2. Section 6 describes related work. In Section 7 we conclude this paper.

[3] http://dcslab.hanyang.ac.kr/nvramos08/EthanMiller.pdf

2 Problem Definition

2.1 Traces Analyzed

There are three real traces we analyze as shown in Table 1. *Microsoft* means Microsoft Windows build server production traces [5] from BuildServer00 to BuildServer07 within 24 hours, and its data size is 223.7GB (including access pattern information). *Harvard* is a research and email NFS trace used by a large Harvard research group [6], and its data size is 158.6GB (including access pattern information). We implemented a metadata crawler that performs a recursive walk of the file system using *stat()* to extract file/directory metadata. By using the metadata crawler, the *Linux* trace is fetched from 22 Linux servers in our data center, and it is different from and much bigger than the *Linux* trace in [7]. Its file system metadata size is 4.53GB, and data size is 3.05TB.

Table 1. Traces

Trace	# of files	Path metadata	Max. length
Harvard	7,936,109	176M	18
Microsoft	7,725,928	416M	34
Linux	10,271,066	786M	21

2.2 Load Balancing

Distributed metadata server cluster must guarantee good load balancing in such a way that they can meet their throughput and latency goals, and both partitioning and replication can be combined to make them scalable. They have to balance two kinds of loads: 1) storage load and 2) request load. The storage load is static for requiring constant storage capacity in each node. Capacity is typically load-balanced by using a hashing-based approach [8]. The request load is dynamic for handling queries from users. Metadata should be distributed as uniformly as possible among nodes, and no node should cope with much more query requests than another node. Although some schemes can balance the utilization of storage space, they do not balance the request load, in which hot spots often occur, i.e., some items are requested more than others. Many real-world workloads have uneven request distributions. Distributed systems typically balance the request load with the following ways. Some systems dynamically move data from overloaded servers to underloaded servers to make the request load uniform. Others rely upon replication to direct queries to the underloaded ones with a number of replicas, substantially improving load balancing [9].

2.3 DROP with Caching

DROP [7] leverages pathname-based locality-preserving hashing (*LpH*) for metadata distribution and location, avoiding the overhead of hierarchical directory traversal. To access data, a client hashes the pathname of the file with the same *LpH* function to locate which MDS contains the metadata of the file, and then contacts the appropriate

MDS. The process is extremely efficient metadata access, typically involving a single message to a single MDS. With losing negligible metadata locality, DROP uses an efficient histogram-based dynamic load balancing mechanism to balance storage load. We can leverage the namespace locality in keys by caching metadata items within the same domain in lookup results, reducing total metadata lookup traffic. DROP maintains namespace locality in metadata placement, so clients do not need to require data from many nodes, and repetitive lookups are avoided because of the lookup cache mechanism. Large amounts of localities exist in distributed systems, e.g., file access locality in P2P systems [10], and they are the basis for distributed caching techniques.

Metadata server architecture is shown in Figure 1. DROP is a SSD/NVM-based [11] key-value store, where *key* is pathname, and *value* is its *inode* information. C^2 is used to deal with request load balancing, in which the lookup cache stores metadata items in recent query results, so future query requests that access keys in cached key ranges entirely bypass the lookup step. Clients could also explore a lookup cache in DirHash or FileHash that is to randomly distribute directories or files according to their pathnames, each of which is assigned to a metadata server, but it would be less effective since future queries may not request keys in recently accessed key ranges. Cache entries may become stale because of potential dynamic system membership. DROP falls back to a normal lookup when a metadata item is not found. It does not affect correctness with a stale cache entry, but it impairs retrieval latency.

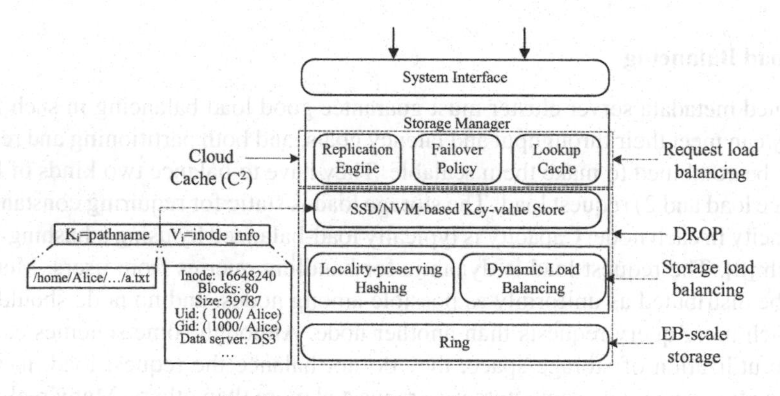

Fig. 1. Metadata Server Architecture

When a file/directory is updated, DROP is responsible to insert its metadata's new versions along the entire path to the root. It makes sure that each read must have a consistent view of the metadata, and it implies that each write must update all the metadata along the full path. When writing temporary files, DROP avoids this overhead with a t-second write-back cache, which is also explored as a buffer. Due to this buffer, multiple reads of the same metadata occurring within a t-second window only require it to be retrieved once. Metadata items seen by clients may be stale by up to t seconds because of this cache, but incomplete writes will never been seen.

2.4 Problem Formulation

Given a set of nodes $\mathbb{S}$ ($\mathbb{S} = \{S_i, i = 1, \cdots, n\}$), with each storing a subset of metadata items $\mathbb{D}$ ($\mathbb{D} = \{D_j, j = 1, \cdots, m\}$) and a specified set of move operations, each of which specifies which item needs to be moved from one node to another one. A question we face and address is how to schedule these move operations. For each metadata item d, there is a subset of source MDSs S_d and target MDSs T_d. In the beginning, only the MDSs in S_d have metadata item d, and all the MDSs in T_d want to receive it. A MDS in T_d becomes a source of item d after it receives item d. Our goal is to find a metadata migration plan using the minimum number of rounds, where there is a constraint that is each MDS just takes part in the transfer of only one item either as a sender or receiver. It is a NP-hard problem [12].

There are a set of nodes $\mathbb{S}$ and a set of metadata items $\mathbb{D}$. Initially, each MDS stores a subset of items. A transfer graph $G = (V, E)$ is built, in which each node represents a virtual node and an edge $e = (u, v)$ represents a metadata item to be moved from a node u to v. Over time, metadata items may be moved to another MDS for load balancing. Note that the transfer graph can be a multi-graph, in which there are multiple edges between two nodes, when multiple metadata items are moved from one node to another.

There are two situations: 1) the request load of an item is smaller than a given request load threshold l_t for a node, and 2) the request load of an item is bigger than l_t for a node. The *Microsoft* trace shows that the hottest file is accessed over 2.5% of total requests and the the combined CDF of hottest 125 files is close to 90% [5]. It tells that the hottest file is much more popular than one of other files that are not in the hottest 125 files. Suppose that there are 20 metadata servers, each of which has five virtual nodes, and there are 100 virtual nodes in total. Any virtual node that maintains the hottest file will be overloaded. For the first situation, we can use adaptive cache diffusion discussed in Section 3, while for the second one, we can use adaptive replication scheme described in Section 4.

3 Adaptive Cache Diffusion

We first present an adaptive cache diffusion approach that leads to low migration overhead and fast convergence. Load-stealing and load-shedding are used to achieve this goal. Cache space is used for retrieval operations of DROP, in which a cached metadata item is placed at a virtual node to accelerate subsequent retrievals. It might be replaced via LRU very soon after it is created.

3.1 System Model

A physical metadata server might have a set of virtual nodes $\mathbb{N} = \{n_1, n_2, \cdots, n_d\}$ with a set of loads $\mathbb{L} = \{l_1, l_2, \cdots, l_d\}$. Load is applied to metadata servers via their virtual nodes, i.e., metadata server S might have load $L_S = \sum_i^d l_i$. A MDS is said to be load-balanced when it satisfy Definition 1, i.e., the largest load is less than t^2 times the smallest load in the DROP system. According to Definition 1, a MDS has an upper target L_u ($L_u = t \times \overline{L}$) and a lower target L_l ($L_l = 1/t \times \overline{L}$). If a MDS finds itself receiving more load than L_u, it considers itself overloaded. Otherwise, it considers itself underloaded if it finds itself receiving less load than L_l. MDSs may want to operate below their capacities to prevent variations in workload from temporary overload.

Definition 1 (**MDS_i is load balancing**). *MDS_i is load balancing if its load satisfies $1/t \leq L_i/\overline{L} \leq t\,(t \leq 2)$.*

File popularity [6] follow Zipf request distributions. The Zipf property of file access patterns is a basic fact of nature. It states that a small number of objects are greatly popular, but there is a long tail of unpopular requests. A Zipf workload means that destinations are ranked by popularity. The Zipf law states that the popularity of the ith-most popular object is proportional to $i^{-\alpha}$, in which α is the Zipf coefficient. Usually, Zipf distributions look linear when plotted on a log-log scale. Figure 2 shows the popularity distribution of file/directory metadata items in the *Microsoft* and *Harvard* traces. Like the Internet, the metadata request distribution as observed in both traces also follows Zipf distributions.

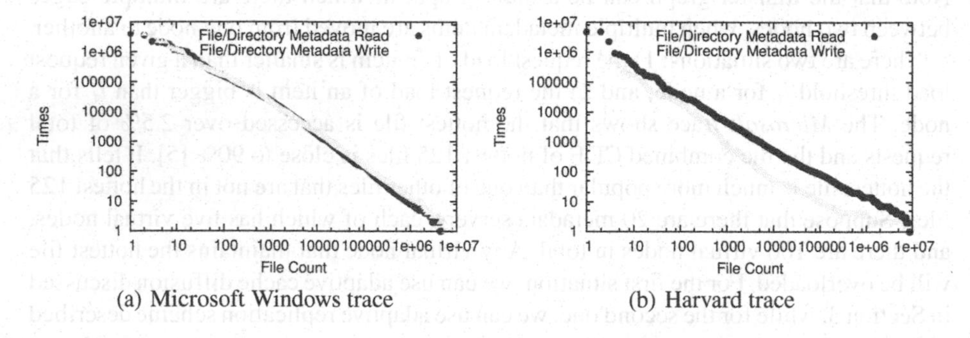

(a) Microsoft Windows trace (b) Harvard trace

Fig. 2. Read and write distribution

3.2 Load Shedding

Load-shedding means that an overloaded node attempts to offload requests to one or more underloaded ones. It may be well suited to the DROP MDS cluster. An overloaded node n_1 has to transfer an item x to another node n_2, and simultaneously create a *redirection pointer* to n_2. The item x also could be replicated at n_2, increasing redundancy and allowing n_1 to control how much load would be shed. In Section 4, we will explain how to effectively place multiple replicas using a multiple-choice scheme.

There are m metadata items in a node, with a tuple of loads $\langle l_1, l_2, \cdots, l_m\rangle$ and a tuple of probabilities $\langle p_1, p_2, \cdots, p_m\rangle$. When this node has a cache of size $c > 0$, the c most frequently requested items will all hit the cache of this node, with two tuples of positive numbers $\langle l_1, l_2, \cdots, l_c\rangle$ and $\langle p_1, p_2, \cdots, p_c\rangle$ respectively. Let L be $t \times \overline{L}$, and this node is overloaded if $\sum_i^c l_i > L$. Therefore, it can be formulated as a 0-1 Knapsack

Problem that is NP-hard, i.e., it is to determine how to reassign some items to other nodes in a way that minimizes metadata migration from this node as follows:

$$\text{maximize} \quad z = \sum_{i=1}^{c} p_i x_i \tag{1a}$$

$$\text{s.t.} \quad \sum_{i=1}^{c} l_i x_i \le L \tag{1b}$$

$$x_i \in \{0,1\}, i \in \{1,2,\cdots,c\} \tag{1c}$$

Constraint (1b) ensures that the total load of metadata items kept in this node is less than L. Constraint (1c) states if an item x_i is kept or not.

3.3 Load Stealing

Load-stealing states that a underloaded node n_1 seeks out load to take from one or more overloaded nodes. The load-stealing node finds such a node n_2, and it makes a replica of an item x in the node n_2, which creates a redirection pointer to n_1 for the item x. A natural idea is to have n_1 attempt to steal metadata items, for which n_1 has a redirection pointer. A metadata item can be placed using multiple choices, and it is associated with one of its r hash locations, which is further explained in Section 4.

There are a number of metadata items from previous nodes with two tuples of positive numbers $\langle l'_1, l'_2, \cdots, l'_{c'} \rangle$ and $\langle p'_1, p'_2, \cdots, p'_{c'} \rangle$ respectively. If $\sum_i^c l_i < L$, this node is in load balancing, and it can take some items with its cache space of $L - L_e$. Therefore, it also can be formulated as a 0-1 Knapsack Problem, i.e., it is to determine how to take some items from other overloaded nodes in a way that maximizes the cache utilization of this node as follows:

$$\text{maximize} \quad z = \sum_{i=1}^{c'} p'_i x_i \tag{2a}$$

$$\text{s.t.} \quad \sum_{i=1}^{c'} l'_i x_i \le L - L_e \tag{2b}$$

$$x_i \in \{0,1\}, i \in \{1,2,\cdots,c'\} \tag{2c}$$

where

$$L_e = \sum_{i=1}^{c} l_i \tag{3}$$

3.4 Traffic Control

During load balancing, a metadata item may be migrated multiple times. DROP uses metadata pointers to minimize metadata migration overhead. For a metadata pointer, a node retrieves the metadata when it has held the pointer for longer than the stabilization time of the pointer. Using metadata pointers only temporarily hurts metadata locality when balancing the load. Besides reducing load balancing overhead, pointers also can make writes succeed even when the target node is at capacity, pointers can be utilized to divert metadata items from heavy nodes to light nodes. However, the node at full capacity will eventually shed some load when balancing the load, just causing temporary additional indirection. Suppose that a node X is heavily loaded, and a node Y takes some items of X to reduce some of X's load. Now X must transfer some of its metadata items to Y. Instead of having X immediately shed some of its metadata items to Y when Y gets some items from X, Y initially maintains metadata pointers to X. Later Y transfers the pointers to Z, and Z ultimately retrieves the actual metadata from X and deletes the pointers.

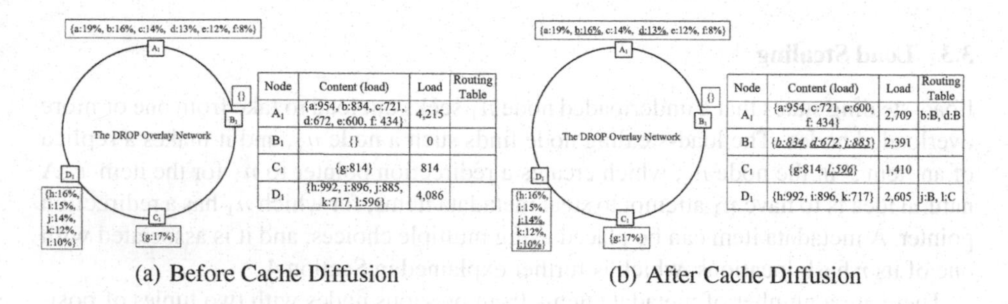

(a) Before Cache Diffusion (b) After Cache Diffusion

Fig. 3. Cache Diffusion. Each metadata server has only one virtual node for illustration.

There is an example of cache diffusion as shown in Figure 3, in which there is load imbalance before cache diffusion as shown in Figure 3(a). There are four nodes, where A_1 and D_1 are overloaded, while B_1 and C_1 are underloaded. After running our cache diffusion approach, we can see there is a good load balancing as shown in Figure 3(b), where the loads of four nodes are all in $[1/t \times \overline{L}, t \times \overline{L}]$, i.e., $[1139.375, 4457.5]$ ($t = 2$). The migrated items are found via the routing tables of the nodes that are responsible for the items.

4 Adaptive Replication Scheme

Replication distribution is a key technique to balance request load [13]. We propose a novel metadata replication mechanism to further balance request workload by placing multiple replicas of popular metadata items in different nodes. In DROP, a ZooKeeper-based linearizable consistency mechanism is proposed in [4] to keep excellent metadata replica consistency among MDSs.

4.1 Random Node Selection

We first present an effective random node selection strategy to achieve coarse load balancing. Let h_0 denote a hash function that maps virtual nodes onto the ring, and $\mathbb{H}$ ($\mathbb{H} =$

$\{h_1, h_2, \cdots, h_k\}$) denote a set of hash functions mapping metadata items onto the ring. The number of replicas r is calculated by $r = \lceil \frac{f}{\theta} \rceil$, where f is the access frequency of a metadata item, and θ is a given threshold. An item x is inserted as a primary replica using the h_0 hash function, and its $r-1$ replicas should be placed in the nodes selected by using k hash functions. Lookups are initiated to find the nodes associated with each of these k hash values by calculating $h_1, h_2, \cdots, h_k$. According to the mapping given by h_0, k lookups can be executed in parallel to find the virtual nodes $n_1, n_2, \cdots, n_k$ in charge of these hash values. After querying the loads of nodes, the underloaded nodes are chosen.

To decrease the overhead of searching for additional nodes, *redirection pointers* are explored. In addition to storing replicas at the $r-1$ underloaded nodes $\mathbb{N}_u^{r-1}$, other nodes $\{\mathbb{S} - \mathbb{N}_u^{r-1}\}$ store a pointer $x \rightarrow \mathbb{N}_u^{r-1}$. To search for the item x, a single query is performed by choosing a hash function h_j at random in an effort to locate one of nodes in $\mathbb{N}_u^{r-1}$. If n_j does not have x, n_j forwards the query request with a pointer $x \rightarrow \mathbb{N}_u^{r-1}$. Query requests take at most one more step. The extra step is necessary with probability $(k-r+1)/k$ if h_j is chosen uniformly at random from the k choices. It incurs the overhead of maintaining the additional pointers, but storing actual items and any associated computation dominate the stored pointers. In addition, we need to answer how to select $r-1$ nodes from $\mathbb{N}_u$ to place x's replicas.

4.2 Topology-Aware Replica Placement

To select the $r-1$ underloaded nodes $\mathbb{N}_u^{r-1}$, we first need to consider the network topological characteristics of nodes so that we place the replicas of an item on the nodes topologically adjacent to the node in charge of the item in DROP. In this way, we can reduce the network bandwidth consumption and the query latency when achieving better load balancing. We employ an effective topology-aware replica placement scheme by introducing a technique that discovers the topological information of nodes. In this technique, the key is how to represent and keep the network topology information so that the topologically close nodes are easily discovered for a given node. Thus, we must have a mechanism that is able to represent the topological location information of nodes. The distributed binning scheme [14] is a simple approach for this purpose.

For example, there is a topology table for node 700, as shown in Figure 4. The landmark node ordering information is employed as part of the node identification information. There are three landmark nodes L_1, L_2 and L_3 that are used, and the link latencies from node 700 to the three landmark nodes are within [0,20), [20,80) and greater than 80ms. The nodes with the same or similar ordering information are topologically close, e.g., node 700:011 is topological closer to node 206:012 than node 124:212. It means the link latency to the node 206:012 is much smaller than to the node 124:212. For each entry in the table, the first item is the order information, and the second one consists of several records, each of which includes the node ID and its workload. For an entry with a tuple $[o, (id, w), \cdots]$, it represents that workloads came from the nodes with the common order information o in the past given period. To choose a node with adequate workload capacity to store a replica, node n_1 in charge of a popular item has to contact those nodes by sending a message. The nodes being selected to store the replicas reply

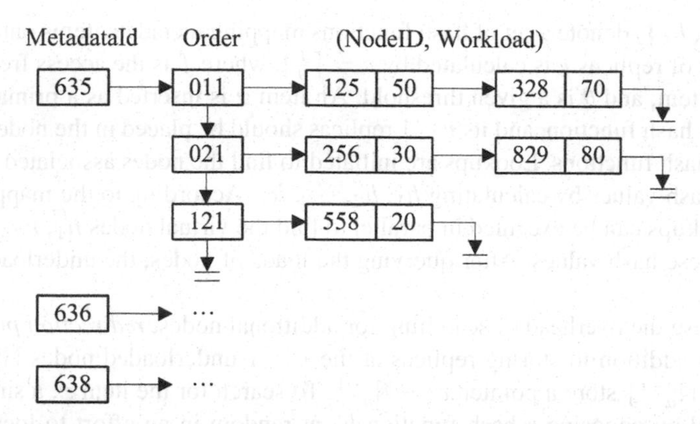

Fig. 4. A sample topology table on node 700:011. When there are three replicas of MetadataId 635 with access times 45 to be placed in DROP, the nodes will be 125, 256 and 558.

with their order information and estimated workload. Meanwhile, direct links to the replicas are created on node n_1.

4.3 Directory-Based Replica Diffusion

We present an efficient directory-based replica diffusion technique in DROP. A replica, as a copy of cached metadata item, is placed in the DROP overlay by its insertion operation. DROP stores directories of pointers to metadata item replicas that are stored in virtual nodes, but their locations are not related to the structure of locality-preserving hashing. When a node has a metadata item whose request times exceed a given threshold, it creates a directory for the item, chooses $r-1$ virtual nodes with the topology-aware replica replacement scheme and stores the item replicas at the nodes, recording them in its directory. When the directory receives a request for the item, it returns directory entries pointing to individual replicas of the item with a single response message. The directory node monitors the request rate for the item to determine if a new replica is created. When the request rate reaches a given threshold, the directory node creates a new replica along with the list of pointers to replicas of the item.

In chain-based replica diffusion, the r replicas of an item are placed on its primary node and its $r-1$ followers. In both replica diffusion techniques, a node has to serve a request if it holds a replica of the requested metadata item. In the chain-based replica diffusion, a node pushes out a replica of the item one overlay hop closer to the source node of the last request if the request rate has exceeded its capability. It also aims to offload some of the demand to more nodes that serve requests. Compared to the chain-based replica diffusion, the directory-based one has three advantages: 1) faster replica transmission speed, 2) higher query parallelism, and 3) better load balancing because of $r-1$ nodes chosen with k random hash functions.

5 Performance Evaluation

In this section, we evaluate the performance of C^2 using one synthetic workload-based simulation and two detailed trace-driven simulations. We have developed a detailed event-driven simulator to validate and evaluate our design decisions and choices. In the first part, we empirically evaluate the convergence rate of C^2. We second measure the metadata migration overhead of C^2. Lastly, we measure the scalability of our adaptive replication scheme.

We define *Load Factor* as follows: $LoadFactor = \frac{Max.Load}{Min.Load}$. Each MDS has five virtual nodes, and the *Linux* trace follows the Zipf distribution with $\alpha = 1.2$. All simulation experiments are conducted on a Linux Server with four Dual-Core AMD Opteron(TM) 2.6GHz processors and 8.0GB of RAM, running 64-bit Ubuntu 12.04. All the experiments are repeated three times, and average results are reported.

5.1 Convergence Rate

In this section, we measure convergence rate using the three approaches with C^2 on the three traces. Convergence rate is critically significant in distributed systems. It includes two metrics: 1) number of rounds that measures how many rounds should be taken to reach load balancing, and 2) time cost that means how fast to achieve load balancing. Figure 5 depicts the number of rounds on the three traces in the cluster of metadata servers when using all the three methods with C^2. We can see that there are at most five rounds to converge to load balancing on the *Linux* trace, and there are at most four rounds to converge to load balancing on both the *Microsoft* trace and the *Harvard* trace. This is because there are much more access frequencies with more metadata items in the *Linux* trace than in both the *Microsoft* and *Harvard* traces, so the system is harder to reach load balancing when using the *Linux* trace than the other two traces. As shown in Figure 5(b) and Figure 5(c), the system is in a state of load balancing before running DirHash with C^2 or FileHash with C^2 in the cluster of ten MDSs.

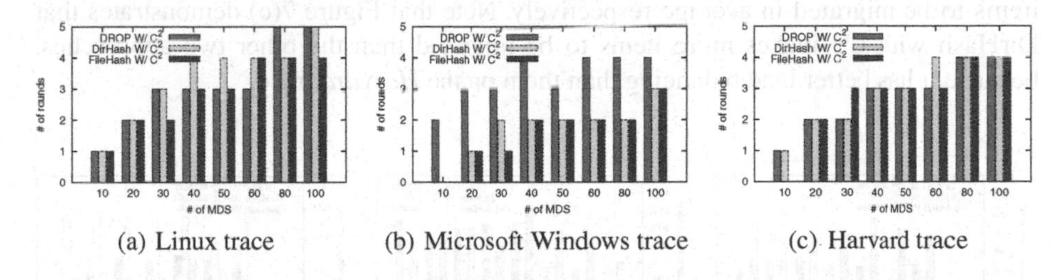

(a) Linux trace (b) Microsoft Windows trace (c) Harvard trace

Fig. 5. Number of Rounds with Varying the Number of Metadata Servers

Figure 6 shows how long it takes using all the three approaches with C^2 to reach load balancing. Figure 6(b) illustrates that DROP with C^2 has much longer time cost than the other two approaches because the *Microsoft* trace has only three first-level directories, and it has more obvious locality than the other two traces. Figure 6(c) shows that DROP with C^2 is close to the other two approaches in time cost because the *Harvard* trace has the most first-level directories among the three traces, and it is the worst in locality among them. Figure 6(a) demonstrates that DROP with C^2 is somewhat longer in time cost than the other two approaches. This is because the *Linux* trace has more first-level

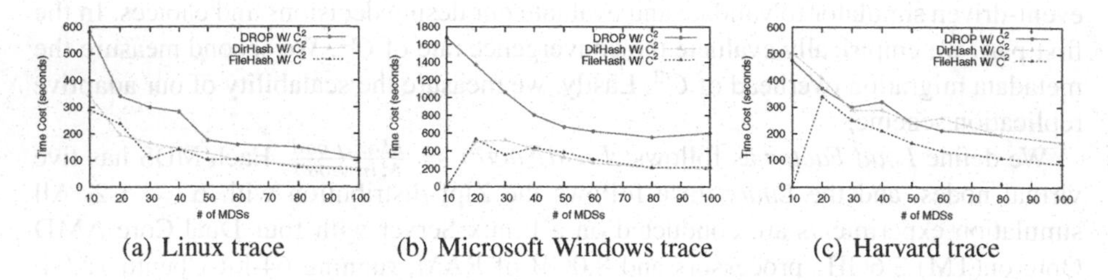

(a) Linux trace (b) Microsoft Windows trace (c) Harvard trace

Fig. 6. Time Cost with Varying the Number of Metadata Servers

directories than the *Microsoft* trace, and much fewer first-level ones than the *Harvard* trace, and it has worse than *Microsoft* and better than *Harvard* in locality. Figure 5 and 6 illustrate that the deployed techniques have excellent efficiency.

5.2 Migration Overhead

As file and directory metadata items are accessed more/less frequently, the request workload distribution in the system changes, and the system may have to migrate cached metadata to maintain request load balancing. Figure 7 shows the metadata migration overhead with excellent scalability. We perform this experiment as follows. Due to skew query requests, all the MDSs in the DROP system are not in a satisfactory load balancing state at the beginning. Metadata items in the *Microsoft* and *Harvard* traces are accessed according to their real-world history access information, while those in the *Linux* trace are requested according to the Zipf-like distribution. Figure 7(a) shows that the three methods cause 24.67%, 17.23% and 16.10% of items to be migrated in average respectively, and Figure 7(b) illustrates that they cause 27.68%, 21.13% and 23.59% of items to be migrated in average respectively. Note that Figure 7(c) demonstrates that DirHash with C^2 makes more items to be migrated than the other two approaches, because it has better load balancing than them on the *Harvard* trace.

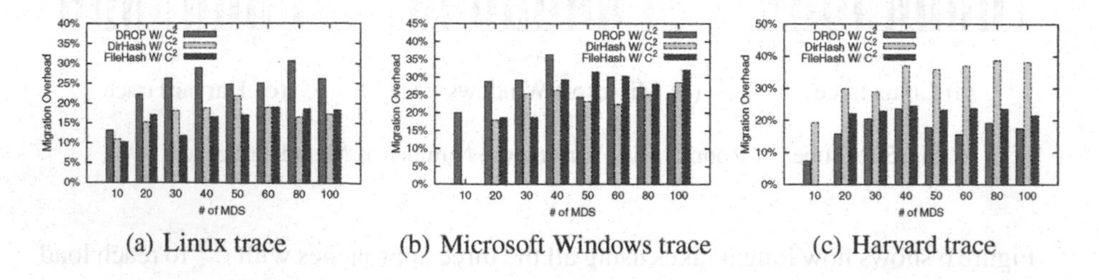

(a) Linux trace (b) Microsoft Windows trace (c) Harvard trace

Fig. 7. Migration Overhead with Varying the Number of Metadata Servers

5.3 Replication Overhead

In this section, we present how many replicas are necessary for metadata items requested much heavily to keep good load balancing. Note that we do not count a main replica. Figure 8 shows that our adaptive replication scheme has excellent scalability with different numbers of metadata servers. When running our adaptive replication scheme on the *Microsoft* trace, the number of replicas varies slightly as the MDS cluster size increases, where its maximum value is 2.96, and its minimum value is 1.0. When running the scheme on the other two traces, the number of replicas rises somewhat more obviously than that on the *Microsoft* trace as the MDS cluster size increases, but the scalability is still excellent on the two traces. The maximum numbers of replicas are 7.6 and 6.68 respectively, while the minimum numbers of replicas are 3.8 and 3.17 respectively on the *Harvard* trace and the *Linux* trace.

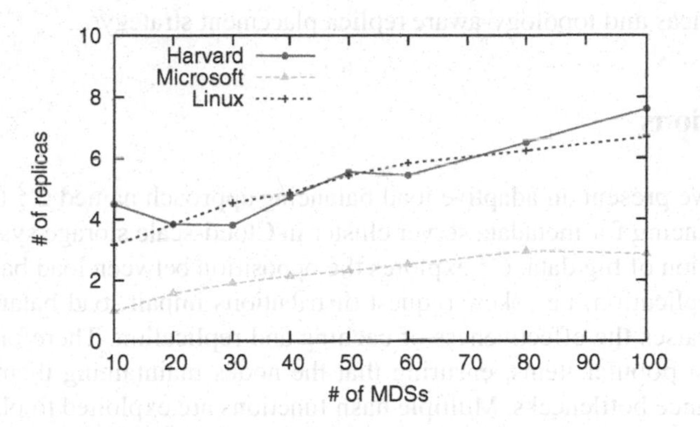

Fig. 8. Replicas for metadata items accessed frequently

6 Related Work

Cloud-scale storage is critical to Cloud-scale data backup [15], and even Cloud-scale data management [16]. In recent years, many load balancing schemes were proposed in distributed metadata organization and management for Cloud-scale storage systems.

Online Migration. The "virtual node" approach [8] was proposed to cope with the imbalance of the key distribution due to hash function. A number of virtual nodes are generated with random IDs in a physical server, so reducing the load imbalance. However, the usage of virtual nodes enormously boosts the amount of routing metadata in each server, therefore causing more maintenance overhead and increasing the number of hops per lookup. In addition, it does not take item popularity into account. On the contrary, the "dynamic ID" approach explores only a single ID per server [17]. The load of a server can be adjusted with a more suitable ID in the namespace. However, the solution requires IDs to be reassigned to maintain load balancing, resulting in a high overhead because of transferring items and updating overlay links. Our motivation for studying the online migration problem lies in how to efficiently migrate metadata for MDS cluster in Cloud-scale storage systems.

Caching and Replication. Hot spots are handled with caches to store popular items in the network, and query requests are considered to be resolved whenever cache hits occur along the entire path. Solutions addressing the uneven popularity of objects are based on caching and replication. *Path replication* replicates objects on all nodes along the full lookup path, e.g., DHash [18] replicates objects in k successors with caching on the lookup path. In the k-choice [9] load balancing approach, multiple hashes are employed to generate a set of IDs in a node, and one of the IDs is chosen at join time to minimize the differences between capacity and load for itself and other nodes affected by its join time. Unfortunately, the last several hops of a lookup are precisely the ones that can least be optimized [19]. Furthermore, a fixed number of replicas do not work well since the request load is dynamic: resources may be wasted if the number is set too high, while the replicas may not be enough to support a high request load if it is set too low. Our replication-based solution is similar to the k-choice approach, with a flexible number of replicas and topology-aware replica placement strategy.

7 Conclusions

In this paper, we present an adaptive load balancing approach named C^2 to handle request load balancing for metadata server cluster in Cloud-scale storage systems, which are the foundation of big data. C^2 explores the opposition between load balancing, and caching and replication, i.e., skew request distributions impair load balancing but simultaneously raises the effectiveness of caching and replication. Therefore, the cache serves the most popular items, ensuring that the nodes maintaining them do not become performance bottlenecks. Multiple hash functions are exploited to place multiple replicas, so balancing the load caused by most frequently accessed items. Our approach enables the system good load balancing even when query request workload is heavily skewed. Extensive simulation results show significant improvements in maintaining a more balanced distributed metadata management system, leading to the improved system with excellent scalability and performance.

Acknowledgement. The authors would like to thank Garth Gibson from Carnegie Mellon University and Jun Wang from University of Central Florida for their help. This work is supported by A*STAR Thematic Strategic Research Programme (TSRP) Grant No. 1121720013.

References

1. Raicu, I., Foster, I.T., Beckman, P.: Making a case for distributed file systems at Exascale. In: LSAP, pp. 11–18 (2011)
2. Amer, A., Long, D., Schwarz, T.: Reliability Challenges for Storing Exabytes. In: International Conference on Computing, Networking and Communications (ICNC), CNC Workshop (2014)
3. Ousterhout, J.K., Costa, H.D., Harrison, D., Kunze, J.A., Kupfer, M.D., Thompson, J.G.: A trace-driven analysis of the UNIX 4.2 BSD file system. In: SOSP, pp. 15–24 (1985)
4. Xu, Q., Arumugam, R., Yong, K.L., Mahadevan, S.: Efficient and Scalable Metadata Management in EB-scale File Systems. IEEE Transactions on Parallel and Distributed Systems 99, 1 (2013) (PrePrints)

5. Kavalanekar, S., Worthington, B.L., Zhang, Q., Sharda, V.: Characterization of storage workload traces from production windows servers. In: Christie, D., Lee, A., Mutlu, O., Zorn, B.G. (eds.) 4th International Symposium on Workload Characterization (IISWC 2008, September 14-16, pp. 119–128. IEEE, Seattle (2008)
6. Ellard, D., Ledlie, J., Malkani, P., Seltzer, M.I.: Passive NFS tracing of email and research workloads. In: Chase, J. (ed.) Proceedings of the FAST 2003 Conference on File and Storage Technologies, March 31 - April 2. USENIX, Cathedral Hill Hotel (2003)
7. Xu, Q., Arumugam, R.V., Yang, K.L., Mahadevan, S.: DROP: Facilitating distributed metadata management in EB-scale storage systems. In: IEEE 29th Symposium on Mass Storage Systems and Technologies, MSST 2013, Long Beach, CA, USA, May 6-10, pp. 1–10 (2013)
8. Stoica, I., Morris, R., Karger, D.R., Kaashoek, M.F., Balakrishnan, H.: Chord: A scalable peer-to-peer lookup service for internet applications. In: SIGCOMM, pp. 149–160 (2001)
9. Ledlie, J., Seltzer, M.I.: Distributed, secure load balancing with skew, heterogeneity and churn. In: 24th Annual Joint Conference of the IEEE Computer and Communications Societies, INFOCOM 2005, March 13-17, pp. 1419–1430. IEEE, Miami (2005)
10. Gummadi, P.K., Dunn, R.J., Saroiu, S., Gribble, S.D., Levy, H.M., Zahorjan, J.: Measurement, modeling, and analysis of a peer-to-peer file-sharing workload. In: Scott, M.L., Peterson, L.L. (eds.) Proceedings of the 19th ACM Symposium on Operating Systems Principles 2003, SOSP 2003, October 19-22, pp. 314–329. ACM, Bolton Landing (2003)
11. Cai, Q., Arumugam, R.V., Xu, Q., He, B.: Understanding the Behavior of Solid State Disk. In: The 18th Asia Pacific Symposium on Intelligent and Evolutionary Systems (to appear, 2014)
12. Khuller, S., Kim, Y.A., Wan, Y.J.: Algorithms for data migration with cloning. In: Neven, F., Beeri, C., Milo, T. (eds.) Proceedings of the Twenty-Second ACM SIGACT-SIGMOD-SIGART Symposium on Principles of Database Systems, June 9-12, pp. 27–36. ACM, San Diego (2003)
13. Xu, Q., Shen, H.T., Cui, B., Hou, X., Dai, Y.: A novel content distribution mechanism in DHT networks. In: Fratta, L., Schulzrinne, H., Takahashi, Y., Spaniol, O. (eds.) NETWORKING 2009. LNCS, vol. 5550, pp. 742–755. Springer, Heidelberg (2009)
14. Ratnasamy, S., Handley, M., Karp, R.M., Shenker, S.: Topologically-aware overlay construction and server selection. In: INFOCOM (2002)
15. Xu, Q., Zhao, L., Xiao, M., Liu, A., Dai, Y.: YuruBackup: A Space-Efficient and Highly Scalable Incremental Backup System in the Cloud. International Journal of Parallel Programming, 1–23 (2013)
16. Cao, Y., Chen, C., Guo, F., Jiang, D., Lin, Y., Ooi, B.C., Vo, H.T., Wu, S., Xu, Q.: Es^2: A cloud data storage system for supporting both OLTP and OLAP. In: Proceedings of the 27th International Conference on Data Engineering, ICDE 2011, April 11-16, pp. 291–302 (2011)
17. Naor, M., Wieder, U.: Novel architectures for P2P applications: The continuous-discrete approach. ACM Transactions on Algorithms 3(3) (2007)
18. Dabek, F., Kaashoek, M.F., Karger, D.R., Morris, R., Stoica, I.: Wide-Area Cooperative Storage with CFS. In: SOSP, pp. 202–215 (2001)
19. Gopalakrishnan, V., Silaghi, B.D., Bhattacharjee, B., Keleher, P.J.: Adaptive replication in peer-to-peer systems. In: 24th International Conference on Distributed Computing Systems (ICDCS 2004), March 24-26, pp. 360–369. IEEE Computer Society, Hachioji (2004)

A Multi-agent Simulation Framework to Support Agent Interactions under Different Domains

Moath Jarrah[1], Bernard P. Zeigler[2], Chi Xu[3], and Jie Zhang[1]

[1] School of Computer Engineering, Nanyang Technological University, Singapore
[2] RTSync Corporation, Arizona, USA
[3] Singapore Institute of Manufacturing Technology, 71 Nanyang Drive, Singapore
{hmoath,zhangj}@ntu.edu.sg, berniezeigler@rtsync.com, cxu@simtech.a-star.edu.sg

Abstract. The ability to study complex systems has become feasible with the new intensive computing resources such as GPU, multi-core, clusters, and Cloud infrastructures. Many companies and scientific applications use multi-agent modeling and simulation platforms to study complex processes where analytical approach is not feasible. In this paper, we use two negotiation protocols to generalize the interaction behaviors between agents in multi-agent environments. The negotiation protocols are enforced by a domain-independent marketplace agent. In order to provide the agents with flexible language structure, a domain-dependent ontology is used. The integration of the domain-independent marketplace with the domain-dependent language ontology is accomplished through an automatic code generation tool. The tool simplifies deploying the framework for a specific domain of interest. Our methodology is implemented in FD-DEVS simulation environment and SES ontological framework.

Keywords: Multi-agent modeling and simulation, business process, negotiation, ontology, automate, FD-DEVS, SES.

1 Introduction

Agent based simulation platforms have been in research in the last decades with the promising that they will give understandings to natural phenomena. However, the computing power was not feasible to tackle the ever-increasing complexity of systems. With the current development in high performance clusters, cloud environment, multi-core, and graphical processing unit, scientists and decision makers are even more ambitious that multi-agent modeling and simulation field will help them study complex systems such as in social sciences, command and control, business processes, chemical reactions, forecast, and many others. Simulating and understanding a system helps decision makers to be proactive and not reactive. Also, it helps companies to make proper decisions to increase their profit and lower their risks. A system consists of many components and each component has its own behavior and interacts with other components in the system. A natural modeling for systems is to represent each component as

H. Handa et al. (eds.), *Proc. of the 18th Asia Pacific Symp. on Intell. & Evol. Systems – Vol. 1*,
Proceedings in Adaptation, Learning and Optimization 1, DOI: 10.1007/978-3-319-13359-1_17

an agent. Agents interact according to rules of interactions. Some agents are intelligent and go through evolution as in the artificial life proposed by Epstein and Axtell in their famous Sugarscape model of artificial societies [6]. Multi-agent based modeling and simulation is needed because systems are becoming more and more complex in their heterogeneity, intelligence, and interactions. In this regard, SciDAC and Argonne lab are collaborating to build a high end framework that demands huge processing power (where IBM Blue gene is to be used) to tackle the most challenging scientific phenomena in the universe through simulation, visualization, and analysis [12] [13] [18].

Researchers in the domain of industry have been using agent based modeling and simulating for their business processes in order to achieve better performance. One application of a business process is in stock markets. In this application, investors try to buy stocks at low prices and sell them when the stock price is high. Investors try to maximize their profits. Hence, taking risks and studying historical data profiles are very useful in making predictions. Examples of such models as in [2] and [16], where heterogeneous agents buy, sell, and hold stocks and bonds.

Supply chain is another suitable domain for applying the multi-agent simulation framework where a network of retailers, distributors, factories or warehouses, and suppliers interact to achieve profits. Every entity is represented by an agent in the simulation and the output results are used to analyze the supply chain performance to better plan and predict future rules. Many industrial companies have focused on this domain of research such as the work by Boeing on automation of their supply chain interactions[11].

The ability to manage and utilize software and/or hardware products and services in current complex distributed system has become increasingly difficult. The complexity results from the fact that there are many aspects and factors that represent the characteristics of these systems. Many attempts exist in literature where solutions were proposed to exploit these resources such as in [4] and [5]. However, the development of methods to experiment and study complex systems is still far from being sophisticated.

Complex systems are dynamic and seldom static in the sense that new components emerge and some components disappear. An example on complex systems can be the Web services. Web Services developments are growing dramatically and millions of resources are being added every day to the World Wide Web. The success in e-commerce, e-learning, online auctions, online marketplaces, information discovery and retrieval has encouraged more and more companies to provide Web Services either to satisfy customer requirements or to manage their distributed computing resources. Hence, a natural modeling and simulation solution is to map a complex system into a multi-agent simulation environment. More on multi-agent design issues and challenges can be found in [19]. Manual software management is not feasible in such dynamic behaviors because of the number of service providers and the heterogeneity in their information management. In this work, we aim at providing a generic automated integration of negotiation protocols with application-specific messaging capabilities. The framework provides a flexible and easy to implement, design, tailor, and deploy a modeling and simulation platform under different applications.

Negotiation rules are mechanisms to allow agents' interaction in order to achieve their goals. However, in designing negotiation systems, the designer needs to address three main issues [1]. First, negotiation techniques should provide brokering in managing loosely coupled service providers. Secondly, the engineering design of management tools should provide enough expressive capabilities for various behaviors or when different domains are encountered. Thirdly, lack of interaction between different requesters and providers yields inefficient and very costly agreements. In summary, negotiation systems should provide an interaction environment where many parties (agents) can be engaged in the negotiation using flexible rules and powerful language capabilities.

In order to reach to a successful modeling and simulation framework design, we identified the following issues to be supported:

- The framework should provide flexible brokering and negotiation capabilities.
- The framework should provide transparency to its subscribers whenever they need.
- New resources should be able to subscribe in a simple and efficient way.
- The framework should provide decision making capabilities on behalf of the agents whenever a user's agent requires it.
- The framework must provide simple and rich expressive primitives and be able to specialize them under different domains in a simple and automated approach.
- The design of the framework must be easy to develop it for a specific domain to shorten the development time which is vital for adding and removing new heterogeneous components to the simulation environment.

Hence, in this work, we show how we developed a flexible, efficient, and automated agent-interaction model that can be utilized by different engineering domains [8]. The model defines different concepts and principles in the negotiation process. Our method consists of an automated domain-independent marketplace architecture that allows agents to interact using two simple and yet powerful negotiation protocols which define the rules of interactions in multi-agent environments. Having a third party "marketplace" supports privacy and transparency among interacted agents. The marketplace agent is implemented using FD-DEVS [7]. In order to provide negotiation in different domains, a dynamic message structuring capability is needed where a message can have different formats based on the selected application. We develop an ontology that contains specialization relations between the different domains of interest [9]. We focus in this paper on the automation and integration of the domain-dependent message structure ontology with the domain-independent marketplace architecture. This paper shows how a designer of a multi-agent platform will have a powerful tool where systems can be tailored based on the operational purpose and objectives.

This paper is organized as follows: section 2 provides a summary of the negotiation protocols for agents interactions. Section 3 provides a summary of the ontology design and message specialization. Section 4 discusses in details the automation phase in developing a tailored platform for a specific application. Section 5 shows a running example for digital photo and printing markets; and finally, we conclude our paper in section 6.

2 Agents Interactions

In most of business and distributed systems, managing the resources and services manually is impossible and autonomous agents are needed to act on behalf of system users. Agents interact using a negotiation process, which is a methodology that is applied to provide bargaining and brokering capabilities between the different agents in a multi-agent environment. In multi-agent domain, agents are not just capable of making decisions in predictable situations, but also they can be intelligent to act in dynamic interaction. The agents need to communicate with each other, exchange data, and share same semantic language (can be an ontology). The objective of interactions is to reach to some agreements that are acceptable by those agents who are engaged in the negotiation. Game theory is a branch of economics that is concerned with interactions between agents to reach to decisions [10] [14] [17]. It imposes mathematical models (functions) that describe each agent's utility function in multi-agent systems in which strategically each agent tries to maximize its individual profit or preference. Autonomous agents have been used in many areas such as search engines, where they crawl the Web to find data or information.

In this framework, we have designed the interaction between agents through a marketplace that insures two negotiation scenarios: one-to-one rule, and service

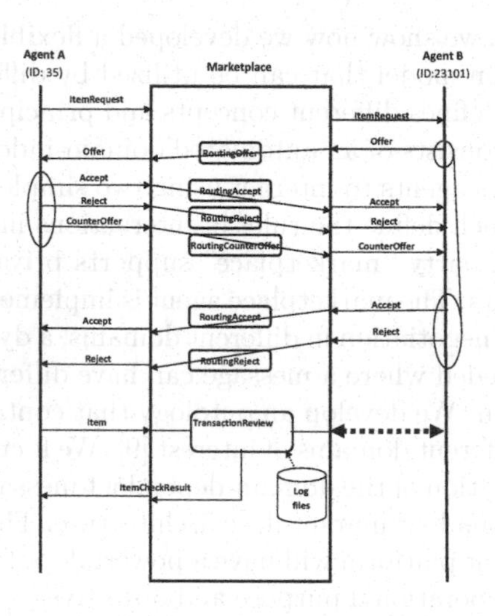

Fig. 1. One-to-One Negotiation Protocol

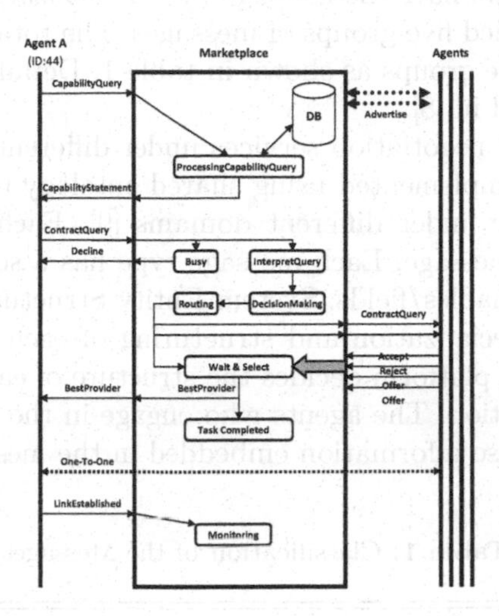

Fig. 2. Service Discovery Negotiation Protocol

discovery rule. Figures 1 and 2 show the two negotiation scenarios. In one-to-one rule, the marketplace forwards a message received from an agent to the corresponding agent using the destination ID field. Hence, in the one-to-one rule, each agent should know the corresponding agent ID. An agent can send to many agents using their IDs. In the service discovery, an agent sends a query to the marketplace which has access to the database of all subscribers, to find service providers. The marketplace responds to the request with a group of service providers or agents (IDs) who potentially can fulfill the request. An agents uses the list of the available service providers (agents) and their capabilities to engage in one-to-one negotiation to reach to an agreement on a specific contract. Any agent can decide on whether to proceed with the negotiation process or not. If the agent chooses to proceed with the interaction, it sends a contract query message to the marketplace agent, and then the marketplace in turn forwards that message to the appropriate agents and wait for responses from them. For more details on the negotiation protocols, refer to [8]. DEVS formalism was used to implement the states of the agents. Coupling and de-coupling is used between agents to reflect the interaction. The messages are sent through the in ports and out ports of the agents.

3 Language of Interaction

Agents interact with each other according to the rules of interaction (negotiation protocols). During an interaction, they send and receive different messages. Messages (for example, *Offer*) carry different information according to their structure (fields). For example, you can have the message *Offer* to contain three fields of information and have the message *Link Established* to have 7 fields, and so on. We have identified five groups of messages. The total number of messages are 17 divided into five groups as shown in table 1. Detailed description on the messages can be found in [8].

In order to support negotiation services under different domains, a dynamic message structure is implemented using shared ontology that defines each message and its structure under different domains [9]. Each domain would be a specialization of the message. Each message type has a separate structural ontology defining its variables/fields. System Entity Structure (SES) formalism is used to define the specialization and structuring of each of the messages [20]. Hence, the user of our platform decides the structure of each of the messages to be used in the interaction. The agents who engage in the interaction should be designed to utilize these information embedded in the messages.

Table 1. Classification of the Messages

Abort	Initiators	Reactors	Completers	Informative
Terminate	ContractQuery	Offer	Reject	Busy
NotMet	CapabilityQuery	CounterOffer	Accept	LinkEstablished
	ItemRequest	Decline		Item
		CapabilityStatement		ItemCheckResult
				BestProvider
				ProvidersChosen

3.1 Specialization and Structuring

In this subsection, we discuss how the message (language of encounter) is designed and implemented in order to support many application in a simple and automated way. Each message has a SES ontology that defines its structure under different domains. In this paper, we consider three applications as examples which are: Supply chain, photo development market [3], and a generic online store (e-commerce site). We will explain how to build the ontology for one message (*ContractQuery*) and the rest follow the same methodology. In supply chain domain, supplier and demand agents interact on contract variables that usually include: inventory level, price, product quantity, and lead time. Hence, a natural *ContractQuery* message will contain the aforementioned fields. In the domain of photo printing/development market, a contract usually consists of variables such as: print job ID, customer name, technology type, paper quality, number of copies, deadline, color. Hence, autonomous agents in a multi-agent environment exchange (interact) offers and counter offers in order to reach to an agreement that is acceptable by the engaged agents. Figure 3 shows how the ontology of the *ContractQuery* looks like.

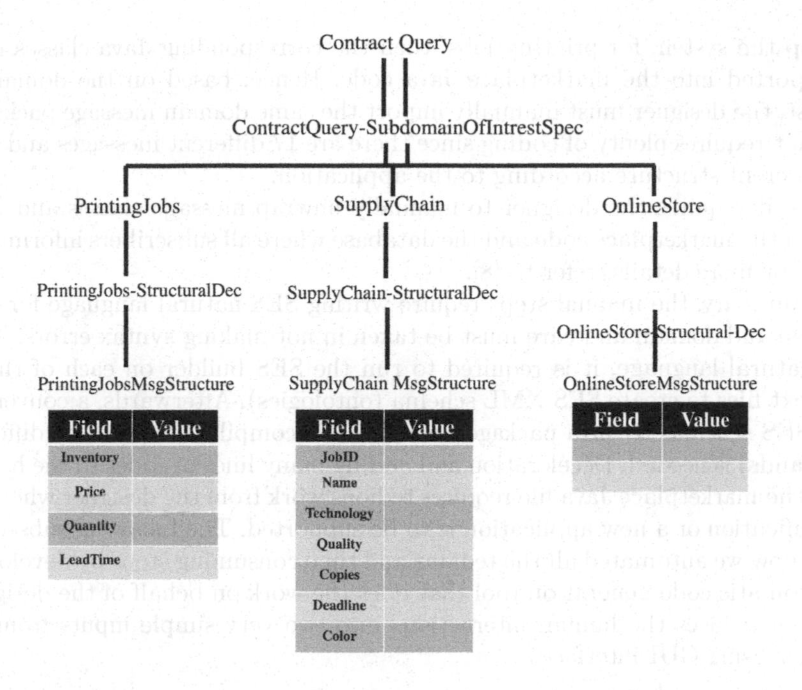

Fig. 3. Contract Query Specialization and Decomposition under Different Domains

4 Automatic Tailored Framework Generation for a Specific Application

In this section, we show the integration details and the automatic code generation capabilities. The integration is required to attach the interaction rules with the message structure in order to result in a working platform where agents can communicate, understand, and negotiate to accomplish their tasks. The implementation and proof-of-the-concept are done in DEVS environment along with the SES ontological framework.

4.1 Integration of the Domain-Independent Negotiation Protocols with the Domain-Dependent Ontology

The marketplace that enforces generic (domain-independent) negotiation rules is created using FD-DEVS GUI tool [9], which is a useful tool to generate Java templates [21]. The output of the tool is a Java file which is a domain-independent generic marketplace template that enforces the negotiation protocols.

In order to integrate the message structures into the domain-independent marketplace, the system designer must create Java classes for each message type consisting of the fields that are defined under the domain of interest. For example, if the designer is developing a framework for supply chain, then all messages specialization for supply chain must be imported. If another designer wants to develop the system for printing jobs, then the corresponding Java classes must be imported into the marketplace Java code. Hence, based on the domain of interest, the designer must manually import the same domain message packages, and that requires plenty of coding since there are 17 different messages and each has different structure according to the application.

Also, it requires the designer to manually unwrap message classes and wrap them in the marketplace code and the database where all subscribers information exist. For more details, refer to [8].

In summary, the manual steps require writing SES natural language for each message and domain and care must be taken in not making syntax errors. After SES natural language, it is required to run the SES builder on each of the 17 SES text files to create SES XML schema (ontologies). Afterwards, a conversion from SES schemas to Java packages using JAXB compiler (requires 17 different commands) is needed. Deceleration and adding many lines of codes to the header file of the marketplace Java file requires tedious work from the designer whenever a modification or a new application is to be supported. The following subsection shows how we automated all the tedious and time consuming steps by developing an automatic code generation tool that does the work on behalf of the designer. The tool reduces the human interactions into two very simple inputs from the designer using GUI interface.

4.2 Automatic Generation and Integration of the Negotiation Marketplace

In order to help the designer in defining the message structures, we have developed a simple and easy to use Graphical User Interface which is shown in figure 4. The user of the GUI can add any domain (for example supply chain) to a message ontology; and define the structure of that message under the domain. For example, The user inputs how many fields the message *ContractQuery* has, and the name of each field in the structure. The user or the designer might select four fields with the name: Inventory, Price, Quantity, and Lead time. If a

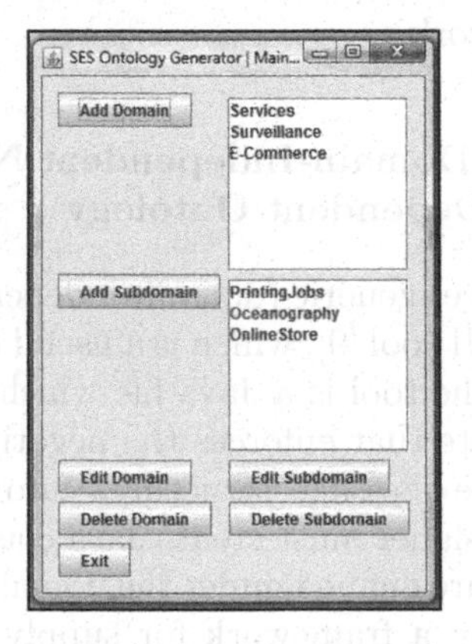

Fig. 4. Domain-Dependent Ontology Creation GUI

message type is not used in agent interactions, then the designer can select that the number of fields for that specific message is zero.

The output of the GUI tool is a collection of SES natural language text files (17 files), one for each message type. Those files are automatically converted into XML representation of the SES ontology in the automatic code generation tool. The tool then converts the XML representation (String sesinxml) into an XML schema by executing the line of code:

String schema = XmlToSes.getSchema(sesinxml);

The returned value of the above code is a schema, where the tool writes automatically 17 schema files for each of the messages into files with the extension .xsd to prepare them for the JAXB compiler. The SES schema is the representation of a master ontology that contains all domains that were defined. A Java method (*ExecJAXBSchemaCompiler*) is called to execute the JAXB compiler to translate the schema files into Java classes using the compiler command (with the appropriate parameters):

xjc schema.xsd d dirName p PackageName

The method iterates on each file and executes the command. Then another Java method named (*PostProcessingJavaClasses*) is called to extend (derived class) of type (*entity*) which is the base class for exchanging messages in DEVSJAVA. Also, the Java method imports the package (*import GenCol.*;*). At this point, the Java packages are complete and can be used by the marketplace generic model to declare the appropriate messages for the specific domain of interest.

4.3 Tailoring of the Marketplace Agent for a Specific Application

The first step in designing the framework is simple as mentioned in the previous subsection, which is to use the GUI to define the message structures. The second step that needs human interaction is very simple as well and all what it needs is to call a Java method (namely *CreateFDDEVSModelFor*) with the domain of interest as a string input such as Supply Chain" or "PrintingJobs. The Java method *CreateFDDEVSModelFor* uses the marketplace XML file that was created in FD-DEVS which defines the negotiation protocols, the in ports, and out ports of the marketplace agents. The in ports and out ports are through which messages are received and sent respectively. Hence, if the designer executes the code:

CreateFDDEVSModelFor("Supply Chain");

Then it generates a tailored marketplace agent model for supply chain domain. The model enforces the negotiation protocols and the message structures defined using the GUI interface for supply chain. The Java method *CreateFD-DEVSModelFor* is responsible of performing the following code generation steps:

1. Import the required Java classes for the negotiation process to take place, and the appropriate message package for a specific application.
2. Declare an instance of each of the negotiation primitives (messages).
3. When receiving a message, it stores it in the corresponding local instance produced in step 2. Then it generates the appropriate code to unwrap the message to get the domain message structure class that has the *get* and *set* methods to allow the designers to access the data received or to set variables to be sent in the message. The objective of storing the messages into local variables provides the capabilities for future data access and processing.
4. Create the JAXB Unmarshaller code to provide the marketplace agent to access its database when search for subscribers services.
5. Prepare domain message structure classes and wrapping them into the corresponding primitive; and then send it through the appropriate out port. This step provides the designer with the flexibility of adding *setV* methods to marshal the messages with data as needed.

In summary, this section showed how we automated the process of generating the marketplace agent given the language of interaction and the domain of interest. For example, if the message is ContractQuery and the domain is PrintingJobs, the tool will select the pruned SES of the ContractQuery ontology that defines the message structure under the domain "PrintingJobs".

5 Experiments and Proof of the Concept

The application of the negotiation activity can be applied into many multi-agent disciplines where a user or an agent initiates the process by asking a query or a request to be fulfilled. The user seeks to find either the best provider for the request or just any provider that can meet the requirements. In this section we show how agents interact to reach an agreement on a contract for online printing jobs; similar to the photo development market where an agent tries to get the service, within specific deadline, price, quality, and color. The marketplace agent helps all negotiating parties to reach to an agreement. For example, if a customer is concerning with printing business cards, the customer might choose thermography, Engraving or Letterpress technology. Also, other aspects for paper could be quality, deadline, color and duplex.

We have designed a user agent model that is searching for a provider who has Business Cards printing capabilities. The user would accept an offer if a contract with the following conditions occurs:

1. If the paper quality is medium or high, the color is RGB and deadline is less than 30.
2. If the paper quality is medium or high, the color is full HD and the deadline is less than 80.

3. If the paper quality is medium or high, the color is grayscale and the deadline is less than 20.

If the offer does not match any of the previous conditions, the user sends back a counter offer asking for the first contract or a modified one based on the history of the offers that were received. In our model, we chose that the user sends the first preference back again.

For the service provider agents, we have arbitrary assigned different photo and printing capabilities using the following pool of technologies: Digital printing can be : Brochures, Journals, Booklets, photos. Embossing Printing can be: Greeting Cards, Metals, Garments. Flexography Printing can be: Milk and Beverage Cartons, Disposable Cups, Containers, Adhesive Tapes, Envelopes, Newspapers, Food and Candy Wrappers. Letterpress Printing can be: Business Cards, Company Letterhead, Proofs, Billheads, Forms, Posters, Embossing, Hot-leaf Stamping. Engraving Printing can be: Stationery, Wedding Cards, Business Cards, Letterhead. Gravure Printing can be: Label, Flexible Packaging, Cartoning. Thermography Printing can be: Fax Printers, Business Cards, Letter Head, Invitation.

Some of the service agents are designed to update their deadline to accomplish the job using some constant value. Others do not and stick to the same deadline value through the whole simulation. The marketplace agent enforces the rules of interaction via reacting and routing the messages accordingly. We have used one agent to represent the user demand and seven agents to represent the service providers. After we ran the simulation in DEVS environment on a single machine, we found that an agreement has been reached with the following contract terms: Customer : Customer, Job Type : Business Cards, Print Server : Print Server 6, Color : FullHDColor, Paper Quality : High, Deadline : 78, Duplex : Yes, Number of Copies : 1, Technology Type : Thermography.
The contract terms that was reached actually satisfy the conditions in the second item of the user agent decision making which is:

– If the paper quality is medium or high, the color is full HD and the deadline is less than 80.

In order to provide a proof of the concept and deploy the framework in a distributed cluster; we have created the model agents and uploaded them on five machines in a DEVS/Service Oriented Architecture cluster. DEVS Service Oriented Architecture is a web services multi-server environment to support DEVS simulator. The system consists of two services, namely *MainService* and *Simulation* Service. For more details on DEVS/SOA system specifications and services, refer to [15].

The same logical behaviors as in the single machine simulation is used for DEVS/SOA. One user agent is used and seven service provider agents were used, and distributed on five machines in the cluster according to table 2. As expected, agent (Print Server 6) established the agreement with the user agent.

Table 2. Agent Distribution in DEVS/SOA Cluster

Machine IP	Agent
150.135.218.200	Customer, Print Server 2, Print Server 4, Print Server 5, and Print Server 7
150.135.218.201	Print Server 1
150.135.218.203	Print Server 3
150.135.218.204	SOAMarketPlace and the Coupled model (ServicesSOAEnv)
150.135.218.206	Print Server 6

6 Conclusion

Two powerful and yet flexible negotiation protocols are used to enforce the rules of interactions in multi-agent modeling and simulation platform. The rules are implemented through a third party marketplace agent, which supervises the negotiation process while preserving privacy and transparency among the system users. Discrete event modeling and simulation environment (DEVS formalism) is used to implement the generic marketplace model. In order to accompany the negotiation protocols with flexible messaging capabilities to handle different complex domains and applications, a dynamic structure of the language of interaction is implemented in SES ontological framework. Each negotiation message has a separate ontology that defines its structure under different domain specialization.

The domain-independent marketplace model integrated with the domain-dependent language of interaction ontology gives system designers a very powerful and easy to use tool. Given the language of interaction structures under a specific domain of interest and the domain name as inputs to the automated code generation tool, it produces a tailored negotiation marketplace agent that is ready to be used. The automated marketplace code generation deceases significantly the time spent to tailor the platform for a specific domain. Different applications in industry and academia can utilize our framework to study their processes and phenomenon while reducing the development time and changing the messages contents in a flexible and simple way.

Acknowledgment. This work is supported by the SIMTech-NTU Joint Lab on Complex Systems.

References

1. Addis, M.J., Allen, P.J., Surridge, M.: Negotiating for Software Services. In: Eleventh International Workshop on Database and Expert Systems Applications, DEXA 2000 (September 2000)
2. Arthur, W.B., Holland, J.H., LeBaron, B., Palmer, R., Tayler, P.: Asset pricing under endogenous expectations in an artificial stock market. In: The economy as an evolving, complex system II, pp. 15–44. Addison Wesley, Redwood City (1997)

3. Chan, Y., Chen, X., Chou, M., Goh, B.H., Haw, C.S., Koh, S., Lee, H.K., Ye, H.Q., Yuan, X.M.: Analysis of a Software Focused Supply Chain in Photo Development Market. In: IEEE International Conference on Industrial Informatics, pp. 759–764 (August 2006)
4. Cooper, T.: Case studies of four industrial meta-applications. In: Sloot, P.M.A., Hoekstra, A.G., Bubak, M., Hertzberger, B. (eds.) HPCN-Europe 1999. LNCS, vol. 1593, pp. 1077–1086. Springer, Heidelberg (1999)
5. DISTAL, Distributed Software On-Demand For Large Scale Engineering Applications, `http://cordis.europa.eu/esprit/src/26386.htm`
6. Epstein, J.M., Axtell, R.: Growing artificial societies: social science from the bottom up. Brookings Institution Press (1996)
7. Hwang, M.H., Zeigler, B.P.: Reachability Graph of Finite and Deterministic DEVS Networks. IEEE Transactions on Automation Science and Engineering 6, 468–478 (2009)
8. Jarrah, M., Zeigler, B.P.: A Modeling and Simulation-based Methodology to Support Dynamic Negotiation for Web Service Applications. Journal Simulation 88, 315–328 (2012)
9. Jarrah, M., Zeigler, B.P.: Ontology-based marketplace for supporting negotiation in different scientific applications. In: IEEE Conference on Systems, Man, and Cybernetics (SMC), pp. 667–672 (October 2012)
10. Krishna, V., Ramesh, V.: Intelligent Agents for Negotiations and Market Games, Part 1: Model. IEEE Transaction on Power Systems 13, 1103–1108 (1998)
11. Kruse, S., Brintrup, A., McFarlane, D., Sanchez, L.T., Owens, K., Krechel, W.E.: Designing Automated Allocation Mechanisms for Service Procurement of Imperfectly Substitutable Services. IEEE Transactions on Computational Intelligence and AI in Games 5, 15–32 (2013)
12. Macal, C.M., North, M.J.: Agent-Based Modeling and Simulation: Desktop ABMS. In: 39th Conference on Winter Simulation, WSC 2007, pp. 95–106. IEEE Press, NJ (2007)
13. Macal, C.M., North, M.J.: Agent-based modeling and simulation. In: Conference on Winter Simulation, WSC 2009, pp. 86–98 (2009)
14. Mahajan, R., Rodrig, M., Wetherall, D., Zahorjan, J.: Experiences Applying Game Theory to System Design. In: Proceeding SIGCOMM PINS Workshop, pp. 183–190 (2004)
15. Mittal, S., Risco-Mart, J.L., Zeigler, B.P.: DEVS/SOA: A Cross-Platform Framework for Net-centric Modeling and Simulation in DEVS Unified Process. Simulation Journal 85, 419–450 (2009)
16. Palmer, R.G., Arthur, W.B., Holland, J.H., LeBaron, B., Tayler, P.: Artificial economic life: a simple model of a stock market. J. Physica D. 75, 264–274 (1994)
17. Persons, S., Wooldridge, M.: Game Theory and Decisions Theory in Multi-Agent Systems. Journal on Autonomous Agents and Multi-Agent Systems 5, 243–254 (2002)
18. Scientific Discovery through Advanced Computing, agent-based modeling and simulation for exascale computing (2014), `http://www.scidacreview.org/0802/html/abms.html`
19. Susan, E.L.: Issues in Multi agent Design Systems. Journal IEEE Expert: Intelligent Systems and Their Applications 12, 18–26 (1997)
20. System Entity Structure, SES (2014), `http://www.ms4systems.com/pages/devs/ses.php`
21. W3C XML Schema for Finite Deterministic(FD) DEVS Models (2014), `http://www.duniptechnologies.com/research/xfddevs/`

Automatic Ultrasound Image Segmentation Framework Based on Darwinian Particle Swarm Optimization

Vedpal Singh[1,*], Irraivan Elamvazuthi[1], Varun Jeoti[1], and John George[2]

Centre for Intelligent Signal and Imaging Research (CISIR)
Department of Electrical and Electronic Engineering
[1] Universiti Teknologi PETRONAS
Bandar Seri Iskandar
31750 Tronoh, Perak Darul Ridzuan, Malaysia
[2] Research Imaging Centre, University of Malaya,
Kuala Lumpur, Malaysia
vedpalsiet101@gmail.com

Abstract. Accurate medical diagnosis and treatment necessitate the application of optimal segmentation of images. Although manual segmentation is easy, it has many problems such as time complexity and error sensitivity. On the other hand, automatic segmentation is fast with less probability of errors. However, it has many problems like low contrast image, unclear boundaries and less accurate. To overcome these problems, optimization methods like Particle Swarm Optimization (PSO), genetic algorithm (GA), etc. can provide more accurate and efficient outcomes. Thus, for the achievement of optimized results, the current study proposes a more optimized 'Singh-Elamvazuthi' ultrasound segmentation framework based on Darwinian Particle Swarm Optimization (D-PSO) for ankle Anterior Talofibular Ligament.

Keywords: ATFL ligament, ultrasound Image segmentation, optimization, PSO, D-PSO, time elapse.

1 Introduction

The key aim of the segmentation is to split images into informative multiple segments. It follows the labeling of every pixel in entire image to utilize the graphic properties. The obtained segmented part holds more meaningful information than complete image in a significant visualization manner. For accurate analysis and image understanding, image segmentation works as a precious approach in categorization and detection of objects [1], [2], [3].

Image classification is playing an important role for accurate segmentation. Figure 1 represents the classification of images based on their shape, color, intensity, texture, coordinate information, boundary, region and structure based information [4].

Image segmentation can be categorized into 4 categories such as histogram thresholding, texture based segmentation, clustering and split and merging of region based methods [6]. In all these segmentation methods, image thresholding is a popular

* Corresponding author.

H. Handa et al. (eds.), *Proc. of the 18th Asia Pacific Symp. on Intell. & Evol. Systems – Vol. 1*,
Proceedings in Adaptation, Learning and Optimization 1, DOI: 10.1007/978-3-319-13359-1_18

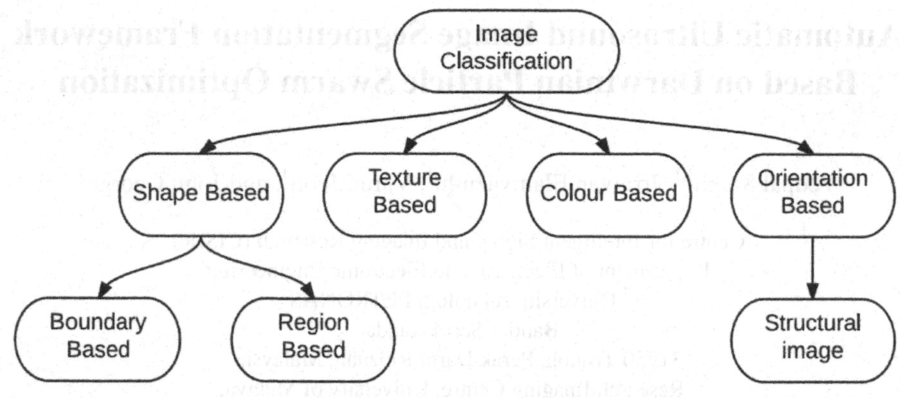

Fig. 1. Image Classification [5]

approach. The main target of the thresholding method is to provide an efficient segmentation regarding two or more clusters [1], [7], [8]. It can be classified into some categories based on property and optimal thresholding approach. Around the world, many researchers are working on image segmentation and they are trying to find out optimal and global method which provides all desirable properties. The property based thresholding segmentation approaches outcomes are reasonable and speedy which would be more helpful in case of multiple thresholding. The optimal thresholding method provides more accurate and optimized segmented outcomes [1], [9].

Medical image segmentation is a more difficult task because of complications in anatomical structures. This study focuses on ankle ATFL ligament segmentation which is presented by Figure 2.

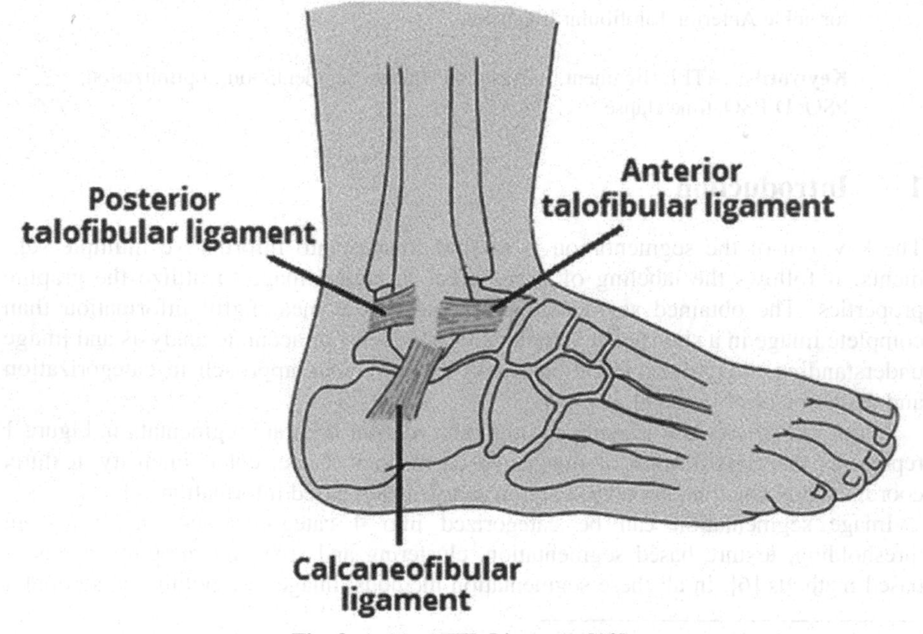

Fig. 2. Ankle ATFL Ligament [10]

Through the normal ultrasound imaging, this ligament is not easily detected by radiologists which could be the main cause of unsatisfied diagnosis. It is a challenging research domain to obtain a segmented ligament area with accurate measurements for proper treatments. Manual segmentation can be an option but it needs expert every time. It is time consuming, painful and error sensitive. Generally, radiologists are using invasive approach for the treatment. To overcome above defined problems, this paper proposes an automatic segmentation 'Singh-Elamvazuthi' framework using D-PSO algorithm [4], [11], [12], [13], [14], [15].

2 Motivation

The current image processing and pattern recognition techniques have many limitations due to lack of optimal segmentation method. Thus, image segmentation needs a global and optimal solution. So, it is an open research problem still because of unsatisfied outcomes [16], [17].

The local search methods are used to solve the complex situations in image processing and object detection. However, some set of segmentation problems can be solved easily but some of them are more difficult to solve. These problems could be solved through optimized algorithms like Particle Swarm Optimization (PSO), Genetic Algorithm (GA) etc. PSO algorithm is a stochastic algorithm which is based on the concept of bird flocks' social behavior. Due to the numerous beneficial features such as efficiency, straight forwardness, more robust and improved accuracy [16], [18], [19], [20], [21], an enhanced version of PSO named D-PSO algorithm was used in this study.

2.1 Introduction to Particle Swarm Optimization

The particle Swarm Optimization (PSO) is a popular optimization approach which used particle swarm as a search space. According to this algorithm, each image pixel is called image particle and a complete set of particles is named as swarm or population. Each image particle has their own velocity and position in population named as search space which traverses in direction of optimal solution [22], [25], [26].

Let us consider N is population and M is a search space in which position of ith image particle is $X_i(x_{i1}, x_{i2}, \ldots\ldots, x_{iM})$ and velocity is denoted by $V_i(v_{i1}, v_{i2}, \ldots\ldots, v_{iM})$. Optimal spatial information of every image particle is $P_i(p_{i1}, p_{i2}, \ldots\ldots, p_{iM})$ with optimized last position is represented as $P_g(p_{g1}, p_{g2}, \ldots\ldots, p_{gM})$. Based on individual's velocity and position [1], [2], [6], [22], each image particle is traversed according to Equations (1), (2) and (3).

$$v_{im}^{k+1} = w^k * v_{im}^k + c_1 * rand() * (p_{im} - x_{im}^k) / \Delta t + c_2 * rand() * (p_{gm} - x_{im}^k) / \Delta t \tag{1}$$

$$x_{im}^{k+1} = x_{im}^{k} + v_{im}^{k} * \Delta t \tag{2}$$

$$w^{k} = w_{\max} - k * (w_{\max} - w_{\min}) / k_{\max} \tag{3}$$

Where, $1 \le m \le M$ and rand () is a random number which performs uniformity distribution $U(0,1)$, acceleration coefficients are c_1 and c_2 used in equation 1. Inertia weight is w in which $w_{\max}$ and $w_{\min}$ represents the maximum and minimum values, k is present iterative time and $k_{\max}$ shows its maximum value. Δt is a unit time. v_{im}^{k+1} and x_{im}^{k+1} used the Equations (4), (5) and (6).

$$v_{im}^{k+1} = \begin{Bmatrix} v_{im}^{k+1}, -v_{\max} \le v_{im}^{k+1} \le v_{\max} \\ v_{\max}, v_{im}^{k+1} > v_{\max} \\ -v_{\max}, v_{im}^{k+1} < -v_{\max} \end{Bmatrix} \tag{4}$$

$$x_{im}^{k+1} = \begin{Bmatrix} x_{im}^{k+1}, x_{\min} \le x_{im}^{k+1} \le x_{\max} \\ x_{init}, x_{im}^{k+1} > x_{\max} \\ x_{init}, x_{im}^{k+1} < x_{\min} \end{Bmatrix} \tag{5}$$

$$x_{init} = x_{\min} + rand() * (x_{\max} - x_{\min}) \tag{6}$$

where, $v_{\max}$ represents the maximum value of v and $x_{\max}$ and $x_{\min}$ maximum value of x [22], [25].

2.2 Darwinian Particle Swarm Optimization Algorithm

Enhancements were introduced to the original PSO to overcome some of the limitations. In 2005, Tillet et al. [27] have done an interesting improvement in PSO based on natural selection approach. This proposed method is named as Darwinian Particle Swarm Optimization (D-PSO) [27]. The main advantage of D-PSO is that it can work with multiple swarms at a time. Each swarm works discretely as a normal PSO with novel natural selection capability. This natural selection is called Darwinian principle of survival of the fittest, where avoidable to local minima. During D-PSO execution, if image particles are searching for local optima, then it could be avoided and then alternative region is chosen for search. In addition, this method provides reward to optimal image particle like particle life and provides a new descendent. However, for image particles that are not doing well, they get punishment like swarm life reduction or complete removal from the swarm [26].

In swarm, each image particle tries to maintain fitness which is calculated by fitness function. It is used for analysis of the performance of each image particle. Fitness value of any image particle can change the position and velocity of neighbors. If global solution is found, then new image particle is produced and particles with bad fitness are deleted from the swarm. In D-PSO, removal and production of image particles and swarm is a leading step. However, it is not a more complex process with some protocols. In the initial check, if number of image particles in swarm less than the predefined criteria, then swarm will be deleted. Subsequently, if any image particle is performing not well, then it is deleted [28]. After the deletion of particles, the reinitiating values are done through the tuning of a specified thresholding number as given in Equation (7).

$$SC_c(N_{kill}) = SC_c^{\max}\left[1 - \frac{1}{N_{kill}}\right] \tag{7}$$

where N_{kill} is indicating the removal of particles from a swarm within a particular time interval in which no particle fitness improvement is reported. Any swarm cannot survive if numerous image particles are removed or does not have large number of image particles. However, any swarm can be introduced with probability, *p, as* represented by Equation (8).

$$p = f \,/\, NS \tag{8}$$

Where, f is a random number, $0 < f < 1$ and NS indicates number of swarms. In Equation (8), generation of more swarms cannot be permitted due to existent of many swarms. Sometimes, parent swarm remains ideal but mostly image particles are elected randomly to design a child swarm. If initial numbers of image particles are not known then remaining image particles are used in initialization. These initialize image particles are merged to newly introduced swarm. Image particles are spawned, when a swarm achieved a global best solution and in swarm maximum demarcated population has not been reached [26]. The Darwinian Particle Swarm Optimization (D-PSO) algorithm is demonstrated by Table 1.

Table 1. Darwinian Particle Swarm Optimization Algorithm

Main Program Loop		*D-PSO Algorithm*
For each swarm in the collection Evolve the swarm Allow the swarm to spawn Delete "failed" swarms	→	For each particle in the swarm Update Particles' Fitness Update Particles' Best Move Particle If swarm gets better Reward swarm: spawn particle: extend swarm life If swarm has not improved

Main Program Loop ***D-PSO Algorithm***

Punish swarm: possibly delete
particle: reduce swarm life

Similar to PSO, the D-PSO algorithm requires some parameters adjustment for efficient program execution such as population initialization, need to define population range of minimum to maximum, and threshold [26], [29].

3 The Proposed Segmentation Framework: Singh-Elamvazuthi Automatic Segmentation Framework

The main aim of this study is to provide a novel automatic framework for efficient segmentation of ultrasound images. It proposed a D-PSO based optimized segmentation framework which consists of some steps shown by Figure 3.

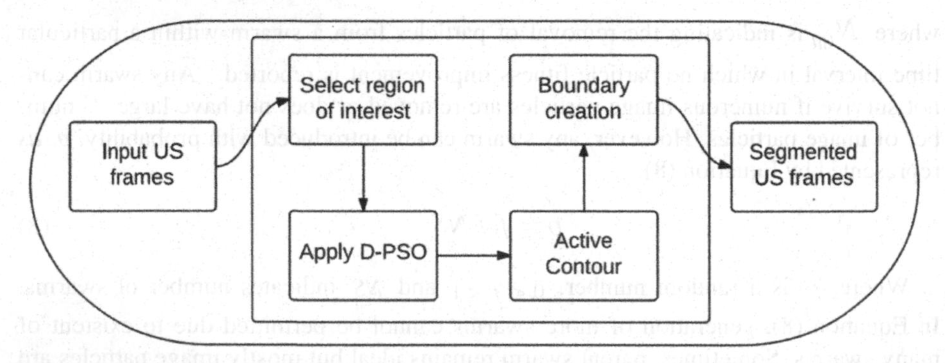

Fig. 3. The proposed 'Singh-Elamvazuthi' automatic ultrasound image segmentation framework

The proposed 'Singh-Elamvazuthi' framework is presented by Figure 3 initially tries to find out exact region of interest from an image. This region can be defined as rectangular, circle or random shape. In this study, it defined in a random shape to provide to reduce the number of iterations and computational complexity. Proposed framework applies the D-PSO algorithm on selected region of interest to obtain more optimized region than traditional particle swarm optimization method. To get more optimized and smoother region with minimum surface energy uses an energy minimization method active contour. The active contour method stretched the surface of segmented image edges until minimum energy level. This step provides a segmented image but its boundary is unclear. Thus, the next step is boundary creation, where it gives a great help in case of blurred image to get a strong boundary for better visualization analysis [25], [30].

4 Results and Discussion

To evaluating the performance of the proposed 'Singh-Elamvazuthi' framework, numerous ultrasound images were used. Current study applied the proposed segmentation framework on ultrasound images for the purpose of accurate segmentation outcomes. This study used a specific approach 'natural selection' for better optimization than the existing segmentation method. For the precise optimization need to established a significant initialization strategy presented by Figure 4.

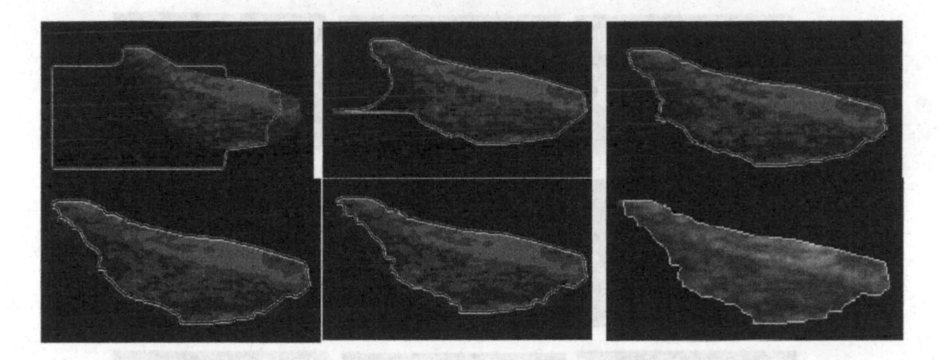

Fig. 4. Initialization processing during the 'Singh-Elamvazuthi' framework execution

Initialization would help in the determination of exact region extraction during image segmentation. Initialization processing of the proposed framework is approximately similar to the existing method. But proposed approach created a difference during the image particle movements which effects optimization ratio in a good way. One more beneficial feature of this framework is to preserve the properties of the ultrasound images without any loss of valuable information for better visualization. The segmented region boundaries are much clearer as shown in Figure 5. It consists of 15 segmented ultrasound images outcomes. Each segmented image represents the ATFL ligament in an accurate visualized manner with smoother boundary. It segments ligament in accurate size which helps in truthful volume determination.

The key aim of this study is to provide the accurate segmented results of ultrasound images in fast and efficient way. Another main target is to make the segmentation process less parametric dependent because it's using only pixel values for the whole process. However, existing method using many features at the same time such as intensity, image texture, and color which makes segmentation process more computationally inefficient.

Figure 5 have shown segmented ultrasound ligament images resulted by the 'Singh-Elamvazuthi' segmentation framework. It represents the images more accurately with appropriate boundaries in a green color to differentiate from image surface. For quantitative analysis of this framework need to analysis some parameters like intensity value of each image with their fitness value and execution time. But this study mainly concern on comparative on the basis of execution time of the proposed framework and existing particle swarm optimization method which is demonstrated by Table 2.

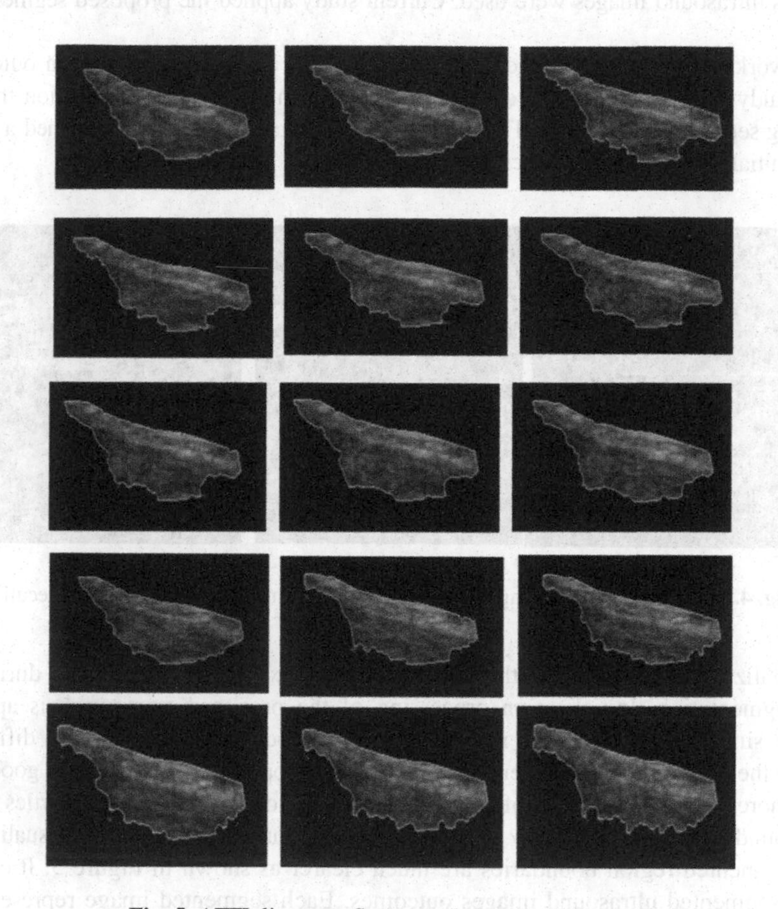

Fig. 5. ATFL ligament Segmented ultrasound Images

Table 2. Comparative analysis of the proposed 'Singh-Elamvazuthi' framework with existing Particle Swarm optimization

Execution Time (sec.)	
The Proposed 'Singh-Elamvazuthi' Framework	Particle Swarm Optimization
2.9472	4.2849
2.6717	2.8209
2.6991	2.8490
2.5557	2.8140
2.5945	2.8524

2.7343	2.8350
2.7328	2.8067
2.7303	2.8436
2.7755	2.8516
2.7001	2.8357

The proposed framework provides the intensity of each image like [27 69 104]. It consists of three values due to RGB image; in case of binary image it provides only one value. Intensity, fitness and time of each image are calculated for performance evaluation of utilized approaches. The fitness capability of image particles is determined by the fitness function for the confirmation of any particle's performing strength. But fitness and intensity values of each image are approximately similar in case of both proposed and existing methods. So, here, these values cannot be used for the performance evaluation. Thus, only execution time is used for quantitative analysis which is represented by Table 2. It provides a demonstration about performance of proposed framework with PSO algorithm. It shows the execution time of the proposed framework is lower than the traditional PSO algorithm.

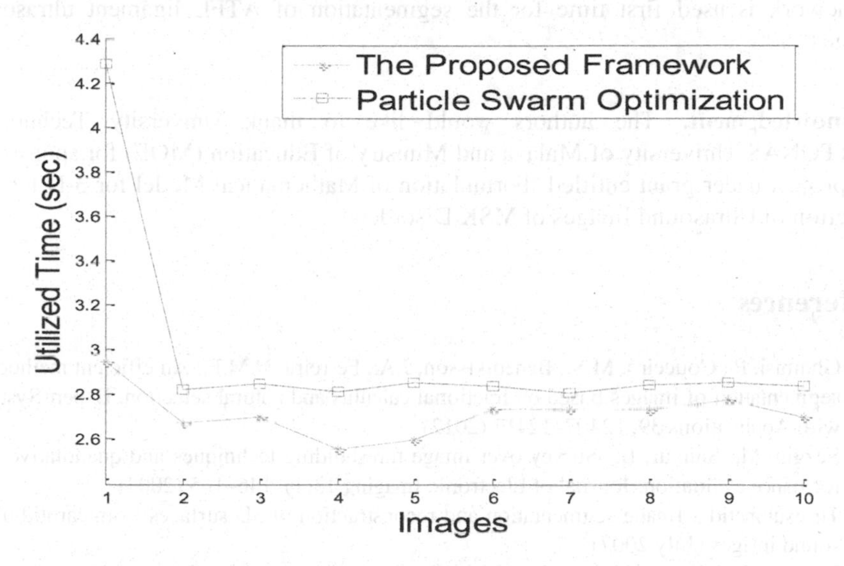

Fig. 6. Utilized time ratio between the proposed 'Singh-Elamvazuthi' framework and Particle Swarm Optimization algorithm

On the basis listed outcomes in Table 2, the D-PSO based proposed approach is faster than the standard PSO algorithm. This approach performs better in terms of fitness values and time taken during processing. From the above segmented resultants and quantitative analysis, it can be concluded that the 'Singh-Elamvazuthi' framework is more beneficial in terms of good image quality with lower execution time shown by Figure 5 and Figure 6 respectively. However, the comparative analysis of fitness and intensity variance would be a precise future footstep for detail analysis of big databases. It would be interesting to companion this research with the large databases of several types of images. It would leads to platform independency of this method. Finally, the proposed 'Singh-Elamvazuthi' segmentation framework demonstrated by this study would provide to help in the detection of tears in injured ligament and volume of tears. Clinically, it is more beneficial information which in the evaluation of healing rate of particular ligament injury.

5 Conclusion

This study proposed a novel and automatic 'Singh-Elamvazuthi' framework for the segmentation of ATFL ligament from ultrasound images. It is based on the Darwinian Particle Swarm Optimization (D-PSO) algorithm. The current study provides the comparison between the performances of the proposed framework and standard Particle Swarm Optimization method based on visualization and computational complexity. This analysis produce the outcomes which have shown about that proposed framework have less computational complexity and better image quality than the existing approach. It should be noted that optimization algorithm D-PSO based framework is used first time for the segmentation of ATFL ligament ultrasound images.

Acknowledgment. The authors would like to thank University Technology PERTONAS, University of Malaya and Ministry of Education (MOE) for sponsoring the project under grant entitled 'Formulation of Mathematical Model for 3-D Reconstruction of Ultrasound Images of MSK Disorders'.

References

1. Ghamisi, P., Couceiro, M.S., Benediktsson, J.A., Ferreira, N.M.F.: An efficient method for segmentation of images based on fractional calculus and natural selection. Expert Systems with Applications 39, 12407–12417 (2012)
2. Sezgin, M., Sankur, B.: Survey over image thresholding techniques and quantitative performance evaluation. Journal of Electronic Imaging 13(1), 146–165 (2004)
3. Tesesubmetida, Image segmentation and reconstruction of 3D surfaces from carotid ultrasound images (July 2007)
4. Ibrahim, S., Khalid, N.E.A., Manaf, M.: Empirical Study of Brain Segmentation using Particle Swarm Optimization. IEEE (2010) ISBN: 978-1-4244-5651
5. Jaiswal, V., Tiwari, A.: A Survey of Image Segmentation based on Artificial Intelligence and Evolutionary Approach. IOSR Journal of Computer Engineering (IOSR-JCE) 15(3), 71–78 (2013) E-ISSN: 2278-0661, ISSN: 2278-8727
6. Brink, A.D.: Minimum spatial entropy threshold selection. IEE Proceedings on Vision Image and Signal Processing 142, 128–132 (1995)
7. Kulkarni, R.V., Venayagamoorthy, G.K.: Bio-inspired algorithms for autonomous deployment and localization of sensor nodes. IEEE Transactions, SMC-C 40(6), 663–675 (2010)
8. Booth, B., Li, X.: Boundary Point Detection For Ultrasound Image Segmentation Using Gumbel Distributions. In: Second International Conference on Signal Processing and Multimedia Applications, Barcelona, Spain, July 28-31 (2007)

9. Akbari, H., Fei, B.: 3D ultrasound image segmentation using wavelet support vector machines. Med. Phys. 39(6) (2012)
10. OLIVER JONES, The Ankle Joint Teach Me Anatomy (May 26, 2014)
11. Souza, J.G., Costa, J.A.F.: Natural Computing Techniques for Data Clustering and Image Segmentation (2007)
12. Guo, Y., Cheng, H.D., Tian, J., Zhang, Y.: A Novel Approach to Breast Ultrasound Image Segmentation Based on the Characteristics of Breast Tissue and Particle Swarm Optimization. In: Proceedings of the 11th Joint Conference on Information Sciences (2008)
13. Alamelumangai, N., DeviShree, J.: PSO Aided Neuro Fuzzy Inference System for Ultrasound Image Segmentation. International Journal of Computer Applications (0975 – 8887) 7(14) (October 2010)
14. Al-Faris, A.Q., Ngah, U.K., Isa, N.A.M., Shuaib, I.L.: Breast MRI Tumour Segmentation using Modified Automatic Seeded Region Growing Based on Particle Swarm Optimization Image Clustering. In: Soft Computing in Industrial Applications. AISC, vol. 223, pp. 49–60. Springer, Heidelberg (2011)
15. Forghani, N., Forouzanfar, M., Eftekhari, A., Mohammad-Moradi, S., Teshnehlab, M.: Application of Particle Swarm Optimization in Accurate Segmentation of Brain MR Images. In: Lazinica, A. (ed.) Particle Swarm Optimization, InTech (2009) ISBN: 978-953-7619-48-0
16. Raju, N.G., Rao, P.A.N.: Particle Swarm Optimization Methods for Image Segmentation Applied In Mammography. Int. Journal of Engineering Research and Applications 3(6), 1572–1579 (2013) ISSN : 2248-9622
17. Talebi, M., Ayatollahi, A., Kermani, A.: Medical ultrasound image segmentation using genetic active contour. J. Biomedical Science and Engineering 4, 105–109 (2011)
18. Saini, K., Dewal, M.L., Rohit, M.: Ultrasound Imaging and Image Segmentation in the area of Ultrasound: A Review. International Journal of Advanced Science and Technology 24 (November 2010)
19. Kaur, R., Kaur, M.: Image Segmentation- A Review. International Journal of Engineering And Computer Science 3(4), 5457–5461 (2014) ISSN:2319-7242
20. Qi, C.: Maximum Entropy for Image Segmentation based on an Adaptive Particle Swarm Optimization. Appl. Math. Inf. Sci. 8(6), 3129–3135 (2014)
21. Spiller, J.M., Marwala, T.: Medical Image Segmentation and Localization using Deformable Templates. Computer Vision and Pattern Recognition (2007)
22. Li, L., Li, D.: Fuzzy entropy image segmentation based on particle swarm optimization. Progress in Natural Science 18, 1167–1171 (2008)
23. Zheng, L., Pan, Q., Li, G., Liang, J.: Improvement of Grayscale Image Segmentation Based On PSOAlgorithm. In: Fourth International Conference on Computer Sciences and Convergence Information Technology (2009)
24. Tandan, A., Raja, R., Chouhan, Y.: Image Segmentation Based on Particle Swarm Optimization Technique. International Journal of Science, Engineering and Technology Research (IJSETR) 3(2) (February 2014)
25. Hakl, H., Uguz, H.: A novel particle swarm optimization algorithm with Levy flight. Applied Soft. Computing 23, 333–345 (2014)
26. Ghamisi, P., Couceiro, M.S., Ferreira, N.M.F., Kumar, L.: Use Of Darwinian Particle Swarm Optimization Technique For The Segmentation of Remote Sensing Images. In: IGARSS IEEE 2012 (2012)
27. Tillett, J., Rao, T.M., Sahin, F., Rao, R., Brockport, S.: Darwinian Particle Swarm Optimization. In: Proceedings of the 2nd Indian International Conference on Artificial Intelligence, pp. 1474–1487 (2005)

28. Lee, C.-Y., Leou, J.-J., Hsiao, H.-H.: Saliency-directed color image segmentation using modified particle swarm optimization. Signal Processing 92, 1–18 (2012)
29. Chang, C.-Y., Lei, Y.-F., Tseng, C.-H., Shih, S.-R.: Thyroid Segmentation and Volume Estimation in Ultrasound Images. IEEE (2008)
30. Khan, W.: Image Segmentation Techniques: A Survey. Journal of Image and Graphics 1(4) (December 2013)

Chinese Text Similarity Computation via the 1D-PW CNN*

Luping Ji**, Xiaorong Pu, and Guisong Liu

School of Computer Science and Engineering,
University of Electronic Science and Technology of China
Chengdu 611731, P.R. China
jlp0813@gmail.com

Abstract. Cellular neural network is one of the classic neural networks. This paper designs a one-dimensional pairwisc CNN, and then develops a global alignment algorithm for the Chinese texts using this CNN. The method mainly includes these processing steps, namely the initialization of the CNN, the generation of the alignment path, and the global alignment of the texts in accordance with the comments. The experiments show that the developed method is efficient, and compared to the other three methods, it could obtain the alignment result of two Chinese texts with the higher similarity and the less time.

Keywords: Chinese text similarity, cellular neural network, cell state matrix, global alignment path.

1 Introduction

In the world, there are more than 20% people using Chinese, and Chinese is becoming an increasingly important information-carrying tool. For any language such as English and Chinese, the text similarity computation is often a very important research field [1].

Text similarity has many significant applications, such as the intellectual property protections, the anti-plagiarism of papers, and the natural language understanding, etc. Currently, more and more Chinese universities are utilizing text similarity detection technology to prevent students from copying text materials from the other papers or dissertations. Almost all of the publishing houses in China take the Chinese text similarity detection technology [2] to check and evaluate the originality of the paper manuscript submitted by the authors.

Chinese text similarity research is so important that it has attracted the great enthusiasms from the researches. For instance, Xu presents a Chinese text similarity computation method by machine translation to translate Chinese text into English text [1]. Xu also researches the text similarity of Chinese papers using

* This work is supported by the National Science Foundation of China (NSFC) under the Grant number 61175061.

** Corresponding author.

H. Handa et al. (eds.), *Proc. of the 18th Asia Pacific Symp. on Intell. & Evol. Systems – Vol. 1*,
Proceedings in Adaptation, Learning and Optimization 1, DOI: 10.1007/978-3-319-13359-1_19

map-reduce technology [2]. Wang explores the Chinese text similarity on some broad topics of word, sentence and document [3]. Li investigates the performance of different similarity measures [4] in KNN algorithm on Chinese texts.

In general, the key step in the text similarity computation is the text alignment. By now, there have been many methods developed for the global alignment of sequences, such as the mixed-integer linear programming [5], the super pairwise alignment [6], and the global visualization alignment [7]. In recent years, some special methods for English texts have also been proposed, such as the Chinese-English word similarity measure [8], the sememe vector space model [9], and the IR lists approach [10]. These algorithms could often achieve the good results on the Latin texts, however they could no longer be equally effective on the Chinese texts.

Moreover, the cellular neural network (CNN) is also a workable tool for the global alignment of the text sequences. As one of the classic neural networks, CNN was originally proposed by Chua [11]. In the past decades, the exponential stability problems [12,13] of the various CNNs were deeply researched. Moveover, the two-dimensional CNN models have been widely used in the applications, such as the image processing [14,15], the chaotic Hash function construction [16]. Besides the two-dimensional CNNs, some one-dimensional models [17, 18] for the CNN also have been proposed. However, the special applications of one-dimensional CNN models are seldom paid enough attention to.

This paper designs a one-dimensional pairwise CNN (1D-PW CNN), and then develop a global alignment algorithm for the Chinese texts. The 1D-PW CNN consists of two one-dimensional CNNs, one is the master, and another is the slave. In the 1D-PW CNN, the master keeps immoveable all the time, while the slave will move a distance forward at each time. By the computation of the cell state matrix, the optimal state matrix can be obtained, an then the global alignment path is generated through back-tracing this matrix. The two Chinese texts are globally aligned in accordance with the path, and then the global similarity computation is conducted.

The remainder of this paper is organized as follows. In Section 2, the 1D-PW CNN is designed and its cell dynamics is described and analyzed. In Section 3, the global alignment algorithm based on the 1D-PW CNN and the similarity computation for two Chinese text sequences are exhibited. For evaluating this algorithm, some comparison experiments on some Chinese text samples are given in Section 4. Finally, Section 5 summarizes this paper.

2 The Proposed 1D-PW CNN Model

The new 1D-PW CNN model proposed in this paper to globally align the Chinese texts and then compute the similarity between them is similar to, but a little different from the one-dimensional models respectively presented in [17] and [18], and it is briefly illustrated in Fig.1. In the model, the 1D-PW CNN consists of two 1-D CNNs, one is the master CNN_1 which keeps itself immoveable all the time, and another is the slave CNN_2 which will move a step forward with the

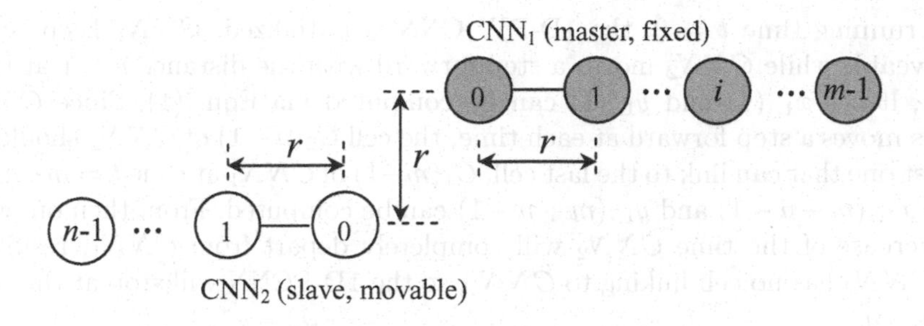

Fig. 1. The proposed 1D-PW CNN model

fixed distance $r = 1$ at each time. Furthermore, $r = 1$ is the distance between two nearest cells and also the critical distance more than which any two cells cannot receive the link from each other. Moreover, CNN_1 contains m cells, and CNN_2 contains n cells. $C_1(i)$, $i = 0, 1..., m-1$, represents the cell of CNN_1, similarly $C_2(j)$, $j = 0, 1, ..., n-1$ represents the cell of CNN_2.

Based on the models presented in [11,17,18], the cell dynamics equations of the 1D-PW CNN are defined as:

$$\begin{cases} x_{1,i}(t+1) = (1 - \frac{1}{CR_x})x_{1,i}(t) + \frac{1}{C}\Big(\sum_{C_l(k) \in N_{1,i}(r,t)} A_k y_{l,k}(t) \\ \qquad + \sum_{C_l(k) \in N_{1,i}(r,t)} B_k u_{l,k} + I_{1,i} \Big) \\ y_{1,i}(t) = f(x_{1,i}(t), I_{1,i}) = \begin{cases} 1, \text{ if } x_{1,i}(t) = I_{1,i}; \\ 0, \text{ else.} \end{cases} \end{cases} \quad (1)$$

where $x_{1,i}(t)$ and $y_{1,i}(t)$ represent the cell state and the output of $C_1(i)$ at the running time t, respectively. $C_l(k)$ represents the neighbor cell in the neighbor domain, $N_{1,i}(r,t)$ of $C_1(i)$. $y_{l,k}(t)$ and $u_{l,k}$ represent the output of $C_l(k)$ at time t and the link input to $C_l(k)$. **A** and **B** are the feedback template to modulate the output of $C_l(k)$ and the control template to modulate the link input to $C_l(k)$, respectively. C and R_x are the two circuit parameters. $I_{1,i}$ is the cell state bias (or threshold) of $C_1(i)$.

Moreover in Eqn. (1), the output function of $C_1(i)$, $f(...)$ is a little different from the output functions in [11,17,18]. In the 1D-PW CNN, $f(...)$ is redefined as a two-valued function, namely $y_{1,i}(t) = 1$ if $x_{1,i}(t)$ is equal to the state threshold $I_{1,i}$ at time t, otherwise $y_{1,i}(t) = 0$.

In the 1D-PW CNN, the slave CNN_2 is merely regarded as the link supplier to CNN_1, so the cell dynamics of CNN_2 is not considered in this paper. For the cell $C_1(i)$ of CNN_1, its neighbor domain will constrain the two types of cells, one is the cells of CNN_1 itself, another is the cells of CNN_2. Meanwhile, $C_1(i)$ will receive both the link input, $u_{l,k}$ and the cell output, $y_{l,k}(t)$ from all the neighbor cells at time t.

At running time $t = 0$, the 1D-PW CNN is initialized. CNN_1 keeps itself immoveable while CNN_2 moves a step forward with the distance $r = 1$ at time $t = 1$, hence $x_{1,i}(1)$ and $y_{1,i}(1)$ can be computed via Eqn. (1). Since CNN_2 always moves a step forward at each time, the cell $C_2(n-1)$ of CNN_2 should be the last one that can link to the last cell, $C_1(m-1)$ of CNN_1 at time $t = m+n-1$, hence $x_{1,i}(m+n-1)$ and $y_{1,i}(m+n-1)$ can be computed. From then on, with the increase of the time CNN_2 will completely depart from CNN_1, resulting that CNN_2 has no cell linking to CNN_1, so the 1D-PCNN will stop at the time $t = m+n$.

By running the 1D-PW CNN at time t, $t = 0, ..., m+n-1$, a series of the cell state $x_{1,i}(t)$ and the cell output $y_{1,i}(t)$ can be obtained. And then these state and output can be used for solving the global alignment problem for the fast similarity computation of the Chinese texts.

3 The Global Alignment and Similarity Computation

To compute the global similarity between two Chinese texts, it is usually necessary to transform the two texts into the two corresponding character sequences. Moreover, another necessary processing step is to globally align the two sequences [8, 9] by moving the characters. In this paper, based on the proposed 1D-PW CNN model, the global alignment method is developed to do the fast alignment of the two Chinese texts.

3.1 The 1D-PW CNN Initialization

To express easily, these symbols are defined as follows. Using S_1 and S_2 to represent the two given Chinese character sequences. The lengths of S_1 and S_2 are L_1 and L_2, respectively. Moreover, $S(i')$, where $0 \le i' \le L_1 - 1$, dedicates the character of S_1 at the location index i', and $S(j')$, where $0 \le j' \le L_2 - 1$, dedicates the character of S_2 at the location index j'.

And then, at time $t = 0$ the 1D-PW CNN is initialized as follows. Firstly, initialize CNN_1 and CNN_2 with $m = L_1 + 1$ and $n = L_2 + 1$ cells, respectively. The cell index i satisfies $0 \le i \le L_1$, and j satisfies $0 \le j \le L_2$. The original link input, $u_{1,i}$ of CNN_1 and $u_{2,j}$ of CNN_2 are respectively initialized as follows:

$$u_{1,i} = \begin{cases} null, & \text{if } i = 0; \\ S_1(i-1), & \text{if } 1 \le i \le L_1. \end{cases} \tag{2}$$

$$u_{2,j} = \begin{cases} null, & \text{if } j = L_2; \\ S_2(L_2 - j - 1), & \text{if } 0 \le j \le L_2 - 1. \end{cases} \tag{3}$$

where $'null'$ represents the blank space character, which will be inserted into the Chinese texts when the Chinese texts are globally aligned.

Moreover, both the parameter C and R_x are set to 1, and $I_{1,i}$ is set to the constant 2. The initial state $x_{1,i}(0) = 0$, where $0 \le i \le L_1$. The two templates, **A** and **B**, are set to $\mathbf{A} = [0, 0, 0]$ and $\mathbf{B} = [0, 1, -1]$, respectively.

3.2 The Computation of the Optimal State Matrix

After the initialization of the 1D-PW CNN, the next step is to compute the optimal cell state, $\widehat{x}_{1,i}(t)$ of CNN_1 at each time t, where $0 \leq i \leq m-1$ and $1 \leq t \leq m+n-1$.

Furthermore, define O_1, O_2 and O_3 respectively as $O_1 = x_{1,i-1}(t-2) + 2I_{1,i} \times y_{1,i}(t)$, $O_2 = x_{1,i}(t-1) - I_{1,i}$ and $O_3 = x_{1,i-1}(t-1) - I_{1,i}$, where both $x_{1,i}(t)$ and $y_{1,i}(t)$ are computed via Eqn. (1). Based on these definitions, at each time t the $\widehat{x}_i(t)$ is calculated by

$$\widehat{x}_i(t) = \max\left\{O_1, O_2, O_3\right\} \tag{4}$$

Then replace $x_{1,i}(t)$ with the optimal cell state $\widehat{x}_{1,i}(t)$. Continue to iteratively compute all the optimal cell state, $x_{1,i}(t)$ by the rule presented above until the 1D-PW CNN stops at the time $t = m+n-1$.

By the iterative computation, a series of cell state values can be obtained, such as $x_{1,0}(0)$ at $t=0$, $x_{1,0}(1)$ and $x_{1,1}(1)$ at $t=1$, $x_{1,0}(2)$, $x_{1,1}(2)$ and $x_{1,2}(2)$ at $t=2$, ..., and the last one $x_{1,m}(m+n-1)$ at $t=m+n-1$. And then, build the match matrix, $O_{m,n}$ as the following pattern

$$O_{m,n} = \begin{pmatrix} x_0(0) & x_1(1) & x_2(2) & ... & x_m(m) \\ x_0(1) & x_1(2) & x_2(3) & ... & x_m(m+1) \\ x_0(2) & x_1(3) & x_2(4) & ... & x_m(m+2) \\ x_0(3) & x_1(4) & x_2(5) & ... & x_m(m+3) \\ ... & ... & ... & ... & ... \\ x_0(n) & x_1(n+1) & x_2(n+2) & ... & x_m(m+n-1) \end{pmatrix} \tag{5}$$

Based on the built match matrix, the global alignment path, P (the initial $P = \varnothing$) can be determined by backtracking $O_{m,n}$. The generation rule of the alignment path is described as the following steps:

1. start from $x_m(m+n-1)$, the right-bottom element of $O_{m,n}$, let $P = \left\{x_m(m+n-1)\right\}$, and $i = m, t = m+n-1$.
2. while $i \neq 0$ and $t \neq 0$, do the loops:
   - let $i_1 = i_2 = i-1$, $i_3 = i$, $t_1 = t_3 = t-1$, $t_2 = t-2$;
   - decide parameters $i' = i_{i_{k'}}$ and $t' = t_{t_{k'}}$ by
   $$k' = \arg \max_{k \in \{1,2,3\}} x_{i_k}(t_k)$$
   - find the node state $x_{i'}(t')$;
   - insert $x_{i'}(t')$ into the path P, as followed by $P = \left\{x_{i'}(t')\right\} \bigcup P$;
   - let $i = i'$ and $t = t'$.
3. obtain the optimal path node set P;
4. the algorithm completes.

In fact, for the global alignment path P, because the back-tracing process starts from $x_m(t = m+n-1)$ and completes at $x_0(t=0)$, the first element of P is just $P_0 = x_0(0)$, and the last one should be just $P_m = x_m(t = m+n-1)$. Similarly, the second element, P_1 is just the one which is obtained at the time $t=1$, and the ith $(1 < i < m+n-1)$ one should be $P_i = x_{k_i}(t=i)$, the cell state obtained at the time $t=i$, and k_i can be directly seen from P.

3.3 The Global Alignment of Two Chinese Texts

After the optimal path P has been obtained, the two Chinese text character sequences, S_1 and S_2, can be aligned under the guidance of P. Suppose, the alignment path, P is presented as $P = \{P_0, P_1, P_2, ...P_k\}$, and so the length of P is equal to $k+1$. Just as mentioned above, $P_0 = x_0(0)$ and $P_k = x_m(m+n)$. The two sequences will be aligned as follows.

Firstly, define the element location index k' for P, where $0 \leq k' \leq k$, i' as the character location index of S_1, and j' as the character index of S_2.

1. let i'=0, $j' = 0$, $k' = 0$;
2. based on P, determine $P_{k'} = x_a(t_a)$, $P_{k'+1} = x_b(t_b)$;
3. compare a, b, t_a and t_b, then S_1 and S_2 are processed on the three different decision conditions:
   - if $b - a = t_b - t_a = 1$, a character '$*$' is inserted into S_2 at the location j', and then let $S_2'(j'+1) = S_1(j')$, where $j' = 0, 1, 2, ..., n-1$;
   - if $b = a$ and $t_b - t_a = 1$, a character '$*$' is inserted into S_1 at the location index i', and then let $S_1'(i'+1) = S_1(i')$, where $i' = 0, 1, 2, ..., m-1$;
   - if $b - a = 1$ and $t_b - t_a = 2$, nothing is done;
4. Let $k' = k'+1$, $i' = i'+1$ and $j' = j'+1$;
5. If $k' < k$, go to step 3, else go to step 7;
6. The global alignment of the two text sequences completes.

By now, these two Chinese texts have been aligned. By the global alignment, S_1 and S_2 have become the two texts with the same sequence lengths.

3.4 Text Similarity Computation

Once the global alignment of two texts completes, the global similarity between the two text sequences will easily be computed. To express conveniently, define $GS(S_1, S_2)$ to represent the numerical value of the global similarity between the text sequences S_1 and S_2.

In general, there are many criteria which could be used for the numerical representation of the text similarity. In this paper, based on the classic Euclidean distance, the global similarity of texts is computed as:

$$GS(S_1, S_2) = \frac{2 \times N_{match}}{Len(S_1) + Len(S_2)} \times 100\% \tag{6}$$

where N_{match} is the number of the character pairs which have been correctly aligned, $Len(S_1)$ represents the text length of S_1, and $Len(S_2)$ represents the text length of S_2.

4 Experiments and Comparisons

To evaluate the performance and the merit of the proposed global similarity computation algorithm for the two Chinese Texts, some simulation and comparison experiments have be carried out. In these experiments, all these compared algorithms are simulated using the MATLAB software (version: R2011a), and then executed on the same computer with 3.2GHz CPU, 2GB memory and Windows XP system.

All the Chinese texts for the experiments, more than 100 Chinese text pairs, come from the Baidu search website and the CNKI (China National Knowledge Infrastructure) database, including the Micro-blog (like twitter), news, reviews, papers, and so on. Furthermore, the length range of these Chinese texts is between 10 and 10 thousands.

To evaluate the performance of our algorithm, the other three methods have been chosen for the efficiency comparisons, and they are the MCALIGN2 by Wang [19] , the global alignment computation using GSA tree by Zhao [20], and the global alignment using GASSST by Guillaume [21].

4.1 Similarity Comparisons

In the global similarity experiments, the 100 chosen Chinese text pairs are tested one by one, and the obtained similarity results by the four algorithms are saved. In order to intuitively exhibit these comparison results and then analyze them, the 100 Chinese text pairs have been sorted according to the similarity detected by the proposed 1D-PW CNN in an ascending order (labeled with '#').

Table 1. The partial experiment results on the similarity

Text Pairs(#)	5#	16#	39#	65#	76#	88#
Lengths	$Len(S_1)=566$	$Len(S_1)=135$	$Len(S_1)=6454$	$Len(S_1)=54$	$Len(S_1)=3886$	$Len(S_1)=1447$
	$Len(S_2)=490$	$Len(S_2)=116$	$Len(S_2)=7129$	$Len(S_2)=32$	$Len(S_2)=4025$	$Len(S_2)=2523$
GSAT	5.68%	15.94%	37.61%	58.14%	64.52%	75.06%
GASSST	5.87%	16.73%	37.74%	58.14%	64.87%	75.62%
MCALIGN2	6.44%	16.73%	37.90%	58.14%	65.07%	75.92%
1DPW-CNN	6.82%	16.73%	38.25%	58.14%	65.43%	76.12%

Table 1 displays a part of the experiment results for the global similarity on the 100 text pairs, including pairs 5#, 16#, 39#, 65#, 76# and 88#. In the table, it can be seen that these four algorithms obtain the same similarity of 58.14% on the text pair 65#. And, on the pair 16#, 1DPW-CNN, MCALIGN2 and GASSST can also get the same similarity of 16.73%, however for the GSAT, the similarity obtained is only 15.94%, a little smaller than those by the others. On the other text pairs, these similarity values are completely different from each other.

In general, from table 1, it can also be seen that, for each text pair, the 1DPW-CNN algorithm often achieves the same similarity as the others, and it obtains the bigger similarity than the four methods sometimes. If the total length of S_1 and S_2 is small, for examples, $16^{\#}$ or $65^{\#}$, there is no significant similarity difference between the proposed method and the other three. However, with the increase of the total length of the text pairs, when $Len(S_1) + Len(S_2)$ is more than one thousand, such as $5^{\#}$, $39^{\#}$, $76^{\#}$ and $88^{\#}$, the 1DPW-CNN method will get significantly higher similarity than the other three.

4.2 Execution Time Comparisons

Besides the global similarity, just as mentioned before, the execution time of the algorithm is also another important evaluation criterion. Thus in experiments, the execution time of these algorithms to compute the similarity on the 100 Chinese text pairs is also tested and saved.

Table 2 exhibits a part of execution time in the experiments with the four methods, including the six Chinese text pairs. As a whole, for each one of the four computation methods, their execution time will become longer and longer with the increase of the total text length of S_1 and S_2.

Table 2. Partial Experiment Results of Execution Time

Text Pairs(#)	$5^{\#}$	$16^{\#}$	$39^{\#}$	$65^{\#}$	$76^{\#}$	$88^{\#}$
Lengths	$Len(S_1)$=566	$Len(S_1)$=135	$Len(S_1)$=6454	$Len(S_1)$=54	$Len(S_1)$=3886	$Len(S_1)$=1447
	$Len(S_2)$=490	$Len(S_2)$=116	$Len(S_2)$=7129	$Len(S_2)$=32	$Len(S_2)$=4025	$Len(S_2)$=2523
GSAT	85.21ms	76.15ms	674.15ms	58.22ms	399.61ms	170.55ms
GASSST	88.46ms	78.43ms	712.62ms	59.54ms	376.27ms	186.39ms
MCALIGN2	81.53ms	72.67ms	502.06ms	56.35ms	330.40ms	160.82ms
1DPW-CNN	78.16ms	71.35ms	353.71ms	56.08ms	279.62ms	152.69ms

According to the execution time of the four methods in Table 2, it is easy to see that, for each of the six exhibited Chinese text pairs, the needed time by 1DPW-CNN is often the shortest. In contrast, the computation time by the GASSST is the longest, and the computation time by the GSAT or the MCALIGN2 is usually more than that by the 1DPW-CNN and less than that by the GASSST.

Moreover, it could also be seen that if the total length of S_1 and S_2 is very small, such as pair $5^{\#}$, the difference of the execution time is also very small, only a few milliseconds. However, if the total length is more than one thousand, such as pair $88^{\#}$, the average difference of the execution time between the 1DPW-CNN and the others almost reaches to 20 milliseconds, approximately 13.1% of the total time by 1DPW-CNN. With the increase of the total length, the time difference becomes bigger and bigger, such as pairs 76# and $39^{\#}$. Especially, when the total length reaches to 10 thousands, such as pair 39#, the longest pair in the table, the time by the GASSST is 712.62 ms, which is approximately twice the length of the time by 1DPW-CNN, 353.71 ms.

5 Conclusions

The global similarity detection is an every important application technology in the Chinese information processing fields. This paper designs a new 1D-PW CNN, which consists of the immoveable master and the movable slave, to develop the global alignment algorithm for the two Chinese text sequences. Via the parallel computation of the cell states, the optimal state matrix is generated, and then the global alignment path is obtained by back-tracing the state matrix. Under the guidance of the alignment path, the two Chinese texts are globally aligned by inserting the blank characters into the texts at the appropriate locations, and then the numerical similarity between these two texts is computed. By the comparison experiments on some Chinese text samples, it shows that the method with the 1D-PW CNN developed in this paper could be efficient to globally align the text sequences. It could often obtain the higher similarity with the less computation time than the other three.

Acknowledgment. This work is supported by the National Science Foundation of China (NSFC) under the Grant number 61175061.

References

1. Yu, X., Jianxun, L., Mingdong, T., Yiping, W.: Empirical Study of Chinese Text Similarity Computation Based on Machine Translation. In: Proceedings of the Seventh International Conference on Semantics Knowledge and Grid (SKG 2011), pp. 156–159 (2011)
2. Fan, X., Qiaoming, Z., Peifeng, L.: Detecting Text Similarity over Chinese Research Papers Using MapReduce. In: Proceedings of 2011 12th ACIS International Conference on Software Engineering, Artificial Intelligence, Networking and Parallel/Distributed Computing (SNPD 2011), pp. 197–202 (2011)
3. Xiuhong, W., Shiguang, J., Shengli, W.: Challenges in Chinese Text Similarity Research. In: Proceedings of 2008 International Symposiums on Information Processing (ISIP 2008), pp. 297–302 (2008)
4. Xiangdong, L., Hangyu, L., Han, J., Huang, L.: Research on the categorization accuracy of different similarity measures on Chinese texts. In: Proceedings of 2011 International Conference on Business Management and Electronic Information (BMEI 2011), vol. 4, pp. 224–227 (2011)
5. McAllister, S.R., Rajgaria, R., Floudas, C.A.: Global pairwise sequence alignment through mixed-integer linear programming: a template-free approach. Optimization Methods and Software 22(1), 127–144 (2007)
6. Shiyi, S., Jun, Y., Adam, Y., Pei, H.: Super Pairwise Alignment (SPA): An Effecient Approach to Global Alignment for Homologous Sequences. Journal of Computational Biology 9(3), 77–486 (2002)

7. Pak, C.W., Kwong-kwok, W., Harlan, F., Jim, T.: Global visualization and alignments of whole bacterial genomes. IEEE Transactions on Visualization and Computer Graphics 9(3), 361–377 (2003)
8. Wu, S., Wu, Y.: Chinese and English Word Similarity Measure Based on Chinese WordNet. Journal of zhengzhou university (Natural Science Edition) 2, 66–69 (2010)
9. Ke, Z., Jun, L., Xilin, C.: Text similarity computing based on sememe Vector Space. In: Proceedings of 2013 4th IEEE International Conference on Software Engineering and Service Science (ICSESS 2013), pp. 208–211 (2013)
10. Metin, S.K., Kisla, T., Karaoglan, B.: Text similarity analysis using IR lists. In: Proceedings of Signal Processing and Communications Applications Conference (SIU 2013), pp. 1–4 (2013)
11. Chua, L.O., Yang, L.: Cellular neural networks: Theory. IEEE Trans. on Circuits and Systems 35, 1257–1272 (1988)
12. Yonggui, K., Cunchen, G.: Global exponential stability analysis for cellular neural networks with variable coefficients and delays. Neural Computing and Applications 17(3), 291–295 (2008)
13. Guowei, Y., Yonggui, K., Wei, L., Xiqian, S.: Exponential stability of impulsive stochastic fuzzy cellular neural networks with mixed delays and reaction Cdiffusion terms. Neural Computing and Applications 23(3-4), 1109–1121 (2013)
14. Amanatidis, D., Tsaptsinos, D., Giaccone, P., Jones, G.: Optimizing motion and colour segmented images with neural networks. Neurocomputing 62, 197–223 (2004)
15. Shukai, D., Xiaofang, H., Lidan, W., Shiyong, G.: Hybrid memristor/RTD structure-based cellular neural networks with applications in image processing. Neural Computing and Applications 25(2), 291–296 (2014)
16. Yangtao, L., Di, X., Huaqin, L., Shaojiang, D.: Parallel chaotic Hash function construction based on cellular neural network. Neural Computing and Applications 21(7), 1563–1573 (2012)
17. Manganaro, G., Gyvez, G.P.: One-dimensional discrete-time CNN with multiplexed template-hardware. IEEE Transactions on Circuits and Systems I: Fundamental Theory and Applications 5, 764–769 (2000)
18. Takahashi, N., Nagayoshi, M., Kawabata, S., Nishi, T.: Stable Patterns Realized by a Class of One-Dimensional Two-Layer CNNs. IEEE Trans. on Circuits and Systems: Regular Papers 11, 3607–3620 (2008)
19. Jun, W., Peter, D.K., Toby, J.: MCALIGN2: Faster, accurate global pairwise alignment of non-coding DNA sequences based on explicit models of indel evolution. BMC Bioinformatics 7, 292 (2006)
20. Zhao-hui, Q., Xiao-Qin, Q., Chen-chen, L.: New method for global alignment of 2DNA sequences by the tree data structure. Journal of Theoretical Biology 263, 227–236 (2010)
21. Guillaume, R., Dominique, L.: GASSST: global alignment short sequence search tool. Bioinformatics 26(20), 2534–2540 (2010)

A New Approach for Product Design by Integrating Assembly and Disassembly Sequence Structure Planning

Samyeon Kim[1], Jung Woo Baek[1], Seung Ki Moon[1,*], and Su Min Jeon[2]

[1] School of Mechanical and Aerospace Engineering,
Nanyang Technological University, Singapore
{samyeon001,JWBaek,skmoon}@ntu.edu.sg
[2] Planning and Operation Management,
Singapore Institute of Manufacturing Technology, Singapore
smjeon@simtech.a-star.edu.sg

Abstract. Assembly and disassembly sequences were planned and assessed independently. However, as eco-friendly product design has been enforced by government regulation, it is needed to plan assembly and disassembly sequence at the same time in early product design stage. This paper aims to propose a sequence structure design method by integrating both assembly and disassembly sequence planning. First, graphs represent feasible assembly sequence with components. Second, adjacency matrices are developed to accumulate the sequence information for computational analysis. Third, a black box diagram is applied to understand and consider the functions of a product during assembly and disassembly operations. Finally, factors are determined to overcome gaps between each assembly and disassembly sequence structure. By considering product architecture with Design Structure Matrix (DSM), we can determine modular design for selective disassembly and interface type for assembly. A case study is performed with two coffee makers to demonstrate the effectiveness of the proposed method.

Keywords: Assembly sequence structure planning, Disassembly sequence structure planning, Product recovery.

1 Introduction

Optimal product design affects the improvement of manufacturing efficiency and the ease of use for customers as well. Product design based on a systematic assembly sequence enables firms to reduce manufacturing lead-time, and thus, to improve the efficiency in production process. On the other hand, by considering the disassembly and reassembly sequences, the firms can develop the product design that provides ease of use to attract consumer preferences.

There are many studies in assembly sequence planning and disassembly sequence planning independently to minimize operation time and cost together with government regulation. The production process in the companies is developed based on the design of each component and assembly sequence. The assembly process consists of

* Corresponding author.

H. Handa et al. (eds.), *Proc. of the 18th Asia Pacific Symp. on Intell. & Evol. Systems – Vol. 1*,
Proceedings in Adaptation, Learning and Optimization 1, DOI: 10.1007/978-3-319-13359-1_20

preparing all parts and subassemblies, deciding interface types for assembly, performing functional test and inspection, and packaging the product. Assembly would have influences on manufacturing time and cost together with product quality. Therefore, assembly-oriented design is proposed to reduce manufacturing cost [1].

On the other hand, a company has started to consider the disassembly-oriented design for maintenance and product recovery to decrease the amount of wasted product to landfill. This is because the company can rapidly launch a new product and lifecycle of the new product is getting shorten because of changing customer needs. Then, the quantity of the wasted products to landfill is increasing. For several decades, governments have enforced manufacturers to produce environmental friendly product design and manufacturing including product recovery. Therefore, the manufacturers have to follow the regulations from the conceptual design to the disposal, i.e., during the product life cycle. Product recovery is needed to minimize the amount of landfill disposal of waste and the usage of natural resources by recovering obsolete parts to serviceable parts [2]. There are several recovery options: recycling for recovering raw materials by carrying out disassembly, reuse to use an item for its own original purpose, reconditioning to use an item for its original purpose with value added, and refurbishment to restore its original purpose [2]. To perform all these recovery options, disassembly operations should be performed and then it is needed to satisfy a desired level of quality for customers after reassembly operations. Disassembly for the product recovery is not always same as reverse sequences of assembly but is able to separate demanded parts selectively [3]. Therefore, disassembly sequence planning is significant to minimize disassembly time and cost by maximizing easiness of disassembly and the quality of reassembled products in terms of customer perspective.

In the previous research, the assembly and disassembly sequence planning were performed independently. Disassembly carries out between usage step and the end of product life. However, the disassembly planning should be considered for easiness to maintain and product recovery at the beginning of product design with assembly planning. In this paper, we propose a novel product design framework from both the company and customer perspectives while considering both assembly and disassembly sequence structure planning. The proposed framework can help to reduce assembly time, maintain products, and reuse end of life products.

In this paper, Section 2 presents a literature review. A method for assembly and disassembly sequence planning is described in Section 3. To demonstrate the effectiveness of the proposed method, a case study is performed with coffee makers in Section 4. The conclusion and future work are presented in Section 5.

2 Literature Review

Assembly sequence planning plays important role in assembly cost [1]. Assembly sequence planning is to deploy components according to sequence to produce a final product based on companies' strategies. The purpose of the assembly planning is to minimize assembly time and cost [4]. The starting point of assembly sequence planning is to model relationships among components by a liaison graph [5], AND/OR graph [6], and assembly precedence diagram which is a directed and acyclic graph

[7]. Based on these graph representations, the optimal assembly sequence is identified by using artificial intelligence algorithms such as genetic algorithm [8] and particle swarm optimization [4].

For disassembly sequence planning, the needs of product recovery and maintenance demands disassembly. The research for disassembly is mainly focused on disassembly sequence planning by minimizing disassembly cost. Also, as complete disassembly is not cost effective and practical, a disassembly sequence planning emphasizes on selective disassembly for product recovery and maintenance. The disassembly sequence before optimization is represented by graphical networks based on precedence relations. Lambert [9] described three types of disassembly representation: tree representation, state representation, and AND/OR graph representation. These representations are starting point of disassembly sequence planning by visualizing feasible disassembly operations. The representations contain nodes which are states of disassembly and arcs which describe disassembly operation based on precedence relationship. Based on graph representation, Behdad et al. [10] determined an optimal disassembly sequence planning by applying stochastic modeling and immersive computing technology. The stochastic modeling resulted in minimizing disassembly time and damage to components during the operations, while immersive computing technology was applied to identify solutions based on human intuition. Askiner and Gupta [3] proposed an approach to generate disassembly sequence plan by using hierarchical tree representation and branch and bound algorithm. They also resulted in the optimal disassembly sequence with the lowest cost. Smith and Chen [11] presented a rule-based recursive method for finding the lowest cost of selective disassembly sequence planning. Sung and Jung [12] developed a heuristic algorithm for disassembly planning and mathematical modeling to find minimum disassembly cost.

Based on literature review, in traditional sequence planning, the assembly sequence planning and disassembly planning were designed separately. This is, disassembly sequence was not considered when planning assembly sequence. Although products can be produced with lowest assembly cost according to companies' plan, customers might spend heavy charges while disassembly operations because a product is not considered to add disassembly strategies for maintenance. The disassembly sequence and the reverse order of assembly sequence are not always same. Consequently, assembly and disassembly sequence planning should be considered at the same time in product design as shown in Fig. 1.

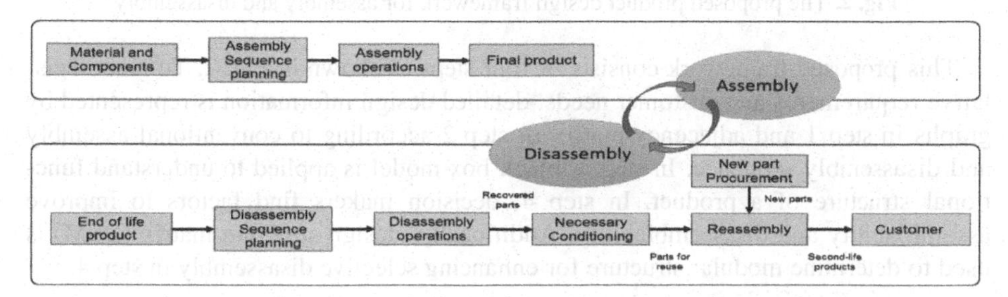

Fig. 1. Sequence planning by integrating assembly and disassembly (Modified from Kwak[13, p.6])

3 A Product Design Framework for Assembly and Disassembly

A novel product design framework is introduced to achieve the production efficiency from company's perspective and to meet the customer's need. From the company's perspective, the assembly sequence is needed to produce the final product from the stage of the material and component. The assembly sequence is closely related to improve production efficiency in the beginning of product lifecycle [14, 15, 16].

On the other hand, the customer satisfaction is related to how to maintain their end of life products. So, a disassembly sequence is required to dismantle a product to components for upgrade or maintenance of the product and then the components are reassembled to the product as shown in Fig. 1 [4, 10].

As shown in Fig. 2, we propose a product design framework by integrating assembly and disassembly planning. The proposed method is related to design for manufacturing (DFM) which is mind-set of decision makers in conceptual design stage in order to produce products more easily and economically [17].

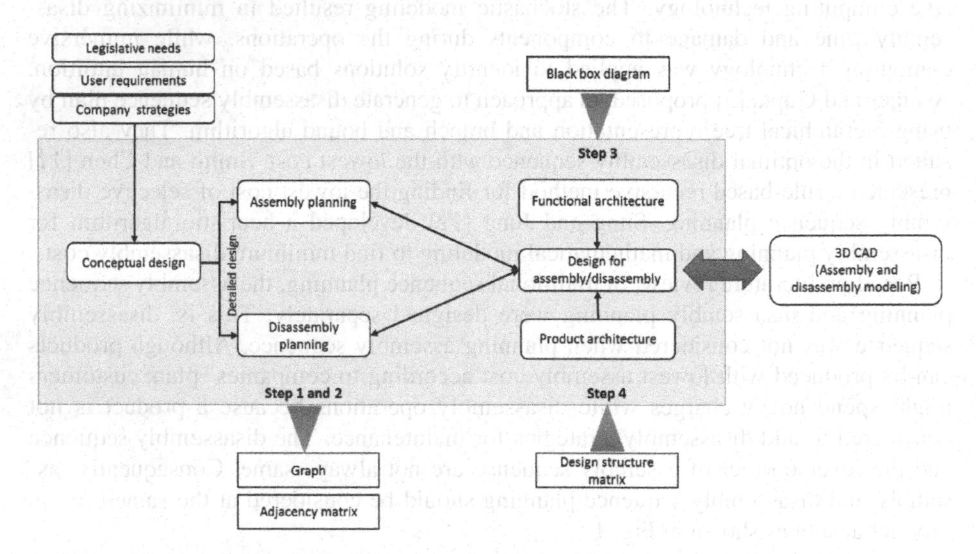

Fig. 2. The proposed product design framework for assembly and disassembly

This proposed framework consists of four steps as shown in Fig. 2. To meet legislative requirements and customer needs, detailed design information is represented by graphs in step 1 and adjacency matrix in step 2 according to conventional assembly and disassembly sequence. In step 3, black box model is applied to understand functional structure of a product. In step 4, decision makers find factors to improve assemblability and disassemblability. Additionally, design structure matrix (DSM) is used to determine modular structure for enhancing selective disassembly in step 4.

Step 1. Graph representation for assembly and disassembly sequence

The conceptual design according to customer needs and government regulation for environment is translated to graph representation for assembly and disassembly sequence planning. The each graph contains assembly and disassembly sequence from experts' knowledge. In the graph, each node represents component and arcs link between nodes with direction and precedence. After developing the graph for assembly sequence, the graph for disassembly is followed reverse order of the assembly sequence.

Table 1. Assembly type, graph, and matrix information (Source: Demoly et al. [14, p.40])

Assembly type	Directed graph	Adjacency matrix	Assembly time	Work content	Example
Interconnected serial (1)	t_{12}, t_{13}, t_{23}	$\begin{bmatrix} 0 & 1 & 1 \\ -1 & 0 & 1 \\ -1 & -1 & 0 \end{bmatrix}$	$t_{12}+t_{23}$	$wc_{12}+wc_{23}$	
Serial (2)	t_{12}, t_{23}	$\begin{bmatrix} 0 & 1 & 0 \\ -1 & 0 & 1 \\ 0 & -1 & 0 \end{bmatrix}$	$t_{12}+t_{23}$	$wc_{12}+wc_{23}$	
Constrained serial (3)	t_{12}, t_{13}	$\begin{bmatrix} 0 & 1 & 1 \\ -1 & 0 & \lambda \\ -1 & -\lambda & 0 \end{bmatrix}$	$t_{12}+t_{13}$	$wc_{12}+wc_{13}$	
Parallel (4)	t_{12}, t_{13}	$\begin{bmatrix} 0 & 1 & 1 \\ -1 & 0 & 0 \\ -1 & 0 & 0 \end{bmatrix}$	$\mathrm{Max}(t_{12}, t_{13})$	$wc_{12}+wc_{13}$	

Step 2. Develop the adjacency matrix for assembly and disassembly

As an adjacency matrix contains directional nature and dependency relationships among parts, the graph representation in step 1 can be translated to the adjacent matrix. The each adjacent matrix contains assembly and disassembly sequence respectively. For disassembly, the adjacent matrix for disassembly is developed based on the transpose of the DSM for assembly. It means that the direction of arcs in the graph is changed to the opposite direction. These matrices will be helpful to save these information on the products in the company's database and then utilize the information to determine optimal sequence planning by using simulation tools and/or computational algorithms.

In steps 1 and 2, the graph and adjacency matrix are developed based on information in Table 1 [14].

$$C_{ij} = \begin{bmatrix} 0 & c_{12} & \cdots & c_{1n} \\ c_{21} & 0 & \cdots & c_{2n} \\ \vdots & \vdots & 0 & \vdots \\ c_{n1} & c_{n2} & \cdots & 0 \end{bmatrix} \tag{1}$$

In the adjacency matrix, C_{ij}, if there are dependency relationship between components i and j, the value in c_{ij} is equal to one and it means that component j is assembled after assembling component i. When there are no physical contact but is precedence relationships, the value of c_{ij} is λ. If there are no relationships, the value in c_{ij} is equal to zero.

There are four assembly types in Table 1: an interconnected serial type that part 1 passes through parts 2 and 3, a serial type, a constrained serial type that part 2 is assembled before part 3 in spite of no relationship, and a parallel type. Therefore, after steps 1 and 2, decision makers can decide assembly types from characteristics of the graph and adjacency matrix.

Step 3. Black box diagram for identifying functional priority
When designing a product with assembly and disassembly sequence planning, designers should understand the functional structure of the product. Therefore, a black box diagram is applied to represent functional structure of a product in the proposed method. The diagram provides a way to understand comprehensive material, energy, and signal flows together with function of modules or parts. Since this step supports a product information in detail, decision makers would make a plan for how to enhance assemblability and disassemblability with keeping function of the product. Also, it would be helpful for a decision maker to decide the functional priority of components while detaching parts.

Step 4. Determine factors to overcome gaps between assembly and disassembly sequences
In this step, decision makers need to determine and improve how to disassemble and assemble a product for maintenance. Assembly and disassembly sequence are related to the functional structure in step 3 as well as physical interrelationship among components. Therefore, design structure matrix is applied to support design for assembly and disassembly sequence by considering the relationship among components. If a company considers that disassembly order is the same as the reverse order of assembly order, customers might spend heavy charges for disassembly and the reassembled product could not guarantee the quality of the product. Therefore, factors improving disassembly operations are determined in this step. Then, these factors are applied to improve assembly and disassembly structures for verification.

4 Case Study

To demonstrate the effectiveness of the proposed method, a case study carried out with two coffee makers, Philips HD7450 [18] and HD7458 [19] in Table 2. These coffee makers in the same manufacturer have simple functions to brew coffee for home use. There are four main parts for the coffee maker like a water reservoir (upper casing), heater, decanter and filter. The parts are mainly performed to achieve a main function of the coffee maker. The main function is to brew coffee with consistent steam supplement. Three sub-functions, which is water/coffee capacity, heating water, and the durability related to product life, are needed to achieve the main function for the coffee maker.

Table 2. Specifications of two coffee makers

Model	HD7450 (Philips)	HD7458 (Philips)
Price	S$49	S$89
Specification	220V, 50Hz, 650W 4~6 cups	220V, 50Hz, 1000W 10~15 cups

Step 1. Graph representation for assembly and disassembly
In this case study, it is difficult to know exact assembly order of their company by product dissection. Accordingly, we assume that the assembly sequence is the reverse order when reassembling component by component of coffee makers. Assembly and disassembly sequence structure graph are determined in Fig. 3. The dash arrows in the graph give precedence relationships among non-connected components. The starting point of the dash arrow has high precedence to operate. Since these two coffeemakers have different design but have the same functional and physical structure, the graphs are the same in the coffee makers.

Step 2. Develop the adjacency matrix for assembly and disassembly
Based on the graph representations, the adjacency matrices are developed as shown in Fig. 4. These adjacency matrices are not organized, so that it is hard to determine assembly types without graph. Accordingly, these matrices would be organized by using the product architecture in Step 4.

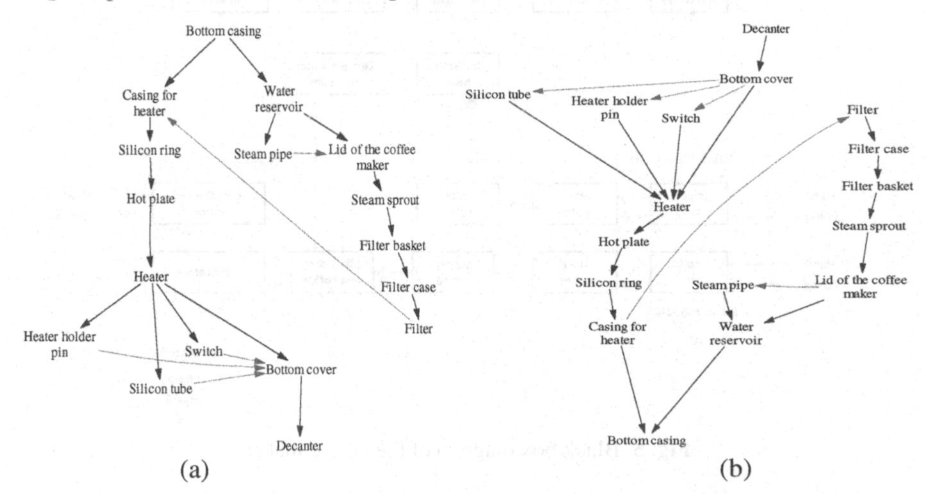

Fig. 3. Sequence graph of the coffee maker for (a) assembly and (b) disassembly

	Assembly	1	2	3	4	5	6	7	8	9	10	11	12	13	14	15	16	17
1	Decanter	0									-1							
2	Switch		0	-1							λ							
3	Heater		1	0							1		-1	1	1			
4	Lid of the maker				0		-λ		-1							1		
5	Filter basket					0										-1	1	
6	Steam pipe				λ		0		-1									
7	Filter							0		λ							-1	
8	Water reservior				1		1		0									-1
9	Casing for heater							-λ		0		1						-1
10	Bottom cover	1	-λ	-1							0			-λ	-λ			
11	Silicon ring									-1		0	1					
12	Hot plate			1								-1	0					
13	Heater holder pin			-1							λ			0				
14	Silicon tube			-1							λ				0			
15	Stem sprout				-1	1										0		
16	Filter case					-1		1									0	
17	Bottom casing								1	1								0

(a)

	Disassembly	1	2	3	4	5	6	7	8	9	10	11	12	13	14	15	16	17
1	Decanter	0									1							
2	Switch		0	1							-λ							
3	Heater		-1	0							-1		1	-1	-1			
4	Lid of the maker				0		λ		1							-1		
5	Filter basket					0										1	-1	
6	Steam pipe				-λ		0		1									
7	Filter							0		-λ							1	
8	Water reservior				-1		-1		0									1
9	Casing for heater							λ		0		-1						1
10	Bottom cover	-1	λ	1							0			λ	λ			
11	Silicon ring									1		0	-1					
12	Hot plate			-1								1	0					
13	Heater holder pin			1							-λ			0				
14	Silicon tube			1							-λ				0			
15	Stem sprout				1	-1										0		
16	Filter case					1		-1									0	
17	Bottom casing								-1	-1								0

(b)

Fig. 4. The adjacency matrix for (a) assembly and (b) disassembly

Step 3. Black box diagram for identifying functional priority
Steps 1 and 2 are focused on assembly and disassembly sequence based on product architecture. However, to improve these sequence structure, decision makers should understand function of each component or module of the coffee makers. The diagram is utilized to describe functional structure of these coffee makes in Fig. 5. This diagram shows five main components with their functions as well as flow of material, energy, and signal. For example, if we check water flow, the cold water in a water reservoir pass through a heater. In the heater, the cold water is changed to steam and then mixed with coffee in filter. Finally, coffee is brewed and then stored in a decanter.

Decision makers can easily understand which parts are disassembled for maintenance or upgrade. Also, decision makers can make plans to reduce probability to incurring damage during disassembly and reassembly.

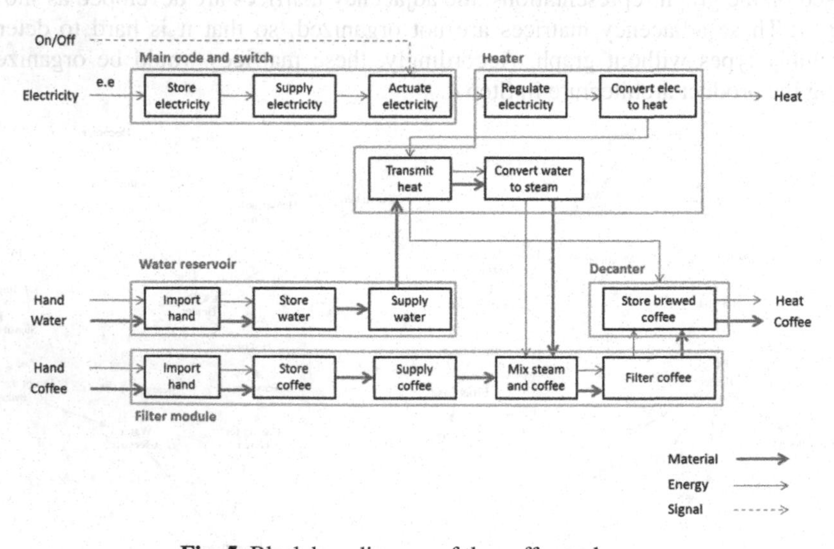

Fig. 5. Black box diagram of the coffee maker

Step 4. Determine factors to overcome gaps between assembly and disassembly sequences

To overcome gaps between assembly and disassembly sequence structures, it is needed to understand the product architecture of the coffee makers. As the adjacency matrices for assembly and disassembly includes information on the assembly and disassembly order, information on product architecture is needed to develop the orders. Therefore, design structure matrix (DSM) is applied to represent interrelationships among components in detail and modularize the components. Consequently, we consider that the first factor to overcome the gaps is disassembility with modularization. In the case study, based on DSM in Fig. 6, three modules are generated by DSM clustering algorithm by Thebeau [20]. In Fig. 7, the first module is related to filter parts. The second module is for heater parts and the third module includes components in the upper casing part of the coffee maker, such as a steam pipe, silicon tube, water reservoir, stream sprout, and lid of the coffee maker. The modularization that is clustering components into chunks is to make high internal coupling within modules and less coupling between modules. Therefore, the assembly and disassembly sequence structures are changed by using the modules as shown in Fig. 8. The structure with modules would improve selective disassembility by changing series structure of components to parallel structure of modules. For example, when removing the water reservoir based on existing disassembly graph in Fig. 3, the water reservoir can be removed after detaching casing for the heater and filter parts. However, the water reservoir can be disassembled selectively without detaching the casing for the heater and filter by considering modular product structure.

	DSM	1	2	3	4	5	6	7	8	9	10	11	12	13	14	15	16	17
1	Bottom cover	0																1
2	Silicon ring		0	1	1													1
3	Hot plate		1	0	1	1											1	1
4	Casing for heater	1	1	1	0	1		1										1
5	Heater			1	1	0	1	1		1							1	1
6	Switch					1	0											1
7	Heater holder pin				1	1		0										
8	Steam pipe								0	1								
9	Silicon tube					1			1	0	1	1						1
10	Water reservoir								1	1	0	1	1	1		1		
11	Steam sprout								1	1		0	1	1		1		
12	Filter basket										1		0	1	1		1	
13	Filter												1	0	1			
14	Filter case												1	1	0			
15	Lid of coffee maker										1	1	1			0		1
16	Decanter		1	1		1											0	1
17	Bottom Casing	1		1	1	1	1			1	1						1	0

Fig. 6. Product architecture representation based on DSM

	Assembly	1	2	3	4	5	6	7	8	9	10	11	12	13	14	15	16	17
1	Decanter	0																-1
2	Filter basket		0	1												-1		
3	Filter case		-1	0	1													
4	Filter			-1	0	λ												
5	Casing for heater				-λ	0	1										-1	
6	Silicon ring					-1	0	1										
7	Hot plate						-1	0	1									
8	Heater							-1	0	1	1	1						1
9	Heater holder pin								-1	0								λ
10	Switch								-1		0							λ
11	Silicon tube								-1			0						λ
12	Water reservior												0	1	1		-1	
13	Steam pipe												-1	0	λ			
14	Lid of the maker												-1	-λ	0	1		
15	Stem sprout		1												-1	0		
16	Bottom casing					1							1				0	
17	Bottom cover	1							-1	-λ	-λ	-λ						0

	Disassembly	1	2	3	4	5	6	7	8	9	10	11	12	13	14	15	16	17
1	Decanter	0																1
2	Filter basket		0	-1												1		
3	Filter case		1	0	-1													
4	Filter			1	0	-λ												
5	Casing for heater				λ	0	-1										1	
6	Silicon ring					1	0	-1										
7	Hot plate						1	0	-1									
8	Heater							1	0	-1	-1	-1						-1
9	Heater holder pin								1	0								-λ
10	Switch								1		0							-λ
11	Silicon tube								1			0						-λ
12	Water reservior												0	-1	-1		1	
13	Steam pipe												1	0	-λ			
14	Lid of the maker												1	λ	0	-1		
15	Stem sprout		-1												1	0		
16	Bottom casing					-1							-1				0	
17	Bottom cover	-1							1	λ	λ	λ						0

Fig. 7. The adjacency matrix with modules for (a) assembly (b) disassembly

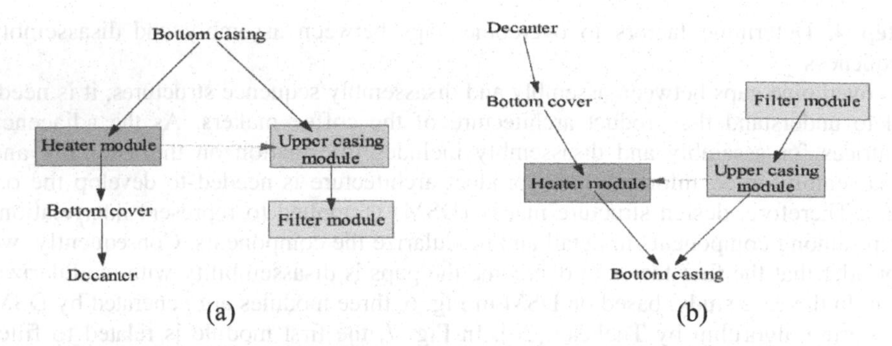

Fig. 8. Sequence structure based on modules for (a) assembly and (b) disassembly

The second factor is interfaces between modules and components. The interfaces between parts and/or modules play pivotal roles in product quality after reassembly. In the HD7450 coffee maker, many screws are used to combine components, compared to the HD7458 coffee maker. In the HD7458, unlike the HD7450 with screws, snap fits are used to combine the water reservoir and bottom casing. Also, snap fits with a steel stick are used to fix the casing for the heater to the bottom casing. The snap fit have a role to minimize the number of parts like screw fasteners and improve efficiency of assembly and reassembly operations. However, it spends more time to disassemble than time to disassemble parts fastened by screws. After recursive assembly and disassembly of two coffee makers, the parts in HD7458 are still strongly mated, while the parts fastened by screws in the HD7450 is not strongly assembled because of abrasion in plastic parts and screw heads.

5 Closing Remarks and Future Work

During assembly and disassembly sequence planning, most of literatures focus on minimum time and cost. In this paper, the proposed framework aims to design assembly sequence structure integrated with disassembly sequence structure simultaneously. Therefore, in the case study, two factors which are modularization of components and interface design are identified to consider simultaneous design for assembly and disassembly. We figured out that the modularization of components can improve selective disassembly and the interface design would help for customer to enhance mating quality of the reassembled product. Analyzing assembly/disassembly interface can link to components inventory tracking system for efficient supply and delivery.

Based on the proposed sequence structures, a possible further research issues is to develop method for choosing and evaluating inventory management policy for multiple components assembly /disassembly manufacture system in supply chain.

Acknowledgements. This work was supported by SIMTech-NTU joint research project and an AcRF Tier 1 grant from Ministry of Education, Singapore.

References

1. Boothroyd, G., Dewhurst, P., Knight, W.A.: Product design for manufacture and assembly. CRC Press (2010)
2. Kwak, M., Kim, H.: Market Positioning of Remanufactured Products With Optimal Planning for Part Upgrades. Journal of Mechanical Design 135, 011007 (2013)
3. GÜngÖr, A., Gupta, S.M.: Disassembly sequence plan generation using a branch-and-bound algorithm. International Journal of Production Research 39, 481–509 (2001)
4. Tseng, Y.-J., Yu, F.-Y., Huang, F.-Y.: A green assembly sequence planning model with a closed-loop assembly and disassembly sequence planning using a particle swarm optimization method. The International Journal of Advanced Manufacturing Technology 57, 1183–1197 (2011)
5. De Fazio, T., Whitney, D.E.: Simplified generation of all mechanical assembly sequences. IEEE Journal of Robotics and Automation 3, 640–658 (1987)
6. Homem de Mello, L.S., Sanderson, A.C.: Representations of mechanical assembly sequences. IEEE Transactions on Robotics and Automation 7, 211–227 (1991)
7. Lin, A.C., Chang, T.C.: An integrated approach to automated assembly planning for three-dimensional mechanical products. THE International Journal of Production Research 31, 1201–1227 (1993)
8. Tseng, Y.-J., Lin, C.-H., Lin, Y.-H.: Integrated assembly and machining planning for electronic products using a genetic algorithm approach. The International Journal of Advanced Manufacturing Technology 36, 140–155 (2008)
9. Lambert, A.J.D.: Disassembly sequencing: A survey. International Journal of Production Research 41, 3721–3759 (2003)
10. Behdad, S., Berg, L.P., Thurston, D., Vance, J.: Leveraging Virtual Reality Experiences With Mixed-Integer Nonlinear Programming Visualization of Disassembly Sequence Planning Under Uncertainty. Journal of Mechanical Design 136, 041005 (2014)
11. Smith, S.S., Chen, W.-H.: Rule-based recursive selective disassembly sequence planning for green design. Advanced Engineering Informatics 25, 77–87 (2011)
12. Sung, J., Jeong, B.: A heuristic for disassembly planning in remanufacturing system. The ScientificWorldJournal 2014, 949527 (2014)
13. Kwak, M.: Green profit design for lifecycle. University of Illinois at Urbana-Champaign (2012)
14. Demoly, F., Yan, X.-T., Eynard, B., Rivest, L., Gomes, S.: An assembly oriented design framework for product structure engineering and assembly sequence planning. Robotics and Computer-Integrated Manufacturing 27, 33–46 (2011)
15. Dong, T., Tong, R., Zhang, L., Dong, J.: A knowledge-based approach to assembly sequence planning. The International Journal of Advanced Manufacturing Technology 32, 1232–1244 (2006)
16. Ou, L.-M., Xu, X.: Relationship matrix based automatic assembly sequence generation from a CAD model. Computer-Aided Design 45, 1053–1067 (2013)
17. Poli, C.: Design for manufacturing [electronic resource]: a structured approach / Corrado Poli, c2001. Butterworth-Heinemann, Boston (2001)
18. `http://www.philips.com.sg/c/coffee-makers/daily-collection-with-glass-jug-black-hd7450_20/prd/`
19. `http://www.philips.com.sg/c/coffee-makers/viva-coffee-maker-hd7458_00/prd/`
20. Thebeau, R.E.: Matlab macro for clustering DSMs (2001), `http://www.dsmweb.org/?id=121`

Community Detection from Signed Social Networks Using a Multi-objective Evolutionary Algorithm

Yujie Zeng and Jing Liu

Key Laboratory of Intelligent Perception and Image Understanding of Ministry of Education
Xidian University, Xi'an 710071, China

Abstract. In this paper, we propose a method for detecting communities from signed social networks with both positive and negative weights by modeling the problem as a multi-objective problem. In the experiments, both real world and synthetic signed networks whose size ranges from 100 to 1200 nodes are used to validate the performance of the new algorithm. A comparison is also made between the new algorithm and an effective existing algorithm, namely FEC. The experimental results show that our algorithm obtains a good performance on both real world and synthetic data, and outperforms FEC clearly.

Keywords: Signed social networks, Community detection problems, Multi-objective evolutionary algorithms.

1 Introduction

The original community detection problems (CDPs)[1] are only designed for networks with positive links. But in social networks, there exist positive links as well as negative links. The positive links may denote friendship, trust, and like between two individuals, while the negative ones may represent hostility, mistrust, and dislike. The networks with both positive and negative relationships are named as signed social networks[2] or signed networks (SNs). Although many community detection methods have been proposed, most of them can only handle networks without negative links, namely unsigned networks[3-6, 12].

For SNs, communities are defined not only by the density of links but also by the signs of links. That is, within communities, the links should be positive and dense, and between communities, the links should be negative or positive and sparse. But this problem is by no means straightforward since it is natural to have some negative links within groups and, at the same time, some positive links between groups. Also, nodes connected by positive links do not belong to the same community, either. Thus, more robust community partitions should properly disregard and retain some positive and negative links so as to identify more natural communities[7]. There are also a few studies focusing on CDPs from SNs[13, 14, 26].

With the intrinsic properties of detecting communities from SNs in mind, we first model this problem as a multi-object problem (MOP). Evolutionary algorithms (EAs) are the most popular method for solving MOPs[8-11]. Therefore, we propose a

H. Handa et al. (eds.), *Proc. of the 18th Asia Pacific Symp. on Intell. & Evol. Systems – Vol. 1*,
Proceedings in Adaptation, Learning and Optimization 1, DOI: 10.1007/978-3-319-13359-1_21

multi-objective evolutionary algorithm (MOEA) for this problem, namely MOEA-SN. In the experiments, both real world and synthetic networks with different properties are used to test the performance of MOEA-SN, and a comparison is also made between MOEA-SN and FEC[7]. The results show that MOEA-SN obtains a good performance and clearly outperforms FEC.

The rest of this paper is organized as follows. Section 2 formally describes the CDPs from SNs and the objective functions we designed. The details of MOEA-SN are given in Section 3. Section 4 presents the experiments. Finally, Section 5 concludes the work by highlighting the contributions.

2 Objective Functions

Given a signed network $G=(V, E, w)$, where $V=\{v_1, v_2, ..., v_n\}$ is the set of nodes, $E\subseteq V\times V=\{(v_i, v_j) \mid v_i, v_j\in V \text{ and } i\neq j\}$ is the set of edges, and w_{ij} is the weight of the edge between nodes v_i and v_j which can be larger than 0 (positive relationship) or smaller than 0 (negative relationship). Let $C=\{C_1, C_2, ..., C_m\}$ be a set of communities in G; that is, $C_i\subset V$ for $i=1, 2, \ldots, m$. The problem of CD from SNs can be accurately expressed through the following conditions,

$$\begin{cases} w_{ij} > 0, & (v_i, v_j)\in E\wedge(v_i\in C_l)\wedge(v_j\in C_l) \\ w_{ij} < 0, & (v_i, v_j)\in E\wedge(v_i\in C_l)\wedge(v_j\in C_k)\wedge(l\neq k) \end{cases}, \quad l, k=1, 2, ..., m \tag{1}$$

The modularity Q proposed by Newman and Girvan[15] is a well known measure that can evaluate the goodness of a partition based on the comparison between the graph at hand and a null model, which is a class of random graphs with the same expected degree sequence of the original graph. Let e_{ij} be the fraction of edges in the network connecting vertices from community i to those of community j, then

$$Q=\sum_{i=1}^{m}\sum_{j=1}^{m}\left(e_{ij}-a_i^2\right) \tag{2}$$

where $a_i=\sum_{j=1}^{m} e_{ij}$.

If a network has n nodes, the weight of edges can be represented by an $n\times n$ adjacency matrix W, which can reflect the structure information of networks. However, it is difficult to extract structure information which can be used for community detection from the original adjacency matrix directly. Thus, we divide the adjacency matrix into two independent submatrices, namely W^+ and W^-, by separating the positive and negative links. In W^+,

$$w_{ij}^{+} = \begin{cases} w_{ij} & w_{ij} \geq 0 \\ 0 & w_{ij} < 0 \end{cases} \quad i, j\text{=1, 2, …, } n \tag{3}$$

In W,

$$w_{ij}^{-} = \begin{cases} 0 & w_{ij} \geq 0 \\ -w_{ij} & w_{ij} < 0 \end{cases} \quad i, j\text{=1, 2, …, } n \tag{4}$$

Then, the positive and negative strengths of node i are given by

$$w_i^{+} = \sum_{j=1}^{n} w_{ij}^{+} \quad i\text{=1, 2, …, } n \tag{5}$$

$$w_i^{-} = \sum_{j=1}^{n} w_{ij}^{-} \quad i\text{=1, 2, …, } n \tag{6}$$

Since the original Q is designed for unsigned networks, to deal with negative links, we separated evaluate the community structure represented by each submatrix,

$$Q^{+} = \frac{1}{w^{+}} \sum_{i=1}^{n} \sum_{j=1}^{n} \left(w_{ij}^{+} - \frac{w_i^{+} w_j^{+}}{w^{+}} \right) \delta(C_i, C_j) \tag{7}$$

$$Q^{-} = \frac{1}{w^{-}} \sum_{i=1}^{n} \sum_{j=1}^{n} \left(w_{ij}^{-} - \frac{w_i^{-} w_j^{-}}{w^{-}} \right) \delta(C_i, C_j) \tag{8}$$

where $w^{+} = \sum_{i=1}^{n} w_i^{+}$ and $w^{-} = \sum_{i=1}^{n} w_i^{-}$, and

$$\delta(C_i, C_j) = \begin{cases} 1, & C_i = C_j \\ 0, & C_i \neq C_j \end{cases} \tag{9}$$

is the Kronecker delta function, which takes the value 1 if nodes i and j belong to the same community, and 0 otherwise. In fact, Q^{+} takes into account the deviation of actual positive weights against a null case random network, and Q^{-} is its counterpart for negative weights[16].

Since we should maximize Q^{+} and minimize Q^{-}, we change Q^{-} to

$$Q^{-} = -\frac{1}{w^{-}} \sum_{i=1}^{n} \sum_{j=1}^{n} \left(w_{ij}^{-} - \frac{w_i^{-} w_j^{-}}{w^{-}} \right) \delta(C_i, C_j) \tag{10}$$

Thus, we have two objectives to maximize, which contradict with each other. Therefore, we can treat the CDP from SNs as a MOP.

3 MOEA-SN

3.1 Initialization

We use a traditional encoding method for community detection, namely symbol encoding [5, 17, 18]. An n-dimensional vector is used as the chromosomes and n is the number of vertices. In a chromosome, each gene represents the corresponding vertex, and the value of the gene is the label of community to which this vertex belongs. Thus, each chromosome corresponds to a partition of the network.

In general, we can initialize the first population at random, but it can not reflect some remarkable densely connected groups of nodes and the algorithm can not convergence to global optima quickly. Thus, we employ the initialization strategy called individual generation via label propagation (IGLP) proposed in [19]. In IGLP, every node is initialized with a unique label and at every step each node adopts the label that most of its neighbors currently have. In this iterative process, densely connected groups of nodes form a consensus on a unique label to form communities. Given a network with n nodes, let C_1, C_2, ..., C_m be the communities found. $C_x(t)$ is the x-th community in the t-th iteration. The initialization algorithm is described in Algorithm 1.

Algorithm 1. Initialization.

1. Initialize the labels of all nodes; that is, for each node x, $C_x(0) \leftarrow x$;
2. $t \leftarrow 1$;
3. Sort all nodes in a random order labeled as X;
4. For each $x \in X$, let $C_x(t) = f\left(C_{x_1}(t), \cdots, C_{x_{im}}(t), C_{xi_{(m+1)}}(t\text{-}1), \cdots, C_{x_{ik}}(t\text{-}1)\right)$, where f returns the label occurring with the highest frequency among neighbors and ties are broken uniformly randomly;
5. If every node has a label that the maximum number of their neighbors has, then stop the algorithm; otherwise, $t \leftarrow t+1$ and go to step 3.

3.2 Evolutionary Operators

There exists a problem in the representation [20]; that is, different chromosomes may correspond to the same network partition. For instance, given a graph of 5 nodes, {3, 1, 2, 3, 1} and {2, 3, 1, 2, 3} denote the same partition {{1, 4}, {2, 5}, 3}. In this case, the popular one-point, two-point, or multiple-point crossover operators used in the EA community may not work effectively. Therefore, the one-way crossover operator introduced in [6] is employed. In the one-way crossover operator, two chromosomes are chosen to be parent, and one of them is treat as a source chromosome and the other as a destination chromosome. We chose a community label randomly in the source chromosome as a destination label, and mark the vertices that belong to this community in the source chromosome as chosen vertices whose values of genes in destination chromosome will be changed. Then, replace the community labels of chosen vertices in the destination chromosome by the labels of chosen vertices in the source chromosome.

The mutation operator with a local search (LMA) [21] is employed. The main idea of LMA comes from the definition of Q,

$$Q = \frac{1}{2E}\sum_{i} f_i, \quad f_i = \sum_{j \in c_{r(i)}} \left(A_{ij} - \frac{k_i k_j}{2E} \right) \tag{11}$$

where E is number of edges, $r(i)$ is the label of the community that node v_i belongs to, and k_i is the degree of node v_i. In (11), the function f can be understood in this way: from the local viewpoint of any vertex, f is the fraction of edges inside communities minus the expected value of the fraction of edges if edges fall at random without regard to the community structure. So f of every vertex can evaluate the quality of network community from the local aspect.

Since the modularity we use is not the original one, here we make some improvement to make the definition of f function be suitable for SNs.

$$f_i = \sum_{j} \left[w_{ij} - \left(\frac{w_i^+ w_j^+}{w^+} - \frac{w_i^- w_j^-}{w^-} \right) \right] \delta(C_i, C_j) \tag{12}$$

From the definition of f we know that if the maximum value of f is found then the maximum value of Q can be found at the same time. Therefore, we search in neighbors of the mutate vertex i, and calculate the values of $f(i)$ if the community label of the mutate vertex is changed into its neighbors' community label. Search for the maximum value of all the $f(i)$ values and change the community label of the mutate vertex into the corresponding neighbor's community label. Algorithm 2 describes the mutation operator, where R is the chromosome and p_m is the mutation rate.

Algorithm 2. Mutation operator with a local search (LMA).

1. for i=1:n
2. if rand() < p_m then
3. $\sigma(i) \leftarrow$ the set of neighbors of node v_i;
4. $L \leftarrow$ the set of community labels of nodes in $\sigma(i)$;
5. $max \leftarrow -\infty$;
6. for each $r \in L$
7. Calculate f_i according to (12) when node v_i is in community r;
8. if $f_i > max$ then
9. $max \leftarrow f_i$;
10. Replace the community label of v_i by r;
11. end if;
12. end for;
13. end if;
14. end for;

3.3 Implementation of MOEA-SN

MOEA-SN borrows the main framework of NSGA-II, which is summarized in Algorithm 3. N is the population size, p_m is the mutation rate, p_c is the crossover rate, and *max_g* is the maximum number of generations. MOEA-SN works as follows: the first part is the initialization with IGLP. The second part is non-domination sorting of the parent population. The third part is performing evolutionary operators such as selection, crossover, and mutation. The last part is non-domination sorting of both parent population and child population and chose the next generation by the crowding distance.

Algorithm 3. MOEA-SN

1. Initialize the population with IGLP, and get the parent population P_t, $t \leftarrow 1$;
2. $F \leftarrow$fast-non-dominated-sort(P_t); //non-domination sorting of the parent population
3. $P_t \leftarrow$select(P_t); // generate the next generation
4. $Q_t \leftarrow$crossover(P_t, p_c);
5. $Q_t \leftarrow$mutation(Q_t, p_m);
6. $R_t \leftarrow P_t \cup Q_t$;
7. $F \leftarrow$fast-non-dominated-sort(R_t);
8. $P_{t+1} \leftarrow \emptyset$ and $i \leftarrow 1$; // chose the next generation by the crowding distance
9. while $|P_{t+1}|+|F_i| \leq N$
10. crowding-distance-assignment(F_i);
11. $P_{t+1} \leftarrow P_{t+1} \cup F_i$;
12. $i \leftarrow i+1$;
13. end while;
14. sort($F_i \prec n$); //Sort individuals in the *i*th layer according to the crowding distance
15. $P_{t+1} \leftarrow P_{t+1} \cup F_i[1:(N-|P_{t+1}|)]$;
16. $t \leftarrow t+1$;
17. if $t \leq max\_g$, go to step 2; otherwise stop.

4 Experiments

In this section, various networks are used to validate the performance of MOEA-SN, and a comparison is also made between MOEA-SN and FEC [7]. The crossover and mutate rates are set to 0.8 and 0.2, respectively. The maximum number of generations is set to 50, and the population size is set to 200. To evaluate the performance of these algorithms, the *Normalized Mutual Information* (*NMI*) metric[22] is adopted to estimate the similarity between the true partitions and the detected ones. Although the multi-objective algorithm can find a Pareto front which includes a set of solutions, we only need the best solution. Therefore, we choose the individual with maximum ($Q^+ + Q^-$) in the last generation as the final result.

4.1 Experiments on Two Illustrative Networks

The two illustrative networks come from [7]. As shown in Fig.1, solid lines denote positive links, and dashed lines denote negative links. The network shown in Fig.1(a) can be divided into three groups, namely (6, 7, 22, 23, 24, 25, 13, 14, 15, 16, 4, 5), (8, 9, 26, 27, 17, 18), and (20, 21, 10, 11, 12, 1, 2, 3, 19, 28). In the three-way partition, the weights of links within groups are all positive, and the weights of links between groups are all negative. There also exists a reasonable two-way partition, namely (6, 7, 22, 23, 24, 25, 13, 14, 15, 16, 4, 5) and (8, 9, 26, 27, 17, 18, 20, 21, 10, 11, 12, 1, 2, 3, 19, 28). The network shown in Fig.1(b) has the same three-way partition as that of Fig.1(a) but there is no two-way partition that can bipartition it.

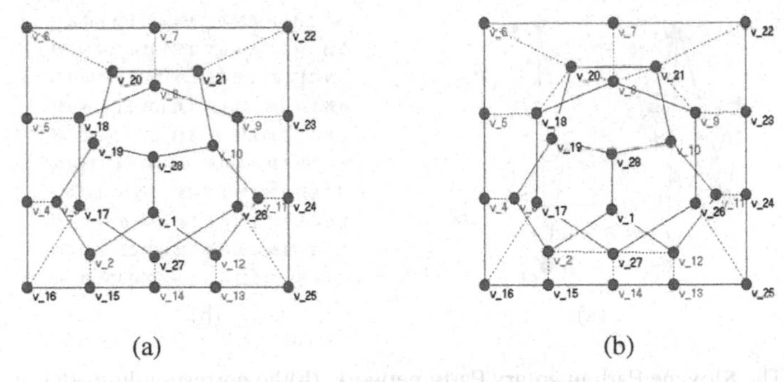

Fig. 1. Two illustrative networks

The communities found by our algorithm are shown in Fig. 2. As can be seen, our algorithm found 2 communities for the network in Fig.1(a) (see Fig.2(a)), but sometimes it can found 3 communities like Fig.2(b). As for the network in Fig.1(b), three communities have been found like Fig.2(b), which are (20, 19, 21, 3, 28, 10, 2, 1, 11, 12), (4, 5, 16, 6, 15, 7, 14, 22, 13, 23, 25, 24), and (8, 26, 17, 9, 18, 27). For these two networks, *NMI* achieves the maximum value of 1.

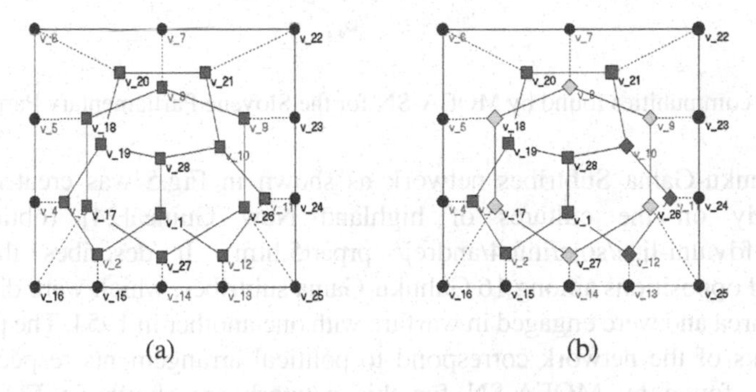

Fig. 2. The communities found by MOEA-SN for the two illustrative networks, where squares, circles, and diamonds stand for nodes in different communities.

4.2 Experiments on Real World Networks

Fig.3 shows the relation network of 10 parties of the Slovene Parliamentary in 1994[23], which was established by a group of experts on Parliament activities. The weights of links in this network were estimated by the average distance between each pair of parties on the scale from -3 to 3, where {-3, -2, -1} mean that the parties are "very dissimilar", "quite dissimilar", and "dissimilar", respectively. 0 means that parties are neither dissimilar nor similar (somewhere in between). {+3, +2, +1} mean that parties are "very similar", "quite similar", and "similar", respectively. The experimental results of MOEA-SN are shown in Fig.4. As we can see, our algorithm find two communities, which are identical to the results given by Kropivnik and Mrvar [23]. The maximum value of *NMI* found by MOEA-SN is 1.

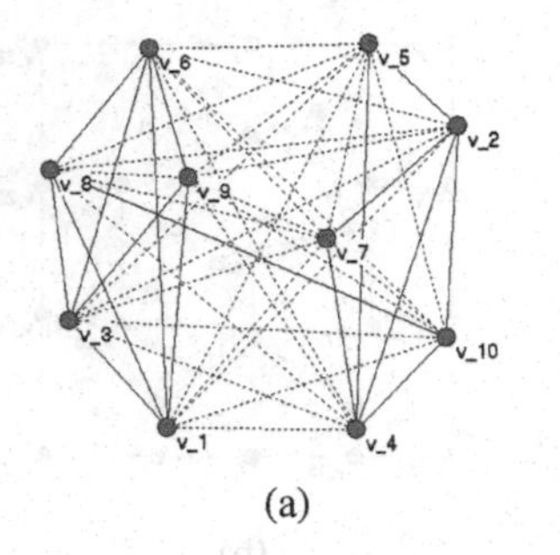

(a)

0	-2.15	1.14	-0.89	-0.77	0.94	-1.70	1.76	1.17	-2.10
-2.15	0	-2.17	1.34	0.77	-1.50	0.57	-2.53	-2.30	0.49
1.14	-2.17	0	-2.03	-0.80	1.38	-1.09	1.77	1.80	-1.74
-0.89	1.34	-2.03	0	1.57	-1.42	1.73	-2.41	-2.54	0.23
-0.77	0.77	-0.80	1.57	0	-1.88	-1.7	-1.2	-1.60	-0.09
0.94	-1.50	1.38	-1.42	-1.88	0	-0.97	1.4	1.16	-1.06
-1.7	0.57	-1.09	1.73	1.70	-0.97	0	-1.84	-1.91	-0.06
1.76	-2.53	1.77	-2.41	-1.20	1.40	-1.84	0	-1.91	-0.06
1.17	-2.30	1.80	-2.54	-1.60	1.16	-1.91	2.35	0	-1.64
-2.1	0.49	-1.74	0.23	-0.09	-1.06	-0.06	-1.32	-1.64	0

(b)

Fig. 3. (a)The Slovene Parliamentary Party network, (b)the corresponding adjacency matrix

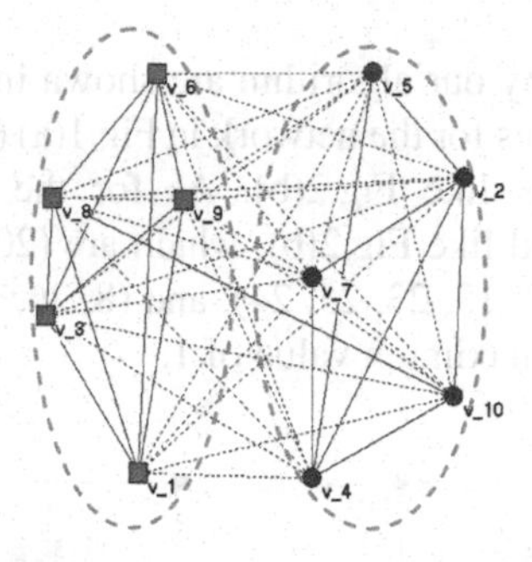

Fig. 4. The communities found by MOEA-SN for the Slovene Parliamentary Party network

The Gahuku-Gama Subtribes network as shown in Fig.5 was created based on Read's study on the cultures of highland New Guinea[24] (obtained from http://mrvar.fdv.uni-lj.si/sola/info4/andrej/ prpart5.htm). It describes the political alliances and oppositions among 16 Gahuku-Gama subtribes, which were distributed in a particular area and were engaged in warfare with one another in 1954. The positive and negative links of the network correspond to political arrangements respectively. The communities found by MOEA-SN for this network are shown in Fig.6, and the maximum value of *NMI* is 1. The three communities found were identical to the three-way partition reported by Doreian and Mrvar[2].

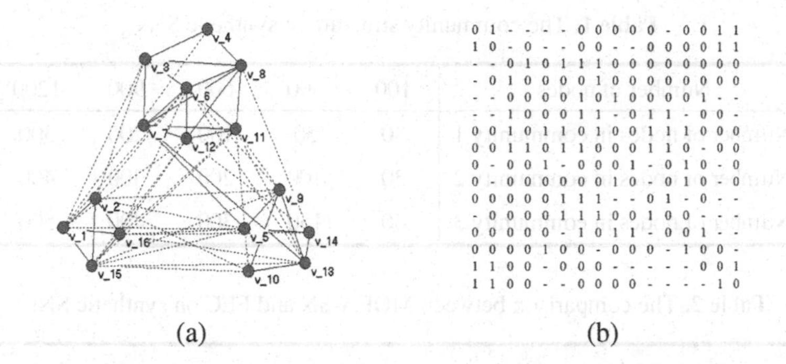

Fig. 5. (a) The Gahuku-Gama Subtribes network, (b) the corresponding adjacency matrix

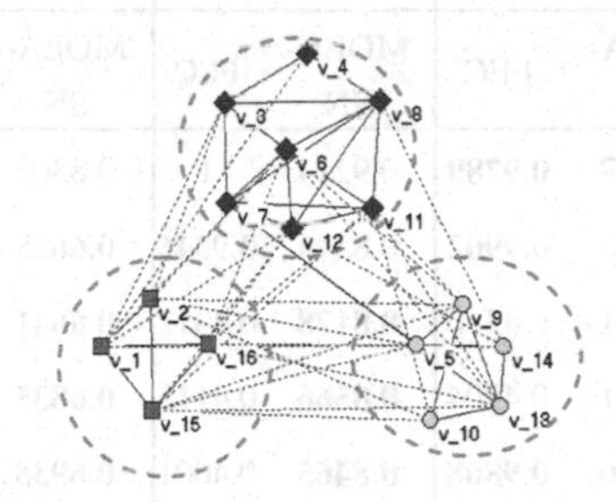

Fig. 6. The communities found by MOEA-SN for the Gahuku-Gama Subtribes network

4.3 Experiments on Synthetic Networks

To further investigate the performance of MOEA-SN, we generate synthetic networks with different properties using the method in [13]. The network is generated based on the BA model[25]. At first, 5 fully-coupled nodes are generated, then, at each time a new node which connects to 4 existing nodes is added. In this network generator, there is a parameter *p*, which is used to adjust the ratio of inter-community edges to inner-community edges. If *p*=0, there are no inter-community edges. If *p*=1, inner- and inter-community edges have the same number. Thus, *p* controls the cohesiveness of the communities inside the generated SNs. That is, the higher the value of *p* is, the more ambiguous the community structure is.

To generate balanced SNs, all inner-community edges are set to positive and inter-community edges are set to negative. Since it is easy to detect communities in balanced SNs, we generate unbalanced SNs based balanced ones as follows. In the generated balanced SNs, a fraction of edges, determined by the parameter *q*, is randomly selected, and then signs of these edges are flipped. In fact, *q* is used to adjust the noise level. Being the same with *p*, the larger the values of *q* are, the more ambiguous the community structure is. In the following experiments, synthetic SNs with 5 different sizes are generated. The number of nodes ranges from 100 to 1200, *p* ranges from 0.1 to 1, and *q* ranges from 0 to 0.2. Their community structures are given in Table 1.

Table 1. The community structure of synthetic SNs

Number of nodes	100	300	600	900	1200
Number of nodes in community 1	30	50	100	200	300
Number of nodes in community 2	30	100	200	300	400
Number of nodes in community 3	40	150	300	400	500

Table 2. The comparison between MOEA-SN and FEC on synthetic SNs

#Nodes	p=0.1									
	q=0		q=0.05		q=0.1		q=0.15		q=0.2	
	MOEA-SN	FEC	MOEA-SN	FEC	MOEA-SN	FEC	MOEA-SN	FEC	MOEA-SN	FEC
100	**1**	**1**	0.9257	**0.9789**	0.9214	**1**	0.8306	**0.8424**	0.7391	**0.8426**
300	**1**	**1**	0.8540	**0.9902**	0.8544	**0.9346**	**0.6628**	0.4936	**0.5763**	0.5336
600	**1**	**1**	**0.8751**	0.4360	**0.8178**	0.4312	**0.6641**	0.4172	**0.5406**	0.3928
900	**1**	**1**	**0.9511**	0.4204	**0.8566**	0.4115	**0.6835**	0.4037	**0.5770**	0.3823
1200	**1**	**1**	0.9636	**0.9868**	**0.8465**	0.4001	**0.6938**	0.3896	**0.5810**	0.3748
#Nodes	p=0.5									
	q=0		q=0.05		q=0.1		q=0.15		q=0.2	
	MOEA-SN	FEC	MOEA-SN	FEC	MOEA-SN	FEC	MOEA-SN	FEC	MOEA-SN	FEC
100	**1**	**1**	**1**	0.9040	**0.9534**	0.6555	**0.9487**	0.5495	**0.9171**	0.6347
300	**1**	**1**	**1**	0.4429	**0.9340**	0.7123	**0.6628**	0.5022	**0.5763**	0.3373
600	**1**	**1**	**0.9806**	0.8755	**0.8635**	0.4385	**0.7171**	0.4410	**0.5560**	0.3068
900	**1**	**1**	**1**	0.5092	**0.9042**	0.5935	**0.8046**	0.4274	**0.5968**	0.3143
1200	**1**	**1**	**0.9897**	0.4946	**0.8722**	0.4156	**0.6938**	0.4439	**0.5810**	0.3190
#Nodes	p=1									
	q=0		q=0.05		q=0.1		q=0.15		q=0.2	
	MOEA-SN	FEC	MOEA-SN	FEC	MOEA-SN	FEC	MOEA-SN	FEC	MOEA-SN	FEC
100	**1**	**1**	**1**	0.7685	**1**	**1**	**0.8197**	0.5336	**0.8240**	0.5699
300	**1**	**1**	**0.9469**	0.7663	**0.8705**	0.5455	**0.6951**	0.5758	**0.6986**	0.4889
600	**1**	**1**	**0.9624**	0.7884	**0.8354**	0.5453	**0.6557**	0.4393	**0.4183**	0.3846
900	**1**	**1**	**0.9226**	0.7141	**0.8610**	0.5883	**0.6932**	0.4295	**0.5034**	0.3655
1200	**1**	0.7993	**0.9905**	0.5842	**0.9336**	0.4733	**0.7142**	0.4091	**0.5084**	0.3286

The performance of MOEA-SN on these synthetic SNs is reported in Table 2, which is also compared with that of FEC, an effective existing algorithm for SNs. As can be seen, *NMI* of MOEA-SN reaches 1 for all balanced networks (q=0). *NMI* of FEC also reaches 1 except for the network with p=1 and 1200 nodes. Thus, both MOEA-SN and FEC can effectively detect communities in balanced SNs. When q increases, communities are more mixed and harder to be detected, therefore, both the performance of MOEA-SN and FEC decreases. However, the performance of MOEA-SN is always much better than that of FEC when p=0.5 and 1. When p=0.1, FEC outperforms MOEA-SN on a couple of networks, but MOEA-SN still outperforms FEC on most of networks.

5 Conclusions

In this paper, we studied CDPs in SNs from the viewpoint of multi-objective problems. The proposed method MOEA-SN adopts a popular multi-objective evolution algorithm NSGA-II's framework. The performance of MOEA-SN is validated on both real world networks and synthetic networks with different properties. The experimental results showed that MOEA-SN can find accurate community structures for both real world network and various synthetic networks, and outperforms an effective existing algorithm, FEC. For unbalanced signed networks with high noises, both the performance of MOEA-SN and FEC decreases. Therefore, in the future work, we will improve the performance of MOEA-SN on networks with high noises.

Acknowledgements. This work is partially supported by the National Natural Science Foundation of China under Grants 61271301 and 61103119, the Research Fund for the Doctoral Program of Higher Education of China under Grant 20130203110010, and the Fundamental Research Funds for the Central Universities under Grant K5051202052.

References

1. Girvan, M., Newman, M.E.J.: Community structure in social and biological networks. Proc. Natl. Acad. Sci. 99(12), 7821–7826 (2002)
2. Doreian, P., Mrvar, A.: A partitioning approach to structural balance. Social Networks 18(2), 149–168 (1996)
3. Clauset, A., Newman, M.E.J., Moore, C.: Finding community structure in very large networks. Phys. Rev. E. 70(6) (2004)
4. Blondel, V.D., Guillaume, J.L., Lambiotte, R., Lefebvre, E.: Fast unfolding of communities in large networks. J. Stat. Mech. Theory Exp. 2008(7) (2008)
5. Tasgin, M., Herdagdelen, A., Bingol, H.: Community detection in complex networks using genetic algorithms. arXiv, 2007(11) (2007)
6. Pizzuti, C.: GA-Net: A genetic algorithm for community detection in social networks. Parallel Problem Solving from Nature 5199, 1081–1090 (2008)
7. Yang, B., Cheung, W.K., Liu, J.: Community mining from signed social networks. IEEE Trans. Knowl. Data Eng. 19(10), 1333–1348 (2007)

8. Zitzler, E., Thiele, L.: Multi-objective evolutionary algorithms: a comparative case study and the strength Pareto approach. IEEE Trans. on Evolutionary Computation 3(4), 257–271 (1999)
9. Zitzler, E., Laumanns, M., Thiele, L.: SPEA2: Improving the strength Pareto evolutionary algorithm. Evolutionary Methods for Design, Optimization and Control with Applications to Industrial Problems, Vol. TIK-Report, No. 103 (2002)
10. Deb, K., Pratap, A., Agarwal, S., Meyarivan, T.: fast and elitist multiobjective genetic algorithm: NSGA-II. IEEE Trans. on Evolutionary Computation 6(26), 182–197 (2002)
11. Goh, C.K., Tan, K.C.: An investigation on noisy environments in evolutionary multiobjective optimization. IEEE Trans. on Evolutionary Computation 11(3), 354–381 (2007)
12. Pizzuti, C.: A multiobjective genetic algorithm to find communities in complex networks. IEEE Trans. Evolutionary Computation 16(3), 418–430 (2012)
13. Wu, L., Ying, X., Wu, X., Lu, A., Zhou, Z.: Examining spectral space of complex network with positive and negative links. International Journal of Social Network Mining 1(1), 91–111 (2012)
14. Li, Y., Liu, J., Liu, C.: A comparative analysis of evolutionary and memetic algorithms for community detection from signed social networks. Soft Computing 18(2), 329–348 (2014)
15. Newman, M.E.J., Girvan, M.: Finding and evaluating community structure in networks. Phys. Rev. E 69(2) (2004)
16. Gòmez, S., Jensen, P., Arenas, A.: Analysis of community structure in networks of correlated data. Phys. Rev. E. 80(1) (2009)
17. He, D.X.: Research on intelligent algorithms for network community mining, Master's thesis, Jilin University, China (2010)
18. Jin, D., Liu, J., Yang, B., He, D.X., Liu, D.Y.: A genetic algorithm with local search strategy for improved detection of community structure. Acta Automatica Sinica 37(7) (2011)
19. Raghavan, U.N., Albert, R., Kumara, S.: Near linear time algorithm to detect community structures in large-scale networks. Phys. Rev. E. 76(3), 53–60 (2007)
20. Hruschka, E.R., Campello, R.J.G.B., Freitas, A.A., de Carvalho, A.C.P.L.F.: A survey of evolutionary algorithms for clustering. IEEE Transactions on Systems, Man, and Cybernetics - Part C: Applications and Reviews 39(2), 133–155 (2009)
21. Shi, C., Yan, Z., Wang, Y., Cai, Y., Wu, B.: Genetic algorithm with local search for community detection in large-scale complex networks. Advances in Complex Systems 13(1), 873–882 (2010)
22. Danon, L., Diaz-Guilera, A., Duch, J., Arenas, A.: Comparing community structure identification. J. Stat. Mech. Theory Exp. 2005(9) (2005)
23. Kropivnik, S., Mrvar, A.: An analysis of the Slovene Parliamentary Parties network. Developments in Statistics and Methodology, pp. 149–168 (1996)
24. Read, K.E.: Cultures of the central highlands, New Guinea. Southwestern J. Anthropology 10(1), 1–43 (1954)
25. Barabási, A.L., Albert, R.: Emergence of scaling in random networks. Science 286(5439), 509–512 (1999)
26. Liu, C., Liu, J., Jiang, Z.: A multi-objective evolutionary algorithm based on similarity for community detection from signed social networks. IEEE Trans. on Cybernetics (2014)

Solving Multimode Resource-Constrained Project Scheduling Problems Using an Organizational Evolutionary Algorithm

Lixia Wang and Jing Liu

Key Laboratory of Intelligent Perception and Image Understanding of Ministry of Education, Xidian University, Xi'an 710071, China

Abstract. In this paper, a new evolutionary algorithm, namely organizational evolutionary algorithm for multimode resource-constrained project scheduling problems (OEA-MRCPSPs), is proposed. In OEA-MRCPSPs, the population is composed of organizations, and the number of organizations is adjusted by the splitting operator and annexing operator. In the evolutionary process of OEA-MRCPSPs, the global and local searches are combined efficiently. In the experiments, the performance of OEA-MRCPSPs is validated on benchmarks. The results show that OEA-MRCPSPs obtains a good performance in terms of both optimal solutions found and average deviations from optimal solutions or critical path lower bounds. The comparison results show that OEA-MRCPSPs outperforms six other existing algorithms.

Keywords: Organization, Evolutionary algorithm, Multimode, Resource-constrained project scheduling.

1 Introduction

The multimode resource-constrained project scheduling problem (MRCPSP) is an extension of the RCPSP [1]. The most important difference between them is that there are several execution modes for each activity in the MRCPSP and each execution mode may need different types of and amounts of resources. Researchers tried to solve it with different kinds of methods. Jozefowska *et al.* [2] and Bouleimen et al. [3] proposed the algorithms based on simulated annealing. Slowiński *et al.* [4] proposed a single-pass approach, a multi-pass approach as well as a simulated annealing algorithm. A biased random sampling approach was proposed by Drexl and Grünewald [5]. A local search strategy was proposed by Kolisch and Drexl [6]. Hartmann *et al.* used genetic algorithms to the MRCPSP [7, 8]. Sprecher *et al.* [9] proposed an exact algorithm based on the branch-and-branch strategy. A new mathematical formulation for the MRCPSP is proposed by Maniezzo and Mingozzi and two new lower bounds were derived [10]. Boctor presented heuristics to solve the MRCPSP without the nonrenewable resources [11-13]. Tseng *et al.* proposed a two-phase genetic local search algorithm in [1].

In our previous work, we proposed an organizational coevolutionary algorithm for classification [14], which achieved a higher predictive accuracy and lower computational

H. Handa et al. (eds.), *Proc. of the 18th Asia Pacific Symp. on Intell. & Evol. Systems – Vol. 1*,
Proceedings in Adaptation, Learning and Optimization 1, DOI: 10.1007/978-3-319-13359-1_22

cost, and an organizational evolutionary algorithm (OEA) for numerical optimization problems [15], which showed a good performance in terms of both the solution quality and the computational cost. In [16], we also applied OEA to general floorplanning problems, which can obtain high quality solutions for various and large-scale problems. All our previous work shows that OEA has a huge potential in solving complex problems. Therefore, in this paper, we apply the OEA to solve MRCPSPs, and the algorithm is named as OEA-MRCPSPs. The performance of OEA-MRCPSPs is tested upon the benchmark problem sets J10, J12, J14, J16, J18, J20, and J30, and the results show that OEA-MRCPSPs obtains a good performance not only in the optimal solutions found but also in the average deviations from optimal solutions or critical path lower bounds.

The remaining parts of this paper are organized as follows. Section 2 describes the formal definition of MRCPSPs. Section 3 presents the details of the OEA-MRCPSPs. Experiments on benchmark problems are shown in Section 4. Finally, Section 5 presents the conclusion of the work in this paper.

2 Problem Definition

In MRCPSPs, the activities of a project are marked by 1, 2, …, n. For an arbitrary activity i, its execution modes are represented by a set EM_i={1, 2, …, M_i}, and a mode means a group of different kinds and quantities of resources, and it costs a related duration. If activity i executes in mode m_i, then it needs $r_{im_ik}^{RR}$ units of renewable resource k, and $r_{im_ik}^{NR}$ units of nonrenewable resource k, d_{im_i} is the related duration.

As indicated above, there are two kinds of constraints in the MRCPSP, one is the precedence constraint, and the other is the resource constraint. The precedence constraint can be described by a graph in Figure 1, where each node represents an activity, and if there is an arrow from node i to j, it means that activity i is the immediate predecessor of activity j, similarly, activity j is called the immediate successor of activity i. An activity can not be arranged before all of its immediate predecessors have been completed. Image resource k belongs to renewable resources, and its gross is a constant Q_k^{RR}, which means no matter at which moment, the total amount of resource k is equal to Q_k^{RR}, while if k is a nonrenewable resource and its gross is a constant Q_k^{NR}, which means in the overall process, the total amount of resource k is stationary.

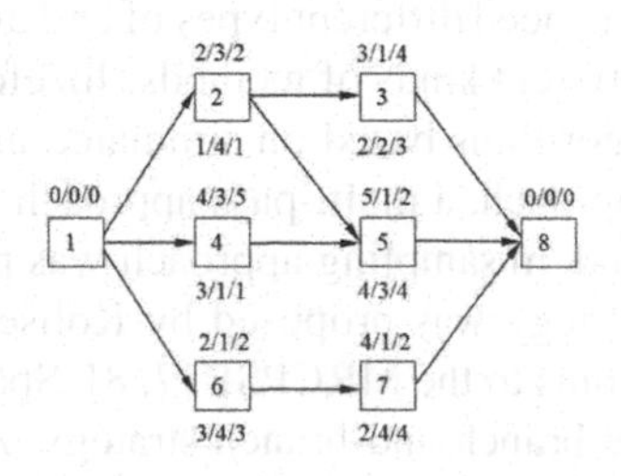

Fig. 1. An instance of the MRCPSP

The objective of MRCPSPs is to optimize the start time and the execution mode for each activity so that the makespan of the whole project is minimized. At the same time, the precedence and resource constraints must be satisfied. In Figure 1, there are two dummy activities called fictitious initial activity and fictitious finish activity that do not need any resources and their durations are zeros. Suppose each activity has two execution modes except the dummy activities, and there are one type of renewable resources and one type of nonrenewable resources. The data above node i mean the duration, the renewable resource cost, and the nonrenewable resource cost for activity i when it is executed under mode 1. Similarly, the data below node i mean the duration, the renewable resource cost, and the nonrenewable resource cost for activity i when it is executed under mode 2.

3 Organizational Evolutionary Algorithm for MRCPSPs

3.1 Organizations for MRCPSPs

In OEA-MRCPSPs, an organization is composed by members and a member is a candidate solution for the problem need to be solved.

Definition 1: A member $\boldsymbol{M}$ consists of two components, namely the activity component and the execution mode component, which are labeled as $\boldsymbol{M}\langle \boldsymbol{A}\rangle$ and $\boldsymbol{M}\langle \boldsymbol{O}\rangle$, respectively. $\boldsymbol{M}\langle \boldsymbol{A}\rangle$ is a permutation of all activities, and all activities in this permutation are arranged according to the precedence constraints,

$$\boldsymbol{M}\langle \boldsymbol{A}\rangle=\{\pi_1, \pi_2, \ldots, \pi_n\} \tag{1}$$

where $(\pi_1, \pi_2, \ldots, \pi_n)$ is a permutation of $(1, 2, \ldots, n)$, and for π_i, i=1, 2, …, n-1, no immediate predecessors of π_i is in $\{\pi_{i+1}, \pi_{i+2}, \ldots, \pi_n\}$. $\boldsymbol{M}\langle \boldsymbol{O}\rangle$ is the set of execution modes for each activity,

$$\boldsymbol{M}\langle \boldsymbol{O}\rangle=\{o_1, o_2, \ldots, o_n\} \tag{2}$$

where $o_i \in EM_{\pi_i}$, i=1, 2, …, n, and stands for the execution mode of activity π_i. □

In order to reduce the search space, the preprocessing procedure in [17] is used. After the preprocessing, the execution modes that violate renewable resource constrains are already removed, and the activity component in a member strictly obeys the precedence constraints. Therefore, a member is an infeasible solution only if it violates the nonrenewable resource constraints. In this paper, we use the fitness function proposed in [8] to evaluate the quality of a member, which is defined as follows,

$$f(\boldsymbol{M})=\begin{cases} mak(\boldsymbol{M}) & \text{If } \boldsymbol{M} \text{ is feasible} \\ max\_fea\_pop\_mak+mak(\boldsymbol{M}) & \text{Others} \\ -min\_project\_CC+SFT(\boldsymbol{M}) & \end{cases} \tag{3}$$

where $mak(\boldsymbol{M})$ indicates the makespan of $\boldsymbol{M}$. $max\_fea\_pop\_mak$ expresses the maximum makespan among all feasible solutions in current population, which should be updated in every generation. $min\_project\_CC$ is the critical path using the minimum durations of activities, which is smaller than the makespan of any infeasible scheduling. $SFT(\boldsymbol{M})$ represents the amount of the nonrenewable resources that exceeds the capacities. For an infeasible solution, $SFT(\boldsymbol{M})$ is undoubtedly larger than zero, and the length of critical path is smaller than any of the makespans of feasible solutions. Therefore, the makespans of infeasible solutions are greater than that of feasible ones.

Based on the members, an organization is defined as follows,

Definition 2: An organization $\boldsymbol{org}$ is a set of members, and the best member is called as the leader of $\boldsymbol{org}$, which is labeled as $\boldsymbol{Leader_{org}}$. That is, $\boldsymbol{org}$ and $\boldsymbol{Leader_{org}}$ satisfy the following conditions,

$$\boldsymbol{org}=\{\boldsymbol{M}_1, \boldsymbol{M}_2, \ldots, \boldsymbol{M}_{|\boldsymbol{org}|}\} \text{ and } \boldsymbol{org}\neq\varnothing \quad (4)$$

$$\boldsymbol{Leader_{org}}\in \boldsymbol{org} \text{ and } f(\boldsymbol{Leader_{org}})\leq f(\boldsymbol{M}_i),\ i=1, 2, \ldots, |\boldsymbol{org}| \quad (5)$$

where $|\boldsymbol{org}|$ denotes the number of members in $\boldsymbol{org}$. The fitness value of an organization is equal to that of its leader. When two organizations are compared, one organization is better than another when its leader is better than another one's leader.

3.2 Evolutionary Operators for Members

There are three types of operators which can be conducted on members directly, namely crossover, mutation, and local search operators.

3.2.1 Crossover Operators

Two crossover operators, labeled as $Crossover_1$ and $Crossover_2$, are employed here, which are also used in [8] and [17]. Both of them are based on two-point crossover operations, operate on two parent members $\boldsymbol{M}_{p1}$ and $\boldsymbol{M}_{p2}$, and generate two child members $\boldsymbol{M}_{c1}$ and $\boldsymbol{M}_{c2}$.

In $Crossover_1$, two cut-points r_1 and r_2 are randomly drawn first. Then, the first r_1 elements of $\boldsymbol{M}_{c1}\langle \boldsymbol{A}\rangle$ are inherited from the first r_1 elements of $\boldsymbol{M}_{p1}\langle \boldsymbol{A}\rangle$. The following r_2-r_1 elements of $\boldsymbol{M}_{c1}\langle \boldsymbol{A}\rangle$ are inherited from the first r_2-r_1 elements of $\boldsymbol{M}_{p2}\langle \boldsymbol{A}\rangle$ which do not appear in the first section of $\boldsymbol{M}_{c1}\langle \boldsymbol{A}\rangle$. The remaining elements of $\boldsymbol{M}_{c1}\langle \boldsymbol{A}\rangle$ are inherited from the remaining elements of $\boldsymbol{M}_{p1}\langle \boldsymbol{A}\rangle$ which do not appear in the first two sections of $\boldsymbol{M}_{c1}\langle \boldsymbol{A}\rangle$. All elements of $\boldsymbol{M}_{c1}\langle \boldsymbol{A}\rangle$ inherited from the parents must keep the order as they appear in $\boldsymbol{M}_{p1}\langle \boldsymbol{A}\rangle$ or $\boldsymbol{M}_{p2}\langle \boldsymbol{A}\rangle$. The mode assignment $\boldsymbol{M}_{c1}\langle \boldsymbol{O}\rangle$ is inherited from the mode assignment of $\boldsymbol{M}_{p1}\langle \boldsymbol{O}\rangle$ or $\boldsymbol{M}_{p2}\langle \boldsymbol{O}\rangle$ along with the corresponding activity. $\boldsymbol{M}_{c2}$ is constructed in the same way with the roles of $\boldsymbol{M}_{p1}$ or $\boldsymbol{M}_{p2}$ being exchanged.

In $Crossover_2$, $\boldsymbol{M}_{c1}\langle \boldsymbol{A}\rangle$ and $\boldsymbol{M}_{c2}\langle \boldsymbol{A}\rangle$ are constructed in the same way as in $Crossover_2$, but $\boldsymbol{M}_{c1}\langle \boldsymbol{O}\rangle$ and $\boldsymbol{M}_{c2}\langle \boldsymbol{O}\rangle$ are constructed differently. To construct the mode assignment of offspring, we first randomly draw a sequence of 0s and 1s. Then for each activity, if the corresponding element in the sequence is 1, then the mode of the corresponding element in $M_{c1}\langle O\rangle$ is equal to that in $M_{p1}\langle O\rangle$ and that in $M_{c2}\langle O\rangle$ is equal to that in $M_{p2}\langle O\rangle$; otherwise, that in $M_{c1}\langle O\rangle$ is equal to that in $M_{p2}\langle O\rangle$ and that in $M_{c2}\langle O\rangle$ is equal to that in $M_{p1}\langle O\rangle$.

3.2.2 Mutation Operators

Two mutation operators, $Mutation_1$ and $Mutation_2$, which are similar to those in [1], are employed. Both $Mutation_1$ and $Mutation_2$ consist of two parts. One is to change the positions of activities; the other is to change the execution modes. In $Mutation_1$, n_1 activities and n_2 execution modes are randomly selected, where both n_1 and n_2 are smaller than one thirds of the total number of activities. First, each activity of the n_1 ones is randomly moved to another position without violating the precedence constrains. The corresponding execution mode is also moved to the destination position, too. Then, each execution mode of the n_2 ones is randomly changed.

$Mutation_2$ works similarly to $Mutation_1$. Instead of selecting activities and execution modes from all available ones, $Mutation_2$ selects n_1 activities and n_2 execution modes from a critical path, since the makspan will not be reduced if the critical path is not changed, where both n_1 and n_2 here are not larger than the total number of activities in the critical path. Then, the following operations are the same to those in $Mutation_1$.

3.2.3 Local Search Operator

In order to improve the search efficiency, a local search operator is designed based on $Mutation_1$; that is, each one among the l_{max} best members in current population will perform T times $Mutation_1$. Each time, if the new member generated by $Mutation_1$ is better than the previous one, it is used to replace the previous one, and $Mutation_1$ is performed again until the upper bound times T is reached.

3.3 Evolutionary Operators for Organizations

Three evolutionary operators which are performed on organizations directly are designed.

3.3.1 Splitting Operator

In OEA-MRCPSPs, if one organization is too large, it will affect the search efficiency. Therefore, an upper bound size, labeled as max_os (>1), is set to prevent an organization being too large. When the size of an organization exceeds the upper bound or satisfies the other condition in (6), the organization will be split.

$$(|\mathbf{org}|>max\_os) \text{ or } ((|\mathbf{org}|\leq max\_os) \text{ and } (U(0,1)<|\mathbf{org}|/N_0)) \tag{6}$$

where $U(0, 1)$ is a uniformly distributed random number between 0 and 1, and N_0 is the initial number of organizations. Then, randomly generate an integer M satisfying $|\mathbf{org}|/3<M<2\times|\mathbf{org}|/3$, then divide the individuals randomly into two sub-organizations until one of them has M members, and the other one has $|\mathbf{org}|-M$ members. After the two child organizations are generated, their best members are selected to be their leaders, and the original organization is removed from the population.

3.3.2 Annexing Operator

The annexing operator reflects the competition between two organizations. The organization with stronger strength will defeat the weak one and annex it to construct a larger organization. Suppose the annexing operator is performed on two parent organizations, $\boldsymbol{org}_{p1}$ and $\boldsymbol{org}_{p2}$, and $f\left(\boldsymbol{Leader}_{\boldsymbol{org}_{p1}}\right) < f\left(\boldsymbol{Leader}_{\boldsymbol{org}_{p2}}\right)$. Then, $\boldsymbol{org}_{p2}$ will be annexed by $\boldsymbol{org}_{p1}$ in the following way to generate a new organization $\boldsymbol{org}_{\text{new}}$. First, all members of $\boldsymbol{org}_{p1}$ are moved to $\boldsymbol{org}_{\text{new}}$ without any change. Then, If $U(0, 1)<AS$ where AS is a predefined parameter in the range of (0, 1), $\boldsymbol{Mutation}_1$ is performed on $\boldsymbol{Leader}_{\boldsymbol{org}_{p1}}$ to generate $|\boldsymbol{org}_{p2}|$ new members for $\boldsymbol{org}_{\text{new}}$; otherwise, $\boldsymbol{Mutation}_2$ is used. Finally, $\boldsymbol{org}_{\text{new}}$ is added to the population and $\boldsymbol{org}_{p1}$ and $\boldsymbol{org}_{p2}$ are removed. After the annexing operator, the two organizations become a larger organization and the leader of the new organization needs to be selected again.

3.3.3 Cooperating Operator

The cooperating operator reflects the cooperative relationship between two organizations. Suppose the cooperating operator is performed on two parent organizations, $\boldsymbol{org}_{p1}$ and $\boldsymbol{org}_{p2}$. If $U(0, 1)<CS$ where CS is a predefined parameter in the range of (0, 1), then $\boldsymbol{Crossover}_1$ is performed on $\boldsymbol{Leader}_{\boldsymbol{org}_{p1}}$ and $\boldsymbol{Leader}_{\boldsymbol{org}_{p2}}$; otherwise, $\boldsymbol{Crossover}_2$ is used. After the cooperating operator, two new members are generated, and the two old leaders are replaced by the new members, as the fitness values of new members either better or worse than those of old leaders, it needs to select new leaders for the two organizations.

3.4 Implementation of OEA-MRCPSPs

In the OEA-MRCPSPs, a population is composed of organizations, and operators are first performed on organizations, while some of them are based on the crossover and mutation operators for members. In the process of initialization, for each member, the activity component and the mode component are generated simultaneously where the method in [4] is used to generate the activity component. Also, when an activity is arranged into the activity component, at the same time, its execution mode is randomly selected and placed into the mode component. In each generation, the size of each organization is first checked. If it satisfies (6), the organization will be split. Then two organizations are randomly selected from the population, the annexing operator or the cooperating operator will conduct on them with the same probability. After that, the local search operator is executed. The population evolves generation by generation. Finally, the member with the best fitness value is output as the result. The details are shown in Algorithm 1.

Algorithm 1. Organizational evolutionary algorithm for MRCPSPs.

1. Conduct the preprocessing;
2. Initialize the population with N_0 organizations, and each has only one member;
3. Calculate the fitness of each member and each organization;
4. while (the termination criteria are not satisfied) do
5. begin
6. For each organization, if it satisfies (6), then conduct the splitting operator on it;
7. Randomly select two organizations from the current population, and conduct the annexing operator or the cooperating operator on them with the same probability;
8. Update the fitness of each member and each organization in the current population;
9. Conduct the local search operator on the current population;
10. end.

4 Experiments

4.1 Configuration Test

In this section, benchmark problem sets J10, J12, J14, J16, J18, J20, and J30, totally 3842 problem instances, from the PLPSIB [18] are used to test the performance of OEA-MRCPSPs. Table 1 lists the number of problem instances in each set. For J10, there are 10 non-dummy activities in each instance, for J12, there are 12 non-dummy activities in each instance, and so on. Each instance has two kinds of renewable resources and two kinds of nonrenewable resources. The optimal solutions for the instances in J10, J12, J14, J16, J18, and J20 are known, while in J30, the optimal solutions of some instances have not been found. Table 2 shows the parameter setting in our experiments.

The OEA-MRCPSPs is tested upon J10~J30 with the maximum schedules evaluated being set to 5000, 10000, 20000, and 50000. For each instance in J10 to J20, 30 independent runs of OEA-MRCPSPs are conducted, while for each instance in J30, 10 independent runs are conducted. For J10 to J20, two indexes are mainly used to evaluate the performance of the algorithm, namely the percentage of the optimal solution found and the average deviation from the optimal solutions which is calculated according to (7). For J30, as parts of the optimal solutions are not known, the only index is the average deviation from the critical path lower bounds in percentage which is calculated by (8). Tables 3 and 4 report the performance of OEA-MRCPSPs.

Table 1. The number of feasible instances of each problem set

Problem set	J10	J12	J14	J16	J18	J20	J30
Number of feasible instances	536	547	551	550	552	554	552

Table 2. The parameters setting of the OEA-MRCPSPs

Parameter	N_0	*max_os*	*AS*	*CS*	l_{max}	*T*
Value	50	10	0.5	0.5	5	5 for J10, J12, J14; 10 for J16; 50 for J18 to J30

$$\text{Average deviation}=\frac{\sum_{\text{instance}_i}\left(\frac{\text{best makespan}_i}{\text{optimal makespan}_i}-1\right)}{\text{number of feasible instances}} \quad (7)$$

$$\text{Average deviation}=\frac{\sum_{\text{instance}_i}\left(\frac{\text{best makespan}_i}{\text{critical path lowerbound}_i}-1\right)}{\text{number of feasible instances}} \quad (8)$$

Table 3. The percentage of optimal solutions found (%) for J10 to J20

Problem set	Maximum number of schedules evaluated			
	5000	10000	20000	50000
J10	97.61	99.00	99.54	99.85
J12	95.17	97.99	99.08	99.79
J14	84.39	90.97	95.26	97.99
J16	79.76	87.44	92.96	96.70
J18	75.44	82.97	88.90	93.96
J20	66.76	75.24	82.71	90.39

From Tables 3 and 4 we can see that with the increasing in the number of schedules evaluated, the performance of OES-MRCPSPs is getting better. Especially, when the number of schedules evaluated increases to 50000, the percentage of optimal solutions found is larger than 90% and the average deviation from optimal solutions is smaller than 0.4% for all instances in J10 to J20.

Table 4. The average deviation (%) from optimal solutions for J10 to J20 and average deviation (%) from critical-path lower bound for J30

Problem set	Maximum number of schedules evaluated			
	5000	10000	20000	50000
J10	0.131	0.052	0.024	0.008
J12	0.250	0.118	0.043	0.010
J14	0.738	0.398	0.200	0.083
J16	0.910	0.523	0.275	0.160
J18	1.223	0.787	0.508	0.283
J20	1.652	1.083	0.681	0.343
J30	17.159	16.029	15.165	14.423

4.2 Comparison with Existing Methods

From existing literatures on MRCPSPs, we find that most of them show the performance by calculating the average deviation from optimal solutions or the critical path lower bounds under the maximum schedules evaluated being a fixed number, while majority of these set the maximum schedules to 5000 or 6000, and they did not give the performance under larger schedules such as 20000 and 50000. Therefore, Table 5 compares the performance of OEA-MRCPSPs with that of other methods for J10 to J20 under 5000 schedules. From the results we can see that the OEA-MRCPSPs outperforms all the other algorithms for J10 to J20.

Table 6 shows the comparison of average deviation from optimal solutions and percentage of optimal solutions found for J10 between OEA-MRCPSPs and other methods under 6000 schedules. It can be seen clearly that the OEA-MRCPSPs also outperforms the other algorithms under this condition. This indicates that the OEA-MRCPSPs obtains a high search efficiency although the number of schedules evaluated is kept in a low level.

Tables 7, 8, and 9 compare the performance of OEA-MRCPSPs and two other methods with the maximum schedules evaluated fixed to 5000, 10000, 20000, and 50000. From these three tables we can see that the OEA-MRCPSPs performs better than the two other methods in most of the conditions.

Table 5. Comparison in terms of average deviation (%) from optimal solutions between OEA-MRCPSPs and other methods under 5000 schedules

Method	Problem set					
	J10	J12	J14	J16	J18	J20
OEA-MRCPSPs	**0.13**	**0.25**	**0.74**	**0.91**	**1.22**	**1.65**
Two-phase GLS [1]	0.33	0.52	0.92	1.09	1.30	1.71
Alcaraz *et al.* [8]	0.24	0.73	1.00	1.12	1.43	1.91
Jozefowska *et al.* [2]	1.16	1.73	2.6	4.07	5.52	6.74

Table 6. OEA-MRCPSPs v.s. other methods for J10 under 6000 schedules

Method	J10	
	Average deviation (%)	Optimal solution found (%)
OEA-MRCPSPs	**0.10**	**98.2**
Hartmann [9]	0.10	98.1
Alcaraz *et al.* [8]	0.18	96.5
Two-phase GLS [1]	0.28	95.8
Kolisch and Drexl [6]	0.50	91.8
Ozdamar [19]	0.86	88.1

Table 7. The comparison in terms of optimal solutions found (%) for J10 to J20 between OEA-MRCPSPs and two other methods

Problem set	Method	Maximum number of schedules being evaluated			
		5000	10000	20000	50000
J10	OEA-MRCPSPs	**97.61**	**99.00**	**99.54**	99.85
	Two-phase GLS [1]	95.16	98.24	99.48	**99.94**
	Jozefowska *et al.* [2]	85.60	93.70	96.60	97.20
J12	OEA-MRCPSPs	**95.17**	**97.99**	**99.08**	**99.79**
	Two-phase GLS [1]	90.57	95.24	98.05	99.44
	Jozefowska *et al.* [2]	80.30	91.60	96.70	97.60
J14	OEA-MRCPSPs	**84.39**	**90.97**	**95.26**	**97.99**
	Two-phase GLS [1]	82.03	89.35	94.07	97.18
	Jozefowska *et al.* [2]	66.40	79.70	89.10	95.10
J16	OEA-MRCPSPs	**79.76**	**87.44**	**92.96**	**96.70**
	Two-phase GLS [1]	77.39	86.07	91.89	95.61
	Jozefowska *et al.* [2]	54.70	68.50	81.80	95.50
J18	OEA-MRCPSPs	**75.44**	**82.97**	**88.90**	**93.96**
	Two-phase GLS [1]	73.38	82.19	88.42	93.21
	Jozefowska *et al.* [2]	43.50	60.10	71.90	87.70
J20	OEA-MRCPSPs	**66.76**	75.24	82.71	**90.39**
	Two-phase GLS [1]	66.66	**75.93**	**83.56**	89.83
	Jozefowska *et al.* [2]	35.70	51.10	62.60	81.00

Table 8. The comparison in terms of average deviation (%) from optimal solutions for J10 to J20 between OEA-MRCPSPs and two other methods

Problem set	Method	Maximum number of schedules being evaluated			
		5000	10000	20000	50000
J10	OEA-MRCPSPs	**0.131**	**0.052**	**0.024**	0.008
	Two-phase GLS [1]	0.330	0.106	0.026	**0.003**
	Jozefowska *et al.* [2]	1.160	0.470	0.270	0.230
J12	OEA-MRCPSPs	**0.250**	**0.118**	**0.043**	**0.010**
	Two-phase GLS [1]	0.524	0.258	0.094	0.025
	Jozefowska *et al.* [2]	1.730	0.740	0.420	0.370
J14	OEA-MRCPSPs	**0.738**	**0.398**	**0.200**	**0.083**
	Two-phase GLS [1]	0.917	0.492	0.253	0.114
	Jozefowska *et al.* [2]	2.600	1.430	0.590	0.240
J16	OEA-MRCPSPs	**0.910**	**0.523**	**0.275**	**0.160**
	Two-phase GLS [1]	1.087	0.624	0.342	0.171
	Jozefowska *et al.* [2]	4.070	2.300	1.060	0.280
J18	OEA-MRCPSPs	**1.223**	0.787	0.508	0.283
	Two-phase GLS [1]	1.301	**0.784**	**0.472**	**0.261**
	Jozefowska *et al.* [2]	5.520	3.310	1.790	0.560
J20	OEA-MRCPSPs	**1.652**	**1.083**	0.681	**0.343**
	Two-phase GLS [1]	1.713	1.084	**0.678**	0.386
	Jozefowska *et al.* [2]	6.740	4.330	2.540	0.800

Table 9. The comparison in terms of average deviation (%) from critical path lower bounds for J30 between OEA-MRCPSPs and the two phase GLS [1]

Problem set	Methods	Maximum schedules that being evaluated			
		5000	10000	20000	50000
J30	OEA-MRCPSPs	**17.159**	**16.029**	**15.165**	**14.423**
	Two-phase GLS [1]	18.332	16.786	16.193	15.683

5 Conclusion

The OEA-MRCPSPs, a new organizational evolutionary algorithm for solving project scheduling problems with multiple modes is proposed in this paper. In the OEA-MRCPSPs, the population is composed by organizations and organizations are composed by members, where each member is composed by two components, namely the activity component and the execution mode component. Based on such a kind of representation for members, the splitting operator, the annexing operator, and the cooperating operator are designed to realize the global search, and the local search operator is designed to realize the local search.

We have tested the OEA-MRCPSPs upon benchmark problem sets J10, J12, J14, J16, J18, J20, and J30 with the maximum schedules evaluated setting to 5000, 10000, 20000, and 50000, respectively. The experimental results show that when the number of maximum schedules evaluated increases, the percentages of optimal solution found increase obviously and the average deviations from optimal solutions (for J10 to J20) or critical path lower bounds (for J30) decrease obviously, which illustrates that the good performance of OEA-MRCPSPs. In the comparison with six other existing methods, the OEA-MRCPSPs obtains a good ranking, and outperforms all the other algorithms.

Acknowledgements. This work is partially supported by the National Natural Science Foundation of China under Grants 61271301 and 61103119, the Research Fund for the Doctoral Program of Higher Education of China under Grant 20130203110010, and the Fundamental Research Funds for the Central Universities under Grant K5051202052.

References

1. Tseng, L.Y., Chen, S.C.: Two-phase genetic local search algorithm for the multimode resource-constrained project scheduling problem. IEEE Transactions on Evolutionary Computation 13(4), 848–857 (2009)
2. Jozefowska, J., Mika, M., Rozycki, R., Waligora, G., Weglarz, J.: Simulated annealing for multimode resource-constrained project scheduling. Annals of Operations Research 102(1-4), 137–155 (2001)
3. Bouleimen, K., Lecocq, H.: A new efficient simulated annealing algorithm for the resource-constrained project scheduling problem and its multiple mode version. European Journal of Operational Research 149(2), 268–281 (2003)
4. Slowinski, R., Soniewicki, B., Weglarz, J.: DSS for multiobjective project scheduling subject to multiple-category resource constraints. European Journal of Operational Research 79, 220–229 (1994)
5. Drexl, A., Grünewald, J.: Nonpreemptive multi-mode resource-constrained project scheduling. IIE Transactions 25(5), 733–750 (1993)
6. Kolisch, R., Drexl, A.: Local search for nonpreemptive multi-mode resourch-constrained project scheduling. IIE Transactions 29, 987–999 (1997)
7. Hartmann, S.: Project scheduling with multiple modes: A genetic algorithm. Annals of Operations Research 102(1-4), 111–135 (2001)
8. Alcaraz, J., Maroto, C., Ruiz, R.: Solving the multimode resource-constrained project scheduling problem with genetic algorithms. Journal of the Operational Research Society 54, 614–626 (2003)
9. Sprecher, A., Drexl, A.: Multimode resource-constrained project scheduling by a simple, general and powerful sequencing algorithm. European Journal of Operational Research 107(2), 431–450 (1998)

10. Maniezzo, V., Mingozzi, A.: A heuristic procedure for the multi-mode project scheduling problem based on Bender's decomposition. In: Project Scheduling International Series in Operations Research & Management Science, pp. 179–196 (1999)
11. Boctor, F.F.: Heuristics for scheduling projects with resource restrictions and several resource-duration modes. International Journal of Production Research 31(11), 2547–2558 (1993)
12. Boctor, F.F.: A new and efficient heuristic for scheduling projects with resource restrictions and multiple execution modes. European Journal of Operational Research 90(2), 349–361 (1996)
13. Boctor, F.F.: An adaption of the simulated annealing algorithm for solving the resource-constrained project scheduling problems. International Journal of Production Research 34, 2335–2351 (1996)
14. Jiao, L., Liu, J., Zhong, W.: An organizational coevolutionary algorithm for classification. IEEE Transactions on Evolutionary Computation 10(1) (2006)
15. Liu, J., Zhong, W., Jiao, L.: An organizational evolutionary algorithm for numerical optimization. IEEE Transactions on Systems, Man and Cybernetics, Part B 37(4) (2007)
16. Liu, J., Zhong, W., Jiao, L.: Moving block sequence and organizational evolutionary algorithm for general floorplanning with arbitrarily shaped rectilinear blocks. IEEE Transactions on Evolutionary Computation 12(5) (2008)
17. Sprecher, A., Hartmann, S., Drexl, A.: An exact algorithm for project scheduling with multiple modes. Operations-Research-Spektrum 19(3), 195–203 (1997)
18. Kolisch, R., Sprecher, A.: PSPLIB: a project scheduling problem library. European Journal of Operational Research 96(1), 205–216 (1996)
19. Özdamar, L.: A genetic algorithm approach to a general category project scheduling problem. IEEE Transactions on Systems, Man, and Cybernetics, Part C 29(1), 44–59 (1999)

Cyber-Physical Systems: The Next Generation of Evolvable Hardware Research and Applications

Garrison Greenwood[1], John Gallagher[2], and Eric Matson[3]

[1] Portland State University, Portland, OR 97201 USA
[2] Wright State University, Dayton, OH 45435 USA
[3] Purdue University, West Lafayette, IN 47907 USA

Abstract. Since the late 1990s the sales of processors targeted for embedded systems has exceeded sales for the PC market. Some embedded systems tightly link the computing resources to the physical world. Such systems are called *cyber-physical systems*. Autonomous cyber-physical systems often have safety-critical missions, which means they must be fault tolerant. Unfortunately fault recovery options are limited; adapting the physical system behavior may be the only viable option. Consequently, autonomous cyber-physical systems are a class of adaptive systems. The evolvable hardware field has developed a number of techniques that should prove to be useful for designing cyber-physical systems although work along those lines has only recently begun. In this paper we provide an overview of cyber-physical systems and then describe how two evolvable hardware techniques can be used to adapt the physical system behavior in real-time. The goal is to introduce cyber-physical systems to the evolvable hardware community and encourage those researchers to begin working in this emerging field.

Keywords: adaptive systems, cyber-physical systems, evolvable hardware, metamorphic system.

1 Introduction

Nature has successfully used evolution to find solutions to difficult problems. It should therefore come as no surprise that researchers would attempt to use Neo-Darwinistic methods to solve hardware design problems. For more than 20 years the *evolvable and adaptive hardware* (EAH) field has investigated how evolutionary algorithms such as genetic algorithms or evolution strategies can evolve hardware solutions [1]. The concept is quite straightforward. Candidate hardware solutions are encoded as individuals in a population. The fitness of an individual tells how well the design operates and the better the performance, the higher the fitness. Fitness is determined either *extrinsically* with simulators or *intrinsically* by constructing the design and then conducting physical tests. Highly fit individuals are then subjected to reproduction operators that create new solutions for evaluation whereas poorly fit individuals die out. The population continues this evolutionary process until an acceptable solution is found. Other biologically inspired algorithms (BIAs) and agent-based approaches are also beginning to surface in EAH research.

H. Handa et al. (eds.), *Proc. of the 18th Asia Pacific Symp. on Intell. & Evol. Systems – Vol. 1*,
Proceedings in Adaptation, Learning and Optimization 1, DOI: 10.1007/978-3-319-13359-1_23

EAH problems fall into two categories: original design and adaption. In original design problems the objective is to evolve a hardware configuration that satisfies a design specification. Configurations that meet all specifications have maximal fitness. In original design problems fitness can be found either intrinsically or extrinsically. Moreover, there is usually few (if any) time constraints so the evolutionary process can run as long as needed to produce an acceptable hardware configuration. Conversely, in adaption problems a system already exists but suddenly no longer performs correctly due to internal faults and/or a changed operational environment. Online EAH methods must reconfigure the system to restore proper behavior. Extrinsic fitness evaluation is impractical unless the exact nature of the fault is known a priori. Fault recovery also often has deadlines so the running time of an evolutionary algorithm may be limited.

The overwhelming majority of prior EAH research has focused on original design problems [2]. Nevertheless, the real promise and ultimate payoff for EAH methods lies in adaption—particularly adapting behavior in autonomous systems. Autonomous systems often operate in harsh environments for extended time periods. In most cases these systems cannot depend on human support to keep them operational. Consequently, autonomous systems can survive only if they are aware and responsive. They must be aware of their surroundings and be able to react in time to counter any threats or performance losses due to internal faults. Adapting behavior in response to changing circumstances is essential to survival.

The interest in adaption has dramatically increased with the recent, rapid growth of *cyber-physical systems*. A cyber-physical system (CPS) tightly integrates computing resources, communications and physical systems. A significant percentage of cyber-physical systems are autonomous, which means the need for new tools and methods for designing adaptive systems is intensifying. New methods for online adaption of cyber-physical systems is also of increasing interest. We believe EAH techniques can play a prominent role in these areas.

In this paper we provide a basic introduction to cyber-physical systems and discuss the disciplines involved in designing such systems. Two EAH-based methods for adapting behavior in cyber-physical systems are described. We begin with a CPS overview in Section 2. This overview will be followed with a description of issues related on online adaption of CPS behavior in Section 3. Section 4 presents two EAH techniques that could be used for CPS behavioral adaption. Finally in Section 5 some issues related to EAH support for CPS design will be mentioned.

2 CPS Overview

We begin with the definition of an embedded system. Definitions vary, but essentially it is an information processing system where the end user is not aware a computer is present. Examples include photocopiers, microwave ovens, engine control in automobiles and price scanners in markets and department stores. More formally,

Definition: (*embedded system*)
An information processing system embedded into an enclosing product [3].

The above definition implies a link between computing resources and physical systems. We will make that connection stronger shortly. But first it is interesting to compare embedded systems against general purpose computing systems such as in a conventional laptop computer. This comparison is shown in the following table:

	Embedded Sys.	General Purpose Computer
clock speed	10-100 MHz	> 2 GHz
memory	KBytes	GBytes
number of tasks	1-5	∞
constraints	size, weight, power	none
focus	correctness	performance

Clearly embedded systems run much slower and have fewer resources. This is not a limiting factor, however, because embedded system computers are asked to only do a small number of dedicated tasks. The interesting difference is the focus. In general purpose computing performance, such as speed or virtually unlimited memory are major selling factors. Conversely, in embedded systems correctness is most important. Embedded systems often perform safety-critical operations where incorrect behavior can have dire consequences.

Embedded systems are ubiquitous. Applications include automobiles, commercial and military aircraft, weapon systems, medical equipment, smart power grids and transportation systems. They are becoming increasingly complex often including multiple processors, sophisticated communication networks and elaborate sensor and actuator systems. Sales of low-end microcontrollers suited for embedded applications exceed that of PC microprocessor sales and have done so for nearly 15 years.

So what exactly is a "cyber-physical system"? Is it just another term for an embedded system? The short answer is no. The term CPS came into popular use as early as 2006 in large part via the efforts of Helen Gill at the U.S. National Science Foundation. A CPS is not a traditional embedded system or sensor net. The term CPS emphasizes the fact that computer resources (the cyber portion) are tightly integrated with a physical system (the physical portion). Cyber capabilities could be incorporated into every physical component. A CPS could have elaborate networks and may be reconfigurable. Control loops can be continuous or discrete. Cyber-physical systems exist at all scales from hand-held devices to power grids spanning large geographical areas. The commonly accepted definition of a CPS is as follows:

Definition: *(cyber-physical system)*
A cyber-physical system is the integration of computation and physical processes [4].

Figure 1 shows the abstract architecture of a CPS. Using the term "cyber-physical" emphasizes the strong link between the cyber and the physical worlds. In a CPS the cyber portion affects the physical system and the physical system affects the cyber portion. The integration of the cyber with the physical is extremely tight. In fact, this integration is so tight it may be impossible to identify whether the system behavior is due to

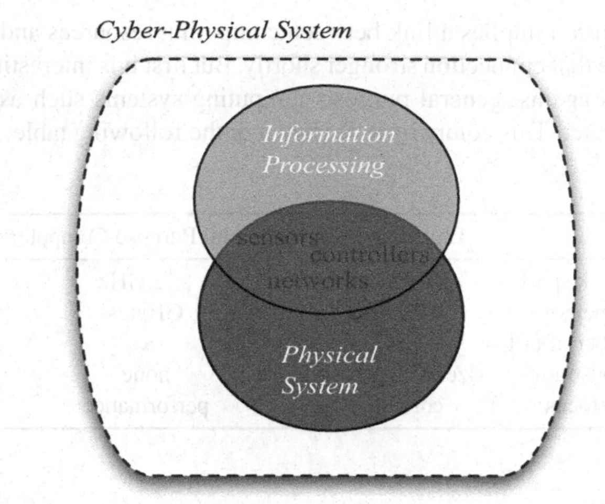

Fig. 1. An abstract view of the CPS architecture. The information processing system typically consists of one or more low-end microcontrollers. Sensors observe the physical system state while controllers provide inputs that alter the physical system state. Networks interconnect the physical and the cyber portions. The physical system can be electronic, mechanical or electromechanical.

computing or physical laws! For example, it may not be possible to tell if an unmanned aerial vehicle maneuver was caused by computer commands or resulted from the natural governing dynamics of the vehicle's airframe. A CPS is not the *union* of the cyber with the physical but rather the *intersection* of the two.

Certain CPS properties appear quite frequently. The list below describes some of those properties although the intended system function and operational environment dictates which ones are relevant in a given application.

– *dependability*

Dependable systems are reliable, maintainable and available. Reliability means the system has a low probability of failure. Maintainable systems are easy to repair. These two properties make the system available, which means it is usable whenever needed.

– *efficiency*

Efficiency covers several diverse topics. Systems that operate under battery power must be energy efficient. Limited memory requires small code sizes and judicious use of data memory. Rugged enclosures needed for high shock and vibration environments must still comply with size and weight constraints.

– *safety*

Some systems operate in proximity to other systems. Safe systems are completely predictable and cannot behave in ways that might cause damage to itself, any system it interfaces to or anything that operates in its neighborhood.

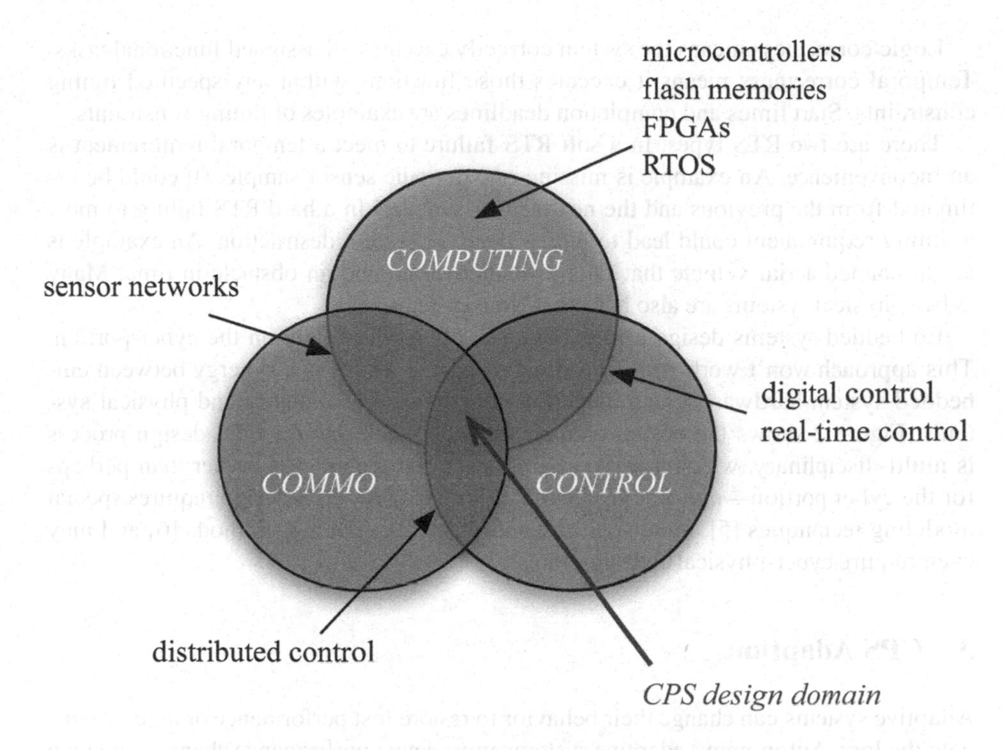

Fig. 2. The design domain for cyber-physical systems is the intersection of a computing design space, a control design space and a communications design space.. The computing design space contains microcontroller circuit designs, real-time operating systems (RTOS), FPGAs, flash memory, ADC and DAC convertors—i.e., the tools, devices and methods typically used in embedded system design work. Sensor networks and inter-processor bus or message-passing systems are common in cyber-physical systems so networking design tools are needed. The control design space involves both discrete and continuous control system design techniques, DSP algorithms, real-time control techniques and so forth. Some cyber-physical systems are distributed, which means distributed computer system and distributed control techniques may also be required.

– *security*

Systems may manipulate and/or store confidential data. Secure systems limit access to authorized users. Any communication with other systems is properly authenticated and messages are encrypted.

– *real-time operation*

Sometimes a system must perform all assigned tasks within specified timeframes. This does *not* necessarily mean fast. Actually the term "real-time" has nothing to do with speed because even an agonizingly slow system could qualify as a real-time system. The formal definition of a real-time system is as follows:

Definition: (*real-time system* (RTS))
A system that is both logically and temporally correct.

Logic correctness means the system correctly executes all assigned functional tasks. Temporal correctness means it executes those functions within any specified timing constraints. Start times and completion deadlines are examples of timing constraints.

There are two RTS types. In a soft RTS failure to meet a temporal requirement is an inconvenience. An example is missing one periodic sensor sample. (It could be estimated from the previous and the next sensor sample.) In a hard RTS failing to meet a timing requirement could lead to injury, death or system destruction. An example is an unmanned aerial vehicle that fails to maneuver around an obstacle in time. Many cyber-physical systems are also hard real-time systems.

Embedded systems design efforts have usually focused only on the cyber portion. This approach won't work with CPS design because a CPS is a synergy between embedded system hardware, communication networks, control theory and physical systems. Figure 2 shows the design domain for a CPS. Clearly the CPS design process is multi-disciplinary, which means existing tools and techniques—other than perhaps for the cyber portion—may not be adequate. For instance, CPS design requires special modeling techniques [5], more sophisticated formal verification methods [6] and may even require cyber-physical codesign tools [7].

3 CPS Adaption

Adaptive systems can change their behavior to restore lost performance or at least mitigate the loss. Autonomous adaptive systems must detect performance changes and then automatically initiate corrective actions. In other words, they must be self-adaptive. Before discussing adaption in cyber-physical systems it is useful to explain why system performance changes. There are two primary reasons: system faults or changes in the operational environment.

Safety-critical cyber-physical systems—particularly autonomous ones—must be *fault tolerant*. Unfortunately, many people do not fully understand what it takes to build a fault tolerant system. Fault tolerant systems must perform two independent functions: *fault detection and isolation* (FDI) and *fault recovery* (FR). FDI operations determine that a fault has occurred and attempt to isolate the fault to a particular subsystem or (hopefully) component within a subsystem. FR methods attempt to correct, mitigate, or in the worst case contain the failure. It is important to realize that FDI and FR operations are usually real-time operations because they have deadlines—i.e., there is a finite amount of time to detect, isolate and repair a system to prevent further damage.

Redundancy is the most widely used FR method. Redundancy can, in principle, recover from any fault because the failed subsystem is replaced with an identical spare. Unfortunately redundancy never works if the fault was caused by a changed operational environment. To fix ideas, consider the MOSFET switch circuit shown in Figure 3. Under normal operation the transistor is ON and delivers current to the load when $V_{GS} > 0$. This happens if $V_{in} = V_{DD}$. The transistor turns OFF when $V_{GS} = 0$ by making $V_{in} = 0$. Thus under normal operation V_{in} assumes only two values: V_{DD} to turn the transistor ON and 0V to turn it OFF. Now suppose this MOSFET switch circuit is in a space probe near the asteroid belt where it could be subjected to nuclear radiation. When a MOSFET is exposed to radiation it now takes $V_{GS} < 0$ to fully turn it off.

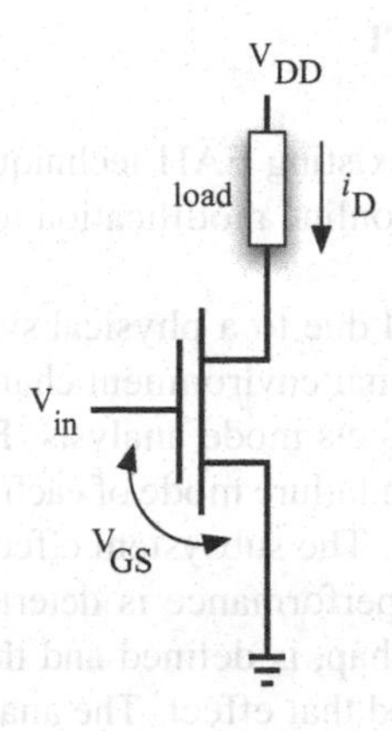

Fig. 3. A MOSFET switch circuit

Thus replacing a failed MOSFET circuit with an exact duplicate won't fix the problem because it was the operational environment that caused the circuit failure.

In many cyber-physical systems size and weight restrictions simply won't accommodate spare circuitry; redundancy as a FR method is not an option. An example of such a system is an unmanned aerial vehicle. In such systems, the only viable FR option may be to not repair the fault or not block the environment but instead directly force the observed behavior to change. We call this forced behavioral change *adaption*.

What is the best way to do this adaption? The obvious answer is reconfiguration. The basic idea behind reconfiguration is the faulty subsystem is re-organized in some way to restore some functionality or, if that won't work, at least isolate it from the rest of the system. Reconfiguration may only partially restore lost performance but it has the advantage of not taking up valuable space with spare hardware. The problem is what happens if the fault is in the physical system? In most (virtually all?) cases the physical system can't be reconfigured. But there is something else that *can* be reconfigured.

Every CPS has a control strategy that dictates how the physical system should react in the current operational environment. Controller circuitry executes the control strategy by generating specific inputs to the physical system to produce a desired behavior. However, the physical system may not respond properly to those inputs if it becomes faulty or the operational environment changes. The solution is to reconfigure the controller itself. A reconfigured controller generates a new set of inputs specifically designed to counteract the lost physical system performance.

It is important to realize the controller is implemented in the cyber portion of a CPS and so it can always be reconfigured. For instance, FPGAs can easily be reprogrammed to assume different functions. New firmware can be downloaded into a fuzzy logic controller. The fact that the controller is in the cyber portion is key to understanding how EAH techniques can support CPS adaption.

4 EAH Adaption Support

In this section we describe two existing EAH techniques that appear in the literature. Both techniques can be used for online modification to a CPS controller and their description will be in that context.

We assume adaption is needed due to a physical system fault arising from either a component failure or an operational environment change. FDI can be achieved using either a bottom-up failure and effects mode analysis (FMEA) or a top-down fault tree analysis (FTA). In the FMEA each failure mode of each component is analyzed to determine how it impacts a subsystem. The subsystem effects are then propagated upwards until the effect on system level performance is determined. In a FTA a catastrophic system-level failure, called a mishap, is defined and then all subsystems are analyzed to see if they could have produced that effect. The analysis proceeds downwards until a particular component or set of components failure modes are identified as the root cause.

From a practical standpoint the FTA is more efficient for several reasons. First, of all the FMEA is far more involved because every failure mode for every component must be analyzed. Second, some component failure modes are highly unlikely so their analysis is not warranted. Conversely, the FTA concentrates only on the most disastrous and likely system failures and targets the specific component failure modes that would produce that specific mishap.

EAH techniques evolve controller designs using evolutionary algorithms. The fitness of each one should be determined intrinsically and in a laboratory environment before deploying the CPS. This is done for several reasons. Extrinsic evaluation requires simulation and it may be difficult to accurately incorporate a system failure into a model, particularly if analytic models are used. Conversely, intrinsic evolution does not require models; the fault is physically injected into the system so the performance effects can be accurately observed and measured. It should also be remembered that evolutionary algorithms conduct a stochastic search. Evolution is directionless, which means some controller designs might produce undesirable physical system behaviors. Doing the evolution in a laboratory environment means these undesirable behaviors can be monitored and contained to prevent physical system damage.

4.1 Metamorphic Systems

The *metamorphic system* was introduced by Greenwood and Tyrrell in 2010 [8]. Figure 4 shows a block diagram. The substrate contains predesigned configurations. These configurations define a unique controller configuration. For example, each configuration could be a fuzzy logic controller rulebase. Or, each configuration could be a parameter set for a linear quadratic regulator or PID controller. The substrate does not contain duplicates. Since each configuration is different, each one defines a unique controller with unique characteristics. Thus each controller configuration produces different CPS behavior; switching controllers adapts the behavior.

The detection mechanism looks at the immediate (short-term) behavior while the assessment module looks at trends (long-term) behaviors to predict if the behavior is

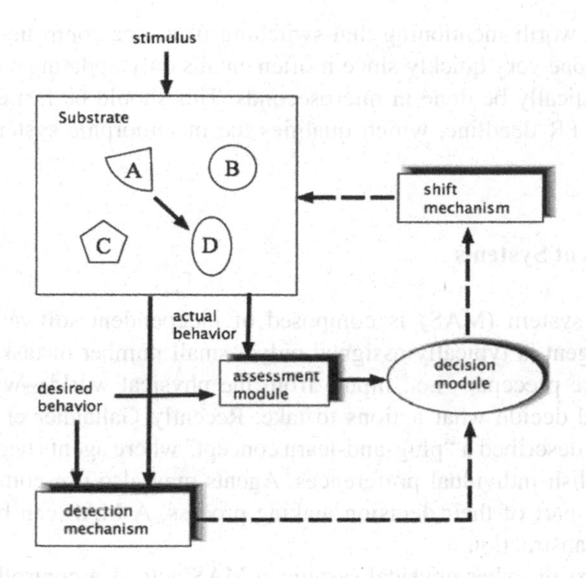

Fig. 4. The metamorphic system block diagram. The substrate contains different controller designs. The arrow from A to D indicates configuration A is replaced with configuration D. Note that at some future time configuration D may be replaced by configuration A.

deteriorating albeit still acceptable. Both compare existing behavior against desired behavior. The decision module is the "brains" and does require some computing capability. Any deviation from the desired behavior causes an event trigger to be sent to the decision module, which determines if a controller change is required. The detection mechanism and assessment module thus provide a FDI capability. If a change is necessary, the decision module tells the shifting mechanism which controller to bring on line.

No assumption is made about the design method used to create the configurations inside the substrate. The designs could come from conventional design formulas or be reused from previous projects. They could even be evolved using evolutionary algorithms[1]. The substrate could contain hardware devices if the controller configuration is implemented entirely in hardware (e.g., in an ASIC). It will not contain any executable code for a microprocessor. Most likely the substrate will contain firmware.

A metamorphic system is ideally suited for CPS adaption. Under laboratory conditions specific faults could be physically injected into the physical system. A controller that restores as much behavior as possible could then be intrinsically evolved. Each evolved controller—one per injected fault—is then placed into the substrate and the decision module is informed of which event triggers from the assessment module or detection mechanisms should select which controller. After a controller has been evolved for each injected fault the metamorphic system is ready for deployment as part of the cyber portion of a CPS.

[1] At this stage of the design process either extrinsic or intrinsic evolution could be used.

Finally, it is worth mentioning that switching from one controller to another controller can be done very quickly since it often entails only updating a memory pointer; FR could realistically be done in microseconds. This should be fast enough to satisfy virtually every FR deadline, which qualifies the metamorphic system as a real-time system.

4.2 Multi-agent Systems

A multi-agent system (MAS) is composed of independent software entities called *agents*. Each agent is typically assigned only a small number of tasks to accomplish. Agents compare precepts—i.e., inputs from the physical world—with their internal preferences and decide what actions to take. Recently Gallagher et al. [9] extended this concept by described a "plug-and-learn concept" where agents negotiate with other agents to establish individual preferences. Agents may also use communication with other agents as part of their decision making process. A MAS can be defined at any design level of abstraction.

In the context of cyber-physical systems a MAS acts as a controller. It consists of agents that sense the operational environment and physical system state, share information with other agents, and make decisions about what signals to send to actuators. Collectively they cooperate to achieve a specified physical system behavior.

Toolkits are normally used to design a MAS. These toolkits provide support for necessary design tasks such as agent creation, scheduling, simulation visualization and so forth. The more popular ones, like REPAST [10] and MASON [11], are open sourced and well documented. However, the design engineer is responsible for coding the individual agent behavior.

Each agent is responsible for executing specific tasks. These tasks are often abstractly expressed as IF-THEN-ELSE rules. This suggests a finite state machine (FSM) structure for encoding agent behavior. Changing the rules—equivalently, changing the FSM structure—adapts the agent's behavior. Since agents act as the CPS controller, the physical system behavior will also adapt. Evolutionary algorithms have a long history of evolving FSM structures. In fact, this is the precisely one application proposed for evolutionary programming almost 50 years ago [12].

MAS-based CPS adaption parallels that of the metamorphic system based adaption. In a laboratory environment FSMs that encode agent behavior are intrinsically evolved for every fault injected into the physical system. Thus each agent stores multiple potential behaviors, each corresponding to a unique controller configuration. The appropriate controller can be brought online once FDI procedures—executed by other agents—identify the specific fault. Alternatively one could evolve a hierarchal FSM that implements a subsumption architecture, which has been widely used in robot control [13].

Some agents can be assigned FDI tasks. They can communicate their findings to other agents to ensure the appropriate FSM controller is brought online to adapt the physical system's behavior.

As in the metamorphic system switching between independent FSMs can be typically in microseconds so the MAS method of adaption should also be able to meet any FR deadline.

5 Final Remarks

In this paper we provided an overview of cyber-physical systems with an emphasis on autonomous systems. These type of systems must be fault tolerant but unfortunately FR methods are limited especially when the fault is in the physical system itself. In such cases the only viable FR option is to adapt the behavior of the physical system by modifying the controller implemented in the cyber portion of the CPS. We offered two existing EAH methods to illustrate how to adapt the behavior of a CPS.

But not all cyber-physical systems are autonomous or even safety-critical. This means not all cyber-physical systems need behavior adaption. Nevertheless, all cyber-physical systems need some form of controller. It was previously mentioned that the predominant research coming from the EAH community involves original design, not adaption. There is no reason why those EAH design tools and methods cannot be exploited for CPS design and development. Indeed, the tight integration of the cyber with the physical makes the design of such systems—even those without some adaptive capabilities—quite challenging. New tools and methods are needed that are tailored to CPS design and test.

In this paper we described two EAH techniques that could be exploited for developing future cyber-physical systems but there are certainly others that can also be used. For instance, some methods used for intrinsic evolution may provide insight into new testing methods for cyber-physical systems. We believe cyber-physical systems is the next generation of evolvable hardware research and we encourage the EAH community to begin actively working in this emerging field.

Readers interested in learning more about cyber-physical systems should see two recent textbooks on the topic [3,14]. At the present time both can be found online as PDFs.

Acknowledgement. This material is based upon work supported by the National Science Foundation under Grant Numbers CNS-1239196, CNS-1239171, and CNS-1239229.

References

1. Greenwood, G., Tyrrell, A.: Introduction to Evolvable Hardware: A Practical Guide for Designing Self-Adaptive Systems. Wiley-IEEE Press (2006)
2. Haddow, P., Tyrrell, A.: Challenges of evolvable hardware: past, present and the path to a promising future. Genetic Prog. & Evol. Mach. 12(3), 183–215 (2011)
3. Marwedel, P.: Embedded system design: embedded systems foundations of cyber-physical systems, 2nd edn. Springer (2011)
4. Lee, E.: Computing foundations and practice for cyber-physical systems: a preliminary report, Tech. Report UCB/EECS-2007-72, EECS Dept., UC Berkeley (2007)
5. Derler, P., Lee, E., Vincentelli, A.: Modeling cyber-physical systems. Proc. of the IEEE 100(1), 13–28 (2012)
6. Sanwal, M.U., Hasan, O.: Formal verification of cyber-physical systems: Coping with continuous elements. In: Murgante, B., Misra, S., Carlini, M., Torre, C.M., Nguyen, H.-Q., Taniar, D., Apduhan, B.O., Gervasi, O. (eds.) ICCSA 2013, Part I. LNCS, vol. 7971, pp. 358–371. Springer, Heidelberg (2013)

7. Bradley, J., Atkins, E.: Toward continuous state-space regulation of coupled cyber-physical systems. Proc. of the IEEE 100(1), 60–74 (2012)
8. Greenwood, G., Tyrrell, A.: Metamorphic systems: a new model for adaptive system design. In: Proc. 2010 IEEE Cong. on Evol. Comp., pp. 3261–3268 (2010)
9. Gallagher, J., Matson, E., Greenwood, G.: On the implications of plug-and-learn adaptive hardware components: toward a cyberphysical systems perspective on evolvable and adaptive hardware. In: Proc. 2013 Int'l Conf. on Evol. Sys., pp. 59–65 (2013)
10. `http://repast.sourceforge.net/`
11. `http://cs.gmu.edu/~eclab/projects/mason/`
12. Fogel, L., Owens, A., Walsh, M.: Artificial Intelligence through Simulated Evolution. John Wiley, NY (1966)
13. Brooks, R.: A robust layered control system for a mobile robot. IEEE J. Robotics & Auto. RA 2, 14–23 (1986)
14. Lee, E., Seshia, S.: Introduction to Embedded Systems: A Cyber-Physical Systems Approach (2011), `http://LeeSeshia.org`

Searching for Agents' Best Risk Profiles

Robert E. Marks

Economics, UNSW Australia,
Sydney, NSW 2052, Australia
robert.marks@gmail.com
http://www.agsm.edu.au/bobm

Abstract. The purpose of this research is to seek the best (highest performing) risk profiles of agents who successively choose among risky prospects. An agent's risk profile is his attitude to perceived risk, which can vary from risk preferring to risk neutral (an expected-value decision maker) to risk averse, or even a dual-risk attitude. We use the Genetic Algorithm to search in the complex stochastic space of repeated lotteries. We examine three families of utility (or value) functions: wealth-independent CARA and wealth-dependent CRRA, in which an agent's risk profile is unchanging, and the Dual-Risk-Profile (DRP) functions from Prospect Theory, in which the agent can be risk-averse (for gains) or risk preferring (for losses). Statistical analysis of the simulation results suggests that the best (profit-maximizing) CRRA functions are risk neutral, while the other functions remain slightly risk-averse. The most profitable are slightly risk-averse DRP functions.

Keywords: decision making under risk, comparing utility functions, Prospect Theory, Genetic Algorithms, risk aversion, risk neutrality.

1 Introduction

Informally, it is widely held that in an uncertain world, with the possibility of the discontinuity of bankruptcy, the most prudent risk profile is risk aversion.[1] Indeed, "Risk aversion is one of the most basic assumptions underlying economic behavior" [2], perhaps because "a dollar that helps us avoid poverty is more valuable than a dollar that helps us become very rich" [3]. But is risk aversion the best risk profile? Even with bankruptcy as a possibility?

To answer this question, we use three kinds of utility function: the wealth-independent exponential utility function, or Constant Absolute Risk Aversion CARA; the Constant Relative Risk Aversion CRRA function, which is sensitive to the agent's level of wealth; and the DRP functions of Prospect Theory, where an agent's risk profile can vary depending on prospects of losing or gaining. We run computer experiments in which each agent chooses among three lotteries, and is then awarded with the outcome of the chosen lottery k.

[1] An preliminary version of this paper was presented at the IEEE Computational Intelligence for Finance Engineering & Economics 2014, London, March 29 [1].

H. Handa et al. (eds.), *Proc. of the 18th Asia Pacific Symp. on Intell. & Evol. Systems – Vol. 1*,
Proceedings in Adaptation, Learning and Optimization 1, DOI: 10.1007/978-3-319-13359-1_24

Repetition of these choices by many agents allows us to use a technique from machine learning – the Genetic Algorithm or GA [4] – to search for the best function from each utility family, where "best" means the highest average payoff when choosing among lotteries.

Modelling the agent's utility directly allows us to avoid the indirect inference of Szpiro [2], who argues that the evolutionary learning technique of the GA does two things: it allows wealth-maximizing agents to succeed even in highly stochastic environments, and it allows the emergence of risk aversion. Indeed, Szpiro argues that risk aversion is the best risk profile to adopt in such an environment. We compare the cumulative winnings (fitnesses) of our agents to see whether this is so.

2 Decisions under Risk and Risk Profiles

The von Neumann-Morgenstern formulation of the decision-maker's attitude to risk is based on the observation that individuals are not always expected-value decision makers. That is, there are situations in which people apparently prefer a lower certain outcome to the higher expected (or probability-weighted) outcome of an uncertain prospect (where the possible outcomes and their possibly subjective, or Bayesian, probabilities are known). An example is paying an insurance premium that is greater than the expected loss without insurance. On the other hand, people will sometimes "gamble" by apparently preferring a lower uncertain outcome to a higher sure thing: this is risk-preferring.

We can formalise this by observing that, by definition, the utility of a lottery is its expected utility, or

$$U(L) = \sum p_i U(x_i), \tag{1}$$

where each (discrete) outcome x_i occurs with probability p_i, and $U(x_i)$ is the utility of outcome x_i. It is useful to define the Certainty Equivalent $\tilde{x}$ (or C.E.), which is a certain outcome which has the identical utility as the lottery:

$$U(\tilde{x}) = U(L) = \sum p_i U(x_i). \tag{2}$$

We can use the C.E. to describe the decision-maker's risk profile [5]. Define the Expected Value $\bar{x}$ of the Lottery as:

$$\bar{x} = \sum p_i x_i. \tag{3}$$

When $\tilde{x} = \bar{x}$, then the decision-maker's utility function exhibits risk neutrality; when $\tilde{x} < \bar{x}$, then risk aversion; and when $\tilde{x} > \bar{x}$, then risk preferring.

2.1 Approximating the Certainty Equivalent

Expand utility $U(.)$ about the expected value $\bar{x}$.

$$U(x_0) \approx U(\bar{x}) + (x_0 - \bar{x})U'(\bar{x}) + \frac{1}{2}(x_0 - \bar{x})^2 U''(\bar{x}).$$

The C. E. $\tilde{x}$ of a continuous lottery is obtained by integration over the probability density function (p.d.f.) $f_x(.)$:

$$U(\tilde{x}) = \int dx_0 U(x_0) f_x(x_0).$$

$$\therefore U(\tilde{x}) \approx U(\bar{x}) + 0 + \frac{1}{2}\sigma^2 U''(\bar{x}), \tag{4}$$

where σ^2 is the variance. But, by expansion,

$$U(\tilde{x}) \approx U(\bar{x}) + (\tilde{x} - \bar{x}) U'(\bar{x}). \tag{5}$$

Therefore, from (4) and (5),

$$\tilde{x} - \bar{x} \approx \frac{1}{2}\sigma^2 \frac{U''(\bar{x})}{U'(\bar{x})}.$$

$$\therefore \tilde{x} \approx \bar{x} + \frac{1}{2}\sigma^2 \frac{U''(\bar{x})}{U'(\bar{x})}. \tag{6}$$

2.2 Risk Aversion

Risk aversion is not indicated by the slope of the utility curve: it's the *curvature* (U''/U'): if the utility curve is locally –

- linear (say, at a point of inflection, where $U'' = 0$), then the decision maker is locally risk neutral;
- concave (its slope is decreasing – Diminishing Marginal Utility), then the decision maker is locally risk averse;
- convex (its slope is increasing), then the decision maker is locally risk preferring.

3 Utility Functions

We consider three types of utility function:

1. those which exhibit constant risk preference across all outcomes (so-called wealth-independent utility functions, or Constant Absolute Risk Aversion CARA functions);
2. those where the risk preference is a function of the wealth of the decision maker (the Constant Relative Risk Aversion CRRA functions); and
3. those in which the risk profile is a function of the prospect of gaining (risk averse) or losing (risk preferring): the DRP Value Functions from Prospect Theory.

3.1 CARA Utility Functions

If an increase of all outcomes in a lottery by an equal amount Δ increases the C.E. of the lottery by Δ, then the decision maker exhibits wealth independence:

$$U(\tilde{x} + \Delta) = U(L') = \sum p_i U(x_i + \Delta).$$

Acceptance of this property restricts possible utility functions to be linear (risk neutral) or exponential – constant-absolute-risk-aversion (CARA) functions.

CARA utility functions charactise risk preference by a single number, the *risk aversion coefficient*, γ. Since CARA utility functions are wealth-independent, any aversion to bankruptcy is thus precluded, by definition. Whether a decision maker exhibits a wealth-independent utility function is an empirical question.

When utility is linear in outcomes, the decision maker is risk-neutral, across all outcomes, but such a simple constant-risk-profile utility function is of no further interest. Instead, we consider the exponential CARA functions, where utility U is given by

$$U(x) = 1 - e^{-\gamma x}, \tag{7}$$

where $U(0) = 0$ and $U(\infty) = 1$, and where γ is the *risk aversion coefficient*:

$$\gamma = -\frac{U''(x)}{U'(x)}. \tag{8}$$

From (6) and (8), for exponential utility,

$$\tilde{x} \approx \bar{x} - \frac{1}{2}\sigma^2\gamma,$$

which indicates that when $\gamma = 0$, then $\tilde{x} \approx \bar{x}$ (risk neutrality), when $\gamma > 0$, then $\tilde{x} < \bar{x}$ (risk averse), and when $\gamma < 0$, then $\tilde{x} > \bar{x}$ (risk preferring), with positive variance.

3.2 CRRA Utility Functions

We want utility functions which are *not* wealth-independent, to see whether such functions will result in risk-averse agents doing best.

The Arrow-Pratt measure of relative risk aversion (RRA) ρ is defined as

$$\rho(w) = -w\frac{U''(w)}{U'(w)} = w\gamma. \tag{9}$$

This introduces wealth w into the agent's risk preferences, so that lower wealth can be associated with higher risk aversion. The risk aversion coefficient γ is as in (8).

The Constant Elasticity of Substitution (CES) utility function:

$$U(w) = \frac{w^{1-\rho}}{1-\rho}, \tag{10}$$

with positive wealth, $w > 0$, exhibits constant relative risk aversion CRRA, as in (9).

Risk Aversion with CES Utility. In the CRRA simulations, we use the cumulative sum of the realisations of payoffs won (or lost, if negative) in previous lotteries chosen by the agent plus the possible payoff in this lottery as the wealth w in (10). Each agent codes for ρ.

From (6), the C.E. with CES utility is approximated by

$$\tilde{x} \approx \bar{x} - \frac{1}{2}\frac{\rho}{w}\sigma^2.$$

Iff $\frac{1}{2}\frac{\rho}{w}\sigma^2 > 0$ (or $\rho/w > 0$), then then C.E. $\tilde{x} <$ the expected mean $\bar{x}$, and the decision maker is risk averse.

With $w > 0$, $\rho > 0$ is equivalent to risk aversion. With $w > 0$ and $\rho = 1$, the CES function becomes the (risk-averse) logarithmic utility function, $U(w) \approx \log(w)$. With $w > 0$ and $\rho < 0$, it is equivalent to risk preferring.

3.3 The Dual-Risk-Profile DRP Function from Prospect Theory

From Prospect Theory [8], we model the DRP Value Function, which maps from quantity X to value V with the following two-parameter equations (with $\beta > 0$ and $\delta > 0$):

$$V(X) = \frac{1 - e^{-\beta X}}{1 - e^{-100\beta}}, 0 \leq X \leq 100, \tag{11}$$

$$V(X) = -\delta \frac{1 - e^{\beta X}}{1 - e^{-100\beta}}, -100 \leq X < 0. \tag{12}$$

The parameter $\beta > 0$ models the curvature of the function, and the parameter $\delta > 0$ the asymmetry associated with losses. The DRP function is not wealth independent. This function (in Fig. 1, with $\delta = 1.75$, for prizes between $\pm\$100$) exhibits the S-shaped asymmetry postulated by Kahneman and Tversky [8]. It exhibits risk seeking (loss aversion) when X is negative with respect to the reference point $X = 0$, and risk aversion when X is positive. We use here a linear probability weighting function (hence no weighting for smaller probabilities). As Fig. 1 suggests, as $\delta \to 1$ and $\beta \to 0$, the value function asymptotes to a linear, risk-neutral function (in this case with a slope of 1).

We use the GA to search the joint plane (β, δ) as the agents (each characterised by a point in the (β, δ) plane) choose the lottery that has the greatest expected value of the three. Each lottery has two known prizes in the interval of [$-\$100$, $+\$100$] of known probabilities, p_i. So the agent chooses the lottery k with the highest expected value:

$$U_k = \sum_{i=1}^{2} p_{ki} V(X_{ki}, \beta, \delta).$$

The GA jointly searches for points in (β, δ) that result in high payoffs after the payoff of each chosen lottery is subsequently realised, based on the probabilities of the possible outcomes.

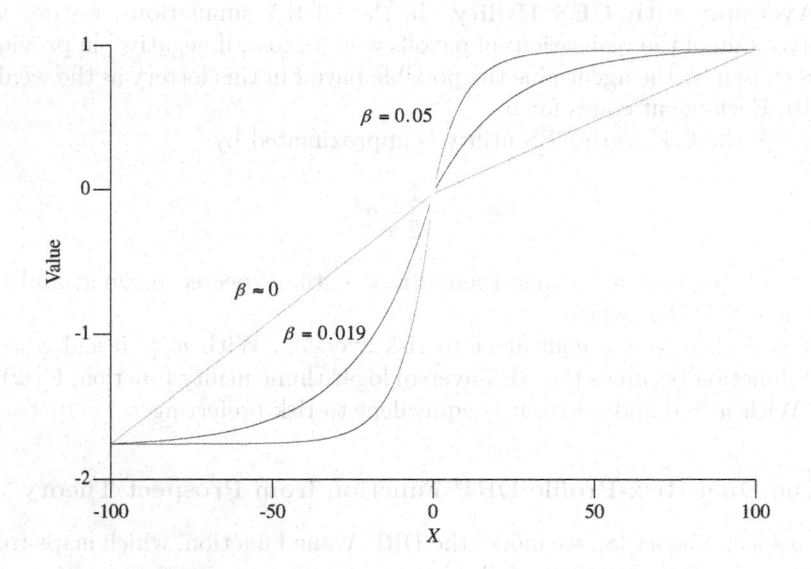

Fig. 1. A Prospect Theory (DRP) Value Function

4 The Simulations

Each lottery is randomly constructed: the two payoffs ("prizes") are randomly chosen in the interval between − and + MAP, (where the Maximum Absolute Prize, MAP, is $100); and the probability is also chosen randomly. (Each lottery has, of course, a single degree of freedom for probability). Each agent calculates the expected utility of each of the three lotteries, using its utility or value function (a function of its γ or ρ/w or (β, δ)), and chooses the lottery k with the highest expected utility. Doing this, agents know the prizes and probabilities of all three lotteries.

Then the actual (simulated) outcome of the chosen lottery k is randomly realised, using its probability. The winnings of the agent (that is, the wealth of the CRRA agent) is incremented accordingly. Each agent successively chooses 1000 lotteries.

Calculate the three expected utilities for lotteries X, Y, and Z, functions of γ (or ρ and w, or (β, δ)):

$$U(X) = p_x U(x_1) + (1 - p_x) U(x_2)$$

$$U(Y) = p_y U(y_1) + (1 - p_y) U(y_2)$$

$$U(Z) = p_z U(z_1) + (1 - p_z) U(z_2)$$

Choose the lottery I with the highest expected utility or value. Win (or lose) whichever prize (i_1 or i_2) is realised in that lottery, based in the lottery's probability p_i.

4.1 Searching with the Genetic Algorithm

We use a population of 100 agents, each of which has an average winnings or a cumulative level of wealth, based on its risk profile and the successive outcomes of its choices among the lotteries. The GA's mutation rate is controllable by the simulator, on-screen.

We use an implementation [6] of the GA to search for the best risk profile. That is, we select the best-performing agents to be the "parents" of the next generation of agents, which is generated by "crossover" and "mutation" of the chromosomes of the pairs of parents. Each of the new generation of agents chooses the lottery k with highest expected utility a thousand times. Again, the best are selected to be the parents of the next generation.

We use the GA simulation in this search as an empirical alternative to solving for the best (highest performing) risk profile analytically. Note that Rabin [3] asserts that "theory actually predicts virtual risk neutrality." We return to this in the Discussion below.

4.2 Simulations with Utility-Maximizing (or Value-Maximizing) Agents

Using NetLogo [7], we model each agent as a binary string which codes to its risk-aversion coefficient/s, (γ for CARA agents, ρ for CRRA agents, and (β, δ) for DRP agents) in the interval ± 1.048576. The DRP agents search for $0 \leq \beta < 0.21$ and for δ in the interval ± 10.48.

Each lottery is a two-prize lottery, where each prize is chosen from a uniform distribution, between − and + MAP (Maximum Absolute Prize), where MAP can be set up to \$100 by the simulator, and the single probability is chosen randomly from uniform [0,1].

Each agent chooses the lottery k with the highest expected utility from (7) and (10), based on its value of γ (respectively, ρ and wealth w), or from (11) and (12), based on its value of β and δ. Then a realised outcome is calculated for that lottery, based on its probability.

Each agent faces 1000 lottery choices, and the cumulative winnings that agent's "fitness" for the GA. Because all agents face the same probabilistic lotteries, we can also compare the the three utility families based on their average winnings or fitnesses. The processes are stochastic. For each model we perform a number n of Monte Carlo simulation runs to obtain sufficient data to analyse the results statistically.

4.3 The CARA Results

The on-line simulations[2] show three things clearly:

1. The mean (black) fitness (cumulative winnings) grows quickly to a plateau after 20 generations or so;

[2] See `http://www.agsm.edu.au/bobm/teaching/SimSS/NetLogo4-models/RA-CARA-EU-312p.html` for a Java aplet and the Netlogo code.

2. the mean, maximum, and minimum risk-aversion coefficients γ converge to close to zero (risk neutrality) over the same period, and
3. Any γ deviation from zero up (more risk-averse) or down (more risk-preferring) leads to the minimum fitness in that generation collapsing from close to the mean fitness.

These observations show that CARA agents perform best (in terms of their lottery winnings) who are closest to risk neutral ($\gamma = 0$). Too risk averse, and they forgo fair lotteries; too risk preferring and they choose too many risky lotteries.

Eye-balling single output plots, however, is not sufficient to reach clear conclusions about the best utility functions. We have preferred 55 independent Monte Carlo runs using the GA to search for better CARA utility functions.

The correlation between γ and Fitness in these MC simulations (where Fitness is the winnings averaged across 100 agents, each of which chooses the expected "best" of three lotteries 1000 times per generation, for 200 generations) is 0.7148. This suggests that the larger the value of γ, the higher the value of Fitness. The *p*-value for γ against a null hypothesis of $H_0 : \mu_\gamma = 0$ is 0.0006, which provides a very strong presumption against the null, that is, although it is close to the risk-neutrality of $\gamma = 0$, the CARA function does not converge to exact risk neutrality.[3] The final means of γ in the 55 runs are significantly positive, suggesting weak risk-aversion.

The wealth-independent CARA utility function precludes bankruptcy. What of utility function that does not exclude this possibility?

4.4 The CRRA Results

We could, of course, put a floor on agent wealth, below which is oblivion, but better to use a utility formulation that is not wealth independent and repeat the search. We use the CES utility functions (10) that exhibits CRRA.

The results are surprising.[4] We have performed 109 independent Monte Carlo runs using the GA to search for better CRRA utility functions.

The correlation between ρ and Fitness is 0.0907. This suggests that ρ and Fitness are not correlated. The *p*-value for ρ against a null hypothesis of $H_0 : \mu_\rho = 0$ is 0.2996, which provides no presumption against the null, that is, the data suggest that the mean $\mu_\rho = 0$, or the CRRA function converges to risk neutrality, despite our prior expectations for this function.

Remember: $\gamma = \frac{\rho}{w}$, so dividing the ρ values by the high w values attained implies corresponding minute values of γ here.

---

[3] For the detailed statistics of this and the other four analyses, see the Appendix in the paper at `http://www.agsm.edu.au/bobm/papers/singapore14.pdf`.

[4] See `http://www.agsm.edu.au/bobm/teaching/SimSS/NetLogo4-models/DRA-CRRA-EU-revCD-312p.html` for a Java aplet and the NetLogo code.

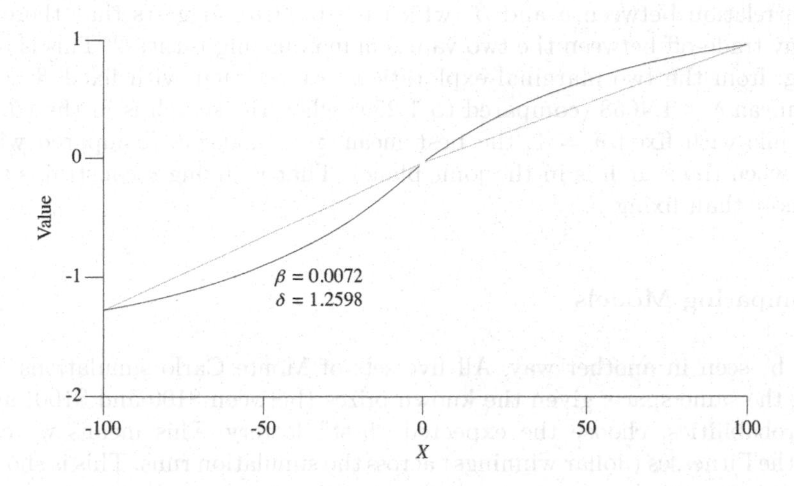

Fig. 2. The Best Prospect Theory (DRP) Value Function

4.5 The Dual-Risk-Profile Results

We considered first the marginal results: holding β constant at zero, asking what values of δ emerge as conditionally best, then, holding δ constant at unity, asking what values of β emerge as conditionally best. That is, first considering a kinked function, possibly linear, and then a symmetric DRP function, possibly linear.

We performed 55 independent Monte Carlo runs using the GA to search for better δ, while holding $\beta = 0$. The correlation between δ and Fitness in these MC simulations (where Fitness is the winnings averaged across the 100 agents, each of which chooses the expected "best" of three lotteries 1000 times per generation, for 200 generations) is 0.5602. This suggests that the larger the value of δ, the higher the value of Fitness. The p-value for δ against a null hypothesis, $H_0 : \mu_\delta = 1$, is effectively zero (< 0.00001), which provides a very strong presumption against the null, that is, the data suggest that the mean $\mu_\delta \neq 1$.

We then performed 50 independent Monte Carlo runs using the GA to search for better β, while holding $\delta = 1$. The correlation between β and Fitness in these MC simulations is 0.0941. This suggests that β and Fitness are not correlated. The p-value for β against a null hypothesis, $H_0 : \mu_\beta = 0$ is 0.0012, which provides a strong presumption against the null, that is, the data suggest that the mean $\mu_\beta \neq 0$.

Then we undertook a search in the (β, δ) plane. We performed 54 independent Monte Carlo runs using the GA to search for better β and δ jointly. Fig. 2 shows the result with the two variables' final means across these runs, where $\beta = 0.007186$ and $\delta = 1.2598$: almost risk neutral (in which an expected-value decision maker becomes an expected-outcome decision maker).[5]

[5] See `http://www.agsm.edu.au/bobm/teaching/SimSS/NetLogo4-models/RA-PTh-EV-both.html` for a Java aplet and the NetLogo code of a DRP Value Function model.

The correlation between δ and β (which is -0.0015) suggests that there is little if any trade-off between the two values in maximizing Fitness.[6] This is not surprising: from the two marginal explorations, we see that with fixed $\beta = 0$, the best mean $\delta = 1.4658$ (compared to 1.2598 when the search is in the (β, δ) plane), while with fixed $\delta = 1$, the best mean $\beta = 0.003846$ (compared with 0.007186 when the search is in the joint plane). That is, fixing δ constrains the Fitness more than fixing β.

4.6 Comparing Models

This can be seen in another way. All five sets of Monte Carlo simulations are searching the same space: given the known prizes (between \$100 and \$100) and known probabilities, choose the expected "best" lottery. This means we can compare the Fitnesses (dollar winnings) across the simulation runs. This is shown in Table 1.

Table 1. Mean Fitnesses of the Five Models

Model	Mean Fitness (\$)
CARA	37,650
CRRA	29,403
DRP with $\beta = 0$	37,666
DRP with $\delta = 1$	38,879
DRP joint β, δ	37,721

It is clear that CRRA, the only model whose "best" parameter reflects risk-neutrality, performs worst at maximizing Fitness, while the best model is the DRP model from Prospect Theory with symmetric ($\delta = 1$) loss-averting and gain-preferring (its utility is convex for losses and concave for gains). This is strange: a constrained optimization outperforming an unconstrained optimisation, when the constrained value is available to the unconstrained. The GA search of the joint model could/should find that Fitness is higher when $\delta \approx 1$ but hasn't. This suggests that the runs be lengthened, perhaps because the joint search in (β, δ) is hard. Indeed, from Table 1, the apex of the hill of optimal fitness is quite flat. It is likely that this anomaly suggests that the GA optimisations have been prematurely terminated; would longer runs change our risk profile results? Further work will tell.

At any rate, Table 1 shows that CARA, DRP with $\beta = 0$, and DRP with joint β, δ search are very close in terms of best Fitness.

---

[6] Another formalization of Prospect Theory Value Functions is to model separate curvature parameters for gains ($X > 0$) and losses ($X < 0$), but this does not really capture the full asymmetry between gains and losses that our model includes. And $\delta \neq 1$ implies a different slope (given $\beta = 0$) for losses.

5 Discussion

Like the GA simulations of Szpiro [2], we find that the best-performing CARA agents are risk-averse, not risk-neutral. Because of the indirect way in which Szpiro modelled the risk profiles of his agents (unlike a referee's suggestion, footnote 3, Szpiro's model "only distinguishes between risk-averse automata and all others"), while our models allow any CARA risk profile to emerge, we argue that our results are more general than Szpiro's.

Rabin [3] suggests a reason why risk-neutral functions will not do better than risk-averse functions, at least for small-stakes lotteries. He argues that von Neumann-Morgenstern expected-utility theory is inappropriate for reconciling actual human behaviour as revealed in risk attitudes over large stakes and small stakes. If there is risk aversion for small stakes, then expected-utility theory predicts wildly unrealistic risk aversion when the decision maker is faced with large stakes. Or risk aversion for large stakes must be accompanied by virtual risk neutrality for small stakes.

But we do not appeal to empirical evidence or even to prior beliefs of what sort of risk profile is best. Whereas there has been much research into reconciling actual human decision making with theory (see [9]), we are interested in seeing what is the best (i.e. most profitable) risk profile for agents faced with risky choices.

We find that for wealth-independent CARA utility functions (exponential) agents do not learn to become risk-neutral decision makers in order to maximise their returns when choosing among risky propositions. But for wealth-dependent CRRA utility functions (CES) our agents do learn to be risk neutral, despite the possibility (even if small) of bankruptcy, or the loss of all accumulated wealth.

Rabin [3] argues that *loss aversion* [8], rather than risk aversion, is a better (i.e. more realistic) explanation of how people actually behave when faced with risky decisions. This is captured in our DRP Value Function.

An analytical study of Prospect Theory Value Functions [10] posits an adaptive process for decision-making under risk such that, despite people being seen to be risk averse over gains and risk seekers over losses with respect to the current reference point [8] – the so-called dual risk attitude, with utility convex for losses and concave for gains – the agent eventually learns to make risk-neutral choices. Their result appears consistent with our results for the CRRA model, although the learning in their model is not that of the GA, but rather agents observing how their choices result in systemic undershooting (or overshooting) of their targets, which then results in more realistic targets and choices. Their lotteries are symmetrical (for tractability), unlike ours. Our results suggest that their results might generalise to asymmetric lotteries, such as ours, at least for CRRA utility.

A simulation study [11] examines the survival dynamics of investors with different risk preferences in an agent-based, multi-asset, artificial stock market and finds that investors' survival is closely related to their risk preferences. Examining eight possible risk profiles, the paper finds that only CRRA investors with relative risk aversion coefficients close to unity (log-utility agents) survive in the long run (up to 500 simulations). This does not appear consistent with our results.

Our last finding is obtained by comparing the mean fitnesses (accumulated winnings) of the five models. We find that a symmetric DRP model (with $\delta = 1$) does better than any of the other models, while the only model which learns to be risk-neutral (the CRRA model) does worst.

6 Conclusion

Using a demonstrative agent-based model – which demonstrates principles, rather than tracking historical phenomena – we have used the Genetic Algorithm to search the complex, stochastic space of decision making under risk, in which agents successively choose among three (asymmetric) lotteries with randomly allocated probabilities and outcomes (two per lottery), in order to maximize their expected utilities. The GA searches for the best-performing utility function, among CARA (or wealth-independent), CRRA (when wealth, and hence bankruptcy, matters), or for the best-performing Value Function, which exhibits the DRP of Prospect Theory, although we use the same parameter β to describe the curvature of both risk averse (gains) and risk preferring (losses), which is a restriction that could be relaxed with further study.

Consistent with our prior belief that a risk-averse agent does best in these circumstances, we find that only one of our three models – CRRA – converges to risk neutrality. Our findings are therefore only partly consistent with analytical work that proves that with symmetric lotteries, and agents with dual risk attitude, risk-neutral decisions are the eventual outcome of agents adjusting their aspirations and targets in response to the realisations of their choices. But the other two models – CARA and the DRP from Prospect Theory – converge on (slightly) risk-averse parameters, when we search using accumulated winnings as the Fitness.

Comparing the mean accumulated winnings across our models, we find that the best performing model is a symmetric DRP model. This might prove of use to future simulators.

Simulations, of course, can not prove necessity, only sufficiency [12], so our results for each of the three functions – CARA, CRRA, and DRP Value Functions – are existence proofs only: the best (highest performing) functions, in choosing among lotteries of known prizes and known probabilities, do not generally tend to risk neutral (linear). The results suggest relaxing the assumption of known probabilities might also be of interest. These results therefore tend to confirm the common knowledge that a small amount of risk aversion is best in a risky world.

Acknowledgment. I should like to thank Simon Grant, Luis Izquierdo, the participants of the Complex Systems Research Summer School 2007 at Charles Sturt University, the participants at the 26th Australasian Economic Theory Workshop 2008 at Bond University, the participants at the 2014 IEEE CIFEr meetings in London, Jasmina Arifovic, James Andreoni, Seth Tisue, Marco LiCalzi, Shu-Heng Chen, Arthur Ramer, David Midgley, and Nigel Gilbert for his implementation of the Genetic Algorithm in NetLogo. Several anonymous reviewers have also been helpful.

Note: Java aplets of the simulation models and the NetLogo code are available online, together with graphical output of the simulation results, as referenced in the three footnotes above. These models will also generate real-time results, including graphs of their performance, when one's computer's Java security allows. Moreover, one can explore the impact of the GA mutation rate on the simulation evolution.

References

1. Marks, R.E.: Learning to be Risk Averse? In: Serguieva, A., Maringer, D., Palade, V., Almeida, R.J. (eds.) Proc. of the 2014 IEEE Computational Intelligence for Finance Engineering & Economics (CIFEr), London, March 28-29, pp. 1075–1079. IEEE Computational Intelligence Society (2014)
2. Szpiro, G.G.: The Emergence of Risk Aversion. Complexity 2, 31–39 (1997)
3. Rabin, M.: Risk Aversion and Expected-Utility Theory: a Calibration Theorem. Econometrica 68, 1281–1292 (2000)
4. Holland, J.H.: Adaptation in Natural and Artificial Systems, 2nd edn. MIT Press, Cambridge (1992)
5. Howard, R.A.: The Foundations of Decision Analysis. IEEE Trans. on Systems Science and Cybernetics ssc-4, 211–219 (1968)
6. Gilbert, N.: Axelrod's Iterated Prisoners' Dilemma Tournament. In: Gilbert, N., Troitzsch, K.G. (eds.) Simulation for the Social Scientist, 2nd edn. Open University Press, Maidenhead (2005), http://cress.soc.surrey.ac.uk/s4ss/code/NetLogo/axelrod-ipd-ga.html
7. Wilensky, U.: NetLogo. Center for Connected Learning and Computer-Based Modeling. Northwestern University, Evanston (Version 4) (1999), http://ccl.northwestern.edu/netlogo
8. Kahneman, D., Tversky, A.: Prospect Theory: an Analysis of Decision Under Risk. Econometrica 47, 263–291 (1979)
9. Arthur, A.B.: Designing Economic Agents that Act like Human Agents A Behavioral Approach to Bounded Rationality. American Economic Review Papers & Proceedings 81, 353–360 (1991)
10. DellaVigna, S., LiCalzi, M.: Learning to Make Risk Neutral Choices in a Symmetric World. Mathematical Social Sciences 41, 19–37 (2001)
11. Chen, S.-H., Huang, Y.C.: Risk Preference, Forecasting Accuracy and Survival Dynamics: Simulation Based on a Multi-Asset Agent-Based Artificial Stock Market. Journal of Economic Behavior and Organization 67(3-4), 702–717 (2008)
12. Marks, R.E.: Analysis and Synthesis: Multi-Agent Systems in the Social Sciences. The Knowledge Engineering Review 27(2), 123–136 (2012)

MA-Net: A Reliable Memetic Algorithm for Community Detection by Modularity Optimization

Leila Moslemi Naeni[1,2], Regina Berretta[1,2], and Pablo Moscato[1,2]

[1] The Priority Research Centre in Bioinformatics, Biomarker Discovery and Information-based Medicine, Hunter Medical Research Institute, Australia
[2] School of Electrical Engineering and Computer Science, The University of Newcastle, Australia
Leila.Mosleminaeni@uon.edu.au,
{Regina.Berreta,Pablo.Moscato}@newcastle.edu.au

Abstract. The information that can be transformed in knowledge from data in challenging real-world problems follows the accelerated rate of the advancement of technology in many different fields from biology to sociology. Complex networks are a useful representation of many problems in these domains One of the most important and challenging problems in network analysis lies in detecting community structures. This area of algorithmic research has attracted great attention due to its possible application in many fields. In this study we propose the MA-Net, memetic algorithm to detect communities in network by optimizing modularity value which is fast and reliable in the sense that it consistently produces sound solutions. Experiments using well-known real-world benchmark networks indicate that in comparison with other state-of-the-art algorithms, MA-Net has an outstanding performance on detecting communities.

Keywords: community detection, modularity, memetic algorithm.

1 Introduction

A variety of real-world complex systems in the fields of biology, sociology and physics can all be represented as complex networks. Patterns of connections or interactions between elements of a given system can be represented as a network, which in its simplest form, can mathematically be modelled as an undirected graph. In a graph representation, the components of the network are typically encoded as nodes and their interactions are represented as edges. For instance, chemical reactions between molecules and proteins in the cell [1-3], hyperlinks between webpages [4, 5], physical connection between neurons [6-8] in the brain, and even consumer behavior models [9] all can be represented as graphs. An underlying hypothesis of these studies is that the pattern of interactions and the structure of a graph greatly affect the behavior of the associated system.

One of the important and challenging problems in graph analysis lies in finding groups hidden in a graph, which is generically called the community detection problem. While there is no universally agreed definition for a community [10], Girvan and

H. Handa et al. (eds.), *Proc. of the 18th Asia Pacific Symp. on Intell. & Evol. Systems – Vol. 1*,
Proceedings in Adaptation, Learning and Optimization 1, DOI: 10.1007/978-3-319-13359-1_25

Newman [11] defined communities as groups of vertices, where the connections are dense within the group, but the connections are sparse between the groups (See Fig.1). With this definition, the community detection problem can be formalised as an optimization problem in undirected graphs, however, due to the huge number of possible ways for partitioning the set of vertices of a graph, even when the graph is small, the community detection problem, like many other clustering problems defined in graphs, naturally lead to formulations of the problems which give rise to NP-hard computational problems [12].

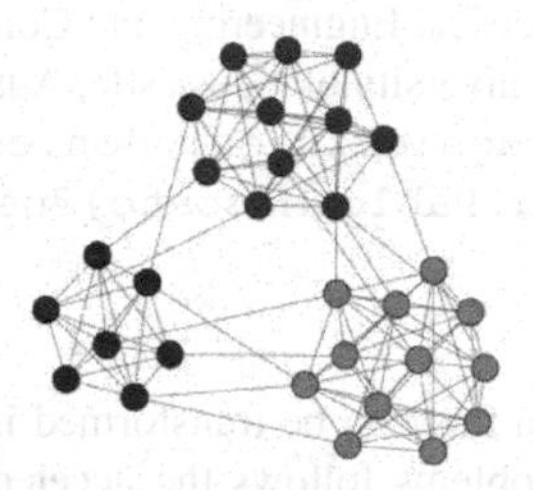

Fig. 1. A small network with three communities that have highly intra-group interactions and sparse inter-group interactions

In order to identify a good community structure in a network, many outstanding community detection algorithms have been proposed following different approaches during the last decade (see a recent comprehensive review in [13]). Optimization-based algorithms are among the most popular methods and the modularity function that was introduced by Newman and Girvan [14] has played a great role in optimization algorithms. Modularity is a quality measure that compares the number of edges in a detected community with the number of edges that are expected to be observed in a random graph. Therefore, higher levels of modularity indicate a greater difference between the detected partition and a random graph, thus modularity optimization acts as a proxy for detecting a better community structure. In this manner, the community detection problem can be considered as a combinatorial optimization problem with the objective function of maximizing the modularity. In a given graph $G = (V, E)$ with the adjacency matrix A, the modularity can be written as follows:

$$Q = \sum_{c=1}^{k} \left[\frac{l_c}{m} - \left(\frac{d_c}{2m} \right)^2 \right], \quad (1)$$

where, m is the total number of edges in the graph and the summation runs over the k communities of the partition; and l_c stands for the number of edges inside the community c. The summation of degrees of all vertices in the community is denoted by d_c. Instead of running modularity equation over communities, it can be reformulated to run over all pairs of nodes. Modularity function can easily be modified to measure the quality of partitions in weighted graphs [15]. It has been proven that modularity optimization gives rise to decision problems which are NP-complete [16].

Although some exact algorithms have also been proposed to solve the modularity optimization problem in small size networks by using column generation algorithms [17, 18], heuristic algorithms are always in need to find good solutions in a reasonable time, particularly in large-scale networks. Many different heuristic algorithms have been proposed for modularity optimization with different approaches, such as label propagation algorithms (LPA) [19] and agglomerative hierarchical clustering [20-24] that iteratively join pairs of nodes to increase the level of modularity. To enhance the performance of heuristic algorithms a variety of metaheuristic algorithms have been proposed to detect communities, for instance, simulated annealing [25, 26], conformational space annealing [27], genetic algorithm [28-30] and memetic algorithm [31-34]. To reduce the running time of the algorithm for large-scale networks some parallel metaheuristics have been proposed [35-37].

While modularity is the most popular objective function for community detection, there are some proven limitations in detecting communities by optimizing modularity [38, 39]. For instance, modularity optimization has a tendency to merge small clusters even when their individual size is small in comparison with the whole network. Moreover, the number of local maxima increases dramatically when the size of the network grows, thus it is easy to get trapped in a local optimal solution. Some various objective functions have been proposed to overcome modularity limitations. For instance, *modularity density* [40], *community score* and *community fitness* [41]. In addition, a few multi-objective algorithms have been proposed to optimize more than a single objective function rather than only optimizing the modularity[42].

In this study we develop a reliable memetic algorithm, MA-Net, to optimize modularity and to detect community structure of the given network. We compare our results with state-of-the-art algorithms. As a future research direction, we aim to extend the proposed algorithm to a multi-objective optimization algorithm.

2 Proposed Algorithm

The proposed algorithm is designed in the framework of Memetic Algorithms (MA). We have taken this approach due to proven successful results and its effectiveness to address many NP-hard combinatorial optimization problems [43]. The modularity function can then be used to evaluate the quality of the communities. Thus, we have chosen it as an objective function for the first implementation of MA-Net, and to allow us to compare the performance of MA-Net against existing state-of-the-art algorithms. Algorithm 1 shows the main framework of our proposed MA-Net. A graph is the main input of the algorithm. In MA-Net, to facilitate and speed up the computations we use the adjacency list which uses less memory compared to the adjacency matrix. We will describe the *initialize_population* procedure, genetic operators (modularity based recombination and mutation), local search strategy and *update_population* procedure in the following sections.

Algorithm 1. MA-Net framework

Input:
Graph $G = (V, E)$;
N_p:population size;
N_r:acceptable number of generation without improvement;
P_m: mutation probability
Result:
I^*: A partition of the graph vertices that aims to achieve maximum modularity:

1: $pop = \{I_1, I_2, ..., I_{N_p}\} \leftarrow initialize\_population(G, N_p)$
2: **repeat**
3: $offspring \leftarrow modularity\_based\_crossover(pop)$
4: $offspring \leftarrow adaptive\_mutation(offspring, P_m)$
5: $offspring \leftarrow local\_search(offspring)$
6: $pop \leftarrow update\_population(offspring)$
7: $I^* \leftarrow$ fittest member of pop
8: **until** $termination\_criterion(pop, N_r)$
9: **return** I^*

2.1 Representation and Initialization

To represent a partition of a given graph G with n nodes, we use string-coding representation as $P = [C_1, C_2, ..., C_n]$. Here, C_i is an integer number, shows the community label, and refers to the community that node v_i belongs to. So by having k communities in partition P, C_i can be any integer number between 1 and k (i.e. $1 \leq C_i \leq k$). Obviously, nodes having the same community label are considered in the same community. A simple example of string-coding representation is illustrated in Fig. 2.

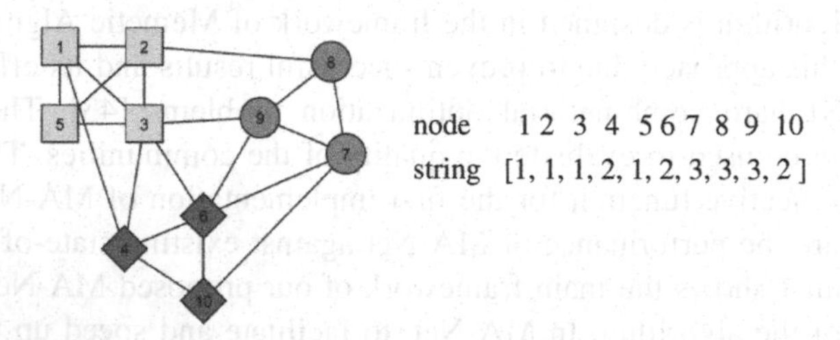

Fig. 2. Example of the solution representation structure. Left: a graph of 10 nodes and partitioned in three communities shown by different colors. Right: the string coding represents this partition.

The *initialize_population* procedure generates the initial population. We first assign a random community label to each node. The random label is an integer number between 1 and the total number of nodes, therefore there is no previous assumption for the number of the communities. Ours is an unsupervised algorithm and it aims at discovering the best number of communities for the graph to maximize modularity. These random solutions are of very low quality as even pairs unconnected nodes can be assigned to the same community. In the second step, to speed up the convergence of the algorithm, we improve the quality of the solution with another procedure. For each solution, we generate a random sequence $\{r_1, r_2, \dots, r_n\}$ of nodes then for each node v_{r_i} in the sequence, move the node to one of its neighbors' community that results in the greatest improvement in the modularity score. We repeat this operation for all solutions of the initial population. To reduce the computational cost of the improvement procedure, we calculate ΔQ by moving one node v_i into one of its neighbors' community [22]. The improvement procedure avoids unnecessary communities by considering only connected neighbors in moving of nodes. This heuristic improvement procedure is easy to implement and very effective to improve the quality of the initial population.

2.2 Genetic Operators

MA-Net explores the search space by two specially designed genetic operators: modularity-based recombination and an adaptive mutation operator. These operators play important roles, working together, in preventing the algorithm to getting trapped into a local optimal solution and exploring the configuration space.

Modularity-Based Recombination Operator

The recombination operator generates an offspring that inherits some characteristic from its two parents, therefore it plays an important role in the global search of the solution space [43]. Traditional recombination operators, including uniform crossover, one-point and two-point crossover can hardly convey community structure of the parents to the offspring and seem less suitable for this case. We propose a modularity-based recombination operator which is an efficient operator specifically designed for the community detection problem by which descendants can inherit useful communities of their parents. The main idea of this operator is to take the communities as the genetic material and try to preserve the best communities of parents for the offspring. The modularity-based recombination procedure is described next.

Firstly, two random members of the population are selected as parents. Since we can safely assume that all members of the population are of a relatively good quality (after the improvement step of the population initialization procedure), a randomly selection strategy could work well in maintaining the population diversity. Let P and Q be the parents and p and q represent the number of communities in each parent, respectively. Then, we sort $(p + q)$ communities in the list L according to their fitness. According to the modularity function (equation 1), the solution fitness is the total sum of the modularity of all communities in the partition. Therefore, we have already computed the fitness of each community from the fitness of the solution. Next, we choose the fittest community from list L and form a same community in the offspring. Then for the second good community in L, we try to form a similar community in the offspring with the nodes that are not assigned to any community before.

We repeat this procedure till all nodes in offspring solution have been assigned to a community.

Fig. 3 illustrates the modularity-based crossover procedure applied to a small graph with 13 nodes. Two random partitions of the graph are selected as parents (Fig. 3 (a) and (b)). The underlined number close to each community shows the ranking of the community according to its modularity, for instance, a community with 5 nodes in parent (a) has the highest fitness and its rank is 1. Communities are transferred to the offspring according to their rank, so in first step nodes of the community rank 1 in parent (a), {1, 2, 3, 4, 5} are assigned in a same community in the offspring. The second priority is for nodes {9, 12, 13} in parent (b) that are not assigned to a community in the offspring in previous step; they form the second community in the offspring. The third priority is for nodes {10, 11, 12, 13} in parent (a). But nodes 12 and 13 were assigned before, so only nodes 10 and 11 are put in a community. The forth priority is for nodes {6, 7, 8} in parent (b) that can be exactly formed a same community in the offspring. In this stage we can see that all nodes in the offspring are partitioned. As shown in Fig. 3, the offspring has 4 communities that are inherited from parents.

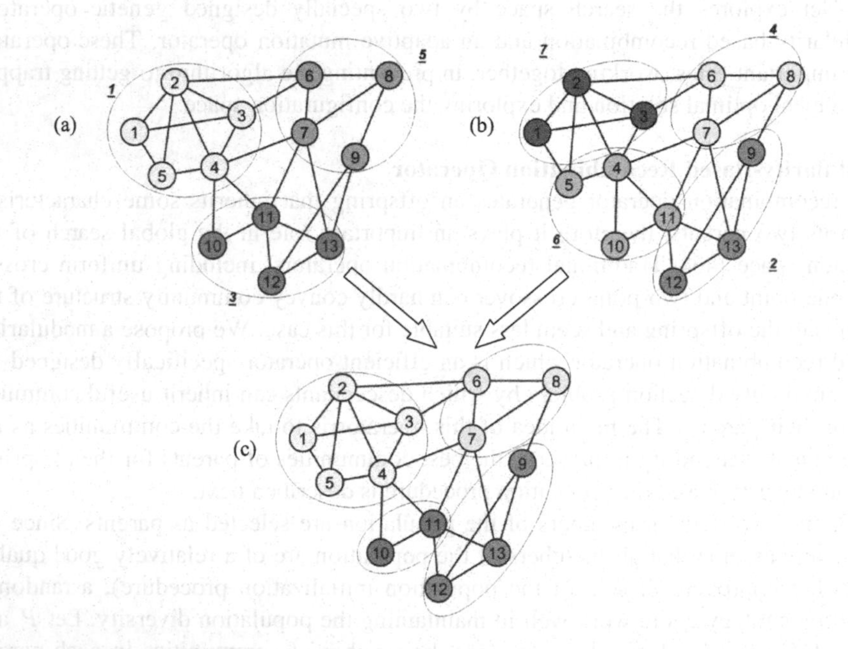

Fig. 3. A schematic illustration of the modularity-based crossover. (a) and (b) are two partition of a simple graph with 12 nodes that are selected as parents. (c) shows the offspring generated from parents by modularity-based crossover.

The modularity-based recombination operator's aim is that offsprings inherit the best structures observed in both originating solutions. While this recombination operator can increase the number of communities, we proposed a local search procedure that can merge communities to improve the modularity. As stated, one of the weaknesses of some modularity optimization algorithms [20, 21] is their tendency to merge the small communities and create huge communities with a large fraction of nodes, but after applying the modularity-based recombination operator the number of communities in offspring is generally more than the number of communities in parents. Therefore, this operator leads the algorithm to explore small communities.

Adaptive Mutation Operator

In order to maintain the diversity of the population and to avoid useless explorations of the search space, we use the neighbour-based mutation strategy [44]. The mutation operator changes the community label of node $\boldsymbol{v_i}$ to the community label of one of its neighbours randomly. Considering only neighbours' communities for changing the community label of each node helps the algorithm to explore search space more wisely. Here we apply the adaptive mutation operator to the existing offspring and generate a new offspring.

The mutation probability $\boldsymbol{P_m}$ indicates the likelihood of a mutation to be performed for each node. Here, we proposed an adaptive mutation operator where $\boldsymbol{P_m}$ is modified by the ability of the algorithm to find a better solution. It means that mutation probability is increasing when the algorithm could not improve the solution. The higher $\boldsymbol{P_m}$, the more changes in the offspring and the broader exploration of the search space.

One of the algorithm's parameters is N_r that indicates the acceptable number of generations without improvement and we used this parameter as the termination criterion, so the algorithm is terminated when the number of generations without improvement in the best solution exceed the value N_r. While the number of generations without improvement is increasing, the mutation probability will increase to expand the capacity of diversification of the algorithm by increasing the amount of changes in the mutated individual. $\boldsymbol{P_m}$ is an algorithm parameter and it will growth linearly to $\boldsymbol{2P_m}$ according to how close the algorithm is to reaching its termination criteria. For instance, in experimental tests (Section 3) $\boldsymbol{P_m}$ is set to 0.05 and the termination criterion is set to 30 generations without improvement. Therefore in the first generation, $\boldsymbol{P_m}$ is 0.05 and when the algorithm is run for 15 generations without improvement, the mutation probability will increase to $\boldsymbol{1.5 * P_m (i.e\ \frac{15}{30} P_m + P_m)}$.

2.3 Local Search

In this study, we employ the vertex movement heuristic. Different implementations of vertex movement heuristics have been used in existing algorithms [22, 23, 45, 46]. In MA-Net, we use a vertex movement heuristic with a stochastic hill climbing strategy to exploit the neighborhood of each solution. In a given graph G with n nodes, we define the neighbour solution of a partition $P = [C_1, C_2, \ldots, C_n]$ with k communities as a partition where a single node v_i is reassigned to another community. Therefore, a movement of a node from its own community to any other community generates a neighbour solution. The local search procedure works as follows:

Firstly, a random sequence of n nodes $L = \{r_1, r_2, \dots, r_n\}$ is generated. Then, for each node v_{r_i} in the sequence, the modularity gain ΔQ of moving this node to the community of one of its random neighbour v_j (i.e. $(v_{r_i}, v_j) \in E$) is computed. If this movement improves the fitness of the partition, the node v_{r_i} obtains the community label of v_j . If the movement does not improve the fitness, another random neighbour of v_{r_i} is selected to compute ΔQ. This process stops when node v_{r_i} is moved or we found that there is no movement for v_{r_i} that improves the fitness of the partition. The same procedure is repeated for all nodes in the random sequence L.

It should be noted that the stochastic hill climbing algorithm does not examine all the neighbours before deciding where to move the node. Therefore, the random neighbour selection strategy of the local search procedure helps to avoid getting stuck in local optima which is the limitation of deterministic hill climbing techniques. Due to the sensitivity of the local search procedure to the nodes sequence, we generate a random sequence each time to assure that we start the local search from different nodes. In addition, to speed up the local search procedure we compute a variation of the modularity ΔQ [34] when moving a node into its neighbour community.

2.4 Updating Strategy

In MA-Net, we use an elitism strategy for the *update_population* procedure. It means that the population is updated for the next generation by replacing the new individual with the least fitted member. But if the new individual is worse than the least fitted member of the population, the population will not change. This strategy guarantees that the better solutions are always retained in the population and remain eligible to be selected as parents in the next generation.

3 Experimental Results

In this section we evaluate the proposed algorithm (MA-Net) on five well-known real-world networks. The proposed algorithm is implemented in Python 2.7 and the experiments are performed on a machine with Intel(R) Xeon(R) CPU E5-1620 at a clock speed of 3.60GHz (4 cores and 8 logical processors) and 16 GB of memory. After initial experiments and applying Wilcoxon singed-rank test [47] to set the parameters, we tuned the algorithm's parameters as follows: population size, N_p, is set to 40, the mutation probability, P_m, is set to 0.05 and the termination criterion is set to 30 generations without improvement in the best solution.

We choose five well-known benchmark networks to compare MA-Net performance with the state-of-the-art algorithms: *Zachary's karate club network* (Karate) [48], the *bottlenose dolphin social network* (Dolphin) [49], *American political books* (Polbook) [50], *American college football network* (Football) [11] and the *jazz musicians network* (Jazz) [51]. Some configurations data of these networks are shown in Table 1.

We ran MA-Net a total of 50 times on each network and we report the average modularity, maximum modularity and the standard deviation of the modularity (runs are independent of each other, that is, we do not use the results of the previous runs to guide the algorithm in the future runs). To evaluate the performance of the MA-Net, we compare the results given by MA-Net with those of five well-known existing algorithms for modularity maximization: GN (a greedy heuristic algorithm) [11], CNM (an improved heuristic algorithm) [20], GATHB (a genetic algorithm) [28] and Meme-Net (a memetic algorithm) [32] and MA-COM (a memetic algorithm) [31]. The MA-Net results in Table 1 include maximum modularity (Q_{max}), average modularity (Q_{avg}) and the standard deviation of modularity (Q_{std}) in 50 runs.

Table 1. Benchmark networks configurations

	Network	Nodes	Edges	Average degree
1	Karate	34	78	4.59
2	Dolphin	62	159	5.13
3	Polbook	105	441	8.40
4	Football	115	613	10.66
5	Jazz	199	2742	27.70

Table 2. The maximum, average and standard deviation of the modularity ($Q_{avg}, Q_{max}, Q_{std}$) obtained by MA-Net in comprasion with the maximal modularity achived by GN, CNM, GATHB, Meme-Net and MA-COM

Network	GN	CNM	GATHB	Meme-Net	MA-COM	MA-Net		
						Q_{max}	Q_{avg}	Q_{std}
1 Karate	0.401	0.381	0.402	0.402	0.420	0.420	0.419	0.002
2 Dolphin	0.519	0.515	0.522	0.518	0.529	0.529	0.523	0.004
3 Polbook	0.510	0.502	0.518	0.523	0.527	0.527	0.526	0.002
4 Football	0.599	0.565	0.551	0.604	0.605	0.605	0.601	0.003
5 Jazz	0.439	0.439	0.445	0.438	0.445	0.445	0.445	0.000

Firstly, the comparison between the maximum modularity in Table 2 shows that for all benchmark networks the maximum modularity (Q_{max}) obtained by MA-Net is superior to the maximum modularity achieved by GN, CNM, GATHB and Meme-Net and it is equal to MA-COM results. We need to highlight that the MA-COM has indeed obtained the best solution for community detection problem in these five networks. Thus considering Q_{max} and the MA-COM results, this implies that MA-Net also finds the best partition with the highest modularity value in all studied networks.

Moreover, comparing Q_{avg} with the best results of others shows that across all networks, even the average results obtained through 50 runs of MA-Net are better than the best results of GN, CNM and GATHB. Only for the Jazz network the best result of GATHB is same as average result of MA-Net. The comparison between the two memetic algorithms of MA-Net and Meme-Net, show that the average results of MA-Net (Q_{avg}) in four networks (Karate, Dolphin, Polbook and Jazz network) are better than the maximal results of Meme-Net and only in the football network the Q_{avg} is less than Meme-Net result.

Finally, to show the reliability of MA-Net in community detection, we reported the Q_{std} that means the standard deviation of modularity values obtained after 50 runs of the algorithm in each network. Across all benchmark networks Q_{std} obtained by MA-Net are very small and always less than 0.005. Surprisingly, in all trials MA-Net obtained the optimal solution for the Jazz network. Therefore the standard deviation Q_{std} in Jazz network is equal to zero.

To conclude, the experimental results demonstrate that although the proposed algorithm is a memetic algorithm with a randomized behavior and no performance guarantee, in compares very well with state-of-the-art algorithms, it is highly reliable in finding good partitions and it is robust in handling different networks and detecting high quality community structures.

4 Conclusion and Future Work

Nowadays, the growth of complex systems research brings the need of novel sets of analytical methods aimed at understanding the underling properties and mechanisms by decomposing it in tightly coupled subsystems. One of the challenging problems in complex network analysis is detecting the community structure in large-scale networks. In this study, we proposed a novel approach to reveal community structure of the network. The proposed strategy is a memetic algorithm with problem-specific recombination and mutation operators. The experimental comparisons on real-world benchmark networks illustrate that our proposed MA-Net performs better than traditional algorithms on the same set of instances and is highly competitive with state-of-the-art methods. In addition, our experiments have shown that the MA-Net approach can always discover good solutions for the community detection problem with a small deviation from the optimal solution for modularity optimization. Future work will aim at investigating the algorithm's performance for larger networks and converting the single modularity optimization problem into a multi-objective problem, which can overcome the weakness of modularity in detecting small communities.

References

1. Gavin, A.-C., et al.: Proteome survey reveals modularity of the yeast cell machinery. Nature 440, 631–636 (2006)
2. Krogan, N.J., et al.: Global landscape of protein complexes in the yeast Saccharomyces cerevisiae. Nature 440(7084), 637–643 (2006)
3. Lee, J., Hidden, L.J.: information revealed by optimal community structure from a protein-complex bipartite network improves protein function prediction (2013), http://www.ncbi.nlm.nih.gov/pubmed/23577106
4. Brin, S., Page, L.: The anatomy of a large-scale hypertextual Web search engine. Computer Networks and ISDN Systems 30(1-7), 107–117 (1998)

5. Smith, M.A., Kollock, P.: Communities in cyberspace, vol. 1, p. 1999. Routledge, London (1999)
6. Ahn, Y.-Y., Bagrow, J.P., Lehmann, S.: Link communities reveal multiscale complexity in networks. Nature 466(7307), 761–764 (2010)
7. Stam, C.J., Reijneveld, J.C.: Graph theoretical analysis of complex networks in the brain. Nonlinear Biomedical Physics 1(1), 3 (2007)
8. Bullmore, E., Sporns, O.: Complex brain networks: graph theoretical analysis of structural and functional systems. Nature Reviews Neuroscience 10(3), 186–198 (2009)
9. de Vries, N.J., Carlson, J., Moscato, P.: A Data-Driven Approach to Reverse Engineering Customer Engagement Models: Towards Functional Constructs. PloS One 9(7), e102768 (2014)
10. Gong, M., et al.: An Improved Memetic Algorithm for Community Detection in Complex Networks. In: WCCI 2012 IEEE World Congress on Computational Intelligence (2012)
11. Girvan, M., Newman, M.E.J.: Community structure in social and biological networks. Proceedings of the National Academy of Sciences 99(12), 7821–7826 (2002)
12. Brandes, U., et al.: On Modularity Clustering. IEEE Transactions on Knowledge and Data Engineering 20(2), 172–188 (2008)
13. Fortunato, S.: Community detection in graphs. Physics Reports 486(3-5), 75–174 (2010)
14. Newman, M.E.J., Girvan, M.: Finding and evaluating community structure in networks. Physical Review E 69(2), 026113 (2004)
15. Newman, M.E.J.: Analysis of weighted networks. Physical Review E 70(5), 056131 (2004)
16. Brandes, U., et al.: Maximizing Modularity is hard. eprint arXiv:physics/0608255 (2006)
17. Xu, G., Tsoka, S., Papageorgiou, L.G.: Finding community structures in complex networks using mixed integer optimisation. The European Physical Journal B 60(2), 231–239 (2007)
18. Aloise, D., et al.: Column generation algorithms for exact modularity maximization in networks. Physical Review E 82(4), 046112 (2010)
19. Barber, M.J., Clark, J.W.: Detecting network communities by propagating labels under constraints. Physical Review E 80(2), 026129 (2009)
20. Clauset, A., Newman, M.E.J., Moore, C.: Finding community structure in very large networks. Physical Review E 70 (2004)
21. Newman, M.E.J.: Fast algorithm for detecting community structure in networks. Physical Review E 69(6), 066133 (2004)
22. Blondel, V.D., et al.: Fast unfolding of communities in large networks. Physics and Society (2008)
23. Waltman, L., Eck, N.: A smart local moving algorithm for large-scale modularity-based community detection. The European Physical Journal B 86(11), 1–14 (2013)
24. Shiokawa, H., Fujiwara, Y., Onizuka, M.: Fast Algorithm for Modularity-based Graph Clustering. In: Twenty-Seventh AAAI Conference on Artificial Intelligence (2013)
25. Liu, J., Liu, T.: Detecting community structure in complex networks using simulated annealing with -means algorithms. Physica A: Statistical Mechanics and its Applications 389(11), 2300–2309 (2010)
26. Rosvall, M., Bergstrom, C.T.: An information-theoretic framework for resolving community structure in complex networks. Proceedings of the National Academy of Sciences 104(18), 7327–7331 (2007)
27. Lee, J., Gross, S.P., Lee, J.: Modularity optimization by conformational space annealing. Physical Review E 85(5), 056702 (2012)
28. Tasgin, M., Bingol, H.: Community Detection in Complex Networks Using Genetic Algorithm. Cornell University Library (2007)

29. Li, J., Song, Y.: Community detection in complex networks using extended compact genetic algorithm. Soft Computing 17(6), 925–937 (2013)
30. Pizzuti, C.: GA-Net: A Genetic Algorithm for Community Detection in Social Networks. In: Rudolph, G., Jansen, T., Lucas, S., Poloni, C., Beume, N. (eds.) PPSN X. LNCS, vol. 5199, pp. 1081–1090. Springer, Heidelberg (2008)
31. Gach, O., Hao, J.-K.: A Memetic Algorithm for Community Detection in Complex Networks. In: Coello, C.A.C., Cutello, V., Deb, K., Forrest, S., Nicosia, G., Pavone, M. (eds.) PPSN 2012, Part II. LNCS, vol. 7492, pp. 327–336. Springer, Heidelberg (2012)
32. Gong, M., et al.: Memetic algorithm for community detection in networks. Physical Review E 84(5) (2011)
33. Liu, D., et al.: Genetic Algorithm with a Local Search Strategy for Discovering Communities in Complex Networks. International Journal of Computational Intelligence Systems 6(2), 354–369 (2013)
34. Ma, L., et al.: Multi-level learning based memetic algorithm for community detection. Applied Soft Computing 19(0), 121–133 (2014)
35. Song, Y., et al.: Community detection using parallel genetic algorithms. In: 2012 IEEE Fifth International Conference on Advanced Computational Intelligence (ICACI) (2012)
36. Riedy, J., Bader, D.A., Meyerhenke, H.: Scalable Multi-threaded Community Detection in Social Networks. In: 2012 IEEE 26th International Parallel and Distributed Processing Symposium Workshops & PhD Forum (IPDPSW) (2012)
37. Riedy, E.J., Meyerhenke, H., Ediger, D., Bader, D.A.: Parallel Community Detection for Massive Graphs. In: Wyrzykowski, R., Dongarra, J., Karczewski, K., Waśniewski, J. (eds.) PPAM 2011, Part I. LNCS, vol. 7203, pp. 286–296. Springer, Heidelberg (2012)
38. Good, B.H., de Montjoye, Y.-A., Clauset, A.: Performance of modularity maximization in practical contexts. Physical Review E 81(4), 046106 (2010)
39. Fortunato, S., Barthélemy, M.: Resolution limit in community detection. In: PNAS, USA (2007)
40. Li, Z., et al.: Quantitative function for community detection. Physical Review E 77(3), 036109 (2008)
41. Pizzuti, C.: A Multi-objective Genetic Algorithm for Community Detection in Networks. In: 21st International Conference on Tools with Artificial Intelligence, ICTAI 2009 (2009)
42. Gong, M., et al.: Community detection in networks by using multiobjective evolutionary algorithm with decomposition. Physica A: Statistical Mechanics and its Applications 391(15), 4050–4060 (2012)
43. Neri, F., Cotta, C., Moscato, P.: Handbook of Memetic Algorithms. SCI, vol. 379. Springer, Heidelberg (2011)
44. Pizzuti, C.: A multiobjective genetic algorithm to find communities in complex networks. IEEE Transactions on Evolutionary Computation 16(3), 418–430 (2012)
45. Liu, X., Murata, T.: Advanced modularity-specialized label propagation algorithm for detecting communities in networks. Physica A: Statistical Mechanics and its Applications 398(7) (2010)
46. Rotta, R., Noack, A.: Multilevel local search algorithms for modularity clustering. J. Exp. Algorithmics 16, 2.1–2.27 (2011)
47. Derrac, J., et al.: A practical tutorial on the use of nonparametric statistical tests as a methodology for comparing evolutionary and swarm intelligence algorithms. Swarm and Evolutionary Computation 1(1), 3–18 (2011)
48. Zachary, W.W.: An Information Flow Model for Conflict and Fission in Small Groups. Journal of Anthropological Research 33(4), 452–473 (1977)

49. Lusseau, D., et al.: The bottlenose dolphin community of Doubtful Sound features a large proportion of long-lasting associations. Behavioral Ecology and Sociobiology 54(4), 396–405 (2003)
50. Krebs, V.: A network of books about US politics sold by Amazon.com (2008), http://www.orgnet.com/
51. Duch, J., Arenas, A.: Community detection in complex networks using Extremal Optimization. Physical Review E 72, 027104 (2005)

Identifying the High-Value Social Audience from Twitter through Text-Mining Methods

Siaw Ling Lo[1,2], David Cornforth[1], and Raymond Chiong[1]

[1] School of Design, Communication and Information Technology, The University of Newcastle, Callaghan, NSW 2308, Australia
[2] School of Information Technology, Nanyang Polytechnic, Singapore
siawling.lo@uon.edu.au,
{david.cornforth,raymond.chiong}@newcastle.edu.au

Abstract. Doing business on social media has become a common practice for many companies these days. While the contents shared on Twitter and Facebook offer plenty of opportunities to uncover business insights, it remains a challenge to sift through the huge amount of social media data and identify the potential social audience who is highly likely to be interested in a particular company. In this paper, we analyze the Twitter content of an account owner and its list of followers through various text mining methods, which include fuzzy keyword matching, statistical topic modeling and machine learning approaches. We use tweets of the account owner to segment the followers and identify a group of high-value social audience members. This enables the account owner to spend resources more effectively by sending offers to the right audience and hence maximize marketing efficiency and improve the return of investment.

Keywords: Twitter, topic modelling, machine learning, audience segmentation.

1 Introduction

Social media has not only transformed the way we share our personal life, it has also transformed the way business is carried out. A recent study [1] found that nearly 80% of consumers would more likely be interested in a company due to its brand's presence on social media. It is therefore not a surprise that 77% of the Fortune 500 companies have active Twitter accounts and 70% of them maintain an active Facebook account to engage with their potential customers [2]. With more companies doing business on social media, how can one stand out from the increasingly crowded social space to find prospective customers from different audiences in social media?

It is no longer feasible for a company to depend on gimmicks (such as incentive referrals) to boost the social media business as that may only provide short-term gain. While a company can adopt approaches like mass marketing to all the "fans" or contacts available, the return may not be justified by the effort and amount of money spent. Furthermore, there is a thin line between broadcasting a general message and spamming, so instead of attracting a greater audience, there is a high probability of

H. Handa et al. (eds.), *Proc. of the 18th Asia Pacific Symp. on Intell. & Evol. Systems – Vol. 1*,
Proceedings in Adaptation, Learning and Optimization 1, DOI: 10.1007/978-3-319-13359-1_26

losing current customers. Hence, it makes sense to identify a target audience to maximize marketing efficiency and improve the return of investment.

Traditionally, an understanding of customers is obtained through customer surveys so that information such as customer preferences can be known. This set of information can be merged with internal company data, for example, product purchase data or transactional data, so that segmentation of customers can be done to better understand the customers and manage offerings according to their interests. However, with the recent proliferation of social media activities, more and more companies are putting in efforts to ensure that their presence is felt in the crowded social space. Even though there is a rich source of customer information to be mined, the real-time nature and free-form expression of social media content poses a challenge in extracting commercially viable information from the vast amount of conversations.

While there are many guidelines or tips on the Web about how to find a target audience on social media, most of these concentrate on searching specific keywords related to products or brands. However, while using this approach can retrieve lists of information using different keywords, it is not capable of determining the relationship among the keywords and providing a more comprehensive view on the subject matter without the help of domain experts. Furthermore, deciding which keywords to use may not be obvious to a non-expert and this may lead to inaccurate information extraction and hence a misunderstood market analysis. On top of this, there is a need to manually consolidate the list of social audience found and to ensure that contents shared by the audience match with the keywords.

Prior work [3][4] has proposed various approaches such as translating both social networks and semantic information into Resource Description Framework (RDF) formats and using RDF methods for correlation, or making use of semantic tagging to correlate current social tagging approaches to make sense of social media data. These approaches, however, require additional efforts of translating and tagging of current social media data, which can be a daunting task considering the huge amount of data and the possible manual effort.

In this paper, we investigate several different methods in order to make use of available resources to identify a group of high-value social audience members without utilizing a considerable amount of human annotation effort. These include text mining methods such as fuzzy keyword matching using Dice coefficient [5] of string similarity, statistical topic modeling with Twitter Latent Dirichlet Allocation (LDA) [6], and machine learning using the Support Vector Machine (SVM) [7]. The hypothesis is based on the idea that followers are interested in the content posted by an owner, and hence they choose and take action to follow the account owner. If that is the case, some of the tweets shared by the followers should be of similar nature to the account owner. In other words, tweets of the account owner (of a similar period of time) can be used to select or identify the group of followers who are interested in the content that the owner has been tweeting. Hence, these followers are more likely to comprise the target audience compared to others who are not sharing similar contents.

In order to achieve this, we use a list of seed words (derived from the owner tweets using term frequency analysis) to generate a baseline using Direct Keyword Match.

This set of seed words is also used in Fuzzy Keyword Match and identification of suitable topic numbers from Twitter LDA. In contrast to how things are done in a traditional machine learning approach, tweets from the account owner are used to build the positive training data instead of tweets extracted from the list of followers. This eliminates the need to manually annotate the vast amount of tweets from the followers and it is more practical if the approach is to be adopted in a real-world application.

The major contributions of this work are enumerated as follows:

i. To the best of our knowledge, our work in this paper is the first attempt to identify the target audience from the list of followers of a Twitter owner's tweets through various methods. It is assumed that those who have tweeted similar contents are more likely to be interested in the owner's tweets, compared to others who have not been sharing similar contents.
ii. From the result observation, it is likely that half or less followers are tweeting similar contents as the owner. This implies that it may not be sensible to try to engage every follower, as not everyone is interested in the content or topic shared. Instead, it makes sense to be selective and target specific groups of followers to maximize the use of allocated marketing expenses and reach out to potential customers in social media.

2 Related Work

As the aims of any business are to increase profit, build a long lasting brand name, and to grow its customer base or engage current customers, it is essential to understand the needs and behaviors of the customers. This understanding can be achieved through different means and at different levels of detail. Most companies define a set of segments that reflect the companies' knowledge of the customers and their traits or behaviors. All other marketing activities, such as customer engagement activities, are targeted and measured according to this segmentation.

However, the segmentation is typically restricted to customer relationship management (CRM) or transaction data obtained either through customer surveys or tracking of product purchases to understand the customer demand. Demographic variables, RFM (recency, frequency, monetary) and LTV (lifetime value) are the most common input variables used in the literature for customer segmentation and clustering [8, 9]. While CRM or organizational transaction data can be coupled with geographical data to obtain additional information, the segmentation remains limited to within an organization's system and does not leverage on shared contents and activities on social media where customers tend to reveal about themselves – life events, personal and business preferences, perception of brands and more.

There have been efforts in deriving or estimating demographics information [10, 11] from available social media data, but this set of information may not be suitable to be used directly in targeted marketing, as temporal effects and types of products to be targeted are usually not considered. Besides that, demographic attributes such as age, gender and residence areas may not be updated and hence may result in a misled conclusion. Recently, eBay has expressed that, due to viral campaigns and major social media activities, marketing and advertising strategies are evolving. Although targeting specific demographics through segmentation still has its value, eBay is focusing on "connecting people with the things they need and love, whoever they are" [12].

Other research on predicting purchase behavior from social media has shown that using Facebook categories, such as likes and n-grams, is better than using demographic features shared on Facebook [13]. Due to the privacy policy of Facebook profiles, this work focuses on Twitter, where most of the contents and activities shared online are open and available. It would be interesting to see if other factors (such as the content shared on social media) can be used to derive alternative approaches to identify the target or high-value social audience for a company or a product.

3 Methods

The focus of this research is to establish an approach that makes use of contents and activities shared on social media platforms to profile and segment the social audience of a Twitter account owner. This account owner can be a business or a government body and the online social audience we are interested to profile or segment is the list of followers of the Twitter account. The architecture of our system is given in Fig. 1.

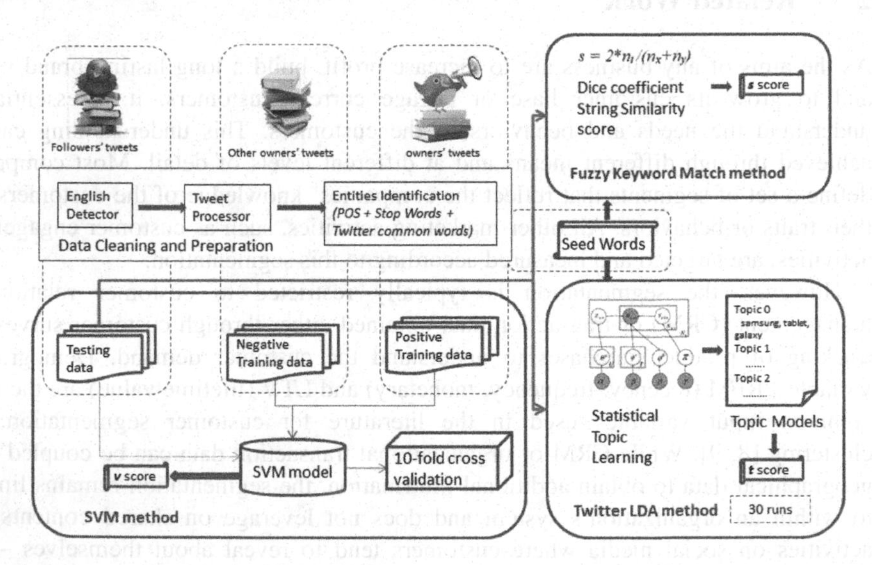

Fig. 1. The system architecture

The tweets from various parties - owners', followers', owners' from other domains are cleaned and preprocessed before preparing for seed words generation and SVM training and testing datasets. The owners' tweets are used as the positive training data while tweets of owners' from other domains are extracted as the negative training data. 10 fold cross-validation is applied on both the positive and negative training datasets for the SVM before the classification of followers' tweets (or the testing data) is conducted. The seed words generated are used in both Fuzzy Keyword Match and Twitter LDA methods. A string similarity score derived from Dice coefficient is calculated through a fuzzy comparison with the seed words on the testing data. A list of topics is learnt from testing data using Twitter LDA and followers with relevant topic numbers are identified. Details of each component are described in the following sections.

3.1 Data Collection

We use the Twitter Search API [14] for our data collection. As the API is constantly evolving with different rate limiting settings, our data gathering is done through a scheduled program that requests a set of data for a given query. The subject or brand selected for this research is Samsung Singapore or "samsungsg" (its Twitter username). At the time of this work, there were 3,727 samsungsg followers. In order to analyze the contents or tweets of the account owner, the last 200 tweets by samsungsg have been extracted. The time of tweets ranges from 2 Nov 2012 to 3 Apr 2013. For each of the followers, the API is used to extract their tweets, giving a total of 187,746 records, and 2,449 unique users having at least 5 tweets in their past 100 tweets of the same period. We reasoned that those with fewer than 5 tweets were inactive in Twitter, as it implied that the user was tweeting an average of less than one tweet in a month (since the period was of 6 months).

3.2 Data Cleaning and Preparation

Tweets are known to be noisy and often mixed with linguistic variations. It is hence very important to clean up the tweet content prior to any content extraction:

- Non-English tweets are removed using the Language Detection Library for Java [15];
- URLs, any Twitter's username found in the content (which is in the format of @username) and hashtags (with the # symbol) are removed;
- Each tweet is pre-processed to lower case.

As tweets are usually informal and short (up to 140 characters), abbreviation and misspelling are often part of the content and hence the readily available Named Entity Recognition package may not be able to extract relevant entities properly. As such, we derive an approach called Entities Identification, which uses Part-of-Speech (POS) [16] tags to differentiate the type of words. All the single nouns are identified as possible entities. If the tag of the first fragment detected is 'N' or 'J' and the consecutive word(s) is of the 'N' type, these words will be extracted as phrases. This approach is then complemented by another process using the comprehensive stop words list used by search engines (http://www.webconfs.com/stop-words.php) in addition to a list of English's common words (preposition, conjunction, determiners) as well as Twitter's common words (such as "rt", "retweet" etc.) to identify any possible entity. In short, the original tweet is sliced into various fragments by using POS tags, stop words, common words and punctuations as separators or delimiters. For example, if the content is "Samsung is holding a galaxy contest!", two fragments will be generated for the content as follows: (samsung) | (galaxy contest).

3.3 Seed Words Generation

All the tweets extracted from samsungsg are subjected to data cleaning and preparation mentioned in the previous section. Each tweet is now represented by the identified fragments or words and phrases. This set of data is further processed using term frequency analysis to obtain a list of seed words (which include "samsung", "galaxy s iii", "galaxy camera" etc.). The words in a phrase are joined by '_' so that they can be identified as a single term but the '_' is filtered in all the matching processes. These seed words are used to generate results for Direct Keyword Match, Fuzzy Keyword Match and identification of suitable topic numbers in the Twitter LDA method.

3.4 Direct Keyword Match

This is the most common method used to find the relevant or suitable social audience for a specific content or product. The list of seed words generated is used to match the tweets from the list of followers. As long as there is a direct word or phrase match with any of the seed words, the follower will be considered as a potential member of a high-value social audience, who is likely to be interested in the content shared by the account owner. The result of this approach is set as the baseline for the rest of the methods.

3.5 Fuzzy Keyword Match

It is not uncommon for Twitter users to use abbreviations or interjections or a different form of expression to represent similar terms. For example, "galaxy s iii" can be represented by "galaxy s 3", which is understandable by a human but cannot be captured by the Direct Keyword Match baseline method. As such, a Fuzzy Keyword Match method using the seed words derived is implemented.

The comparison here is based on a Dice coefficient string similarity score [5] using the following expression,

$$s = 2*n_t/(n_x+n_y) \tag{1}$$

where n_t is the number of characters found in both strings, n_x is the number of characters in string x and n_y is the number of characters in string y. For example, to calculate the similarity between "process" and "proceed":

x = process bigrams for x = {pr ro oc ce es ss}

y = proceed bigrams for y = {pr ro oc ce ee ed}

Both x and y have 6 bigrams each, of which 4 of them are the same. Hence, the Dice coefficient string similarity score is 2*4/(6+6) = 0.67.

Similar to the Direct Keyword Match method, each of the tweets of every follower is compared with the seed words and the highest score of any match is maintained as the *s* score of the follower.

3.6 Twitter LDA

Recently, LDA [17], a renowned generative probabilistic model for topic discovery, has been used in various social media studies [6][18]. LDA uses an iterative process to build and refine a probabilistic model of documents, each containing a mixture of topics. However, standard LDA may not work well with Twitter as tweets are typically very short. If one aggregates all the tweets of a follower to increase the size of the documents, this may diminish the fact that each tweet is usually about a single topic. As such, we have adopted the implementation of Twitter LDA [6] for unsupervised topic discovery among all the followers.

As the volume of the tweet set from all the followers is within 200,000, only a smaller number of topics (from 10-50, with an interval of 10) from Twitter LDA are used. These 5 different topic models were run for 100 iterations of Gibbs sampling, while the other model parameters or Dirichlet priors were kept constant: $\alpha = 0.5$; $\beta_{word} = 0.01$, $\beta_{background} = 0.01$ and $\gamma = 20$. Suitable topics are chosen automatically via comparison with the list of seed words. The result or the list of audience identified by each topic model is a consolidation of 30 runs where a score is assigned to each follower using the following calculation:

$$t = n_m/n_r \tag{2}$$

where n_m is the total number of matches and n_r is the total number of runs. If a particular follower is found in 5 runs then the *t* score assigned is 5/30 = 0.17.

3.7 The SVM

The SVM is a supervised learning approach for two- or multi-class classification, and has been used successfully in text categorization [7]. It separates a given known set of {+1, -1} labeled training data via a hyperplane that is maximally distant from the positive and negative samples respectively. This optimally separating hyperplane in the feature space corresponds to a nonlinear decision boundary in the input space. More details of the SVM can be found in [19].

The positive dataset is generated using processed tweets from the account owner (i.e., samsungsg). The negative dataset is randomly generated from account owners of 10 different domains (online shopping deals, food, celebrities, parents, education, music, shopping, politics, Singapore news, traffic), which are ilovedealssg, hungrygowhere, joannepeh, kiasuparents, MOEsg, mtvasia, tiongbahruplaza, tocsg (TheOnlineCitizen), SGnews and sgdrivers respectively. These domains have been chosen as they are the main topics discovered from the list of tweets of all the followers using Twitter LDA. The respective account owners have been selected as they are the popular Twitter accounts in Singapore according to online Twitter analytic tools such as wefollow.com.

LibSVM implementation of RapidMiner [20] is used in this study and the sigmoid kernel type is selected as it produces higher precision prediction than other kernels, such as RBF (Radial Basis Function) and polynomial.

As the input of the SVM is through a matrix of feature vectors, the feature vectors used in this study are created using term frequency analysis from the content of the tweets after the data cleaning process specified in Section 3.2 and word stemming using Porter [21]. In order to capture the features from the tweets of the followers, three different approaches are used in extracting features for testing data construction in this study. The details are listed below. As a result, there are three matrices of feature vectors accounting for the three approaches respectively. Since the number of tweets shared by each follower is different, the three approaches for representing the followers' tweets are:

- Extract topical representation features of all the tweets from each follower using the top topical words from Twitter LDA;
- Extract word representation features of all the tweets from each follower using term frequency;
- Extract word representation features from each follower's tweets and treat each set as individual testing data, where each tweet will be classified as either positive or negative. The final assignment of the v score is based on the following representation:

$$v = n_p/n_a \tag{3}$$

where n_p is the total number of tweets classified as positive and n_a is the average number of tweets shared by all the followers (71 tweets per follower for this study). If 5 tweets of a particular follower are classified as positive, then the v score assigned is 5/71 * *normalized factor* so that the score range is within [1, 0].

4 Experiments and Results

The results obtained from the various methods were compared with a random annotated sample of the followers. The contents of a total of 300 followers (which were randomly sampled) were annotated manually as either a potential high-value social audience according to the content shared by the account owner or not a target audience. This set of data was used in the evaluation of the various methods described in Sections 3.4, 3.5, 3.6 and 3.7.

4.1 Numbers of High-Value Audience Identified

Out of the 2,449 "active" followers (excluding those tweeted less than 5 tweets), the numbers of the followers who were tweeting similar contents measured by various methods are listed in Table 1.

Table 1. Numbers of high-value audience identified by various methods

Methods	Audience Numbers	% within active followers (2,449)	% within all the followers (3,727)
Direct Keyword Match	321	13%	9%
Fuzzy Keyword Match	1115	46%	30%
Twitter LDA 10 topics*	760	31%	20%
Twitter LDA 20 topics*	582	24%	16%
Twitter LDA 30 topics*	527	22%	14%
Twitter LDA 40 topics*	414	17%	11%
Twitter LDA 50 topics*	424	17%	11%
SVM	736	30%	20%

*The results are consolidated from 30 runs.

4.2 Results of Twitter LDA

As shown in Table 1, in general, the size of high-value audience decreases with the increase of the topic numbers. A further analysis was done and the group of audience members identified with topic numbers greater than 30 remained the same. Hence further result analysis has been carried out on topic models from 10, 20 and 30.

Table 2 presents some sample topic groups and their topical words. The table shows that using seed words derived from the account owner can identify relevant contents from the list of followers.

Table 2. Sample topic groups and their topical words (IDs are the topic group IDs)

Models	IDs	Top topical words
Twitter LDA 10 topics	3	google, android, apps, mobile, galaxy, tablet
	4	samsung, galaxy, mobile, phone, android, tv, camera, smartphone
Twitter LDA 20 topics	8	galaxy, samsung, android, phone, mobile, apps, smartphone
	17	samsung, galaxy, app, tablet
	19	samsung, tv, led, mobile, smart, phone, laptop
Twitter LDA 30 topics	3	samsung, galaxy
	18	samsung, galaxy, android, google, app, phone, mobile, tablet, smartphone
	25	samsung, tv, led, camera, lcd, smart, hd

4.3 Results of SVM

The 10 fold cross-validation of the training data yields an accuracy of 88%, with class precision and recall as presented in Table 3.

Table 3. SVM 10 fold cross-validation results

	True samsungsg	True others	Class precision
Predicted samsungsg	165	13	92.7%
Predicted others	35	187	84.2%
Class recall	82.5%	93.5%	

The results of the testing data from various approaches using the SVM as compared to the baseline method – Direct Keyword Match – are showed using Receiver Operator Characteristic (ROC) curves in Fig. 2. There are 3 approaches:

1. SVM_LDA: all the tweets of each follower are represented as a single feature using top topical words from Twitter LDA.
2. SVM_TF: all the tweets of each follower are represented by top frequency terms.
3. SVM_Ind: a v score is generated through the classification of each tweet of the follower.

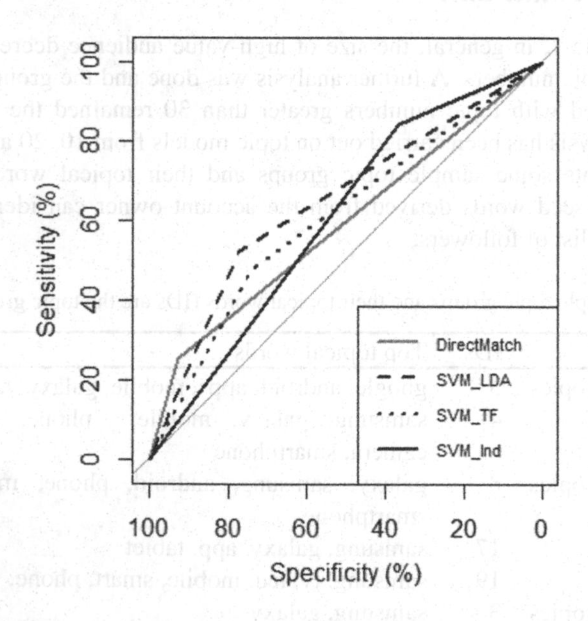

Fig. 2. ROC curves based on testing data of various approaches using the SVM

All the three approaches have performed better than the baseline Direct Keyword Match method, with the third approach (SVM_Ind), which classifies individual tweets instead of combining all the tweets in a single feature, having the higher sensitivity.

This is essential as it is more capable of identifying the true high-value social audience for the account owner.

4.4 Comparison of Various Methods

To compare the various methods, ROC curves, as shown in Fig. 3, are plotted on all the results. It is observed that Fuzzy Keyword Match has the best result (the largest area under the curve), followed by the Twitter LDA topic modeling methods. The SVM or machine learning method has a higher sensitivity as compared to the baseline method, Direct Keyword Match, but it has not performed as well as the other methods.

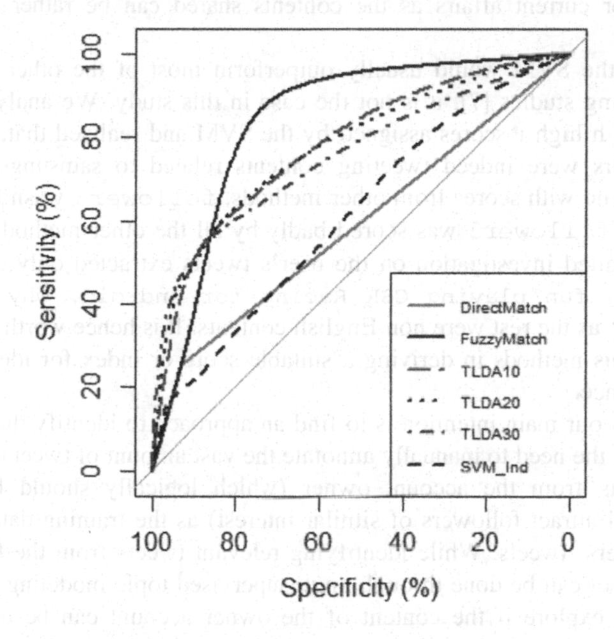

Fig. 3. ROC curves of various methods

5 Discussion

It is interesting to observe from the results that, while most of the account owners may think that their followers would be truly interested in their contents, this may not be the case as shown in Table 1. It is likely that half or less than half of all the followers are actually tweeting similar contents to them.

One possible reason Fuzzy Keyword Match has emerged as the top performer may be due to the account chosen. "samsungsg" being a technology and mobile company, tends to tweet contents with specific terms such as products or events. It is therefore likely that the target audience who are also interested in the similar content will be tweeting similar terms or text. For example, the *s* score (generated by the Fuzzy

Keyword Match method) is the highest for both Twitter users, `follower1` and `follower2`, as shown in Table 4, and a detailed study on their tweets indeed showed that they have tweets related to Samsung. `follower1` shared a lot of tweets on technology and mobile news, such as "A new galaxy is born, follow @SamsungMobile for updates on the Samsung S III.", "Let the Smart TV experience begin | Samsung Smart TV". While `follower2` did share one tweet on "3 galaxy, 2 xp, 1 iphone, 1 mac and latest 1 wins 8, under 1 roof. That wld make me? Complicated. :)", this user mainly tweeted about daily chores. This may explain why Twitter LDA methods did not generate a high score for `follower2`. While the Fuzzy Keyword Match method seems to perform well, this may not be the case for more generic accounts such as parents groups or current affairs as the contents shared can be rather diverse and conceptual.

Even though the SVM would usually outperform most of the other methods in various text mining studies [7], it is not the case in this study. We analyzed the top few followers with high v scores assigned by the SVM and realized that, while most of these followers were indeed tweeting contents related to samsungsg and their scores were in-sync with scores from other methods, `follower3` wasn't. As can be seen in Table 4, `follower3` was scored badly by all the other methods except for the SVM. A detailed investigation on the user's tweets extracted only one relevant tweet – "Having fun playing CSR Racing for Android, why not join me for FREE?" as the rest were non-English contents. It is hence worth considering combining various methods in deriving a suitable score or index for identifying the high-value audience.

In addition, as our main intention is to find an approach to identify the high-value audience without the need to manually annotate the vast amount of tweet contents, we have used tweets from the account owner (which logically should be tweeting contents that will attract followers of similar interest) as the training data instead of using the followers' tweets. While identifying relevant tweets from the followers as the training dataset can be done through an unsupervised topic modeling method, we are interested to explore if the content of the owner account can be used for this purpose. Analyses using followers' tweets will be studied in the future, which we expect would provide better results.

Table 4. Interesting followers identified. The highest score of each user is bolded. The s score is generated by Fuzzy Keyword Match, TLDA10 t score is generated by Twitter LDA 10 topics, TLDA20 t score is generated by Twitter LDA 20 topics, TLDA30 t score is generated by Twitter LDA 30 topics, and v score is generated by the SVM.

Twitter name	s score	TLDA10 t score	TLDA20 t score	TLDA30 t score	v score
follower1	**1.0**	**1.0**	**1.0**	**1.0**	0.18
follower2	**0.9**	0.1	0.03	0.0	0.2
follower3	0.35	0.2	0.13	0.03	**1.0**

The various scores generated, such as the *s* score from Fuzzy Keyword Match, can be used to segment followers into groups of high-value social audience members, which a company or organization can use to engage depending on the resources available. For example, if the company or organization only has a limited amount of budget to reach out to 100 followers, the top 100 scorers would have a higher possibility of being interested than a randomly generated list of 100. In fact, a preliminary result using an average value that is built from the combination of the various scores has shown to identify 86% of the 63 high-value audience members from the 300 randomly annotated users. In other words, the scores derived from the various methods have high potential to be customized for segmentation of followers for social media marketing and engagement.

6 Conclusion and Future Work

In this study, we have investigated the use of various text mining methods to identify the high-value social audience from a list of followers using the contents of a Twitter account owner, "samsungsg". It is assumed that those who have tweeted similar contents are more likely to be interested in the owner's tweets as compared to those who have not been sharing similar contents.

Our results show that the Fuzzy Keyword Match method has produced the best performance in identifying the high-value social audience. It should be noted that achieving an accuracy of 100% for the application area of targeted marketing is unnecessary as any improvement of mass marketing is going to be beneficial for business companies.

From the result observation, it is likely that half or less of the followers are sharing similar contents as the owner, hence it makes sense to segment or identify a group of social audience members who are the target audience for further engagement. Our approach in identifying this group of high-value audience members enables companies or organizations of any Twitter account owner to devise their marketing or engagement plan according to the segment or group of social audience members so as to maximize the use of allocated budgets and successfully reach out to customers in the crowded social media space.

We have used "samsungsg" as a case study in this paper. For future work, we plan to extend it to include other account owners to verify if the observation is consistent across Twitter or if there are other features that can play a role in identifying the high-value audience. Furthermore, tweets related to a product or a company are usually opinionated, and hence it is important to identify the context through semantic knowledge base enrichment [22], in order to provide additional analysis in terms of sentiment for high-value social audience profiling. We would also like to see if the use of biologically inspired natural language processing methods [23] such as the Extreme Learning Machine [24], which has gained increasing popularity recently, would achieve good results in unstructured text analysis. It will be of interest to explore this area and improve on the results.

References

1. Unlocking the power of social media | IAB UK, http://www.iabuk.net/blog/unlocking-the-power-of-social-media
2. 2013 Fortune 500 - UMass Dartmouth, http://www.umassd.edu/cmr/socialmediaresearch/2013fortune500/
3. Breslin, J.G., Passant, A., Vrandečić, D.: Social semantic web. In: Handbook of Semantic Web Technologies, pp. 467–506. Springer (2011)
4. Torres, D., Diaz, A., Skaf-Molli, H., Molli, P.: Semdrops: A Social Semantic Tagging Approach for Emerging Semantic Data. In: 2011 IEEE/WIC/ACM International Conference on Web Intelligence and Intelligent Agent Technology (WI-IAT), pp. 340–347. IEEE (2011)
5. Kondrak, G., Marcu, D., Knight, K.: Cognates can improve statistical translation models. Presented at the Proceedings of the 2003 Conference of the North American Chapter of the Association for Computational Linguistics on Human Language Technology: Companion Volume of the Proceedings of HLT-NAACL 2003–short papers, vol. 2 (2003)
6. Zhao, W.X., Jiang, J., Weng, J., He, J., Lim, E.-P., Yan, H., Li, X.: Comparing twitter and traditional media using topic models. In: Clough, P., Foley, C., Gurrin, C., Jones, G.J.F., Kraaij, W., Lee, H., Mudoch, V. (eds.) ECIR 2011. LNCS, vol. 6611, pp. 338–349. Springer, Heidelberg (2011)
7. Joachims, T.: Text categorization with support vector machines: Learning with many relevant features. In: Nédellec, C., Rouveirol, C. (eds.) ECML 1998. LNCS, vol. 1398, pp. 137–142. Springer, Heidelberg (1998)
8. Mo, J., Kiang, M.Y., Zou, P., Li, Y.: A two-stage clustering approach for multi-region segmentation. Expert Systems with Applications 37, 7120–7131 (2010)
9. Namvar, M., Khakabimamaghani, S., Gholamian, M.R.: An approach to optimised customer segmentation and profiling using RFM, LTV, and demographic features. International Journal of Electronic Customer Relationship Management 5, 220–235 (2011)
10. Mislove, A., Viswanath, B., Gummadi, K.P., Druschel, P.: You are who you know: inferring user profiles in online social networks. In: Proceedings of the Third ACM International Conference on Web Search and Data Mining, pp. 251–260. ACM (2010)
11. Kosinski, M., Stillwell, D., Graepel, T.: Private traits and attributes are predictable from digital records of human behavior. Proceedings of the National Academy of Sciences 110, 5802–5805 (2013)
12. How Ebay Uses Twitter, Smartphones and Tablets to Snap Up Shoppers, http://www.ibtimes.co.uk/how-ebay-uses-twitter-smartphones-tablets-snap-shoppers-1443441
13. Zhang, Y., Pennacchiotti, M.: Predicting purchase behaviors from social media. In: Proceedings of the 22nd International Conference on World Wide Web, pp. 1521–1532. International World Wide Web Conferences Steering Committee (2013)
14. Using the Twitter Search API | Twitter Developers, https://dev.twitter.com/docs/using-search
15. Nakatani, S.: language-detection - Language Detection Library for Java - Google Project Hosting, http://code.google.com/p/language-detection/
16. Toutanova, K., Manning, C.D.: Enriching the knowledge sources used in a maximum entropy part-of-speech tagger. Presented at the Proceedings of the 2000 Joint SIGDAT Conference on Empirical Methods in Natural Language Processing and Very Large Corpora: Held in Conjunction with the 38th Annual Meeting of the Association for Computational Linguistics, vol. 13 (2000)

17. Blei, D.M., Ng, A.Y., Jordan, M.I.: Latent dirichlet allocation. The Journal of Machine Learning Research 3, 993–1022 (2003)
18. Yang, M.-C., Rim, H.-C.: Identifying Interesting Twitter Contents Using Topical Analysis. Expert Systems with Applications 41, 4330–4336 (2014)
19. Burges, C.J.: A tutorial on support vector machines for pattern recognition. Data Mining and Knowledge Discovery 2, 121–167 (1998)
20. Predictive Analytics, Data Mining, Self-service, Open source - RapidMiner, `http://rapidminer.com/`
21. Willett, P.: The Porter stemming algorithm: then and now. Program: Electronic Library and Information Systems 40, 219–223 (2006)
22. Weichselbraun, A., Gindl, S., Scharl, A.: Enriching semantic knowledge bases for opinion mining in big data applications. Knowledge-Based Systems (in press, 2014)
23. Cambria, E., Mazzocco, T., Hussain, A.: Application of multi-dimensional scaling and artificial neural networks for biologically inspired opinion mining. Biologically Inspired Cognitive Architectures 4, 41–53 (2013)
24. Cambria, E., Huang, G.-B., Kasun, L.L.C., Zhou, H., Vong, C.-M., Lin, J., Yin, J., Cai, Z., Liu, Q., Li, K.: Extreme Learning Machines. IEEE Intelligent Systems 28, 30–59 (2013)

Understanding the Behavior of Solid State Disk

Qingchao Cai[1], Rajesh Vellore Arumugam[2], Quanqing Xu[2], and Bingsheng He[1]

[1] School of Computer Engineering, Nanyang Technological University, Singapore
{qccai,bshe}@ntu.edu.sg
[2] Data Storage Institute, A*STAR, Singapore
{Rajesh_VA,Xu_Quanqing}@dsi.a-star.edu.sg

Abstract. In this paper, we develop a family of methods to characterize the behavior of new-generation Solid State Disks (SSDs). We first study how writes are handled inside the SSD by varying request size of writes and detecting the placement of requested pages. We further examine how this SSD performs garbage collection and flushes write buffer. The result shows that the clustered pages must be written and erased simultaneously, otherwise significant storage waste will arise if such clustered pages are partially written.

We then conduct two case studies to analyze the storage efficiency when an SSD is used for server storage and the cache layer of a hybrid storage system. In the first case, we find that a moderate storage waste exists, whereas in the second case, the number of written pages caused by a write request can be as much as 4.2 times that of pages requested, implying an extremely low storage efficiency. We further demonstrate that most of such unnecessary writes can be avoided by simply delaying the issuance of internal write requests, which are generated when a read request cannot be serviced by the cache layer. We believe that this study is helpful to understand the SSD performance behavior for data-intensive applications in the big-data era.

Keywords: Storage, Solid State Disk, Hybrid storage system, Algorithm.

1 Introduction

NAND-flash based Solid state disks (SSDs)[1] have been incorporated into the computer storage architecture over the past several years, and now have become an important supplementary to traditional rotational hard disk drives (HDDs). Compared with their rotational counterparts, SSDs have a much higher read/write throughput, and due to the absence of moving mechanical components, SSDs are able to sustain an order of magnitude less random access latency.

The layout of data in SSDs is much more complicated than in HDDs. The storage space of an SSD can be partitioned into multiple domains, each containing a number of flash memory pages that share some specific resources [6]. Due

[1] We restrict our discussion to flash based SSDs, as most SSDs in the market are of this kind.

H. Handa et al. (eds.), *Proc. of the 18th Asia Pacific Symp. on Intell. & Evol. Systems – Vol. 1*,
Proceedings in Adaptation, Learning and Optimization 1, DOI: 10.1007/978-3-319-13359-1_27

to resource contention, an access requesting two pages within a same domain might have a longer latency than that requesting two pages placed in different domains. The difference in the internal structure and access latency between SSD and HDD can also lead to the different way in which access requests are serviced. We explore the service of access requests inside SSDs, as it can be used to reveal how SSD realizes its specific internal structure and assist the incorporation of SSD into storage systems.

In this paper, we develop a family of methods to characterize the behavior of a representative SSD. First, we carry out an investigation on how write requests are serviced inside this SSD. To this end, we issue multiple writes with varying request size to the SSD, and then detect the placement of requested pages via comparing the latencies among a set of carefully designed read requests. The result implies there exist clustered pages which must be written simultaneously, and pages for servicing write requests are chosen such that there are least number of partially written clustered pages. Second, we study how garbage collection is performed by overwriting the certain page of clustered pages that have been completely written and then measuring the resulted page placement, and find that the constituting pages of a clustered page must also be erased at same time. In addition, we extract the length of flush periods, defined as the interval between two flushes of SSD write buffer, using a method similar to that of investigating the service of write requests. The difference is that the varying parameter is no longer the request size of writes, but the interval between two consecutive write requests instead.

The characteristics of clustered pages that the four pages must be written and erased simultaneously implies there will be a waste of storage if a clustered page is partially written. In order to quantify storage efficiency, we conduct two case studies in which the SSD is used for different purposes. In the first case, we analyze the block access traces of ten server applications, and find that if the same sequence of write requests are issued to the SSD, hundreds of thousands of wasted pages, i.e., the unwritten pages of partially written clustered pages, will be produced. In the second case, we use Flashcache [24] to deploy a hybrid storage system with SSD serving as cache layer, and collect the traces of accesses to the SSD cache for multiple IO access patterns. The result shows that due to the long inter-arrival interval of internal write requests which are generated when a read request cannot be serviced by cache layer, the wasted pages can be up to 1.2 times more than those requested when reads account for the majority of IOs, which in turn leads to a write amplification up to 4.2. We show that most of such wasted pages can be eliminated by simply first delaying the issuance of internal write requests and then flushing them simultaneously. We believe that our findings can guide the design and implementation of data-intensive applications on SSDs in the big data era.

The remainder of paper is organized as follows. Section 2 describes the background and related works. The methods of capturing SSD behavior and the corresponding results are presented in Section 3 in detail. Two case studies are presented in Section 4 to quantify SSD storage efficiency. We finally conclude this work in Section 5.

2 Background and Related Works

2.1 Solid State Disk

Data is accessed in page granularity in SSDs, and in this sense a flash memory page can be viewed as a block of hard disks. The difference between them is that flash pages do not support in-place update, and can be overwritten only after being erased. The erase operation of SSDs, however, is not page-based, but in a granularity of erase blocks. An erase block is comprised of a number (usually 64 or 128) of consecutive flash pages, and as a result, each time when a block is to be erased, the valid pages in it should first be copied to the free pages of other blocks, which leads to the notorious write amplification problem of SSDs.

To hide these behavioral differences, a flash translation layer (FTL) [7] [10] [14] [17] is employed in SSDs. To support out-of-place update, FTL provides the map of logical page number to physical page number, giving an illusion of in-place update to the host. Another important function of FTL is to perform garbage collection (GC). GC erases one or more blocks when there are no sufficient free pages to service write requests or when device is idle, and generally the blocks with least valid pages are selected for GC so that write amplification is minimized. In addition, since each SSD block can only withstand a limited number of erase cycles, FTL also implements wear-leveling to evenly spread the writes to each block and hence extend SSD lifetime.

FTL is usually implemented as a firmware run by an *SSD controller*. SSD controller translates incoming read/write requests into flash memory operations and issues commands to flash memory through a flash controller. Besides the SSD controller, there are three other major components inside an SSD. The *host interface logic* connects device to the host via an interface connector such as SATA. A *RAM buffer* is also commonly deployed in SSDs to improve access performance by temporarily storing data accessed and buffering write requests. Data is persistently stored in an array of *flash memory packages* which are connected to the flash controller via multiple channels. Each flash memory package is composed of multiple dies, and each die further contains multiple planes, each with a number of flash blocks inside. The design issues of SSD architecture are discussed in detail in [4] [8].

The four-level hierarchy of flash memory corresponds to four levels of parallelism: channel-level, package-level, die-level and plane-level. Flash pages across different channels, packages or dies can be operated independently, and can thus support parallel operations over them natively. However, the plane-level parallelism is not activated in general, unless there are multiple operations of same type simultaneously accessing flash pages across different planes of the same die, in which case the plane-level parallelism can be exploited through *n-plane command* which enables n (typically 2 or 4) planes of a same die to work simultaneously. There have been many studies [11] [19] [23] toward effectively exploiting the rich parallelism inside SSDs for better IO performance. Since the parallelism of SSDs can be exploited effectively by sequential writes, some studies [15] [16] [18] tailor up-level applications to make SSD writes as sequential as possible.

2.2 The Extraction of SSD Parameters

Since SSD parameters can substantially affect the performance of device, it is thus of practical meaning to extract them as it can guide the design of systems and applications to exploit SSD performance more effectively.

While part of these parameters such as page size and block size are well documented, there also exist some implicit parameters, e.g., parallel degree and size of clustered page, hiding inside SSD internals. Chen et al. [6] probe the size of chunks, which consist of pages that are continuously allocated within a single domain, parallel degree and page mapping policy of several SSDs, and give a detailed discussion on the influence of parallelism on SSD performance.

Another work of SSD parameter extraction is [12], which develops a set of micro benchmarks to extract the size of clustered page/block and read/write buffer, and modifies Linux block layer such that the incoming reads/writes are aligned with the boundary of clustered pages and then split into pieces with the same size of read/write buffer. Although the term "clustered page" is also used in this work to represent an internal storage unit of SSD, it has a different meaning from the counterpart used in our work. By its definition in [12], a clustered page is actually composed of pages across different SSD domains, and hence closely related to the degree of parallelism inside SSD. On the contrary, the clustered page defined in our work consists of flash pages that are placed in the same domain and must be written and erased simultaneously.

3 The Measurement of SSD Parameters

3.1 Experimental Enviorment

The experiment is conducted on an HP xw6600 workstation, which is equipped with an Intel Quad Core Xeon(R) E5420 2.5GHz processor and 4GB main memory. For the OS, we use Ubuntu 12.04 with Kernel 3.2.0 and install it in in a 250GB Seagate 7200RPM hard disk. The device for measurement is a 128GB SSD produced by a mainstream SSD manufacturer. It is built upon multi-level cells (MLC) flash memories, and the 4KB random read and write latencies of this device are 33 μs and 12.5 μs, respectively. To avoid the interference from the OS (e.g., page cache and file system), we perform the measurement directly on the raw block device. Following the previous study [6], we choose *noop* as the IO scheduler for this SSD, leaving the optimization for access requests handled by the device itself. For the sake of expression, this SSD will be referred to as "SSD-A" in the following text.

3.2 Characterizing SSD Behaviors

We adopt the generalized model presented in [6] to profile SSD internals. As described in this model, an SSD consists of multiple *domains*, each of which is a set of flash memories that share some specific resources; the pages continuously allocated within one domain comprise a *chunk*.

Table 1. Description of the parameters and functions

name	description
cycle	length of variation cycles of stride_read latency
dev_size	SSD device size
max_offset	max. offset in page unit
max_rq_size	max. write size in page unit
pg_size	page size of SSD
range	size of address space initialized
read_buf_size	SSD read buffer size
rand_pos(pos, size)	randomly choose an address within $[0, pos)$ aligned to size
SSD_read(pos, size)	read size bytes against pos
SSD_write(spos, epos, size)	sequentially fill range $[spos, epos)$ with size-byte write requests, during which OS page cache and SSD write buffer are both disabled
single_write(pos)	write one page at pos with SSD write buffer enabled
stride_read(pos, offset)	read two pages with two concurrent threads, each for one page; the 1st thread reads against pos, and the 2nd one skips offset over the 1st one

Servicing Write Requests. We intend to investigate how flash pages get written under different write patterns, thereby revealing how write requests are serviced. To this end, we first initialize several disjoint address ranges of SSD-A with writes of varying request sizes. We disable the page cache of operating system and the write buffer of SSD-A so that each write requests will be directly handled by flash memories. Since due to resource contention, the pages inside a domain will experience a longer read latency compared with those across multiple domains, we issue a set of read requests that are able to realize this difference in

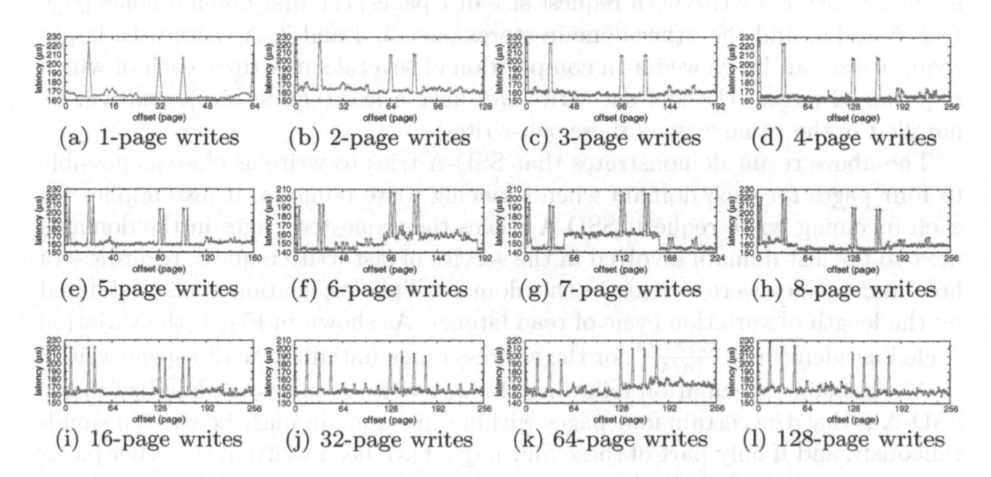

(a) 1-page writes (b) 2-page writes (c) 3-page writes (d) 4-page writes
(e) 5-page writes (f) 6-page writes (g) 7-page writes (h) 8-page writes
(i) 16-page writes (j) 32-page writes (k) 64-page writes (l) 128-page writes

Fig. 1. The variation of page placement with the request size of writes

Procedure 1. Measuring SSD page placement (I)

```
for i ← 1 to max_rq_size do
    spos ← (i − 1) × range;   epos ← i × range
    rq_size ← i × pg_size
    SSD_write(spos, epos, rq_size)
    for j ← 1 to max_offset do
        latency ← 0;    offset ← j × pg_size
        for k ← 0 to 1000 do
            latency ← stride_read(spos, offset) + latency
            //polute SSD read buffer
            SSD_read(spos + max_offset × pg_size, read_buf_size);
        end
        print rq_size, offset, latency/1000
    end
end
```

read latency, and compare their service time to derive how pages are placed. The whole process is shown in Procedure 1 in detail. The parameters and functions used in this paper are summarized in Table 1.

In Procedure 1, we create two concurrent threads for each read. Each thread reads only one page. The first thread reads against the start address of current address range, and the second thread skips `offset` pages over the first one. Figure 1 shows how latency of 2-thread read varies with `offset` in several address ranges initialized with writes of different request size. Since the latency varies periodically with `offset`, we only plot the first two cycles in Fig. 1.

From Fig. 1, we can make the following derivation regarding the service of write requests inside SSD-A. *For a write with a request size of* n $(n \leq 32)$ *pages, numbered from 1 to* n, $\lceil n/4 \rceil$ *domains will be used to handle this request such that domain* i $(0 < i < \lceil n/4 \rceil)$ *holds pages* $\{2i-1, 2i, 2(i+\lceil n/4 \rceil)-1, 2(i+\lceil n/4 \rceil)\}$, *and domain* $\lceil n/4 \rceil$ *holds remainder pages.* For instance, 2 domains will be involved in the service of a write with request size of 7 pages; the first domain holds page 1, 2, 5 and 6, and the other domain stores page 3, 4 and 7. A write with larger request size can be viewed as a composition of several sub-writes, each of which requests 32 pages (the last sub-write can have a less request size), and will be handled in the same way as these sub-writes.

The above result demonstrates that SSD-A tries to write as close as possible to four pages for each domain when servicing write requests. It also implies for each incoming write request, SSD-A places the requested pages in the domains next to the last domain involved in the service of last write request, regardless of how many pages were written in this domain. This implication is also validated by the length of variation cycle of read latency. As shown in Fig. 1, the variation cycle has a length of $\frac{128 \times n/4}{\lceil n/4 \rceil}$ for the address range initialized with n-page writes.

A reasonable speculation following the way write requests are handled inside SSD-A is that the certain four pages within same domain must be written simultaneously, and if only part of these four pages have been written, the other pages can be programmed (written) only after the written ones have been erased.

Procedure 2. Measuring SSD page placement (II)

```
len ← cycle × pg_size        //cycle is measured in Procedure 1
for i ← 0 to 4 do
    SSD_write(0, device_size, 256 × pg_size)
    rq_size ← (i + 1) × pg_size
    SSD_write(0, range, rq_size)
    for j ← 1 to max_offset do
        latency ← 0;   offset ← j × pg_size
        for k ← 0 to 50000 do
            pos ← rand_pos(range, len) + rand_pos(len/4, rq_size)
            latency ← stride_read(pos, offset) + latency
        end
        print rq_size, offset, latency/50000
    end
end
```

To verify the above speculation, we carry out another experiment in a similar way to Procedure 1. We first sequentially fill the whole address space of SSD-A with writes of a large request size so that the placement of pages is the same as that shown in Fig. 1j-1l. Then, starting from address 0, we sequentially write SSD-A with a request size of one page, during which the OS page cache and on-device write buffer are disabled.

We use the same method as Procedure 1 to detect the placement of pages within the address range filled in the second write phase. This time we allow the first thread to read addresses other than the start position of address range, i.e., address 0. Specifically, each time the first thread reads against an address randomly selected from the address set $\{i \times \texttt{cycle} \times \texttt{pg\_size} + j \times \texttt{rq\_size} | 0 \leq i < \frac{\texttt{range}}{\texttt{cycle} \times \texttt{pg\_size}}, 0 \leq j < 8\}$, where cycle is the cycle length measured in Procedure 1 (32 in this case, as shown in Fig. 1a), and the meanings of other variables can be found in Table 1. We do this because, as can be inferred from Fig. 1, the latency of 2-thread reads keeps almost unchanged when the first thread reads against different addresses of this set. The detailed implementation is shown in Procedure 2.

We repeat this experiment for four times. Each time we choose a different write size in the second write phase, and adjust the candidate address set for the first read thread accordingly. The experiment result is shown in Fig. 2.

Comparing Fig. 2 with Fig. 1a - 1d, it is intuitive to observe that when write request size of the second write phase is less than four pages, there will be a substantial difference in page placement between the two scenarios with/without the first write phase, and such difference disappears when the request size of second-phase writes increases to four pages. Therefore, we can infer that pages written in the second phases of the first three runs (corresponding to Fig. 2a, 2b and 2c, respectively) have been relocated and compacted to provide more available flash pages, otherwise the corresponding page placement should keep unchanged as writes of seconde phase are gradually serviced. This result also confirms our speculation made above: the certain four pages within same domain

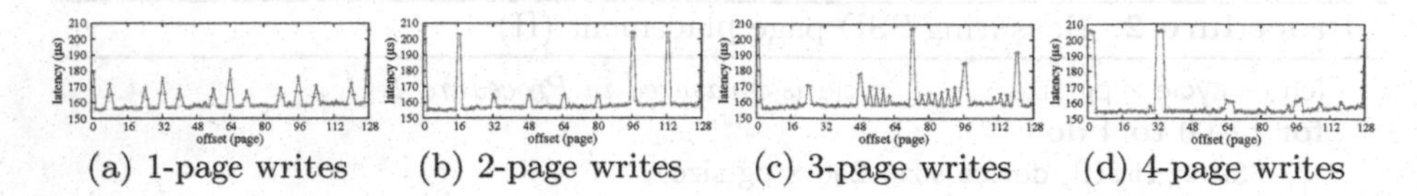

(a) 1-page writes (b) 2-page writes (c) 3-page writes (d) 4-page writes

Fig. 2. The variation of page placement with the size of write request and the availability of free space

must be programmed simultaneously; if only part of these four pages have been written, the other pages are left unable to service write requests until the written ones have been erased, leading to a waste in storage. For the sake of expression, we call each such four pages *a clustered page.*

From the definition of clustered pages, it is not difficult to see that storage waste will raise from the service of writes with a request size that is not a multiple of four pages, since at least one involved clustered page will be partially written. Such feature of clustered pages helps to understand our result regarding the service of write requests. As we have mentioned, a write requesting n pages will be handled such that page $2i-1$, $2i$, $2(i+\lceil n/4 \rceil)-1$ and $2(i+\lceil n/4 \rceil)$ requested in this write will be placed in the same domain and occupy a full clustered page. In this way, for each write request, there is at most one clustered page, i.e., the last one involved, that might will be partially written, and the storage waste is thus minimized. In the mean time, the parallelism among domains can be effectively exploited.

Garbage Collection. The SSD erase granularity has been studied in [12] based on the assumption that after the whole SSD has been sequentially written, the speed of following random writes with a request size equal to the size of erase unit must be same as that of sequential writes, as there is no page relocation in both cases. In this work, we are more interested in whether the blocks with pages in same clustered pages must be erased simultaneously or not, for which a more intuitive result can be obtained by carefully overwriting the clustered pages that have been completely filled.

As shown in Procedure 3, we first sequentially fill the whole space of SSD-A with writes of a request size of four pages so that each involved clustered page is fully written and page placement inside SSD-A is same as in Fig. 2d. After that, from address 0, we gradually overwrite the first page of clustered pages. As write process progresses, garbage collection will be revoked to reclaim the overwritten pages, and we are then able to answer the question concerned by examining whether the non-overwritten pages have been relocated.

Figure 3b - 3d present the placement of non-overwritten pages after garbage collection, and their counterpart before garbage collection is shown in Fig. 3a for comparison purpose. It can be easily derived from Fig. 3 that the three pages of clustered pages that did not get overwritten during the write process have been relocated after garbage collection. We are thus able to conclude that the four

Procedure 3. Measuring page placement after garbage collection

```
chunk_size ← 4 × pg_size
SSD_write(0, device_size, chunk_size)
for i ← 0 to range/chunk_size do
    single_write(i × chunk_size)
end
len ← cycle × pg_size
for i ← 1 to 4 do
    for j ← 0 to max_offset do
        latency ← 0;   offset ← j × pg_size
        for k ← 0 to 50000 do
            /* each time the first thread reads the (i+1)-th page of a clustered
            page                                                              */
            pos ← rand_pos(range, len) + rand_pos(len/4, chunk_size) +
            i × pg_size
            latency ← stride_read(pos, offset) + latency
        end
        print i, j, latency/50000
    end
end
```

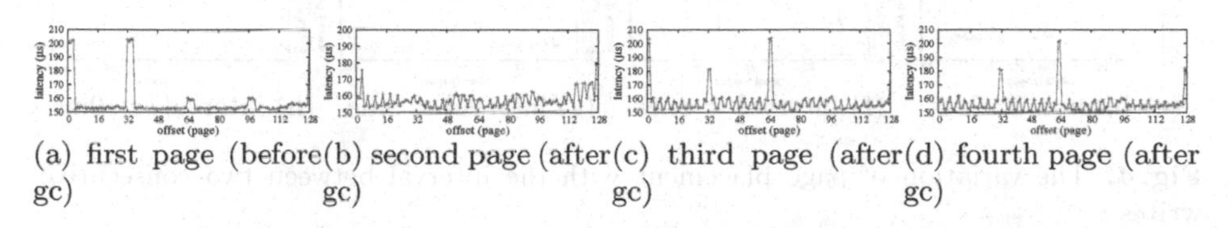

(a) first page (before gc) (b) second page (after gc) (c) third page (after gc) (d) fourth page (after gc)

Fig. 3. Page placement inside clustered pges before/after garbage collection

pages of each clustered page must be erased simultaneously, and a write with a request size of n pages thus actually leads to $4 \times \lceil n/4 \rceil$ pages being written in the sense that there is no difference between a written page and an unwritten one within the same clustered page, both of which cannot be programmed until being erased. As a result, the write amplification resulted from a write requesting n pages is $4 \times \lceil n/4 \rceil / n$, and this number may be further increased by 1 in the case where there are many partially written clustered pages that can be relocated and compacted to generate a large number of free pages.

Flushing Write Buffer. The problem of partially written clustered pages can be alleviated by the existence of write buffer in most SSDs. When a write request arrives, the SSD first buffers it and later flushes all the buffered write requests to the flash memory for persistent storage. As such, the number of pages written each time is increased, reducing the ratio of partially written clustered pages.

The flush of write buffer will be triggered when the buffer is full, or after a certain time period, which we call *flush period*. The corresponding two parameters associated with SSD write buffer are thus the size of write buffer and the length of flush period. As the measurement of the former parameter has been conducted in [12], we are more interested in the latter parameter.

We have revealed how request size of writes affects page placement for the case with write buffer disabled. Conversely, we can also infer from observed page placement the request size of writes and the number of buffered pages for flush, both of which are in principle the same. In this regard, we follow the same way as we did in Procedure 1 with the exception that the parameter varying across address ranges is no longer the request size of writes, which is kept constant at 1 page, but the interval between two consecutive writes instead. In addition, the write buffer is no longer disabled during the initialization phase. Due to page limit and its similarity to Procedure 1, the detailed implementation is not presented in this paper.

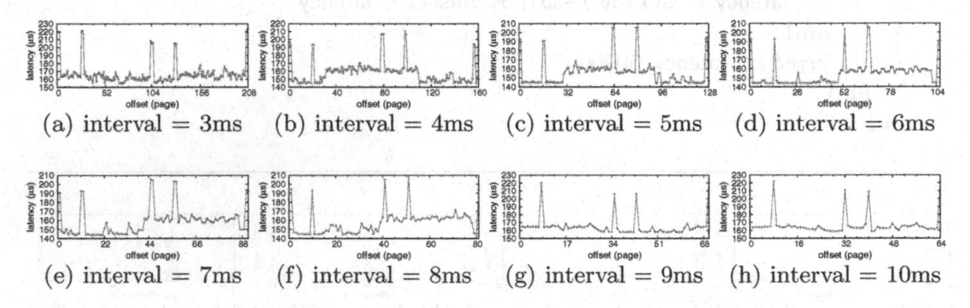

(a) interval = 3ms (b) interval = 4ms (c) interval = 5ms (d) interval = 6ms

(e) interval = 7ms (f) interval = 8ms (g) interval = 9ms (h) interval = 10ms

Fig. 4. The variation of page placement with the interval between two consecutive writes

Figure 4 gives the measured latency of two-thread reads for different inter-arrival interval of write requests. For demonstration purpose, we only provide in this figure the result for cases in which the buffered pages for each flush occupy only one clustered page. Comparing Fig. 4 with Fig. 1, we can find that Fig. 4h and 4c are, respectively, the same as Fig. 1a and 1b, which means the numbers of buffered pages for flush in corresponding two cases are 1 and 2, respectively, and the length of flush period is thus 10 milliseconds. Moreover, as can be observed from Fig. 4, the read latency in each case exhibits a periodical pattern with a cycle length reversely proportional to the inter-arrival interval. It can be inferred from this observation that the flush of write buffer is performed once every 10 milliseconds, rather than in 10 milliseconds after the arrival of the earliest buffered write, in which case there would be two buffered pages for each flush if the inter-arrival interval of writes is within the range $(5ms, 10ms)$, and Fig. 4d-4g thus must have the same cycle length as Fig. 4c.

4 SSD Storage Efficiency

4.1 Server Storage

We first study the case in which SSD-A is used as storage device for server applications. To this end, we download the HDD block access traces of two OLTP applications (Financial 1 and Financial 2) [3] and eight other server applications [1] [25]. Table 2 gives the general information of these block traces.

Table 2. Description of server traces

name	no. WR records	name	no. WR records
Financial 1	4,099,354	Home 4	2,354,032
Financial 2	653,082	Online	4,211,728
Home 1	8,882,821	Web Mail	6,381,984
Home 2	4,901,076	Web Research	2,413,936
Home 3	908,835	Web Users	5,127,100

For each trace, we intend to investigate the storage efficiency when the same sequence of writes are issued to SSD-A. To simplify the investigation, we assume that two consecutive writes with an interval less than a certain threshold will be flushed simultaneously; the threshold is chosen to be longer (20ms in our investigation) than the time to access data in most modern HDDs so that two consecutive writes with an interval longer than the threshold are likely to be independent, and thus will be issued with the same interval in SSD case. Under this assumption, the number of buffered pages for flush will be an over-estimation to the real value as pages for flush are those that have been buffered over a time period longer than the flush period.

Figure 5 presents the number of wasted pages, i.e., unwritten flash pages in clustered pages, that will be incurred if the same sequence of writes of each trace are issued to SSD-A. It can be seen from this figure that each trace will generate hundreds of thousands of wasted pages, which means a moderate degree of storage waste, as compared with the number of write records in Table 2.

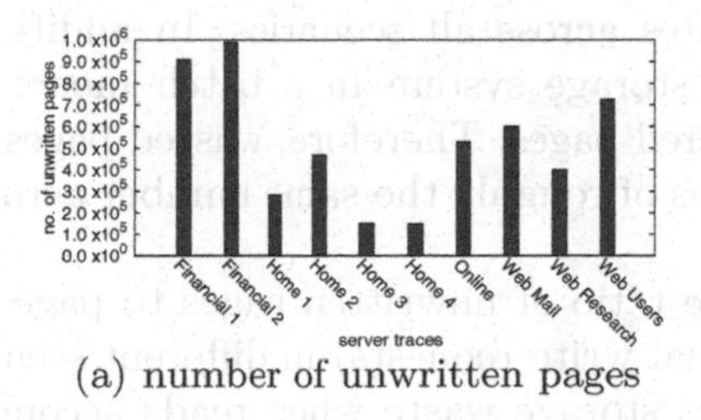

(a) number of unwritten pages

Fig. 5. Storage waste in different server traces

4.2 Hybrid SSD/HDD Storage System

Due to their superior access performance but relatively high cost, SSDs have been extensively used as an additional cache layer on the top of HDDs to form hybrid SSD/HDD storage systems with improved storage performance [9] [13] [21] [22]. In such systems, if a read request cannot be serviced by cache layer, it will generate a write request to the cache, and the inter-arrival interval of such write requests, which we will call *internal write requests* in the sense that they are issued inside the storage system, is thus no less than the service time of read requests of hard disks, which can be as much as tens of milliseconds due to high position delay [2]. As a result, if SSD-A is employed as the cache layer, the internal write requests can lead to a significant waste of cache storage because of their slow arrival rate and the existence of clustered pages inside cache.

We carry out several experiments to investigate the storage efficiency when SSD-A is used as the cache layer of hybrid storage systems. The experimental platform is Flashcache [24], a popular open source solution to hybrid SSD/HDD systems, and the tool for I/O test is fio [5]. For experiment, we issue a set of random reads and writes, each requesting one page, to storage system, and explore the resulted storage waste. We run the experiment five times, each lasting five minutes and with a varied fraction of reads. The page cache of operating system and SSD write buffer are both enabled during the experiment.

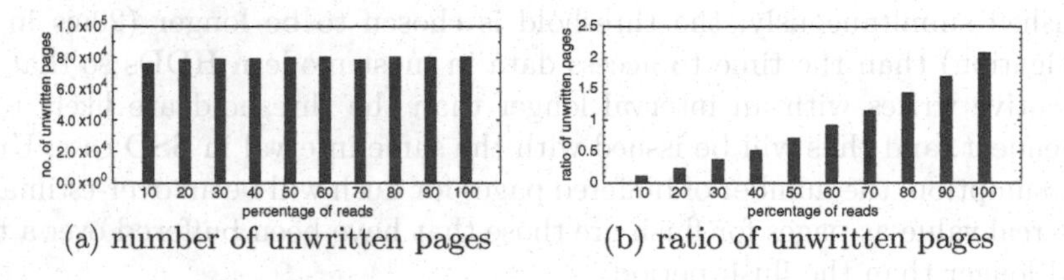

(a) number of unwritten pages (b) ratio of unwritten pages

Fig. 6. Storage waste in hybrid storage system

The result of storage efficiency is demonstrated in Fig. 6. Since the OS page cache is enabled, writes are first handled in the memory and thus completed much faster than reads. Consequently, there are roughly the same number of reads and internal writes across all scenarios. In addition, writes buffered in memory are issued to storage system in a batch mode, and thus incur little partially written clustered pages. Therefore, wasted pages are mostly caused by internal writes, and thus of roughly the same number across all scenarios, which is verified in Fig. 6a.

Fig. 6b describes the ratio of unwritten pages to pages requested (including those involved in internal write requests) in different scenarios. This figure also shows there is a serious storage waste when reads account for a large fraction of total IOs. For instance, in the scenario with all IOs being reads, each write request leads to 2.2 wasted pages on average, implying a write amplification up to 4.2, which will be achieved after page relocation.

When OS page cache is enabled, the data read from storage device will be stored in memory, i.e., page cache, so that future reads requesting the same data can be satisfied without disk access. Therefore, for hybrid storage systems, the issuance of internal write requests can be safely delayed without performance loss, as long as there is a copy of the corresponding data in memory. Consequently, we can suspend the issuance of internal write requests until there are a certain number of such write requests accumulated, and then simultaneously issue them to the cache layer to improve storage efficiency.

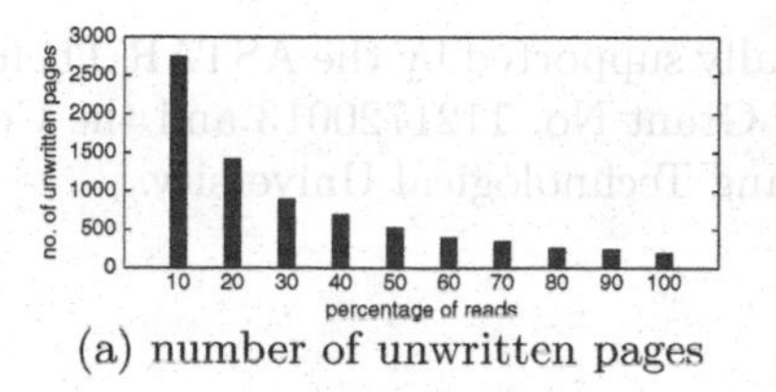

(a) number of unwritten pages

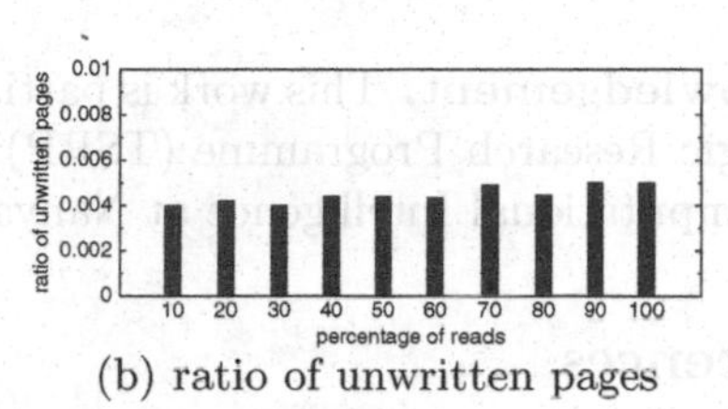

(b) ratio of unwritten pages

Fig. 7. Storage waste in hybrid storage system with the isssue of internal write requests delayed

We implement the above method in Flashcache [24]. In our implementation, the issuance of delayed internal write requests takes place when a certain time period has passed by after last issuance, or the number of delayed requests exceeds a specific threshold. We repeat the above experiments and show the corresponding results in Fig. 7. We can draw from this figure that there will be little storage waste when the issuance of internal write requests is delayed: in all cases, the number of wasted pages is no more than 0.5% that of pages requested. In addition, it is worth noting that this method of reducing cache writes is orthogonal to those presented in [21] which reduce the writes to cache by neglecting the data that have been requested only a limited times.

Besides improving storage efficiency, there are some other advantages that can arise from delaying the issuance of internal write requests. First, issuing multiple writes requests at same time can effectively exploit the rich parallelism inside SSD. In addition, read/write interference inside SSD, which has been reported to be able to significantly hamper access performance of SSDs [6][20], can also be alleviated as a result of reduced number of flushes of SSD write buffer.

5 Conclusion

In this paper, we carry out an extensive investigation on the behavior of new-generation SSDs, and obtain two major findings. First, the investigation exposes the existence of clustered pages, each of which consists of certain four flash pages that must be programmed and reclaimed simultaneously. Second, pages are placed inside the SSD such that the parallelism of the SSD can be effectively exploited, on the premise that the number of partially written clustered pages, which are the source of storage waste, is minimized.

In order to quantify the impact of clustered pages on storage efficiency, we then conduct two case studies, i.e., server storage and cache layer of a hybrid storage system. For the latter case, we find that when most IOs are reads, the storage efficiency is extremely low due to the long inter-arrival interval of internal write requests which are generated when a read cannot be serviced by cache layer. By delaying the issuance of such internal write requests, we can substantially reduce the number of wasted pages, thereby enhancing the storage efficiency and extending the lifetime of SSD.

Acknowledgement. This work is partially supported by the ASTAR Thematic Strategic Research Programme (TSRP) Grant No. 1121720013 and the Center for Computational Intelligence at Nanyang Technological University.

References

1. FIU Traces, `http://iotta.snia.org/traces/390` (retrieved September 11, 2014)
2. Hard disk drive, `http://en.wikipedia.org/wiki/Hard_disk_drive` (retrieved September 11, 2014)
3. UMassTraceRepository, `http://traces.cs.umass.edu/index.php/Storage/Storage`
4. Agrawal, N., Prabhakaran, V., Wobber, T., Davis, J.D., Manasse, M.S., Panigrahy, R.: Design tradeoffs for ssd performance. In: ATC, Boston, Massachussetts, USA, pp. 57–70 (2008)
5. Axboe, J.: fio, `https://github.com/axboe/fio` (retrieved September 11, 2014)
6. Chen, F., Lee, R., Zhang, X.: Essential roles of exploiting internal parallelism of flash memory based solid state drives in high-speed data processing. In: HPCA, San Antonio, Texas, USA, pp. 266–277 (2011)
7. Chen, F., Luo, T., Zhang, X.: Caftl: A content-aware flash translation layer enhancing the lifespan of flash memory based solid state drives. In: FAST, San Jose, California, USA (2011)
8. Dirik, C., Jacob, B.: The performance of pc solid-state disks (ssds) as a function of bandwidth, concurrency, device architecture, and system organization. In: ISCA, Austin, TX, USA, pp. 279–289 (2009)
9. Guerra, J., Pucha, H., Glider, J., Belluomini, W., Rangaswami, R.: Cost effective storage using extent based dynamic tiering. In: FAST, San Jose, CA, USA (2011)
10. Gupta, A., Kim, Y., Urgaonkar, B.: Dftl: A flash translation layer employing demand-based selective caching of page-level address mappings. In: ASPLOS XIV, Washington, DC, USA, pp. 229–240 (2009)
11. He, B., Yu, J.X., Zhou, A.C.: Improving update-intensive workloads on flash disks through exploiting multi-chip parallelism. IEEE Transactions on Parallel and Distributed Systems (2014)
12. Kim, J., Seo, S., Jung, D., Kim, J.S., Huh, J.: Parameter-aware i/o management for solid state disks (ssds). IEEE Transactions on Computers 61(5), 636–649 (2012)
13. Koltsidas, I., Viglas, S.D.: Flashing up the storage layer. Proceedings of the VLDB Endowment 1(1), 514–525 (2008)
14. Lee, S.W., Park, D.J., Chung, T.S., Lee, D.H., Park, S., Song, H.J.: A log buffer-based flash translation layer using fully-associative sector translation. ACM Transactions on Embedded Computing Systems 6(3), article No. 18 (2007)

15. Li, Y., He, B., Luo, Q., Yi, K.: Tree indexing on flash disks. In: ICDE, Shanghai, China, pp. 1303–1306 (2009)
16. Li, Y., He, B., Yang, R.J., Luo, Q., Yi, K.: Tree indexing on solid state drives. Proceedings of the VLDB Endowment 3(1-2), 1195–1206 (2010)
17. Ma, D., Feng, J., Li, G.: Lazyftl: A page-level flash translation layer optimized for nand flash memory. In: SIGMOD, Athens, Greece (2011)
18. Min, C., Kim, K., Cho, H., Lee, S.W., Eom, Y.I.: Sfs: Random write considered harmful in solid state drives. In: FAST, San Jose, CA, USA (2012)
19. Park, C., Seo, E., Shin, J.Y., Maeng, S., Lee, J.: Exploiting internal parallelism of flash-based ssds. Computer Architecture Letters 9(1), 9–12 (2010)
20. Park, S., Shen, K.: Fios: a fair, efficient flash i/o scheduler. In: FAST, San Jose, CA, USA (2012)
21. Pritchett, T., Thottethodi, M.: Sievestore: A highly-selective, ensemble-level disk cache for cost-performance. In: ISCA, Saint-Malo, France, pp. 163–174 (2010)
22. Saxena, M., Swift, M.M., Zhang, Y.: Flashtier: A lightweight, consistent and durable storage cache. In: EuroSys, Bern, Switzerland, pp. 267–280 (2012)
23. Seol, J., Shim, H., Kim, J., Maeng, S.: A buffer replacement algorithm exploiting multi-chip parallelism in solid state disks. In: CASE, Grenoble, France, pp. 137–146 (2009)
24. Srinivasan, M.: Flashcache, `https://github.com/facebook/flashcache` (retrieved September 11, 2014)
25. Verma, A., Koller, R., Useche, L., Rangaswami, R.: Srcmap: energy proportional storage using dynamic consolidation. In: FAST, San Jose, California, USA (2010)

Supply Chain Risk Mitigation and the Application Potential of Complex Systems Approaches

Zhengping Li and Rajpreet Kaur Gulati

Singapore Institute of Manufacturing Technology
71 Nanyang Drive, Singapore 638075
zpli@simtech.a-star.edu.sg, rajpreetgulati@gmail.com

Abstract. Supply chains are becoming more complex and vulnerable. How to mitigate supply chain risks (SCRs) becomes a critical issue in supply chain management (SCM). This paper reviews the researches on SCR mitigation and identifies the current status and limitations. A new framework on SCR mitigation is proposed which intends to draw a clear roadmap for identifying, prioritizing and mapping SCRs to mitigation policies, evaluating and implementing related mitigation policies. Considering the complexities of SCR issues, the paper also elaborates the application potential of complex systems (CS) approaches for SCR mitigation.

Keywords: supply chain, risk mitigation, complex systems, mitigation policy.

1 Introduction

With the increasing scales and complexities of supply chains as a result of globalization, outsourcing and lean initiatives, today's supply chains become more vulnerable [1] [2] and risk management has become a critical issue in SCM.

A risk management project can be divided into the phases of risk identification, assessment, mitigation, and monitoring [3, 4]. Among which, risk mitigation is the process of designing mitigation policies and algorithms to reduce related loss with minimum cost. It intends to design, select, prioritize and execute measures which reduce risk exposure with least cost [3, 5].

The classical mitigation policies in SCR management (SCRM) are the policies on inventory, buffer capacity, dual sourcing, distribution and transportation alternatives arrangements etc. [6] [7]. The policies help companies to protect against recurrent, low-impact risks in supply chains but ignore high-impact, low-likelihood risks. Recently, disruptions, both natural and man-made, have made organizations to rethink on SCRM as they may rely on their partners spanning several nations and continents with the globalization of supply chains. Disasters, such as floods and Tsunami in Asia, had big impacts on the economy of the regions where they occurred. However, these disasters also impact organizations in Europe and USA as their suppliers are in countries such as China, India or South East Asia [8].

H. Handa et al. (eds.), *Proc. of the 18th Asia Pacific Symp. on Intell. & Evol. Systems – Vol. 1*,
Proceedings in Adaptation, Learning and Optimization 1, DOI: 10.1007/978-3-319-13359-1_28

Researchers and supply chain managers have proposed many policies & techniques on mitigating SCRs. However, an elusive question is how to design, select, match and apply the right mitigation policy and technique to deal with a particular risk and this is related to a number challenges.

First, SCRs are evolving with new uncertainties and disruptions. The most dangerous risk may not be the one with high impact and probability which can be detected and identified with limited efforts, but the risks with low probability and detectability [9]. The traditional mitigation policies by adjusting inventory, capacity, sourcing policy, and logistics alternatives become not so valid in dealing the risks. The design of new mitigation policies and the selection of the right policies become challenging.

Second, lacking of professional knowledge with the complexities of SCRs causes difficulties for companies to design and select the right mitigation policies. For example, companies have never experienced risks such as the 2008 Financial Crisis and 2011 Thailand Flood before. They lack of the knowledge on these risks and related propagations, not to mention selecting the right mitigation policies.

Third, the capabilities on designing effective mitigation policies for handling complex SCRs are still lacking. Traditional analytical approaches have been used to provide exact solutions on optimizing supply chain strategies and operations. However it is difficult to understand how a complex supply chain can work as a whole and what are the interplays of various factors and components in complex risk scenarios. In addition, the traditional approaches assume the structure and components of supply chains are fixed and actually this is not true and supply chains are dynamic all the time. To break the limitations, new approaches are needed to provide better choices on investigating the collective behaviours of supply chains and enables observing emergent behaviours in complex supply chain networks (SCNs) [10].

This paper conducts a comprehensive literature review, proposes a new framework on SCR mitigation and elaborates on the potential of CS approaches for SCR mitigation. It is organized as follows: Section 2 reviews researches on SCR mitigation; Section 3 proposes a new framework for SCR mitigation; Section 4 discusses on the potential of CS approaches for SCR mitigation and Section 5 concludes the paper.

2 Literature Review on SCR Mitigation

SCR is about any threat of interruption to the operations of supply chains. SCRs can be classified as supply, demand, operational, disruption risks based on the sources of the risks [7, 11]. Many literatures have analysed and modelled these risks and some go further on the mitigation of the risks.

2.1 Supply Risk and Mitigation Policies

Supply risks are caused by multiple risk sources including supply yield-capacity uncertainty [12], lead time uncertainty [12], price uncertainty [6, 12, 13], commitment [6], shipment disruptions[9], lack of ownership [7] etc.

Resilience is considered as an important policy for dealing with supply risk. Tang [12] reviewed works on supply management and suggested using multi-supplier strategy to handle supply risk and using flexible supply contract to improve resilience. Tang and Tomlin [6] proposed flexible contracts to mitigate supply risks and the inability to change order quantity once submitted. The paper provided a stylized model for mitigating supply risk. Tang [13] suggested "flexible supply based" and "supply alliance network" as mitigation strategy for supply risk. Hua Lee,1993 [14] proposed mathematical model considering flexibility to deal with material shortage issues.

Supply chain redesign is used as a mitigation policy. Blackhurst [15] discussed flexible and real time supply chain reconfiguration which takes effect once a disruption occurs. Agent-based P-Trans-Nets Models are used for task decomposition, distribution, and resource allocation. Blackhurst et al [16] proposed a 3-stage processes for SCRM: disruption discovery, disruption recovery and supply chain redesign.

Redundancy has been applied in mitigating supply risk. 'Redundancy' approaches such as the using of safety stocks or multiple sourcing are suggested by Thun [17] who proposed empirical analysis models based on a survey of German automotive industry to conclude that redundancy policies are effective means to deal with supplier quality and unreliability issues. Carvalho [18] proposed alternative transport links as policy for mitigating supply delay. They presented a simulation study for a case of a Portuguese automotive supply chain. Kull et al [19] suggested network reconfiguration to mitigate supply risk. Their exploratory findings suggest that increased inventory in a tiered supply chain can sometimes increase supply risk rather than decrease it.

Other approaches have also been proposed. Jüttner et al [7] identified "Lack of Ownership" as a supply risk and suggested policies including vertical integration, buffer inventory and capacities, and contract requirements on suppliers. Oke et al [20] discussed collaborative planning with supplier to mitigate the risk of supplier shutdown and resilience policies for mitigating the risk of losing key suppliers. Benyoucef, M. and M. Canbolat [21] quantified a comprehensive set of local and global sourcing risk factors. A process failure mode effects analysis is used to rank causes of failures, to calculate total risks in terms of dollars and to evaluate optimum risk mitigation strategies. Swink & Zsidisin [22] hypothesized that the relationship between focused commitment strategies to suppliers and buyer's performance is nonlinear, taking the form of inverted u-shaped curves, with the exception of 'quality' which exhibits a positive linear relation. Sirivunnabood and Kumara [23] proposed a multiple agent based approach to evaluate mitigation policies for supply risks. Behdani et al. [24] suggested conceptual policies of excess inventory, alternative suppliers and order reassignment for mitigating supplier disruption in a chemical supply chain by agent-based modelling.

2.2 Demand Risk and Mitigation Policies

Demand risks are caused by uncertainties in demand volume fluctuations, changes in preferences, cancellations of order, inflexibility and wrong order forecast. To mitigate this type of risk, different mitigation policies have been proposed.

Resilience has been identified as policy for handling demand risks. Doege et al [25] investigated risk management of power supply market and discussed on the hedging value of production flexibility. Tang & Tomlin [6] suggested that when supply is inflexible, one can use a pricing mechanism to influence customer demand so as to reduce demand risks. Swafford [26] discussed supply chain flexibility which covers procurement, distribution, manufacturing and product development functions and represents abilities to reduce supply chain lead times, ensure production capacity and provide product variety to improve customer responsiveness. Cucchiella and Gastaldi [27] proposed resilience as a mitigation policy for managing uncertainties in supply chains by utilizing real options theory to cover one or more risks inside the supply chain. Ben-Tal et al. [28] considered resource flexibility for handling demand uncertainties. They proposed a methodology to generate a robust logistics plan that mitigated demand uncertainty in humanitarian relief supply chains.

In addition, postponement has been used as a policy in face of increasing product variety in demand. It is useful in handling demand variations in markets where a single product may have multiple derivatives due to different language, culture, government or technological requirements, and greatly reduces inventory carrying and transportation costs. Whang and Lee [29], Jüttner [7], Tang & Tomlin [6], and Tang [12] addressed postponement as a demand risk mitigation strategy. Yang and Yang [30] proposed postponement policy and draw insights from NAT (normal accident theory) that addresses catastrophic accidents and apply them to supply chain disruptions.

Supply chain redesign is used as policy for mitigating demand risks. Klibi [31] proposed 'SCN design' as a policy to protect against fluctuations in the prices of finished products, energy costs, labour costs and exchange rates. Klibi [32] studies modelling approaches to design resilient supply networks with location–transportation problem under demand uncertainty and disruptions. They proposed several stochastic programming models incorporating alternative resilience formulations. Pearson and Masson [33] proposed an equilibrium model for handling demand variation and forecast errors in a fashion SCN. The model can be used to examine contractual details as well as strategic and game theoretical concepts of a network and agent-based simulation was used in their solution. Baghalian and Rezapour [34] used network redesign as mitigation policy and developed a stochastic mathematical model for designing multi-product SCNs under uncertainty. The resultant model incorporates the cut-set concept in reliability theory and robust optimization concept. Kumar [35] illustrates that the ability to react to risk factors offers potential solutions to robust supply chain design. Computational intelligence techniques such as genetic algorithms, particle swarm optimisation and artificial bee colony are applied in solution evaluation.

Product substitution has been proposed as policy for mitigating demand risks. Tang [12] show a firm can increase product substitutability by selecting a specific combination of products with similar features. Rajaram and Tang [36] presented a single period stochastic model that sells two substitutable products. They showed that the optimal order quantity of each product and the retailer's profit increase as product substitutability increases. Bitran and Dasu [37] presented models for determining the optimal production quantities when higher-grade chips can be used as substitutes for lower-grade chips. Pricing mechanism was used to entice customers to shift their demand from one product to another, so the demand uncertainties can be mitigated ([38]).

Besides the above, other policies were proposed for mitigating demand risks. Tang [12] studied supply contracts to mitigate demand uncertainties. The strategies on supply contract include flexible prices ([39], [40]), buy back contract ([41]), revenue sharing contracts ([42]), and contracts with quantity flexibility (QF) and minimum order ([43]). Li and Marlin [44] proposed a robust Model Predictive Control (MPC) model for real-time supply chain optimization under uncertainties, which provides the solution of a constrained, bi-level stochastic optimization problem. Ben-Tal et al [45] proposed multi-echelon multi-period inventory control as an policy for handling demand uncertainty.

2.3 Operational Risk and Mitigation Policies

Operational risks are caused by multiple risk sources, including uncertainty in product safety, machinery / equipment breakdowns, production capacity problems [6, 27], material delivery / processing delays [9] and lead time uncertainty [15] etc. Different types of mitigation policies have been proposed in mitigating operational risks.

Quality control is identified as a mitigation policy for operational risks. Six Sigma and regulatory compliance and analytical methods are discussed for operational risk control [46]. Sun et al.[47] presented a quality risk management model for a supplier–assembler supply chain. They develop a P-chart solution model to find out the optimal due-date that minimizes the total cost.

Resilience is also used as mitigation policy operational risk. Sheffi & Rice [9] argued that ‘conversion flexibility’, which involves the use of standard processes across facilities with interoperability, allows a firm to operate in another facility when one is disrupted or to replace sick or otherwise unavailable operators. Tang & Tomlin [6] and Thun & Hoenig [17] proposed flexible process strategies which allows the firm to produce multiple products to compete on product variety and cost. Cucchiella & Gastaldi [27] addressed risks such as insufficient production capacity and delays in receiving critical information and examine ‘real options’ in risk avoidance policies, such as deferring investment, outsourcing, scaling down and abandoning current operations. They suggested that the real option theory allows increasing flexibility and enable the selection of possible options to protect the firm against operational risks.

Supply chain redesign is used as mitigation policy for operational risks. Speier et al. [48] develops a framework to examine the threat of potential disruptions on supply chain processes and security. It focuses on supply chain design strategies to mitigate these risks by integrating three theoretical perspectives—normal accident theory, reliability theory, and situational crime prevention. In addition, Blackhurst & O’Grady [15] proposed P-Trans-net model to identify those nodes along the supply chain that contribute to the longest lead times and delays. Wang [49] discussed on node overload problem in scale-free network by simulation and he suggested use transhipment of workload in (supply chain) network.

2.4 Disruption Risk and Mitigation Policies

Disruption risk is due to natural disasters (fire, earthquake, flood, rock fall, landslide, avalanche, etc.), weather (iciness, storm, heat) and political instability (strike, war, terrorist attacks, embargo, political labour conflicts, and industrial disputes).

Resilience has been used as important policy for mitigating disruptions. Klibi [31] suggested flexibility such as 'resource flexibility' and 'shortage response actions' into SCN design for handling disruptions such as social and political risks and natural disasters. Braunscheidel and Suresh [50] studied disruption risk and suggested resilience as the main mitigation policy with market orientation and learning orientation model based on structural equation modelling technique – partial least squares (PLS). Sawik [51] proposed a mixed integer programming model to determine resilient supply portfolios in case of disruptions. Ji and Zhu [52] developed supply chain disruption risk management strategies with the properties of efficiency and resilience analysed, and related with actual practice.

Redundancy is used as mitigation policy for disruption risk. Schmitt [53] studied disruption risk in the production and distribution sections of a supply chain. Their strategies take the advantage of the networks and include satisfying demand from alternative locations in the network, procuring material or transportation from alternative sources or routes, and holding strategic inventory reserves in network. Schmitt [54] studied risks from both supply disruptions and demand uncertainty and compare their impacts and mitigating strategies. They analyse inventory placement and back-up methodologies in a multi-echelon network and view their effect on reducing SCR.

Supply chain redesign has been used as policy for mitigating disruption risks. Klibi [31] gave a review on design robust value-creating supply chain for tackling disruption risks and pointed out that most of SCN design literatures consider simplified static and deterministic models. Ratick et al [55] suggested a 'geographical dispersion' strategy to spread risks associated with single point of failure events, natural and anthropogenic events affecting the value stream or a node.

In addition, Kleindorfer and Saad [56] underlines the importance of collaborative information sharing to manage disruption risk in supply chains under a cost-efficient manner. Zhao et al [57] investigated environment risk and suggested manufacturers to select mitigation strategies based on government rules. They proposed game theory model for environment risk reduction by using the concept of 'tolerability of risk'. Zhu and Dou [58] also uses game theory to analyse the strategies selected by manufacturers to reduce life cycle environmental risk. They suggested subsidiaries and penalties as mitigation measures in their evolutionary game model. [59] investigates the methods for mitigating bankruptcy propagation through coordination and proposed contract schemas as mitigation policies for a two-stage supply chain network.

2.5 Summary

As reviewed, the current works on SCR mitigation policies are mainly based on analysis of the features of risks and most of them focus on traditional issues of supply chain structure, supply contract, inventory policies, capacity buffers, and visibility through IT and data analysis techniques. Traditional OR approaches are applied to design the mitigation policies and algorithms. However, we observed a few limitations on current SCR mitigation research:

1) Lack of consistent approach focusing on risk mitigation from risk identification, ranking, matching, to the design, evaluation and implementation of mitigation policies.
2) Lack of the approach on mapping/matching SCRs to mitigation policies.
3) Quite some of the researches just proposed conceptual models and the models with algorithms are mainly based on traditional OR and simulation approaches that are with limitations in handling complex risk issues.
4) Although there are a number of researches on mitigation policies and algorithms, there is still a gap in evaluating and validating the mitigation policies / algorithms before implementing them in practices.

This research caters for the above issues and proposes a new SCR mitigation framework and then study on the potential of CS approaches for mitigating SCRs in complex supply chain scenarios.

3 IRMEI Framework on SCR Mitigation

3.1 A New Framework for SCR Mitigation - IRMEI

As discussed, mitigation is the most critical process in SCRM since it reduces the loss of SCRs directly. While there are many researches on risk mitigation, a consistent framework focusing on the mitigation of complex SCRs is still lacking. An IRMEI (Identification, Ranking, Matching, Evaluation and Implementation) SCR mitigation framework is hence proposed (Figure 1). As shown in the Figure, 5 major steps are included in the framework.

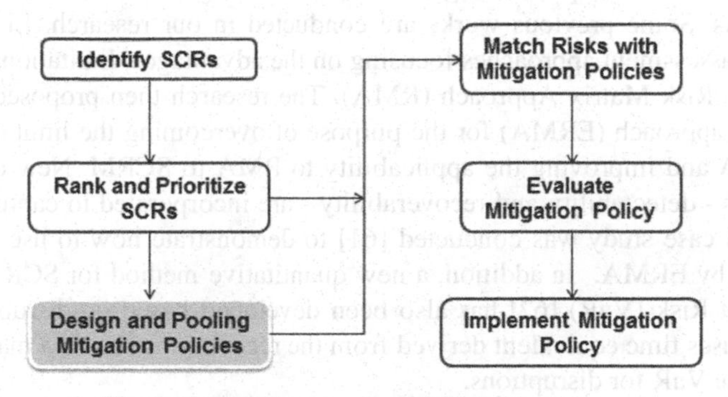

Fig. 1. IRMEI Framework of SCR Mitigation

Step 1 –Identification: a supply chain is mapped and SCRs are identified with communication to the people interviewed. Risk analysis and propagation need to be investigated to identify critical risks and nodes in a SCN.

Step 2 - Ranking: SCRs are ranked and prioritized based on risk assessment approaches. Risks are ranked and prioritized based on their impacts (severity), probability, detectability and other features.

Step 3 – Matching: the selected SCRs are matched to the right mitigation policies. Mitigation policies should be designed and verified before adding them to a pool ofpolicies. This step should also include the work of designing, categorizing and pooling of mitigation policies based on research review and industry case studies.

Step 4 – Evaluation: to evaluate the effectiveness of mitigation policies. The evaluation of the policies is mainly based on the efficiency of the policies and the cost of implementation.

Step 5 – Implementation: This is the step to realize the mitigation policy in a SC with setting the necessary parameters and make resources available for mitigation.

3.2 SCR Identification

Developing a supply chain map which clearly shows the products (components), nodes (partners) and connections (flows) can help to identify risks in a SCN. Given the complexity of supply chains, the mapping can have the perspective of focus and exclude non-critical entities. The processes of both developing and disseminating the map should lead to a common understanding of the SCN that would include what was deemed important to managing or monitoring the chain [60]. Available mapping approach such as Value Stream Mapping can add the risk hotspot to the map. The identification can involve various organization units and provides the opportunity for the managers to be involved in discussing the risks which need to be highlighted. The managers will also have the chance to tell why a risk is more critical than another.

3.3 SCR Ranking

Risk assessment is for prioritizing risks and allocating limited resources to mitigate critical risks. Some previous works are conducted in our research. Li et al [61] reviews risk assessment approaches focusing on the advantages, limitations and applications of the Risk Matrix Approach (RMA). The research then proposed an extended risk matrix approach (ERMA) for the purpose of overcoming the limitations of traditional RMA and improving the applicability to RMA to SCRM. New dimensions of risk metrics - detectability and recoverability - are incorporated to capture SCR complexities. A case study was conducted [61] to demonstrate how to use user data and rank SCRs by ERMA. In addition, a new quantitative method for SCR measurement by Value at Risk (VaR) [62] has also been developed based on disruption recovery models. It uses time equivalent derived from the recovery modes as a main measure to calculate the VaR for disruptions.

3.4 Matching SCRs to Mitigation Policies

Mitigation policies can be categorized into major types based on the risk issues related to Structure, Visibility, Resilience and Buffer etc. Structure issue refers to weak and risky supply chain topology and SCN redesign is a strategy about changing SC structures. Visibility issue mainly refers to the risks due to the un-availability of necessary information to aware the risks and the insight for sensing risks which are produced based on basic risk information. Buffer issue mainly refers to putting strategic redundancy in inventory and capacities in key locations of a supply chain for dealing with possible uncertainty and risks in future. Resilience issue is about improving flexibility and agility of the supply chain with existing resources. With the growing complexities of SCNs, resilience is considered as a more and more important strategy for SCR mitigation and CS approaches may play a more active role on investigating the resilience and survivability of SCNs.

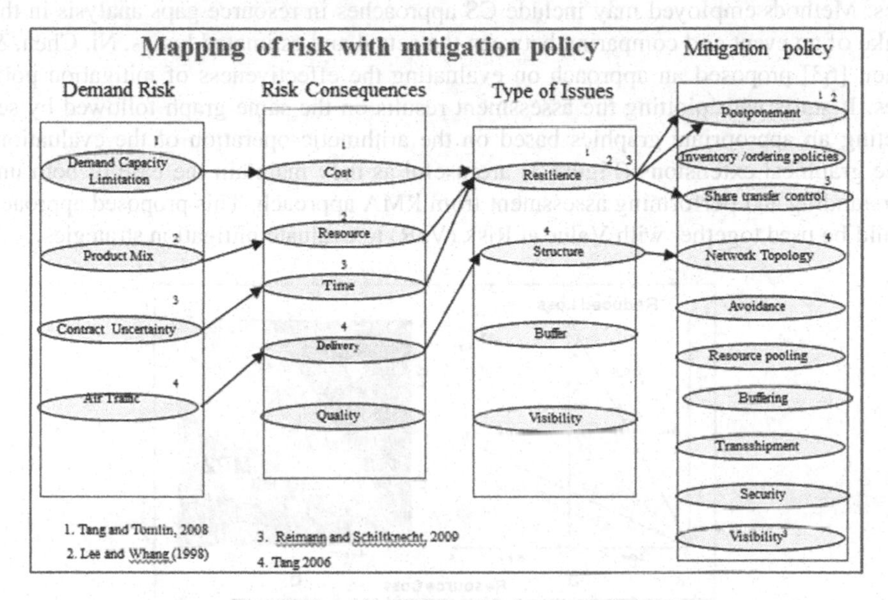

Fig. 2. Matching SCR to Mitigation Policies

The matching of risks to mitigation policies should be based on the consequences of the risks and not the scenario ([7]). A risk may cause negative consequences on performances, such as on cost, resource, time and product quality etc., and the selected mitigation policies should be applied to reduce the impacts. For example, if responsiveness is critical for a supply chain and the focus of risk mitigation should be mainly the impact to lead-time and the measure on reducing lead-time should be the main mitigation policies considered.

Based on the consequences of the risks, Figure 2 shows examples on mitigation policy matching considering risk type, consequence, issues and SCR mapping to policies. In case 1 (Tang and Tomlin [6]), the main risk is the delay of order confirmation, and it may incur high cost on discarding products that may be outdated. The mitigation policy is to postpone the final customization until the demand is confirmed. This requires the resilience policy in product customization. In case 4, Tang [13] introduces a case that air-traffic problem causes late delivery due to the problem in the airport.

The issue is a structure problem of the distribution network. So the mitigation policy can be multi-modal and alternative routes in the distribution network.

3.5 Evaluation of Mitigation Policy

Evaluation of the effectiveness of mitigation policies should go beyond economic losses and include also, but not restricted to, speed of recovery, extent of recovery and reduction in losses. One of the simplest and easily understood methods is to consider the cost-benefit ratio of the risk mitigation strategy. Rapid modeling and evaluation methods for assessing complex SCN scenario are needed in assessing mitigation policies. Methods employed may include CS approaches in resource gaps analysis in the wake of an event and comparing between the actual and estimated losses. Ni, Chen, & Chen [63] proposed an approach on evaluating the effectiveness of mitigation policies. It starts with plotting the assessment results on the same graph followed by selecting an appropriate graphics based on the arithmetic operation of the evaluation. The graphical extensions (Figure 3) are useful as they maintain the ease of both understanding and performing assessment from RMA approach. This proposed approach could be used together with Value at Risk (VaR) to evaluate mitigation strategies.

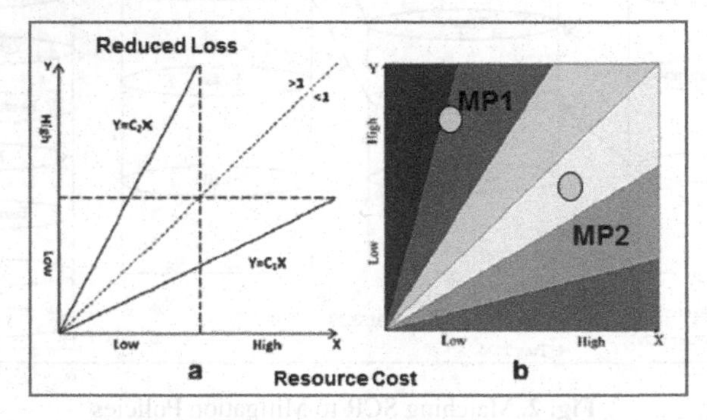

Fig. 3. Evaluation of risk mitigation policies

4 The Potential of CS Approaches for SCR Mitigation

CS focuses on how parts at a micro-level in a complex system affect emergent behaviour and overall outcome at the macro-level [64]. It is a bottom-up approach that attempts to understand how changes at the micro level and in the number of individuals can have emergent behaviours at the macro level. The increasing interests in CS are driven predominantly by new trends, challenges and demands in practical systems, e.g. economics and supply chain systems, where industries have the interests on how to design, build and manage systems as they increase in scale and connectivity.

Tang & Nurmaya [65] reviewed more than 200 papers on SCRM and identified that the trends on SCRM to be more on managing complexity, propagation, integration and networking with increasing scale and connectivity of the supply chains since 2006. So, new technologies for SCRM based on system complexities are more needed by industry for the purpose of preparing the supply chains for unforeseen crises and improving the flexibility and robustness of the supply chains. Li et al. [10] identified 5 major CS technologies for SCRM based on an investigation of the ScienceDirect Database, include dynamic systems, agent-based modelling and simulation, network theory, game theory and evolutionary computation.

Due to the limitation of traditional mathematical and simulation approaches in meeting the needs for modelling and analysing the complex supply chains and designing sustainable and self-adaptive supply chains, researchers have begun to use CS approaches for SCRs mitigation and relief. Agent-based approaches are one of the most used CS technologies. Giannakis & Louis [66] and Behdani et al. [24] proposed multi-agent based decision support systems for risk management of manufacturing and chemical supply chains. Sirivunnabood & Kumara [23] used an agent-based simulation to evaluate risk mitigation strategies for supplier risks. There are also applications of other CS approaches on SCR mitigation. Kumar [35] applied genetic algorithms, particle swarm optimisation and artificial bee colony in robust supply chain design and evaluation. Schmitt [53] used network theory approach. Zhao et al [57] and Zhu and Dou [58] proposed game theory model for risk mitigation.

However, the researches on applying CS to SCRM are still on its starting stage. There are still a lot of potentials on applying it to SCR mitigation.

First, how to make a supply chain resilient is a critical issue so that the SCN can adapt to uncertainties and disruptions with the available resources and acceptable costs. A potential to apply CS is to manage the increasing complexities in scale, connectivity, range, system and risks in supply chains. Traditional SCR mitigation approaches should be enhanced to increase supply chain resilience through the management of multiple variables.

Second, rapid modelling and evaluation of mitigation policies is a potential area for CS. Companies currently lack of technologies & tools for rapid modelling and evaluation of mitigation policies in complex SCNs and hence cannot take immediate and effective measures to mitigate the disruptions. Near real-time technologics are needed and CS technologies with rapid inference and dependency analysis capabilities would be good candidates for this purpose. In addition, supply chain with different structures, parameters and policies will present different performance behaviours in facing disruptions and CS technologies could play an extensive role for evaluating the effectiveness of the mitigation policies in these scenarios.

Third, CS technologies are needed to quantify disruption impacts and mitigate the variations to hasten the recovery from disruptions, for example, technologies responding to "catastrophe" events and recovery based on catastrophe theory and multi-objective optimization technologies. These technologies are also useful for contingency planning where leading companies deal with the SCRs by holding reserves. A challenge is how to mitigate risk by intelligently positioning and sizing the reserves without decreasing profits.

Finally, the validation of CS models for SCR mitigation is important for the deployment of the technologies. Although there is a growing awareness of SCRM by CS approaches and a number of conceptual models were proposed, effective validation of the models for practical application are still lacking.

5 Conclusions

The paper proposed a SCR mitigation framework based on a comprehensive review of SCR mitigation policies and studied the application potential of CS technologies to SCR mitigation.

Firstly it elaborated the challenges in SCR mitigation and clarified the objective of the research. Secondly, literatures on researches of SCR mitigation approaches and policies were reviewed. Thirdly, based on the reviews, a new SCR mitigation framework -IRMEI was proposed. Issues on risk identification, ranking, matching to mitigation policies, and the evaluation of mitigation policies were elaborated. Fourthly, the paper investigated the current states of CS technologies for SCR mitigation and identified the potentials of using CS technologies for SCR mitigation.

In future work, CS technologies will be developed and applied to SCR mitigation policy design, evaluation and validation. Especially, CS approaches will be used to capture the dynamic and complex nature of SCNs for the purpose of designing effective techniques to improve the resilience of the supply chains.

Acknowledgement. This work is supported by A*STAR TSRP funding, Singapore Institute of Manufacturing Technology and Nanyang Technological University.

References

1. Christopher, M.: Logistics and supply chain management. Financial Times/Prentice Hall (2010)
2. Wagner, S.M., Neshat, N.: Assessing the vulnerability of supply chains using graph theory. International Journal of Production Economics 126(1), 121–129 (2010)
3. De Souza, R., Goh, M., Meng, F.: A risk management framework for supply chain networks. TLI - Asia Pacific White Papers Series (2007)
4. Naraharisetti, P.K., et al.: From PSE to PSE2—Decision support for resilient enterprises. In: FOCAPO 2008 – Selected Papers from the Fifth International Conference on Foundations of Computer-Aided Process Operations, vol. 33(12), pp. 1939–1949 (2009)
5. Christopher, M.: Logistics and supply chain management: creating value-adding networks. Pearson Education (2005)
6. Tang, C., Tomlin, B.: The power of flexibility for mitigating supply chain risks. International Journal of Production Economics 116(1), 12–27 (2008)
7. Jüttner, U., Peck, H., Christopher, M.: Supply chain risk management: outlining an agenda for future research. International Journal of Logistics: Research and Applications 6(4), 197–210 (2003)
8. Trkman, P., McCormack, K.: Supply chain risk in turbulent environments—A conceptual model for managing supply chain network risk. International Journal of Production Economics 119(2), 247–258 (2009)
9. Sheffi, Y., Rice, J.: A supply chain view of the resilient enterprise. MIT Sloan Management Review 47(1) (2005)
10. Li, Z.P., Tan, P.S., Yee, Q.M., et al.: A Review of Complex Systems Technologies for Supply Chain Risk Management. In: IEEE International Conference on Systems, Man, and Cybernetics 2013, Manchester, UK (2013)

11. Christopher, M., Peck, H.: Building the resilient supply chain. The International Journal of Logistics Management 15(2), 1–14 (2004)
12. Tang, C.S.: Perspectives in supply chain risk management. International Journal of Production Economics 103(2), 451–488 (2006)
13. Tang, C.S.: Robust strategies for mitigating supply chain disruptions. International Journal of Logistics: Research and Applications 9(1), 33–45 (2006)
14. Lee, H.L., Billington, C.: Material management in decentralized supply chains. Oper. Res. 41(5), 835–847 (1993)
15. Blackhurst, J., Wu, T., O'grady, P.: Network-based approach to modelling uncertainty in a supply chain. International Journal of Production Research 42(8), 1639–1658 (2004)
16. Blackhurst, J., et al.: An empirically derived agenda of critical research issues for managing supply-chain disruptions. International Journal of Production Research 43(19), 4067–4081 (2005)
17. Thun, J.-H., Hoenig, D.: An empirical analysis of supply chain risk management in the German automotive industry. International Journal of Production Economics 131(1), 242–249 (2011)
18. Carvalho, H., et al.: Supply chain redesign for resilience using simulation. Computers & Industrial Engineering 62(1), 329–341 (2012)
19. Kull, T., Closs, D.: The risk of second-tier supplier failures in serial supply chains: Implications for order policies and distributor autonomy. European Journal of Operational Research 186(3), 1158–1174 (2008)
20. Oke, A., Gopalakrishnan, M.: Managing disruptions in supply chains: A case study of a retail supply chain. International Journal of Production Economics 118(1), 168–174 (2009)
21. Benyoucef, M., Canbolat, M.: Fuzzy AHP-based supplier selection in e-procurement. International Journal of Services and Operations Management 3(2), 172–192 (2007)
22. Swink, M., Zsidisin, G.: On the benefits and risks of focused commitment to suppliers. International Journal of Production Research 44(20), 4223–4240 (2006)
23. Sirivunnabood, S., Kumara, S.: Comparison of mitigation strategies for supplier risks: A multi agent-based simulation approach. In: IEEE/INFORMS International Conference on Service Operations, Logistics and Informatics, SOLI 2009. IEEE (2009)
24. Behdani, B., Adhitya, A.: Mitigating supply disruption for a global chemical supply chain - Application of agent-based modeling. In: Proceedings of the 11th International Symposium on Process Systems Engineering, Singapore (2012)
25. Doege, J., et al.: Risk management in power markets: The hedging value of production flexibility. European Journal of Operational Research 199(3), 936–943 (2009)
26. Swafford, P.M., Ghosh, S., Murthy, N.: The antecedents of supply chain agility of a firm: scale development and model testing. Journal of Operations Management 24(2), 170–188 (2006)
27. Cucchiella, F., Gastaldi, M.: Risk management in supply chain: a real option approach. Journal of Manufacturing Technology Management 17(6), 700–720 (2006)
28. Ben-Tal, A., et al.: Robust optimization for emergency logistics planning: Risk mitigation in humanitarian relief supply chains. Transportation Research Part B: Methodological 45(8), 1177–1189 (2011)
29. Whang, S., Lee, H.: Value of postponement. In: Product Variety Management, pp. 65–84. Springer (1998)
30. Yang, B., Yang, Y.: Postponement in supply chain risk management: a complexity perspective. International Journal of Production Research 48(7), 1901–1912 (2010)

31. Klibi, W., Martel, A., Guitouni, A.: The design of robust value-creating supply chain networks: A critical review. European Journal of Operational Research 203(2), 283–293 (2010)
32. Klibi, W., Martel, A.: Modeling approaches for the design of resilient supply networks under disruptions. International Journal of Production Economics 135(2), 882–898 (2012)
33. Pearson, M., Masson, R., Swain, A.: Process control in an agile supply chain network. International Journal of Production Economics 128(1), 22–30 (2010)
34. Baghalian, A., Rezapour, S., Farahani, R.Z.: Robust supply chain network design with service level against disruptions and demand uncertainties: A real-life case. European Journal of Operational Research 227(1), 199–215 (2013)
35. Kumar, S.K., Tiwari, M.K., Babiceanu, R.F.: Minimisation of supply chain cost with embedded risk using computational intelligence approaches. International Journal of Production Research 48(13), 3717–3739 (2009)
36. Rajaram, K., Tang, C.S.: The impact of product substitution on retail merchandising. European Journal of Operational Research 135(3), 582–601 (2001)
37. Bitran, G.R., Dasu, S.: Ordering policies in an environment of stochastic yields and substitutable demands. Operations Research, 999–1017 (1992)
38. Chod, J., Rudi, N.: Resource flexibility with responsive pricing. Operations Research 53(3), 532–548 (2005)
39. Lariviere, M.A., Porteus, E.L.: Selling to the newsvendor: An analysis of price-only contracts. Manufacturing & Service Operations Management 3(4), 293–305 (2001)
40. Cachon, G.P.: Supply chain coordination with contracts. In: Handbooks in Operations Research and Management Science, vol. 11, pp. 229–340 (2003)
41. Emmons, H., Gilbert, S.M.: Note. The role of returns policies in pricing and inventory decisions for catalogue goods. Management Science 44(2), 276–283 (1998)
42. Dana Jr., J.D., Spier, K.E.: Revenue sharing and vertical control in the video rental industry. The Journal of Industrial Economics 49(3), 223–245 (2001)
43. Tsay, A.A., Lovejoy, W.S.: Quantity flexibility contracts and supply chain performance. Manufacturing & Service Operations Management 1(2), 89–111 (1999)
44. Li, X., Marlin, T.E.: Robust supply chain performance via Model Predictive Control. Computers & Chemical Engineering 33(12), 2134–2143 (2009)
45. Ben-Tal, A., Boaz, G., Shimrit, S.: Robust multi-echelon multi-period inventory control. European Journal of Operational Research 199(3), 922–935 (2009)
46. Tarantino, A.: Governance, Risk, and Compliance Handbook: Technology, Finance, Environmental, and International Guidance and Best Practices. John Wiley & Sons (2008)
47. Sun, J., Matsui, M., Yin, Y.: Supplier risk management: An economic model of P-chart considered due-date and quality risks. International Journal of Production Economics 139(1), 58–64 (2012)
48. Speier, C., et al.: Global supply chain design considerations: Mitigating product safety and security risks. Journal of Operations Management 29(7-8), 721–736 (2011)
49. Wang, J.: Mitigation strategies on scale-free networks against cascading failures. Physica A: Statistical Mechanics and its Applications 392(9), 2257–2264 (2013)
50. Braunscheidel, M.J., Suresh, N.C.: The organizational antecedents of a firm's supply chain agility for risk mitigation and response. Journal of Operations Management 27(2), 119–140 (2009)
51. Sawik, T.: Selection of resilient supply portfolio under disruption risks. Omega 41(2), 259–269 (2013)

52. Ji, G., Zhu, C.: Study on supply chain disruption risk management strategies and model. In: 2008 International Conference on Service Systems and Service Management. IEEE (2008)
53. Schmitt, A.J.: Strategies for customer service level protection under multi-echelon supply chain disruption risk. Transportation Research Part B: Methodological 45(8), 1266–1283 (2011)
54. Schmitt, A.J., Singh, M.: A quantitative analysis of disruption risk in a multi-echelon supply chain. International Journal of Production Economics 139(1), 22–32 (2012)
55. Ratick, S., Meacham, B., Aoyama, Y.: Locating backup facilities to enhance supply chain disaster resilience. Growth and Change 39(4), 642–666 (2008)
56. Kleindorfer, P.R., Saad, G.H.: Managing disruption risks in supply chains. Production and Operations Management 14(1), 53–68 (2005)
57. Zhao, R., et al.: Using game theory to describe strategy selection for environmental risk and carbon emissions reduction in the green supply chain. Journal of Loss Prevention in the Process Industries 25(6), 927–936 (2012)
58. Zhu, Q.-H., Dou, Y.-J.: Evolutionary game model between governments and core enterprises in greening supply chains. Systems Engineering-Theory & Practice 27(12), 85–89 (2007)
59. Sun, Y., Xu, X., Hua, Z.: Mitigating bankruptcy propagation through contractual incentive schemes. Decision Support Systems 53(3), 634–645 (2012)
60. Barroso, A., Machado, V., Machado, V.C.: Supply Chain Resilience Using the Mapping Approach. In: Li, P. (ed.) Supply Chain Management. InTech (2011)
61. Li, Z.P., Yee, Q.M.G., Tan, P.S., Lee, S.G.: An Extended Risk Matrix Approach for Supply Chain Risk Assessment. In: IEEE international Conference on Industrial Engineering and Engineering Management (IEEM), Bangkok, December 10-13 (2013)
62. Zhang, A.N., Wagner, S.M., Goh, M., Terhorst, M., Ma, B.: Quantifying Supply Chain Disruption Risk Using VaR. In: The IEEE International Conference on Industrial Engineering and Engineering Management, Hong Kong (2012)
63. Ni, H., Chen, A., Chen, N.: Some extensions on risk matrix approach. Safety Science 48(10), 1269–1278 (2010)
64. McElroy, M.W.: Integrating complexity theory, knowledge management and organizational learning. Journal of Knowledge Management 4(3), 195–203 (2000)
65. Tang, O., Nurmaya Musa, S.: Identifying risk issues and research advancements in supply chain risk management. International Journal of Production Economics 133(1), 25–34 (2011)
66. Giannakis, M., Louis, M.: A multi-agent based framework for supply chain risk management. Journal of Purchasing and Supply Management 17(1), 23–31 (2011)

An Adaptive Clustering and Re-clustering Based Crowding Differential Evolution for Continuous Multi-modal Optimization

Soham Sarkar[1], Rohan Mukherjee[2], Subhodip Biswas[2], Rupam Kundu[2], and Swagatam Das[3]

[1] Department of Electronics and Communication Engineering, RCC Institute of Information Technology, Kolkata-15, WB, India
sarkar.soham@gmail.com

[2] Department of Electronics and Telecommunication Engineering, Jadavpur University, Kolkata-32, WB, India
{rohan.mukherjii,sub17was,rupam2422}@gmail.com

[3] Electronics and Communication Sciences Unit, Indian Statistical Institute, Kolkata-108, WB, India
swagatam.das@isical.ac.in

Abstract. In real-life a particular system, operating under a given set of conditions, may need to switch other set of conditions due to change in physical condition or failure in its existing state. Niching techniques facilitates in such situations by tracking multiple optima (solutions). When integrated with Evolutionary Algorithms (EAs), they seek parallel convergence of population members to find multiple solutions to a problem (landscape) without loss in optimality. In this paper an effective new grouping strategy namely adaptive clustering and re-clustering (ACaR) is proposed based on Fuzzy c-means clustering technique and is integrated with a hybrid of crowding niching technique and a real-parameter optimizing algorithm called Differential Evolution (DE). The performance of the proposed ACaR-CDE algorithm has been evaluated on different niching benchmark problems with diverse characteristics ranging from simple objectives to complex composite problems and compared with other published state-of-the-art niching algorithms. From experimental observation, we observe that the proposed strategy is apt in restraining solutions within its local environment, typically applicable to niching environment.

Keywords: Clustering, Differential Evolution, Fuzzy C-Means, niching.

1 Introduction

The generic term "niching" [1] [2] refers to a method of finding and preserving multiple stable niches, or favorable parts of the solution space possibly around multiple solutions (global and local optima), so as to prevent convergence to a single solution [3]. niching methods rely on population diversity since it is highly critical in multimodal landscapes and multi-criteria cases [1]. Popular niching techniques includes *crowding* [6], *speciation* [7], *fitness sharing* [8] [9] , *clearing* [10] and Restricted Tournament Selection

H. Handa et al. (eds.), *Proc. of the 18th Asia Pacific Symp. on Intell. & Evol. Systems – Vol. 1*, Proceedings in Adaptation, Learning and Optimization 1, DOI: 10.1007/978-3-319-13359-1_29

[11]. For more details on niching and related applications interested readers can refer to the survey by [12].

DE [5] has gained wide acceptance in the EA community owing to its simple yet robust nature and has been applied to solve a plethora of real-life problems. This performance can be attributed to the stochastic differential nature of mutation i.e. using difference vector between randomly selected population samples. This leads to basin transfer of individuals where members with higher fitness tend to attract other inferior members, thereby inducing hill climbing behavior [13]. Thus an external implementation of similarity based grouping techniques is cardinal to restrict the distance between offspring and the parents in niching algorithms.

A popular modification is Crowding DE (CDE) as proposed by Thomson *et al.* [14] by which the offspring, as generated by DE, competes with the nearest member of the population to preserve diversity and positional information within the population. Though CDE is an improvement upon classical DE, it fails to adapt itself when the fitness landscape has multiple global peaks, or involves high dimensionality. It is to be noted CDE does nothing to restrict the internal stochastic nature of mutation in DE thereby disrupting localization.

Motivated by these observation, we propose a *Adaptive Clustering and Re-clustering* theory for improved selection of solution groups. The grouping of individuals not only ensures close proximity (Euclidean distance) among each other, but also assures generation of off-springs within the domain of the parent pool. Selection of proper clustered local environment for each solution is possible if an efficient grouping can be done. Our analysis is based on an adaptive setting of the entire population. It‘s proper implementation requires determination of a grouping strategy in which each solution selects the best group to which it may belong for maximum favorable neighbors. We have used *Fuzzy c-means* algorithm [16] with decisive cluster selection by the solutions on basis of membership function and variable cluster size for neighbourhood determination. This decisively restricts the population to a more efficient local search.

A robust clustering used in ACaR-CDE adaptively sets the required number of clusters at each stage of the algorithm. The clustered populations are lent to efficient optimization using CDE for tracking multiple peaks. Later it has been experimentally verified that ACaR-CDE is efficient in retaining diversity in population, maintaining an adaptive cluster system, performing a restricted local search and is an efficient optimizer for complex niching landscapes.

The remainder of the paper is organized as follows. Section 2 revisits various niching techniques and their applications in evolutionary computation. In Section 3 classical DE and CDE have been discussed. Section 4 introduces the proposed clustering and re-clustering theory. In Section 5 the Experimental Setup is shown and the numerical results are discussed in Section 6. Lastly Section 7 concludes the paper.

2 Clustered Neighborhood Based Crowding Differential Evolution

2.1 Previous Approaches of Neighborhood Selection and Comparison

Earlier techniques of locality selection focused on fixed parameters for restricting mutation of parent vectors within its local environment. This confines the effectiveness of the algorithm to regulation of its parametric set. It is impossible to presume the nature of an unknown function and such sets can seldom be effective as a black-box optimizer. Since the neighbourhood of each solution being static, the algorithm cannot vary with the progress of the search algorithm. Consequently poor efficiency results in varied settings of data problems. The eventual purpose of niching to retain solutions in restricted local search domains is thus compromised. For complex landscapes with high dispersion metric (high concentration of local optima around global ones) it is difficult to reach the optimum as peak density and its corresponding specifications like height, width vary randomly over space. The individuals are in general attracted to positions in the basin of wider optimum, inhibiting multimodal optimization. Hence the problem of mutation over larger distance persists. The idea is depicted in Fig. 1.

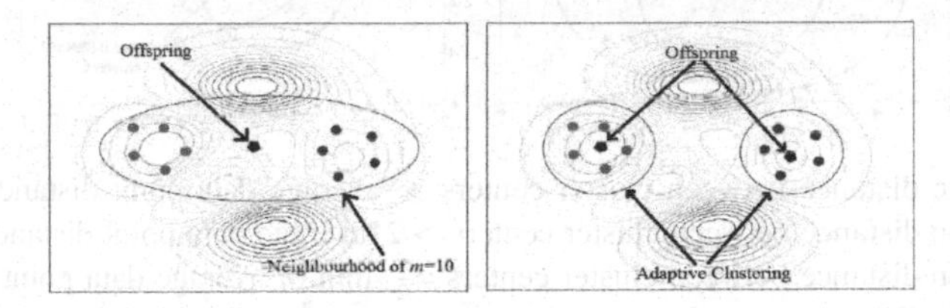

Fig. 1. Advantage of clustered neighborhood selection over neighborhood

This scheme of generating mutant groups ensures that local environments are selected for every individual adaptively in each stage by considering the current state of other members in the population. To determine the membership of a solution in local neighborhood of other, comparison with all members of the population is observed to determine the current stage of optimization. For example let us assume a test solution X whose clustered locality is to be considered. Let another solution Y is considered to be in close proximity to X and thus in its neighborhood. This proximity is determined by comparison with other solutions which are at relatively greater distance from X as compared to Y. This comparative analysis is critical for optimization in multimodal surfaces with high dispersion metrics where adaptive setting of parameters ensure proper locality selection.

2.2 Adaptive Clustered Neighborhoods

The population has to be grouped such that, at any instance it can have an adjustable number of clusters with variable number of constituting solutions. In ACaR-CDE the selection of a cluster for any individual is based on a comparative analysis with all other solutions. The population is congregated with the optimum number of clusters suited for distributed optimization which constitutes solutions best suited for it. Thus the population can adaptively select clustered neighborhoods. The member of the clusters can then independently mutate among themselves and operate within itself for tracking multiple optima within its territory.

Fuzzy c-Means [16] is a data clustering technique in which each member has a definite membership function to belong to a certain group or cluster and is preferred for high dimensional search space. Each data point can belongs to multiple data clusters

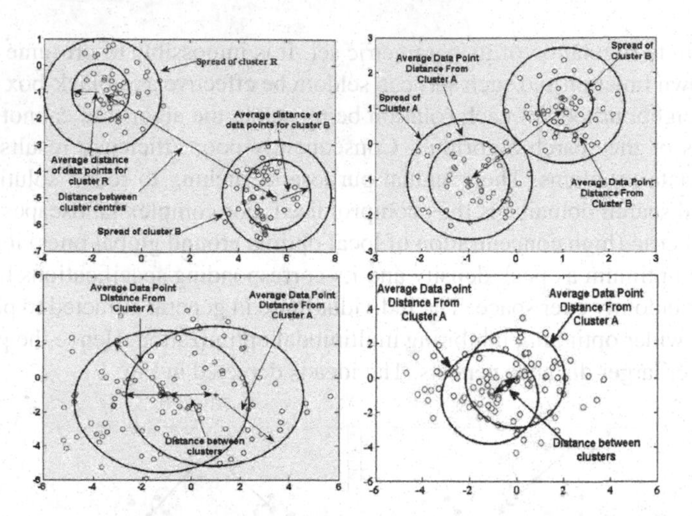

Fig. 2. (a): When distance between cluster centers $\gg$ average data point distance from cluster center, (b): When distance between cluster centers > 2*average data point distance from cluster center, (c): When distance between cluster centers $=\frac{1}{2}*$min of average data point distance from cluster center, (d): When distance between cluster centers $< \frac{1}{2}*$min of average data point distance form cluster center.

but with a different degree of belonging or membership to each. A data point in close proximity to a cluster-centroid will have a high degree of belonging to belong to that cluster than a solution that is far away from that cluster-centroid. Thus selection of a data point in data clusters follow a comparative analysis with rest of its members. Consequently membership function for the data points are set for all clusters. The clustering technique is implemented such that a data point gets selected into the cluster with which it shares highest degree of membership. Maximum number of clusters *NC* is predefined and adaptation over this cluster number is implemented.

Proper adaptation is effectuated via two strategies- *a cluster survival by virtue of maximum fuzzy membership function* and *a distance based cluster merging technique* . Initially data points are distributed in different regions of search space. In later stages of the algorithm (that follows a niching techniques) data points are found to congregate as groups in proximity to various basins of possible optimum. Some cluster centers defined by fuzzy c-means may be redundant. That is the total number of clusters defined is more than the number of optima located by the algorithm. The fuzzy membership function defined for that cluster may not be maximum for any data point. In that case that cluster is identified to be discarded and NC (number of clusters) is decreased by one. The next stage of the algorithm proceeds with this reduced cluster number.

Another scenario that can arise in a niching landscape. A redundant cluster center can have some data points belonging to it by membership criterion. In other words, two cluster centers occupy the same position in the landscape and share data points that would have actually been in a single cluster. The data points get distributed between two cluster centers hampering the local search. The distance based cluster merging technique helps in merging these two data clusters that should have constituted a single cluster. The decision is made on a comparison of distance between two cluster centers and average distance of constituting data points from each cluster center. Fig 2 depicts the various cases possible. Only when distance between two cluster centers is less than half of the average distance between cluster center and constituting data points, the two clusters are regarded redundant and hence merged. Fig 2(a) shows the ideal case when data cluster center distance is much greater, Fig 2(b) shows when two clusters are close but clusters are unique, Fig 2(c) shows the critical case where average data point distance is half the cluster center distance. Populations can then be suitably merged as clear from the picture. Fig 2(d) shows a more apt case for cluster merging. Thus though a maximum number of possible clusters NC_{max} are declared at the start of the algorithm, the number of clusters NC at any iteration may adapt to a value below the maximum i.e. $NC < NC_{max}$.

By ACaR-CDE algorithm the data cluster number gradually decreases and adapts to the number of possible global optima. Ideally actual cluster number at any instant should be the number of optima actually present in the landscape. There are also some additional advantage inherent to this algorithm. Clusters may have variable density of

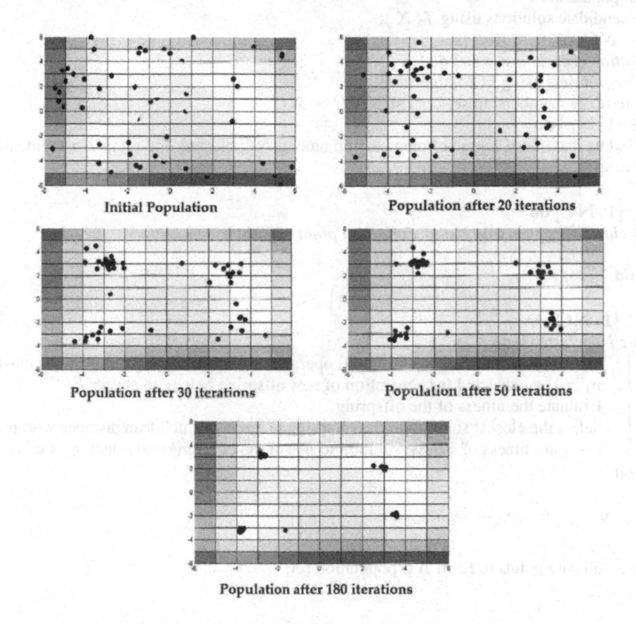

Fig. 3. Solution distribution with increasing iteration for Himmelblau's function

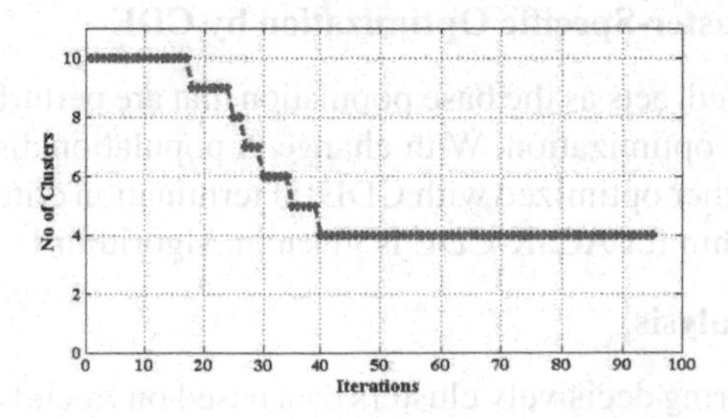

Fig. 4. Decreasing cluster size with iterations for Himmelblau's Function

points i.e. some clusters might be dense whereas some clusters may have very few members. This adds to the adaptive nature of cluster. At the end of each generation the solutions are de-clustered to belong in the data space independently. The beginning of the next generation is marked by re-clustering of these data point set. This eliminates any chance of solutions to diverge away from clusters and consequently hampering mutation in CDE with larger distance between parents. A re-clustering strategy can ensure maintenance of the local search process throughout the algorithm. The convergence profile along with the varying cluster number are depicted for Himmelblau's Function [3] [15] in Fig 3 and Fig 4. These clusters handle different cluster size and hence an adaptive clustering is achieved that has suited this problem accurately.

Algorithm 1. Basic Structure Algorithm for ACaR-CDE

```
input :
i) Control parameters: Maximum cluster number (NC_max), Crowding Factor (CF = 1) [14], threshold level (ε).
ii) Objective function f(X_i), dimensionality (n), maximum FEs (max_F Es) and population size Np.
output: Final population X and functional value f(X).
1  begin
2      Initialize trial population ;
3      Evaluate the candidate solutions using f(X);
4      Set N_C ⟵ NC_max;
5      while termination criteria is not met do
6          Form N_C clusters using FCM algorithm;
7          Compute fuzzy membership set U of size NP × NC;
8          for i ∈ [1, Np] do
9              Find its maximum degree of membership among NC clusters and set its cluster identity to that
               cluster;
10         end
11         for i ∈ [1, NC] do
12             if cluster i is empty || contains one data point only then
13                 N_C ⟵ N_C − 1 ;
14             end
15         end
16         for k ∈ [1, NC] do
17             for i ∈ cluster k do
18                 Generate an offspring u_i using DE operators with parents selected from the cluster i.e.
                   r1, r2, r3 selected for generation of new offspring belong to cluster k;
19                 Evaluate the fitness of the offspring;
20                 Select the closest solution to the offspring in terms of Euclidean distance within the cluster;
21                 Compare fitness of closest solution to that of the offspring and select u_i if it is better;
22             end
23         end
24         X' ⟵ X' ∪ x';
25         Evaluate x';
26         De-cluster all data points to form Np population set;
27     end
28 end
```

2.3 Multimodal Cluster-Specific Optimization by CDE

The clusters thus formed, acts as the base population that are perturbed by Crowding DE [14] for efficient local optimization. With change in population distribution population is re-clustered and further optimized with CDE till termination criterion is satisfied. The basic structure algorithm for ACaR-CDE is given in Algorithm 1.

2.4 Complexity Analysis

Fuzzy C-means clustering decisively clusters data based on Euclidean distance from the cluster centroids. Thus the order of complexity [16] is $\mathbf{O}(n \times c^2 \times d \times i)$ where i is the number FCM over the entire dataset, n is the number of data points, c is the number to be formed and d is the dimension of the problem.

Clustering in ACaR-CDE forms data groups which are further optimized by CDE and the complexity of the problem stands out to be $\mathbf{O}({n^2}_{max}d)$ where n_{max} denotes the maximum cardinality of the clusters formed and $n_{max} < n$ and usually it is less than n_{max}/c where c is the number of clusters formed. Thus the total complexity of ACaR-CDE is $\mathbf{O}(nc^2di+{n^2}_{max}d)$.

3 Experimental Setup

3.1 Test Functions

For evaluating the performance of ACaR-CDE and other niching algorithms, two different tests are conducted. The problem formulation is adapted from [15], [17] and [18]. Experiment 1 constitutes two sets of benchmark problems widely used for niching problems. The benchmarks used for this test is shown in Table 1. Set1 comprises of 14 basic functions of varied nature while set 2 has 15 challenging composite functions.

Experiment two has different sets of benchmark problems with different mode of performance evaluation. The test functions, as shown in Table 4, along with results, experimental procedure, experimental objective and performance measure can be found in details in [18].

Population Size, Maximum Number of Evaluations and Parameter Setting. The required level of accuracy, the allowed maximum number of functional evaluations and population size for function Set 1 are shown below in Table 3. Maximum number of

Table 1. Test Functions for Experiment One (CF: Composition Function)

Test Function Set 1 [15]		Test Function Set 2 [15], [18]	
Test Function name	No. of global optima / Dimension	Test Function name	No. of global optima/ Dimension
E1-F1: Two-peak trap	1/1	E1-F15: CF 1	8/10
E1-F1: Central two-peak trap	1/1	E1-F16: CF 2	6/10
E1-F3: Five-uneven-peak trap	2/1	E1-F17: CF 3	6/10
E1-F4: Equal Maxima	5/1	E1-F18: CF 4	6/10
E1-F5: Decreasing Maxima	1/1	E1-F19: CF 5	6/10
E1-F6: Uneven Maxima	5/1	E1-F20: CF 6	6/10
E1-F7: Uneven Decreasing Maxima	1/1	E1-F21: CF 7	6/10
E1-F8: Himmelblau's function	4/2	E1-F22: CF 8	6/10
E1-F9: Six-hump camel back	2/2	E1-F23: CF 9	6/10
E1-F10: shekels foxholes	1/2	E1-F24: CF 10	6/10
E1-F11: 2D inverted Shubert function	18/2	E1-F25: CF 11	8/10
E1-F12: 1D inverted Vincent function	6/1	E1-F26: CF 12	8/10
E1-F13: 2D inverted Vincent function	36/2	E1-F27: CF 13	10/10
E1-F14: 3D inverted Vincent function	216/3	E1-F28: CF 14	10/10
		E1-F29: CF 15	10/10

clusters can be set at approximately $NP/10$ but as far as optimization for a certain level of accuracy within a fixed number of functional evaluations is concerned, for best results maximum number of clusters can be set as shown in Table 3. It can be tuned by running algorithm once or twice by an approximate knowledge on number of peaks in the landscape. DE parameters are set at:

Performance Measure. To compare the performance of different niching algorithms on the above mentioned test functions for Experiment 1 three strategies of performance evaluation are considered. Results are averaged over 25 independent runs:

1. Success Rate or percentage of runs in which an algorithm can detect all the global peaks within the specified level of accuracy (ϵ) as given in Table 1. Detecting a peak within a level of accuracy signifies Euclidean distance between each global peak and a solution to be less than ϵ.
2. Average number of peaks found for each environment. Level of accuracy remains as section (B.I).
3. Success Rate in detecting local optima. $E1 - F1, E1 - F2, E1 - F3, E1 - F5, E1 - F10$ are considered for this experiment as in [15]. $\epsilon$and other conditions same as in section (B.I).
4. Average number of peaks detected for simultaneous detection of local and global peaks. Functions same as in section (B.III) and criterion same as section (B.II).

Table 2. Test Functions for Experiment Two [15]

Test Function name	Number of Global peaks/ Dimension	Test Function name	Number of Global peaks/ Dimension
E2-F1: Waves	10/2	E2-F8: Shubert	18/2
E2-F2: Six-hump camel back	2/ 2	E2-F9: Ackley	1/2
E2-F3: Sphere	1/ 2	E2-F10:Michalewicz	1/2
E2-F4: Shifted Rastrigin	1/ 2	E2-F11: Ursem F1	1/2
E2-F5: Rotated Hybrid Composition Function	1/ 2	E2-F12: Ursem F3	1/2
E2-F6: Rescaled Six-hump camel back	2/2	E2-F13: Ursem F4	1/2
E2-F7: Branin RCOS	3/2		

Algorithms compared are NCDE [15], NSDE [15], NShDE [15], CDE [14], SDE [3], ShDE [3], FERPSO [17], SPSO [18], r2pso [17], r3pso [17], r2pso-lhc [17], r3pso-lhc [17], CMA-ES [18], SCMA-ES [19].

For the set of functions considered in Experiment 2, performance evaluation criterion is based on peak accuracy and distance accuracy [3], [15], [18].

1. *Peak accuracy*: It is defined as the accuracy of an algorithm in simultaneous detection of multiple peaks in terms of difference in fitness of peaks and closest fitness of a solution for all peaks to be considered [15]. The measure can be erroneous if multiple peaks are located close enough in the Euclidean space or they are of same fitness values. It is given by:

$$peak\_accuracy = \sum_{i=1}^{NOP} |f(peaks_i) - f(x)|, \text{ where } NOP{=}number\ of\ peaks$$

Table 3. Parameter Setting for Experiment 1

Function	ϵ	Maximum cluster number	Population size	No. of function evaluations
E1-F1	0.05	5	50	10000
E1-F2	0.05	5	50	10000
E1-F3	0.05	5	50	10000
E1-F4	10^{-6}	5	50	10000
E1-F5	10^{-6}	5	50	10000
E1-F6	10^{-6}	5	50	10000
E1-F7	10^{-6}	5	50	10000
E1-F8	0.0005	5	50	10000
E1-F9	10^{-6}	5	50	10000
E1-F10	10^{-6}	5	50	10000
E1-F11	0.05	30	250	100000
E1-F12	10^{-4}	10	100	20000
E1-F13	10^{-3}	40	500	200000
E1-F14	0.001	150	1000	400000

2. *Distance accuracy*: It is a more reliable measure of niching algorithms. It is given by the sum of the distance of closest solutions to different peaks in the niching environment. It is calculated by:

$$distance\_accuracy = \sum_{i=1}^{NOP} \| \boldsymbol{peaks}_i - \boldsymbol{x}) \|$$
$$\text{, where } NOP{=}number\ of\ peaks$$

Algorithms compared are NCDE [15], NSDE [15], NShDE [15], CDE [14], SCMA-ES [19],TSC [18], SCGA [7], DFS [9], TSC2 [18].

4 Experimental Results

4.1 Experiment 1

Results and Discussions: Test Function Set 1. Table 4 and 5 show the result for average number of peaks detected and success rates of contending algorithms for set 1 problems respectively. Ranks are denoted along with the results and the calculated total ranks have also been shown. Results indicate that algorithms NCDE, NShDE and ACaR-CDE can most accurately detect maximum global peaks. NSDE suffers in cases where there are numerous global optima in close proximity. All neighborhood based schemes as well as ACaR-CDE have considerable success rate accuracy. NShDE enjoys superiority in cases where there is significant number of global optima because of its sharing scheme. ACaR-CDE in both the above tests has been ranked better than all other state-of-the-arts including Neighborhood techniques. ACaR-CDE has best results followed by NCDE. NShDE has equally good characteristics in these experiments on $2-D$ landscapes. NSDE on the other hand suffers in last two test functions with large number of closely spaced optima mainly because species seeds remove redundant solutions in close proximity marked by radius parameter. Evidently the results conclude that CDE variants are best optimizers in 2-D space followed by SDE and ShDE variants.

Table 4. Average number of peaks found and Rank for Experiment 1: Set 1

Fnc	NCDE	NShDE	NSDE	CDE	ShDE	SDE	FER-PSO	SPSO	r2pso	r3pso	r2pso-lhc	r3pso-lhc	SCMA-ES	ACaR-CDE
E1-F1	1	1	1	1	1	1	0.72	0.48	0.76	0.84	0.56	0.6	1	1
	4.5	4.5	4.5	4.5	4.5	4.5	11	14	10	9	13	12	4.5	4.5
E1-F2	1	1	1	1	1	1	1	0.44	0.88	0.96	0.44	0.56	1	1
	5	5	5	5	5	5	5	13.5	11	10	13.5	12	5	5
E1-F3	2	2	2	2	2	1.96	0.8	0.24	0.48	0.6	0.48	0.6	2	2
	4	4	4	4	4	8	9	14	12.5	10.5	12.5	10.5	4	4
E1-F4	5	5	5	3.84	3.28	4.72	4.84	4.88	4.92	4.88	5	4.92	0.04	5
	3	3	3	12	13	11	10	8.5	6.5	8.5	3	6.5	14	3
E1-F5	1	1	1	0.72	0.44	1	1	1	1	1	1	1	1	1
	6.5	6.5	6.5	13	14	6.5	6.5	6.5	6.5	6.5	6.5	6.5	6.5	6.5
E1-F6	5	5	5	3.96	3.28	4.6	5	4.92	4.88	4.72	4.92	4.88	0	5
	3	3	3	12	13	11	3	6.5	8.5	10	6.5	8.5	14	3
E1-F7	1	1	1	0.6	0.4	1	1	1	1	1	1	1	0.96	1
	6	6	6	13	14	6	6	6	6	6	6	6	12	6
E1-F8	4	3.92	4	0.32	0.16	3.72	3.68	0.84	2.92	2.76	3	3.12	3.44	4
	2	4	2	13	14	5	5	12	10	11	9	8	7	2
E1-F9	2	2	2	0.04	0.04	2	1.96	0.08	1.44	1.56	1.56	1.48	2	2
	3.5	3.5	3.5	13.5	13.5	3.5	7	12	11	8.5	8.5	10	3.5	3.5
E1-F10	1	0.96	1	0.52	0.96	0.32	1	0.56	0.88	0.76	0.72	0.6	0.04	1
	2.5	5.5	2.5	12	5.5	13	2.5	11	7	8	9	10	14	2.5
E1-F11	18	18	18	17.7	16.56	12.4	17.4	8.52	15.2	15.6	15.1	16.2	2.16	18
	2.5	2.5	2.5	5	7	12	6	13	10	9	11	8	14	2.5
E1-F12	5.8	5.88	5.84	5.56	5.6	4.88	5.6	5.52	5.16	5.36	5.28	1.52	6	
	3	2	4	7	5.5	13	9.5	5.5	8	12	9.5	11	14	1
E1-F13	35.9	35.96	30.6	33.8	35.92	22.8	23.6	25.7	21.8	22.2	22.5	23.1	1.4	33.96
	3	1	6	5	2	10	8	7	13	12	11	9	14	4
E1-F14	179	198.9	84.28	152	197.8	50.6	68.6	70.1	40.6	45.4	42.2	43.3	0.04	185.4
	4	1	6	5	2	9	8	7	13	10	12	11	14	3
Total Rank	52.5	51.5	58.5	124.5	117	117	96.5	126.5	133	131	130	129	140.5	50.5

4.2 Results and Discussions: Test Function Set 2

Benchmark problems for set 2 are highly complex with higher dimension hybrid composite functions and poses a challenge. Average number of peaks detected for these complex benchmarks are shown in Table 6. Results establish the difference between SOAs and our proposed algorithm. SOAs other than neighborhood variants fail considerably in these cases. Except $E1-CF3$, results of ACaR-CDE mark that adaptive clustering and re-clustering collectively restrict local mutation even when problem demands parallel convergence. Unlike previous case, NSDE ranks second in these problem which indicate that SDE variant was better suited for these $10-D$ cases.

However use of our clustering technique improvises CDE such that it even enjoys superiority over SDE optimizers.On the other hand, NShDE fail in these complex optimizing scenarios and results are far behind other Neighborhood techniques. NCDE also fail to perform well mainly because restriction of mutation in these high dimensional problems was not totally apt and narrow optima were missed. ACaR-CDE could locate more optima as cluster formation ensured more exploitation in different domains and narrow spaces were covered. As these problems even constitute more challenging peaks and some optima were even narrower and located in remote scenarios they could not be located under given accuracy. However relaxing accuracy limit we observed that some optimum in some scenarios were discovered by the solutions. Other SOAs also fail to locate these optima. As evident from the results again ACaR-CDE proves to be more

Table 5. Success rate (in %) and Rank found for Experiment 1: Set 1

Fnc	NCDE	NShDE	NSDE	CDE	ShDE	SDE	FER-PSO	SPSO	r2pso	r3pso	r2pso-lhc	r3pso-lhc	SCMA-ES	ACaR-CDE
E1-F1	100 4.5	100 4.5	100 4.5	100 4.5	100 4.5	100 4.5	72 11	48 14	76 10	84 9	56 13	60 12	100 4.5	100 4.5
E1-F2	100 5	100 5	100 5	100 5	100 5	100 5	100 5	44 13.5	88 11	96 10	44 13.5	56 10	100 5	100 5
E1-F3	100 4	100 4	100 4	100 4	100 4	96 8	20 9	4 13.5	8 11	8 11	4 13.5	8 11	100 4	100 4
E1-F4	100 3	100 3	100 3	28 12	4 13	72 11	84 10	88 8.5	92 6.5	88 8.5	100 3	92 6.5	0 14	100 3
E1-F5	100 6.5	100 6.5	100 6.5	72 13	44 14	100 6.5	100 6.5	100 6.5	100 6.5	100 6.5	100 6.5	100 6.5	100 6.5	100 6.5
E1-F6	100 3	100 3	100 3	28 12	8 13	60 11	100 3	92 7	88 9	72 10	92 7	92 7	0 14	100 3
E1-F7	100 6	100 6	100 6	60 13	40 14	100 6	100 6	100 6	100 6	100 6	100 6	100 6	96 12	100 6
E1-F8	100 2	92 4	100 2	0 13	0 13	72 5.5	72 5.5	0 13	28 8.5	24 10.5	28 8.5	24 10.5	44 7	100 2
E1-F9	100 3	100 3	100 3	0 13	0 13	100 3	96 7	0 13	56 9.5	60 8	56 9.5	52 11	100 3	100 3
E1-F10	100 2.5	96 5.2	100 2.5	52 12	96 5.5	32 13	100 2.5	56 11	88 7	76 8	72 9	60 10	4 14	100 2.5
E1-F11	100 2.5	100 2.5	100 2.5	72 5	28 7	0 13	52 6	0 13	4 10	4 10	4 01	20 8	0 13	100 2.5
E1-F12	84 3.5	88 2	84 3.5	56 9.5	68 6.5	48 12.5	60 8	72 5	68 6.5	56 9.5	52 11	48 12.5	0 14	100 1
E1-F13	88 3	96 1	24 4.5	8 6	92 2	0 10.5	0 10.5	0 10.5	0 10.5	0 10.5	0 10.5	0 10.5	0 10.5	24 4.5
E1-F14	0 7.5	0 7.5	0 7.5	0 7.5	0 7.5	0 7.5	0 7.5	0 7.5	0 7.5	0 7.5	0 7.5	0 7.5	0 7.5	0 7.5
Total Rank	62.5	58	57.5	129.5	122	117	97.5	151	128.5	133	136	129	129	55

apt to handle complex landscapes as defined by this benchmark suite than NCDE and CDE. It ranks best among all other state-of-the arts.

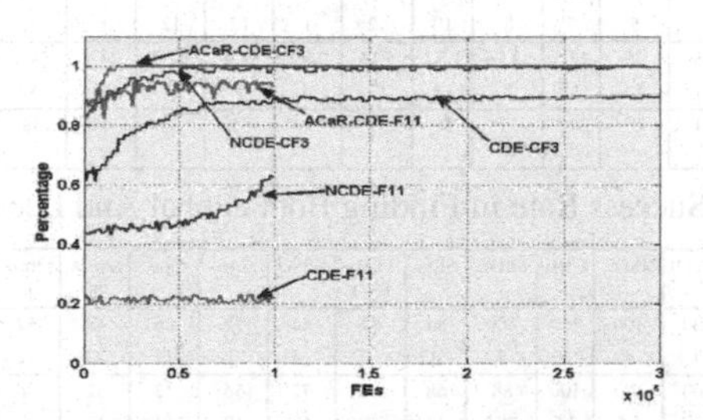

Fig. 5. Neighborhood test on test Functions E1-F11 and E1-CF3

4.3 Accuracy in Detecting Global as Well as Local Peaks

Results to compare the performance of ACaR-CDE with other state-of-the-arts for simultaneous detection of global as well as local peaks in multimodal landscapes are tabulated in Table 7 and 8.

Table 6. Average number of peaks found and Rank for Experiment 1: Set 2

Fnc	NCDE	NShDE	NSDE	CDE	ShDE	SDE	FER-PSO	SPSO	r2pso	r3pso	r2pso-lhc	r3pso-lhc	SCMA-ES	ACaR-CDE
E1-CF1	5.18	3.7	6.7	0	0	1.79	1.08	0	0	0	0	0	2	6
	3	4	1	11	11	6	7	11	11	11	11	11	5	2
E1-CF2	3.6	2.8	4	1.2	1.1	1.2	2	0	0	0	0	0	1.9	4
	3	4	1.5	7.5	9	7.5	5	10	10	10	10	10	6	1.5
E1-CF3	5.8	4	6	0.7	1.11	1.5	2.5	0	0	0	0	0	2.7	4
	2	3.5	1	8	7	6	5	11	11	11	11	11	4	3.5
E1-CF4	4.8	4.5	5.4	0	0	0	0	0	0	0	0	0	0.2	4.8
	2.5	4	1	10	10	10	10	10	10	10	10	10	5	2.5
E1-CF5	5.2	3.6	5.9	1.1	1.3	1.3	2	0	0	0	0	0	1.9	4.78
	2	3	1	8	7	6	4	11	11	11	11	11	5	3
E1-CF6	3	3	3	0	0	1.4	1.2	0	0	0	0	0	2.6	4
	3	3	3	11	11	6	7	11	11	11	11	11	5	1
E1-CF7	1.80	1	1.9	0	0	1	0.5	0	0	0	0	0	1	
	3	4	2	11	11	4	6	11	11	11	11	11	11	4
E1-CF8	3	3	3	0	0	1.4	1.5	0	0	0	0	0	2.3	4
	3	3	3	11	11	6	5	11	11	11	11	11	4	1
E1-CF9	3	3	3	0	0	1.8	1.5	0	0	0	0	0	1.7	4
	3	3	3	11	11	5	7	11	11	11	11	11	6	1
E1-CF10	1.3	1	2	0	0	1.1	1.1	0	0	0	0	0	1.2	2
	3	7	1.5	11	11	5.5	5.5	11	11	11	11	11	4	1.5
E1-CF11	2.8	2.2	4	0	0	1.3	0	0	0	0	0	0	0.7	6.5
	3	4	2	10.5	10.5	5	10.5	10.5	10.5	10.5	10.5	10.5	6	1
E1-CF12	2.5	2	2.9	0	0	1.6	1.6	0	0	0	0	0	1.7	4
	3	4	2	11	11	6.5	6.5	11	11	11	11	11	5	1
E1-CF13	2.3	1	3.8	0	0	0.9	0.3	0	0	0	0	0	1.4	4
	3	5	2	11	11	6	7	11	11	11	11	11	4	1
E1-CF14	1	1	1	0	0	1	1	0	0	0	0	0	1	1
	4	4	4	11	11	4	4	11	11	11	11	11	4	4
E1-CF15	3.8	2.4	4	0	0	1.6	1.2	0	0	0	0	0	2	5.8
	3	4	2	11	11	6	7	11	11	11	11	11	5	1
Total Rank	43.5	59.5	30	150	149.5	89.5	96.5	162.5	162.5	162.5	162.5	162.5	71	26

Table 7. Average Number of Peaks Found (Global And Local)

Func	NCDE	NShDE	NSDE	CDE	ShDE	SDE	FER-PSO	SPSO	r2pso	r3pso	r2pso-lhc	r3pso-lhc	SCMA-ES	ACaR-CDE
E1-F1	2	2	2	2	2	1.84	1.48	1.44	1.72	1.48	1.48	1.52	2	2
	4	4	4	4	4	8	12	14	9	12	12	10	4	4
E1-F2	2	2	2	2	1.84	1.68	1.88	1.72	1.36	1.24	1.52	1.76	2	2
	3.5	3.5	3.5	3.5	8	11	7	10	13	14	12	9	3.5	3.5
E1-F3	5	5	3.76	4.44	3.6	3.04	0.64	3.08	0.8	0.4	3	2.16	2	5
	2	2	5	4	6	8	13	7	12	14	9	10	10	2
E1-F5	5	5	4.76	4.28	3.12	1.52	1	5	1	1	4.52	2.8	0.48	5
	2	2	5	7	8	10	12	1	12	12	6	9	14	2
E1-F10	25	25	25	12.5	24.96	1.32	5.16	24.9	24.4	24.3	24.8	24.6	0.88	25
	2.5	2.5	2.5	11	5	13	12	6	9	10	7	8	13	2.5
Total Rank	14	14	20	28.5	31	50	56	38	55	62	46	46	44.5	14

Table 8. Success Rate in Finding Both Global And Local Peaks

Fnc	NCDE	NShDE	NSDE	CDE	ShDE	SDE	FER-PSO	SPSO	r2pso	r3pso	r2pso-lhc	r3pso-lhc	SCMA-ES	ACaR-CDE
E1-F1	100	100	100	100	100	84	64	44	72	56	48	52	100	100
	4	4	4	4	4	8	10	14	9	11	13	12	4	4
E1-F2	100	100	100	100	84	68	88	72	56	32	52	76	100	100
	3.5	3.5	3.5	3.5	8	11	7	10	12	14	13	9	3.5	3.5
E1-F3	100	100	40	44	12	4	0	0	0	0	8	0	0	100
	2	2	5	4	6	9	12	12	12	12	7	12	12	2
E1-F5	100	100	76	48	4	0	0	100	0	0	64	4	0	100
	2.5	2.5	5	7	8.5	12	12	2.5	12	12	6	8.5	12	2.5
E1-F10	100	100	100	0	96	0	0	92	60	52	84	76	0	100
	2.5	2.5	2.5	12.5	5	12.5	12.5	6	9	10	7	8	13	2.5
Total Rank	14.5	14.5	20	31	31.5	52.5	53.5	44.5	54	59	46	49.5	44	14.5

4.4 Percentage of Neighborhood Mutation

The percentage of total offspring that are being restricted in the local environment of each cluster is of utmost importance in Neighborhood based algorithms for niching environments. A niching technique is more stable and efficient if more local mutations are possible. The percentage of local mutation in ACaR-CDE is shown in Fig 5 and the average distance between parent and offspring in every iteration is shown in Fig 6.

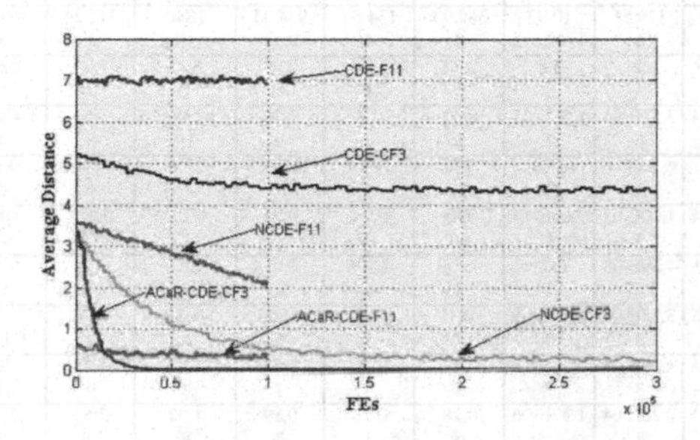

Fig. 6. Average parent-offspring distance test on test Functions E1-F11 and E1-CF3

Here it is evident that NShDE and NCDE perform equally well as an optimizer for locating local and global optima. NSDE cannot attain 100% success rate in all instances as a result of which optimizing for local peaks still remains a problem. Note that *E1-F10* and *E1-F5* constitutes 25 and 5 peaks respectively with equal height with no existence of local optima. *E1-F1* and *E1-F2* has only one local optimum but spaced at good distance apart. *E1-F3* however has 5 optima of which 3 are local in a 1-D space. NSDE however fails to restrict solutions to local optima under influence of global seeds that are close enough. Overall comparison concludes ACaR-CDE to be the best optimizer along with the NCDE, NShDE for this test instance.

The nearest peak of the parent is calculated and if the same happens to be the nearest peak of the offspring they it is concluded that both are searching for the same optima. Similar analysis is also carried out for NCDE [15] and CDE [14]. If parent and offspring are within same niche it is considered to be a successful local mutation. The percentage of offspring with successful local mutation is noted and plotted. The plot shows that ACaR-CDE has more successful local mutation within limited FEs than NCDE and CDE. Significant difference is clearly visible for E1-F11 i.e. 18 peak Shubert function that needs to maintain stable niches and is a challenge to multimodal optimization. Moreover average distance between parent and offspring produced per iteration is plotted along with the same for NCDE and CDE.

Graph clearly depicts better performance of ACaR-CDE in terms of local mutation within the neighborhood as compared to other schemes. It can be observed that ACaR-CDE outputs do fluctuate with iteration. This is desired and causes due to adaptive clustering. A solution that is initially away from all multimodal peaks keeps selecting different clusters for mutation till it finds the best cluster suited for it and remains there to track the optima within that locality.

Table 9. Peak Accuracy

Func	NCDE	NShDE	NSDE	TSC2	CDE	TSC	NCMA-ES	SCGA	DFS	ACaR-CDE
E2-F1	0.10 2	0.14 3	6.72 6	1.84 5	1.59 4	7.7 7	8.89 8	18.59 9	20.93 10	2.03E-07 1
E2-F2	2.75E-04 2	0.14 3	5.14 7	2.91 4	3.3 5	6.18 8	3.9 6	6.37 9	7.27 10	2.05E-09 1
E2-F3	3.01E-33 2	6.85E-30 3	6.56E-47 1	1.81E-07 4	4.48E-04 9	4.90E-06 8	3.92E-06 6	2.86E-07 5	4.17E-06 7	1.13E-12 4
E2-F4	0 2	0.0028 4	0 2	1.63 9	0.11 7	1.73 10	0.19 8	0.01 5	0.02 6	0 2
E2-F5	51.10 3	179.56 6	10.92 2	369.93 8	134.64 4	934.45 9	1840 10	317.24 7	164.85 5	3.89E-06 1
E2-F6	1.95E-04 2	0.21 3	5.8 8	2.77 5	1.78 4	2.99 6	5.12 7	7.06 10	6.89 9	2.07E-08 1
E2-F7	1.70E-04 3	3.51E-04 5	1.58E-05 2	0.02 6	0.1 7	1.79 9	1.96 10	0.73 8	3.42E-04 4	6.25E-09 1
E2-F8	4.99 4	6.09 5	2.96 3	727.9 7	115.4 8	1628.46 10	52.6 6	1381.05 9	0.11 2	3.49E-02 1
E2-F9	9.73E-04 2	1.10E-03 3	7.60E-14 1	0.85 10	0.24 9	0.23 8	0.01 7	0.003 5.5	0.003 5.5	1.00E-03 4
E2-F10	4.39E-05 2.5	4.39E-05 2.5	0.02 7	0.009 5	0.006 4	0.01 6	0.07 8	0.48 10	0.37 9	3.82E-07 1
E2-F11	3.18E-05 3	3.18E-05 3	3.18E-05 3	0.1 7	0.002 5	0.005 6	0.64 8	0.76 9	0.92 10	1.03E-09 1
E2-F12	1.69E-04 2	4.26E-04 3	0.40 7	0.32 5	0.1 4	1.68 7	1.89 8	4.28 9	4.3 10	8.81E-11 1
E2-F13	1.37E-06 2.5	9.20E-04 4	1.37E-06 2.5	0.28 6	0.12 5	0.89 7	1.02 8	2.53 10	2.11 9	0 1
Total Rank	32	47.5	50.5	81	75	101	100	95.5	96.5	20

4.5 Peak Accuracy: Results and Discussions

Results on deviation of best located solution from all peaks is given in Table IX. ACaR-CDE outmatches all other algorithms including Neighborhood techniques in peak accuracy measurement. It clearly depicts that besides its explorative nature and efficient handling of multiple peaks it is more exploitative in optimization than any other methods. The results prove that when fitness of the current best individual is a concern in any optimization problem, ACaR-CDE is the best algorithm to opt for. Neighborhood

Table 10. Distance Accuracy

Fnc	NCDE	NShDE	NSDE	TSC2	CDE	TSC	NCMA-ES	SCGA	DFS	ACaR-CDE
E2-F1	0.02 2	0.037 3	1.55 6	0.79 5	0.41 4	3.26 7	3.87 8	11.56 10	11.52 9	4.78E-05 1
E2-F2	4.80E-03 2	0.10 3	2.23 6	2.09 5	1.99 4	6.18 7	3.19 8	7.02 10	6.22 9	1.05E-05 1
E2-F3	2.63E-17 2	6.36E-016 3	1.87E-24 1	9.32E-05 6	5.25E-03 10	9.08E-05 5	5.48E-04 8	1.65E-04 7	8.12E-04 9	1.07E-06 4
E2-F4	8.00E-05 3	7.72E-04 4	0 1.5	0.05 8	0.03 7	0.07 9	0.14 10	0.01 5	0.02 6	0 1.5
E2-F5	0.02 3	0.0304 4	6.30E-03 2	0.49 6	0.07 5	1.07 8	0.71 7	2.51 9	3.05 10	2.50E-03 1
E2-F6	3.30E-03 2	0.03 4	0.28 5	0.54 6	0.22 3	3.94 8	1.48 7	5.31 10	4.55 9	6.84E-06 1
E2-F7	0.01 3	0.02 4	1.80E-03 2	0.45 6	0.21 5	5.48 9	4.56 8	6.04 10	3.63 7	5.41E-05 1
E2-F8	0.16 3	0.18 4	0.03 2	33.2 8	3.12 5	59.2 9	31.0 7	22.1 6	88.2 10	2.12E-02 1
E2-F9	3.42E-04 2	3.76E-04 4	2.70E-14 1	0.21 10	0.05 8	0.06 9	2.96E-03 7	9.78E-04 6	8.18E-04 5	3.69E-04 3
E2-F10	1.00E-05 2	3.20E-05 3	0.05 7	0.02 5	0.01 4	0.03 6	0.15 8	0.92 10	0.72 9	9.06E-06 1
E2-F11	1.32E-06 3	6.80E-04 4	1.28E-09 1	0.21 7	9.00E-03 5	0.01 6	1.42 8	1.55 9	1.87 10	6.40E-08 2
E2-F12	6.81E-04 2	1.00E-03 3	0.43 6	0.34 5	0.12 4	1.77 7	1.99 8	4.32 10	4.29 9	2.38E-05 1
E2-F13	4.93E-10 3	6.93E-04 4	1.05E-16 2	0.95 6	0.36 5	7 8	5.42 7	7.68 10	7.36 9	0 1
Total Rank	32	45	40.5	88	69	108	101	112	91	19.5

techniques deteriorate in performance maintenance because selection of proper neighbors and subsequent optimization in some instances may lead to a slower optimization. ACaR-CDE thus proves itself to be a robust optimizer. However if multiple peaks have same functional values, peak accuracy may give faulty indication of multimodality though in many practical life solutions peak accuracy is the type of solution opted for. To remove ambiguity we also include distance accuracy test that is more scientific towards testing a niching technique.

4.6 Distance Accuracy: Results and Discussions

The results on distance accuracy of ACaR-CDE along with other algorithms are shown in Table X. Like all other previous cases ACaR-CDE proves to be the best algorithm in discovering all multimodal peaks in best accuracy level in terms of distance from the peaks. The results indicate superiority of ACaR-CDE in getting the closest solutions to multiple peaks in the environment, both global and local. Neighborhood techniques lag far behind in distance accuracy results.NCDE is the sole algorithm that obtains solutions close to that obtained by ACar-CDE. However incorporation of adaptive clustering and re-clustering clearly justifies a better performance in optimization in multimodal fields. NSDE though a strong optimizer, cannot perform well in terms of exploitation of known solutions. The problem reflects more when functional landscape is complicated and may be a concern in practical life landscapes.

5 Conclusion

In this paper an effective optimizer called ACaR-CDE with a unique grouping strategy has been proposed. The effectiveness and feasibility of ACaR-CDE has been validated on different niching benchmark problems with diverse characteristics from simple objectives to complex Composite Problems and compared with other state-of-the-art EAs. ACaR-CDE has outperformed competitive algorithms in all cases as evident from the calculated ranks in all test cases considered.

References

1. Mahfoud, S.: Niching Method for Genetic Algorithms, Doctoral Dissertation. Technical Report, Department of Computer Science, University of Illinois at Urbana-Champaign, Urbana, IL, USA, Illinois Genetic Algorithms Laboratory, IlliGAL, Report No. 95001 (1995)
2. Mahfoud, S.W.: A comparison of parallel and sequential niching methods. In: Proceedings of 6th International Conference on Genetic Algorithms, Pittsburg, USA, July 15-19, pp. 136–143 (1995) ISBN 155860-370-0
3. Biswas, S., Das, S., Kundu, S.: Inducing Niching Behavior in Differential Evolution through Local Information Sharing. TEVC (2014), doi:10.1109/TEVC.2014.2313659
4. Singh, G., Deb, K.: Comparison of multimodal optimization algorithms based on evolutionary algorithms, pp. 1305–1312. ACM Press, Seattle (2006)
5. Storn, R., Price, K.: Differential Evolution- A Simple and Efficient Heuristic for Global Optimization over Continuous Spaces. Journal of Global Optimization 11, 341–359 (1997)

6. De Jong, K.A.: An Analysis of The Behavior of a Class of Genetic Adaptive Systems. Ph. D. Thesis, University of Michigan (1975)
7. Li, J.-P., Balazs, M.E., Parks, G.T., Clarkson, P.J.: A species conserving genetic algorithm for multimodal function optimization. Evolutionary Computation 10(3), 207–234 (2002)
8. Goldberg, D.E., Richardson, J.: Genetic algorithms with sharing for multimodal function optimization. In: GECCO, pp. 41–49 (1987)
9. Cioppa, A.D., Stefano, C.D., Marcelli, A.: Where are the niches? dynamic fitness sharing. TEVC 11(4), 453–465 (2007)
10. Deb, K.: Genetic algorithms in multimodal function optimization, the Clearing house for Genetic Algorithms. M.S thesis and Rep. 89002, Univ. Alabama, Tuscaloosa (1989)
11. Harik, G.R.: Finding multimodal solutions using restricted tournament selection. In: Proceedings of the 6th International Conference on Genetic Algorithms, San Francisco, pp. 24–31 (1995)
12. Das, S., Maity, S., Qu, B.-Y., Suganthan, P.N.: Real-parameter evolutionary multimodal optimization- A survey of the state-of-the-art. Swarm and Evolutionary Computation 1, 71–88 (2011)
13. Das, S., Suganthan, P.N.: Differential evolution A survey of the state-of-the-art. TEVC 15(1), 4–31 (2011)
14. Thomsen, R.: Multimodal optimization using Crowding-based differential evolution. In: CEC, pp. 1382–1389 (2004)
15. Qu, B.Y., Suganthan, P.N.: Differential Evolution with Neighborhood Mutation for Multimodal Optimization. TEVC 16(5), 601–611 (2012)
16. Bezdek, J.C.: Pattern Recognition with Fuzzy Objective Function Algorithms. Plenum Press (1981)
17. Parrott, D., Li, X.: Locating and Tracking Multiple Dynamic Optima by a Particle Swarm Model Using Speciation. TEVC 10(4), 440–458 (2006)
18. Stoean, C., Preuss, M., Stoean, R., Dumitrescu, D.: Multimodal optimization by means of a topological species conservation algorithm. TEVC 14(6), 842–864 (2010)
19. Shir, O.M., Bäck, T.: Niching in Evolution Strategies. In: GECCO, New York (2005)

A Network Connectivity Embedded Clustering Approach for Supply Chain Risk Assessment

Xiao Feng Yin, Xiuju Fu, Loganathan Ponnambalam, and Rick Siow Mong Goh

Computing Science Department, Institute of High Performance Computing,
1 Fusionopolis Way, #16-16 Connexis North, Singapore 138632
{yinxf,fuxj,ponnaml,gohsm}@ihpc.a-star.edu.sg

Abstract. In recent years, increased attention has been shown to the supply chain risk management due to the occurrences of several high profile disruptions which had resulted in significant social, economic and political impact globally. However, there aren't direct and easy ways of understanding the risk of an entire supply chain. In this paper, a network connectivity embedded k-means clustering approach has been proposed to determine at-risk clusters of nodes which share similar risk profiles and linkages with the focal company. The proposed approach uses a multiple dimensional feature vector to represent the risks that nodes are facing, their geographical locations, supply chain attributes and network connectivity attributes. The clustering approach is able to reduce the complexity of a large supply chain network to facilitate in-depth targeted analysis and simulations. The effectiveness of the proposed approach has been illustrated by experiments that successfully identify the risk clusters and critical risk zones.

Keywords: Supply Chain Risk Management, Supply chain risk clustering, k-means clustering.

1 Introduction

The supply chain risk management has been shown an increasing global attention in recent years owning to huge impacts from occurrences of some high profile disruptions all over the world. Risks and disruptions including earthquakes, economic crises, strikes and terrorist attacks have repeatedly hit the supply chain and its operations. In addition, due to the globalization and constantly adoption of practices such as outsourcing, drop shipment and vendor management inventory, it is important for a company to keep track of happenings of potential risks in its entire supply chain network including disasters and social unrest around the regions of its key suppliers as they may have eventually an impact on the focal company itself. The appropriate management of constraints of the supply, manufacturing, and demand and relevant risks is essential for the sustainable development and business success of a company [1, 2, 3].

The recent catastrophic events like flood in Thailand, earthquake in Japan demonstrated that the effects of the natural disruption not only realised at the place of

H. Handa et al. (eds.), *Proc. of the 18th Asia Pacific Symp. on Intell. & Evol. Systems – Vol. 1*,
Proceedings in Adaptation, Learning and Optimization 1, DOI: 10.1007/978-3-319-13359-1_30

its origination but also propagated to the other members of the supply chain. Therefore, the risk management and mitigation process should consider the whole supply chain network while more attentions shall be paid to the risk zones. Disaster-related risks and disruptions normally are with low probability but low predictability. This leaves companies insufficiently time buffer to react which may cause widespread business disruptions [4] like: i) damaging facilities; ii) upending shipping schedules; iii) interfering with production; and iv) impairing the ability to meet customers' expectations for high quality and timely services.

Taking the 2010 volcanic eruption of Eyjafjallajökull in Iceland as an example, it caused enormous disruption to air travel across western and northern Europe over an initial period of six days in April 2010. Big global company such as Infineon's DC in Europe (DCE) was also badly affected as delivery of products to customers is usually through air. DCE had to resort to land transportation by trucks to clear the backlogs while the customer service had to be compromised due to longer delivery time. Meanwhile, more security measures had to be deployed as goods stolen or lost is happening much more easily and frequently by land transportation than by air.

2 Supply Chain Risk Clustering

The empirical analysis in automotive industry conducted by Thun and Hoenig [5] revealed that the rising complexity and increasing uncertainties and dependencies among companies fostered supply chain risk and resulted in increased supply chain vulnerability and pooper visibility. Tang [1] had reviewed various quantitative models for managing supply chain risks. He concluded that those quantitative models were designed for managing operational risks primarily, not disruption risks, due to inaccurate measures of the probability of an occurrence of a major disruption and the potential impact of a disruption in a large supply chain network. There aren't direct and easy ways of understanding the risk of an entire supply chain.

The Thailand flood during 2011 resulted in a global shortage of hard disk supply due to the overly centralised hard disk manufacturers over the area that was badly affected. If such a risk cluster of a supply chain network can be identified earlier, precautions can be taken to source more suppliers from other regions or mitigation plans can be studied and prepared. In this paper, a supply chain risk clustering approach is proposed to handle supply chain risk management. The essence of the clustering problem is to develop a methodology for determining groups of "similar" nodes which can be treated as a basic unit for tactic and strategic analysis of supply chain network. The risk clustering is capable of reducing the complexity of a large supply chain network and identify topological risks in supply chain networks. It is to facilitate the evaluations of potential at-risk facilities and clusters hidden in supply chain networks that cannot be easily detected. Besides, targeted measurements, analysis and simulations can be designed and employed to a specific cluster of supply chain after the clusters have been determined. By doing so, better results and performances can be expected.

Clustering technology had been utilised in many researches to simplify the supply chain network and assist the decision making. Irfan *et al.* [6] proposed a k-means algorithm to find the cluster centres of different supply chain tiers, such as customers, retailers, distribution centres and manufacturers, to assist the business decisions. In order to simplify the supply chain and production network, Doring *et al.* [7] worked on a k-means clustering approach for grouping of state spaces of production network. Hu et.al. [8] applied neural network based fuzzy clustering to study the supply chain quality management while considering macro variables such as political influence and law and regulation and micro variables such as price and quantity. In order to reduce the complexity of an extended supply chain network for planning and scheduling, Yin *et al.* [9] proposed a hybrid evolutionary approach to the clustering of supply chain by considering material flow and the total cost. Tabrizi and Razmi [10] applied fuzzy set theory to understand the extant uncertainties and risks during the phase of supply chain network design.

There were few studies addressing the supply chain risk clustering. Hallikas *et al.* [11] explored the network risks, risk-management measures and supplier classification to achieve collaborative risk management and learning. A work had also been carried out by Reniers *et al.* [12] to understand the systemic risk that taken into consideration the safety and security index and the supply chain index of a typical supply chain cluster in chemical industry. It built upon an established cluster network. The holistic analysis that takes into consideration connectivity, geographical lactation and supply chain risk in supply chain risk management context to facilitate the evaluations of potential at-risk facilities and clusters and assess the overall risk that a company is facing is lacking.

In the rest of the paper, a network connectivity embedded k-means clustering algorithm was depicted in Section 3, followed by the experiments and discussions in Section 4. The main conclusions reached in this work are then summarised in the last section, Section 5.

3 A Network Connectivity Embedded k-Means Clustering

3.1 Problem Description

Given a supply chain network and a set of nodes of the network $\mathfrak{X} = \{X_1, X_2, \ldots, X_n\}$ where each node X_i is a multiple dimensional feature vector representing the supply chain and related attributes to be considered, the clustering algorithm needs to partition the n nodes into k clusters with centroids of clusters $\mathcal{C} = \{C_1, C_2, \ldots, C_k\}$ so as to minimise the squared error as expressed in Eq 1.

$$\operatorname{argmin} \sum_{j=1}^{k} \sum_{i=1, y_i=j}^{n} \left\| X_i - C_j \right\|^2 \tag{1}$$

Where

y_i cluster assignment of node X_i, $i = 1,2,\ldots,n$
C_j centroid of cluster j, $j = 1,2,\ldots,k$

Cluster assignment y_i can be established given the centroids of the clusters using Eq 2 as follows.

$$y_i = argmin_j \|X_i - C_j\|^2 \quad (2)$$

3.2 Mathematical Model - A Generic Model

As mentioned in Section 3.1, node X_i is a multiple dimensional feature vector. It represents the risks of a node facing, the geographical location, supply chain attributes and network connectivity attributes. By considering the above supply chain and risk specific features, Eqs 3 and 4 can be derived to represent X_i and C_j respectively.

$$X_i = \{X_i^G, X_i^R, X_i^T, X_i^C\} \quad (3)$$

$$C_j = \{C_j^G, C_j^R, C_j^T, C_j^C\} \quad (4)$$

Where

$X_i^G, X_i^R, X_i^T, X_i^C$ feature vectors of node i representing geographical location, risks facing, supply chain attributes and network connectivity attributes, respectively

$C_j^G, C_j^R, C_j^T, C_j^C$ centroids of cluster j representing geographical location, risks facing, supply chain attributes and network connectivity attributes, respectively.

Eqs 1 and 2 are extended to incorporate different ways of evaluating centroids and squared errors for geographical location, risks facing, supply chain attributes and network connectivity attributes.

$$y_i = argmin_j \left[w^G\|X_i^G - C_j^G\|^2 + w^R\|X_i^R - C_j^R\|^2 + w^T\|X_i^T - C_j^T\|^2 + w^C\|X_i^C - C_j^C\|^2\right] \quad (5)$$

Where

w^G, w^R, w^T, w^C weightages for geographical location, risks facing, supply chain attributes and network connectivity attributes, respectively

The optimisation criteria and updating of centroids are given by Eqs 6 to 10 respectively.

$$argmin \sum_{j=1}^{k} \sum_{i=1, y_i=j}^{n} \left[w^G\|X_i^G - C_j^G\|^2 + w^R\|X_i^R - C_j^R\|^2 + w^T\|X_i^T - C_j^T\|^2 + w^C\|X_i^C - C_j^C\|^2\right] \quad (6)$$

$$C_j^G = \sum_{i=1, y_i=j}^{n} X_i^G \Big/ n_j \quad (7)$$

$$C_j^R = \sum_{i=1, y_i=j}^{n} X_i^R \Big/ n_j \quad (8)$$

$$C_j^T = \sum_{i=1, y_i=j}^{n} X_i^T \Big/ n_j \quad (9)$$

$$C_j^C = \sum_{i=1, y_i=j}^{n} X_i^C \Big/ n_j \quad (10)$$

Where

n_j number of nodes in cluster j

4 Experiments and Discussions

Two experiments are used to illustrate the effectiveness of the proposed algorithm. As shown in Figure 1, risks of a node facing consist of city risk and country risk; geographical location is represented latitude and longitude; supply chain attributes include tier information. Transactional information and relative importance level between nodes can also be included as part of the supply chain attributes. As for risk scores of a city and its country, they can be found from public domains. The Economist Intelligence Unit (http://viewswire.eiu.com/) provides updated and relatively reliable risk scores for some major cities and countries. Figure 2 illustrates how the connectivity of a network can be translated into multiple two dimensional matrix to be considered by k-means. By incorporating a scenario simulator which is able to define different sets of parameters, various supply chain clusters can be generated which emphases different concerned factors. It is able to form risk clusters of an entire supply chain which summarise the risks the supply chain of a company is exposing. It can also analyse the risk zones in respect to the geographical locations and raise alerts if critical facilities are all located in similar high risk zone.

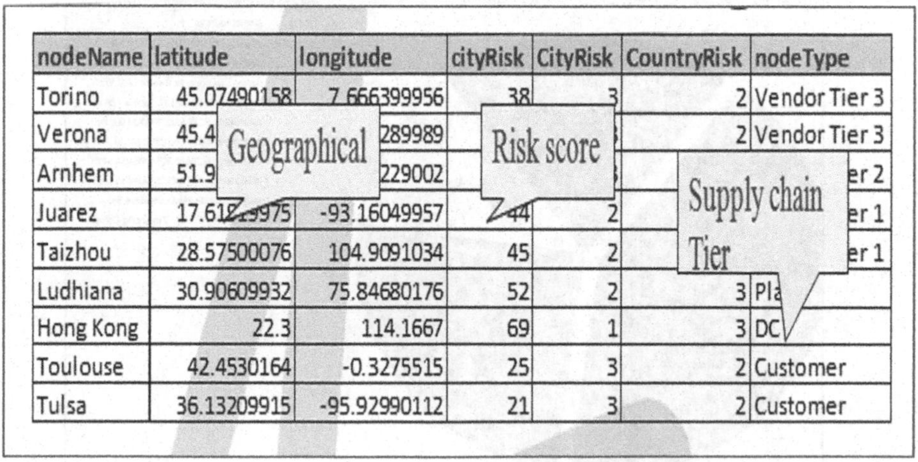

nodeName	latitude	longitude	cityRisk	CityRisk	CountryRisk	nodeType
Torino	45.07490158	7.666399956	38	3	2	Vendor Tier 3
Verona	45.4[illegible]	[illegible]289989	[illegible]	[illegible]	2	Vendor Tier 3
Arnhem	51.9[illegible]	[illegible]229002	[illegible]	[illegible]	[illegible]	[illegible]er 2
Juarez	17.61[illegible]975	-93.16049957	44	2	[illegible]	[illegible]er 1
Taizhou	28.57500076	104.9091034	45	2	[illegible]	[illegible]er 1
Ludhiana	30.90609932	75.84680176	52	2	3	Pl[illegible]
Hong Kong	22.3	114.1667	69	1	3	DC
Toulouse	42.4530164	-0.3275515	25	3	2	Customer
Tulsa	36.13209915	-95.92990112	21	3	2	Customer

Fig. 1. Illustration of risks of a node facing, the geographical location, and supply chain attributes

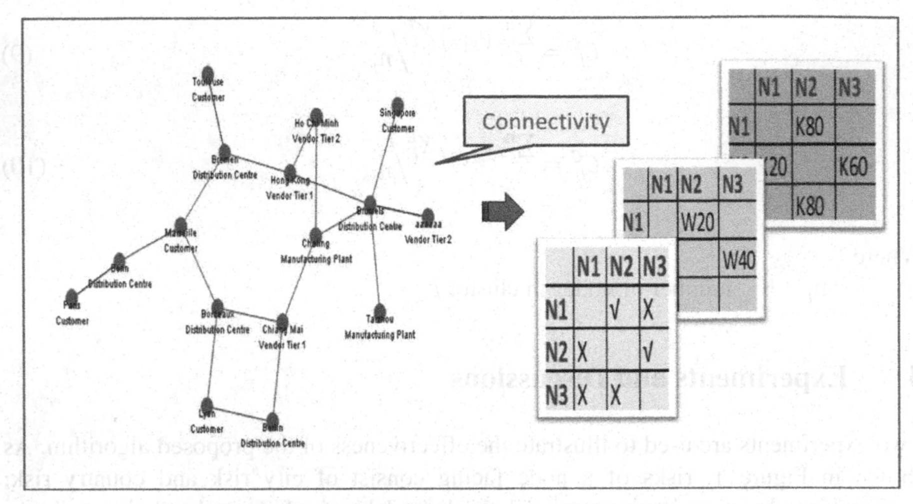

Fig. 2. Illustration of network connectivity attributes

4.1 Study 1: Risk Cluster Identification

One of the purposes of the study is to find the high risk zone in respect to the risks nodes are exposing and the network connectivity. In the experiment, Thailand and its cities such as Bangkok and Chiang Mai are given high risk scores due to the past flooding issue and political unrest. Relatively high risk scores are configured for China and Vietnam owning to recent diplomatic and territorial disputes. As shown in Figure 3, the clustering algorithm partitioned the supply chain nodes into five categories that are identified in advance. The details in clusters of very high risk and high risk (Figure 4) show the interconnected nodes located in Thailand, Vietnam and China. This illustrates that the clustering algorithm can work on not only the risk scores but also network connectivity that is critical in the context of supply chain management.

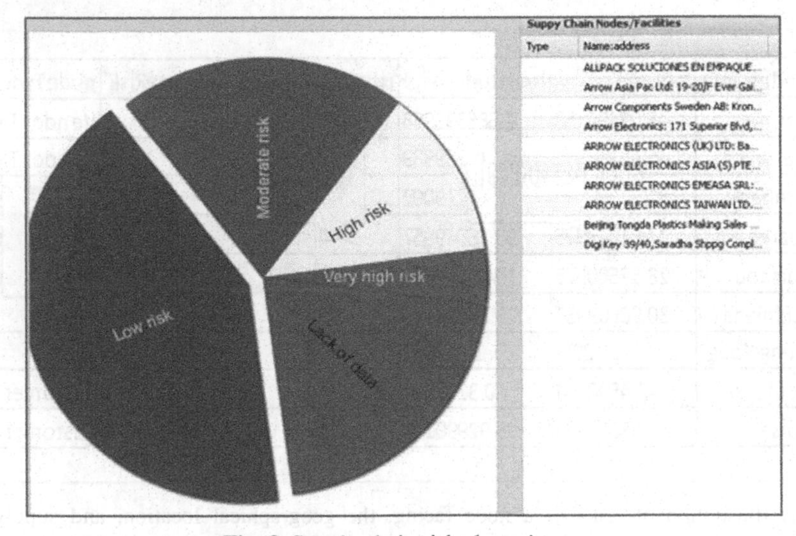

Fig. 3. Supply chain risk clustering

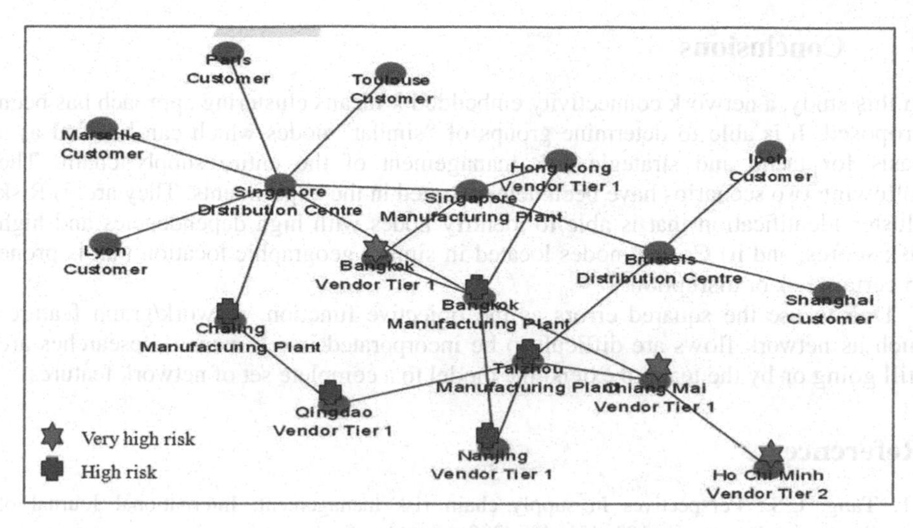

Fig. 4. High risk nodes

4.2 Study 2: Critical Zone Cluster Identification

In this experiment, parameters are tuned to emphasis more on the geographical locations besides the risk factors and the connectivity. The result is very similar as that from Study 1 except that the manufacturing plant in Bangkok has now been shifted into the very high risk cluster (Figure 5). It is due to the reason that Bangkok geographically is very close to other nodes in the very high risk cluster such as Ho Chi Minh and Chiang Mai. The change demonstrates that the geographical location plays a bigger role in this experiment. It is important to understand the risks a company is facing due to centralised suppliers and manufacturing facilities in different industrial sectors.

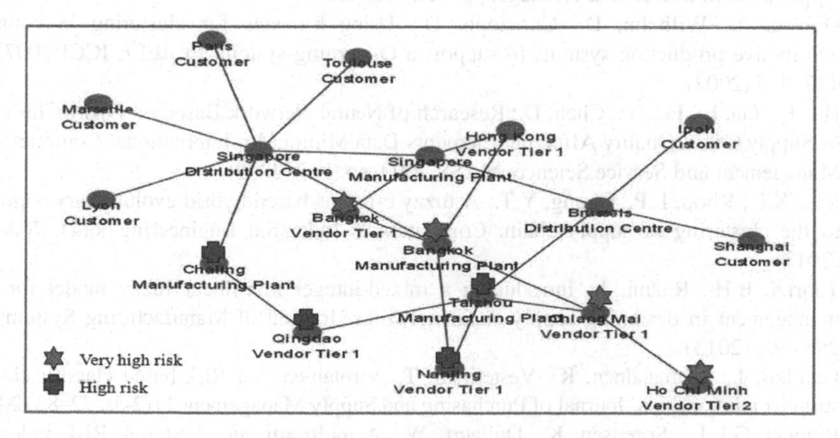

Fig. 5. Critical nodes cluster

5 Conclusions

In this study, a network connectivity embedded k-means clustering approach has been proposed. It is able to determine groups of "similar" nodes which can be used as a basis for tactic and strategic risk management of the entire supply chain. The following two scenarios have been demonstrated in the experiments. They are: i) Risk cluster identification that is able to identify nodes with high dependencies and high risk scores; and ii) Critical nodes located in similar geographic location that is prone to certain risk or disruption.

Due to use the squared errors as the objective function, network/graph features such as network flows are difficult to be incorporated into k-means. Researches are still going on by the team to extend the model to a complete set of network features.

References

1. Tang, C.S.: Perspectives in supply chain risk management. International Journal of Production Economics 103, 451–488 (2006)
2. Tang, O., Musa, S.N.: Identifying risk issues and research advancements in supply chain risk management. International Journal of Production Economics 133, 25–34 (2011)
3. Bearzotti, L.A., Salomone, E., Chiotti, O.J.: An autonomous multi-agent approach to supply chain event management. International Journal of Production Economics 135, 468–478 (2012)
4. CFO research services, Preparing for the Worst: Natural Disasters and Supply-Chain Risk Management (2013), `http://www.fmglobal.com/assets/pdf/P09179.pdf` (accessed on May 8, 2013)
5. Thun, J.H., Hoenig, D.: An empirical analysis of supply chain risk management in the German automotive industry. International Journal of Production Economics 131, 242–249 (2011)
6. Irfan, D., Xu, X., Deng, S., Khan, I.A.: Clustering Framework for Supply Chain Management (SCM) System. In: IEEE 2007 Second Workshop on Digital Media and its Application in Museum & Heritage, pp. 422–426 (2007)
7. Doring, A., Wilhelm, D., Christoph, D.: Using k-means for clustering in complex automotive production systems to support a Q-learning-system. In: IEEE ICCI 2007, pp. 487–497 (2007)
8. Hu, J., Hua, E., Fei, Y., Chen, D.: Research of Neural Network Based on Fuzzy Clustering in Supply Chain Quality Affecting Elements Data Mining. In: International Conference on Management and Service Science, MASS 2009, pp. 1–5 (2009)
9. Yin, X.F., Khoo, L.P., Chong, Y.T.: A fuzzy c-means based hybrid evolutionary approach to the clustering of supply chain. Computers & Industrial Engineering 66(4), 768–780 (2013)
10. Tabrizi, B.H., Razmi, J.: Introducing a mixed-integer non-linear fuzzy model for risk management in designing supply chain networks. Journal of Manufacturing Systems 32, 295–307 (2013)
11. Hallikas, J., Puumalainen, K., Vesterinen, T., Virolainen, V.: Risk-based classification of supplier relationships. Journal of Purchasing and Supply Management 11(2-3), 72–82 (2005)
12. Reniers, G.L.L., Sorensen, K., Dullaert, W.: A multi-attribute Systemic Risk Index for comparing and prioritizing chemical industrial areas. Reliability Engineering and System Safety 98, 35–42 (2012)

Proposal and Evaluation of a Robust Pheromone-Based Algorithm for the Patrolling Problem with Various Graph Structure

Shigeo Doi

National Institute of Technology, Tomakomai College,
443 Aza-nishikioka, Tomakomai-shi, Hokkaido, 059-1275, Japan
doi@jo.tomakomai-ct.ac.jp

Abstract. Recently, the urgent necessity to develop an algorithm to resolve patrolling problems has become evident. This problem is modeled using a graph structure and defined as the requirement that an agent or multi-agents patrol each node in the graph at the shortest regular intervals possible. To solve the problem, some central controlled algorithms have been proposed. However, these algorithms require a central control system, and therefore, their reliability strongly depends on the reliability of the central control system. Thus, the algorithm has a lower ability in severe environments, for example, in the case of communication between an agent and the central control system. Instead of a central controlled algorithm, some autonomous distributed algorithms have been proposed. In this paper, we propose an autonomous distributed algorithm, called pheromone and inverse degree-based Probabilistic Vertex-Ant-Walk (pidPVAW), which is an improved version of pheromone-based Probabilistic Vertex-Ant-Walk (pPVAW). pPVAW is based on Probabilistic Vertex-Ant-Walk (PVAW). These algorithms use a pheromone model corresponding to fixed points for agent communication and cooperative patrolling. The difference between pidPVAW and pPVAW is that when an agent determines the neighbor node to which it moves the next time, pidPVAW takes into consideration the degree of each neighbor node, whereas pPVAW does not. This consideration is useful for scale-free or tree-like graphs. It is considered that lower degree nodes cannot easily be visited by agents when they use pPVAW. In contrast, when agents use pidPVAW, they can visit these lower degree nodes with ease. pidPVAW inherits some parts of the useful behavior of pPVAW, such as that agents using pPVAW do not return to the last visited node.

Keywords: patrolling problem, ant colony system, multi-agent systems.

1 Introduction

When security officers patrol a monitored area at midnight as part of their task, for example, usually they visit certain fixed points at regular intervals and check whether an intruder has entered the area. However, these officers can

H. Handa et al. (eds.), *Proc. of the 18th Asia Pacific Symp. on Intell. & Evol. Systems – Vol. 1*,
Proceedings in Adaptation, Learning and Optimization 1, DOI: 10.1007/978-3-319-13359-1_31

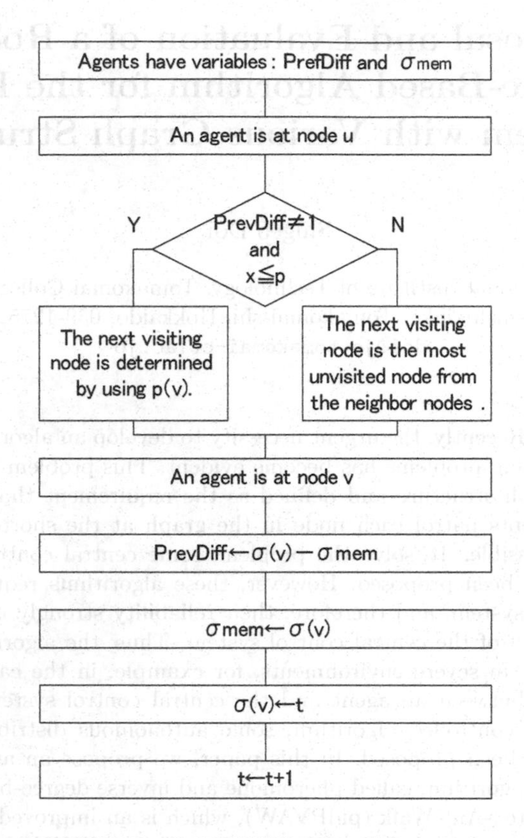

Fig. 1. Flowchart of pPVAW and pidPVAW. The difference between pidPVAW and pPVAW is only in the equation for calculating $p(v)$.

be ambushed or pursued by malicious intruders, which would constitute a very serious incident.

Consequently, security robots have been developed recently and used instead of human security officers. However, their purpose is to check for emergencies by walking along a fixed path, and therefore, they cannot change their behavior if and when conditions change. When the conditions do change, the ability to change the agents' behavior is a particularly important property. Thus, these robots are autonomous and have the ability to change their behavior by themselves. In addition, the introduction of a central control system to control the robots could lead to unreliability, because the robots would then have to communicate with each other and exchange their information through the system.

Therefore, if the communication channels between the system and the robots were disabled, the behavior of the robots would not be effective and they could not accomplish their tasks. Thus, the robots should be able to work effectively and communicate with each other to exchange information without a central control system.

Table 1. Comparison of algorithms

Items \Algorithm	PVAW	pPVAW	pidPVAW
May return to last visited node	Yes	No	No
Uses pheromone information	No	Yes	Yes
Uses degree information	No	No	Yes

In this paper, we design an autonomous, robust, and distributed algorithm to deal with dynamic environments that use security robots in the place of security officers, using pheromone information corresponding to fixed points that one of the robots should visit at regular intervals in the monitored area. This approach is also applicable to room-cleaning robots. In addition, we also take into account the ease of visiting these fixed points in the monitored area.

The monitored area is expressed as a undirected graph $G = (V, E)$, where V is a set of nodes and E is a set of links between two nodes. A node in V corresponds to a fixed point that the robots should visit. We use the degree of each node to represent the ease of visiting the node. A node having less connectivity cannot easily be visited by robots. We developed an algorithm pheromone-based Probabilistic Vertex-Ant-Walk (pPVAW) [5]. We added the feature described above to pPVAW and developed an algorithm called pheromone and inverse degree-based Probabilistic Vertex-Ant-Walk (pidPVAW). This feature is effective for graphs that have a biased degree distribution. In addition, on a grid graph, pidPVAW can achieve at least the same level of performance as pPVAW. A comparison of pidPVAW and pPVAW by means of computer simulations shows that the performance of pidPVAW is better than that of pPVAW on scale-free graphs [2].

2 Problem Definition and Related Work

First, let us define the patrolling problem. Let $G = (V, E)$ be an undirected graph, where V is a set of nodes and E is a set of links. Nodes represent the places that agents should visit at regular intervals. If $(i, j) \in E$, an agent is able to move on the link from i to j using one unit of simulation time. A patrolling problem is defined as the problem of enabling agents to visit all the nodes in intervals that are as short as possible during the simulation time. In general, as the number of agents increases, the visiting interval for all the nodes should be shorter.

Architectures for multi-agent patrolling were discussed in [12]. Spanning-tree based algorithms were proposed in [9] and [6]. Theoretical analyses were proposed in [3] and [10]. In addition, various algorithms, such as graph partitioning [13], reinforcement learning [14], and the ant colony system [11], were proposed under static environments. Various approaches for a patrolling problem were discussed in [1].

Elor et al. proposed Probabilistic Vertex-Ant-Walk (PVAW) [7][8]. In PVAW, when an agent is deciding to which neighbor node it should move, it selects

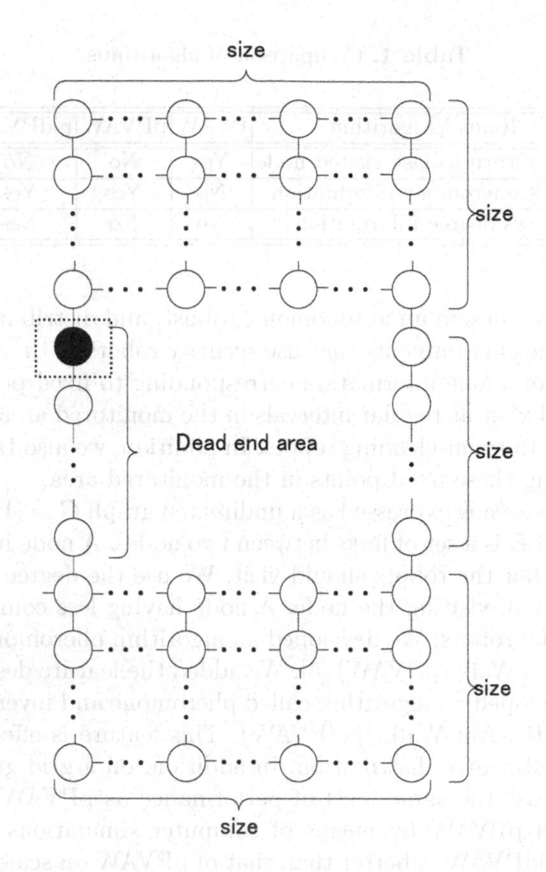

Fig. 2. Graph structure using simulations in Scenario 1

mainly the least visited neighbor node with probability $1 - p$. Otherwise, with probability p, it selects a neighbor node randomly. We developed an improved version of PVAW, called pPVAW [5]. The difference between pPVAW and PVAW is that a robot using pPVAW determines its selection of a next node in proportion to the quantity of attractive pheromones with probability p , while a robot using PVAW determines its selection of a next node randomly. In this paper, we describe the development of and propose an algorithm, pidPVAW, which is based on pPVAW. The selection of a next node by a robot using pidPVAW is inversely proportional to the degree of the neighbor nodes, in addition to the quantity of attractive pheromone, in order to visit nodes that have less connectivity.

2.1 pPVAW

pPVAW [5] is an improved version of the PVAW algorithm. PVAW is an algorithm that conducts patrols on a graph. Let $\sigma(v)$ be the last time when node v was visited by an agent. When an agent is at node v, it determines to visit the

Table 2. The distribution of degree of the graph shown in Fig. 3

Degree	1	2	3	4	5	6	7	8	9	10	11	12	13	14	15	16	17	18
Number of Nodes	6677	1672	638	305	205	142	76	56	43	33	20	24	20	14	4	7	1	4
Degree	19	20	21	22	23	24	25	26	27	28	30	31	32	33	34	35	36	38
Number of Nodes	6	2	2	6	3	3	3	2	1	2	1	1	1	1	2	1	2	1
Degree	39	40	41	43	44	45	47	50	52	53	57	59	63	66	83	109	-	-
Number of Nodes	2	1	1	1	2	1	1	1	2	1	1	1	1	1	1	1	-	-

next node where the value of $\sigma(v)$ is the lowest among the neighbor nodes of v with probability $1 - p$. In other words, it visits the node where the most time has elapsed since its last visit with probability $1-p$. However, it determines the next node randomly with probability p. In addition, agents have internal variables $PrevDiff$ and σ_{mem} to determine whether they move randomly. Thus, the algorithm assumes that an agent at node w is able to acquire $\sigma(v), v \in N(w)$, where $N(w)$ is a set of neighbor nodes of w.

We discovered by means of computer simulations that the performance of pPVAW is better than or equal to that of PVAW. The assumption made in pPVAW is identical to that made in PVAW; thus, agents in pPVAW have to acquire the last time the neighbor nodes were visited from their current node. Assuming that the agent is at node w, the probability that the agent moves to $v \in N(w)$ is given as

$$p(v) = \frac{\max\limits_{n \in N(w)} \sigma(n) - \sigma(v)}{\sum\limits_{m \in N(w)} \left(\max\limits_{n \in N(w)} \sigma(n) - \sigma(m) \right)} \tag{1}$$

where $N(w)$ is a set of neighbor nodes of w. A flowchart for pPVAW is shown in Fig. 1, where $x \in U(0, 1)$, where $U(0, 1)$ is a uniform distribution on [0, 1]. The difference between pPVAW and PVAW is that using PVAW the agent randomly selects the next node with probability p, whereas using pPVAW the agent selects a neighbor node that is proportional to $p(v)$. The value $\max_{n \in N(w)} \sigma(n) - \sigma(m)$ corresponds to the attractive pheromone for the agents.

2.2 pidPVAW

pidPVAW is an improved version of the pPVAW algorithm. In this algorithm, we suppose that it may be difficult for agents to visit lower degree nodes, and focus on this supposition. Thus, we also take into account the degree of each neighbor node so that the lower degree nodes can be more easily visited using pidPVAW than pPVAW. In a grid environment, this algorithm should be similar to pPVAW. In a graph, for example, a scale-free graph, it is easy for agents using pidPVAW to visit not only hub nodes but also leaf nodes. The difference between pidPVAW and pPVAW is only in the equation of $p(v)$. Now, we assume

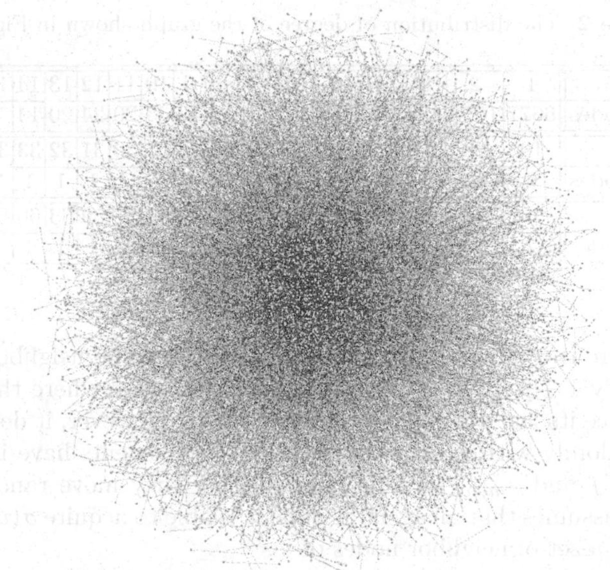

Fig. 3. Graph structure using simulations of Scenario 2. The average degree of all nodes is 2, the diameter of the graph is 27, and the radius of the graph is 14.

that $D(w)$ is the degree of $w \in V$, and the agent is at node w. The probability of the agent moving to $v \in N(w)$ is given as

$$p(v) = \frac{\frac{1}{D(v)}\left(\max\limits_{n \in N(w)} \sigma(n) - \sigma(v)\right)}{\sum\limits_{m \in N(w)} \left(\frac{1}{D(m)}\left(\max\limits_{n \in N(w)} \sigma(n) - \sigma(m)\right)\right)} \tag{2}$$

The equation of $p(v)$ indicates that $p(v)$ is inversely proportional to the degree of the neighbor nodes. In a grid environment, in many cases, an agent using pidPVAW shows behavior similar to that of an agent using pPVAW. When an agent is going to determine which node among neighbor nodes to visit next, in most cases, the degree of all the neighbor nodes is equal, and therefore, $p(v)$ obtained by pidPVAW is equal to $p(v)$ obtained by pPVAW.

With the exception described above, when an agent situated at an edge node in a grid environment determines which node it will visit next, it is easier for it to select an edge node of the grid than the other nodes. In pidPVAW, we assume that an agent can acquire additional information, e.g., the degree of neighbor nodes, when it determines the next node to visit.

Table 3. Simulation Settings Common to Both Scenarios

Item	Setting Value
Simulation Time	1000000
The Number of Agents	2, 4, 6, 8, 10, respectively
Probability p	0.1

2.3 Features of Agents in These Algorithms

We assume that an agent requires one unit of simulation time when it moves between adjacent nodes. From equations (1) and (2), $p(v) = 0$ for the last visited node v, and therefore, one characteristic of a pPVAW or pidPVAW agent is that it does not return to the last visited node. In particular, let us consider a ring graph; the degree of all nodes in the ring graph is two. In pPVAW or pidPVAW, an agent visits each node at regular intervals, whereas in PVAW, the agent may return to the previously visited node with probability p. A comparison of the three algorithms is shown in Table 1.

3 Simulation

3.1 Overview

We used two scenarios for the evaluation of the algorithms. The purpose of the first scenario was to evaluate the algorithms' adaptiveness to a sudden environment change. The purpose of the second scenario was to evaluate performance on a graph that has a biased degree of distribution. In this simulation, we used a scale-free graph. Some parts of the simulation for Scenarios 1 and 2 had the same parameters settings, as shown in Table 3. For example, the simulation time for each run was set to 1000000 units for both scenarios.

3.2 Scenario 1: Sudden Change on the Environment

The first scenario, Scenario 1, used the graph in Fig. 2. In this scenario, we changed the parameter *size* from 10 to 100 in increments of 10 in Fig. 2. The black node in Fig. 2 was alive and connected to its north and south node at the beginning. When half of each simulation time had passed (i.e., after 500000 simulation time units had passed), the black node was no longer alive and was disconnected from its north node and south node. Thus, a dead end emerged; it is not easy to visit the nodes in this dead end. This scenario unveils whether an algorithm is adaptive to a change in environment.

3.3 Scenario 2: Static Scale-Free Graph

The second scenario, Scenario 2, used the graph in Fig. 3. The graph is scale-free. The distribution of degree of all nodes in Fig. 3 is shown in Table 2. In

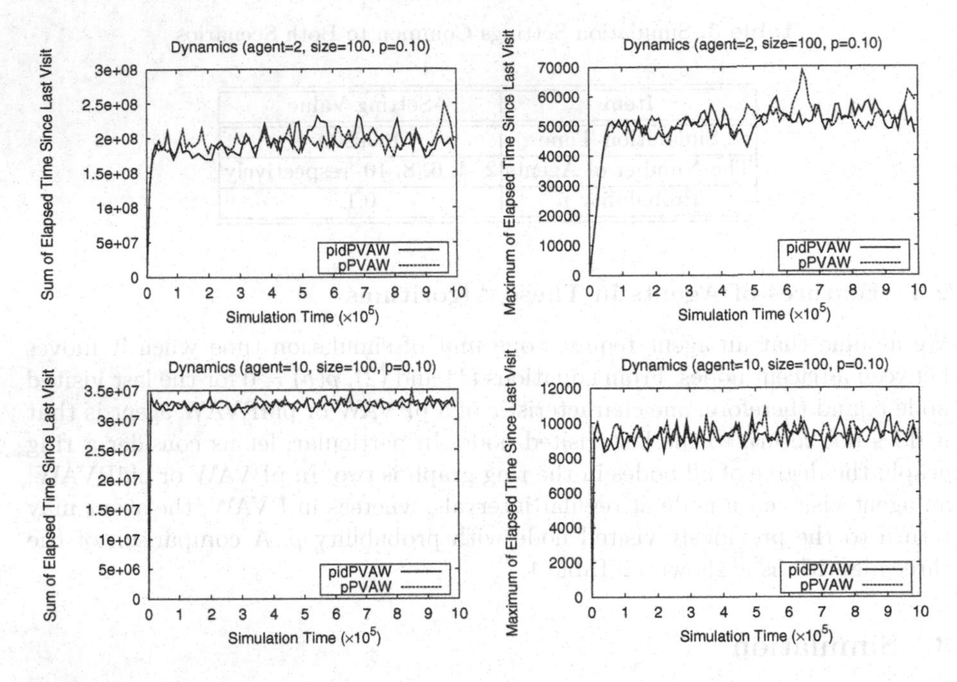

Fig. 4. Simulation result for Scenario 1 with *size*= 100, and p = 0.10. In the two graphs in the first (top) row the setting *agent* = 2 was used and in the two graphs in the second (bottom) row the setting *agent* = 10 was used. For each row, the graph at the left shows s, the sum of the visiting interval for all nodes, and the graph at the right shows m, the maximum visiting interval for all nodes.

this scenario, about 67% of all nodes had only one link and about 17% percent of all nodes had only two links. This graph seems coarse, and therefore, it was difficult for the agents in the graph to visit all the nodes. Scenario 2 ran under a static environment. In other words, the graph structure did not change during simulations in Scenario 2.

3.4 Evaluation Criteria

We used the average visiting interval time of all nodes as a statistical evaluation criterion and the following values of s and m in equations (3) and (4) as our evaluation criteria for dynamic behavior. t is the current time of each simulation, and therefore, $t - \sigma(v)$ represents the elapsed time since the last visit by agents.

$$s = \sum_{v \in V} (t - \sigma(v)) \tag{3}$$

$$m = \max_{v \in V} (t - \sigma(v)) \tag{4}$$

Table 4. Average node visiting interval obtained from Scenario 1. These statistics are the averages over 10 simulations.

	Agent=2		Agent=4		Agent=6	
Algorithm	pidPVAW	pPVAW	pidPVAW	pPVAW	pidPVAW	pPVAW
size=10	1.10×10^2	1.10×10^2	5.49×10^1	5.49×10^1	3.69×10^1	3.66×10^1
size=20	4.20×10^2	4.20×10^2	2.12×10^2	2.10×10^2	1.40×10^2	1.40×10^2
size=30	9.30×10^2	9.30×10^2	4.65×10^2	4.65×10^2	3.10×10^2	3.10×10^2
size=40	1.64×10^3	1.64×10^3	8.20×10^2	8.20×10^2	5.47×10^2	5.47×10^2
size=50	2.55×10^3	2.55×10^3	1.27×10^3	1.27×10^3	8.50×10^2	8.50×10^2
size=60	3.65×10^3	3.66×10^3	1.83×10^3	1.83×10^3	1.22×10^3	1.22×10^3
size=70	4.96×10^3	4.96×10^3	2.48×10^3	2.48×10^3	1.66×10^3	1.66×10^3
size=80	6.46×10^3	6.49×10^3	3.24×10^3	3.23×10^3	2.16×10^3	2.16×10^3
size=90	8.17×10^3	8.17×10^3	4.09×10^3	4.09×10^3	2.73×10^3	2.72×10^3
size=100	1.00×10^4	1.01×10^4	5.04×10^3	5.04×10^3	3.36×10^3	3.36×10^3

	Agent=8		Agent=10	
Algorithm	pidPVAW	pPVAW	pidPVAW	pPVAW
size=10	2.74×10^1	2.74×10^1	2.19×10^1	2.19×10^1
size=20	1.05×10^2	1.05×10^2	8.40×10^1	8.40×10^1
size=30	2.33×10^2	2.33×10^2	1.86×10^2	1.86×10^2
size=40	4.10×10^2	4.10×10^2	3.28×10^2	3.28×10^2
size=50	6.38×10^2	6.38×10^2	5.10×10^2	5.10×10^2
size=60	9.15×10^2	9.15×10^2	7.32×10^2	7.32×10^2
size=70	1.24×10^3	1.24×10^3	9.94×10^2	9.94×10^2
size=80	1.62×10^3	1.62×10^3	1.30×10^3	1.30×10^3
size=90	2.05×10^3	2.05×10^3	1.64×10^3	1.64×10^3
size=100	2.52×10^3	2.52×10^3	2.02×10^3	2.02×10^3

4 Results

The average graphs obtained from 10 simulations for s and m are shown in Figs. 4, and 5, respectively. The statistical results obtained from 10 simulations are shown in 5.

4.1 Scenario 1: Sudden Change on the Environment

In Scenario 1, the results s and m of pidPVAW obtained from Fig. 4 were similar to those of pPVAW and the perspective of s and m did not vary between the two algorithms throughout the simulations. In particular, the performance results, shown in Table 4, showed no difference between the two algorithms while the number of agents changed to 2, 4, 6, 8, and 10, respectively. The reason is that the distribution of degree of the graph used in Scenario 1 is almost constant. Scenario 1 used a graph that consists of two square grids and two corridors, and therefore, all the nodes except those on the edges or corridors had four

Table 5. Average node visiting interval obtained from Scenario 2. These statistics are the averages over 10 simulations. The ratios are based on the visiting interval in the case where the number of agents is 2.

Number of Agents	pidPVAW	Ratio of pidPVAW	pPVAW	Ratio of pPVAW
2	7.81×10^3	1	7.83×10^3	1
4	4.10×10^3	1.91	4.06×10^3	1.93
6	2.74×10^3	2.85	2.71×10^3	2.89
8	2.05×10^3	3.82	2.03×10^3	3.86
10	1.68×10^3	4.65	1.64×10^3	4.77

links. After half of the simulation time passed, the distribution of the degree of nodes was almost the same, while the black node was removed. Therefore, the difference in behavior between these two algorithm is considered to be small, and pidPVAW can achieve a performance equivalent to that of pPVAW.

4.2 Scenario 2: Static Scale-Free Graph

In Scenario 2, the performance of pidPVAW obtained from Fig. 5 is better than that of pPVAW. In both cases where the number of agents was 2 or 10, the difference between pidPVAW and pPVAW in terms of the sum of elapsed time since the last visit, s, and the maximum of elapsed time since the last visit, m, is evident. In particular, in the case where the number of agents was 2, m of pidPVAW varied around 12000 units. On the other hand, m of pPVAW varied around 15000 units. Therefore, m of pidPVAW was about 20% less than that of pPVAW. However, when the number of agents increased, the difference became small. These results imply that agents using pidPVAW visit the nodes that have less connectivity. The reason is that the distribution of the degree of the graph used in Scenario 2 is biased.

When an agent determines the node to which it will move next, the nodes having lower connectivity can easily be selected by the agents by taking the degree of nodes into consideration. Agents using pPVAW can go to hub nodes more easily than agents using pidPVAW. In other words, agents using pPVAW can ignore leaf nodes more easily than agents using pidPVAW, and the average visiting interval of pidPVAW is almost the same as that of pPVAW, as shown in Table 5. The average visiting interval did not change while the number of agents did.

As a result, pidPVAW achieves a better performance than pPVAW, as shown in Fig. 5. In other words, pidPVAW is effective in an environment where the distribution of the degree of nodes in the graph is biased. A pidPVAW agent is apt to visit a lower degree node that has not been visited by an agent for a long time. These results show that pidPVAW can be considered more robust to graph topology than pPVAW.

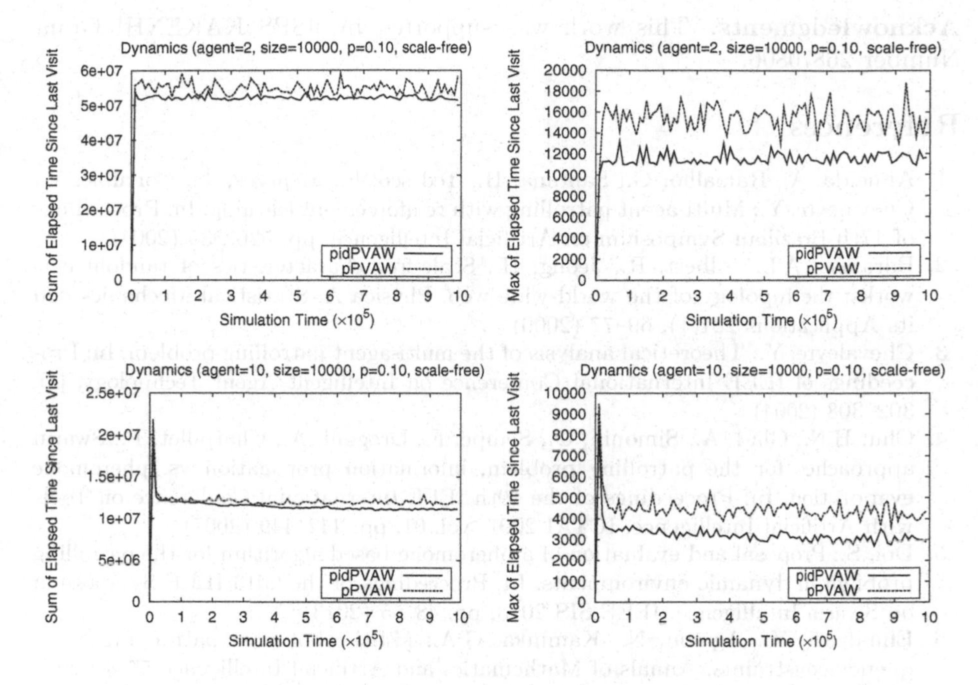

Fig. 5. Simulation result for Scenario 2. The graph had 10000 nodes; the degree distribution of all nodes is shown in Table 2. In the two graphs in the first (top) row the setting *agent* = 2 was used and in the two graphs in the second (bottom) row the setting *agent* = 10 was used. For each row, the graph at the left shows s, the sum of the visiting interval for all nodes, and the graph at the right shows m, the maximum visiting interval for all nodes.

5 Conclusions and Future Work

In this paper, we proposed a patrolling algorithm, pidPVAW, that takes into consideration the degree of neighbor nodes. We confirmed that pidPVAW gives a performance similar to that of pPVAW in the case of environmental changes through computer simulations. In addition, we confirmed that pidPVAW provides a better performance in the case of biased distribution of the degree of nodes in an environment. Thus, it is effective to take the degree of neighbor nodes into consideration. As a result, if an agent can perceive the degree of the neighbor nodes, this should be exploited for effective patrolling. Consequently, pidPVAW, which takes the degree of neighbor nodes into consideration, is a more stable algorithm for resolving patrolling problems than is pPVAW.

In the future, we shall investigate parameter settings and algorithm enhancements for pidPVAW. In particular, we shall also involve the pheromone propagation model [4] in order to visit the nodes that cannot easily be visited by agents.

Acknowledgments. This work was supported by JSPS KAKENHI Grant Number 26870806.

References

1. Almeida, A., Ramalho, G., Santana, H., Tedesco, P., Menezes, T., Corruble, V., Chevaleyre, Y.: Multi-agent patrolling with reinforcement learning. In: Proceedings of 17th Brazilian Symposium on Artificial Intelligence, pp. 526–535 (2004)
2. Barabási, A.L., Albert, R., Jeong, H.: Scale-free characteristics of random networks: the topology of the world-wide web. Physica A: Statistical Mechanics and its Applications 281(1), 69–77 (2000)
3. Chevaleyre, Y.: Theoretical analysis of the multi-agent patrolling problem. In: Proceedings of IEEE International Conference on Intelligent Agent Technology, pp. 302–308 (2004)
4. Chu, H.N., Glad, A., Simonin, O., Sempe, F., Drogoul, A., Charpillet, F.: Swarm approaches for the patrolling problem, information propagation vs. pheromone evaporation. In: Proceedings of the 19th IEEE International Conference on Tools with Artificial Intelligence, ICTAI 2007, vol. 01, pp. 442–449 (2007)
5. Doi, S.: Proposal and evaluation of a pheromone-based algorithm for the patrolling problem in dynamic environments. In: Proceedings of the 2013 IEEE Symposium on Swarm Intelligence, IEEE SIS 2013, pp. 48–55 (2013)
6. Elmaliach, Y., Agmon, N., Kaminka, G.A.: Multi-robot area patrol under frequency constraints. Annals of Mathematics and Artificial Intelligence 57(3), 293–320 (2009)
7. Elor, Y., Bruckstein, A.M.: Autonomous multi-agent cycle based patrolling. Technion CIS Technical Reports CIS-2009-15 (2009)
8. Elor, Y., Bruckstein, A.M.: Autonomous multi-agent cycle based patrolling. In: Dorigo, M., Birattari, M., Di Caro, G.A., Doursat, R., Engelbrecht, A.P., Floreano, D., Gambardella, L.M., Groß, R., Şahin, E., Sayama, H., Stützle, T. (eds.) ANTS 2010. LNCS, vol. 6234, pp. 119–130. Springer, Heidelberg (2010)
9. Gabriely, Y., Rimon, E.: Spanning-tree based coverage of continuous areas by a mobile robot. Annals of Mathematics and Artificial Intelligence 31(1-4), 77–98 (2001)
10. Glad, A., Simonin, O., Buffet, O., Charpillet, F., et al.: Theoretical study of ant-based algorithms for multi-agent patrolling. In: 18th European Conference on Artificial Intelligence Including Prestigious Applications of Intelligent Systems (PAIS 2008) ECAI 2008, pp. 626–630 (2008)
11. Lauri, F., Charpillet, F., et al.: Ant colony optimization applied to the multi-agent patrolling problem. In: IEEE Swarm Intelligence Symposium (2006)
12. Machado, A., Ramalho, G., Zucker, J.D., Drogoul, A.: Multi-agent patrolling: An empirical analysis of alternative architectures. In: Sichman, J.S., Bousquet, F., Davidsson, P. (eds.) MABS 2002. LNCS (LNAI), vol. 2581, pp. 155–170. Springer, Heidelberg (2003)
13. Portugal, D., Rocha, R.: Msp algorithm: multi-robot patrolling based on territory allocation using balanced graph partitioning. In: Proceedings of ACM Symposium on Applied Computing, pp. 1271–1276 (2010)
14. Santana, H., Ramalho, G., Corruble, V., Corruble, V.: Multi-agent patrolling with reinforcement learning. In: AAMAS2004 Proceedings of the Third International Joint Conference on Autonomous Agents and Multiagent Systems, vol. 3, pp. 1122–1129 (2004)

A Choreographer Leads the Pareto Efficient Equilibrium in the Three-Person Network Externality Games: Experimental Evidence*

Hiromasa Takahashi[1], Kazuhito Ogawa[2,**], and Hideo Futamura[3]

[1] Faculty of International Studies, Hiroshima City University
[2] Faculty of Sociology, Kansai University
[3] Faculty of International Studies, Hiroshima City University
kz-ogawa@kansai-u.ac.jp

Abstract. We examine the condition that brings the most efficient outcome under the three-person network externality game. examine three conditions: (1) a role 1 player decides one of the payoff structures (one is the stag-hunt game and the other has the same payoff regardless of player's choice for role 1 player and the dominant strategy for role 0 players) in the first stage and in the second stage role 0 players play the game given by the first stage decision, (2) in the first stage role 0 players play although they do not know the payoff structure, and in the second staeg, a role 1 player observes the first stage decision and decides the payoff structure, (3) cheap talk between role 0 players is introduced into the condition (2). The condition that brings the most efficient outcome is the condition (1): deciding the payoff structure first promotes the probability of realization of the efficient outcome.

Keywords: three-person network externality game, experimental study and equilibrium selection.

1 Introduction

A stag-hunt game is an example of network externality. A stag-hunt game has two pure-strategy Nash equilibria: the risk dominant equilibrium and the payoff dominant equilibrium. According to the prediction from Evolutionary game theory, the risk dominant equilibrium is observed in the long run.(Young, 1998 [12]). Although the payoff dominant equilibrium maximizes the social welfare, it is not observed in the long run.

What realizes the payoff dominant strategy in the stag-hunt game? We introduce the third person, whom we call a "role 1 player" into this game. A role 1

* This is very preliminary version.
** Corresponding author.

H. Handa et al. (eds.), *Proc. of the 18th Asia Pacific Symp. on Intell. & Evol. Systems – Vol. 1*,
Proceedings in Adaptation, Learning and Optimization 1, DOI: 10.1007/978-3-319-13359-1_32

player decides the game structure: one is the stag-hunt game and the other is very trivial game (the payoff level is the same regardless of choice). That is, the game that the rest of two players (we call them a role 0 player) play depends on the choice of a role 1 player. If a role 1 player chooses the stag-hunt game, two role 0 players face with the situation where they offer the risk dominant strategy or the risk dominant strategy.

Does the payoff dominant equilibrium realize in the above mentioned game? If this equilibrium realize, what condition is required? We explore these questions from economic experiments. Adding a role 1 player can make the realization of the payoff dominant equilibrium difficult. However, doing so may make the realization easy. As far as we know, there is no study on this issue.

In the real world, there is the situation wherein a third player select a payoff structure from potential payoff structures and one of them is the stag-hunt. For example, a firm choose the style of cooperate governance and employees work under the cooperate governance. A kind of team production is stag-hunt. On the otherhand, employees can endogeneously establish the cooperate governance in a firm and the firm follow or confirm it. Another example is what happens when establishing an unified standard on something. A commission as the third player, proposes an unified standard on a technology. If a lot of firms which utilize this technology follow this proposal, network externality emerges. On the other hand, firms endogeneously develop the defacto standard and attain network externality, In this case, the commission confirm this.

This paper is organized as follows: In Section 2, we provide the prior experimental studies which investigate how to realize the payoff dominant equilibrium. In Section 3, details of our experiment are introduced. Section 4 describes the outcomes of our experiment. Section 5 discusses the experimental results. Section 6 concludes the paper.

2 Prior Studies

2.1 Experimental Studies on Equilibrium Selection in Coordination Games

Cooper et al.(1990) [3] investigate the experimental equilibrium selection criteria in 3×3 coordination games and find that an equilibrium is realized but that the equilibrium is not always Pareto efficient one. Moreover, they find dominated strategies affect the equilibrium selection. Berninghaus and Ehrhart (2001) [1] experimentally investigate the necessary information to realize the Pareto efficient equilibrium in the weakest link game ($n(\geq 3)$ person coordination game). The realized equilibrium when a participant receives individual decision result of each other participant in the same group and when a participant receives the decision distribution of his or her group is more efficient than the realized equilibrium when a participant receives the amount of the lowest donation.

2.2 Experimental Studies on Human Behavior under Network Externality

Chacravarty(2003) [4] conduct a series of experiments to the network externality model of Katz and Shapiro(1986) [7] and observe the different results from theoretical prediction. Mak and Zwick (2010) [8] conduct the experiment on public good game with network externality. They find that the higher the total amount of donation to the public good is, the higher the probability of realizing the Pareto efficient equilibrium is. Shichijo et al.(2008) [11] establish the theory on the way to familiarize the good with network externality and check this by the laboratory experiments. They find that cheap talk and costly signal improve the coordination problem.

2.3 Experimental Studies on the Effect of Cheap Talk

Let us introduce the brief survey on cheap talk experiments, where players' messages have no direct payoff implications (Crawford, 1998 [5]). Palfrey and Rosenthal (1991) [9] utilize the three-person public good game experiment in which the cost of donation to the public good is private information and invesitigate whether participants behave according to the theoretical prediction and whether cheap talk increases the gained payoff of participants. They find that cheap talk does not improve the gained payoff of participants. Charness and Grosskopf (2004) [2] experimentally investigate when cheap talk promotes participants to the strategy which brings Pareto efficient equilibrium. They find that when a participant receives the information on the choice of othre players, cheap talk promotes him or her to choose the strategy which brings Pareto efficient equilibrium.

Devetag and Ortmann (2007) [10] survey how to decrease the coordination failure in coordination games. According to their survey, decreasing the payoff of secure action, increasing the number of rounds, conducting experiments with fixed matching, increasing the amount of information on the decision result, and introducing pre-play communication (costly or costless) realize the efficient outcome more easily.

3 Experimental Settings

We had conducted a series of experiments in Hiroshima City University and Kansai Univeerisy from November, 2011 to July 2013. A session lasts about 90 minutes.

Let us explain the common settings in the treatments A, AwithCT, and C. These are also common knowledge among participants. 2/3 of the participants are assigned as "Role 0" and the rest of them are assigned as "Role 1" at the beginning of the session. The assigned role remains unchanged throughout the session. A group consists of three participants: two Role 0 players and one Role 1 player. The anonymity of a paricipant is kept.

Table 1. The payoff matrix when a role 1 player chooses **L**

	The strategy of a role 0 player (column player)					
	X			**Y**		
The strategy of a role 0 player (row)	row	columun	role 1	row	columun	role 1
X	200	200	200	80	160	120
Y	160	80	120	160	160	120

Table 2. The payoff matrix when a role 0 player chooses **R**

	The strategy of a role 0 player (column player)					
	X			**Y**		
The strategy of a role 0 player (row)	row	columun	role 1	row	columun	role 1
X	80	80	160	80	160	160
Y	160	80	160	160	160	160

In the treatment SH, all the participatns are assigned to the same role. The anonymity of a paricipant is also kept. These are also common knowledge among participants in the treatment SH.

A session has 20 rounds. Group is randomly made in each round. The payoff structure is common knowledge. At the end of each round, the decision results of a group is informed of the group member as Charness and Grosskopf (2004) —citecharnessbrit2004. These are the common settings among all the treatments.

The experimental payoff of 1 is equal to JPY 1. The sum of the experimental payoff is paid at the end of session. Average reward is 2,500 JPY.

In the treatment A, role 0 players simultaneously and independently choose X or Y in the first stage. A role 1 player observes the decision results of role 0 players and chooses L or R in the second stage. We can interpret the decision procedure in this treatmet as meaning that the payoff structure is determined by a role 1 player (if he or she chooses L (R), the payoff matrix in the table 1 (2) realizes). Therefore, role 0 players make decision before the payoff structure is determined.

Table 3. The payoff matrix in the treatment SH

	Pair's strategy			
	X		Y	
Your strategy	Your payoff	Pair's payoff	Your payoff	pair's payoff
X	200	200	80	160
Y	160	80	160	160

In the treatment C, in the first stage, a role 1 player chooses L or R and the payoff strucuture is determined (tables 1 or 2). In the second stage, role 0 players observe which the payoff structure is determined, and simultaneously and independently choose X or Y.

In the treatment AwithCT, most of the procedure is the same as the treatment A. Only the difference is introduction of cheap talk. Cheap talk is done before the first stage. In the cheap talk stage, one of role 0 players (randomly chosen) send message to the other role 0 players. The content of the message is "what I will choose when I make decision".

In the treatment SH, there are two role 0 players. They play the game in the table 3 simultaneously and independently. The payoff structure is the stag hunt game and the two player version of the table 1.

3.1 Predictions

Let us explain the equilibrium in each treatment. In the treatment A, there are two sub-game perfect Nash equilibrium: the payoff dominant equilibrium wherein role 0 players chooses X and a role 1 player chooses L and the other equilibrium wherein role 0 players chooses Y and role 1 player chooses R. Considering that the risk dominant equilibrium is often observed in the evolutionary game theory, role 0 players often choose Y in the first stage. Thus, we predict the second equilibrium realizes more often than the first.

Prediction I. The risk dominant equilibrium realizes more often than the payoff dominant equilibrium in the treatment A.

In the treatment C, there are two sub-game perfect Nash equilibrium: the payoff dominant equilibrium wherein role 0 players chooses X and a role 1 player chooses L and the other equilibrium wherein role 0 players chooses Y and role 1 player chooses R. A role 1 player who believes that role 0 players choose X with $p \geq 1/2$ chooses L in the first stage and the equilibrium is payoff dominant. Otherwise, a role 1 player chooses R in the first stage and the equilibrium is risk dominant.

Prediction II. In the treatment C, if a role 1 palyer chooses X, then role 0 players choose X.

If this prediction is supported, we observe the payoff dominant equilibrium more than the risk dominant equilibrium.

The equilibrium prediction in the treatment AwithCT is the same as those in the treatment A.

Prediction III. (Y, Y, R) realizes more often than (X, X, L) in the treatment AwithCT.

In the treatment SH, there are two pure strategy Nash equilibirum: the payoff dominant equilibrium (L, L) and the risk dominant equilibrium (R, R). Because our experimental setting is similar to the evolutionary game theory, the latter equilibrium will realize in the long run.

Prediction IV. (Y, Y) realizes more often than (X, X) in the treatment SH.

Which facilitates the realization of the payoff dominant equilibrium, treatments A or C? If the treatment X (A or C) facilitates the realization more, is the share of the equilibrium higher than that in the treatment SH? Although we do not have any referential predictions, these questions are important because we clarify what kind of decision order improves social welfare (the sum of payoffs among three players). All these questions

4 Experimental Results

Table 4 indicates the profile of the participants. The parcentage of female participants seems to be large.

Let us examine four predictions in the section 3.1. The tables 5 and 6 support Predictions I, III, and IV. On Prediciton II, these tables suggest that a role 1 player believes that role 0 players choose X with the probability more than 1/2.

We now explore what kind of game settings leads the payoff dominant equilibrium. In table 5, the percentage of offering the payoff dominant strategy in each role is significantly lower in treatment A than in the treatment C (t−test and

Table 4. Profile of participants

	a role	0		1	
univ	treatment	F	M	F	M
Hiroshima City Univ.	A	10	5	2	1
	AwithCT	9	6	3	0
	C	8	2	4	5
	SH	12	6		
Kansai Univ.	A	18	9	12	6
	AwithCT	5	5	13	4
	C	5	4	7	2
	SH	11	9		

Table 5. The percentage of offering the payoff dominant strategy

	Role 0			Role 1		
treatment	mean	sd	N	mean	sd	N
A	0.294	0.456	840	0.136	0.343	420
AwithCT	0.255	0.436	600	0.113	0.318	300
C	0.777	0.417	480	0.838	0.370	240
C (when choosing L in the fist stage)	0.925	0.013	402			
SH	0.478	0.500	760			

Table 6. The percentage of realizing the payoff dominant equilibrium

treatment	mean	sd	N
A	0.129	0.335	420
AwithCT	0.110	0.313	300
C	0.713	0.453	240
SH	0.392	0.489	760

Kriscal Wallis test, 1% significance). Similarly, table 6 indicates that the payoff dominant equilibrium realizes more in the treatment C than in the treatment A (t−test and Kriscal Wallis test, 1% significance).

The percentage of offering the payoff dominant strategy in role 0 is significantly lower in treatment SH than in the treatment C (t−test and Kriscal Wallis test, 1% significance). Similarly, the payoff dominant equilibrium realizes more in the treatment C than in the treatment SH (t−test and Kriscal Wallis test, 1% significance). These results suggest that the existence of a role 1 player leads the payoff dominant equilibrium in the (second stage) stag hunt game.

The percentage of offering the payoff dominant strategy in role 0 is significantly higher in treatment SH than in the treatment A (t−test and Kriscal Wallis test, 1% significance). Similarly, the payoff dominant equilibrium realizes more in the treatment SH than in the treatment A (t−test and Kriscal Wallis test, 1% significance).

The percentage of offering the payoff dominant strategy in role 0 is not significantly higher in treatment AwithCT than in the treatment A (t−test and Kriscal Wallis test). Similarly, the observation rate of the payoff dominant equilibrium in the treatment AwithCT is not different from that in the treatment A (t−test and Kriscal Wallis test). Thus, the effect of cheap talk is not significant.

4.1 Econometric Analysis

We now explain the econometric analysis on the experimental results. Table 7 indicates that Treatment A dummy is negative and significant in models (2) and (3) and that period×treatment A is negative and significant in all the models. Thus, the results from simple statistical tests is supported by regression models.

Table 8 indicates that treatment ACT dummy is insignificant in all the models and that period×treatment ACT dummy is negative and significant in models

Table 7. Random effect logit regression: A vs. C. Strategy is 1 if a player chooses the payoff dominant strategy. Otherwise, 0. Outcome is 1 if the payoff dominant equilibrium is realized. Otherwise, 0.

	(1)	(2)	(3)
	strategy (role 0)	strategy (role 1)	outcome
period	0.071**	0.095***	0.070**
	(0.031)	(0.036)	(0.029)
treatment A dummy	-0.606	-3.318***	-1.781***
	(0.790)	(1.125)	(0.605)
period×treatment A	-0.296**	-0.232***	-0.206***
	(0.150)	(0.061)	(0.061)
_cons	1.151*	2.051*	0.506
	(0.647)	(1.053)	(0.368)
lnsig2u			
Constant	1.983	1.264	0.694
	(.)	(.)	(.)
N	1320	660	660

robust standard errors clustered by session in parentheses
* $p < 0.1$, ** $p < 0.05$, *** $p < 0.01$

(2) and (3). These results are different from simple statistical tests: introducing cheap talk decreases the possibility of offering the risk dominant strategy in role 1 players and thus the realization of the efficient outcome decreases.

Table 9 indicates that treatment C dummy is significant and positive in both models. This means that role 0 players offer the payoff dominant strategy more in the treatment C (when role 1 player chooses L in the first stage) than in the treatment SH and that the possibility of realizing the efficient outcome is higher in the treatment C than in the treatment SH.

Table 10 indicates that no independent variables are significant when the dependent variable is choice and that treatment A dummy × period is significant and negative when the dependent variable is outcome. Thus, the possibility of observing the efficient outcome decreases as period goes on in the treatment A.

In summary, we attain the following results from econometric analyses.

1. The efficient outcome is observed more in the treatment C than in the treatment A.
2. The effect of cheap talk does not promote the payoff dominant strategy. Therefore, the percentage of the efficient outcome is not different between in the treatment A and in the treatment AwithCT.
3. The efficient outcome is observed more in the treatment C than in the treatment SH.
4. The efficient outcome is observed less in the treatment A than in the treatment SH as period goes on.

Table 8. Random effect logit regression: A vs. AwithCT

	(1) choice (role 0)	(2) choice (role 1)	(3) outcome
period	-0.227	-0.129***	-0.134***
	(0.149)	(0.043)	(0.052)
treatmentACT Dummy	0.390	1.214	1.240
	(0.725)	(0.996)	(0.987)
period×treatment ACT	-0.089	-0.225***	-0.215***
	(0.193)	(0.050)	(0.062)
Constant	0.544	-1.061**	-1.222**
	(0.460)	(0.453)	(0.512)
lnsig2u			
_cons	2.060	0.181	0.466**
	(.)	(0.211)	(0.222)
N	1440	720	720

Robust standard errors clustered by session in parentheses
* $p < 0.1$, ** $p < 0.05$, *** $p < 0.01$

Table 9. Random effect logit regression: C vs. SH. Conditional choice means choice in the second stage when a role 1 player choose L.

	(1) conditional choice (role 0)	(2) outcome
period	-0.025	0.110
	(0.119)	(0.070)
treatment C dummy	5.733*	4.864***
	(3.031)	(1.774)
period×treatment C	0.135	-0.048
	(0.149)	(0.077)
Constant	0.435	-2.787*
	(0.651)	(1.694)
lnsig2u		
_cons	3.143	2.325
	(.)	(.)
N	1162	1162

Robust standard errors clustered by session in parentheses
* $p < 0.1$, ** $p < 0.05$, *** $p < 0.01$

Table 10. A vs. SH

	(1)	(2)
	choice (role 0)	outcome
period	-0.024	0.110
	(0.115)	(0.068)
treatmentA dummy	0.150	0.776
	(0.780)	(1.701)
period×treatmentA	-0.217	-0.283**
	(0.204)	(0.116)
Constant	0.314	-2.789*
	(0.614)	(1.640)
lnsig2u		
_cons	2.702	2.328
	(.)	(.)
N	1600	1600

Robust standard errors in parentheses
* $p < 0.1$, ** $p < 0.05$, *** $p < 0.01$

5 Discussion

Experimental results indicates that the setting in the treatment C is the best to realize the payoff dominant equilibrium in the three person network externality game experiments. For the realization of the payoff dominant equilibrium (X, X, L), it is necessary that a role 1 player believes that each role 0 player chooses X in the second stage after a role 1 player offer L in the first stage. For a role 1 player, doing so is risky because if role 0 players fail to coordinate, the payoff of a role 1 player is at most the payoff of 120. If a role 1 player chooses R in the first stage, he or she always attaines the payoff of 160.

Although theory indicates that choosing L has the risk of decrease of the payoff in the treatment C, the experimental result shows 92% of the observations in the second stage are the payoff dominant strategy if a role 1 player chooses L in the first stage. The expected payoff of a role 1 player is 184, which is higher than when choosing R in the first stage.

This result indicates two possibilities: First possibility is that role 1 players beleive that role 0 players will choose X with high probability if he or she chooses L in the first stage. Second is that role 1 players influence the decision makings of role 0 players through his or her first stage decision. Choosing L is the message to role 0 players that a role 1 player wants role 0 players to choose X because all the players get the highest payoff. According to the second possibility, role 0 players exactly understand what role 1 player want to tell.

Introducing cheap talk into the treatment A does not encourage role 0 players to offer the paoff dominant strategy. This is different from suvey in the Devetag and Ortmann (2007) [10] and the results from Shichijo et al.(2008) [11] but is similar to the results in Palfrey and Rosenthal (1991) [9]. In our cheap talk setting, a role 0 player tells his or her intention to the other role 0 player. Telling the intention all other players will improve the performance of cheap talk and increase the possibility of offering the payoff dominant starategy in the treatment A.

Let us interpret our experimental results. Because a role 1 player decides the payof structure, it is natural to call him or her a a choreographer in the sense of Gintis (2010) [6]. Although our experiments focus on the three-person network externality game, our experimental results suggest that when a choreographer dettermines the payoff structure first, the efficient outcome realizes easily.

6 Conclusion

We conduct a series of experiments to clarify what promotes the realization of the payoff dominant equilibrium in the three-person network externality games. We experimentally find that when a role 1 player, who decides the payoff structure, make decision in the first stage, the most efficient outcome arises the most easily. This is due to the suggestion that the belief of a role 1 player that role 0 players suceed to coordination in the second stage is high and that offering L is a kind of signal that promotes role 0 players to choose the payoff dominant strategy. Our experimental results suggest that the payoff structure should be settled first for increasing the possibility of realizing the efficient outcome.

In our study, we do not conduct the treatment wherein two players play SH but one of them offer X or Y in the first stage and the other player observes the first stage decision and chooses X or Y. If the percentage of the efficient outcome is lower in this treatment than in the treatment C, the importance of a choreographer is confirmed. This will be our future study.

References

1. Berninghaus, S.K., Ehrhart, K.M.: Coordination and Information: Recent Experimental Evidence. Economics Letters 73, 345–351 (2001)
2. Charness, G., Grosskopf, B.: What makes cheap talk effective? Experimental evidence. Economics Letters 83(3), 383–389 (2004)
3. Cooper, R., et al.: Selection Criteria in Coordination Games: Some Experimental Results. American Economic Review 80(1), 218–233 (1990)
4. Chakravarty, S.: Experimental Evidence on Product Adoption in the Presence of Network Externalities. Review of Industrial Organization 23(3), 233–254 (2003)
5. Crawford, V.: A Survey of Experiments on Communication via Cheap Talk. Journal of Economic Theory 78(2), 286–298 (1998)
6. Gintis, H.: Social Norms as Choreography. Politics, Philosophy, and Economics 9(3) (2010)
7. Katz, M.L., Shapiro, C.: Technology Adoption in the Presence of Network Externalities. Journal of Political Economy 94(4), 822–841 (1986)
8. Mak, V., Zwick, R.: Investment decisions and coordination problems in a market with network externalities: an experimental study. Journal of Economic Behavior and Organization 76(3), 759–773 (2010)

9. Palfrey, T., Rosenthal, H.: Testing for Effects of Cheap Talk in a Public Goods Game with Private Information. Games and Economic Behavior 3, 183–220 (1991)
10. Devetag, G., Ortmann, A.: When and why? A critical survey on coordination failure in the laboratory. Experimental Economics 10(3), 331–344 (2007)
11. Shichijo, T., Akai, K., Kusakawa, T., Saijo, T.: Designing a Mechanism to Cope with Coordination Failure in the Goods with Network Externalities: An Experimental Study, mimeo (2008)
12. Young, P.: Individual Strategy and Social Structure: An Evolutionary Theory of Institutions. Princeton University Press, Princeton (1998)

Figure Pattern Creation Support for Escher-Like Tiling by Interactive Genetic Algorithms

Satoshi Ono[1], Megumi Kisanuki[1], Hirofumi Machii[1], and Kazunori Mizuno[2]

[1] Department of Information Science and Biomedical Engineering, Graduate School of Science and Engineering, Kagoshima University 1-21-40, Korimoto, Kagoshima 890-0065, Japan
[2] Department of Computer Science, Faculty of Engineering, Takushoku University 815-1 Tatemachi, Hachioji, Tokyo 193-0985, Japan
ono@ibe.kagoshima-u.ac.jp, mizuno@cs.takushoku-u.ac.jp

Abstract. A tiling design is very significant for mathematical aspect. There has been some studies on producing tiling designs using a computer. Most of them, however, are difficult to support users' creativity because several or various output figure cannot be obtained for one input. In this paper, we have proposed a system that can support users to create tile figures for Escher like tiling designs. The system can create several possible candidate figures from the input one, where repetition of reproducing or varying partial shapes of the figure and evaluating or redrawing the figure by users is executed by employing interactive genetic algorithms. We demonstrate that our system can create various tile figures, which are as close as possible to the input figure and reflected to users' preference.

Keywords: genetic algorithm, interactive evolutionary computation, image processing, tiling, Escher, design support.

1 Introduction

A tiling of the plane is a collection of one or several types of figure shapes, or tiles, that cover the plane without any gaps and overlaps. Designs using tilings are ubiquitous around us, appearing in facade design of buildings, decoration of clothes, accessories, and tablewares, etc. Tilings are composed of wide range of shapes from regularly geometrical to more complicated. Because tilings are also very interesting for mathematical aspects, there has been many discussions as for properties and natures of tiling patterns. However, it is difficult to manually construct complex tiling patterns due to tight restrictions of tilings, i.e., preciously adjusting edges of adjacent tiles while taking account of concavo-convex shapes should be needed.

On the other hand, in solving the optimization problems that seem to be difficult to define an objective, or fitness, function by a numerical formula, e.g., evaluation of individual preference, tastes and senses, etc., an alternative model exploiting human evaluation has been adopted for evolutionary algorithms (EA),

H. Handa et al. (eds.), *Proc. of the 18th Asia Pacific Symp. on Intell. & Evol. Systems – Vol. 1*, Proceedings in Adaptation, Learning and Optimization 1, DOI: 10.1007/978-3-319-13359-1_33

called interactive evolutionary computation (IEC)[9]. A genetic algorithm (GA), which is one of the most general EAs, can also deal with large search spaces and interactivity is the best way to take into account the user preference in solving such kinds of optimization problems.

In this paper, we have developed a system that can create figure patterns forEscher-like tiling pattens in cooperation with users[7]. In our system, interactive GAs are employed to find a figure shape for tiling. Users cannot only choose a possible candidate solution, or a candidate figure, out of ones presented by the system, but also update the goal solution according to users' preference by redrawing the candidate figure. The system thus can create various shapes of figures that completely satisfy restrictions for tilings while supporting divergent thinking of users.

2 Overview

2.1 Tiling

M. C. Escher, the Dutch graphic artist, had left very interesting and ingenious art forms based on the geometry of two- and three-dimensional spaces although he had little mathematical background. The most popular are his created pictures based on tilings, or tessellations, of the two-dimensional plane. These images feature one or several kinds of figures, most of which are animal-like forms such as birds, fish, and reptiles, that complement each other, thus forming a complete coverage of the plane.

In order to automatically discover such tilings, *Escherization problem* has been posed by [5], defined as follows:

Problems ("ESCHERIZATION"): Given a closed plane figure S (the 'goal shape'), find a new closed figure T such that:

1. T is as close as possible to S; and
2. copies of T fit together to form a tiling of the plane.

In other words, an Escher-like tiling is a collection of figures obtained as a solution to the above problem.

We focus on only *isohedral tilings*, which are ones in which a single tile can cover the entire plane through repeated application of rigid motions from the tiling's symmetry group. Isohedral tilings are classified into 93 types, each of which is identified as IH01, IH02, ..., IH93, called a 'tiling pattern (TP)' in this paper.

Each TP has a separate polygon figure satisfying geometric constraints required for tilings. A rough contour of this polygon figure are changed by positions of vertices of the polygon. A point cloud in which some controllable points are contained is put on edges between vertices and the more specific shape of the figure is obtained by changing the position of some points in the point cloud.

2.2 Interactive Genetic Algorithms

A genetic algorithm (GA) is one of the most popular evolutionary computation algorithms, which is based on the principles of the evolution via natural selection, employing a population of individuals that undergo selection by variation inducing operators including recombination, or crossover, and mutation[3]. In GAs, a set, or a candidate solution, of elements in the state space is defined as an individual, each of which is represented by a chromosome composed of genes that corresponds to values for problem variables. A set of individuals is called a population. Individuals are partially taken from the population and used to reproduce a new population for the next generation. Reproductive success for each individual varies with its own fitness value calculated by a fitness function used to evaluate individuals, i.e., the more suitable individuals may probably be reproduced as new individuals, or offsprings, by applying genetic operators. These are repeated until a certain condition is satisfied, e.g., predefined number of generations is attained, enabling to get an optimal or quasi-optimal solution.

Interactive GAs (IGAs) is defined as GAs that use human evaluation instead of evaluation by a fitness function[9]. These are effectively used for the problems where the definition of fitness functions is difficult or not known, e.g. visual appeal or attractiveness including human knowledge, preference, or sensitiveness. In IGAs, the individual is designed according to the problem as well as general GAs, whereas the definition of a fitness function cannot be required. The genetic operations are also applied to the selected individuals in the similar manner to general GAs.

2.3 Related Work

Kaplan has proposed the method to solve *Escherization problems* based on simulated annealing (SA)[6]. This method attempts to obtain the optimal figure T as close as possible to S by repeating 2 processes: evaluating the figure and adjusting edge shapes of the figure. The latter provides a new candidate figure with locally adjusted shapes. Even when the candidate figure has lower fitness value than the previous one, the candidate may be accepted with probability fixed by the temperature parameter of SA and updated for the next search step. He has tried to get the optimal figure for all 93 isohedral tiling patterns and demonstrated that the optimal figure can be comparatively easily obtained for input figures with almost convex shapes

Koizumi and Sugihara have proposed the isohedral tiling design method using matrix operations[8]. In this method, the *Escherization problem* is redefined as an optimization problem and also reduced to calculating a maximum eigenvalue of the symmetry matrix. They have demonstrated that the optimal figure can be obtained for input figures with not only convex shapes but also more complicated shapes. This method, however, can get only one figure for an input figure, meaning that it is difficult to reflect users' preference. Users may also require repetition of putting control points on edge segments until they obtain a satisfied figure.

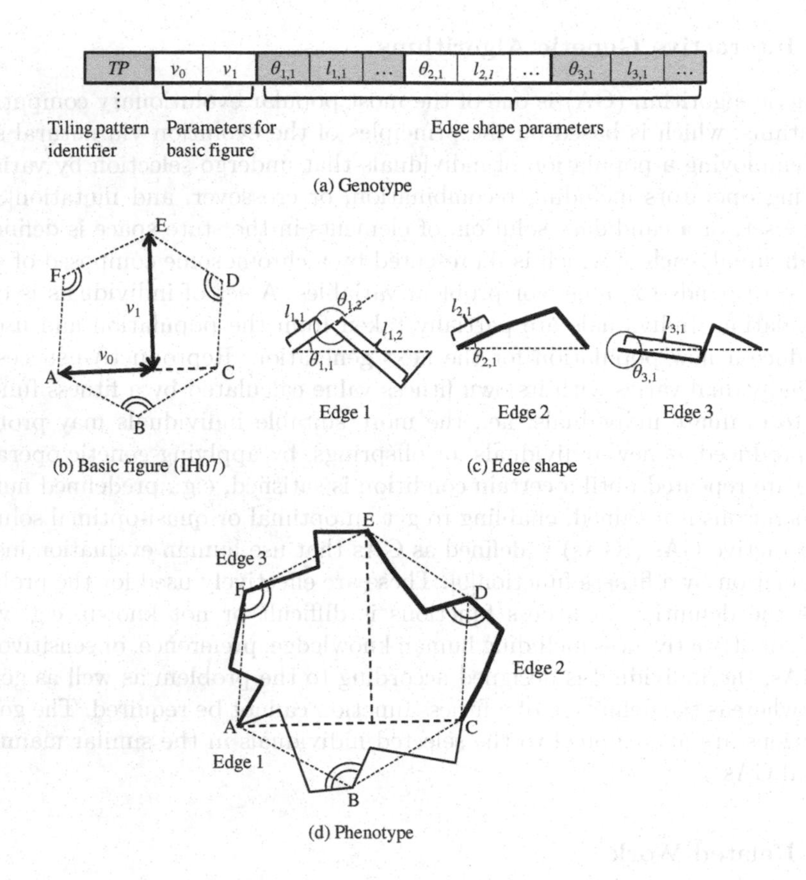

Fig. 1. An example of a chromosome representation, or a genotype and its corresponding phenotype for IH07

3 The Creation Support Method for Escher-Like Tiling

3.1 Basic Principles

The proposed method can create tiling patterns for the isohedral tiling as close as possible to a user input figure. An Escher-like tiling design can also produce based on the obtained figure. Our developed system exploiting this method is based on IGAs, aiming to not only reflect users' preference but also support users' divergent thinking. Our principles of development of the system are as follows:

i) Genetic operations: A solution derived from Koizumi's method[8] is employed as one of individuals, or candidate solutions, in an initial population, in which a superior individual is contained as a result. In order to make the most use of this individual, individuals with a higher fitness value are probably selected as a parent for the crossover operation, where the individual with the best fitness can survive in the population for the next generation.

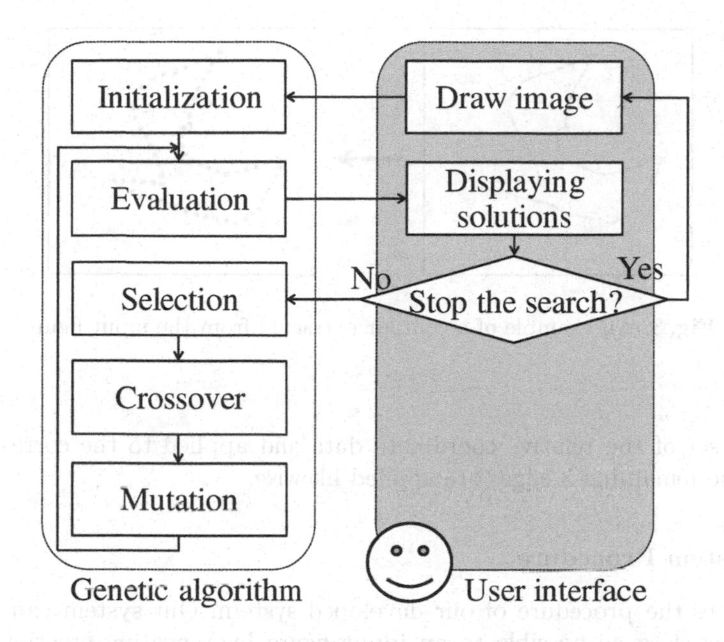

Fig. 2. The system procedure for the proposed system

ii) Iterated IGAs: In general IEC, users choose a promising candidate individual out of several ones generated by the system. However, user preferred figures may be not necessarily contained in the candidates. In our method, users thus cannot only choose a candidate but also adjust and partially redraw the figure so that the figure corresponding to the candidate can be closer to users' preference.

3.2 Encoding

Each chromosome, or individual in GAs, is represented as a set of elements by which local shapes of a figure are fixed. Fig. 1 gives an example of the chromosome representation of GAs and conversion from the genotype to the phenotype, where the TP is IH07 which corresponds to the tiling by applying a rigid motion, i.e., 120° rotation and translation. The genotype (Fig. 1 (a)) is composed of the TP identifier, the set of inherent parameters of the TP (Fig. 1 (b)), and the set of the relative coordinate data, i.e., angles and lengths of line segments, of points allocated between vertices (Fig. 1 (c)). In the set of the relative coordinate data of Fig. 1, if shapes of 3 edges, or AB, CD, and EF, can be defined, ones of the remaining 3 edges, or BC, DE, and FA, can also be represented as well, corresponding to "Edge 1" ~ "Edge 3" in Fig. 1 (c), respectively. As for the conversion from the genotype to the phenotype (Fig. 1 (d)), a contour is first created by inherent parameters of the TP, i.e., the relative positions between shape vertices of the figure are fixed. Then, each shape of 3 edges is created

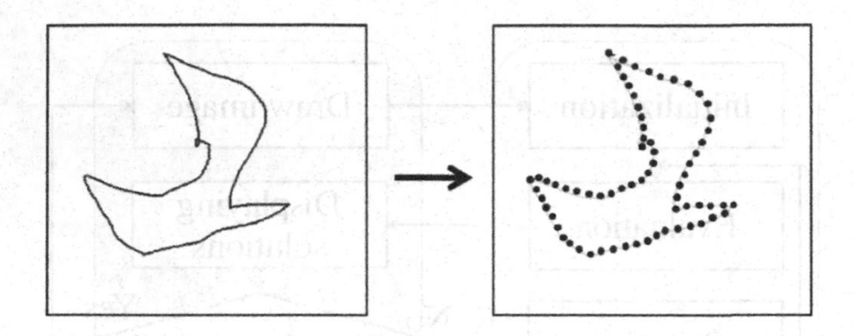

Fig. 3. An example of a contour extracted from the input figure

from the set of the relative coordinate data and applied to the corresponding edges. The remaining 3 edges are applied likewise.

3.3 System Procedure

Fig. 2 gives the procedure of our developed system. Our system can produce a figure as close as possible to an input figure by repeating creation of new candidate figures and evaluation of ones. Even if GAs start the search process, users can interrupt the process at any time to redraw the figure, enabling to easily adopt users' preference as a final output.

(1) **Input an Initial Figure Image**

Users first input a drawn figure image to the system. Then, the system acquires vertex coordinates of a point sequence, or a set of points, from the contour extracted from the input image[2]. A certain number, which is specified by users, of candidate vertex coordinates is selected from the acquired points according to basic shapes of the TP, as shown in Fig. 3.

(2) **Generate the Initial Population**

An initial population of individuals is generated: a candidate figure is automatically created by Koizumi's method[8] from vertex coordinates selected in (1), being added to the initial population. By adding this candidate to the population, we expect that conversing to a superior solution closer to the input image can be facilitated. The other candidates to be added to the population are created based on the figure created by Koizumi's method as well, in which partial shapes of the user input image are randomly applied and arranged.

(3) **Evaluate Fitness Values**

The system evaluates similarity of the shape between the input figure and figures each of which individual generated by GAs forms. Similarity is calculated based on a turning function stated in Sec. 3.4, being defined as a fitness value of each individual.

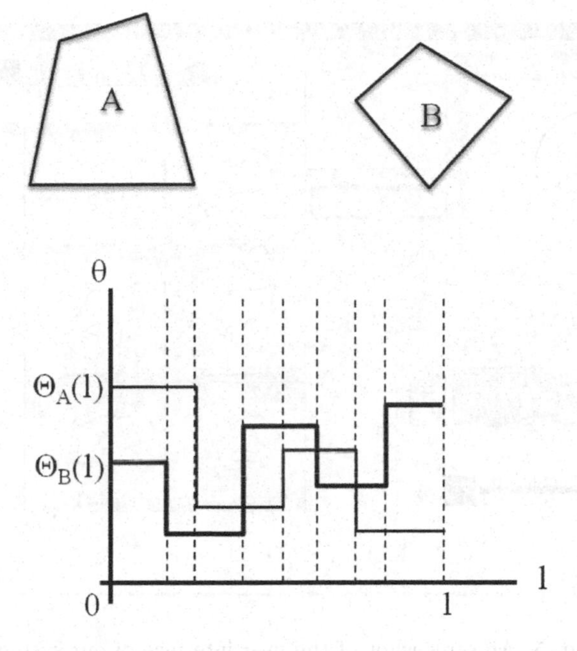

Fig. 4. An example of calculating similarity between different polygons

(4) **Apply Genetic Operators**

Selection: The top-two individuals on fitness values are unconditionally left for the next generation by elitist selection. The four individuals of the lowest fitness values are made die out. To supplement decreasing number of individuals, each individual selected by elitist selection is reproduced for the next generation.

Crossover: Naive fitness proportionate selection, or roulette wheel selection, is used for selecting potentially useful candidates for recombination except for duplication of selected candidates for the elite. 2-point crossover is employed, where the first, second, and third genes, each of which correspond to *tiling pattern ID* and parameters, v_0 and v_1, respectively, defined as the Genotype in Fig. 1, is excluded from the range of crossover.

Mutation: For each gene in all individuals except for elite two ones, whether mutation occurs or not is determined according to mutation rate. Even when a value of a parameter is modified by mutation, the value is replaced so that the value can satisfy restrictions required for TPs.

(5) **Presenting candidate figures to users**

After repeating GA process for a certain number of generations, the system present several possible candidate figures to users. Users can choose one of those according to their own preference. Besides, when there is no candidates satisfying their preference, users can also redraw the figure partially or entirely. Then, GAs restart their main process for the figure renewedly updated as the target one.

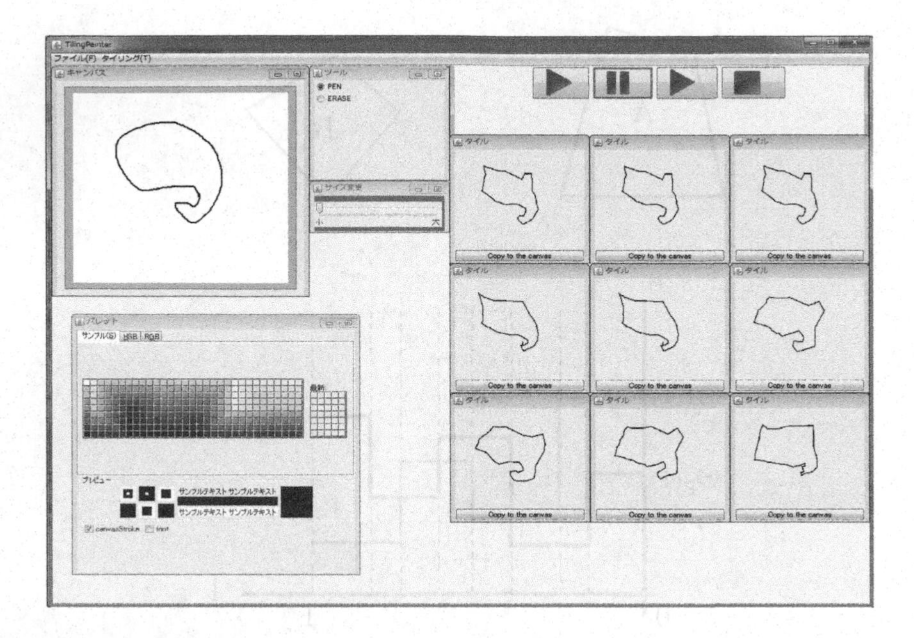

Fig. 5. An screenshot of the user interface of our system

3.4 Turning Function

When a figure closer to the one input or redrawed by users, the individual corresponding to the figure has a higher fitness value. It is thus necessary to compare how similar shapes of two different contours are. To calculate a fitness value of each candidate individual as a evaluation function, we adopt a turning function[1] that is a cumulative measure of the angles through which a polygonal curve turns. As shown in Fig. 4, the turning function $\Theta_A(l)$ of a polygon A gives the angle between the counter-clockwise tangent and the horizontal axis as a function of distance, l, along the polygonal curve, where the distance l is normalized $(0 \leq l < 1)$. Similarity between two polygons, A and B, is defined as

$$D = \int_0^1 \{\Theta_A(l) - \Theta_B(l)\} dl, \qquad (1)$$

$$\text{where } 0 \leq \text{l} \leq 1,$$

showing that two shapes are more similar when the value of D is closer to 0.

3.5 User Interface

Fig. 5 gives the user interface of our system, which is composed of the user draw window (left) and the candidate presentation window (right): the former includes the canvas to draw a figure and the palette tool to adjust the figure. Users can draw the figure without any restrictions on the canvas and, if necessary, adjust

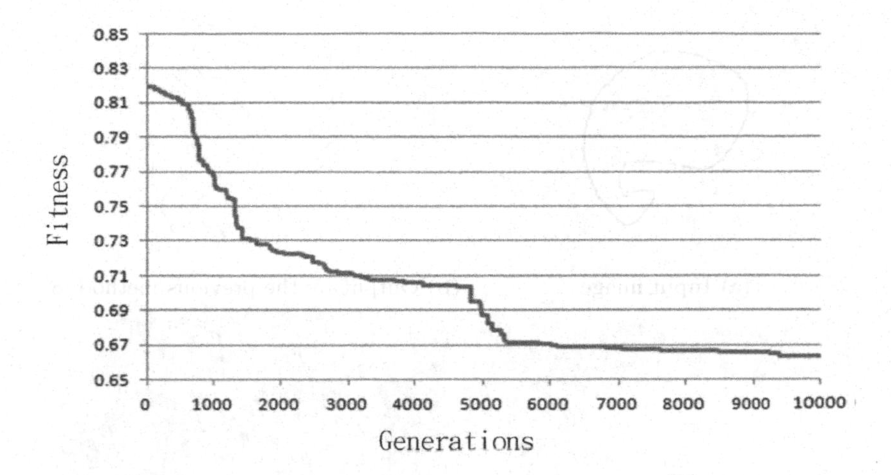

Fig. 6. A chage of the fitness value of the best individual in GAs without user evaluation

the figure by applying some functions of the palette tool including selecting drawn objects, controlling thickness of line segments to draw, and altering colors of line segments.

When users choose a preferred figure out of candidate ones presented in the presentation window, the figure is reflected to the canvas that enables users to directly redraw and edit the figure.

4 Evaluation

4.1 Output Examples by Non-interactive GAs

To confirm ability of GAs without user evaluation, i.e., evaluating fitness values by only similarity based on the turning function, we conduct the experiment for the following settings: the number of individuals, or the population size, and the upper limit of generations are fixed at 100 and 10,000, respectively. The number of elite to retain for the next generation is 2. As for genetic operators, 2-point crossover is employed and the mutation rate is set to 0.5%.

Fig. 6 gives the experimental result on the change of the fitness value of the best individual as a function of the number of generations, where the average value for 5 trials is plotted. As shown in Fig. 6, the fitness value is decreased in accordance with increasing number of generations. Fig. 7 gives one of examples generated by GAs without user evaluation, where (a), (b), (c), and (d) correspond to the input figure, the figure generated by Koizumi's method from the input, the output figure generated by GAs, and the tiling image by embedding the output, respectively. Compared with (b), the figure (c) may be closer to the input figure (a) at the upper right side shape and the lower roughness shape.

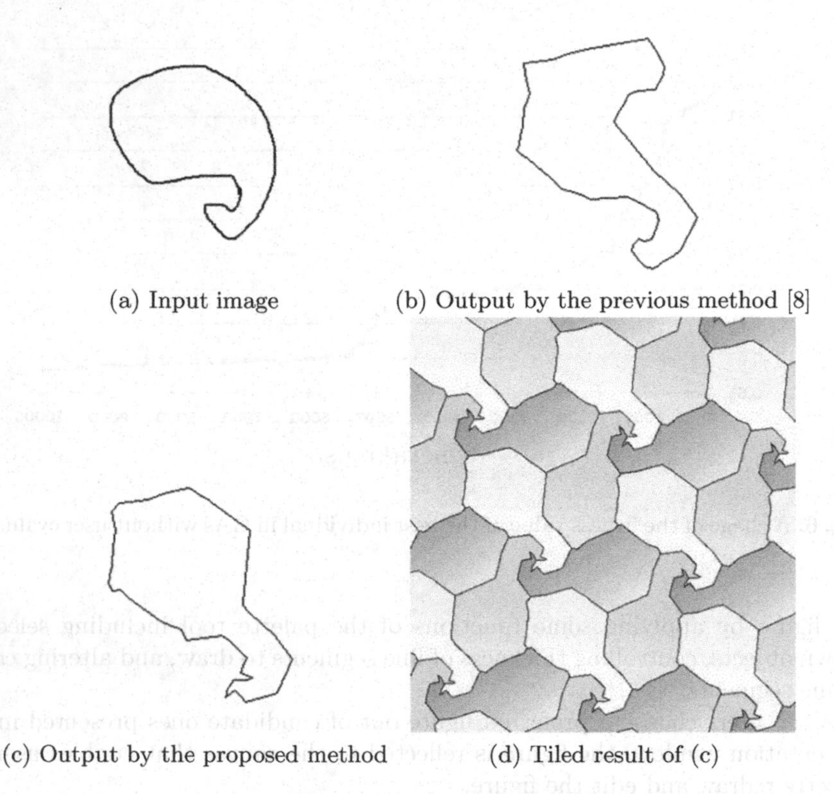

(a) Input image (b) Output by the previous method [8]

(c) Output by the proposed method (d) Tiled result of (c)

Fig. 7. An example of a tiling where a figure output by GAs without user evaluation is used

4.2 Output Examples by Interactive GAs

Fig. 8 gives an example of the execution process of the proposed method, or iterated IGAs, from the input to the output. A dog's face like figure input as a initial figure is gradually varied as progressing IGAs, finally turning into a completely different shape, e.g., looking like a dog's body shape, by changing users' preference.

Fig. 9 gives the examples of a tiling in the proposed method for the initial input figure same as Fig. 8, where (a) and (b) correspond to the output figure and its tiled image and (c) and (d) are the ones for another output figure, respectively. As shown in Fig. 9, the proposed method can create distinct figure shapes, whose copies can fit together to form a tiling, even if the same input is given, enabling to produce various tiling images while supporting the way of thinking of users.

Furthermore, the proposed method can create various and effective figure patterns as shown in Fig. 10.

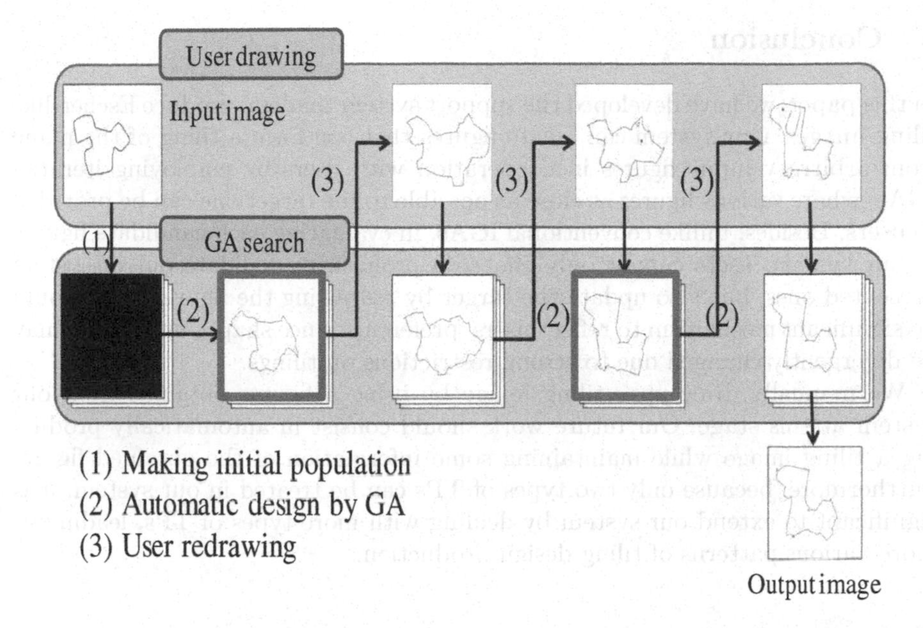

Fig. 8. An example of the execution process of the proposed system

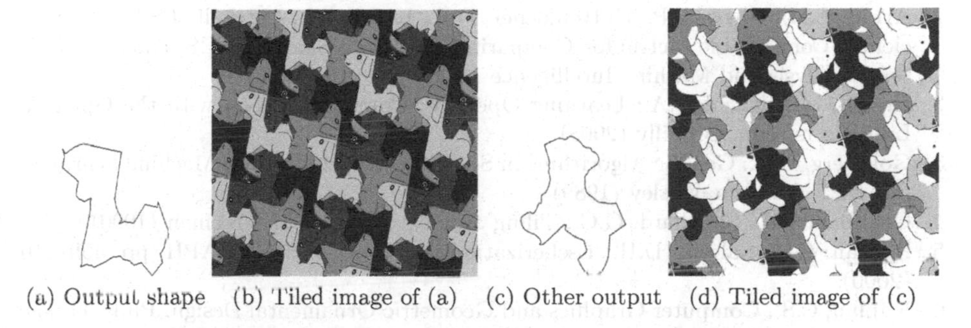

(a) Output shape (b) Tiled image of (a) (c) Other output (d) Tiled image of (c)

Fig. 9. An example of a tiling in the proposed method

Fig. 10. Other examples of tiled images using figures obtained by the proposed method

5 Conclusion

In this paper, we have developed the support system that can produce Escher-like tiling images. Our system can create figures that can form a tiling of the plane from arbitrary input figures in cooperation with users by employing iterated IGAs, where various figures as close as possible to the target one can be provided to users. Besides, unlike conventional IGAs, in evaluating each candidate figures in our system, users cannot only choose a promising candidate out of system presented ones, but also update the target by redrawing the figure. This should be significant mechanism to reflect users' preference since shapes users want may be divergently changed due to strong restrictions on tilings.

We manually draw up a tiling image by using a figures obtained from our system at this stage. Our future work should consist in automatically producing a tiling image while maintaining some information of the obtained figure. Furthermore, because only two types of TPs can be treated in our system, it is significant to extend our system by dealing with more types of TPs, leading to more various patterns of tiling design production.

References

1. Arkin, E.M., Chew, L.P., Huttenlocher, D.P., Kedem, K., Mitchell, J.S.B.: An Efficiently Computable Metric for Comparing Polygonal Shapes. IEEE Trans. on Pattern Analysis and Machine Intelligence 13(3), 209–216 (1991)
2. Bradski, G., Kaehler, A.: Learning OpenCV: Computer Vision with the OpenCV Library, 1st edn. O'Reilly (2008)
3. Goldberg, D.E.: Genetic Algorithms in Search, Optimization and Machine Learning, pp. 1–25. Addison–Wesley (1989)
4. Grunbaum, B., Shephard, G.C.: Tiling and Patterns. W. H. Freeman (1990)
5. Kaplan, C.S., Salesin, D.H.: Escherization. In: Proc. of SIGGRAPH, pp. 499–510 (2000)
6. Kaplan, C.S.: Computer Graphics and Geometric Ornamental Design. Ph.D Thesis, University of Washington, Seattle (2002)
7. Kisanuki, M., Ono, S., Machii, H., Mizuno, K., Nakayama, S.: Escher-Like Tiling Pattern Design by Interactive Iterative Genetic Algorithm, NICOGRAPH International 2013, poster (2013)
8. Koizumi, H., Sugihara, K.: Computer-aided design of escher-like tiling, NICOGRAPH Paper Contest (2009) (in Japanese)
9. Takagi, H.: Interactive evolutionary computation: fusion of the capabilities of EC optimization and human evaluation. Proc. IEEE 89, 1275–1296 (2001)

Angle-Based Outlier Detection Algorithm with More Stable Relationships*

Xiaojie Li, Jian Cheng Lv, and Dongdong Cheng

The Machine Intelligence Laboratory, College of Computer Science, Sichuan University, Chengdu 610065, P.R. China
lvjiancheng@scu.edu.cn

Abstract. Outlier detection is very useful in many applications, such as fraud detection and network intrusion. The angle-based outlier detection (ABOD) method, proposed by Kriegel, plays an important role in identifying outliers in high-dimensional spaces. However, ABOD only considers the relationships between each point and its neighbors and does not consider the relationships among these neighbors, causing the method to identify incorrect outliers. In this paper, we provide a small but consistent improvement by replacing the relationships between each point and its neighbors with more stable relationships among neighbors. Compared with other related methods, which work best in either high or low-dimensional spaces, our method gives significant gains in both high and low-dimensional spaces. Experimental results on both synthetic and real-world datasets demonstrate the effectiveness of our method.

Keywords: Outlier detection, outlier factor, angle-based.

1 Introduction

Outlier detection is an important data mining task for many practical applications, such as fraud detection, public health, industrial damage detection, and network intrusion. According to Hawkins [1], "an outlier is an observation that deviates so much from other observations as to arouse suspicion that it was generated by a different mechanism." This is different from noisy data, because noise involves random errors or variance in a measured variable and should be removed before outlier detection [2].

Unlike most other data mining tasks, such as clustering, classification, and pattern analysis, which aim to find general patterns, outlier detection is generally used to identify observations that deviate significantly from the general data distribution [3, 4]. This is reasonable since the observations are usually generated by certain mechanisms or statistical processes. The distinct deviations from the main distribution are then assumed to originate from a different mechanism

* This work was supported by the National Natural Science Foundation of China under grant 61375065 and the National Program on Key Basic Research Project (973 Program) under grant 2011CB302201.

H. Handa et al. (eds.), *Proc. of the 18th Asia Pacific Symp. on Intell. & Evol. Systems – Vol. 1*,
Proceedings in Adaptation, Learning and Optimization 1, DOI: 10.1007/978-3-319-13359-1_34

such as fraud. Furthermore, detecting outliers is more interesting than detecting normal observations. For example, a liver disorder detection system might consider normal observations to represent healthy patients and outlier observations to denote patients with liver disorders.

Many areas of data mining, machine learning, and pattern recognition require analysis of high-dimensional data. Outlier detection in such cases involves tough challenges. Typically, distance-based discrimination is poor in a high-dimensional space [3, 5]. Thus, distance-based criteria are unsuitable to varying degrees in high-dimensional spaces. Furthermore, irrelevant or redundant attributes interfere with outlier detection in high-dimensional spaces [4, 6–8, 5, 9].

Existing outlier detection methods can be categorized from several different perspectives [2, 3, 10]. According to the underlying approach adopted by each technique, existing approaches can be divided into three categories [3]: statistical models (e.g., depth or deviation-based methods [11–14]), spatial proximity-based models (e.g., distance [15–19] or density-based methods [20–23]), and complex model adaptations of different models to a specific problem. For example, given a univariate data set following a Gaussian distribution, Shewhart [24] used 3σ as the threshold for identifying outliers, where σ is the standard deviation. While most of these methods are distribution-based, standard deviation is highly sensitive to outliers [11]. In general, density-based methods compare the local density of each point with those of its immediate neighbors, with the various methods differing most significantly in their density estimation strategies [20, 21].

However, some methods are infeasible for high-dimensional data owing to their computational complexity [3]. Others that appear practical in a high-dimensional space rely implicitly or explicitly on distances. Because of the poor discrimination of distance in high-dimensional space, the performance of these methods is inadequate. Fortunately, the angle-based outlier detection (ABOD) method used by Kriegel [3] plays an important role in identifying outliers in high-dimensional spaces. A point is defined as an outlier if most other points are located in similar directions. This can be quantified using the angle-based outlier factor (ABOF) of the observation. If the ABOF value is small, the observation is identified as an outlier; otherwise it is a normal point. Nevertheless, since ABOD only considers the relationships between each point and its neighbors, and does not consider the relationships among its neighbors, it can yield incorrect outliers. Refer to Part 2.1 for details.

To solve the problems discussed above, we present a small but consistent improvement by replacing the relationships between each point and its neighbors with more stable relationships among the neighbors. Compared with related methods, which work better in either high- or low-dimensional spaces, our method gives significant gains in both high- and low-dimensional spaces. Experimental results on both synthetic and real-world datasets demonstrate the effectiveness of our method.

The remainder of this paper is organized as follows. Section 2 describes the problem with ABOD and then presents our methods. In Section 3, experimental results on both synthetic and real-world datasets demonstrate the effectiveness of our method. Finally, conclusions are drawn in Section 4.

2 The Model

2.1 The Problem with ABOD

Given data $\mathbf{X} = \{\mathbf{x}_1, \mathbf{x}_2, \cdots, \mathbf{x}_n\} \in R^{m\times n}$, each column vector $\mathbf{x}_i$ denotes an observation, m indicates the number of variables or features, and n indicates the number of observations. Let $\overline{\mathbf{x}_i\mathbf{x}_j}$ denote the difference vector $\mathbf{x}_j - \mathbf{x}_i$. For point $\mathbf{x}_i$, ABOF($\mathbf{x}_i$) is the variance across the angles between the difference vectors of $\mathbf{x}_i$ and its neighbors [3]. Formally,

$$\text{ABOF}(\mathbf{x}_i) = \underset{\mathbf{x}_j,\mathbf{x}_k\in\mathbf{X}}{\text{Var}}\left(\frac{\overline{\mathbf{x}_i\mathbf{x}_j}\bullet\overline{\mathbf{x}_i\mathbf{x}_k}}{\|\overline{\mathbf{x}_i\mathbf{x}_j}\|^2\cdot\|\overline{\mathbf{x}_i\mathbf{x}_k}\|^2}\right), \tag{1}$$

where $i, j, k \in \{1, \cdots, n\}$ and $i \neq j, i \neq k, j \neq k$, and $\bullet$ denotes the inner product. The term within the parentheses can be rewritten as

$$\frac{1}{\|\overline{\mathbf{x}_i\mathbf{x}_j}\|}\cdot\left(\frac{\overline{\mathbf{x}_i\mathbf{x}_j}}{\|\overline{\mathbf{x}_i\mathbf{x}_j}\|}\bullet\frac{\overline{\mathbf{x}_i\mathbf{x}_k}}{\|\overline{\mathbf{x}_i\mathbf{x}_k}\|}\right)\cdot\frac{1}{\|\overline{\mathbf{x}_i\mathbf{x}_k}\|} = cos(\theta)\cdot\frac{1}{\|\overline{\mathbf{x}_i\mathbf{x}_j}\|\|\overline{\mathbf{x}_i\mathbf{x}_k}\|}, \tag{2}$$

where θ denotes the angle between $\overline{\mathbf{x}_i\mathbf{x}_j}$ and $\overline{\mathbf{x}_i\mathbf{x}_k}$. Clearly, (2) involves three points $\{\mathbf{x}_i, \mathbf{x}_j, \mathbf{x}_k\}$. It is important to represent the relationships among three points $\{\mathbf{x}_i, \mathbf{x}_j, \mathbf{x}_k\}$.

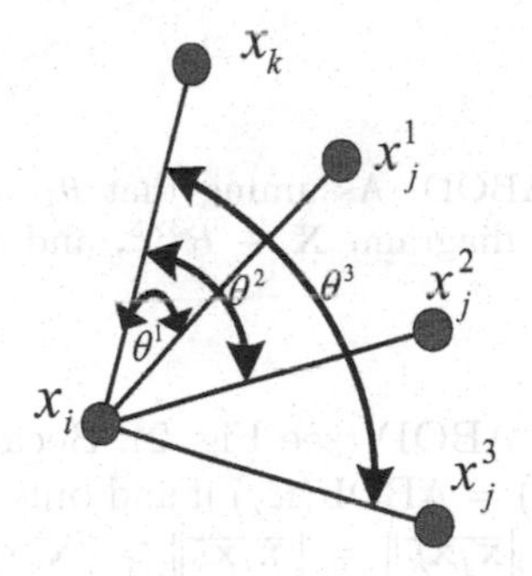

Fig. 1. Assuming $\theta^1 < \theta^2 < \theta^3$ and $\|\overline{\mathbf{x}_i\mathbf{x}_j^1}\| = \|\overline{\mathbf{x}_i\mathbf{x}_j^2}\| = \|\overline{\mathbf{x}_i\mathbf{x}_j^3}\|$

As a general rule, if $cos(\theta)$ dominates (2), then ABOD works better than purely distance-based approaches. Denote by $\{\theta^1, \theta^2, \theta^3\}$ three different angles between $\overline{\mathbf{x}_i\mathbf{x}_k}$ and $\overline{\mathbf{x}_i\mathbf{x}_j^t}, t \in \{1, 2, 3\}$ (see Fig. 1). Assume that $\theta^1 < \theta^2 < \theta^3$ and $\|\overline{\mathbf{x}_i\mathbf{x}_j^1}\| = \|\overline{\mathbf{x}_i\mathbf{x}_j^2}\| = \|\overline{\mathbf{x}_i\mathbf{x}_j^3}\|$. For different angles, it holds that

$$\text{ABOF}(\mathbf{x}_i^t) = \text{Var}\begin{pmatrix} 0 & \frac{cos(\theta^t)}{\|\overline{\mathbf{x}_i\mathbf{x}_j^t}\|\|\overline{\mathbf{x}_i\mathbf{x}_k}\|} \\ \frac{cos(\theta^t)}{\|\overline{\mathbf{x}_i\mathbf{x}_j^t}\|\|\overline{\mathbf{x}_i\mathbf{x}_k}\|} & 0 \end{pmatrix},$$

where $\mathbf{x}_i^t$ corresponds to the angle θ^t. Clearly, the term $\frac{cos(\theta^t)}{\|\overline{\mathbf{x}_i\mathbf{x}_j^t}\|\|\overline{\mathbf{x}_i\mathbf{x}_k}\|}$ determines the value of ABOF($\mathbf{x}_i^t$). The larger this entry, the greater the value of the ABOF. Since $cos(\theta^1) > cos(\theta^2) > cos(\theta^3)$ and $\|\overline{\mathbf{x}_i\mathbf{x}_j^1}\| = \|\overline{\mathbf{x}_i\mathbf{x}_j^2}\| = \|\overline{\mathbf{x}_i\mathbf{x}_j^3}\|$, it follows that

$$\frac{cos(\theta^1)}{\|\overline{\mathbf{x}_i\mathbf{x}_j^1}\|\|\overline{\mathbf{x}_i\mathbf{x}_k}\|} > \frac{cos(\theta^2)}{\|\overline{\mathbf{x}_i\mathbf{x}_j^2}\|\|\overline{\mathbf{x}_i\mathbf{x}_k}\|} > \frac{cos(\theta^3)}{\|\overline{\mathbf{x}_i\mathbf{x}_j^3}\|\|\overline{\mathbf{x}_i\mathbf{x}_k}\|}.$$

Therefore, ABOF($\mathbf{x}_i^1$) > ABOF($\mathbf{x}_i^2$) > ABOF($\mathbf{x}_i^3$). This illustrates that different angles yield different ABOFs under the same relationships between $\mathbf{x}_i^t$ and its neighbors. The smaller the angle, the greater the value of the ABOF. $\{\mathbf{x}_i, \mathbf{x}_j^t, \mathbf{x}_k\}$ are most likely in the same direction when θ^t is small, and ABOD works well in such cases.

However, it is not sufficient to only consider the relationships between $\mathbf{x}_i$ and its neighbors if $cos(\theta^1) = cos(\theta^2) = cos(\theta^3)$. A simple case is presented to

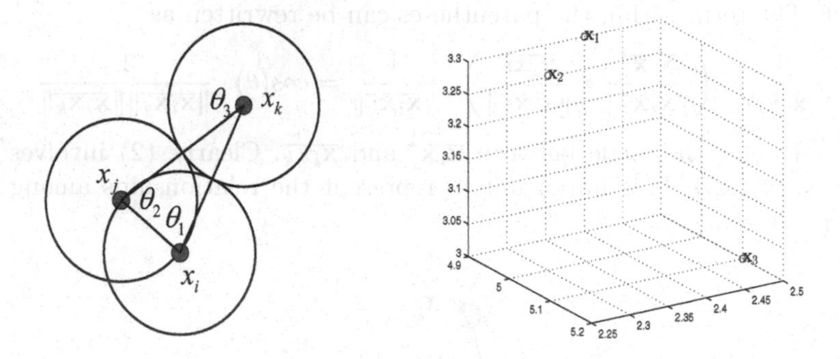

Fig. 2. The problem with ABOD. Assuming that $\theta_1 = \theta_2 > \theta_3$, then $\|\overline{\mathbf{x}_j\mathbf{x}_k}\| = \|\overline{\mathbf{x}_i\mathbf{x}_k}\| > \|\overline{\mathbf{x}_i\mathbf{x}_j}\|$ in the left diagram. $\mathbf{X} \in R^{3\times3}$, and $\mathbf{x}_3$ is an outlier in the right graph.

illustrate the problem with ABOD (see Fig. 2). Because it does not make sense for ABOF($\mathbf{x}_k$) = ABOF($\mathbf{x}_i$) = ABOF($\mathbf{x}_j$) if and only if $\theta_1 = \theta_2 = \theta_3$, we assume that $\theta_1 = \theta_2 > \theta_3$ so that $\|\overline{\mathbf{x}_j\mathbf{x}_k}\| = \|\overline{\mathbf{x}_i\mathbf{x}_k}\| > \|\overline{\mathbf{x}_i\mathbf{x}_j}\|$ (see Fig. 2 (left)). This means that $\mathbf{x}_k$ is an outlier and therefore

$$\text{ABOF}(\mathbf{x}_k) < \text{ABOF}(\mathbf{x}_i) = \text{ABOF}(\mathbf{x}_j). \tag{3}$$

For ABOD, it holds that

$$\text{ABOF}(\mathbf{x}_k) = \text{Var}\begin{pmatrix} 0 & \frac{cos(\theta_3)}{\|\overline{\mathbf{x}_k\mathbf{x}_j}\|\|\overline{\mathbf{x}_k\mathbf{x}_i}\|} \\ \frac{cos(\theta_3)}{\|\overline{\mathbf{x}_k\mathbf{x}_j}\|\|\overline{\mathbf{x}_k\mathbf{x}_i}\|} & 0 \end{pmatrix},$$

$$\text{ABOF}(\mathbf{x}_i) = \text{Var}\begin{pmatrix} 0 & \frac{cos(\theta_1)}{\|\overline{\mathbf{x}_i\mathbf{x}_k}\|\|\overline{\mathbf{x}_i\mathbf{x}_j}\|} \\ \frac{cos(\theta_1)}{\|\overline{\mathbf{x}_i\mathbf{x}_k}\|\|\overline{\mathbf{x}_i\mathbf{x}_j}\|} & 0 \end{pmatrix}.$$

Similarly, the terms $\frac{cos(\theta_3)}{\|\overline{\mathbf{x}_k\mathbf{x}_j}\|\|\overline{\mathbf{x}_k\mathbf{x}_i}\|}$ and $\frac{cos(\theta_1)}{\|\overline{\mathbf{x}_i\mathbf{x}_k}\|\|\overline{\mathbf{x}_i\mathbf{x}_j}\|}$ determine the sizes of ABOF($\mathbf{x}_k$) and ABOF($\mathbf{x}_i$), respectively. The larger this entry, the greater the value of the ABOF. Note that $\|\overline{\mathbf{x}_i\mathbf{x}_j}\| = \|\overline{\mathbf{x}_j\mathbf{x}_i}\|$. The ABOD method works well if

$$\|\overline{\mathbf{x}_k\mathbf{x}_i}\| > \frac{cos(\theta_3)}{cos(\theta_1)}\|\overline{\mathbf{x}_i\mathbf{x}_j}\| = \frac{cos(\theta_3)}{cos(\theta_2)}\|\overline{\mathbf{x}_j\mathbf{x}_i}\|. \tag{4}$$

Since $\|\overline{\mathbf{x}_i\mathbf{x}_k}\| = \|\overline{\mathbf{x}_k\mathbf{x}_i}\| = \|\overline{\mathbf{x}_k\mathbf{x}_j}\| = \|\overline{\mathbf{x}_j\mathbf{x}_k}\|$, Eq. (4) implies

$$\frac{cos(\theta_3)}{\|\overline{\mathbf{x}_k\mathbf{x}_j}\|\|\overline{\mathbf{x}_k\mathbf{x}_i}\|} < \frac{cos(\theta_1)}{\|\overline{\mathbf{x}_i\mathbf{x}_k}\|\|\overline{\mathbf{x}_i\mathbf{x}_j}\|} = \frac{cos(\theta_2)}{\|\overline{\mathbf{x}_j\mathbf{x}_k}\|\|\overline{\mathbf{x}_j\mathbf{x}_i}\|},$$

so that ABOF($\mathbf{x}_k$) < ABOF($\mathbf{x}_i$) = ABOF($\mathbf{x}_j$). Hence, (3) holds. However, if

$$\|\overline{\mathbf{x}_i\mathbf{x}_j}\| < \|\overline{\mathbf{x}_k\mathbf{x}_i}\| < \frac{cos(\theta_3)}{cos(\theta_1)}\|\overline{\mathbf{x}_i\mathbf{x}_j}\| = \frac{cos(\theta_3)}{cos(\theta_2)}\|\overline{\mathbf{x}_j\mathbf{x}_i}\|, \tag{5}$$

then ABOD identifies an incorrect outlier. Since it follows that

$$\frac{cos(\theta_3)}{\|\overline{\mathbf{x}_k\mathbf{x}_i}\|} > \frac{cos(\theta_1)}{\|\overline{\mathbf{x}_i\mathbf{x}_j}\|} = \frac{cos(\theta_2)}{\|\overline{\mathbf{x}_j\mathbf{x}_i}\|}, \tag{6}$$

by $\|\overline{\mathbf{x}_k\mathbf{x}_j}\| = \|\overline{\mathbf{x}_i\mathbf{x}_k}\| = \|\overline{\mathbf{x}_j\mathbf{x}_k}\|$ we have

$$\frac{cos(\theta_3)}{\|\overline{\mathbf{x}_k\mathbf{x}_i}\|\|\overline{\mathbf{x}_k\mathbf{x}_j}\|} > \frac{cos(\theta_1)}{\|\overline{\mathbf{x}_i\mathbf{x}_j}\|\|\overline{\mathbf{x}_i\mathbf{x}_k}\|} = \frac{cos(\theta_2)}{\|\overline{\mathbf{x}_j\mathbf{x}_i}\|\|\overline{\mathbf{x}_j\mathbf{x}_k}\|},$$

so that

$$\text{ABOF}(\mathbf{x}_k) > \text{ABOF}(\mathbf{x}_i) = \text{ABOF}(\mathbf{x}_j).$$

However, this contradicts (3). As mentioned previously, if the value of ABOF($\mathbf{x}_i$) is small, then $\mathbf{x}_i$ is identified as an outlier. Thus, the outlier would be identified as $\mathbf{x}_i$ or $\mathbf{x}_j$ rather than $\mathbf{x}_k$. For example, in Fig. 2 (right), $\mathbf{x}_3$ is considered to be an outlier. Since $\|\overline{\mathbf{x}_3\mathbf{x}_1}\| = 0.3724 < \frac{cos(\theta_3)}{cos(\theta_1)}\|\overline{\mathbf{x}_1\mathbf{x}_2}\| = 0.6913$ and the ABOF values of $\mathbf{X}$ are

$$\begin{cases} \text{ABOF}(\mathbf{x}_1) = 4.2517, \\ \text{ABOF}(\mathbf{x}_2) = 4.2517, \\ \text{ABOF}(\mathbf{x}_3) = 14.6640. \end{cases}$$

This confirms that $\mathbf{x}_1$ or $\mathbf{x}_2$ is an outlier whereas $\mathbf{x}_3$ is not. This contradiction shows that ABOD should not only consider the relationships between each point and its neighbors, but should also use more stable relationships among neighbors. Furthermore, it is unreasonable to measure the variance of the angles. In the following section, an approach to overcome this problem is proposed.

2.2 Proposed Method

Enhanced ABOD (EABOD) Model. In this part, we demonstrate a small but consistent advantage of replacing the relationships between each point and its neighbors with more stable relationships among neighbors. Based on the same definition used in ABOD, the following definition of an outlier is then proposed:

Definition 1. *Given data* $\mathbf{X}$, *an outlier is a point with a small outlier factor. The smaller the outlier factor, the more likely it is that the point is an outlier.*

Define $\overline{\mathbf{x}_i\mathbf{x}_{jk}} = \overline{\mathbf{x}_i\mathbf{x}_j} + \overline{\mathbf{x}_i\mathbf{x}_k}$ and $\beta_{jk} = \|\overline{\mathbf{x}_i\mathbf{x}_j}\| + \|\overline{\mathbf{x}_i\mathbf{x}_k}\|$, chosen to capture more relationships between a pair of neighbors ($\mathbf{x}_j$ and $\mathbf{x}_k$) of $\mathbf{x}_i$. We originally intended to use $\|\overline{\mathbf{x}_i\mathbf{x}_{jk}}\|$ to normalize the angles in (2). However, the triangle inequality implies that

$$0 < \|\overline{\mathbf{x}_i\mathbf{x}_{jk}}\| \le \beta_{jk}. \tag{7}$$

To reduce the influence of distance on the outlier factor, the right side of Eq. (7) is used. In accordance with Definition 1, the *enhanced* angle-based outlier factor (EABOF) of $\mathbf{x}_i$ is defined to be

$$\mathrm{EABOF}(\mathbf{x}_i) = \underset{\mathbf{x}_j,\mathbf{x}_k \in \mathbf{X}}{\mathrm{Var}} \left(\frac{1}{\beta_{jk}} \cdot \frac{\overline{\mathbf{x}_i\mathbf{x}_j} \bullet \overline{\mathbf{x}_i\mathbf{x}_k}}{\|\overline{\mathbf{x}_i\mathbf{x}_j}\|^2 \cdot \|\overline{\mathbf{x}_i\mathbf{x}_k}\|^2} \right), \tag{8}$$

where $i, j, k \in \{1, \cdots, n\}$ and $i \ne j, i \ne k$. This is an *enhanced* version of ABOD (EABOD). Equation (8) puts emphasis on the angles and greatly reduces the role of distance, since the term within the parentheses can be rewritten as

$$\frac{1}{\beta_{jk}} \cdot cos(\theta^i_{jk}) \cdot \frac{1}{(\|\overline{\mathbf{x}_i\mathbf{x}_j}\| \cdot \|\overline{\mathbf{x}_i\mathbf{x}_k}\|)}, \tag{9}$$

where θ^i_{jk} denotes the angle between $\overline{\mathbf{x}_i\mathbf{x}_j}$ and $\overline{\mathbf{x}_i\mathbf{x}_k}$. In general, $\beta_{jk} > 1$, especially in high-dimensional spaces. It follows from $\beta_{jk}(\|\overline{\mathbf{x}_i\mathbf{x}_j}\| \cdot \|\overline{\mathbf{x}_i\mathbf{x}_k}\|) > \|\overline{\mathbf{x}_i\mathbf{x}_j}\|\|\overline{\mathbf{x}_i\mathbf{x}_k}\|$ that

$$\frac{1}{\beta_{jk}(\|\overline{\mathbf{x}_i\mathbf{x}_j}\| \cdot \|\overline{\mathbf{x}_i\mathbf{x}_k}\|)} < \frac{1}{\|\overline{\mathbf{x}_i\mathbf{x}_j}\|\|\overline{\mathbf{x}_i\mathbf{x}_k}\|}. \tag{10}$$

It is clear from Eq. (10) that the expression in (9) places more emphasis on angles than that in (2).

Furthermore, by the inequality in (7), Eq. (8) takes into account more stable relationships between neighbors, which may overcome the problem with the ABOD method (see Fig. 2). By Definition 1, there must be $\mathbf{x}_i$, $\mathbf{x}_j$ and $\mathbf{x}_k$ such that

$$\mathrm{EABOF}(\mathbf{x}_k) < \mathrm{EABOF}(\mathbf{x}_i) = \mathrm{EABOF}(\mathbf{x}_j). \tag{11}$$

From (8), it follows that

$$\mathrm{EABOF}(\mathbf{x}_k) = \mathrm{Var}(\mathbf{M}_k), \quad \mathbf{M}_k = \begin{pmatrix} \frac{1}{2\|\overline{\mathbf{x}_k\mathbf{x}_j}\|^3} & \frac{cos(\theta_3)}{\beta_{ji}\|\overline{\mathbf{x}_k\mathbf{x}_j}\|\|\overline{\mathbf{x}_k\mathbf{x}_i}\|} \\ \frac{cos(\theta_3)}{\beta_{ji}\|\overline{\mathbf{x}_k\mathbf{x}_j}\|\|\overline{\mathbf{x}_k\mathbf{x}_i}\|} & \frac{1}{2\|\overline{\mathbf{x}_k\mathbf{x}_i}\|^3} \end{pmatrix},$$

$$\mathrm{EABOF}(\mathbf{x}_i) = \mathrm{Var}(\mathbf{M}_i), \quad \mathbf{M}_i = \begin{pmatrix} \frac{1}{2\|\overline{\mathbf{x}_i\mathbf{x}_j}\|^3} & \frac{cos(\theta_1)}{\beta_{jk}\|\overline{\mathbf{x}_i\mathbf{x}_j}\|\|\overline{\mathbf{x}_i\mathbf{x}_k}\|} \\ \frac{cos(\theta_1)}{\beta_{jk}\|\overline{\mathbf{x}_i\mathbf{x}_j}\|\|\overline{\mathbf{x}_i\mathbf{x}_k}\|} & \frac{1}{2\|\overline{\mathbf{x}_i\mathbf{x}_k}\|^3} \end{pmatrix},$$

where $\beta_{ji} = \|\overline{\mathbf{x}_k\mathbf{x}_j}\| + \|\overline{\mathbf{x}_k\mathbf{x}_i}\|$ and $\beta_{jk} = \|\overline{\mathbf{x}_i\mathbf{x}_j}\| + \|\overline{\mathbf{x}_i\mathbf{x}_k}\|$. Given $\|\overline{\mathbf{x}_i\mathbf{x}_k}\| = \|\overline{\mathbf{x}_j\mathbf{x}_k}\| > \|\overline{\mathbf{x}_i\mathbf{x}_j}\|$ and $\theta_1 = \theta_2 > \theta_3$, we have

$$1 > cos(\theta_3) > 0, \text{ and } 2\|\overline{\mathbf{x}_k\mathbf{x}_i}\|^3 = 2\|\overline{\mathbf{x}_k\mathbf{x}_j}\|^3 = \beta_{ji}\|\overline{\mathbf{x}_k\mathbf{x}_j}\|\|\overline{\mathbf{x}_k\mathbf{x}_i}\|, \tag{12}$$

$$1 > cos(\theta_1) > 0, \text{ and } 2\|\overline{\mathbf{x}_i\mathbf{x}_j}\|^3 < \beta_{jk}\|\overline{\mathbf{x}_i\mathbf{x}_j}\|\|\overline{\mathbf{x}_i\mathbf{x}_k}\|. \tag{13}$$

From the law of cosines and the triangular trilateral relation theorem, it follows that

$$\frac{cos(\theta_1)}{\beta_{jk}\|\overline{\mathbf{x}_i\mathbf{x}_j}\|\|\overline{\mathbf{x}_i\mathbf{x}_k}\|} < \frac{cos(\theta_3)}{\beta_{ji}\|\overline{\mathbf{x}_k\mathbf{x}_j}\|\|\overline{\mathbf{x}_k\mathbf{x}_i}\|} < \frac{1}{2\|\overline{\mathbf{x}_k\mathbf{x}_j}\|^3} < \frac{1}{2\|\overline{\mathbf{x}_i\mathbf{x}_j}\|^3}. \tag{14}$$

The entries of $\mathbf{M}_k$ are in the interval $A = \left[\frac{cos(\theta_3)}{\beta_{ji}\|\overline{\mathbf{x}_k\mathbf{x}_j}\|\|\overline{\mathbf{x}_k\mathbf{x}_i}\|}, \frac{1}{2\|\overline{\mathbf{x}_k\mathbf{x}_j}\|^3}\right]$, and the entries of $\mathbf{M}_i$ are in the interval $B = \left[\frac{cos(\theta_1)}{\beta_{jk}\|\overline{\mathbf{x}_i\mathbf{x}_j}\|\|\overline{\mathbf{x}_i\mathbf{x}_k}\|}, \frac{1}{2\|\overline{\mathbf{x}_i\mathbf{x}_j}\|^3}\right]$. From (14), we see that $A \subseteq B$, so the variance of $\mathbf{M}_k$ is smaller than that of $\mathbf{M}_i$. For the example in Fig. 2 (left), the EABOFs of $\mathbf{X}$ are as follows:

$$\begin{cases} \mathrm{EABOF}(\mathbf{x}_1) = 7139.8, \\ \mathrm{EABOF}(\mathbf{x}_2) = 7139.8, \\ \mathrm{EABOF}(\mathbf{x}_3) = 0.2. \end{cases}$$

Thus $\mathbf{x}_3$ is an outlier by Definition 1. That is, we can solve this problem with EABOD.

Variant of the EABOD Model. As opposed to $\frac{1}{\beta_{jk}}$, in particular when $\beta_{jk} < 1$, $cos(\cdot)$ returns real values in the interval $[-1, 1]$ for real arguments. To emphasize the role of the angle in EABOFs, we propose an approximation variant (acos(EABOD)) that uses the inverse cosine ($acos(\cdot)$), which returns values in the interval $[0, \pi]$. This is formulated as follows:

$$\mathrm{acos}(\mathrm{EABOD}(\mathbf{x}_i)) = \underset{\mathbf{x}_j, \mathbf{x}_k \in \mathbf{X}}{\mathrm{Var}} \left(\frac{1}{\beta_{jk}} acos\left(\frac{\overline{\mathbf{x}_i\mathbf{x}_j} \bullet \overline{\mathbf{x}_i\mathbf{x}_k}}{\|\overline{\mathbf{x}_i\mathbf{x}_j}\|^2 \cdot \|\overline{\mathbf{x}_i\mathbf{x}_k}\|^2} \right) \right), \tag{15}$$

where $i, j, k \in \{1, 2, \cdots, n\}$ and $i \neq j, i \neq k$.

Algorithm 1. EABOD Algorithm

Input: $\mathbf{X} \in R^{m \times n}$
Output: The top-n outlier
Step 1: Calculate the angle-based outlier factors of $\mathbf{X}$ via (8) or (15)

The Equivalence between EABOD and ABOD. Because of the condition $j \neq k$ in (1), EABOD does not equal ABOD even if $\beta_{jk} = 1$ for all $i, j, k \in \{1, \cdots, n\}, i \neq j, i \neq k$.

Remark 1. Given data $\mathbf{X} \in R^{m \times n}$, EABOD identifies with ABOD if and only if $\beta_{jk} = 1$ for all $i, j, k \in \{1, 2, \cdots, n\}, i \neq j, i \neq k, j \neq k$.

Remark 2. For each point $\mathbf{x}_i \in \mathbf{X}, i \in \{1, 2, \cdots, n\}$ with $n > 2$, the number of neighbors is $n-1$. From the calculations in (1) and (8), we find that the outcome in Remark 1 occurs with probability at most $P_{max} = \frac{1}{\left(2^{\frac{n(n-2)(n-1)}{2}}\right) - 1}$. For example, $(n = 3, P_{max} = \frac{1}{2^3 - 1})$ and $(n = 4, P_{max} = \frac{1}{2^{12} - 1})$. Thus observing Remark 1 is almost non-existent in the real world.

2.3 Algorithm

The whole EABOD algorithm is summarized in Algorithm 1. Several strategies with which we have experimented are:

1. ABOD using (1),
2. EABOD using (8),
3. acos(EABOD) using (15), and
4. acos(ABOD) using the following (16):

$$\text{acos}(\text{ABOF}(\mathbf{x}_i)) = \underset{\mathbf{x}_j, \mathbf{x}_k \in \mathbf{X}}{\text{Var}} \left(acos \left(\frac{\overline{\mathbf{x}_i \mathbf{x}_j} \bullet \overline{\mathbf{x}_i \mathbf{x}_k}}{\|\overline{\mathbf{x}_i \mathbf{x}_j}\|^2 \cdot \|\overline{\mathbf{x}_i \mathbf{x}_k}\|^2} \right) \right), \tag{16}$$

where $i, j, k \in \{1, 2, \cdots, n\}$ and $i \neq j, i \neq k, j \neq k$. We propose (16) to illustrate the effectiveness of acos(EABOD). The second and third strategies have proven to yield better performance through theoretical analysis. We now present experiments to illustrate the effectiveness of our method.

3 Experiments and Discussion

In this section, we evaluate the performances of both EABOD and acos(EABOD) through a number of synthetic and real-world experiments. To verify the effectiveness of our methods, we compare them with other related angle-based methods, namely ABOD and acos(ABOD).

3.1 Effectiveness

Two simple metrics, precision (P) and recall (R), are used to evaluate the performance of each algorithm at retrieving the most likely outlier [25]. Formally, we define

$$\begin{cases} a = Card\{relevant\ records\ retrieved\}, \\ b = Card\{relevant\ records\ not\ retrieved\}, \\ c = Card\{irrelevant\ records\ retrieved\}. \end{cases}$$

Then $\mathrm{P} = \frac{a}{a+c} \cdot 100\%$ and $\mathrm{R} = \frac{a}{a+b} \cdot 100\%$. Both metrics are usually expressed as percentages and are, in general, inversely related. In this paper, we only calculate the precision and recall in the top-n retrieval steps; that is, after retrieving a fixed number of instances ranked by the specific method.

3.2 Experiment on Synthetic Data

For the data $\mathbf{X} = \{\mathbf{x}_1, \mathbf{x}_2, \mathbf{x}_3\}$ (see Fig. 2 (right)), $\mathbf{x}_3$ is an outlier. Table 1 shows the top-1 ranked outlier and the corresponding precision and recall values. Our method correctly identifies the outlier in the first retrieval step. This has been discussed in Sections 2.1 and 2.2.

Table 1. Top-1 ranked outlier, showing the precision and recall for each method on the synthetic dataset (see Fig. 2 (right))

Method	outlier	P(R)(%)
ABOD	$\mathbf{x}_1$	0/0
acos(ABOD)	$\mathbf{x}_1$	0/0
EABOD	$\mathbf{x}_4$	100/100
acos(EABOD)	$\mathbf{x}_4$	100/100

3.3 Experiments on Real Data

There is meaning behind outlier detection ranking [3]. To demonstrate this, a number of experiments on real-world datasets are provided.

Iris Dataset. The Iris datasets consists of 50 samples from each of three species of iris: (1) Iris setosa, (2) Iris virginica, and (3) Iris versicolor, with four features measured in millimeters from each sample: sepal length, sepal width, petal length, and petal width [26, 27]. As shown in [28], the doubtful outliers in the three classes, where we abbreviate $\mathbf{x}_i$ to i, are as follows:

$$\begin{cases} \textit{Outliers in class 1: 42.} \\ \textit{Outliers in class 2: 58, 61, 94, 99.} \\ \textit{Outliers in class 3: 107, 118, 132, 119.} \end{cases}$$

Table 2 shows the top-n ranked outliers and the corresponding precision and recall values for each class. In the second class, acos(ABOD) and acos(EABOD) detect all the outliers in the first n retrieval steps, while ABOD and EABOD detect the normal instances $\mathbf{x}_{84}$ and $\mathbf{x}_{71}$ as outliers. Moreover, these four methods identify all outliers in the first and third classes, respectively.

Table 2. Top-n ranked outliers and the corresponding precision and recall for each method on the Iris dataset

Method	class	*Top-n* outlier	P(R)(%)
ABOD	1	$\mathbf{x}_{42}$	100
	2	$\mathbf{x}_{99}, \mathbf{x}_{61}, \mathbf{x}_{84}, \mathbf{x}_{71}$	50
	3	$\mathbf{x}_{107}, \mathbf{x}_{118}, \mathbf{x}_{132}, \mathbf{x}_{119}$	100
acos(ABOD)	1	$\mathbf{x}_{42}$	100
	2	$\mathbf{x}_{99}, \mathbf{x}_{61}, \mathbf{x}_{58}, \mathbf{x}_{94}$	100
	3	$\mathbf{x}_{107}, \mathbf{x}_{118}, \mathbf{x}_{132}, \mathbf{x}_{119}$	100
EABOD	1	$\mathbf{x}_{42}$	100
	2	$\mathbf{x}_{99}, \mathbf{x}_{61}, \mathbf{x}_{84}, \mathbf{x}_{71}$	50
	3	$\mathbf{x}_{107}, \mathbf{x}_{118}, \mathbf{x}_{132}, \mathbf{x}_{119}$	100
acos(EABOD)	1	$\mathbf{x}_{42}$	100
	2	$\mathbf{x}_{99}, \mathbf{x}_{61}, \mathbf{x}_{58}, \mathbf{x}_{94}$	100
	3	$\mathbf{x}_{107}, \mathbf{x}_{118}, \mathbf{x}_{119}, \mathbf{x}_{132}$	100

AR Dataset. To demonstrate more clearly the meaning behind the outlier detection ranking, we use the AR database, which contains 35 facial images from five different individuals [29]. Figure 3 shows some of the images for the different facial expressions (neutral, smile, anger, and scream) and illumination conditions (left light on, right light on, both lights on). For each individual, images with illumination conditions are considered to be outliers. Note that each image is resized to 60×43, which is represented by a point in 2580-dimensional space.

The top-15 outliers were retrieved to detect illuminated images for each individual. Figure 4 shows the top-15 ranked outliers for each method (top to bottom: EABOD, acos(EABOD), ABOD and acos(ABOD)). We see that ABOD identifies two normal points (the 13th and 14th images in the third row) as outliers, and acos(ABOD) only detects two outliers (the 11th and 13th images in the last row). Table 3 shows the normal points retrieved, and the precision and recall in the first 15 retrieval steps. Notice that EABOD and acos(EABOD) achieve the best performance.

FEI Face Dataset. We also used a subset of the FEI face database [30, 31], consisting of only frontal face images of 100 individuals. There are two frontal images for each individual, one with a neutral (non-smiling) expression and the other with a smiling expression. The 100 images with smiling expressions are considered to be outliers. Figure 5 shows some images from the data used. Note

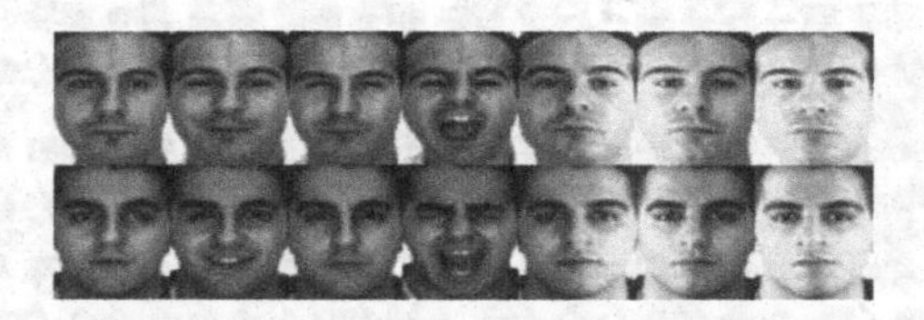

Fig. 3. Some images from the AR database. The conditions (left to right) are (1) neutral, (2) smile, (3) anger, (4) scream, (5) left light on, (6) right light on, (7) both lights on.

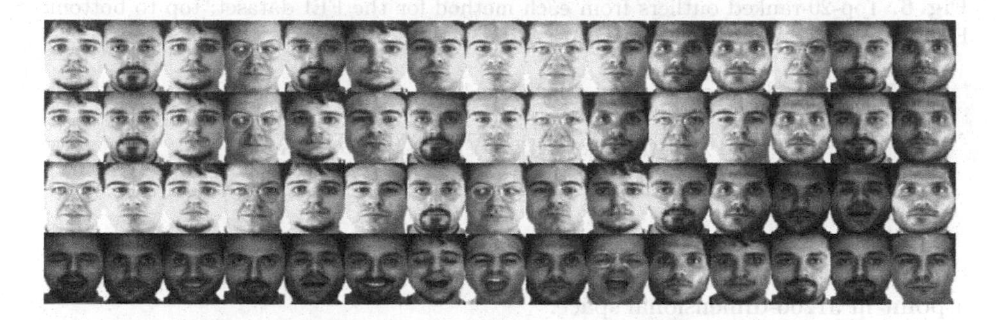

Fig. 4. Top-15 ranked outliers by each method for the AR face dataset: (top to bottom) EABOD, acos(EABOD), ABOD and acos(ABOD).

Table 3. The normal points retrieved, and the corresponding precision and recall in the top-15 retrieving step for the AR dataset

Method	*Normal points*	P(R)(%)
ABOD	*31, 32*	86.67
acos(ABOD)	*too many*	13.33
EABOD		100.00
acos(EABOD)		100.00

Fig. 5. Some frontal images from the FEI face dataset

Table 4. The precisions corresponding to Fig. 6

Method	ABOD	acos(ABOD)	EABOD	acos(EABOD)
P(R)(%)	90.00	0.00	95.00	95.00

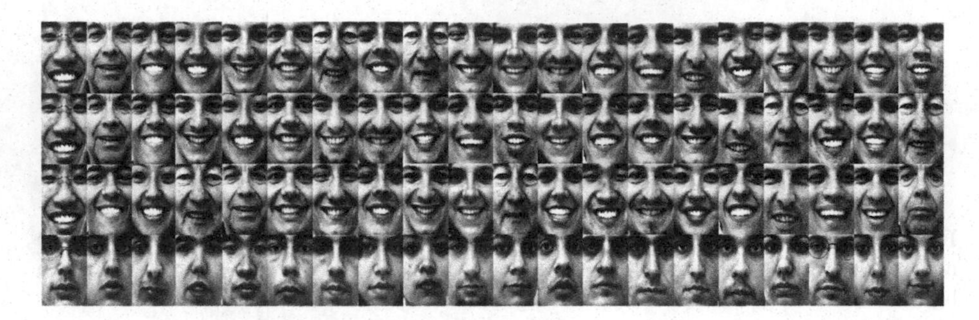

Fig. 6. Top-20-ranked outliers from each method for the FEI dataset: top to bottom: EABOD, acos(EABOD), ABOD, acos(ABOD)

Table 5. The precision and recall for all the outliers for the FEI dataset

Method	ABOD	acos(ABOD)	EABOD	acos(EABOD)
P(R)(%)	74.00	74.00	80.00	84.00

that all these images were cropped to 162×193 pixels, which is represented by a point in 31266-dimensional space.

Figure 6 shows the top-20 ranked outliers from each method (top to bottom: EABOD, acos(EABOD), ABOD, acos(ABOD)). The corresponding precision is shown in Table 4. It is clear that our methods, EABOD and acos(EABOD), achieve the best performance. There are two irrelevant images (the 11th and 20th images in the third row) retrieved by ABOD, while acos(ABOD) has the worst performance. Table 5 shows the precision and recall for the outliers in the first 100 retrieval steps. It would reach the same conclusion.

4 Conclusion

In this paper, we have proposed a small but consistent improvement to outlier detection by replacing the relationships between each point and its neighbors with more stable relationships among the neighbors. Compared with the well-established angle-based method ABOD, which works well in high-dimensional spaces, our method gives significant gains in both high and low-dimensional spaces. Experiments on both synthetic and real-world datasets demonstrate the effectiveness of our method.

Acknowledgments. This work was supported by the National Natural Science Foundation of China under grant 61375065 and the National Program on Key Basic Research Project (973 Program) under grant 2011CB302201.

References

1. Hawkins, D.M.: Identification of outliers, vol. 11. Springer (1980)
2. Han, J., Kamber, M., Pei, J.: Data mining: concepts and techniques. Morgan Kaufmann (2006)
3. Kriegel, H.P., Zimek, A., et al.: Angle-based outlier detection in high-dimensional data. In: Proceedings of the 14th ACM SIGKDD International Conference on Knowledge Discovery and Data Mining, pp. 444–452. ACM (2008)
4. Dang, X.H., Assent, I., Ng, R.T., Zimek, A., Schubert, E.: Discriminative features for identifying and interpreting outliers. In: Proc. ICDE (2014)
5. Kriegel, H.P., Kröger, P., Zimek, A.: Clustering high-dimensional data: A survey on subspace clustering, pattern-based clustering, and correlation clustering. ACM Transactions on Knowledge Discovery from Data (TKDD) 3(1), 1 (2009)
6. Lv, J.C., Tan, K.K., Yi, Z., Huang, S.: Stability and chaos of a class of learning algorithms for ica neural networks. Neural Processing Letters 28(1), 35–47 (2008)
7. Lv, J.C., Yi, Z., Tan, K.K.: Determination of the number of principal directions in a biologically plausible pca model. IEEE Trans. Neural Netw. 18(3), 910–916 (2007)
8. Cheng Lv, J., Yi, Z., Tan, K.: Convergence analysis of xu's lmser learning algorithm via deterministic discrete time system method. Neurocomputing 70(1), 362–372 (2006)
9. Houle, M.E., Kriegel, H.-P., Kröger, P., Schubert, E., Zimek, A.: Can shared-neighbor distances defeat the curse of dimensionality? In: Gertz, M., Ludäscher, B. (eds.) SSDBM 2010. LNCS, vol. 6187, pp. 482–500. Springer, Heidelberg (2010)
10. Kriegel, H.P., Kröger, P., Zimek, A.: Outlier detection techniques. In: Tutorial at the 13th Pacific-Asia Conference on Knowledge Discovery and Data Mining (2009)
11. Anscombe, F.J.: Rejection of outliers. Technometrics 2(2), 123–146 (1960)
12. Chandola, V., Banerjee, A., Kumar, V.: Anomaly detection: A survey. ACM Computing Surveys (CSUR) 41(3), 15 (2009)
13. Barnett, V., Lewis, T.: Outliers in statistical data, vol. 3. Wiley, New York (1994)
14. Rousseeuw, P.J., Driessen, K.V.: A fast algorithm for the minimum covariance determinant estimator. Technometrics 41(3), 212–223 (1999)
15. Hautamäki, V., Kärkkäinen, I., Fränti, P.: Outlier detection using k-nearest neighbour graph. In: ICPR (3), pp. 430–433 (2004)
16. Knox, E.M., Ng, R.T.: Algorithms for mining distancebased outliers in large datasets. In: Proceedings of the International Conference on Very Large Data Bases, pp. 392–403. Citeseer (1998)
17. Knorr, E.M., Ng, R.T.: Finding intensional knowledge of distance-based outliers. In: VLDB, vol. 99, pp. 211–222 (1999)
18. Knorr, E.M., Ng, R.T.: A unified approach for mining outliers. In: Proceedings of the 1997 Conference of the Centre for Advanced Studies on Collaborative Research, p. 11. IBM Press (1997)
19. Fan, H., Zaïane, O.R., Foss, A., Wu, J.: A nonparametric outlier detection for effectively discovering top-N outliers from engineering data. In: Ng, W.-K., Kitsuregawa, M., Li, J., Chang, K. (eds.) PAKDD 2006. LNCS (LNAI), vol. 3918, pp. 557–566. Springer, Heidelberg (2006)
20. Breunig, M.M., Kriegel, H.P., Ng, R.T., Sander, J.: Lof: identifying density-based local outliers. ACM Sigmod Record 29, 93–104 (2000)
21. Breunig, M.M., Kriegel, H.-P., Ng, R.T., Sander, J.: Optics-of: Identifying local outliers. In: Żytkow, J.M., Rauch, J. (eds.) PKDD 1999. LNCS (LNAI), vol. 1704, pp. 262–270. Springer, Heidelberg (1999)

22. Jin, W., Tung, A.K.H., Han, J., Wang, W.: Ranking outliers using symmetric neighborhood relationship. In: Ng, W.-K., Kitsuregawa, M., Li, J., Chang, K. (eds.) PAKDD 2006. LNCS (LNAI), vol. 3918, pp. 577–593. Springer, Heidelberg (2006)
23. Papadimitriou, S., Kitagawa, H., Gibbons, P.B., Faloutsos, C.: Loci: Fast outlier detection using the local correlation integral. In: Proceedings of the 19th International Conference on Data Engineering 2003, pp. 315–326. IEEE (2003)
24. Shewhart, W.A.: Economic control of quality of manufactured product, vol. 509. ASQ Quality Press (1931)
25. Jizba, R.: Measuring search effectiveness. Creighton University Health Sciences Library and Learning Resources Center (2000)
26. Fisher, R.A.: The use of multiple measurements in taxonomic problems. Annals of Eugenics 7(2), 179–188 (1936)
27. Duda, R.O., Hart, P.E., et al.: Pattern classification and scene analysis, vol. 3. Wiley, New York (1973)
28. Acuna, E., Rodriguez, C.: A meta analysis study of outlier detection methods in classification. Technical paper (2004)
29. Martinez, A.M., Benavente, R.: The ar face database. CVC Technical Report 24 (1998)
30. Thomaz, C.E., Giraldi, G.A.: A new ranking method for principal components analysis and its application to face image analysis. Image and Vision Computing 28(6), 902–913 (2010)
31. Tenorio, E.Z., Thomaz, C.E.: Analise multilinear discriminante de formas frontais de imagens 2d de face

An Opposition-based Self-adaptive Hybridized Differential Evolution Algorithm for Multi-objective Optimization (OSADE)

Jin Kiat Chong and Kay Chen Tan

Department of Electrical and Computer Engineering, National University of Singapore,
4 Engineering Drive, 117576, Singapore
{G0900273,eletankc}@nus.edu.sg

Abstract. This paper presents a novel opposition-based self-adaptive hybridized Differential Evolution algorithm termed as OSADE for solving continuous multi-objective optimization problems. OSADE is developed using a modified version of a self-adaptive Differential Evolution variant and hybridizing it with the Multi-objective Evolutionary Gradient Search (MO-EGS) to act as a form of local search. Through the use of a test suite of benchmark problems, a comparative study of this newly developed algorithm and some state-of-the-art algorithms, such as NSGA-II, Non-dominated Sorting Differential Evolution (NSDE), MOEA/D-SBX, MOEA/D-DE and MO-EGS, is being presented by employing the Inverted Generational Distance (IGD) and the Hausdorff Distance (HD) performance indicators. From the simulation results, it is seen that OSADE is able to achieve competitive, if not better, performance when compared to the other algorithms in this study.

Keywords: Differential Evolution, evolutionary multi-objective optimization, self-adaptation, opposition-based learning, continuous multi-objective optimization problems.

1 Introduction

Differential evolution (DE) [5] is an efficient and popular evolutionary algorithm known for its simplicity and ease of use, and it is also recognized as one of the most powerful stochastic real-parameter optimization algorithms in current use. The distinguishing difference between this state-of-the-art algorithm and the other evolutionary algorithms (EAs) is that DE involves the perturbation of the current generation individuals with the scaled differences of other randomly selected and distinct population members. Therefore, there is no separate probability distribution for generating new off-springs. As compared to most standard evolutionary algorithms, DE requires lesser parameters, and has a faster convergence rate in most cases together with stronger global convergence ability and robustness. Despite the strengths witnessed in DE, it is inevitable that there are still certain drawbacks in this evolutionary algorithm when placed under certain problems, applications or environment. These drawbacks [6, 7]

H. Handa et al. (eds.), *Proc. of the 18th Asia Pacific Symp. on Intell. & Evol. Systems - Vol. 1*,
Proceedings in Adaptation, Learning and Optimization 1, DOI: 10.1007/978-3-319-13359-1_35

include stagnation and premature convergence in multimodal problems due to DE being trapped in local optima. In order to overcome these weaknesses, several studies [8, 9] on the hybridization of DE with other evolutionary algorithms or local search have been proposed with promising results. The performance of DE is also highly attributed to the setting of its control parameters which are namely the mutation factor F, crossover probability CR and population size NP. As such, researchers introduced the use of self-adaptation in DE variants [10, 11] to dynamically update the parameter values throughout the evolution process without any preceding knowledge of the relationship between the parameter settings and the properties of the optimization problems given. This approach not only eliminates the need to fix the parameter values at the start of the optimization process, but also improves convergence performance over classic DE in terms of speed and reliability when compared over several benchmark problems.

In this paper, a novel opposition-based self-adaptive hybridized DE variant termed as OSADE is proposed to overcome the drawbacks of DE highlighted above, and the proposed algorithm is applied on continuous unconstrained multi-objective (MO) optimization problems [1], [3], and is compared with some state-of-the-art algorithms.

The rest of the paper is organized as follows. Section 2 describes the details of the proposed algorithm in this work. This is followed by Section 3 which presents the comparative studies with five state-of-the-art algorithms for the selected test benchmark problems. The paper is concluded in Section 4.

2 The Proposed Algorithm: OSADE

OSADE is a hybridization of a novel opposition-based self-adaptive non-dominated sorting differential evolution variant with the multi-objective evolutionary gradient search (MO-EGS). The proposed algorithm incorporates opposition-based learning into a self-adaptive mechanism for the DE control parameters (mutation factor and the crossover probability), and is then hybridized with MO-EGS which acts as a form of local search to enhance the exploitation abilities of the overall algorithm. In OSADE, it follows the mutation and crossover strategies from a frequently used DE variant known as the DE/rand/bin/1. As OSADE is designed to solve multi-objective optimization problems, hence there is no single best solution but rather a set of Pareto-optimal solutions. As such, the non-dominated sorting, ranking and elitism techniques as found in NSGA-II [2] are incorporated into OSADE for the comparison of the quality of the solutions found so as to obtain the Pareto optimal solutions.

In order to improve convergence and eliminate the need to determine optimized parameter values for differential evolution, the approach in the self-adaptive mechanism in [12] is being extended for this study. The approach traces its principles to a self-adaptive mechanism found in evolutionary strategies whereby the mutation factor and the crossover probability of differential evolution are being encoded in every individual of the population. This means that there will be two additional variables in every individual, and they will also undergo the evolutionary process for their values to be adjusted appropriately in a self-adaptive manner for every generation. In order to improve this self-adaptation method, a novel approach inspired by Opposition-Based Learning [13] is proposed in this study to enhance the optimization of the values of the DE parameters during the evolutionary process. As seen from several evolutionary optimization methods, the use of random guesses or estimates are frequently used when we are looking for a solution to a problem. At the same time, the evolutionary optimization methods will also work towards the search for the optimal solution and this incurs computational time which is related to the distance between the estimated solutions and the optimal one. However, the guess might be far from the exact solution as it may be based on past experience or could be totally random. As such, in order to accelerate convergence towards the optimal solution, the opposite number of the estimated solution will be also checked as there will be 50% chance that the estimated solution will be further from the optimal one compared to the opposite solution according to probability theory.

The concept of the Opposite Number in Opposition-Based Learning is being employed in this study and its definition is as follows:

$$\breve{x} = y + z - x \tag{1}$$

Where $x \in [y, z]$is a real number, and $\breve{x}$ represents the Opposite Number.

The pseudo code of OSADE is shown in Algorithm 1. The initial candidate population is randomly generated and evaluated before assigning fitness to the solutions. Binary tournament selection is then performed to choose promising solutions for reproduction. In the reproduction stage, the self-adaptive mechanism in [12] is being extended into a novel opposition-based self-adaptive differential evolution operator in this study for the generation of offspring. The key difference in the self-adaptive mechanism in OSADE is that opposition-based learning is being applied in the updating of the mutation factor. It is also to be mentioned that the adaption of DE parameters in [12] utilizes simple averaging of the encoded parameter values from four different individuals. However in this study, weighted averaging is used instead. The current individual and three randomly selected individuals will be compared based on their Pareto ranks and/or niche counts to determine the best individual which will be awarded a highest weightage for its contribution to the averaging of the parameter values. Individuals with better Pareto ranks are preferred as they are nearer to the optimal Pareto front, and the selection of individuals with lower niche counts promotes diversity [21]. Hence this approach allows fitter individuals to give higher contribution towards the adaptation of the parameter values so as to achieve near-optimal DE parameter values. With this, the DE parameter values will be adapted accordingly within the lower and upper bounds. The Opposite Number of the mutation factor F will also be computed. Two different mutant vectors will be generated using the value of F and its Opposite Number, and they will be compared whereby the non-dominated one will be the offspring. If they are non-dominated to each other, one of them will be randomly selected. Through reproduction, N child solutions will be produced and archived. M solutions will be selected from archive to undergo local search (MO-EGS operator), and the solutions generated will be added to the archive. The updated archive will become the child population of the generation, and will be combined with the parent population. Elitism is performed to select the parent population for the next generation. The process is iterated until the stopping criterion is met.

3 Simulation Results and Discussions

A total of 19 test benchmark problems from [14, 15] are chosen to test the optimization performance of OSADE in terms of converging to the true Pareto front and the ability to maintain a diverse set of solutions. With these unconstrained test problems that may possess two or three objective functions, OSADE is compared against five other state-of-the-art algorithms which are namely NSGA-II-SBX [2], MOEA/D-SBX [18], NSDE [20], MOEA/D-DE [19] and MO-EGS. In Table 1, the parameter settings for the simulations are being presented. In order to draw a fair comparison of the optimization performance by the algorithms, the Inverted Generational Distance (IGD) [16] and the Hausdorff Distance (HD) [17] are being used as the performance metrics as they consider both proximity and diversity. For these two metrics, a lower value indicates better performance.

The simulation results over 30 independent runs in terms of the measurement of the average IGD and HD values with their standard deviations are presented in Tables 2 to 5. The parentheses beside the test problems indicate the number of objective functions (M) and decision variables (D) for the specific test instance. The best IGD and HD values will be in bold for every test instance. As observed from the results in the tables, OSADE achieved the best IGD and HD values in 8 out of the 10 UF problems. As for WFG problems, OSADE managed to achieve the best IGD and HD values in 4 out of 9 WFG problems. Due to space constraint, the evolutions plots and evolved Pareto front curves for only UF1 and WFG1 are being presented in this study. From the HD evolution plots in Fig. 1 and Fig. 2, it can be observed that OSADE is able to achieve faster convergence than the other algorithms compared for these selected test problems. For UF1, OSADE takes about 100 generations to find the approximate Pareto front. As for WFG1, it is observed that the initial convergence is much faster than the other algorithms, but it may take more than 500 generations for full convergence. However, the convergence performance of OSADE for WFG1 is still superior over the other algorithms tested. As seen from the evolved Pareto front curves in Fig. 3 and Fig. 4, it is demonstrated that OSADE is able to achieve better proximity and spread over the other algorithms for both test problems.

The promising results achieved by OSADE can be attributed to the use of the novel opposition-based self-adaptive mechanism for updating the DE parameters, especially for the mutation factor F. Through the use of this mechanism, the value of the mutation factor can be dynamically adapted according to the evolutionary process, and with the help of opposition-based learning, there is also higher chance of the parameter being set to near optimal values. Another contributing factor is the hybridization of the DE variant here with MO-EGS to act as local search. The DE variant is capable of strong global search to find promising solutions in the search space, and the incorporation of MO-EGS may potentially help in the exploitation of solutions as it uses the gradient information of the trajectory of solutions to determine the favourable movements in the search space. In this way, a larger selection pressure can be induced which helps in the overall convergence of the solutions. As such, there can be a balance of exploration and exploitation of the search space by OSADE, and this aids in convergence while maintaining diversity of solutions. In order to solve UF problems well, algorithms need to be able to generate solution sets of higher diversity so as to explore the search space effectively, especially during the earlier stages of the search due to the presence of complicated Pareto Sets in these problems. It has also been indicated that algorithms such as the MOEA/D-SBX may not be suitable in dealing with the test instances in the UF test suite as the population in MOEA/D-SBX may lose diversity and the SBX operator in MOEAs have the shortcomings of producing inferior solutions [19]. As OSADE is inherently a DE variant, hence it also possesses strong exploratory capability that is able to produce a set of diverse solutions, and this makes it suitable and effective for solving UF problems. For the case of WFG problems, these problems are mainly subjected to bias or shift transformations which results in most of the solutions for these problems to be located in certain section of the search space. As such, the use of local search operator in OSADE can also help in the determination of favourable search direction towards the optima, and this enhances the convergence ability of OSADE in handling WFG problems as well.

Table 1. Parameter Settings

Parameter	Value
Population Size, NP	100 or 300 (2 or 3 objectives)
Number of Runs	30
Stopping Criterion	500 generations x NP
Distribution index in SBX, η_c	20
Distribution index in polynomial mutation, η_m	20
Mutation Rate, p_m	1/(Decision Variables)
Crossover Probability for DE operator, CR	0.8
Mutation Factor for DE operator, F	0.5
Probability that parents are selected from neighbourhood (for MOEA/D), δ	0.9
Maximal number of solutions updated, η_r (for MOEA/D)	2
Neighbourhood size, T (for MOEA/D)	20
Fitness assignment approach in MO-EGS	Hypervolume performance indicator
Parameter in MO-EGS progress vector, κ	10
Number of trial solutions in MO-EGS, N	50
Initial F and CR values for OSADE	0
Lower / Upper Bounds of F and CR for OSADE	0.1 (lower) / 0.9 (upper)

Algorithm 1: Pseudo Code for Proposed Algorithm OSADE

Begin

1: **Initialization**: At generation g=0, randomly generate NP initial population, Pop1 (g)

2: **Evaluation**: Evaluate the fitness of every individual (X_i) in the initial population

3: **Do while** ("Stopping criterion is not met")

4: **Fitness Assignment**: Assign fitness to every individual in Pop1 (g) by Pareto Ranking and Crowding Distance.

5: **Selection**: Select NP individuals using binary tournament selection.

6: **Reproduction**: Use the novel opposition-based self-adaptive Differential Evolution (DE) as described here:

7: **For i = 1: NP do**

8: Select the indices r_1, r_2, r_3 randomly under uniform distribution such that $r_1 \neq r_2 \neq r_3 \neq i$

9: Compare the current (i-th) individual and the individuals indexed by r_1, r_2, r_3 in terms of their Pareto ranks and/or niche counts, and select the best individual.

10: Retrieve the values of $F_{i,g}$, $F_{r1,g}$, $F_{r2,g}$, $F_{r3,g}$, and $CR_{i,g}$, $CR_{r1,g}$, $CR_{r2,g}$, $CR_{r3,g}$ which are encoded in the above 4 selected individuals.

11: Assign a weight factor of 0.4 to the best ranking individual and weight factors assigned to the subsequent ranked individuals will be decremented by 0.1.

12: **Self-adaptation of the DE parameters:** Calculate the values of $F_{i,g+1}$ and $CR_{i,g+1}$ using the following formulae:

$$\langle F_g \rangle_i = \frac{w_1 F_{i,g} + w_2 F_{r1,g} + w_3 F_{r2,g} + w_4 F_{r3,g}}{w_1 + w_2 + w_3 + w_4}$$

$$\langle CR_g \rangle_i = \frac{w_1 CR_{i,g} + w_2 CR_{r1,g} + w_3 CR_{r2,g} + w_4 CR_{r3,g}}{w_1 + w_2 + w_3 + w_4}$$

whereby w_1, w_2, w_3 *and* w_4 represent the weight factors assigned to the different individuals as mentioned in Step 11.

13: Calculate the updated F and CR for the next generation as follows:

$$F_{i,g+1} = \langle F_g \rangle_i \; x \; e^{\tau N(0,1)} \qquad CR_{i,g+1} = \langle CR_g \rangle_i \; x \; e^{\tau N(0,1)}$$

whereby $\tau = 1/(8\sqrt{2D})$, D is the dimension of the problem, N (0, 1) denotes a random generated number under Gaussian distribution.

14: Generate the Opposite Number of the mutation scale factor F according to the below formula:

$$F\_opp_{i,g+1} = F_{upper} + F_{lower} - F_{i,g+1}$$

Algorithm 1: Pseudo Code for Proposed Algorithm OSADE (continued)

15: Randomly select an integer from (1,D) to be j_{rand}
16: Create 2 Trial Vectors V1 and V2 as shown in next step.
17: **for** j = 1 to D **do**
  if $rand_j(0,1) \leq CR$ or $j = j_{rand}$ **then**
    $V1_{i,j} = X_{r1,j} + F_{i,g+1} * (X_{r2,j} - X_{r3,j})$
    $V2_{i,j} = X_{r1,j} + F\_opp_{i,g+1} * (X_{r2,j} - X_{r3,j})$
  else
    $V1_{i,j} = X_{i,j}$, $V2_{i,j} = X_{i,j}$
  end if
 end for
18: Let the offspring be represented by $U_{i,g+1}$.
19: Compare the 2 Trial Vectors V1 and V2 as shown below:
 if V1 dominates V2
  $U_{i,g+1} = V1$, $F_{i,g+1} = F_{i,g+1}$, $CR_{i,g+1} = CR_{i,g+1}$
 else if V2 dominates V1
  $U_{i,g+1} = V2$, $F_{i,g+1} = F\_opp_{i,g+1}$, $CR_{i,g+1} = CR_{i,g+1}$
 else if V1 and V2 are non-dominated to each other
  if rand(0,1) < 0.5
    $U_{i,g+1} = V1$, $F_{i,g+1} = F_{i,g+1}$, $CR_{i,g+1} = CR_{i,g+1}$
  else
    $U_{i,g+1} = V2$, $F_{i,g+1} = F\_opp_{i,g+1}$, $CR_{i,g+1} = CR_{i,g+1}$
  end if
 end if
20: **End For**
21: **Evaluation**: Calculate the objective values of all the child solutions.
22: **Archiving**: Store all child solutions into a fixed-size archive tempPop.
23: Select M individuals via binary tournament selection from tempPop for the local search
24: **Local Search**: Perform the MO-EGS on the M individuals to generate offspring. Add the offspring to tempPop if they are not dominated by any of the archived members. If the predetermined archive size is reached, eliminate the most crowded archive member.
25: Let the updated archive tempPop become the child population PopQ.
26: Merge all the solutions in Pop1 (g) and PopQ into a combined population PopR. Perform fitness assignment for all solutions in PopR.
27: **Elitism**: Select NP solutions with the lowest Pareto rank or with largest crowding distance from PopR to form a new parent population Pop1 (g+1).
28: g = g + 1
29: **End Do**
End

Table 2. Comparison of IGD values for UF problems

Algorithm	UF1(2,30)	UF2(2,30)
NSGAII-SBX	0.1200±0.0245	0.0478±0.0102
MOEA/D-SBX	0.1257±0.0497	0.0574±0.0296
NSDE	0.0523±0.0128	0.0450±0.0056
MOEA/D-DE	0.0488±0.0289	0.0342±0.0236
MO-EGS	0.1588±0.0956	0.0510±0.0045
OSADE	**0.0404±0.0051**	**0.0204±0.0012**
Algorithm	**UF3(2,30)**	**UF4(2,30)**
NSGAII-SBX	0.2370±0.0403	0.0538±0.0022
MOEA/D-SBX	0.3092±0.0532	0.0566±0.0047
NSDE	0.1387±0.0318	0.0730±0.0078
MOEA/D-DE	**0.0744±0.0377**	0.0824±0.0078
MO-EGS	0.1748±0.0099	0.1495±0.0076
OSADE	0.2087±0.0101	**0.0408±0.0002**
Algorithm	**UF5(2,30)**	**UF6(2,30)**
NSGAII-SBX	0.3047±0.1036	0.1462±0.0465
MOEA/D-SBX	0.4359±0.0993	0.1744±0.0548
NSDE	0.8773±0.1742	0.0442±0.0088
MOEA/D-DE	0.6670±0.1384	0.0465±0.0291
MO-EGS	1.3705±0.2850	0.0698±0.0091
OSADE	**0.1494±0.0119**	**0.0240±0.0043**
Algorithm	**UF7(2,30)**	**UF8(3,30)**
NSGAII-SBX	0.1683±0.1340	0.2203±0.0047
MOEA/D-SBX	0.3224±0.1380	0.1607±0.0379
NSDE	0.0329±0.0090	0.1478±0.0143
MOEA/D-DE	0.0234±0.0063	**0.0937±0.0082**
MO-EGS	0.0723±0.0040	0.1168±0.0173
OSADE	**0.0178±0.0020**	0.1098±0.0028
Algorithm	**UF9(3,30)**	**UF10(3,30)**
NSGAII-SBX	0.1710±0.0423	0.3274±0.0596
MOEA/D-SBX	0.1220±0.0566	0.3257±0.1810
NSDE	0.1822±0.0671	2.4853±0.2086
MOEA/D-DE	0.1058±0.0485	0.6108±0.0047
MO-EGS	0.1995±0.0622	5.1364±0.7926
OSADE	**0.0805±0.0022**	**0.3197±0.0456**

Table 3. Comparison of IGD values for WFG problems

Algorithm	WFG1(2,30)	WFG2(2,30)
NSGAII-SBX	1.3881±0.0889	0.1017±0.0652
MOEA/D-SBX	1.1814±0.1142	0.1567±0.0374
NSDE	1.2669±0.0118	**0.0306±0.0100**
MOEA/D-DE	1.2308±0.0033	0.0500±0.0064
MO-EGS	1.2441±0.0048	0.3351±0.0549
OSADE	**0.6882±0.0195**	0.0339±0.0003
Algorithm	WFG3(2,30)	WFG4(2,30)
NSGAII-SBX	0.0250±0.0014	0.0184±0.0008
MOEA/D-SBX	0.0299±0.0033	**0.0155±0.0005**
NSDE	**0.0245±0.0011**	0.0994±0.0036
MOEA/D-DE	0.0292±0.0026	0.0791±0.0103
MO-EGS	0.1833±0.0392	0.1585±0.0113
OSADE	0.0338±0.0008	0.0304±0.0024
Algorithm	WFG5(2,30)	WFG6(2,30)
NSGAII-SBX	0.0671±0.0010	0.0482±0.0041
MOEA/D-SBX	0.0665±0.0007	0.0447±0.0070
NSDE	0.0733±0.0011	0.0819±0.0204
MOEA/D-DE	0.0673±0.0002	0.0818±0.0182
MO-EGS	0.0794±0.0102	0.1110±0.0096
OSADE	**0.0608±0.0014**	**0.0354±0.0041**
Algorithm	WFG7(2,30)	WFG8(2,30)
NSGAII-SBX	0.0172±0.0007	0.0802±0.0038
MOEA/D-SBX	**0.0141±0.0001**	0.0766±0.0054
NSDE	0.0332±0.0018	0.1272±0.0121
MOEA/D-DE	0.0180±0.0008	0.1118±0.0112
MO-EGS	0.1083±0.0117	0.2172±0.0095
OSADE	0.0159±0.0003	**0.0705±0.0007**
Algorithm	WFG9(2,30)	
NSGAII-SBX	0.0200±0.0019	
MOEA/D-SBX	**0.0177±0.0013**	
NSDE	0.0335±0.0008	
MOEA/D-DE	0.0348±0.0194	
MO-EGS	0.1594±0.0263	
OSADE	0.0194±0.0003	

Table 4. Comparison of HD values for UF problems

Algorithm	UF1(2,30)	UF2(2,30)
NSGAII-SBX	0.0461±0.0085	0.0214±0.0062
MOEA/D-SBX	0.0511±0.0214	0.0303±0.0161
NSDE	0.0244±0.0186	0.0179±0.0026
MOEA/D-DE	0.0224±0.0154	0.0170±0.0128
MO-EGS	0.0599±0.0295	0.0197±0.0018
OSADE	**0.0169±0.0025**	**0.0082±0.0005**
Algorithm	UF3(2,30)	UF4(2,30)
NSGAII-SBX	0.0897±0.0137	0.0173±0.0009
MOEA/D-SBX	0.1162±0.0198	0.0187±0.0019
NSDE	0.0471±0.0092	0.0235±0.0024
MOEA/D-DE	**0.0300±0.0138**	0.0264±0.0025
MO-EGS	0.1107±0.0188	0.0477±0.0025
OSADE	0.0698±0.0035	**0.0130±0.0001**
Algorithm	UF5(2,30)	UF6(2,30)
NSGAII-SBX	0.0261±0.0111	0.0559±0.0181
MOEA/D-SBX	0.0288±0.0093	0.0681±0.0197
NSDE	0.1310±0.0350	0.0170±0.0043
MOEA/D-DE	0.0649±0.0228	0.0232±0.0136
MO-EGS	0.2439±0.0424	0.0339±0.0041
OSADE	**0.0132±0.0001**	**0.0121±0.0002**
Algorithm	UF7(2,30)	UF8(3,30)
NSGAII-SBX	0.0712±0.0528	0.1828±0.0076
MOEA/D-SBX	0.1326±0.0514	0.0793±0.0216
NSDE	0.0183±0.0088	0.0607±0.0198
MOEA/D-DE	0.0123±0.0043	0.0967±0.0539
MO-EGS	0.0278±0.0015	0.0517±0.0079
OSADE	**0.0083±0.0012**	**0.0464±0.0011**
Algorithm	UF9(3,30)	UF10(3,30)
NSGAII-SBX	0.0756±0.0172	0.2872±0.2685
MOEA/D-SBX	0.0548±0.0290	**0.1326±0.0565**
NSDE	0.0706±0.0332	0.8295±0.0696
MOEA/D-DE	0.1585±0.0534	0.2138±0.0218
MO-EGS	0.0434±0.0115	1.8028±0.1555
OSADE	**0.0296±0.0005**	0.5281±0.2944

Table 5. Comparison of HD Values for WFG problems

Algorithm	WFG1(2,30)	WFG2(2,30)
NSGAII-SBX	0.3344±0.0257	0.0454±0.0335
MOEA/D-SBX	0.2740±0.0314	0.0663±0.0184
NSDE	0.2836±0.0026	**0.0076±0.0022**
MOEA/D-DE	0.2755±0.0007	0.0153±0.0021
MO-EGS	0.2793±0.0012	0.0834±0.0179
OSADE	**0.1542±0.0044**	0.0083±0.0001
Algorithm	WFG3(2,30)	WFG4(2,30)
NSGAII-SBX	0.0062±0.0004	0.0047±0.0003
MOEA/D-SBX	0.0073±0.0009	**0.0039±0.0001**
NSDE	**0.0062±0.0003**	0.0223±0.0008
MOEA/D-DE	0.0071±0.0006	0.0185±0.0024
MO-EGS	0.0416±0.0088	0.0357±0.0026
OSADE	0.0085±0.0002	0.0077±0.0006
Algorithm	WFG5(2,30)	WFG6(2,30)
NSGAII-SBX	0.0152±0.0003	0.0109±0.0009
MOEA/D-SBX	0.0154±0.0002	0.0101±0.0016
NSDE	0.0168±0.0002	0.0185±0.0045
MOEA/D-DE	0.0156±0.0004	0.0184±0.0041
MO-EGS	0.0183±0.0024	0.0255±0.0024
OSADE	**0.0138±0.0003**	**0.0081±0.0009**
Algorithm	WFG7(2,30)	WFG8(2,30)
NSGAII-SBX	0.0046±0.0003	0.0185±0.0009
MOEA/D-SBX	**0.0037±0.0001**	0.0177±0.0012
NSDE	0.0077±0.0004	0.0288±0.0027
MOEA/D-DE	0.0044±0.0002	0.0254±0.0026
MO-EGS	0.0246±0.0027	0.0490±0.0023
OSADE	0.0044±0.0001	**0.0164±0.0002**
Algorithm	WFG9(2,30)	
NSGAII-SBX	0.0053±0.0005	
MOEA/D-SBX	**0.0048±0.0002**	
NSDE	0.0081±0.0002	
MOEA/D-DE	0.0084±0.0084	
MO-EGS	0.0366±0.0059	
OSADE	0.0053±0.0001	

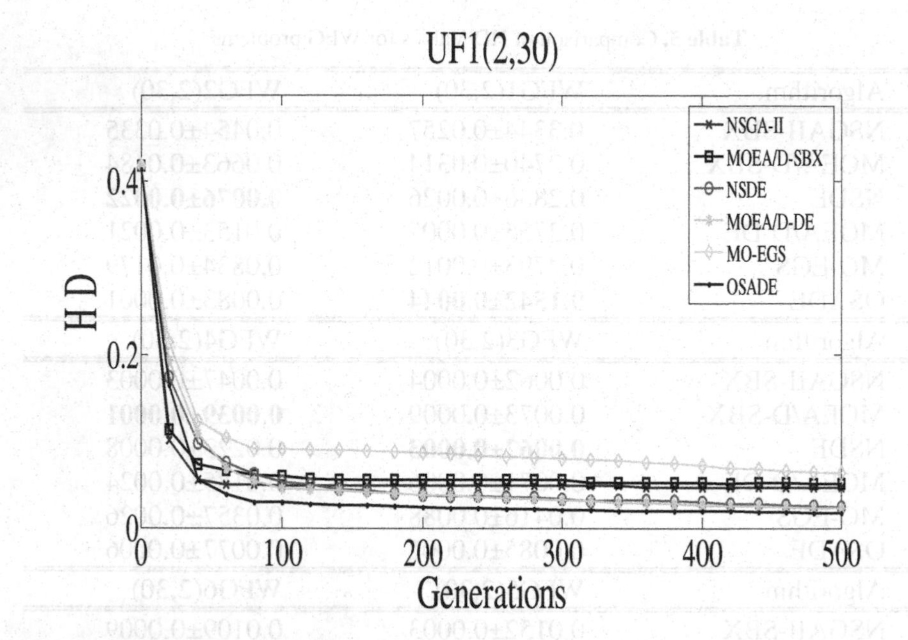

Fig. 1. Evolution of HD average values for UF1 over 30 independent runs

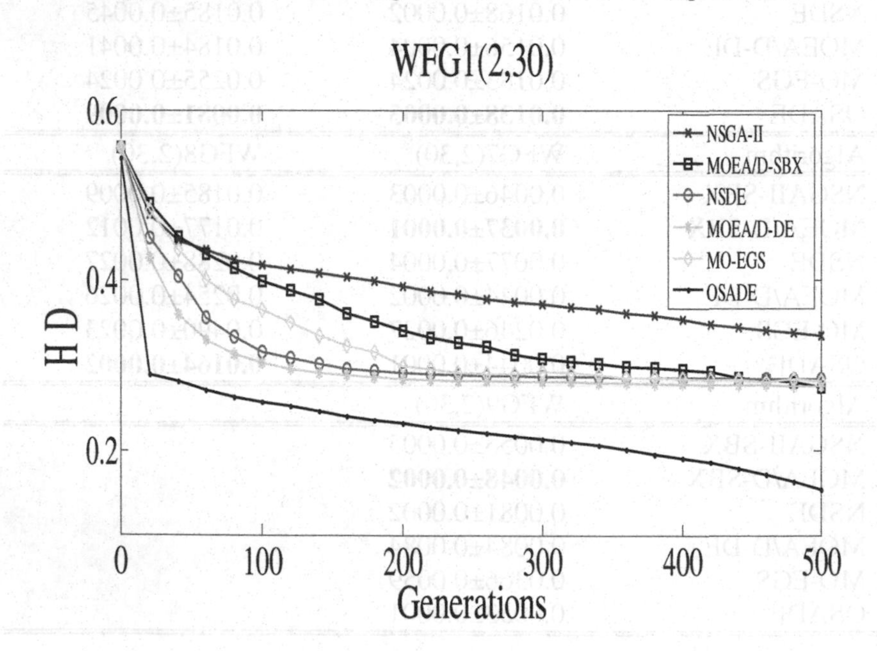

Fig. 2. Evolution of HD average values for WFG1 over 30 independent runs

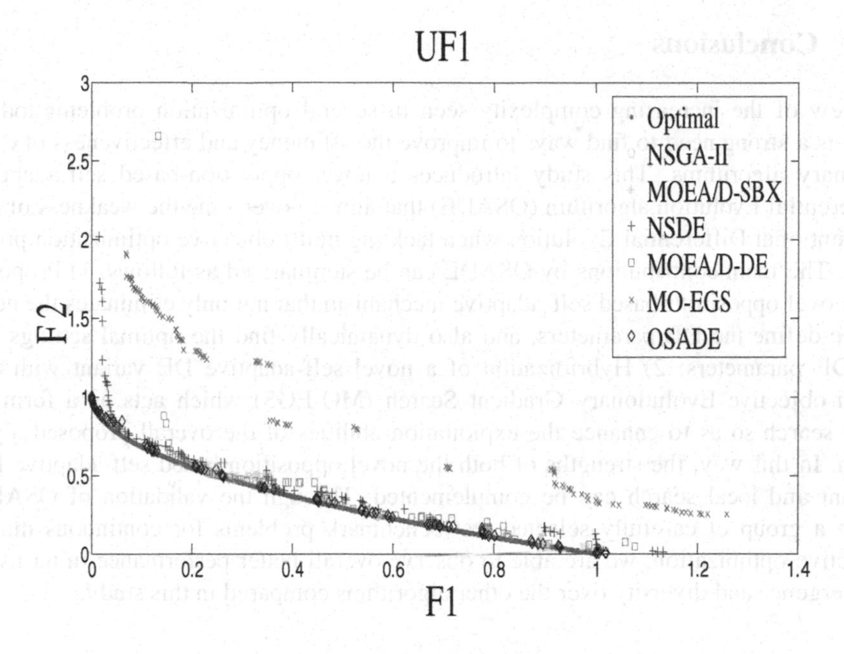

Fig. 3. Evolved Pareto front for UF1

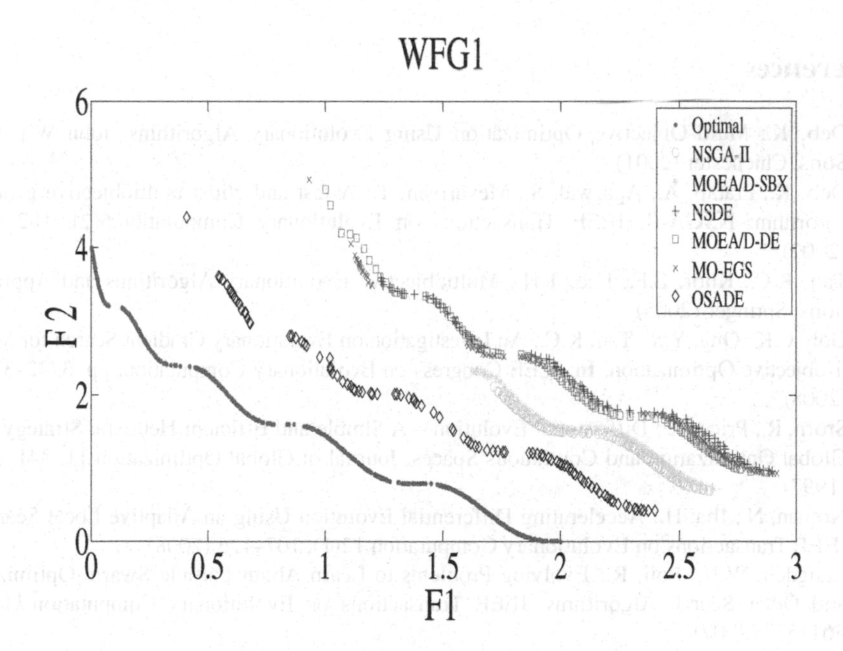

Fig.4. Evolved Pareto front for WFG1

4 Conclusions

In view of the increasing complexity seen in several optimization problems today, there is a strong need to find ways to improve the efficiency and effectiveness of evolutionary algorithms. This study introduces a novel opposition-based self-adaptive Differential Evolution algorithm (OSADE) that aims to overcome the weakness of the conventional Differential Evolution when tackling multi-objective optimization problems. The main contributions by OSADE can be summarized as follows: 1) Proposal of a novel opposition-based self-adaptive mechanism that not only eliminates the need to pre-define the DE parameters, and also dynamically find the optimal settings for the DE parameters; 2) Hybridization of a novel self-adaptive DE variant with the Multi-objective Evolutionary Gradient Search (MO-EGS) which acts as a form of local search so as to enhance the exploitation abilities of the overall proposed algorithm. In this way, the strengths of both the novel opposition-based self-adaptive DE variant and local search can be complemented. Through the validation of OSADE using a group of carefully selected test benchmark problems for continuous multi-objective optimization, we are able to observe overall better performance in terms of convergence and diversity over the other algorithms compared in this study.

Acknowledgement. This work was supported by the Singapore Ministry of Education Academic Research Fund Tier 1.

References

1. Deb, K.: Multi-Objective Optimization Using Evolutionary Algorithms. John Wiley & Sons, Chichester (2001)
2. Deb, K., Pratap, A., Agarwal, S., Meyarivan, T.: A fast and elitist multiobjective genetic algorithm: NSGA-II. IEEE Transactions on Evolutionary Computation 6(2), 182–197 (2002)
3. Tan, K.C., Khor, E.F., Lee, T.H.: Multiobjective Evolutionary Algorithms and Applications. Springer (2005)
4. Goh, C.K., Ong, Y.S., Tan, K.C.: An Investigation on Evolutionary Gradient Search for Multi-objective Optimization. In: IEEE Congress on Evolutionary Computation, pp. 3742–3747 (2008)
5. Storn, R., Price, K.: Differential Evolution – A Simple and Efficient Heuristic Strategy for Global Optimization and Continuous Spaces. Journal of Global Optimization 11, 341–359 (1997)
6. Noman, N., Iba, H.: Accelerating Differential Evolution Using an Adaptive Local Search. IEEE Transactions on Evolutionary Computation 12(1), 107–125 (2008)
7. Langdon, W.B., Poli, R.: Evolving Problems to Learn About Particle Swarm Optimizers and Other Search Algorithms. IEEE Transactions on Evolutionary Computation 11(5), 561–578 (2007)
8. Fan, H.-Y., Lampinen, J.: A Trigonometric Mutation Operation to Differential Evolution. Journal of Global Optimization 27(1), 105–129 (2003)

9. Das, S., Abraham, A., et al.: Differential Evolution Using a Neighbourhood-based Mutation Operator. IEEE Transactions on Evolutionary Computation 13(3), 526–553 (2009)
10. Zhang, J., Sanderson, A.C.: JADE: Adaptive Differential Evolution With Optional External Archive. IEEE Transactions on Evolutionary Computation 13(5), 945–958 (2009)
11. Brest, J., Greiner, S.: Self-Adapting Control Parameters in Differential Evolution: A Comparative Study on Numerical Benchmark Problems. IEEE Transactions on Evolutionary Computation 10(6), 646–657 (2006)
12. Zamuda, A., Brest, J.: Differential Evolution for Multiobjective Optimization with Self-adaptation. In: IEEE Congress on Evolutionary Computation, pp. 3617–3624 (2007)
13. Tizhoosh, H.R.: Opposition-Based Learning: A New Scheme for Machine Intelligence. In: Proceedings of International Conference on Computational Intelligence for Modelling Control and Automation, pp. 695–701 (2005)
14. Zhang, Q., et al.: Multiobjective Optimization Test Instances for the CEC 2009 Special Session and Competition. Technical Report CES-487, University of Essex and Nanyang Technological University (2008)
15. Huband, S., Hingston, P.: A Review of Multiobjective Test Problems and a Scalable Test Problem Toolkit. IEEE Transactions on Evolutionary Computation 10(5), 477–506 (2006)
16. Coello, C.A.C., Cortes, N.C.: Solving Multiobjective Optimization Problems Using an Artificial Immune System. Journal Genetic Programming and Evolvable Machines 6(2), 163–190 (2005)
17. Schutze, O., et al.: Using the Averaged Hausdorff Distance as a Performance Measure in Evolutionay Multiobjective Optimization. IEEE Transactions on Evolutionary Computation 16(4), 504–522 (2012)
18. Zhang, Q., Li, H.: MOEA/D: A Multiobjective Evolutionary Algorithm Based on Decomposition. IEEE Transactions on Evolutionary Computation 11(6), 712–731 (2007)
19. Li, H., Zhang, Q.: Multiobjective Optimization Problems with Complicated Pareto Sets, MOEA/D and NSGA-II. IEEE Transactions on Evolutionary Computation 13(2), 284–302 (2009)
20. Iorio, A.W., Li, X.: Solving Rotated Multi-objective Optimization Problems Using Differential Evolution. In: Webb, G.I., Yu, X. (eds.) AI 2004. LNCS (LNAI), vol. 3339, pp. 861–872. Springer, Heidelberg (2004)
21. Goh, C.K., Tan, K.C., et al.: A Competitive and Cooperative Co-evolutionary Approach to Multiobjective Particle Swarm Optimization Algorithm Design. European Journal of Operational Research 202(1), 42–54 (2010)

Agent-Based Modeling of Economic Volatility and Risk Propagation on Evolving Networks

Yoshito Suzuki[1], Akira Namatame[1], and Yuji Aruka[2]

[1] Dept. of computer science, National Defense Academy,
Yokosuka, Japan
{em52039,nama}@nda.ac.jp
[2] Faculty of Commerce, Chuo University,
Tokyo, Japan
aruka@tamacc.chuo-u.ac.jp

Abstract. Networks increase interdependence, which creates challenges for managing risks. This is especially apparent in areas such as financial institutions and enterprise risk management, where the actions of a single agent (firm or bank) can impact all the other agents in interconnected networks. In this paper, we use agent-based modeling (ABM) in order to analyze how local defaults of supply chain participants propagate through the dynamic supply chain network and interbank networks and form avalanches of bankruptcy. We focus on the linkage dependence among agents at the micro-level and estimate the impact on the macro activities. Combining agent-based modeling with the network analysis can shed light on understanding the primary role of banks in lending to the wider real economy. Understanding the linkage dependency among firms and banks can help in the design of regulatory paradigms that rein in systemic risk while enhancing economic growth.

Keywords: agent-based economics, systemic risk, evolving credit networks, financial networks.

1 Introduction

Macro economy has created well defined approaches and several tools that seemed to serve us for the last decades. However, recent economic fluctuations and financial crises emphasize the need of alternative frameworks and methodologies to be able to replicate such phenomena in order to a deeper understanding the mechanism of eco-nomic crisis and fluctuation. Financial markets are driven by the real economy and in turn they also have a profound effect on it. Understanding the feedback between these two sectors leads to a deeper understanding of the stability, robustness and efficiency of the economic system [8]. Agent-based approaches in Macroeconomics from bottom-up are getting more and more attention recently. Research on this line has been initiated by the series of their works Delli Gatti, et al. [4,5]. Their model simulating the behavior of interacting heterogeneous agents (firms and banks) is able to generate a large number of stylized facts. To jointly account for an ensemble of the facts regarding both micro-macro properties together

H. Handa et al. (eds.), *Proc. of the 18th Asia Pacific Symp. on Intell. & Evol. Systems – Vol. 1*,
Proceedings in Adaptation, Learning and Optimization 1, DOI: 10.1007/978-3-319-13359-1_36

with macro aggregates including GDP growth rates, output volatility, business cycle phases, financial fragility, and bankruptcy cascades, agent-based approaches are getting more and more attention recently.

Historically financial markets were driven by the real economy and in turn they also had a profound effect on it. In recent decades, a massive transfer of resources from the productive sector to the financial sector has been one of the characteristics of global economic systems. This process is mainly responsible for the growing financial instability. In production sectors, there has been dramatic increase in the output volatility and uncertainty. Financial inter-linkages play an important role in the emergence of financial instabilities. Recently many research have focused on the role of linkages along the two dimensions of contagion and liquidity and they suggest that regulators have to look at the interplay of network topology, capital requirements, and market liquidity [9], [14]. In particular for contagion, the position of institutions in the network matters and their impact can be computed through stress tests even when there are no defaults in the system [2].

For the data-driven study using empirical data, many scholars use a collection of daily snapshots of the Italian interbank money market originally provided by the Italian electronic Market for Inter-bank, referred to e-MID in the text [6], [11], [18]. However, even central banks and regulators have only a dim view of the interconnec-tions between banks at a moment in time, and thus the systemic risk in the financial networks, and each bank's contribution to this risk, are poorly known. A natural starting point is to utilize complementary approach to data-driven approach, basing their systemic risk measures on accessing and interpreting data on balance sheets and trading. As understanding of the most critical systemic attributes improves, this network description can be extended to wider jurisdictions and can record more detail: complex transactions such as credit risk derivatives, more complex institutional behavior such as internal risk limit systems and responses to counterparty risk changes.

In this paper, we investigate the effect of credit linkages on the macroeconomic activity by developing the network-based agent model. In particular, we study the linkage dependence among agents (firms and banks) at the micro-level and to estimate their impact on the macro activities such the GDP growth rate, the size and growth rate distributions of agents. We propose the model refinement strategy which validate through the some universal laws and properties based on empirical studies revealing statistical properties of macro-economic time series. The purpose of the network-based model of systemic risks is to build up the dependence among agents (firms and banks) at the micro-level and to estimate their impact on the macro stability.

Phase structure of hypothetical financial systems: the relation between basicnetwork parameters such as connectivity, homogeneity and uncertainty, and macro-scopic systemic risk measures is non-linear, non-intuitive and difficult to predict. Such emergent features will reflect profound properties of real world financial networks that can be understood by first looking at deliberately simplified agent-based simulation models. Simulation studies of complex hypothetical financial networks that map out these types of features will lead to improved understanding of the resilience of net-works, and perhaps ultimately to pragmatic rules of thumb for network participants. This line of inquiry also links systemic network theory strongly to other areas of network science, from which we may draw additional ideas and intuition. We show that three stylized facts: a fat-tailed size distribution of the firm sizes, a tent-shaped growth rate distribution, the scaling relation of the growth rate variance with firm size. We then address the questions of validating and verifying simulations. We validate with the widely acknowledged stylized facts which describe the firm (and bank) growth rates of fat tails, tent distribution, volatility, etc., and recall that some of these properties are directly linked to the way time is taken into account [19]. The growth of firm size, the distribution of firm sizes, the distribution of sizes of the new firms in each year and find it to be well approximated by a log-normal. We validate the simulation results in terms of such as (i) the distribution of the logarithm of the growth rates, for a fixed growth period of one year, and for companies with approximately the same size Sdisplays an exponential form.

In the second part of our work, we investigate the effect of credit linkages on the firms activities to explain some key elements occurred during the recent economic and financial crisis. From this perspective, the network theory is a natural candidate for the analysis of interacting agent systems. The financial sector can be regarded as a set of agents (banks and firms) who interact with each other through financial transactions. These interactions are governed by a set of rules and regulations, and take place on an interaction graph of all connections between agents. The network of mutual credit relations between financial institutions and firms plays a key role in the risk for con-tagious defaults. In particular, we study the repercussions of inter-bank connectivity on agents' performances, bankruptcy waves and business cycle fluctuations. Our findings suggest that there are issues with the role that the bank system plays in the real economy and in pursuing economic growth. Indeed, our model shows that a heavily- interconnected inter-bank system increases financial fragility, leading to economic crises and distress contagion.

2 Risks in a Connected Systems

There is empirical evidence that as the connectivity of a network increases, there is an increase in the network performance, but at the same time, there is an increase in the chance of risk contagion which is extremely large. If external shocks at some agents are propagated to the other connected agents due to failure, the domino effects often come with disastrous consequences. The network is only as strong as its weakest link, and trade-offs are most often connected to a function that models system performance management. The qualification of risks lies in their connections. An interdependent risk to one system may present an opportunity to other systems. Therefore a systemic risk impacts the integrity of the whole system as well as its components.

In a networked world, the risks faced by any one agent depend not only on that agents actions but also on those of others. The fact that the risk one actor faces is often determined in part by the activities of others gives a unique and complex structure to the incentives that agents face as they attempt to reduce their exposure to these inter-dependent risks. The concept of interdependent risks refers to situations in which multiple agents act separately generate common risk. Protective management can reduce the risk of a direct loss to each agent, but there is still some chance of suffering damage from others actions. The fact that the risk is often determined in part by the behavior of others imposes independent risk structures on the incentives that agents face for reducing risk or investing in risk mitigation measures. Kunreuther and Heal were initially led to analyze such situations by focusing on the interdependence of security problems [12]. An interdependent security setting is one in which each individual or firm that is part of an interconnected system must decide independently whether to adopt protective strategies that mitigate future losses. The analysis focused on protection against discrete, low-probability events in a variety of protective settings with somewhat different cost and benefit structures: airline security, computer security, fire protection, and vaccination. Under some circumstances, the interdependent security problem resembles the familiar prisoners dilemma in which the only equilibrium is the decision by all agents not to invest in protection even though everyone would be better off if they had decided to incur this cost. In other words, a protective strategy that would benefit all agents if widely adopted may not be worth its cost to any single agent and it is better off simply taking a free ride on the others' investments.

The financial crises triggered numerous studies on the systemic risks caused by contagion effects via interconnections in the modern banking networks. Systemic risks result in continuous large-scale defaults or systemic failure among the networked banks and financial institutions [9], [14]. There is empirical evidence that as the connectivity of a network increases, there is an increase in the network performance, but at the same time, there is an increase in the chance of risk contagion which is extremely large. The formulation of systemic risk can greatly benefit from a complex network approach. Allen and Gale introduced the use of network theories to enrich our understanding of financial systems and studied how the financial system responds to contagion when financial institutions are connected with different network topologies [1]. They how the banking network topology affect the stability of both finance market and product market by changing the density of connections among banks. While the risk of contagion may be expected to be larger in a highly interconnected banking system, we show that shocks may have complex effects on financial institutions as well as the firms.

Many studies analyze the financial systems such as financial stability and contagion using the network theory and other network analysis methods. In an inter-bank market, banks facing liquidity shortages may borrow liquidity from other banks that have liquidity surpluses. This system of liquidity swapping provides the interbank market with enhanced liquidity sharing. Furthermore, it also brings down the risk of contagion among the interconnected banks when unexpected problems arise. Solvency or liquidity problems of a single bank can travel through the interbank linkages to other banks and become a causality of systemic failure; this highlights the importance of interbank markets for financial stability [15], [17]. A systemic risk is also defined as the risk of a phase transition from one equilibrium condition to another equilibrium. This transition is characterized by self-reinforcing mechanisms that make it difficult to reverse.

3 Agent Based Modeling of Economic Risks

Our work is based on an existing agent-based model [4,5]. Here, we consider multiple banks which can operate not only in the credit market but also in the inter-bank market. In our model, firms may ask for loans from banks to increase their production rate and profit. If contacted banks face liquidity shortage when trying to cover the firms' requirements, they may borrow from a surplus bank in the inter-bank network. In this market, therefore, lender banks share with borrowers bank the risk for the loan to the firm. We model the inter-bank network as preference attachment.

In our model, bankruptcies are determined as financially fragile firms fail, that is their net worth becomes negative. If one or more firms are not able to pay back their debts to the bank, the bank's balance sheet decreases and, consequently, the firms' bad debt, affecting the equity of banks, can also lead to bank failures. As banks, in case of shortage of liquidity, may enter the interbank network, the failure of borrower banks could lead to failures of lender banks. Agents' bad debt, thus, can bring about a cascade of bankruptcies among banks. The source of the domino effect may be due to indirect interaction between bankrupt firms and their lending banks through the credit market, on one side, and to direct interaction between lender and borrower banks through the inter-bank network, on the other side.

3.1 Firms Behavior

The goods market is implemented following the model of Delli Gatti et al. [5] where output is supply-determined, that is firms sell all the output they optimally decide to produce. In the model of Delli Gatti, there are two types of firms: Downstream (D) firms that produce consumption goods using labor and intermediate goods; Upstream (U) firms that produce intermediate goods on demand from D firms.

The D firms demand both for labor $N_{i,t}$ and for intermediate goods $Q_{i,t}$ de-pending on their financial conditions, that is captured by net worth $A_{i,t}$. Respectively, the demands of the i-th firm are given by

$$N_{i,t} = c_1 A_{i,t}^{\beta} \tag{1}$$

$$Q_{i,t} = c_2 A_{i,t}^{\beta} \tag{2}$$

The consumption goods are sold at a stochastic price $u_{i,t}$ that is a random variable extracted from a uniform distribution between $(0,1)$. The U firms produce intermediate goods employing only labor, so that for the j-th the demand is

$$Q_{j,t} = c_3 N_{j,t} \tag{3}$$

Many D firms can be linked to a single U firm but each D firm has only one supplier for intermediate goods among the U firms. The price of intermediate goods is

$$p_{j,t} = 1 + r_{j,t} = \alpha A_{j,t}^{-\alpha} \tag{4}$$

where $r_{j,t}$ is the interest on trade credit, which is assumed to be dependent only on the financial condition of the U firm. In particular, if the j-th firm is not performing well, it will give credit with a less favorable interest rate. While the production of D firms is determined by their worth $A_{i,t}$, the production of U firms is determined by the demand on the part of D firms

$$Q_{j,t} = c_2 \sum A_{i,t}^{\beta} \tag{5}$$

Analogously, the demand for labor will be

$$N_{j,t} = c_1 \sum A_{i,t}^{\beta} \tag{6}$$

Each period a subset of firms enter in the credit market asking for credit. The amount of credit requested by companies is related to their investment expenditure, which is, therefore, dependent on interest rate and firm's economic situation. If the net worth of the firms is not sufficient to pay the wage bill, they will demand credit to a bank. For each firm the credit demand is

$$B_t = c_4 N_t - A_t \tag{7}$$

where the functional form of N_t changes if we are considering U firm or D firm. We assume that many firms can be linked to a single bank but each firm has only one supplier of loans. Without entering in the details, we point out that the interest rate on loans is a decreasing function of the banks net worth and penalizes financially fragile firms.

The profits of the D and U firm are evaluated from the difference between their gains and the costs, and the profit of the j-th U firm at time t is given by

$$\pi_{j,t} = (1 + r_{j,t})Q_{j,t} - (1 + r_{j,t})B_{j,t} \tag{8}$$

At each time step the net worth of each firm is updated according to

$$A_{j,t+1} = A_{j,t} + \pi_{j,t} - D_{j,t} \tag{9}$$

Bankruptcy occurs if the net worth becomes negative. The bankrupt firm leaves the market. Therefore $D_{j,t}$ in Eq.(9) is the bad debt, that takes into account the possibility that a borrower cannot pay back the loan because it goes bankrupt (that is, $A_{j,t} \leq 0$). In this framework the lenders are the U firms and the banks and both U and D firms can be borrowers. The total number of agents (U and D firms) is kept constant over time. Therefore when firms fail, they are replaced by new entrants which are on average smaller than incumbents. So, entrants' size is drawn from a uniform distribution cen-tered around the mode of the size distribution of incumbent firms.

3.2 Banks Behavior

The primary purpose of banks is to channel their funds towards loans to companies. Consulted banks, analyzed their own credit risk and the firm's risk, may grant the requested loan, when they have enough supply of liquidity. The supply of credit is a percentage of banks' equity because financial institutions adopt a system of risk management based upon an equity ratio. When consulted banks do not have liquidity to lend, they can enter in the interbank system, in order not to lose the opportunity of earning on investing firms.

Similar to firms, we have a constant population of competitive banks indexed by $j = 1, \ldots, B$. Each bank has a balance sheet structure defined as $S_{j,t} = E_{j,t} + D_{j,t}$ with $S_{j,t}$ being the credit supply, $E_{j,t}$ the equity and $D_{j,t}$. The primary function of banks activity is to lend their funds through loans to firms, as this is their way to make money via interest rates. Bank j offers its interest rate to the borrower firm i:

$$r^i_{j,t} = \sigma A^{-\beta}_{j,t} + \theta l^{-\theta}_{i,t} \tag{10}$$

So the interest rate is decreasing with the borrower's financial robustness. In a sense, we adopt the principle according to which the interest rate charged by banks incorpo-rates an external finance premium increasing with the leverage and, therefore, inversely related to the borrower's net worth.

When firm i needs loan, it contacts a number of randomly chosen banks. Credit linkages between firms and banks are defined by some bipartite graph Contacted banks, checked the investment risk and their amount of liquidity ($S_{j,t} \geq L^i_{d,t}$), offer an interest rate in Eq.(10). After exploring the lending conditions of the contacted banks, each firm asks the consulted banks for credit starting with the one offering the lowest interest rate. If in the credit market, the contacted financial institutions have not enough supply of liquidity to fully satisfy the firm's loan, then banks consider to use the inter-bank market. As in the credit market, the requiring bank (borrower) asks the lacking fraction of the loan requested by the firm from a number of randomly chosen banks (lenders). Among the contacted banks, the banks satisfying the risk threshold in Eq.(8) and having enough supply of liquidity offer the loan to the asking bank for an inter-bank interest rate, which equals the credit market interest rate in Eq.(9). Among this subset of offering banks, the borrower bank chooses the lender bank, starting with the one offering the lowest interest rate.

At the end of each period t, after trading has taken place, financial institutions update their profits. The bank's profit depends on interests on credit market (first term), on interests on inter-bank market (second term), which can be either positive or negative depending on bank j net position (lender or borrower), on interests payed on deposits and equity (third term). Bank net worth evolves according to:

$$E_{j,t+1} = E_{j,t} - 1 + \pi_{j,t} - B_j^i - B_j^k \quad (11)$$

with the last two terms on the right side being firms and banks' bad debts respectively. Similar to firms, banks go bankrupt when their equity at time t becomes negative, and the failed bank leaves the market.

The total number of banks is kept constant over time. Therefore when banks fail, they are replaced by new entrants which are on average smaller than incumbents.

3.3 Credit Network Formation

We define the network formation dynamics, how D firms look for the linkage with U firm and how D and U firms look for a bank to ask their loans. In order to establish the product supply linkages, the firms take the partner choice rule, that is, they search for the minimum of the prices charged by a randomly selected set of possible suppliers. It can change supplier only if a better partner is found. Similarly in order to establish the credit linkages the frms take the partner choice rule: they search for the minimum of the interest rate charged among the loan offered banks. If contacted banks face liquidity shortage when trying to cover the firms' requirements, they may borrow from a surplus bank in the inter-bank system. In this market, therefore, lender banks share with borrower banks the risk for the loan to the firm.

3.4 Interbank Network Formation

We model the inter-bank network based on some connection rules: (i) random connection (random graph as a bench mark), (ii) net-worth based connection, an agent with higher net worth is selected as a partner, (iii) interest based connection, an agent offering a lower interest is selected as a partner.

Bankruptcies are determined as financially fragile firms fail, that is their net worth becomes negative. If one or more firms are not able to pay back their debts to the bank, the bank's balance sheet decreases and, consequently, the firms' bad debt, affecting the equity of banks, can also lead to bank failures. As banks, in case of shortage of liquidity, may enter the interbank market, the failure of borrower banks could lead to failures of lender banks. Agents' bad debt, thus, can bring about a cascade of bankruptcies among banks. The source of the domino effect may be due to indirect interaction between bankrupt firms and their lending banks through the credit market, on one side, and to direct interaction between lender and borrower banks through the inter-bank system, on the other side. Their findings suggest that there are issues with the role that the bank system plays in the real economy and in pursuing economic growth.

Indeed, their model shows that a heavily-interconnected inter-bank system increases financial fragility, leading to economic crises and distress contagion. The process of contagion gains momentum and spreads quickly to a large number of banks once some critical banks fail.

3.5 Clearing Mechanism

We consider the closed economy and the total number of banks, U and D firms are kept constant over time: bankrupt firms or banks are replaced one to one. In a closed real and financial economy, and there is no sources or sinks of money, except a new entrant with small net worth are only source money. In the case, firms go bankruptcy, bad debt is absorbed by banks. The interbank market is useful for the finance sector. When a bank cannot lend the loan to a firm, it can get necessarily fund through the interbank market. The linkages among banks are also effective for shock absorbing: if some bank goes bankruptcy, other banks which have the credit linkage may partially absorb. This mechanism of risk sharing or shock absorbing will bring positive effects on real economy.

In order to clarify the role of financial networks we investigate the clearing up mechanism explicitly.

(*Bailout mode 1*) A bailout is financed by the central bank or collected from the household and firms. This is the case where there is no interbank network and no other bank pay for the debt of the other bank.

(*Bailout mode 2*) Each defaulting bank is bailed out, and the required fund is col-lected from the other banks with linkage. This is the case where the interbank network exists and other banks pay for the debt.

4 Simulations and Results

The model is studied numerically for different values of the parameter ρ, which drives the inter-bank connectivity. We consider an economy consisting of $N = 1000$ firms and $B = 50$ banks and do simulation over the time period span of $T = 1000$. Each firm is initially given the same amount of capital $K_{i,0} = 100$, net-worth $A_{i,0} = 65$ and loan $L_{i,0} = 35$. We also set other parameters as follows: $\tau = 4$, $\Phi = 0.8$, $c = 1$, $\lambda = 0.3$, $\alpha = 0.1$, $\chi = 0.8$, $\psi = 0.1$ and $\theta = 0.05$. The probability of attachment between firms and banks in the credit market is $x = 0.05$. In this way the number of firm's out-going links is less than three. The reason being that in a highly connected random network synchronization could be achieved via indirect links. The effects of direct contagion among financial institutions are easier to be tested in a network where indirect synchronization is less likely to arise. We repeat simulations 100 times with different random seeds. We start by analyzing the effect of inter-bank linkages on the systemic risk. Then we analyze the correlation between the financial and the real sector of the economy.

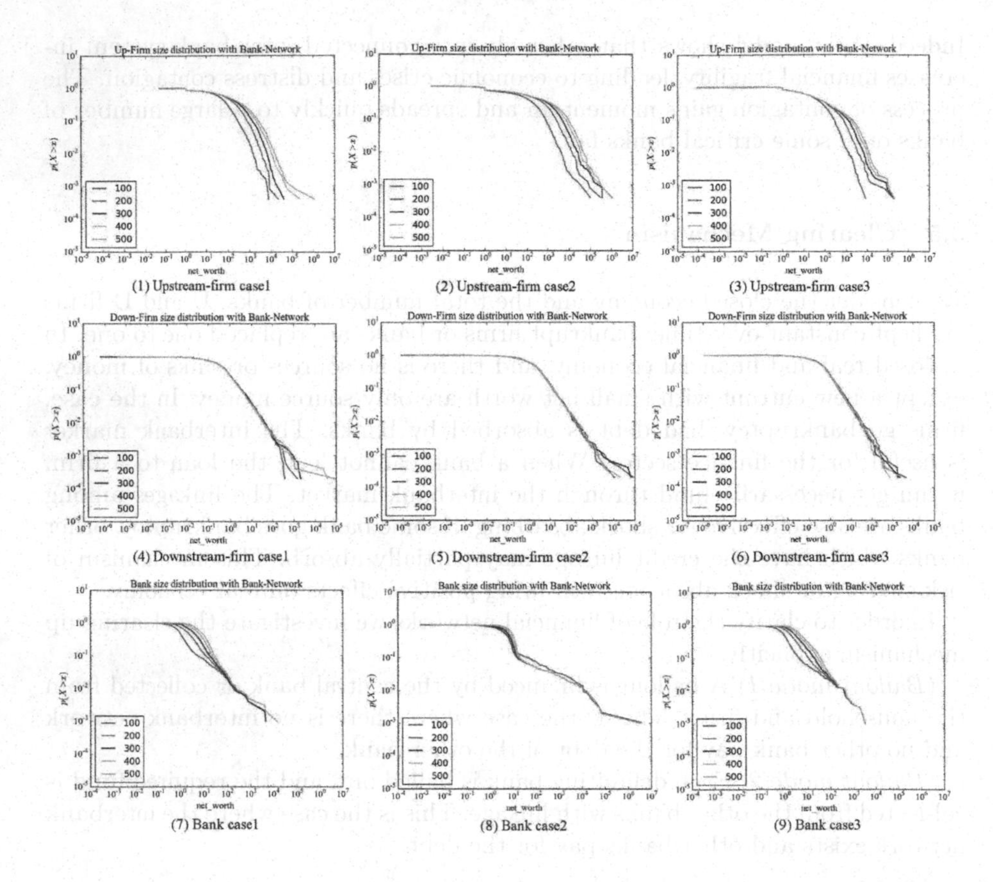

Fig. 1. How the size cumulative distributions in term of the net-worth of firms (Downstream firms, Upstream-firms, and Banks) at the time periods, 100, 200,300,400, and 500

4.1 Economic Growths (Firm and Bank Growth)

Given that the simulated aggregated output time series show an upward trend, we extract the trend component. By developing simple interactive structures among agents and feedback effects, we can reproduce the source of business fluctuations: on one side, due to indirect interactions between bankrupt firms and their lending banks through the credit market and, on the other side, due to direct interactions between lender and borrower banks through the interbank system. In particular, we analyzed the effect of inter-bank linkages on the GDP expansion and contractions. Our model displays quite realistic features, starting from the three-sector structure of the credit network. Also the bankruptcy avalanches mechanism is realistic. Moreover, the model is able to reproduce the main empirical facts about the statistical properties of firms. Fig.1 show the size cumulative distributions in term of the net-worth of firms (Down-stream firms,

Upstream-firms, and Banks) at the time periods, 100, 200,300,400, and 500. The topics on economic growth processes have then attracted much interest. Recent empirical studies suggest that the firm size distribution follows a power law distribution [19]. Regarding other empirical evidences, the results shown in Fig.1 shows reasonable agreement with the known data: the distribution of firm size is a power law and the one of aggregate growth rates is a double exponential.

4.2 Contagion Effects on the Real Economy

In our model, bankruptcies are determined as financially fragile firms fail, that is their net worth becomes negative. If one or more firms are not able to pay back their debts to the bank, the bank's balance sheet decreases and, consequently, the firms' bad debt, affecting the equity of banks, can also lead to bank failures. As banks, in case of shortage of liquidity, may enter the interbank market, the failure of borrower banks could lead to failures of lender banks. Agents' bad debt, thus, can bring about a cascade of bankruptcies among banks. The source of the domino effect may be due to indirect interaction between bankrupt firms and their lending banks through the credit market, on one side, and to direct interaction between lender and borrower banks through the inter-bank system, on the other side.

One of the goals of our work is the study of bankruptcy avalanches and their connection with the dynamics of the credit networks. Fig. 2 shows some conceptual framework of default cascade on the credit network. Suppose that a random price fluctuation causes the bankruptcy of some U firms. Consequently, the loans they took will not be fulfilled and the worth of the lenders (banks and D firms) will decrease. Eventually, this will result in a bankruptcy of some of them and, more importantly, in an increase of the interest rates charged on their old and new borrowers. This, in turn, will increase the probability of a bankruptcy of a D firm, and so on. The credit network has a scale free structure and then the default of a highly connected agent may provoke an avalanche of bankruptcy.

The question we address here is whether phenomena of collective bankruptcies are related to the initial setting of parameters. Usually simulation starts with homogeneous firms and banks. Early stage of the simulation (before 200 period) many firms and banks default. However after that, especially banks and D-firms grow as extremely heterogeneous agents and the size distribution obeys power low, then a few banks and firms default occasionally. In order to answer this mysterious observation, we inves-tigate the effects of inter-bank linkages on contagion phase in the financial market. In particular, we focus on one of the most extreme examples of systemic failure, namely bank bankruptcies. Fig.3 shows the average number of failed banks over all times. Collective bankruptcies arise from the complex nature of agent interaction. To better analyze this mystery, we further need data-driven analysis of simulated data.

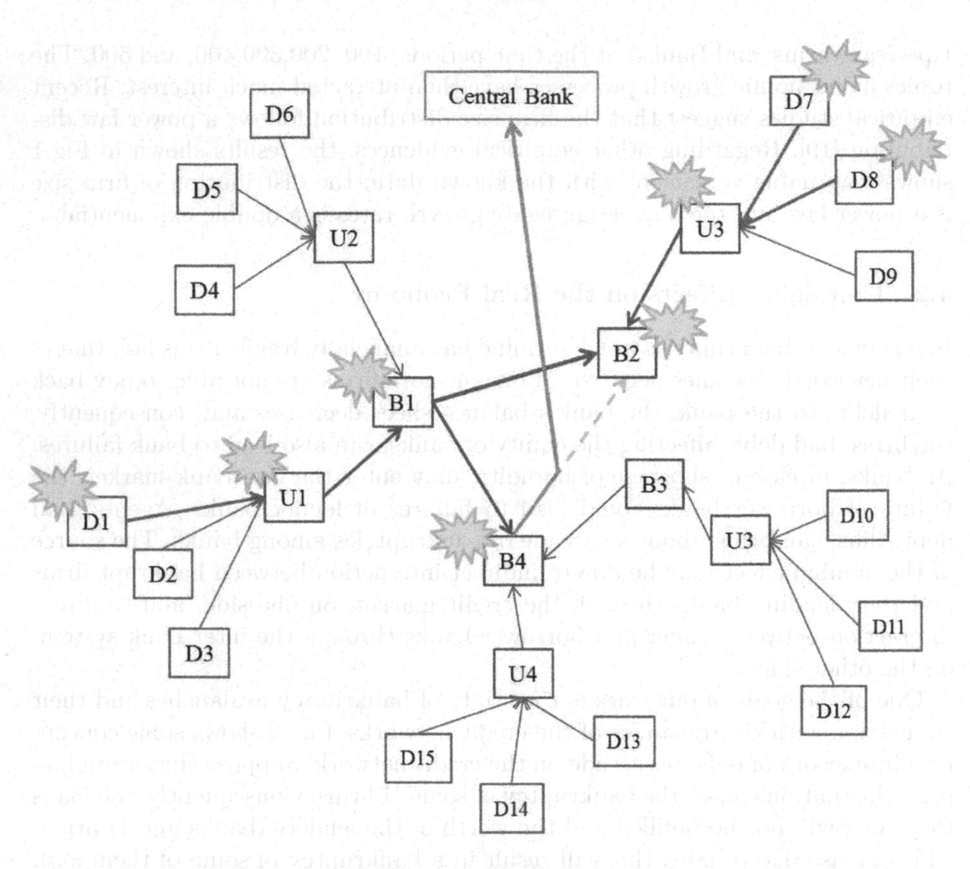

Fig. 2. shows some conceptual framework of default cascade on the credit network

4.3 Economic Growths (Firm and Bank Growth)

We explicitly model the agents' interaction as credit networks. In our model, firms may ask for loans from banks to increase their production rate and profit. If contacted banks face liquidity shortage when trying to cover the firms' requirements, they may borrow from a surplus bank in the inter-bank system. In this market, therefore, lender banks share with borrowers bank the risk for the loan to the firm. Some snapshot of the credit linkages among firms and banks are shown in Fig.4 and the interbank network is shown in Fig.5 and they are always evolving by changing the connectivity.

Addressing the question of how firms and banks form networks is not just of theoretical interest, but also informs understanding of how effectively the financial sector performs its economic functions. Combining as agent-based modeling and a financial network creation model can shed light on understanding the primary role of banks in lending to the wider economy, how banks are likely to interact, as well as the optimal design of new institutions such as central clearing houses.

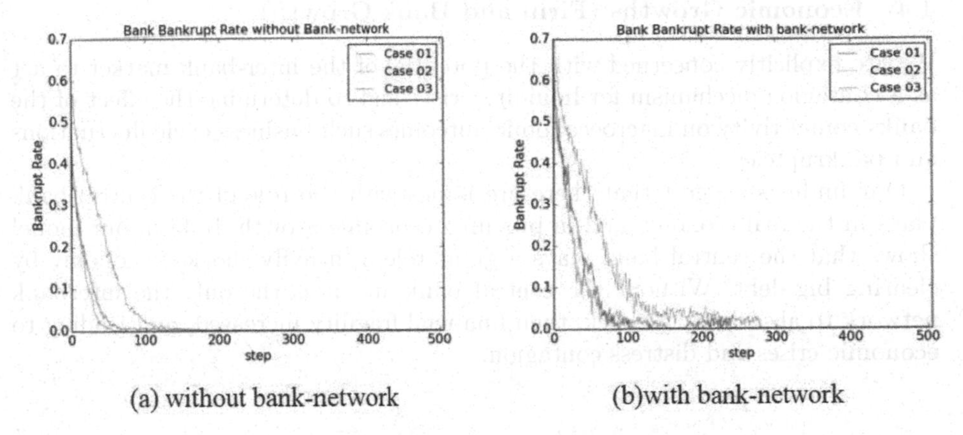

(a) without bank-network (b)with bank-network

Fig. 3. shows the average number of failed banks over all times. Collective bankruptcies arise from the complex nature of agent interaction.

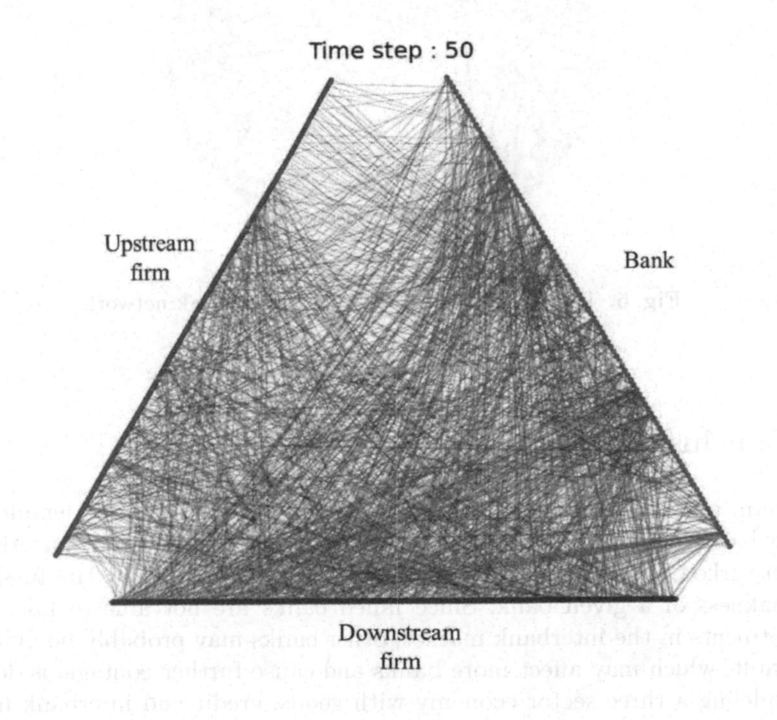

Fig. 4. Some snapshot of the evolving credit linkages among firms and banks

4.4 Economic Growths (Firm and Bank Growth)

We are explicitly concerned with the potential of the inter-bank market to act as a contagion mechanism for liquidity crises and to determine the effect of the banks connectivity on macroeconomic outcomes such business cycle fluctuations and bankruptcies.

Our findings suggest that there are issues with the role of the central bank plays in the real economy and in pursuing economic growth. Indeed, our model shows that the central bank plays a great role in heavily shocked economy by clearing big debt. Without the central bank and it is the only the interbank network to absorb a big shock, then financial fragility increased, and leading to economic crises and distress contagion.

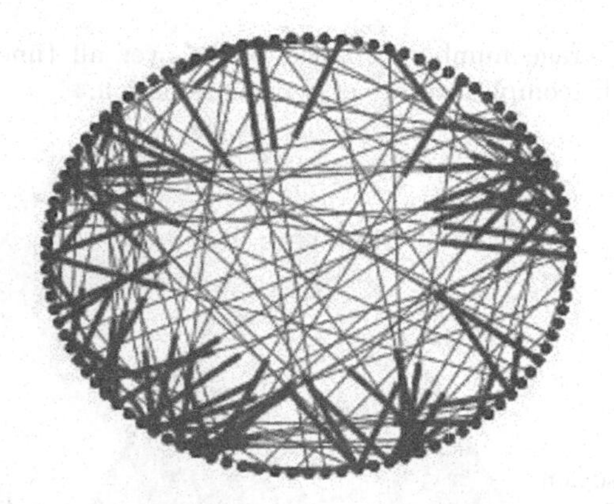

Fig. 5. The snapshot of the evolving interbank network

5 Conclusions

The main purpose of this paper is to study the contagious bank defaults, that is, the bank defaults that influence other banks through interconnectivity of the networked banking system, and not the defaults caused by the fundamental weakness of a given bank. Since failed banks are not able to honor their commitments in the interbank market, other banks may probably be influenced to default, which may affect more banks and cause further contagious defaults. By modeling a three sector economy with goods, credit and interbank market, we have been able to analyze the role of financial network on the agents performance and macro dynamics in terms of the growth and stability. Our results also support that the interaction among market participants (firms and banks) is a key element to reproduce important stylized facts about bankruptcy waves and business cycle fluctuations. In particular, we have shown that the existence the strong linkages among banks generates larger bankruptcy cascades due to the larger systemic risk. When our inter-bank network reaches the phase transition, the presence of many interconnected banks suggests that the credit network is more susceptible to the domino effect. In this case, in fact, when failures occur, many agents are potentially compromised. However, our model has shown that the relationship between risk propagation and risk sharing cannot be clearly defined. Our findings suggest that there are issues with the role that the bank system plays in the real economy and in pursuing economic growth. Indeed, our model shows that a heavily-interconnected inter-bank system increases financial fragility, leading to economic crises and distress contagion if the role of the central bank is weak.

References

1. Allen, F., Gale, D.: Financial contagion. J. Political Econ. 108(1), 1–33 (2000)
2. Battiston, S., et al.: DebtRank: Too Central to Fail? Financial Networks, the FED and Systemic Risk. Scientific Reports, `http://www.nature.com/srep/2012/120802/srep00541/full/srep00541.html`
3. Bramoull Ye, Y., Kranton, R.: Risk-sharing networks. Journal of Economic Behavior and Organization 64(3-4), 275–294 (2007)
4. Delli Gatti, D., Di Guilmi, C., Gallegati, M., Palestrini, A.: A new approach to business fluctuations: heterogeneous interacting agents, scaling laws and financial fragility. Journal of Economic Behavior and Organization 56(4), 489–512 (2005)
5. Delli Gatti, D., Gallegati, M., Greenwal, B., Stiglitz, J.: The financial accelator in an evolving credit network. Journal of Economic Dynamic and Control 34, 1627–1650 (2010)
6. Fricke, D., Lux, T.: Core-Periphery Structure in the Overnight Money Market: Evidence from the E-mid trading platform. Kiel Working Paper, No.1759 (2012)
7. Gai, P., Kapadia, S.: Contagion in financial networks. Proc. R. Soc. Interface, 466–480 (2010)
8. Gallegati, M., Greenwald, B., Richiardi, M.G., Stiglitz, J.E.: The asymmetric effect of diffusion processes: Risk sharing and contagion. Global Economy Journal 8(3), 1–20 (2008)
9. Haldane, A., May, R.: Systemic risk in banking ecosystems. Nature 469, 351–355 (2011)
10. Ide, K., Namatame, A.: A mesoscopic approach to modeling and simulation of systemic risks. IEEE CIFEr (2014)
11. Iori, G., Jafarey, S., Padilla, F.G.: Systemic risk on the interbank market. Journal of Economic Behavior and Organization 61 (2006)
12. Kunreuther, H., Heal, G.: Interdependent Security. Journal of Risk and Uncertainty 26, 231–249 (2003)
13. Legendi, R., Gulyas, L.: Replication of the Macro ABM Model. CRISIS, working paper (2012)
14. May, R.M., Levin, S.A., Sugihara, G.: Ecology for bankers. Nature 451, 893–895 (2008)

15. May, R.M., Arinaminpathy, N.: Systemic risk: the dynamics of model banking systems. Journals of the Royal Society 7, 823–838 (2010)
16. Namatame, A., Tran, H.A.: Enhanceing the resilience of networked agents thorough risk sharing. Advances in Complex Systems 21(2) (2013)
17. Nier, E., Yang, J., Yorulmazer, T., Alentorn, A.: Network models and financial stability. Journal of Economic Dynamics and Control 31(6), 2033–2060 (2007)
18. Soramaki, K., Bech, M.L., Arnold, J., Glass, R.J., Beyeler, W.E.: The topology of interbank payment flows. Physica A 379, 317–333 (2007)
19. Stanley, M., Amaral, L., Sergey, V., Havlin, L., Leschhorn, H., Maass, P., Stanley, E.: Scaling behavior in the growth of companies. Nature 379, 804–806 (1996)

Cuckoo Optimization Algorithm for Optimal Power Flow

Tu Nguyen Le Anh[1], Dieu Ngoc Vo[1,*], Weerakorn Ongsakul[2], Pandian Vasant[3], and Timothy Ganesan[4]

[1] Department of Power Systems, HCMC University of Technology, Ho Chi Minh City, Vietnam
{tunguyen8x,vndieu}@gmail.com
[2] Energy Field of Study, Asian Institute of Technology, Pathumthani 12120, Thailand
ongsakul@ait.ac.th
[3] Department of Fundamental and Applied Sciences, Universiti Teknologi Petronas, 31750 Tronoh, Perak, Malaysia
pvasant@gmail.com
[4] Department of Chemical Engineering, Universiti Teknologi Petronas, 31750 Tronoh, Perak, Malaysia
timothy.andrew@petronas.com.my

Abstract. This paper proposes a cuckoo optimization algorithm (COA) method for solving optimal power flow (OPF) problem. The proposed method is inspired from the life of the family of cuckoo. In the proposed method, there are two main components including mature cuckoos and cuckoo's eggs. During the survival competition, the survived cuckoo societies immigrate to a better environment and restart the process. The cuckoo's survival effort hopefully converges to a state that there is only one cuckoo society with the same maximum profit values. The COA method has been tested on the IEEE 30, 57, and 118-bus systems with three kinds of objective function and the obtained results have been compared to those from conventional particle swarm optimization (PSO) method. The result comparison has shown that the proposed method can obtain better optimal solution than the conventional PSO. Therefore, the proposed COA could be a useful method for implementation in solving the OPF problem.

Keywords: Cuckoo habitat, cuckoo optimization algorithm, cuckoo style, optimal power flow.

1 Introduction

The objective of an optimal power flow (OPF) problem is to find the steady state operation point of generators in the system so as their total generation cost is minimized while satisfying various generator and system constraints such as generator's real and reactive power, bus voltage, transformer tap, switchable capacitor bank, and transmission line capacity limits. In the OPF problem, the controllable variables usually determined are real power output of generators, voltage magnitude at generation buses, injected reactive power at compensation buses, and transformer tap settings. Traditionally, mathematical programming techniques can effectively deal with the problem.

* Corresponding author.

H. Handa et al. (eds.), *Proc. of the 18th Asia Pacific Symp. on Intell. & Evol. Systems – Vol. 1*,
Proceedings in Adaptation, Learning and Optimization 1, DOI: 10.1007/978-3-319-13359-1_37

However, due to the incorporation of FACTS devices to systems, valve point effects or multiple fuels to generators recently, the OPF problem becomes more complicated and the mathematical programming techniques are not a proper selection. Therefore, it requires more powerful search methods for a better implementation. Due to its importance, the OPF problem has been widely studied in the world in [5], [6], [7], [8].

The OPF problem has been solved by several conventional methods such as gradient-based method in [9]; linear programming (LP) in [10], [11]; non-linear programming (NLP) in [12], [13]; quadratic programming (QP) in [14], [15]; Newton-based methods in [16], [17], [18]; semidefinite programming in [19], and interior point method (IPM) in [20], [21], [22]. Generally, the conventional methods can find the optimal solution for an optimization problem with a very short time. However, the main drawback of these methods is that they are difficult to deal with non-convex optimization problems with non-differentiable objective. Moreover, these methods are also very difficult for dealing with large-scale problems due to large search space. Meta-heuristic search methods recently developed have shown that they have capability to deal with this complicated problem. Several meta-heuristic search methods have been also widely applied for solving the OPF problem such as genetic algorithm (GA) in [23], [24], [25]; simulated annealing (SA) in [26]; tabu search (TS) in [27]; evolutionary programming (EP) in [28], [29]; particle swarm optimization (PSO) in [30], and differential evolution (DE) in [31]. These meta-heuristic search methods can overcome the main drawback from the conventional methods with the problem not required to be differentiable. However, the optimal solutions obtained by these methods for optimization problems are near optimum and quality of the solutions is not high when they deal with large-scale problems; that is the obtained solutions may be local optimums with long computational time.

In this paper, a newly Cuckoo optimization algorithm (COA) is proposed for solving optimal power flow (OPF) problem. The proposed COA can quickly move to optimal solution. The proposed method has been tested on benchmark functions, the IEEE 30, 57 and 118-bus systems. The results from the proposed COA are also validated by comparing to those from the conventional method.

2 Problem Formulation

In the OPF problem, the considered variables include control variables and state variables. The control variables include real power injected at generation buses excluding the slack bus, voltage at generation buses, tap ratio of transformers, and reactive power injected by capacitor banks. The state variables include power generation at the slack bus, voltage at load buses, reactive power output of generators, and power flow in transmission lines. In addition, the OPF problem also includes equality constraints which are power flow equations and inequality constraints which are limits control variables and state variables.

Generally, the OPF problem can be formulated as a constrained optimization as follows:

$$Min \sum_{i=1}^{N_g} F_i(P_{gi}) \tag{1}$$

where the fuel cost function $F_i(P_{gi})$ of generating unit i can be expressed in one of the forms as follow:

- *Quadratic Function*: The fuel cost of each thermal generator is represented as a quadratic function of its power output.

$$F_i(P_{gi}) = a_i + b_i P_{gi} + c_i P_{gi}^2 \tag{2}$$

- *Valve Point Effect*: The effect of valve points in boilers of thermal generating units is represented by a sinusoidal component added to the quadratic function.

$$F_i(P_{gi}) = a_i + b_i P_{gi} + c_i P_{gi}^2 + \left| e_i \times \sin\left(f_i \times \left(P_{gi,min} - P_{gi}\right)\right)\right| \tag{3}$$

- *Multiple Fuels*: A generator may have different fuels where each fuel is represented by a piecewise quadratic function as follow:

$$F_i(P_{gi}) = \begin{cases} a_{i1} + b_{i1} P_{gi} + c_{i1} P_{gi}^2, \text{fuel } 1, P_{gi,min} \leq P_{gi} \leq P_{gi1} \\ \cdots \\ a_{ik} + b_{ik} P_{gi} + c_{ik} P_{gi}^2, \text{fuel } k, P_{gik-1} \leq P_{gi} \leq P_{gik} \\ \cdots \\ a_{in} + b_{in} P_{gi} + c_{in} P_{gi}^2, \text{fuel } n, P_{gn-1} \leq P_{gi} \leq P_{gi,max} \end{cases} \tag{4}$$

We have to minimize one of three cost functions and satisfy equality and inequality constraints.

$$P_{gi} - P_d = V_i \sum_{j=1}^{N_b} V_j \left[G_{ij} \cos(\delta_i - \delta_j) + B_{ij} \cos(\delta_i - \delta_j)\right], i = 1, \ldots, N_b \tag{5}$$

$$Q_{gi} + Q_{ci} - Q_{di} = V_i \sum_{j=1}^{N_b} V_j \left[G_{ij} \cos(\delta_i - \delta_j) + B_{ij} \cos(\delta_i - \delta_j)\right], i = 1, \ldots, N_b \tag{6}$$

The real power, reactive power, and voltage at generation buses should be within between their lower and upper bounds:

$$P_{gi,min} \leq P_{gi} \leq P_{gi,max} \; ; i = 1, \ldots, N_g \tag{7}$$

$$Q_{gi,min} \leq Q_{gi} \leq Q_{gi,max} \; ; i = 1, \ldots, N_g \tag{8}$$

$$V_{gi,min} \leq V_{gi} \leq V_{gi,max} \; ; i = 1, \ldots, N_g \tag{9}$$

The switchable Shunt capacitor banks and the tap setting of each transformer should be within their lower and upper limits:

$$Q_{ci,min} \leq Q_{ci} \leq Q_{ci,max} \; ; i = 1, \ldots, N_c \tag{10}$$

$$T_{k,min} \leq T_k \leq T_{k,max} \; ; k = 1, \ldots, N_t \tag{11}$$

The voltage at load buses and power flow in transmission lines should not exceed their limits:

$$V_{li,min} \leq V_{li} \leq V_{li,max} \; ; i = 1, \ldots, N_d \tag{12}$$

$$S_l \leq S_{l,max}\,; l = 1, \ldots, N_l \tag{13}$$

$$S_t = \max \{ |S_{ij}|, |S_{ji}| \} \tag{14}$$

So we have two vectors u and x representing control variables and state variables, respectively.

$$\boldsymbol{u} = \{P_{g2}, \ldots, P_{gN_g}, V_{g1}, \ldots, V_{gN_g}, T_1, \ldots, T_{N_t}, Q_{c1}, \ldots, Q_{cN_c}\}^T \tag{15}$$

$$\boldsymbol{x} = \{P_{g1}, Q_{g1}, \ldots, Q_{gN_g}, V_{l1}, \ldots, V_{lN_d}, S_{l1}, \ldots, S_{lN_t}\}^T \tag{16}$$

3 Proposed Cuckoo Optimization Algorithm

COA is inspired by a special life style of cuckoo bird, there is no cuckoo bird give birth to live young. Mature cuckoos have to find a place to safely place their eggs and hatch the host bird nests. After that, the feed responsibility will belong to host bird. Some of chicks has come out or egg is laid in bad Habitat will be killed. There is only a number of cuckoo's eggs have chance to grow up and become mature cuckoo. All mature cuckoos will move forward to the best Habitat. After some iteration, the cuckoo populations will converge in a Habitat with best profit values.

Like other evolutionary algorithms, the proposed algorithm starts with an initial population of cuckoos. These initial cuckoos have some eggs to lay in some host birds' nests. Some of these eggs which are more similar to the host bird's eggs have the opportunity to grow up and become a mature cuckoo. Other eggs are detected by host birds and are killed. The grown eggs reveal the suitability of the nests in that area. The more eggs survive in an area, the more profit is gained in that area. So the Habitat in which more eggs survive will be the term that COA is going to optimize.

3.1 Generating Initial Cuckoo Habitat

In order to solve an optimization problem, it's necessary that the values of problem variables be formed as an array. In GA and PSO terminologies this array is called "**Chromosome**" and "**Particle Position**", respectively. But here in cuckoo optimization algorithm (COA) it is called "**Habitat**". In an N_{var} dimensional optimization problem, a habitat is an array of $1\times N_{var}$, representing current living Habitat of cuckoo. This array is defined as follows:

$$Habitat = [x_1, x_2, \ldots, x_{Nvar}] \tag{17}$$

Each of the variable values ($x_1, x_2, \ldots, x_{Nvar}$) is floating point number and a Habitat may represent for a vector of control variables in OPF problem. The profit of a habitat is obtained by evaluation of profit function f_p at a habitat of ($x_1, x_2, \ldots, x_{Nvar}$).

$$Profit = f_p(habitat) = f_p(x_1, x_2, \ldots, x_{Nvar}) \tag{18}$$

As it is seen COA is an algorithm that maximizes a profit function. To use COA in cost minimization problems such as minimizing the fitness functions in OPF problem, one can easily maximize the following profit function

$$Profit = -\,Cost\,(habitat) = -\,f_c(x_1, x_2, \ldots, x_{Nvar}) \tag{19}$$

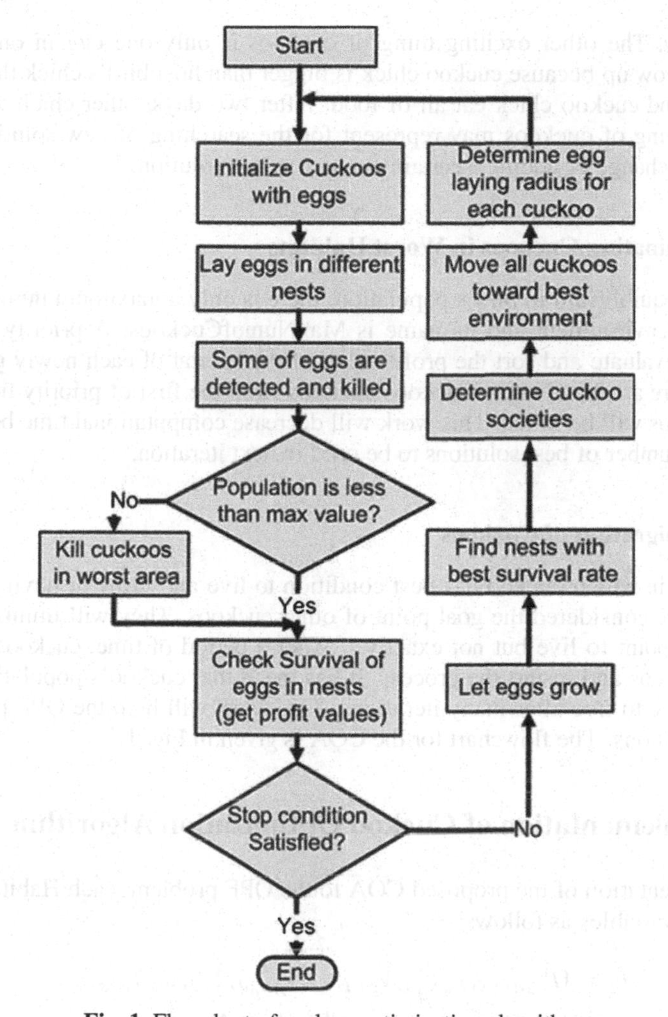

Fig. 1. Flowchart of cuckoo optimization algorithm

3.2 Cuckoo's Style for Eggs Laying

In the first iteration of this optimization algorithm, a candidate habitat matrix of size $N_{Pop} \times N_{Var}$ is generated. Then dedicated some random eggs to each cuckoo and calculate its ELR (Eggs Laying Radius). The ELR is defined as:

$$\mathrm{ELR} = \alpha \times \frac{\text{Number of current cuckoo's eggs} \times (\text{Varhi} - \text{Varlow})}{(\text{Total number of eggs})} \tag{20}$$

where α is an integer to handle the maximum value of ELR and the Var_{hi} and Var_{low} is the up and down limits of optimal variables. This ELR has purposes to determine and limit the searching space in each iteration.

After defining ELR, let cuckoos lay eggs inside their corresponding ELR. Some eggs are recognized by host bird will be killed. The rest eggs hatch host bird nest and chicks grow. The other exciting thing of cuckoos is only one egg in one nest have chance to grow up because cuckoo chick is bigger than host bird's chick three times in body size and cuckoo chick eat all of food. After two days, other chick will be died. The egg laying of cuckoos may represent for the searching of new solution for OPF problem by change or adding a certain value to origin solution.

3.3 Eliminating Cuckoos in Worst Habitats

Due to the equilibrium in bird's population, there is only a maximum number of cuckoos live in environment and its name is MaxNumofCuckoos. A priority list may be created by evaluate and sort the profit value of the habitat of each newly grown cuckoos. So, there are MaxNumofCuckoos cuckoos from the first of priority list alive, and other cuckoos will be killed. This work will decrease computational time because there are only a number of best solutions to be used in next iteration.

3.4 Immigration of Cuckoos

The habitat in which cuckoo has best condition to live and grow or having best profit value will be considered the goal point of other cuckoos. They will immigrate toward to the goal point to live but not exactly. After a period of time, cuckoos grow up to mature cuckoos and restart the process. It has mean that cuckoo's population will find the best place to live after many iterations. This work will help the OPF problem find the best solutions. The flowchart for the COA is given in Fig. 1.

4 Implementation of Cuckoo Optimization Algorithm

For implementation of the proposed COA to the OPF problem, each Habitat represents the control variables as follows:

$$H_d = \{P_{g2d}, \ldots, P_{gN_gd}, V_{g1d}, \ldots, V_{gN_gd}, T_{1d}, \ldots, T_{N_td}, \\ Q_{c1d}, \ldots, Q_{cN_cd}\}^T; \quad d = 1, \ldots, N_H \tag{21}$$

where H_d is a Habitat of cuckoo, N_H is number of cuckoo birds or number of habitat in cuckoo population.

The fitness function to be minimized in COA for the problem is based on the problem objective function and dependent variables including real power generation at the slack bus, reactive power outputs at the generation buses, load bus voltages, and apparent power flow in transmission lines. The fitness function is defined as follows:

$$FT = f(x,u) + K_q \sum_{i=1}^{N_g} (Q_{gi} - Q_{gi}^{lim})^2 \\ + K_v \sum_{i=1}^{N_g} (V_{li} - V_{li}^{lim})^2 + K_s \sum_{l=1}^{N_t} (S_{gi} - S_{l,max})^2 \tag{22}$$

where K_q, K_v, and K_s are penalty factors for reactive power generations, load bus voltages, and power flow in transmission lines, respectively. The limits of the dependent variables or vector $\boldsymbol{x}$ in (14) are generally determined based on their calculated values as follows:

$$\boldsymbol{x}^{lim} = \begin{cases} \boldsymbol{x}_{max} & if\ \boldsymbol{x} > \boldsymbol{x}_{max} \\ \boldsymbol{x}_{min} & if\ \boldsymbol{x} < \boldsymbol{x}_{min} \\ \boldsymbol{x} & otherwise \end{cases} \tag{23}$$

where $\boldsymbol{x}$ and $\boldsymbol{x}^{lim}$ respectively represent the calculated value and limits of Q_{gi}, V_{li}, or S_l.

The overall procedure of the propose COA for solving OPF problem is address as following steps:

Step 1: Set the controlling parameters for COA including:

- NumCuckoos is the number of cuckoos in initial population,
- MinNumberofEggs is Minimum number of eggs,
- MaxNumberofEggs is Maximum number of eggs,
- MaxIter is Maximum number of iterations,
- Motioncoeff is the variable to control the forwarding to goal point process,
- MaxNumofCuckoos is the maximum number of Cuckoos that can live at the same time,
- RadiusCoeff is control parameter of egg laying,
- npar is number of optimal variables. It is set equal to number of variables included in vector H_d,
- varHi and varLow are two matrix sized [1 × npar]. They are set equal to upper and lower limits of vector H_d. They are H_{max} and H_{min} , respectively.
- And penalty factors K_q, K_v, K_s for constraints.

Step 2: Initialize population and determine initial position of habitats (Habitat of each cuckoo) by using the variables given in step 1. Each Habitat of cuckoos and a random goal point are created as follow:

$$H_d^{(0)} = (\, varHi - varLo\,) \times rand1 + varLo \tag{24}$$

$$GoalPoint^{(0)} = (varHi - varLo) \times rand1 + varLo \tag{25}$$

where $rand_1$ is random value in [0,1]. Habitat is a random matrix sized [1 × npar], and it representing living environment of each cuckoo.

Step 3: Calculate the number of eggs for each cuckoo as follow:

$$Number\ of\ current\ cuckoo's\ eggs$$
$$= (MaxNumberofEggs - MinNumberofEggs) \times rand_2 + MinNumberofEggs \tag{26}$$

where $rand_2$ is a random value in [0,1].

Step 4: Calculate the egg laying radiuses ELR and the adding value for laying eggs to new Habitat and update the Habitat:

$$ELR_d^{max} = \frac{Number\ of\ current\ eggs \times (VarHi - VarLo) \times radiuscoeff}{(Total\ number\ of\ eggs)} \quad (27)$$

To lay eggs, we produced some radius values less than maximum of egg laying radius.

$$ELR_d = ELR_d^{max} \times rand_3 \quad (28)$$

where $rand_3$ is a random matrix sized [1× *Number of current eggs*], so ELR_d is a matrix has a number of egg rows and npar columns.

$$Add_d = (-1)^{rand_4} \times ELR_d \times \cos(agles) + LR_d \times \sin(agles) \quad (29)$$

$$H_d^{(1)} = H_d^{(0)} + Add_d \quad (30)$$

where $rand_4$ is a random value , it can be set to 1 or 2 only, and angles is a random line space represent for the flying angles of cuckoo. Each row of matrix $H_d^{(1)}$ is a candidate for vector habitat H_d , then check for limit for each $H_d^{(1)}$ and the egg laying process is done.

Step 5: Kill cuckoo's eggs in the same Habitat because only one egg can go to nest. If there are two or more $H_d^{(1)}$ belong to Habitat$^{(1)}$ equal. We just keep one of those and kill the others.

Step 6: Solve power flow based on the newly obtained value of Habitat for each cuckoos. The rest of $H_d^{(1))}$ after step 5 will be used. Each $H_d^{(1)}$ is an input variable for an iteration of power flow problem.

Step 7: Evaluate fitness function FT in (20) for each $H_d^{(1)}$. Sorting value of all fitness function as a priority list and store the relevant $H_d^{(1)}$.

Step 8: Base on the priority list, keep a number of MaxNumofCuckoos $H_d^{(1)}$ and kill the others at the bottom of list.

Step 9: The $H_d^{(1)}$ at the first of list is the best solution for H_d in (19). It has the best cost and will be the new goal point for others going forward to its habitat. Their new Habitat is determined as:

$$GoalPoint^{(1)} = H_d^{(1)}(best\ cost) \quad (31)$$

$$H_d^{(2)} = \left(GoalPoint^{(1)} - H_d^{(1)}\right) \times rand_5 \times Motioncoeff + H_d^{(1)} \quad (32)$$

where $rand_5$ is random number in [0,1]. Then check for $H_d^{(2)}$ limits.

Step 10: Update new Habitats of population to $H_d^{(2)}$, and store best cost , goal point.

Step 11: Check for stopping critical. If *Iter* < *IterMax*, *Iter* = *Iter* + 1 and return to step 2. Otherwise, stop.

The flowchart of COA implementation for solving the OPF problem is given in Fig. 2.

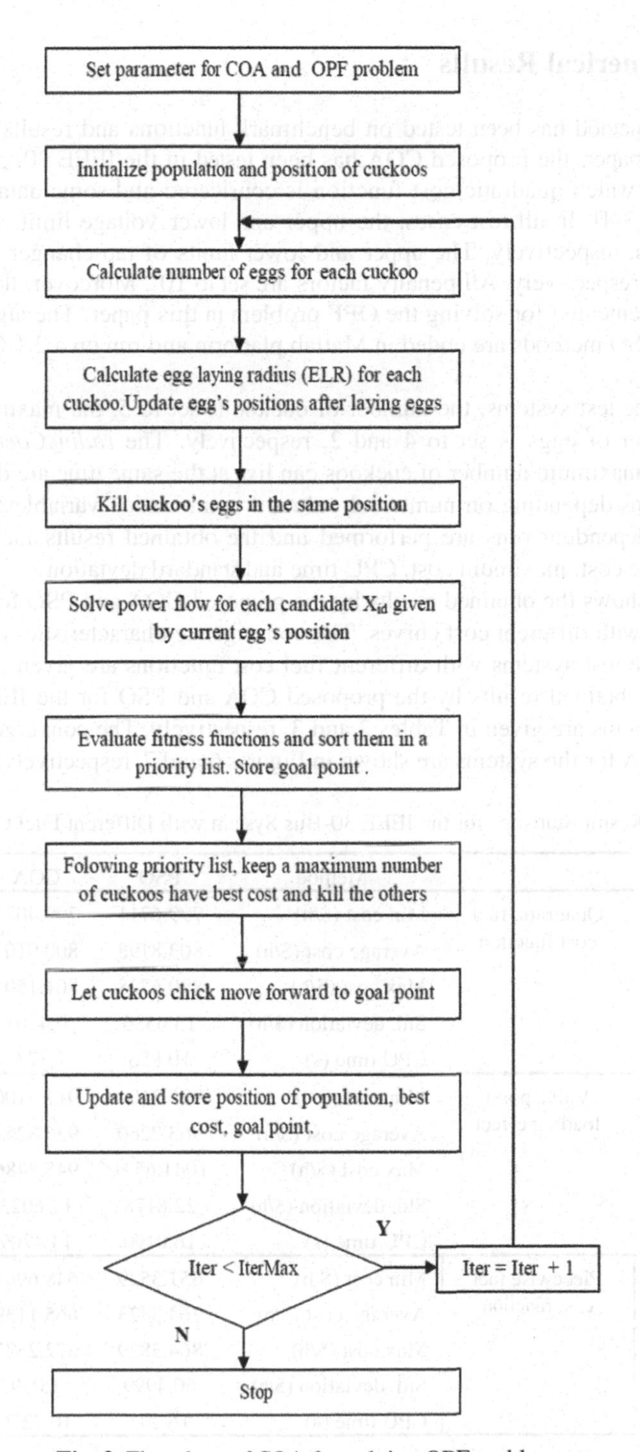

Fig. 2. Flowchart of COA for solving OPF problem

5 Numerical Results

The COA method has been tested on benchmark functions and results were given in [1]. In this paper, the proposed COA has been tested in the IEEE 30, 57 and 118 bus systems, in which quadratic cost function is considered and some data for these systems from [3-4]. In all test cases, the upper and lower voltage limits are set to 1.10 and 0.95 pu, respectively. The upper and lower limits of tap changer are set to 1.10 and 0.9 pu, respectively. All penalty factors are set to 10^6. Moreover, the PSO method is also implemented for solving the OPF problem in this paper. The algorithms of the COA and PSO methods are coded in Matlab platform and run on a 2.1 GHz with 4 GB of RAM PC.

For all the test systems, the number of cuckoo is set to 5, the maximum and minimum number of eggs is set to 4 and 2, respectively. The *radiusCoeff*, *motionCoeff* factors and maximum number of cuckoos can live at the same time are different for the three systems depending on numerical value of optimization variables. For each system, 50 independent runs are performed and the obtained results include minimum cost, average cost, maximum cost, CPU time and standard deviation.

Table 1 shows the obtained results by the proposed COA and PSO for the IEEE 30-bus system with different cost curves. The convergence characteristics of the proposed COA for the test systems with different fuel cost functions are given in Figures 3, 4, and 5. The obtained results by the proposed COA and PSO for the IEEE 57-bus and 118-bus systems are given in Tables 2 and 3, respectively. The convergence characteristics of COA for the systems are shown in Figures 6 and 7, respectively.

Table 1. Result Statistics for the IEEE 30-Bus System with Different Fuel Cost Function

	Method	PSO	COA
Quadratic fuel cost function	Min cost ($/h)	799.6711	799.4027
	Average cost ($/h)	803.8198	800.0103
	Max cost ($/h)	890.6728	801.1501
	Std. deviation ($/h)	13.6536	0.4201
	CPU time (s)	10.856	7.3788
Valve point loading effect	Min cost ($/h)	923.2104	917.3100
	Average cost ($/h)	963.7260	936.3251
	Max cost ($/h)	1041.6531	945.2486
	Std. deviation ($/h)	22.8178	11.6023
	CPU time (s)	16.0190	14.4709
Piecewise fuel cost function	Min cost ($/h)	651.3529	648.6967
	Average cost ($/h)	764.3423	665.1139
	Max cost ($/h)	864.3829	672.2527
	Std. deviation ($/h)	60.4999	11.0392
	CPU time (s)	16.203	10.2276

Table 2. Result Statistic for the IEEE 57-Bus System

Method	PSO	COA
Min cost ($/h)	42109.7231	41901.9977
Average cost ($/h)	44688.4203	42176.3511
Max cost ($/h)	49320.6668	43982.6423
Std. deviation ($/h)	1786.3245	610.1708
CPU time (s)	8.81	7.80

Table 3. Result Statistic for the IEEE 118-Bus System

Method	PSO	COA
Min cost ($/h)	145520.0109	133110.4316
Average cost ($/h)	158596.1725	138260.4028
Max cost ($/h)	184686.8248	153110.4316
Std. deviation ($/h)	9454.4231	4580.9556
CPU time (s)	132.233	110.5267

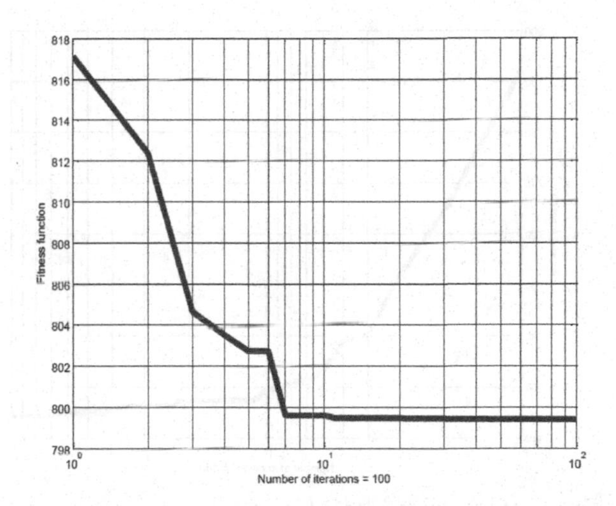

Fig. 3. Convergence characteristic with quadratic fuel function for the IEEE 30-bus system

In the same testing conditions for the IEEE-30, 57, 118 bus systems, COA can obtain better solution than PSO. COA is not only fulfill all constrains of OPF problem but also have result values smaller than PSO in both of minimum cost, maximum cost, average cost and standard deviation. The results show that COA has shorter time of running application and the convergence characteristic show that COA quickly move to the goal point.

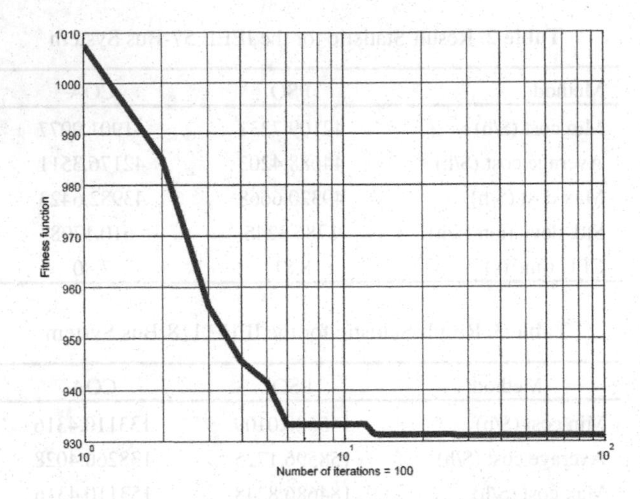

Fig. 4. Convergence characteristic with valve point loading effect for the IEEE 30-bus system

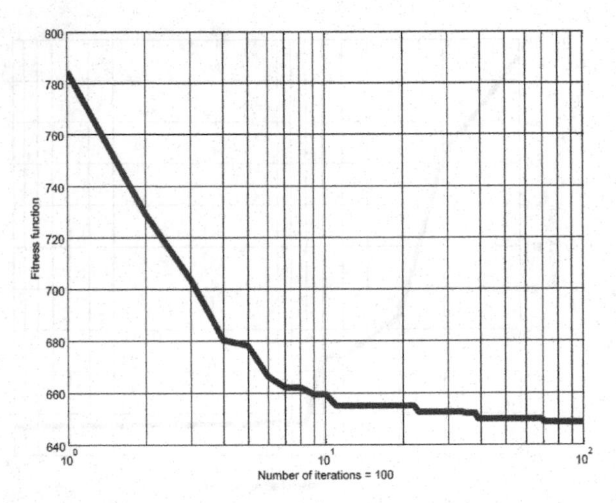

Fig. 5. Convergence characteristic with piecewise fuel cost function for the IEEE 30-bus system

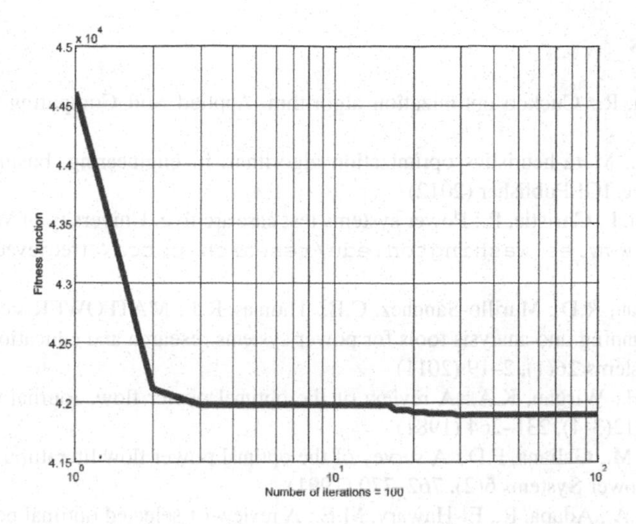

Fig. 6. Convergence characteristic with quadratic fuel function for the IEEE 57-bus system

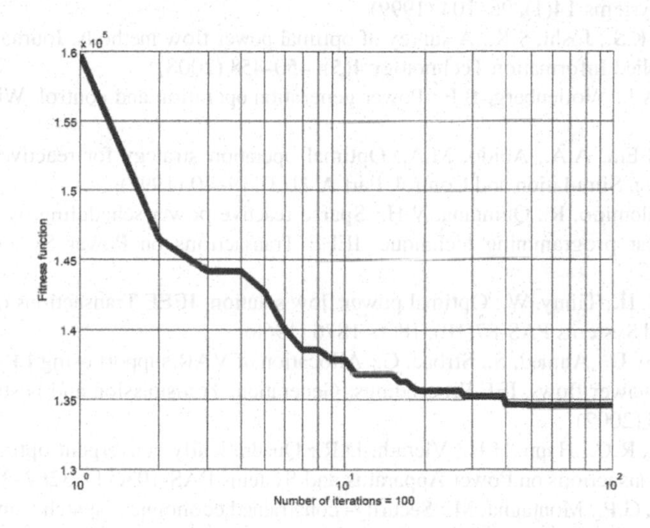

Fig. 7. Convergence characteristic with quadratic fuel function for the IEEE 118-bus system

6 Conclusion

In this paper, the proposed COA has been effectively implemented for solving OPF problem. The proposed COA is a new nature inspired method which can efficiently solve optimization problems. The proposed method has been tested on the large-scale system including the IEEE 30-bus, 57-bus, and 118-bus systems. The obtained result show that COA can perform better than PSO and it can achieve better cost value of fitness function, sorter computational time of running application. Therefore, proposed COA can be a favorable method for solving large-scale OPF problems.

References

1. Rajabioun, R.: Cuckoo optimization algorithm. Applied Soft Computing 11, 5508–5518 (2011)
2. Vasant, P.: Meta-heuristics optimization algorithms in engineering, business, economic, and finance. IGI Publisher (2012)
3. Dabbagchi, I., Christie, R.: Power systems test case archive. University of Washington, http://www.ee.washington.edu/research/pstca/ (retrieved February 20, 2011)
4. Zimmerman, R.D., Murillo-Sánchez, C.E., Thomas, R.J.: MATPOWER steady-state operations, planning and analysis tools for power systems research and education. IEEE Trans. Power Systems 26(1), 2–19 (2011)
5. Happ, H.H., Wirgau, K.A.: A review of the optimal power flow. Journal of the Franklin Institute 312(3-4), 231–264 (1981)
6. Huneault, M., Galiana, F.D.: A survey of the optimal power flow literature. IEEE Transactions on Power Systems 6(2), 762–770 (1991)
7. Momoh, J.A., Adapa, R., El-Hawary, M.E.: A review of selected optimal power flow literature to 1993-I. Nonlinear and quadratic programming approaches. IEEE Transactions on Power Systems 14(1), 96–104 (1999)
8. Pandya, K.S., Joshi, S.K.: A survey of optimal power flow methods. Journal of Theoretical and Applied Information Technology 4(5), 450–458 (2008)
9. Wood, A.J., Wollenberg, B.F.: Power generation operation and control. Wiley, New York (1996)
10. Abou El-Ela, A.A., Abido, M.A.: Optimal operation strategy for reactive power control modelling. Simulation and Control, Part A 41(3), 19–40 (1992)
11. Mota-Palomino, R., Quintana, V.H.: Sparse reactive power scheduling by a penalty function linear programming technique. IEEE Transactions on Power Systems 1(3), 31–39 (1986)
12. Dommel, H., Tinny, W.: Optimal power flow solution. IEEE Transactions on Power Apparatus and Systems PAS-87(10), 1866–1876 (1968)
13. Pudjianto, D., Ahmed, S., Strbac, G.: Allocation of VAR support using LP and NLP based optimal power flows. IEE Proceedings. Generation, Transmission and Distribution 149(4), 377–383 (2002)
14. Burchett, R.C., Happ, H.H., Vierath, D.R.: Quadratically convergent optimal power flow. IEEE Transactions on Power Apparatus and Systems PAS-103(11), 3267–3276 (1984)
15. Granelli, G.P., Montagna, M.: Security- constrained economic dispatch using dual quadratic programming. Electric Power Systems Research 56, 71–80 (2000)
16. Sun, D.I., Ashley, B., Brewer, B., Hughes, A., Tinney, W.F.: Optimal power flow by Newton approach. IEEE Transactions on Power Apparatus and Systems PAS-103(10), 2864–2875 (1984)
17. Santos Jr., A., da Costa, G.R.M.: Optimal power flow solution by Newton's method applied to an augmented Lagrangian function. IEE Proceedings. Generation, Transmission and Distribution 142(1), 33–36 (1995)
18. Lo, K.L., Meng, Z.J.: Newton-like method for line outage simulation. IEE Proceedings - General Transmissions and Distributions 151(2), 225–231 (2004)
19. Bai, X., Wei, H., Fujisawa, K., Wang, Y.: Semidefinite programming for optimal power flow problems. International Journal of Electrical Power & Energy Systems 30(6-7), 383–392 (2008)
20. Yan, X., Quintana, V.H.: Improving an interior point based OPF by dynamic adjustments of step sizes and tolerances. IEEE Transactions on Power Systems 14(2), 709–717 (1999)

21. Wang, M., Liu, S.: A trust region interior point algorithm for optimal power low problems. International Journal of Electrical Power & Energy Systems 27(4), 293–300 (2005)
22. Capitanescu, F., Glavic, M., Ernst, D., Wehenkel, L.: Interior-point based algorithms for the solution of optimal power flow problems. Electric Power Systems Research 77(5-6), 508–517 (2007)
23. Lai, L.L., Ma, J.T.: Improved genetic algorithms for optimal power flow under both normal and contingent operation states. International Journal of Electrical Power & Energy Systems 19(5), 287–292 (1997)
24. Wu, Q.H., Cao, Y.J., Wen, J.Y.: Optimal reactive power dispatch using an adaptive genetic algorithm. International Journal of Electrical Power & Energy Systems 20(8), 563–569 (1998)
25. Osman, M.S., Abo-Sinna, M.A., Mousa, A.A.: A solution to the optimal power flow using genetic algorithm. Applied Mathematics and Computation 155(2), 391–405 (2004)
26. Roa-Sepulveda, C.A., Pavez-Lazo, B.J.: A solution to the optimal power flow using simulated annealing. International Journal of Electrical Power & Energy Systems 25(1), 47–57 (2003)
27. Abido, M.A.: Optimal power flow using tabu search algorithm. Electric Power and Components Systems 30(5), 469–483 (2002)
28. Wu, Q.H., Ma, J.T.: Power system optimal reactive dispatch using evolutionary programming. IEEE Transactions on Power Systems 10(3), 1243–1249 (1995)
29. Yuryevich, J., Wong, K.P.: Evolutionary programming based optimal power flow algorithm. IEEE Transactions on Power Systems 14(4), 1245–1250 (1999)
30. Abido, M.A.: Optimal power flow using particles warm optimization. International Journal of Electrical Power & Energy Systems 24(7), 563–571 (2001)
31. Cai, H.R., Chung, C.Y., Wong, K.P.: Application of differential evolution algorithm for transient stability constrained optimal power flow. IEEE Transactions on Power Systems 23(2), 719–728 (2008)

Nomenclature

a_i, b_i, c_i fuel cost coefficients of generating unit i
e_i, f_i fuel cost coefficients of generating unit i considering valve point effects
a_{ik}, b_{ik}, c_{ik} fuel cost coefficients of generating unit i corresponding to fuel k
G_{ij}, B_{ij} transfer conductance and susceptance between bus i and bus j, respectively
N_b number of buses
N_c number of switchable capacitors
N_d number of load buses
N_g number of generating units
N_l number of transmission lines
N_t number of transformer with tap changing
n_i number of fuels for unit i
P_{di}, Q_{di} real and reactive power demands at bus i, respectively
P_{gi}, Q_{gi} real and reactive power outputs of generating unit i, respectively
Q_{ci} reactive power compensation source at bus i
S_{ij}, S_{ji} apparent power flow from bus i to bus j and from bus j to bus i
S_l maximum apparent power flow in transmission line l connecting between buses i and j
T_k tap-setting of transformer branch k
V_{gi}, V_{li} voltage magnitude at generation bus i and load bus i, respectively
V_i, δ_i voltage magnitude and angle at bus i, respectively

ε Constrained Differential Evolution Algorithm with a Novel Local Search Operator for Constrained Optimization Problems

Wenchao Yi, Xinyu Li*, Liang Gao, and Yinzhi Zhou

State Key Laboratory of Digital Manufacturing Equipment &Technology,
Huazhong University of Science and Technology, Wuhan 430074, PR China
lixinyu@mail.hust.edu.cn

Abstract. Many practical problems can be classified into constrained optimization problems (COPs). ε constrained differential evolution (εDE) algorithm is an effective method in dealing with the COPs. In this paper, ε constrained differential evolution algorithm with a novel local search operator(εDE-LS) is proposed by utilizing the information of the feasible individuals. In this way, we can guide the infeasible individuals to move into the feasible region more effectively. The performance of the proposed εDE-LS is evaluated by the 22 benchmark test functions. The experimental results empirically show that εDE-LS is highly competitive comparing with some other state-of-the-art approaches in constrained optimization problems.

Keywords: Constrained optimization problems, constraint handling technique, ε constrained differential evolution, mutation operator.

1 Introduction

Differential evolution (DE) algorithm is an efficient evolutionary algorithm, which was firstly proposed by Storn and Price [1]. During the past decade, DE algorithm shows competitive performance in solving constrained optimization problems (COPs). In real world applications, a lot of optimization problems are subjected to constraints, which can be categorized into COPs. A general COPs can be stated as follows:

$$
\begin{aligned}
&\min f(\vec{x}) \\
&s.t.\ g_j(\vec{x}) \le 0,\ j = 1,\dots,q \\
&\quad\ h_j(\vec{x}) = 0,\ j = q+1,\dots,m
\end{aligned}
\tag{1}
$$

Where $\vec{x} = (x_1,\dots,x_n)$ is generated within the range $L_i < x_i < U_i$. L_i and U_i denote the lower and upper bound in each dimension. $g_j(\vec{x})$ denotes the *j*th inequality constraint and $h_j(\vec{x})$ denotes the (*j-q*)th equality constraint.

* Corresponding author.

H. Handa et al. (eds.), *Proc. of the 18th Asia Pacific Symp. on Intell. & Evol. Systems – Vol. 1*,
Proceedings in Adaptation, Learning and Optimization 1, DOI: 10.1007/978-3-319-13359-1_38

The research on utilizing DE algorithm to solve COPs has attracted promising attention in recent years. A variety of constraint handling techniques have emerged to deal with the COPs. The most common used one is the penalty function method. Huang et al. [2]proposed a co-evolutionary DE algorithm, in which a special adaptive penalty function was proposed to deal with the constraints. Although the penalty function method is simple and efficient, it is still a difficult task to set a proper penalty parameter for the method.

The multiobjective technique is an efficient method in solving COPs. Wang et al. [3] introduced multiobjective technique based DE algorithm to solve COPs. An infeasible solution replacement mechanism based multiobjective approach is proposed, which mainly focus on guiding the population moving towards the promising and feasible region more efficiently. Gong et. al [4]proposed a multiobjective tenique based DE for COPs, in which multiobejctive technique based constraint handling technique was proposed. However, using the multiobjective technique to tackle the COPs is still difficult to design an effective framework in solving COPs.

The methods by adding extra rules or operator in handling COPs also attract considerable interest from the researchers. Storn [5] proposed a constraint adaptive method, which firstly make all the individuals as feasible ones by relaxing the constraints then decresing the relaxation till reach the original constraints. Lampinen et. al [6] presented three effective rules to handle the constraints, which include the feasible ones always better than the infeasible ones, the one with better fitness function value wins among the feasible ones and the one with less constraint violation wins among the infeasible ones. Among these researches, εDE method is a very effective one. εDE was proposed by Takahama et. al [7] in 2006, in which using ε constraint handling technique to deal with the constraints. The experimental results showed the εDE not only can find the feasible ones rapidly, but can achieve excellent success performance and rate as well. In Takahama et. al [7], a local search based on the information of first-order derivative was proposed. However, it is usually difficult to calculate the first-order derivative. Also, the calculation of first-order derivative is time-cost. In this paper, a local search operator is proposed. It can avoid calculating first-order derivative of constraint functions, but is more effective.

The proposed mutation operator mainly utilizing the information of the feasible and the infeasible individuals, which focus on guiding the infeasible individual move along the direction of the feasible individuals. In order to evaluate the performance of the proposed algorithm, twenty-two benchmark test function collected from the special session on the constrained real-parameter optimization of the 2006 IEEE congress on evolutionary computation are adopted in this paper.

This paper is organized as follows. We firstly will give a general introduction of DE algorithm and εDE as a foundation in Section 2. In Section 3, we will give a detailed introduction of the proposed εDE-LS algorithm, in which the framework is included. The experimental results and the comparison will be presented in Section 4. Finally, conclusions will be given in Section 5.

2 DE and εDE Algorithm

2.1 DE Algorithm

DE algorithm is an efficient but simple EAs, which can be divided into four phase, that is, initialization, mutation, crossover and selection.

During the initialization phase, the *NP* *n*-dimensional individuals $x_i^g = \left(x_{i,1}^g, \ldots, x_{i,n}^g\right), i = 1, \ldots, NP$ are generated. *g* denotes the generation number.

Then the mutation phase will be adopted to generate the mutation vectors. Several mutation operators have been proposed. The DE/rand/1/*exp* proposed by will be utilized in this paper, where *exp* denotes the exponent crossover operator. The mutation vector can be calculated as follows:

$$v_i^g = x_{r1}^g + F * (x_{r2}^g - x_{r3}^g) \tag{2}$$

Where *F* denotes the predefined scale parameter, r_1, r_2, and r_3 are three mutually different generated indexes which should be different from index *i* within the range [1,*NP*]. Then a check will be made to make sure the generated v_i^g are within the boundaries, which can be described as follows:

$$v_{i,j}^g = \begin{cases} \min\{U_j, 2*L_j - v_{i,j}^g\}, \text{if } v_{i,j}^g < L_j \\ \max\{L_j, 2*U_j - v_{i,j}^g\}, \text{if } v_{i,j}^g > U_j \end{cases} \tag{3}$$

Where L_j and U_j denote the lower and upper bound in *j*-th dimension.

The exponent crossover operator makes the trail vector contains a consecutive sequence of the component taken from the mutation vector. The exponent crossover operator can be given as follows:

$$u_{i,j}^g = \begin{cases} v_{i,j}^g, \, if \;\, j \in \{\text{k}, \langle k+1 \rangle_n, \ldots, \langle k+L-1 \rangle_n\} \\ x_{i,j}^g, \, otherwise \end{cases}, \; j = 1, \ldots, \text{n} \tag{4}$$

Where $L, k \in [1, \text{n}]$ are both random indexes. $<j>$ is *j* if *j*<*n* and *j*=*j*-*n* if *j*>*n*.

During the selection phase, a better individual between the trail vector u_i^g and target vector x_i^g will be chosen according to their fitness function value:

$$x_i^{g+1} = \begin{cases} u_i^g, \, if \;\, f(u_i^g) < f(x_i^g) \\ x_i^g, \, else \end{cases} \tag{5}$$

2.2 εDE Algorithm

In the εDE algorithm, the constraint violation $\Phi(x_i^g)$ is defined as the sum of all constraints:

$$\Phi(x_i^g) = \sum_{j=1}^{q} \max\{0, g_j(x_i^g)\} + \sum_{j=q+1}^{m} h_j(x_i^g) \tag{6}$$

After generating the new target vector through the DE algorithm. The ε level comparison is used in εDE algorithm to help decide which individual is better. The comparison can be given as follows:

$$(f_1, \Phi_1) <_{\varepsilon} (f_2, \Phi_2) \Leftrightarrow \begin{cases} f_1 < f_2, \text{if } \Phi_1, \Phi_2 \le \varepsilon \\ f_1 < f_2, \text{if } \Phi_1 = \Phi_2 \\ \Phi_1 < \Phi_2, \text{otherwise} \end{cases} \tag{7}$$

Where f is the objective fitness function value, and the ε level is set as formula given below:

$$\varepsilon(g) = \begin{cases} \Phi(x_\theta), \ g = 0 \\ \varepsilon(0) * (1 - g/T_c)^{cp}, 0 < g < T_c \\ 0, \ g \ge T_c \end{cases} \tag{8}$$

Where x_θ is the top θ-th individual and we set θ=0.2*N in the paper. T_c is a predefined generation number. cp is the control parameter in ε level comparison and we set cp as 5 in the paper.

3 The Proposed εDE-LS Algorithm

Usually, in the COPs, the feasible region is continuous. The surrounding region of feasible individuals has more possibility to be feasible. So the feasible individuals can guide the infeasible ones move towards to the feasible region. Motivated by the interaction between the feasible and infeasible individuals, we design a novel local search operator "DE/current-to- feasible/2" to improve the performance of εDE algorithm. The "DE/current-to- feasible/2" is a transformation version of "DE/current-to-best/2" mutation strategy. It can be presented as follows:

$$v_i^g = x_i^g + a * \left(x_{feas\_r_1}^g - x_i^g\right) + b * \left(x_{feas\_r_2}^g - x_i^g\right) \tag{9}$$

Where $feas\_r_1$ and $feas\_r_2$ are two random indexes chosen from the feasible individual set Q. So the number of feasible individuals must be more than 2. *a* and *b* are two random generated numbers within the range [0,1]. If any dimension in v_i^g exceeds the boundary, then randomly chosen a feasible individual and makes the specific dimension in v_i^g equal to the related dimension in chosen feasible individual.

So in εDE-LS, DE algorithm is used to generate offspring, and ε constrained algorithm is used to choose better individual survive into next generation. For those infeasible solutions, if the number of feasible individuals is less than 2, the local search phase is skipped. If the number of feasible individuals is more than 2, then for each infeasible individual, local search operator is used. Then the ε level comparison is adopted to choose a better one between the individual and offspring generated by local search operator.. The framework of the proposed εDE-LS algorithm can be given as follows:

εDE-LS algorithm
1: Initialize the individuals $\{x_1^0,\ldots,x_{NP}^0\}$
2: Initialize the ε level value, ε=ε(0)
3: **While** *Fes<maxFES*
4: {
5: Mutation phase
6: Crossover phase
7: Using ε level comparison in selection phase
8: *Fes=Fes+NP*;
9: Store the ε-feasible ones in *Q*; Store the other individuals as infeasible ones in *W*
10: **If** the number of feasible individuals in *Q* is bigger than 2
11: **For** *i*=1: *size (W)*
12: using DE/current-to-feasible/2 to generate $V_i'^g$
13: **For** *j*=1:*n*
14: **If** $V_{i,j}'^g$ exceeds the boundary
15: $V_{i,j}'^g = x_{feas\_r_3,j}^{g+1}$
16: **End If**
17: **End For**
18: choose better one between $V_i'^g$ and x_i^{g+1} to survive into next generation
19: *Fes=Fes*+1;
20: **End for**
21: **End If**
22: Update ε level value;
23: }

Fig. 1. The pesudocode of εDE-LS algorithm

4 Experimental Results

4.1 Parameter Settings

The twenty-two benchmark test functions are collected from Liang et. al [8] are adopted in evaluating the performance of the proposed algorithm. The detailed information about the benchmark can be referred to Liang et. al [8]. The parameter settings of the proposed algorithm are as follows:

Table 1. Parameter setings of εDE-LS algorithm

popsize	40,40,100*
MaxFES	$5\times10^3, 5\times10^4, 5\times10^5$
θ	0.2
T_c	0.2* *MaxFES* /*popsize*
cp	5
F	[0.5,1.0]
CR	[0.9,1.0]

*The popsize is 40, 40 and 100 with 5×10^3, 5×10^4, 5×10^5 finess evaluations (FES),respectively.

4.2 Performance of εDE-LS Algorithm

25 independent runs are conducted for the test benchmark functions with 5×10^3, 5×10^4, 5×10^5 FES, respectively.The torlerance value δ for the equality constraints is set as 0.0001. The best, median, worst, mean and standard deviation of the error value ($f(\vec{x})-f(\vec{x}^*)$), where $f(\vec{x}^*)$ is the best objective fitness function value for each benchmark test function that ever known. *c* is the number of the violated constraints at the median solution : the three numbers refers to the constraints bigger than 1, between 0.01 and 1.0 and between 0.0001and 0.01, respectively. *v* is mean value of the violations of all the constraints at the median solution. The number in parentheses after best, median and worst solutions is the number of violated constraints.

Table 2. Function error values achieved when FES= 5×10^3, FES= 5×10^4 and FES= 5×10^5 for function G01-G05

		G01	G02	G03	G04	G05
5×10^3	Best	1.2042E+00	3.5286E-01	-2.4983E-01	8.2965E-01	5.1278E+03
	Median	2.6309E+00	3.9958E-01	9.6055E-01	3.3045E+00	5.2755E+03
	Worst	5.5134E+00	4.7644E-01	1.0005E+00	9.5916E+00	5.7907E+03
	c	0,0,0	0,0,0	0,0,0	0,0,0	0,3,0
	v	0	0	0	0	2.1310E-02
	Mean	2.7307E+00	4.0586E-01	6.4434E-01	4.0203E+00	5.3362E+03
	Std	7.9984E-01	2.9802E-02	4.9011E-01	2.0836E+00	1.8425E+02

Table 2. (*Continued*)

5×10⁴	Best	1.4440E-07	4.1605E-04	2.1900E-06	-3.6380E-12	-1.8190E-12
	Median	6.8327E-07	1.1172E-02	8.9556E-02	-3.6380E-12	-9.0949E-13
	Worst	3.1932E-06	6.2807E-02	2.7449E-01	-3.6380E-12	4.5045E-06
	c	0,0,0	0,0,0	0,0,0	0,0,0	0,0,0
	v	0	0	0	0	0
	Mean	9.9408E-07	1.8072E-02	1.0027E-01	-3.6380E-12	2.0085E-07
	Std	8.2529E-07	1.7363E-02	8.1116E-02	0.0000E+00	9.0155E-07
5×10⁵	Best	0.0000E+00	8.6703E-10	-2.8866E-15	-3.6380E-12	-1.8190E-12
	Median	0.0000E+00	9.3817E-09	-2.6645E-15	-3.6380E-12	-18190E-12
	Worst	0.0000E+00	3.7095E-08	-2.4425E-15	-3.6380E-12	-1.8190E-12
	c	0,0,0	0,0,0	0,0,0	0,0,0	0,0,0
	v	0	0	0	0	0
	Mean	0.0000E+00	1.1279E-08	-2.6024E-15	-3.6380E-12	-1.8190E-12
	Std	0.0000E+00	9.4316E-09	1.6367E-16	0.0000E+00	0.0000E+00

Table 3. Function error values achieved when FES= 5×10^3 , FES= 5×10^4 and FES= 5×10^5 for function G06-G10

		G06	G07	G08	G09	G10
5×10³	Best	6.9833E-04	1.5394E+01	4.1633E-17	3.1226E+00	1.2622E+03
	Median	4.1042E-03	2.1236E+01	5.5511E-17	5.5926E+00	2.1678E+03
	Worst	4.6504E-02	3.6400E+01	6.9389E-17	1.1330E+01	4.8200E+03
	c	0,0,0	0,0,0	0,0,0	0,0,0	0,0,0
	v	0	0	0	0	0
	Mean	8.5336E-03	2.2857E+01	5.4401E-17	5.8105E+00	2.2786E+03
	Std	1.0454E-02	6.3786E+00	5.5511E-18	1.8589E+00	7.5188E+02
5×10⁴	Best	-1.6371E-11	1.1864E-04	2.7756E-17	-1.1369E-13	1.1531E-02
	Median	-1.6371E-11	2.8898E-04	4.1633E-17	1.1369E-13	5.5891E-02
	Worst	-1.6371E-11	1.2395E-03	4.1633E-17	2.2737E-13	7.8631E+01
	c	0,0,0	0,0,0	0,0,0	0,0,0	0,0,0
	v	0	0	0	0	0
	Mean	-1.6371E-11	3.7846E-04	3.7192E-17	6.8212E-14	8.1563E+00
	Std	0.0000E+00	2.7710E-04	6.6071E-18	8.0389E-14	1.7592E+01
5×10⁵	Best	-1.6371E-11	-1.4566E-13	2.7756E-17	-2.2737E-13	-7.2760E-12
	Median	-1.6371E-11	2.8422E-14	2.7756E-17	-1.1369E-13	-4.5475E-12
	Worst	-1.6371E-11	2.9488E-13	2.7756E-17	-1.1369E-13	1.4508E-08
	c	0,0,0	0,0,0	0,0,0	0,0,0	0,0,0
	v	0	0	0	0	0
	Mean	-1.6371E-11	-1.5632E-14	2.7756E-17	-1.5916E-13	2.0654E-09
	Std	0.0000E+00	1.0766E-13	0.0000E+00	5.6843E-14	4.3554E-09

Table 4. Function error values achieved when FES= 5×10^3 , FES= 5×10^4 and FES= 5×10^5 for function G11-G15

		G11	G12	G13	G14	G15
5×10^3	Best	1.6943E-10	8.6228E-05	5.3633E-01	-4.7264E+01	2.2286E-03
	Median	8.1219E-05	7.5430E-03	9.7924E-01	-4.2850E+01	9.6215E+02
	Worst	6.9003E-02	4.2052E-01	1.9416E+00	-3.9904E+01	9.6535E+02
	c	0,0,0	0,0,0	0,3,0	0,3,0	0,0,1
	v	0	0	2.0749E-01	1.1952E-01	3.6741E-04
	Mean	7.7345E-03	5.0767E-06	9.9447E-01	-4.3113E+01	7.7047E+02
	Std	1.5332E-02	1.0478E-05	2.5999E-01	2.0303E+00	3.9265E+02
5×10^4	Best	0.0000E+00	0.0000E+00	1.6771E-02	5.6943E-07	-1.1369E-13
	Median	0.0000E+00	0.0000E+00	5.2714E-01	4.1865E-06	-1.1369E-13
	Worst	0.0000E+00	0.0000E+00	9.2521E-01	2.1939E-04	-1.1369E-13
	c	0,0,0	0,0,0	0,0,0	0,0,0	0,0,0
	v	0	0	0	0	0
	Mean	0.0000E+00	0.0000E+00	5.3646E-01	2.2823E-05	-1.1369E-13
	Std	0.0000E+00	0.0000E+00	3.0524E-01	5.2442E-05	0.0000E+00
5×10^5	Best	0.0000E+00	0.0000E+00	-2.2204E-16	1.4211E-14	-1.1369E-13
	Median	0.0000E+00	0.0000E+00	-2.2204E-16	1.4211E-14	-1.1369E-13
	Worst	0.0000E+00	0.0000E+00	-1.9429E-16	2.1316E-14	-1.1369E-13
	c	0,0,0	0,0,0	0,0,0	0,0,0	0,0,0
	v	0	0	0	0	0
	Mean	0.0000E+00	0.0000E+00	-2.1649E-16	1.7053E-14	-1.1369E-13
	Std	0.0000E+00	0.0000E+00	1.1331E-17	3.5527E-15	0.0000E+00

Table 5. Function error values achieved when FES= 5×10^3 , FES= 5×10^4 and FES= 5×10^5 for function G16-G19 and G21

		G16	G17	G18	G19	G21
5×10^3	Best	3.9394E-03	8.8160E+03	-2.6751E-01	5.3038E+01	4.0133E+02
	Median	9.6029E-03	8.9614E+03	5.1285E-01	1.0054E+02	6.7771E+02
	Worst	2.4616E-02	9.2010E+03	8.0098E-01	1.3151E+02	9.6477E+02
	c	0,0,0	0,4,0	0,0,0	0,0,0	0,3,1
	v	0	9.5126E-02	0	0	1.8268E-03
	Mean	1.1051E-02	8.9672E+03	4.9005E-01	9.6983E+01	6.7307E+02
	Std	5.3806E-03	1.0383E+02	1.8203E-01	2.4519E+01	1.5715E+02
5×10^4	Best	3.7748E-15	7.7749E+00	3.3295E-06	1.2885E-02	2.7546E-05
	Median	3.2196E-14	8.3318E+01	1.8149E-05	3.2261E-02	4.5160E+00
	Worst	6.3349E-13	3.4429E+02	1.2721E-04	1.151E-01	1.3099E+02
	c	0,0,0	0,0,0	0,0,0	0,0,0	0,0,0
	v	0	0	0	0	0
	Mean	1.0176E-13	7.5867E+01	2.8616E-05	3.6846E-02	5.0162E+01
	Std	1.5267E-13	6.6433E+01	2.9068E-05	2.1706E-02	5.7877E+01

Table 5. (*Continued*)

5×10^5	Best	3.7748E-15	-5.8000E-03	1.7130E-11	1.5960E-08	0.0000E+00
	Median	3.7748E-15	-5.8000E-03	1.8764E-10	5.4233E-08	2.2989E-299(6)
	Worst	3.7748E-15	2.9749E+02	7.7298E-10	1.6040E-07	2.7393E-259(6)
	c	0,0,0	0,0,0	0,0,0	0,0,0	0,6,0
	v	0	0	0	0	2.4686E-01
	Mean	3.7748E-15	3.9079E+01	1.9762E-10	6.4420E-08	1.1283E-260
	Std	0.0000E+00	6.5092E+01	1.8283E-10	3.2980E-08	0.0000E-00

Table 6. Function error values achieved when FES= 5×10^3 , FES= 5×10^4 and FES= 5×10^5 for function G23-G24

		G23	G24
5×10^3	Best	-4.9639E+02	3.5212E-09
	Median	-1.3372E+02	6.5883E-08
	Worst	2.3647E+02	4.0692E-07
	c	0,4,0	0,0,0
	v	1.1303E-01	0
	Mean	-9.5647E+01	1.1387E-07
	Std	1.6961E+02	1.1752E-07
5×10^4	Best	2.3201E-01	3.2863E-14
	Median	3.7585E+01	3.2863E-14
	Worst	3.1068E+02	3.2863E-14
	c	0,0,0	0,0,0
	v	0	0
	Mean	4.9361E+01	3.2863E-14
	Std	6.1611E+01	0.0000E+00
5×10^5	Best	7.9471E-08	3.2863E-14
	Median	7.9992E-06	3.2863E-14
	Worst	4.6202E-02	3.2863E-14
	c	0,0,0	0,0,0
	v	0	0
	Mean	1.8593E-03	3.2863E-14
	Std	9.2381E-03	0.000E+00

As shown in Table 2-6, in spite of the test functions G05, G13, G14, G15, G17, G21, G23, for other 15 test benchmark functions, the proposed algorithm can obtain feasible solutions within 5×10^3 FES. All the test benchmark functions can obtain feasible solutions within 5×10^4 FES. Especially, to function G11 and G12, the best known solutions are obtained within 5×10^4 FES. In 5×10^5 FES, 9 out of 22 test benchmark functions (i.e. G03, G04, G05, G06, G07, G09, G10, G13, G15) can obtain a more precisely solutions than the best known solutions. In conclusion, the solutions obtained by the proposed εDE-LS algorithm (except G21) are close to the best known solutions within 5×10^5 FES.

In Table 7, we present the number of FES needed in each run for each test benchmark function when satisfying the success condition: $f(\vec{x})-f(\vec{x}^*)\le 1.0E-04$ and $\vec{x}$ is feasible solution. The best, median, worst, mean and std denote the least, median, most, mean and standard deviation FES when meets the success condition during the 25 independent runs. The feasible rate is the ratio between the feasible solutions and 25 achieved solutions within 5×10^5 FES. The success rate is the ratio between the number of success runs and 25 runs within 5×10^5 FES. The success performance is the mean number of FES for successful runs multiplied by the total runs and divided by the number of successful runs.

In Table 8, a comparison with respect to other state-of-the-art algorithms in terms of the success performance. The related success performance of other state-of-the-art algorithms is can be referred to Wang et al.[3].

Table 7. The success performance, feasible rate and success rate of the εDE-LS algorithm

Pro.b	Best	Median	Worst	Mean	Std	Feasible Rate	Success Rate	Success performance
G01	33360	37480	40680	37290	2012.9	100%	100%	37290
G02	204100	273100	294000	256662	54591.1	100%	100%	256662
G03	97800	102200	106300	101960	2288.4	100%	100%	101960
G04	12440	14600	16040	14565	977.2	100%	100%	14565
G05	16240	29360	43960	28365	8458.6	100%	100%	28365
G06	23160	25200	27800	25426	1077.2	100%	100%	25426
G07	202900	212700	222900	212300	5168.8	100%	100%	212300
G08	320	7240	10080	7162	1745.0	100%	100%	7162
G09	18840	19920	21520	20061	777.61	100%	100%	20061
G10	243600	286900	384300	306412	53413.7	100%	100%	306412
G11	5600	7200	7800	7102	504.4	100%	100%	7102
G12	1120	2960	5040	3038	957.4	100%	100%	3038
G13	82100	85400	87000	84812	1419.3	100%	100%	84812
G14	32360	39720	47720	39766	4893.0	100%	100%	39766
G15	9480	10400	11840	10435	653.7	100%	100%	10435
G16	13920	20800	25080	19832	3037.2	100%	100%	19832
G17	173800	326800	366500	257400	56434	100%	56%	459642
G18	35520	44560	49680	43797	3947.7	100%	100%	43797
G19	287900	314700	335300	314744	10950.7	100%	100%	314744
G21	NA	NA	NA	NA	NA	NA	NA	NA
G23	384200	431300	499800	433073	33406.8	100%	92%	451117
G24	2440	2960	3360	2963	221.5	100%	100%	2963

Table 8. εDE-LS with respect to MDE[9], MPDE[10], GDE[11], jDE-2[12], CMODE[3] in terms of success performance

Prob.	Success performance						
	εDE	MDE	MPDE	GDE	jDE-2	CMODE	εDE-LS
G01	5.9E+04	7.5E+04	4.3E+04	4.1E+04	5.0E+04	1.2E+05	**3.7E+04**
G02	1.5E+05	**6.0E+04**	3.0E+05	1.5E+05	1.5E+05	1.9E+05	2.6E+05
G03	8.9E+04	4.5E+04	**2.5E+04**	3.5E+06	NA	7.5E+04	1.0E+05
G04	2.6E+04	4.2E+04	2.1E+04	1.5E+04	4.1E+04	7.3E+04	**1.5E+04**
G05	9.7E+04	**2.1E+04**	2.2E+05	1.9E+05	4.5E+05	2.9E+04	2.8E+04
G06	7.4E+03	**5.2E+03**	1.1E+04	6.5E+03	2.9E+04	3.5E+04	2.5E+04
G07	7.4E+04	1.9E+05	5.7E+04	1.2E+05	1.3E+05	1.6E+05	**2.1E+04**
G08	1.1E+03	**9.2E+02**	1.5E+03	1.5E+03	3.2E+03	5.9E+03	7.2E+03
G09	2.3E+04	**1.6E+04**	2.1E+04	3.0E+04	5.5E+04	7.1E+04	2.0E+04
G10	1.1E+05	1.6E+05	**4.8E+04**	8.3E+04	1.5E+05	1.8E+05	3.1E+05
G11	1.6E+04	**3.0E+03**	2.3E+04	8.5E+03	5.4E+04	6.0E+03	7.1E+03
G12	4.1E+03	**1.3E+03**	4.2E+03	3.1E+03	6.4E+03	5.0E+03	3.0E+03
G13	3.5E+04	**2.2E+04**	7.4E+05	8.7E+05	NA	3.1E+04	8.5E+04
G14	1.1E+05	2.9E+05	4.3E+04	2.3E+05	9.8E+04	1.1E+05	**4.0E+04**
G15	8.4E+04	1.0E+04	2.0E+05	7.5E+04	2.4E+05	1.3E+04	**1.0E+04**
G16	1.3E+04	**8.7E+03**	1.3E+04	1.3E+04	3.2E+04	2.9E+04	2.0E+04
G17	9.9E+04	**2.6E+04**	7.3E+05	2.1E+06	1.1E+07	1.4E+05	4.6E+05
G18	5.9E+04	1.0E+05	4.4E+04	4.8E+05	1.0E+05	1.1E+05	**4.4E+04**
G19	3.5E+04	NA	**1.2E+05**	2.0E+05	2.0E+05	2.5E+05	3.2E+05
G21	1.4E+05	1.1E+05	2.1E+05	5.8E+05	1.3E+05	1.3E+05	NA
G23	**2.0E+05**	3.6E+05	2.1E+05	1.1E+06	3.6E+05	2.4E+05	4.5E+05
G24	3.0E+03	1.8E+03	4.3E+03	3.1E+03	1.0E+04	2.2E+04	**3.0E+03**

From Table 7-8, we can conclude that the performance of εDE-LS algorithm is highly competitive. 19 out of 22 test benchmark functions can achieve 100% success rate within 5×10^5 FES. εDE-LS algorithm achieves 100% feasible In terms of success performance, εDE-LS algorithm obtained the least FES in test benchmark function G01, G04, G07, G14, G15, G18, and G24 comparing with other six state-of-the-art algorithms. As success performance indicate that the proposed εDE-LS requires less than 1×10^4 FES for 4 test benchmark functions, less than 5×10^4 FES for 14 test benchmark functions, less than 5.0×10^5 FES for 21 test benchmark functions to obtain the require accuracy.

5 Conclusion

This paper proposed the εDE-LS algorithm, in which a novel local search operator designed for COPs are introduced. The feasible and infeasible individuals can interact with each other by applying the proposed mutation operator. By utilizing the novel mutation operator as the local search engine, we can guide the population moving towards the feasible region more effective. The effectiveness of the proposed εDE-LS algorithm is demonstrated by 22 test benchmark functions collected from IEEE CEC2006 special session on constrained real parameter optimization. The experimental results suggest that εDE-LS algorithm is highly competitive in terms of accuracy and convergent speed. εDE-LS algorithm can successfully solve 21 test benchmark functions and can achieve 21 feasible optimal solutions consistently. The success performance of εDE-LS algorithm is highly competitive when compares with other state-of-the-art algorithms. As the effectiveness and efficiency of the proposed algorithm demonstrated above, we can conclude that the εDE-LS is highly competitive one in dealing with COPs and should gain attention from researchers in the future. Besides, the performance of the εDE-LS can be further studied through using other indicators. In the future, more real world applications can be tested by the proposed εDE-LS algorithm. Moreover, as a part of the future direction, the performance of εDE-LS may be further improved by discovering a more efficient mutation operator.

Acknowledgement. This research work is supported by the National Basic Research Program of China (973 Program) under Grant no. 2011CB706804, and the Natural Science Foundation of China (NSFC) under Grant no. 51005088 and 51121002.

References

1. Storn, R., Price, K.: Differential evolution — a simple and efficient heuristic for global optimization over continuous spaces. Journal of Global Optimization 11, 341–359 (1997)
2. Huang, F.Z., Wang, L., He, Q.: An Effective Co-evolutionary Differential Evolution for Constrained Optimization. Applied Mathematics and Computation 286, 340–356 (2007)
3. Wang, Y., Cai, Z.X.: Combining Multiobjective Optimization with Differential Evolution to Solve Constrained Optimization Problems. IEEE Transactions on Evolutionary Computation 16, 117–134 (2012)
4. Gong, W., Cai, Z.: A Multiobjective Differential Evolution Algorithm for Constrained Optimization. In: 2008 Congress on Evolutionary Computation (CEC 2008), pp. 181–188 (2008)
5. Storn, R.: System Design by Constraint Adaptation and Differential Evolution. IEEE Transactions on Evolutionary Computation, 22–34 (1999)
6. Lampinen, J.: A Constraint Handling Approach for Differential Evolution Algorithm. In: Proceedings of the Congress on Evolutionary Computation (CEC 2002), pp. 1468–1473 (2002)
7. Takahama, T., Sakai, S.: Constrained Optimization by the ε-Constrained Differential Evolution with Gradient-Based Mutation and Feasible Elites. In: 2006 IEEE congress on Evolutionary Computation (CEC 2006), pp. 308–315 (2006)
8. Liang, J.J., Runarsson, T.P., Mezura-Montes, E., et al.: Problems Definitions and Evaluation Criteria for the CEC' 2006 Special Session on Constrained Real-parameter Optimization (2006), http://www.ntu.edu.sg/home/EPNSugan/cec2006/technicalreport.pdf

9. Mezura-Montes, E., Velázquez-Reyes, J., CoelloCoello, C.A.: Modified Differential Evolution for Constrained Optimization. In: Proceedings of the Congress on Evolutionary Computation (CEC 2006), pp. 332–339 (2006)
10. Tasgetiren, M.F., Suganthan, P.N.: A Multi-populated DifferentialEvolution Algorithm for Solving Constrained Optimization Problem. In: Proceedings of the Congress on Evolutionary Computation (CEC 2006), pp. 33–40 (2006)
11. Kukkonen, S., Lampinen, J.: Constrained Real-parameter Optimizationwith Generalized Differential Evolution. In: Proceedings of the Congress on Evolutionary Computation (CEC 2006), pp. 207–214 (2006)
12. Brest, J., Zumer, V., Maucec, M.S.: Self-adaptive DifferentialEvolution Algorithm in Constrained Real-parameter Optimization. In: Proceedings of the Congress onEvolutionary Computation (CEC 2006), pp. 215–222 (2006)

Structurization of Design Space for Launch Vehicle with Hybrid Rocket Engine Using Stratum-Type Association Analysis

Kazuhisa Chiba[1], Masahiro Kanazaki[2], Shin'ya Watanabe[3], Koki Kitagawa[4], and Toru Shimada[4]

[1] Hokkaido University of Science, Sapporo 006-8585, Japan
kazchiba@hus.ac.jp
http://www.geocities.jp/thousandleaf_k/index_e.html
[2] Tokyo Metropolitan University, Tokyo 191-0065, Japan
[3] Muroran Institute of Technology, Hokkaido 050-8585, Japan
[4] Institute of Space and Astronautical Science, Japan Aerospace Exploration Agency, Sagamihara 252-5210, Japan

Abstract. A single-stage launch vehicle with hybrid rocket engine has been conceptually designed by using design informatics, which has three points of view, i.e., problem definition, optimization, and data mining. The primary objective of the present design is that the down range and the duration time in the lower thermosphere are sufficiently secured for the aurora scientific observation, whereas the initial gross weight is held down to the extent possible. The multidisciplinary design optimization was performed by using a hybrid evolutionary computation. Data mining was also implemented by using the stratum-type association analysis. Consequently, the design information regarding the tradeoffs has been revealed. Furthermore, the hierarchical dendrogram generated by using the stratum-type association analysis indicates the structure of the design space in order to improve the objective functions. Thereupon, it has been revealed the versatility of the synthetic system as design informatics for real-world problems.

Keywords: Design informatics, Evolutionary computation, Data mining, Stratum-type association analysis, Application to real-world problems, Single-stage launch vehicle for scientific observation, Hybrid rocket engine using solid fuel and liquid oxidizer.

1 Introduction

Design informatics is essential for practical design problems. Although solving design optimization problems is important under the consideration of many disciplines of engineering[1], the most significant part of the process is the extraction of useful knowledge of the design space from results of optimization runs. The results produced by multiobjective optimization (MOO) are not an individual optimal solution but rather an entire set of optimal solutions due to tradeoffs.

H. Handa et al. (eds.), *Proc. of the 18th Asia Pacific Symp. on Intell. & Evol. Systems – Vol. 1*,
Proceedings in Adaptation, Learning and Optimization 1, DOI: 10.1007/978-3-319-13359-1_39

That is, the result of an MOO is not sufficient from the practical point of view as designers need a conclusive shape and not the entire selection of possible optimal shapes. On the other hand, this set of optimal solutions produced by an evolutionary MOO algorithm can be considered a hypothetical design database for design space. Then, data mining techniques can be applied to this hypothetical database in order to acquire not only useful design knowledge but also the structurization and visualization of design space for the conception support of basic design. This approach was suggested as design informatics[7]. The goal of this approach is the conception support for designers in order to materialize innovation. This methodology is constructed by the three essences as 1) problem definition, 2) efficient optimization, and 3) structurization and visualization of design space by data mining. A design problem including objective function, design variable, and constraint, is strictly defined in view of the background physics for several months (problem definition is the most important process for all designers because it directly gives effect on the quality of design space. Since the garrulous objective-function/design-variable space including physics and design information which is not inherently necessary to consider should be performed unnecessary evolutionary exploration and mining, it is conceived to be low-quality design space), then optimization is implemented in order to acquire nondominated solutions (quasi-Pareto solutions) as hypothetical database. Data mining is performed for this database in order to obtain design information. Mining has the role of a postprocess for optimization. Mining result is the significant observations for next design phase and also becomes the material to redefine a design problem.

Intelligent and evolutionary systems including design informatics mentioned above have the significance for not only the systems themselves but also their applications to practical problems in order that science contributes toward the real world. Results themselves do not possess versatility in application problems due to their particularity. The versatility of a system is indeed critical in application problems because it is revealed that application range is expanded. Furthermore, the application results indicate the guidance for the improvement of systems. In the present study, a single-stage launch vehicle with hybrid rocket engine using solid fuel and liquid oxidizer for the scientific observation of aurora will be conceptually designed by using design informatics approach. The final objective is that the advantage of re-ignition in the science mission for aurora observation on hybrid rocket will be quantitatively revealed. As a first step, an optimization problem on single-time ignition, which is the identical condition of the current solid rocket, was defined so as to obtain the design information[5]. As a second step, the implication of solid fuels in the performance of hybrid rocket was revealed because the regression rate is one of the key elements for the performance of hybrid rocket[6]. Finally, the sequence using multi-time ignition, which is the advantage of hybrid rocket, will be investigated in order to reveal the ascendancy of hybrid rocket for aurora observation. This study corresponds to the additional extraction of the design information for the above first step.

This study is a milestone to observe the quantitative difference of performance regarding ignition time.

2 Design Informatics

2.1 Optimization Method

Design informatics after the definition of detailed problem is constructed by two phases as optimization and data mining. Evolutionary computation is used for optimization. Although a surrogate model[17] like as the Kriging model[14], which is a response surface model developed in the field of spatial statistics and geostatistics, can be employed as optimization method, it will not be selected because it is difficult to deal with a large number of design variables. In addition, since the designers require to present many exact optimum solutions for the decision of a compromise one, an evolutionary-based Pareto approach as an efficient multi-thread algorithm, which the plural individuals are parallel conducted, is employed instead of gradient-based methods. The optimizer used in the present study is the hybrid evolutionary method between the differential evolution (DE) and the genetic algorithm (GA)[4]. Moreover, global design information is primarily essential in order to determine a compromise solution. The view of hybridization is inspired by the evolutionary developmental biology[2]. When there is the evolution which the Darwinism cannot explain in the identical species, each individual might have a different evolutionary methodology. When the practical evolution is imitated for the evolutionary computation, the different evolutionary algorithms might ultimately be applied to each individual in population. The making performance of next generation for each methodology depends on not only their algorithms but also the quality of candidate of parent in the archive of nondominated solutions. The present hybridization is intended to improve the quality of candidate of parent by sharing the nondominated solutions in the archive among each methodology. It was confirmed that this methodology had the high performance regarding the convergence and diversity, as well as the strength for noise[4]. Note that noise imitates the error on computational analyses and experiments and is described as the perturbation on objective functions. It is an important factor when the optimization for practical engineering problem is considered.

First, multiple individuals are generated randomly as an initial population. Then, objective functions are evaluated for each individual. The population size is equally divided into sub-populations between DE and GA (although sub-population size can be changed at every generations on the optimizer, the determined initial sub-populations are fixed at all generations in the present study). New individuals generated by each operation are combined in next generation. The nondominated solutions in the combined population are archived in common. It is notable that only the archive data is in common between DE and GA. The respective optimization methods are independently performed in the present hybrid methodology.

The present optimization methodology is a real-coded optimizer[19]. Although GA is based on the real-coded NSGA-II (the elitist nondominated sorting genetic algorithm)[9], it is made several improvements on in order to be progressed with the diversity of solutions. Fonseca's Pareto ranking[10] and the crowding distance[9] are used for the fitness value of each individual. The stochastic universal sampling[3] is employed for parents selection. The crossover rate is 100%. The principal component analysis blended crossover-α (PCABLX)[24] and the confidence interval based crossover using L_2 norm (CIX)[12] are used because of the high performance for the convergence and the diversity as well as the strength for noise[4]. The subpopulation size served by GA is equally divided for these two crossovers. The mutation rate is set to be constant as the reciprocal of the number of design variables. For alternation of generations, the Best-N selection[9] is used. DE is used as the revised scheme[20] for multiobjective optimization from DE/rand/1/bin scheme. The scaling factor F is set to be 0.5. The present optimizer has the function of range adaptation[22], which changes the search region according to the statistics of better solutions, for all design variables. In the present study, the range adaptation is implemented at every 20th generations.

2.2 Data-Mining Technique

The new data mining technique named as the stratum-type association analysis [25] has been applied to analyzing nondominated solutions. The previous study[8] employed the rough set theory in order to obtain the concrete rule regarding the design principles of design variables. Since the rough set theory gives individual rules based on the machine learning, it does not reveal the correlation knowledge among them. The present methodology systematizes individual rules and structurize design space using a hierarchical dendrogram so that methodology obtains a bird's-eye view of it in order to have useful knowledge. The feature of the present methodology is a recursive clustering using association rules and a multi-granular analysis. The results of the recursive clustering can be visualized as a hierarchical dendrogram in which each node is a sub cluster of nondominated solutions. The present system is expected to extract design information from microscopic to macroscopic view points due to the structurization of design space.

Procedure of System. First, the present system discretizes continuous data for logical analysis. Second, association rules are derived from discretized data through logical analysis. Since designer would like to primarily acquire the design knowledge regarding objective function because it corresponds to design requirement. The association rules regarding objective function will be extracted at antecedent and consequent processes. The association rules are gradually integrated into subsets regarding the degree of coincidence at antecedent and consequent processes. Finally, the present methodology constructs a structured hierarchical dendrogram by clustering subsets based on synthetic correlations. It is essential that the methodology generates the subsets of nondominated solutions using not the similarity among them but the inherent characteristics in each nondominated solution.

Clustering Method for Association Rules. A hierarchical dendrogram can be simply generated through clustering the association rules with similarity and a multi-granular analysis. Therefore, the present methodology can be optionally selected regarding clustering manner. The methodology integrates the generated rules into subsets using not similarity but the accordance with the coincidence of the rules generated by the antecedent and consequent operations. A generated hierarchical dendrogram is simply constructed by the inclusive correlation among the subsets. Indeed, all elements which contain all extracted association rules are selected. All combinations of these elements are generated without overlap. Then, the subsets are generated in order to concentrate the elements. Thereupon, all of the subsets have common characteristics. That is, all nodes of a hierarchical dendrogram have one common characteristic at least. The characteristic in a node is useful knowledge for designers.

3 Problem Definition

Single-stage rockets have been researched and developed for the scientific observations and the experiments of high-altitude zero-gravity condition, whereas multi-stage rockets have been also studied for the orbit injection of payload. The Institute of Space and Astronautical Science (ISAS), Japan Aerospace Exploration Agency (JAXA) has been operating K (Kappa), L (Lambda), and M (Mu) series rockets as the representatives of solid rocket in order to contribute to the space scientific research. A lower-cost and more efficient rocket is necessary due to the retirement of M-V in 2008 and in order to promote space scientific research. In fact, E (Epsilon) rocket began to be operated from September 2013. On the other hand, the launch vehicle with hybrid rocket engine using solid fuel and liquid oxidizer has been researched and developed as an innovative technology in mainly Europe and United States[15,23]. The present study will investigate the conceptual design in order to develop a next-generation single-stage launch vehicle with hybrid rocket engine. Since the technologies of hybrid rocket engine for single-stage and multi-stage are not independent, the solution of the fundamental physics regarding single-stage hybrid rocket is diverted to multi-stage one. A hybrid rocket offers the several advantages as higher safety, lower cost, and pollution free flight. The multi-time ignition is the especial ascendancy of hybrid rocket engine[21]. On the other hand, the disadvantage of a hybrid rocket engine is in its combustion. As a hybrid rocket engine has low regression rate of solid fuel due to turbulent boundary layer combustion, the thrust of hybrid rocket engine is less than that of pure solid and pure liquid engines which can obtain premixed combustion[16]. In addition, as the mixture ratio between solid fuel and liquid oxidizer is temporally fluctuated, thrust changes with time. Multidisciplinary design requirements should be considered in order to surmount the

Table 1. Limitation of upper/lower values of each design variable

serial number	*design variable*	*design space*
dv1	initial mass flow of oxidizer	$1.0 \leq \dot{m}_{\rm oxi}(0)$ [kg/sec] ≤ 30.0
dv2	fuel length	$1.0 \leq L_{\rm fuel}$ [m] ≤ 10.0
dv3	initial radius of port	$0.01 \leq r_{\rm port}(0)$ [m] ≤ 0.30
dv4	combustion time	$10.0 \leq t_{\rm burn}$ [sec] ≤ 40.0
dv5	initial pressure in combustion chamber	$3.0 \leq P_{\rm cc}(0)$ [MPa] ≤ 6.0
dv6	aperture ratio of nozzle	$5.0 \leq \epsilon$ [-] ≤ 8.0
dv7	elevation at launch time	$50.0 \leq \phi(0)$ [deg] ≤ 90.0

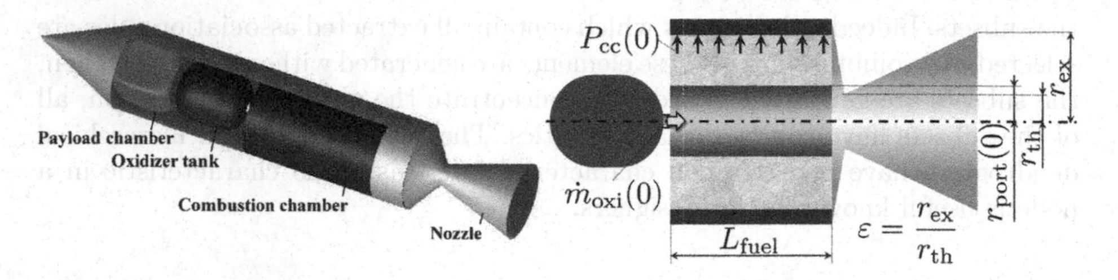

Fig. 1. Conceptual illustrations of hybrid rocket and its design variables regarding the geometry. Aperture ratio of nozzle ϵ is described by using the radius at nozzle exit $r_{\rm ex}$ and the radius at nozzle throat $r_{\rm th}$.

disadvantage of hybrid rocket engine. Moreover, exhaustive design information will be obtained in order to additionally consider productive and market factors for practical problems (it is difficult that optimization deals with them due to the difficulty of the definition).

The conceptual design for a single-stage hybrid rocket[18], simply composed of a payload chamber, an oxidizer tank, a combustion chamber, and a nozzle, is considered in the present study shown in Fig. 1. A single-stage hybrid rocket for aurora scientific observation will be focused because the rocket for more efficient scientific observation is desired for successfully obtaining new scientific knowledge on the aurora observation by ISAS in 2009. In addition, a single-stage hybrid rocket problem fits for the resolution of the fundamental physics regarding hybrid rocket engine and for the improvement of the present design problem due to its simplification.

3.1 Objective Functions

Three objective functions are defined in the present study. First objective is the maximization of the down range in the lower thermosphere (altitude of 90 to 150 [km]) R_d [km] (obj1). Second is the maximization of the duration time in the lower thermosphere T_d [sec] (obj2). It recently turns out that atmosphere has furious and intricate motion in the lower thermosphere due to the energy injection, which leads aurora, from high altitude. The view of these objective functions are to secure the horizontal distance and time for the competent observation of atmospheric temperature and the wind for the elucidation of atmospheric dynamics and the balance of thermal energy. Third objective is the minimization of the initial gross weight of launch vehicle $M_{\mathrm{tot}}(0)$ [kg] (obj3), which is generally the primary proposition for space transportation system.

3.2 Design Variables

Seven design variables are used as initial mass flow of oxidizer $\dot{m}_{\mathrm{oxi}}(0)$ [kg/sec] (dv1), fuel length L_{fuel} [m] (dv2), initial radius of port $r_{\mathrm{port}}(0)$ [m] (dv3), combustion time t_{burn} [sec] (dv4), initial pressure in combustion chamber $P_{\mathrm{cc}}(0)$ [MPa] (dv5), aperture ratio of nozzle ϵ [-] (dv6), and elevation at launch time ϕ [deg] (dv7). Note that there is no constraint except the limitations of upper/lower values of each design variable summarized in Table 1. These upper/lower values are exhaustively covering the region of design space which is physically admitted. When there is a sweet spot (the region that all objective functions proceed optimum directions) in the objective-function space, the exploration space would intentionally become narrow due to the operation of range adaptation on the evolutionary computation.

3.3 Evaluation Method

First of all, the mixture ratio between liquid oxidizer and solid fuel $O/F(t)$ is computed by the following equation.

$$
\begin{aligned}
O/F(t) &= \frac{\dot{m}_{\mathrm{oxi}}(t)}{\dot{m}_{\mathrm{fuel}}(t)}. \\
\dot{m}_{\mathrm{fuel}}(t) &= 2\pi r_{\mathrm{port}}(t) L_{\mathrm{fuel}} \rho_{\mathrm{fuel}} \bar{\dot{r}}_{\mathrm{port}}(t), \\
r_{\mathrm{port}}(t) &= r_{\mathrm{port}}(0) + \int \dot{r}_{\mathrm{port}}(t) dt.
\end{aligned}
\tag{1}
$$

$\dot{m}_{\mathrm{oxi}}(t)$ and $\dot{m}_{\mathrm{fuel}}(t)$ are the mass flow of oxidizer [kg/sec] and the mass flow of fuel [kg/sec] at time t, respectively. $r_{\mathrm{port}}(t)$ is the radius of port [m] at t, L_{fuel} describes fuel length, and ρ_{fuel} is the density of fuel [kg/m^3]. $\dot{r}_{\mathrm{port}}(t)$ describes the regression rate. After that, an analysis of chemical equilibrium is performed by using NASA-CEA (chemical equilibrium with applications) [11], then trajectory, thrust, aerodynamic, and structural analyses are respectively implemented. The present rocket is assumed as a point mass. As the time step is set to be 0.5 [sec] in the present study, it takes roughly 10 [sec] for the evaluation of an individual using a general desktop computer.

A combustion chamber is filled with solid fuel with a single port at the center to supply oxidizer. As the regression rate to the radial direction of the fuel $\dot{r}_{\mathrm{port}}(t)$ [m/sec] generally governs the thrust power of hybrid rocket engine, it is a significant parameter. The following experimental model[13,26] is used in the present study.

$$
\begin{aligned}
\dot{r}_{\mathrm{port}}(t) &= a_{\mathrm{fuel}} \times G_{\mathrm{oxi}}^{n_{\mathrm{fuel}}}(t) \\
&= a_{\mathrm{fuel}} \times \left(\frac{\dot{m}_{\mathrm{oxi}}(t)}{\pi r_{\mathrm{port}}^2(t)} \right)^{n_{\mathrm{fuel}}},
\end{aligned}
\tag{2}
$$

where, $G_{\rm oxi}(t)$ is oxidizer mass flux [kg/m^2/sec]. $a_{\rm fuel}$ [m/sec] and $n_{\rm fuel}$ [-] are the constant values experimentally determined by fuels. In the present study, liquid oxygen as liquid oxidizer and polypropylene as thermoplastic resin for solid fuel are used in order to adopt swirling flow for the supply mode of oxidizer. Therefore, $a_{\rm fuel}$ and $n_{\rm fuel}$ are respectively set to be 8.26×10^{-5} [m/sec] and 0.5500.

4 Results

4.1 Optimization Result

The population size is set to be 18 and evolutionary computation is performed until 3,000 generations when the evolution is roughly converged. The plots of

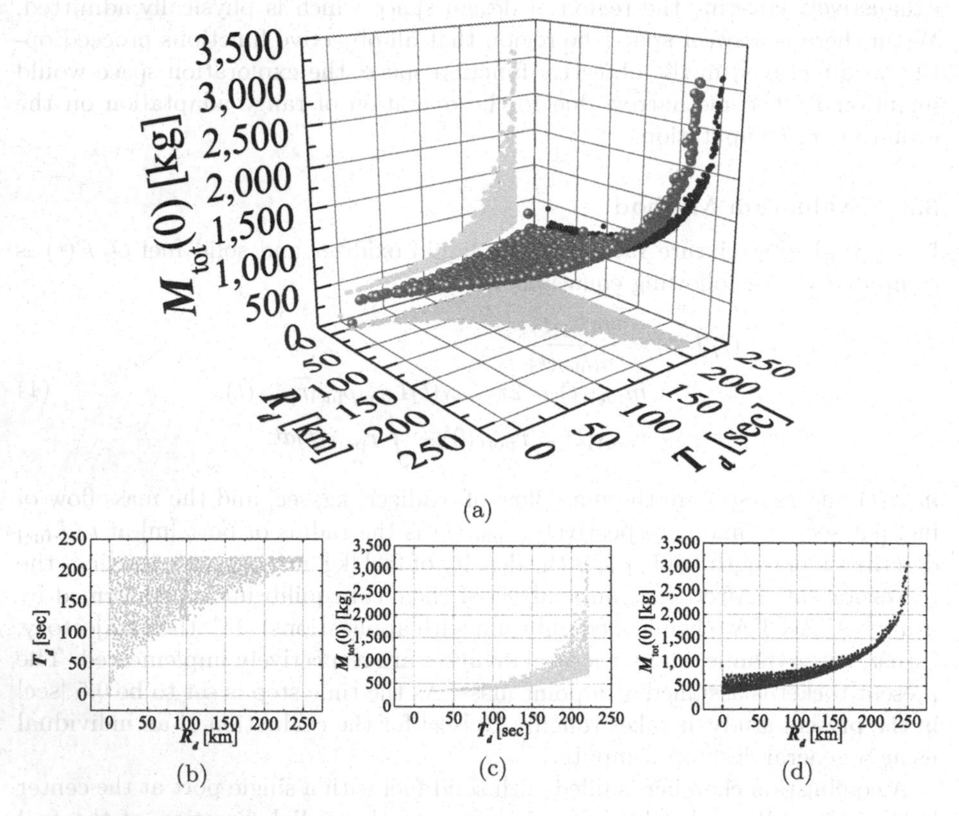

Fig. 2. Plots of nondominated solutions derived by optimization, (a) plotted in three-dimensional objective-function space (red) and their plots projected onto two dimensions, (b) plots projected onto two dimension between down range R_d (obj1) and duration time T_d (obj2) (light green), (c) plots projected onto two dimension between duration time T_d (obj2) and initial gross weight $M_{\rm tot}(0)$ (obj3) (light blue), and (d) plots projected onto two dimension between down range R_d (obj1) and initial gross weight $M_{\rm tot}(0)$ (obj3) (blue).

acquired nondominated solutions are shown in Fig. 2, which reveals that there generates no multimodal and clean convex curved surface.

There is no tradeoff between the down range R_d and the duration time T_d in the lower thermosphere shown in Fig. 2(b). This figure also shows that there are upper limits of roughly 250 [km] for R_d and of roughly 220 [sec] for T_d. Therefore, the projection plots onto two dimension between R_d and T_d do not converge in one point. In the present study, the initial mass flow of oxidizer $\dot{m}_{\text{oxi}}(0)$ (dv1) has the limitation of upper/lower values. Since the regression rate of fuel $\dot{r}_{\text{port}}(t)$ as an empirical model uses the mass flow of oxidizer $\dot{m}_{\text{oxi}}(t)$, $\dot{r}_{\text{port}}(t)$ has constraints. As a result, the upper limits are generated for R_d and T_d.

There is an incomplete tradeoff between T_d and the initial gross weight $M_{\text{tot}}(0)$ shown in Fig. 2(c). The convex nondominated surface to optimum direction with incompleteness is generated due to the upper limit of T_d. As the inclination $\Delta M_{\text{tot}}(0)/\Delta T_d$ is small on the convex curve, T_d can be substantially improved when trifling $M_{\text{tot}}(0)$ would be sacrificed. In addition, Fig. 2(c) shows that the minimum $M_{\text{tot}}(0)$ to reach the upper limit of T_d (roughly 220 [sec]) is approximately 700 [kg]. And also, the minimum $M_{\text{tot}}(0)$ to attain to the lower thermosphere (altitude of 90 [km]) is approximately 350 [kg]. As these values are better than those of the solid rockets which are operated at present for scientific observation, it suggests that hybrid rocket has an advantage even when hybrid rocket does not have a sequence of multi-time ignition.

There is a severe tradeoff between R_d and $M_{\text{tot}}(0)$ shown in Fig. 2(d) (although R_d strictly has the upper limit, it seems that the clean convex curve is generated because the upper limit is on the edge of the nondominated surface). This figure shows that the maximum R_d is roughly 130 [km] when the minimum $M_{\text{tot}}(0)$ to reach the upper limit of T_d (roughly 700 [kg]) is adopted. $M_{\text{tot}}(0)$ should be absolutely increased in order to have more R_d (greater than 130 [km]) despite no increase of T_d (remaining roughly 220 [sec]). This fact suggests that the design strategies for the maximizations of R_d and T_d are different.

4.2 Data-Mining Result

A discretization is necessary for rule generation. The hybrid discretization manner between an equivalent distance and an equivalent frequency methods[25] is utilized. The number of discretization is set to be 10 for the objective functions and the design variables in the present study. The value of minimum support depends on the generated association rule and the number of node. The present minimum support is set to be 10% in order to reduce the number of elements because many nodes with the small influence are generated. As a result, 39 nodes are generated with the minimum confidence of 90%. The hierarchical dendrograms which the first stratum has the association rule regarding the objective functions are constructed for the minimization of all of the three objective functions. Note that there are severe tradeoffs between the initial gross weight and the other two objective functions, which was already revealed in the optimization results shown in Fig. 2. And also, since the initial gross weight is the minimization function while the other two objective functions are maximization functions,

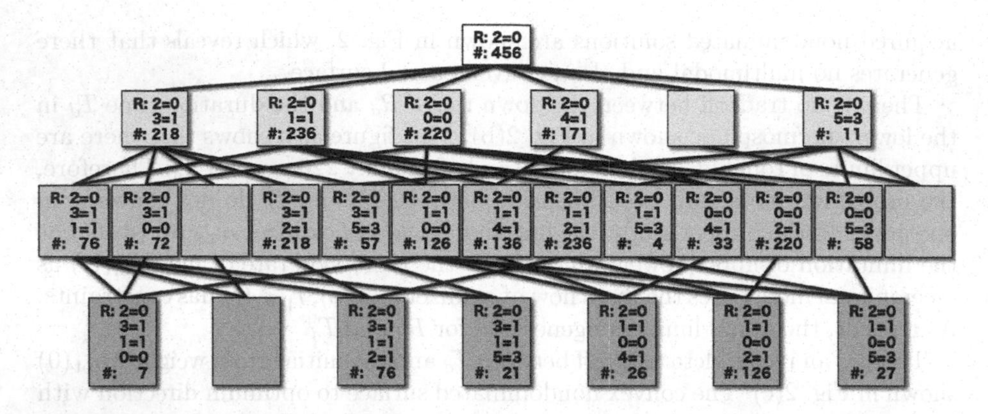

Fig. 3. Hierarchical dendrogram generated by the stratum-type association analysis for the minimization of the initial gross weight $M_{\text{tot}}(0)$. The present dendrogram is constructed by 27 nodes. The two-line explanations show the rule (which is described as 'the variables' = 'its discretized region') from the top line using "R:" and the correspondent number of all nondominated solutions with the rule at the bottom using "#:". The variable numbers of 0 to 9 respectively denote obj1, obj2, obj3, dv1, dv2, $\cdots$, and dv7. Since all variables are discretized into 10 in the present study, the discretized region is described by using the number of 0 to 9. The top node is in the first stratum and the bottom nodes are in the fourth stratum. The node with no information denotes that its minimum support is less than the value set artificially (10% in the present case).

Fig. 3 as the result for the minimization of the initial gross weight is shown as the representative of the generated hierarchical dendrograms. The decoding manner of each node in the present hierarchical dendrogram shown in Fig. 3 is explained in the caption of Fig. 3.

The first stratum has 456 nondominated solutions (because the total number of 999 is obtained for nondominated solutions by the optimization process, the proportion of application is roughly 45.6%) which have the attribute with the minimum-value node of $M_{\text{tot}}(0)$. This result indicates that roughly half number of the nondominated solutions has the small value of $M_{\text{tot}}(0)$.

The second stratum has six nodes with two attributes. Since one attribute is "2=0", i.e., the minimum-value node of $M_{\text{tot}}(0)$, the node indicates the tradeoff information when the other attribute is the rule regarding the other objective functions. In contrast, the node indicates the information regarding the effective design variable to minimize $M_{\text{tot}}(0)$ when the other attribute is the rule regarding the design variables. The third node from the left has the rule for R_d with the lowest values (the number of nondominated solutions is 220). This node reveals the severe tradeoff because R_d for the 48% of the nondominated solutions is in the discretized region for minimum value when $M_{\text{tot}}(0)$ is minimized. The second node from the left has the rule for T_d with low values (the number of nondominated solutions is 236). This node also reveals the severe tradeoff because T_d for 52% of the nondominated solutions is in the discretized region for minimum value when $M_{tot}(0)$ is minimized. Furthermore, the first node from the left has the rule for $\dot{m}_{oxi}(0)$ (dv1) with low values (the number of nondominated solutions is 218). The fourth node from the left is the rule for L_{fuel} (dv2) with low values (the number of nondominated solutions is 171). These nodes reveal that dv1 and dv2 are essential to restrain $M_{tot}(0)$. In addition, dv1 and dv2 are the significant design variables in order to also reduce the other two objective functions. The first node from the right is the rule for $r_{port}(0)$ (dv3) with low values (the number of nondominated solutions is 11 as a small number). $r_{port}(0)$ (dv3) indirectly gives an effect because small $r_{port}(0)$ (dv3) gives an effect on restraining the mass flow of oxidizer $\dot{m}_{oxi}(t)$ although $r_{port}(0)$ (dv3) does not have a direct effect.

The third stratum has 13 nodes with three attributes. 10 nodes of those consist of two rules regarding the objective function and one rule regarding the design variable. Since the knowledge of the second stratum shows the effective design variable for the objective functions, the observation of these 10 nodes can be omitted due to the comprehension into the design knowledge from the second stratum. The crucial point in the third stratum is that merely one node exists which has one rule regarding the objective function and two rules regarding the design variables. This result reveals that the correlation between the objective functions is strong and one design variable to depend on each objective function can be narrowed.

The fourth stratum has seven nodes with four attributes. Since the knowledge of the third stratum reveals that the present problem has severe tradeoffs, five nodes have three rules regarding the objective function and one rule regarding the design variable. The fourth node of the left merely has two rules regarding the objective functions and two rules regarding the design variables. However, since the indicated design variables are $\dot{m}_{oxi}(0)$ (dv1) and $r_{port}(0)$ (dv3), the knowledge from the second stratum can be similarly interpreted.

Although the stratum-type association analysis cannot instruct the optimum and pessimum directions, the severeness of tradeoff can be quantitatively shown by using the number of application. In addition, the magnitude of the influence of the design variables on the objective functions can be quantitatively elucidated. Moreover, since the nondominated solutions to apply to a node are generated, the stratum-type association analysis is also useful for the selection of specific nondominated solutions in order to perform the data mining for the significant local region in the objective-function space.

5 Conclusions

The next-generation single-stage launch vehicle with the hybrid rocket engine of solid fuel and liquid oxidizer in place of the present pure solid-fuel rockets has been conceptually designed by using design informatics in order to contribute to the development of the low cost launch vehicle system and efficient space scientific observation. The objective functions as the design requirements in the design problem is the maximization of the down range and the duration time in the lower thermosphere as well as the minimization of initial gross weight. The hybrid evolutionary computation between the differential evolution and the genetic algorithm is employed for the efficient exploration in the design space. The stratum-type association analysis is used in order to structurize and visualize the design space. As a result, the design information has been revealed regarding the tradeoffs among the objective functions and the behavior of the design variables in the design space. Furthermore, the hierarchical dendrogram indicates the structure of the design space in order to improve the objective functions. Consequently, the design strategy for the maximizations of down range and duration time is different because the duration time can easily attain to the upper limit rather than the down range. Moreover, the structurization generated by the stratum-type association analysis indicates the concrete design strategy regarding the significant design variables for the objective functions. The results show the quantitative data to compare the performances of solid-fuel rocket in present and hybrid rocket with multi-time ignition. Design informatics as the present intelligent and evolutionary system has successfully structured the design space for the present engineering problem and the present study has also revealed the expansion of the application range of design informatics in the real world.

References

1. Arias-Montano, A., Coello, C.A.C., Mezura-Montes, E.: Multiobjective evolutionary algorithms in aeronautical and aerospace engineering. IEEE Transactions on Evolutionary Computation 16(5), 662–694 (2012)
2. Arthur, W.: The emerging conceptual framework of evolutionary developmental biology. Nature 415(6873), 757–764 (2002)
3. Baker, J.E.: Adaptive selection methods for genetic algorithms. In: Proceedings of the International Conference on Genetic Algorithms and their Applications, pp. 101–111. Lawrence Erlbaum Associates (1985)
4. Chiba, K.: Evolutionary hybrid computation in view of design information by data mining. In: Proceedings on IEEE Congress on Evolutionary Computation, pp. 3387–3394. IEEE (2013)
5. Chiba, K., Kanazaki, M., Nakamiya, M., Kitagawa, K., Shimada, T.: Conceptual design of single-stage launch vehicle with hybrid rocket engine for scientific observation using design informatics. Journal of Space Engineering 6(1), 15–27 (2013)
6. Chiba, K., Kanazaki, M., Nakamiya, M., Kitagawa, K., Shimada, T.: Diversity of design knowledge for launch vehicle in view of fuels on hybrid rocket engine. Journal of Advanced Mechanical Design, Systems, and Manufacturing 8(3), JAMDSM0023, 1–14 (2014)
7. Chiba, K., Makino, Y., Takatoya, T.: Design-informatics approach for intimate configuration of silent supersonic technology demonstrator. Journal of Aircraft 49(5), 1200–1211 (2012)
8. Chiba, K., Obayashi, S.: Knowledge discovery in aerodynamic design space for flyback-booster wing using data mining. Journal of Spacecraft and Rockets 45(5), 975–987 (2008)
9. Deb, K., Pratap, A., Agarwal, S., Meyarivan, T.: A fast and elitist multiobjective genetic algorithm: NSGA-II. IEEE Transactions on Evolutionary Computation 6(2), 182–197 (2002)

10. Fonseca, C.M., Fleming, P.J.: Genetic algorithms for multiobjective optimization: Formulation, discussion and generalization. In: Proceedings of the Fifth International Conference on Genetic Algorithms, pp. 416–423. Morgan Kaufmann (1993)
11. Gordon, S., McBride, B.J.: Computer program for calculation of complex chemical equilibrium compositions and applications I. analysis. In: NASA Reference Publication RP-1311. NASA (1994)
12. Hervás-Martínez, C., Ortíz-Boyer, D., García-Pedrajas, N.: Theoretical analysis of the confidence interval based crossover for real-coded genetic algorithms. In: Guervós, J.J.M., Adamidis, P.A., Beyer, H.-G., Fernández-Villacañas, J.-L., Schwefel, H.-P. (eds.) PPSN 2002. LNCS, vol. 2439, pp. 153–161. Springer, Heidelberg (2002)
13. Hirata, K., Sezaki, C., Yuasa, S., Shiraishi, N., Sakurai, T.: Fuel regression rate behavior for various fuels in swirling-oxidizer-flow-type hybrid rocket engines. In: AIAA Paper 2011-5677. AIAA (2011)
14. Jeong, S., Murayama, M., Yamamoto, K.: Efficient optimization design method using kriging model. Journal of Aircraft 42(2), 413–420 (2005)
15. Karabeyoglu, M.A.: Advanced hybrid rockets for future space launch. In: Proceedings on 5th European Conference for Aeronautics and Space Sciences. EUCASS (2013)
16. Karabeyoglu, M.A., Altman, D., Cantwell, B.J.: Combustion of liquefying hybrid propellants: Part 1, general theory. Journal of Propulsion and Power 18(3), 610–620 (2002)
17. Keane, A.J.: Statistical improvement criteria for use in multiobjective design optimization. AIAA Journal 44(4), 879–891 (2006)
18. Kosugi, Y., Oyama, A., Fujii, K., Kanazaki, M.: Multidisciplinary and multi-objective design exploration methodology for conceptual design of a hybrid rocket. In: AIAA Paper 2011-1634, AIAA (2011)
19. Oyama, A., Obayashi, S., Nakamura, T.: Real-coded adaptive range genetic algorithm applied to transonic wing optimization. Applied Soft Computing 1(3), 179–187 (2001)
20. Robič, T., Filipič, B.: DEMO: Differential evolution for multiobjective optimization. In: Coello Coello, C.A., Hernández Aguirre, A., Zitzler, E. (eds.) EMO 2005. LNCS, vol. 3410, pp. 520–533. Springer, Heidelberg (2005)
21. Saraniero, M.A., Caveny, L.H., Summerfield, M.: Restart transients of hybrid rocket engines. Journal of Spacecraft and Rockets 10(3), 215–217 (1973)
22. Sasaki, D., Obayashi, S.: Efficient search for trade-offs by adaptive range multi-objective genetic algorithms. Journal of Aerospace Computing, Information, and Communication 2(1), 44–64 (2005)
23. Simurda, L., Zilliac, G., Zaseck, C.: High performance hybrid propulsion system for small satellites. In: AIAA Paper 2013-3635. AIAA (2013)
24. Takahashi, M., Kita, H.: A crossover operator using independent component analysis for real-coded genetic algorithms. In: Proceedings of IEEE Congress on Evolutionary Computation 2001, pp. 643–649. IEEE (2001)
25. Watanabe, S., Chiba, Y., Kanazaki, M.: A proposal on analysis support system based on associate rule analysis for non-dominated solutions. In: Proceedings on IEEE World Congress on Computational Intelligence, Beijing, China, pp. E–14855. IEEE (2014)
26. Yuasa, S., Shiraishi, N., Hirata, K.: Controlling parameters for fuel regression rate of swirling-oxidizer-flow-type hybrid rocket engine. In: AIAA Paper 2012-4106. AIAA (2012)

Evolving the Parameters of Differential Evolution Using Evolutionary Algorithms

Saber Elsayed and Ruhul Sarker

School of Engineering and Information Technology, University of New South Wales at Canberra (UNSW Canberra), Australia
{s.elsayed,r.sarker}@adfa.edu.au

Abstract. Differential evolution has shown tremendous success in solving different complex optimization problems. However, the performance is highly dependent on the selection of its parameters. Although many techniques have been introduced to adaptively (or self-adaptively) determine the parameters, the task is recognized as a tedious one. In this research, we investigate the use of evolutionary algorithms, such as covariance adaptation matrix evolution strategy, differential evolution and genetic algorithm, to self-adaptively determine the possible values of both the amplification factor and crossover rate. The performances of the algorithms are compared to each other, as well as to a standard differential algorithm, by solving a well-known set of benchmark problems. The experimental results show that such an approach can improve the performance of differential evolution, however further investigation is required to find the appropriate evolutionary algorithm for evolving parameters.

Keywords: differential evolution, covariance adaptation matrix evolution strategy, genetic algorithm, self-adaptation.

1 Introduction

The evolutionary algorithms (EAs), such as genetic algorithms (GA) [1], differential evolution (DE) [2] and evolution strategies (ES) [3], are popular choice to many researchers and practitioners for solving their complex optimization problems. Among EAs, DE has shown its superiority to many other algorithms in solving problems with different mathematical properties. However, it is well-known that DE parameters such as amplification factor (F), crossover rate (Cr) and population size (PS) play a vital role on its success, which led researchers to investigate this research topic, and propose different adaptive and self-adaptive mechanisms to avoid a trial-and-error approach in the selection of parameters. This directs us to the no-free launch theorem [4], which shows that one set of parameters may be well suited for a set of problems that may not work well for another problem, or another class, or range of problems.

While solving an optimization problem, one of the interesting mechanisms to self-adaptively determine the DE parameter values is using an evolution process that may involve DE or any other EAs. This mechanisms dates back to 2002 when Abbass [5] proposed a self-adaptive operator (crossover and mutation) for multi-objective

H. Handa et al. (eds.), *Proc. of the 18th Asia Pacific Symp. on Intell. & Evol. Systems – Vol. 1*,
Proceedings in Adaptation, Learning and Optimization 1, DOI: 10.1007/978-3-319-13359-1_40

optimization problems, where F was generated using a Gaussian distribution $N(0,1)$ and then updated using a DE algorithm. This technique has been modified in [6-9], in which DE variants with more than one difference vectors (DV) were used to evolve parameters, as well as F and Cr were initialized using $N(0.5, 15)$. However, to the best of our knowledge, the use of other EAs to self-adaptively determine DE parameters is rare.

Other mechanisms, which do not depend on EAs, have also been proposed. Qin *et al.* [10] proposed a DE algorithm (SaDE). In it, F was approximated by a normal distribution N (0.5, 0.3), and truncated to the interval (0, 2]. The crossover probabilities were randomly generated according to an independent normal distribution with mean Cr_m and a standard deviation value of 0.1. The Cr_m values remained fixed for five generations before the next re-generation. Cr_m was initialized to 0.5, and it was updated every 25 generations based on the recorded successful Cr values since the last Cr_m update. Using fuzzy logic controllers, Liu and Lampinen [11] presented a fuzzy adaptive DE, whose inputs incorporated the relative function values and individuals of successive generations to adapt the parameters for mutation and crossover. Brest et al. [12] proposed a self-adaptation scheme for the DE control parameters, known as jDE. The control parameters were adjusted by means of evolution of F and Cr. In jDE, a set of F and Cr values was assigned to each individual in the population, augmenting the dimensions of each vector. Zhang et al. [13] introduced an adaptive DE algorithm with optional external memory (JADE). In it, at each generation, the crossover probability Cr_z of each individual x_z was independently generated according to a normal distribution of mean μCr and standard deviation of 0.1. μCr was initialized at a value of 0.5 and updated. Similarly, F_z of each individual x_z was independently generated according to a Cauchy distribution with a location parameter μF and a scale parameter 0. The location parameter μF was initialized to 0.5 and subsequently updated at the end of each generation. Das *et al.*[14] introduced two versions for adapting F in DE. In the first scheme, F was randomly chosen between 0.5 and 1.0, while in the second scheme, F was initialized with a value of 1.0, and then linearly reduced to 0.1 during the evolution process. Generally speaking, such techniques may need adapting other parameters which may affect the performance of DE.

In this paper, we have evolved two DE parameters (such as F and Cr) by using three different algorithms. They are: (1) DE (this variant is recognized as Var1); (2) covariance adaptation matrix evolution strategy (CMA-ES)[15] (Var2); and (3) GA (Var3). That means, we are applying DE to solve the optimization problems, and within DE, we are using one of the above three algorithms to self-adaptively select DE parameters. The performances of these variants are compared to each other as well as to a DE with a single set of parameter values. From the results obtained, it is clear that the self-adaptive mechanism is better than a DE with a fixed set of parameters. Among the three variants, Var2 is the best considering the best fitness values found, while based on the average fitness values, Var3 is the best. However, these two variants are computational expensive in comparison with Var1.

The rest of this paper is organized as follows: section 2 presents and overview of DE. Section 3 discusses the self-adaptive mechanisms used in this paper, while section 4 presents the computational results. Finally, conclusions are elaborated in section 5.

2 Differential Evolution

DE uses the concept of a larger population from a GA and self-adapting mutation from ES [2], and differs from traditional EAs mainly in its generation of new vectors which adds the weighted difference vector (DV) between two individuals to a third individual [16]. It performs well when the feasible patches are parallel to the axes [17] but converges prematurely when dealing with a multi-modal fitness function because it loses its diversity [18, 19].

2.1 Mutation

The simplest form of this operation is that a mutant vector is generated by multiplying an amplification factor, F, by the difference between two random vectors, with the result added to a third random vector (DE/rand/1) [20] as:

$$\vec{x}_{z,t} = \vec{x}_{r_1,t} + F\left(\vec{x}_{r_2,t} - \vec{x}_{r_3,t}\right) \tag{1}$$

where r_1, r_2, r_3 are random numbers (1,2, ..., PS), $r_1 \neq r_2 \neq r_3 \neq z$, x a decision vector, F a positive control parameter for scaling the DV and t the current generation.

This operation enables DE to explore the search space and maintain diversity and there are many strategies for it, such as DE/best/1 [20], DE/current-to-best/1[21]. For more details, readers are referred to [22].

2.2 Crossovers

The DE family of algorithms usually depends on two crossover schemes, exponential and binomial, which are briefly discussed below.

In an exponential crossover, firstly, an integer, l, is randomly chosen within the range [1, D] and acts as the point in the target vector from where the crossover or exchange of components with the donor vector starts. Another integer, L, chosen from interval [1, D] denotes the number of components the donor vector actually contributes to the target. After the generation of l and L, the trial vector is obtained as:

$$u_{z,j,t} = \begin{cases} v_{z,j,t} & for\ j = \langle l \rangle_D, \langle l+1 \rangle_D, \ldots, \langle l+L-1 \rangle_D \\ x_{z,j,t} & for\ all\ other\ j \in [1, D] \end{cases} \tag{2}$$

where $j = 1,2, \ldots, D$, and the angular brackets, $\langle l \rangle_D$, denote a modulo function with a modulus of D and starting index of l.

The binomial crossover is performed on each of the jth variables whenever a randomly chosen number (between 0 and 1) is less than or equal to the crossover rate, Cr. In this case, the number of parameters inherited from the donor has a (nearly) binomial distribution as:

$$u_{zj,t} = \begin{cases} v_{zj,t}, & if\ (rand \leq Cr\ or\ j\ = jrand) \\ x_{zj,t}, & otherwise \end{cases} \tag{3}$$

where $rand \in [0,1]$ and $j_{rand} \in [1,2, \ldots, D]$ is a randomly chosen index which ensures that $\vec{u}_{z,t}$ receives at least one component from $\vec{v}_{z,t}$.

2.3 Selection

An offspring will be selected if it is better than its parent.

3 Self-adaptive DE Variants Based on EAs

To start with, in this paper, Table 1 shows the general framework of DE used in this paper.

Table 1. General framework of DE used in this paper

STEP 1: At generation t = 1, generate an initial random population of size PS. The variables of each individual (z) must be within a range such as:

$$x_{z,j} = \underline{x}_{z,j} + rand \times (\overline{x}_{z,j} - \underline{x}_{z,j})$$

where $\underline{x}_{z,j}, \overline{x}_{z,j}$ are the lower and upper bounds of the decision variable x_j, and *rand* is a random number, $rand \in [0,1]$.

STEP 2: Generate initial values for F and Cr

STEP 3: Evolve F and Cr using an EA, as shown in **3.1, 3.2** and **3.3**

STEP 4: Generate new offspring as follows:

4.1 Generate the offspring vector $\vec{u}_z$, using DE/current-to-best/bin and the corresponding F_z and Cr_z obtained in **Step 3** such as

$$u_{zj,t} = \begin{cases} x_{i,j} + F_z \cdot \left((x_{r_1,j} - x_{r_2,j}) + (x_{best,j} - x_{i,j}) \right), & if\ (rand \le Cr_z\ or\ j = jrand) \\ x_{z,j}, & otherwise \end{cases}$$

where r_1 and r_2 are random integer numbers $\in [1, PS]$ and both are not similar to i

4.2 Update $\boldsymbol{F_z}$ and $\boldsymbol{Cr_z}$, if required, see **3.1**

STEP 8: Stop if the termination criterion is met; **else,** set $t = t + 1$ **and** go to **STEP 3.**

In this paper, three variants are used to self-adaptively generate F and Cr, as described below.

3.1 Var1: Adapting F and Cr Using DE

Here, Cr and F are self-adaptively calculated using a simple DE algorithm, as follows:

- At $t = 1$, each individual in PS is assigned with $\dot{F}_z$ and $\dot{Cr}_z$, where $\dot{F}_z = N(0.5, 0.1)$ and $\dot{Cr}_z = N(0.5, 0.1)$. If the value is less than 0.01 or larger than 1.0, it is reflected back to be between to 0.01 and 1, respectively.
- Then, both parameters are calculated as follows:

$$F_z = \begin{cases} \dot{F}_{r_1} + rand_1 \times (\dot{F}_{r_2} - \dot{F}_{r_3}), & if\ (rand_2 < \tau_1) \\ rand_3 & otherwise \end{cases} \tag{4}$$

$$Cr_z = \begin{cases} \dot{Cr}_{r_1} + rand_4 \times (\dot{Cr}_{r_2} - \dot{Cr}_{r_3}), & if\ (rand_2 < \tau_1) \\ rand_5 & otherwise \end{cases} \tag{5}$$

where $rand_\Gamma \in [0,1]\ \forall \Gamma = 1,2 \ldots,5$ and $\tau_1 = 0.75$. If the value is less than 0.01 or larger than 1, it is truncated to 0.1 and 1, respectively.

- If the new offspring is better than its parent, then $\dot{F}_z = F_z$ and $\dot{Cr}_z = Cr_z$.

3.2 Var2: Adapting F and Cr Using CMA-ES

Here, Cr and F are self-adaptively calculated using CMA-ES, such that

- At $t = 1$, $\vec{x}^t_{mean}$ is initialized as $N(0.5, 0.1, np)$, where np is 2, $\vec{x}^t_{1,1}$ refers to F_z, while $\vec{x}^t_{1,2}$ refers to Cr_z.
- A new population, which represents both F and Cr, is generated using CMA-ES:

$$\overrightarrow{xp}^{t+1}_{z=1:PS} = \vec{x}^t_{mean} + \sigma^t B^t Q^t G_{z=1:PS} \tag{6}$$

  where $G_{k=1:PS}$ are independent realizations of a D-dimensional standard normal distribution with zero-mean and a covariance matrix equal to the identity matrix I. These base points are rotated and scaled by the eigenvectors B^t and the square root of the eigenvalues Q^t of the covariance matrix C^t. The C^t, and the global step-size σ^t are continuously updated after each generation t [15].
- Then, both parameters are calculated as follows:

$$F_z = \begin{cases} xp_{z,1}, & if\ (rand_1 < \tau_1) \\ rand_2 & otherwise \end{cases} \tag{7}$$

$$Cr_z = \begin{cases} xp_{z,2}, & if\ (rand_1 < \tau_1) \\ rand_3 & otherwise \end{cases} \tag{8}$$

- Then $\vec{x}^t_{mean}$, and all other CMA-ES's parameters are updated as suggested in https://www.lri.fr/~hansen/cmaes.m. It is worthy to mention here that, to measure the quality of each $\overrightarrow{xp}_z$, the objective function of the corresponding $\vec{x}_z$ is used instead.

3.3 Var3: Adapting F and Cr Using GA

Here, a multi-parent crossover GA (MPC-GA) [23] is used to evolve DE parameters.

- At $t = 1$, an initial population (xp) of $\dot{F}_z$ and $\dot{Cr}_z$ is initialized, where $xp_{z,1:2} = N(0.5, 0.1)$.
- Then an archive pool (A_{arch}) is filled with the best m individuals (based objective function of the corresponding $\vec{x}_z$.
- Then a tournament selection procedure, with size tc, takes place, from which the best individual is chosen and saved in the selection pool.
- for each three consecutive individuals in the selection pool, three offspring are generated as

$$\vec{y}_1 = \overrightarrow{xp}_1 + \beta \times (\overrightarrow{xp}_2 - \overrightarrow{xp}_3) \tag{9}$$

$$\vec{y}_2 = \overrightarrow{xp}_2 + \beta \times (\overrightarrow{xp}_3 - \overrightarrow{xp}_1) \quad (10)$$

$$\vec{y}_3 = \overrightarrow{xp}_3 + \beta \times (\overrightarrow{xp}_1 - \overrightarrow{xp}_2) \quad (11)$$

where $\beta = N(0.7, 0.1)$ [23].

- On each generated $\vec{y}_z$, a diversity operator is applied. In it, for each individual a uniform random number $\in [0, 1]$ is generated, if it is less than a predefined probability, $p = 0.1$, then $y_z^j = x_{arch}^j$.
- Subsequently, set $xp = y$
- Then, both parameters are calculated as follows:

$$F_z = \begin{cases} xp_{z,1}, & if\ (rand_1 < \tau_1) \\ rand_2 & otherwise \end{cases} \quad (12)$$

$$Cr_z = \begin{cases} xp_{z,2}, & if\ (rand_1 < \tau_1) \\ rand_3 & otherwise \end{cases} \quad (13)$$

4 Experimental Results

In this section, a comparison among all variants is elaborated by solving a set of problems presented in the CEC2014 competition on real-parameter optimization [24], which contains 30 test problems with 30 dimensions, with the following mathematical properties: F01-F03 are unimodal functions, F04-F16 are simple multimodal functions, F17-F22 are hybrid functions, while F23-F30 are composition functions.

To add to this, all variants were compared with a DE, as shown in equation 4, with fixed values of its parameters, such that *F*=0.5, *Cr*=0.5, while *PS* was set to 100 individuals for all variants and τ_1=0.75. All variants were run 51 times for each test problem, where the stopping criterion was to run for up to 10,000D FEs. The algorithm was coded using Matlab R2012b, and was run on a PC with a 3.4 GHz Core I7 processor with 16 GB RAM, and windows 7.

To begin with, the best results obtained by all variants are shown in Table 2. From these results, it is clear that all variants were better than DE for F01, with the consideration that Var1 performs in F01, while all variants were able to obtain the same values for F02 and F03. For the multi-modal test problems, Var2 was the best for most of the test problems. However, Var1 was superior to all other variants for F05 and F13 and F15, while DE was the best in F14. In regards to the hybrid function, Var2 was the best for four test problems, while Var1 was the best for only F22 and Var3 performed best in F21. For the composition functions, Var2 performed best for 6 test functions. DE and Var1 obtained the same best result in F26, while Var3 was the best in F29.

Considering the average results obtained by all variants, see Table 3, it is noticed that Var3 was the best for the unimodal test functions, followed by Var1. However, Var2 was the worst variant for those test problems. For the multi-modal test functions,

Table 2. Best Results obtained by all variants

Prob.	DE1	Var1	Var2	Var3
F01	2.604E+05	**1.276E+03**	2.422E+03	6.483E+03
F02	**0.000E+00**	**0.000E+00**	**0.000E+00**	**0.000E+00**
F03	**0.000E+00**	**0.000E+00**	**0.000E+00**	**0.000E+00**
F04	1.441E+01	1.012E-04	**0.000E+00**	**0.000E+00**
F05	2.074E+01	**2.000E+01**	2.022E+01	2.042E+01
F06	2.259E-04	4.765E-01	**0.000E+00**	9.643E-02
F07	**0.000E+00**	**0.000E+00**	**0.000E+00**	**0.000E+00**
F08	1.333E+01	**0.000E+00**	**0.000E+00**	1.417E+01
F09	1.199E+02	3.283E+01	**2.098E+01**	3.310E+01
F10	1.955E+02	9.155E+00	**1.767E+00**	4.642E+02
F11	5.248E+03	1.942E+03	**1.348E+03**	2.148E+03
F12	1.506E+00	2.782E-01	**2.427E-01**	4.764E-01
F13	1.416E-01	**1.407E-01**	1.744E-01	1.578E-01
F14	**1.065E-01**	1.286E-01	1.503E-01	1.842E-01
F15	1.070E+01	**3.422E+00**	3.698E+00	5.246E+00
F16	1.084E+01	1.065E+01	**8.485E+00**	1.025E+01
F17	1.790E+04	8.806E+02	**2.211E+02**	6.994E+02
F18	1.893E+02	4.807E+01	**3.225E+01**	4.820E+01
F19	4.091E+00	5.492E+00	**2.604E+00**	4.365E+00
F20	4.888E+01	4.376E+01	**1.423E+01**	1.540E+01
F21	2.105E+03	3.068E+02	2.219E+02	**1.471E+02**
F22	5.073E+01	**2.544E+01**	2.592E+01	3.123E+01
F23	**3.152E+02**	**3.152E+02**	**3.152E+02**	**3.152E+02**
F24	2.232E+02	2.235E+02	**2.217E+02**	2.233E+02
F25	2.027E+02	2.033E+02	**2.026E+02**	2.027E+02
F26	**1.001E+02**	**1.001E+02**	1.002E+02	1.001E+02
F27	3.038E+02	3.745E+02	**3.000E+02**	3.004E+02
F28	7.202E+02	7.169E+02	**6.494E+02**	7.500E+02
F29	8.856E+02	5.318E+02	7.195E+02	**4.822E+02**
F30	6.329E+02	9.568E+02	**5.052E+02**	7.080E+02

Var1, Var2, Var3 and DE were able to obtain the best results for 6, 3, 3, 1 test function (s), respectively. For the hybrid functions, Var2 was superior to all other variants for 5 test functions, while Var2 performed best for F22. Considering the composition functions, DE, Var1 and Var3 were able to obtain the same result in F23. Var1 and Var3 obtained the same result in F26, while DE was the best for 3 test functions, while Var3 was the best for F28, F29 and F30.

To continue our analysis, the average computational time for each variant were calculated. The computational time was calculated as the average time consumed to reach the best known solutions with an error 1.0E-08, i.e. the stopping criteria is $[f(\vec{x}) - f(\vec{x^*}) \leq 1.0E-08]$, where $f(\vec{x^*})$ is the best known solution. The summary results are shown in Table 4. From this table, Var1 was the best.

Similarly, the average numbers of fitness evaluations to reach the above mentioned stopping criterion were recorded, see Table 5. From that table, it is found that Var3 performed best.

Table 3. Average Results obtained by all variants

Prob.	DE	Var1	Var2	Var3
F01	1.160E+06	6.398E+04	3.979E+06	**3.865E+04**
F02	2.004E+03	**0.000E+00**	1.746E+07	**0.000E+00**
F03	**0.000E+00**	**0.000E+00**	3.886E+02	**0.000E+00**
F04	1.147E+02	1.719E+01	8.887E+01	**1.578E+00**
F05	2.092E+01	**2.000E+01**	2.046E+01	2.056E+01
F06	**1.132E+00**	1.593E+01	4.088E+00	2.292E+00
F07	1.335E-02	1.090E-02	5.373E-01	**5.987E-03**
F08	6.529E+01	**1.615E-04**	1.023E+01	2.000E+01
F09	1.444E+02	5.552E+01	**5.425E+01**	6.272E+01
F10	2.705E+03	**1.272E+02**	4.348E+02	7.950E+02
F11	6.140E+03	**2.782E+03**	2.974E+03	3.499E+03
F12	2.043E+00	**4.404E-01**	7.345E-01	7.981E-01
F13	2.582E-01	2.625E-01	2.561E-01	**2.535E-01**
F14	2.511E-01	**2.310E-01**	2.465E-01	2.546E-01
F15	1.301E+01	7.306E+00	**6.854E+00**	6.903E+00
F16	1.183E+01	1.130E+01	**1.036E+01**	1.103E+01
F17	8.314E+04	1.810E+03	9.054E+03	**1.403E+03**
F18	3.653E+02	1.323E+02	1.488E+02	**1.265E+02**
F19	6.492E+00	1.149E+01	7.627E+00	**6.340E+00**
F20	8.107E+01	1.743E+02	6.582E+01	**2.909E+01**
F21	3.953E+03	8.281E+02	3.149E+03	**3.964E+02**
F22	2.504E+02	1.941E+02	**1.416E+02**	1.499E+02
F23	**3.152E+02**	**3.152E+02**	3.157E+02	**3.152E+02**
F24	**2.285E+02**	2.339E+02	2.289E+02	2.290E+02
F25	**2.038E+02**	2.088E+02	2.062E+02	2.048E+02
F26	1.081E+02	**1.042E+02**	1.081E+02	**1.042E+02**
F27	**3.935E+02**	4.464E+02	4.192E+02	4.021E+02
F28	8.876E+02	9.502E+02	9.049E+02	**8.631E+02**
F29	2.155E+05	3.839E+05	1.666E+05	**8.067E+02**
F30	1.978E+03	2.388E+03	3.896E+03	**1.786E+03**

Table 4. Computational time, in seconds, of each variant

DE1	Var1	Var2	Var3
18.35	**14.58**	24.81	20.59

Table 5. Average number of fitness evaluations

DE2	Var1	Var2	Var3
289259	281778.3	287584	**280234**

To study the statistical difference between any two stochastic algorithms, a non-parametric test, Wilcoxon Signed Rank Test [25], is chosen. As a null hypothesis, it is assumed that there is no significant difference between the best and/or mean values of two samples. Whereas the alternative hypothesis is that there is a significant difference in the best and/or mean fitness values of the two samples, with a significance level of 5%. Based on the test results, one of three signs (+, −, and ≈) is assigned for the comparison of any two algorithms (shown in the last column), where the "+" sign means the first algorithm is significantly better than the second, the "−" sign means that the first algorithm is significantly worse, and the "≈" sign means that there is no significant difference between the two algorithms. The results are shown in **Table 6**. From this table, considering the best results, it is found that all variants are statistically better than DE. To add to this, Var2 is statistically better than all variants. In regards to the average results, it is interesting to find that Var3 was statically better than all variants, while there were no significant differences among other variants.

Table 6. The Wilcoxon non-parametric test among all variants, based on the best and averge fitness values obtained, where × means that no statisitcal test was applicable

Variants	Best fitness value				Average fitness value			
	DE	Var1	Var2	Var3	DE	Var1	Var2	Var3
DE	×	−	−	−	×	≈	≈	−
Var1 (DE & DE)	+	×	−	≈	≈	×	≈	−
Var2 (DE & CMA-ES)	+	+	×	+	≈	≈	×	−
Var3 (DE & GA)	+	≈	−	×	+	+	+	×

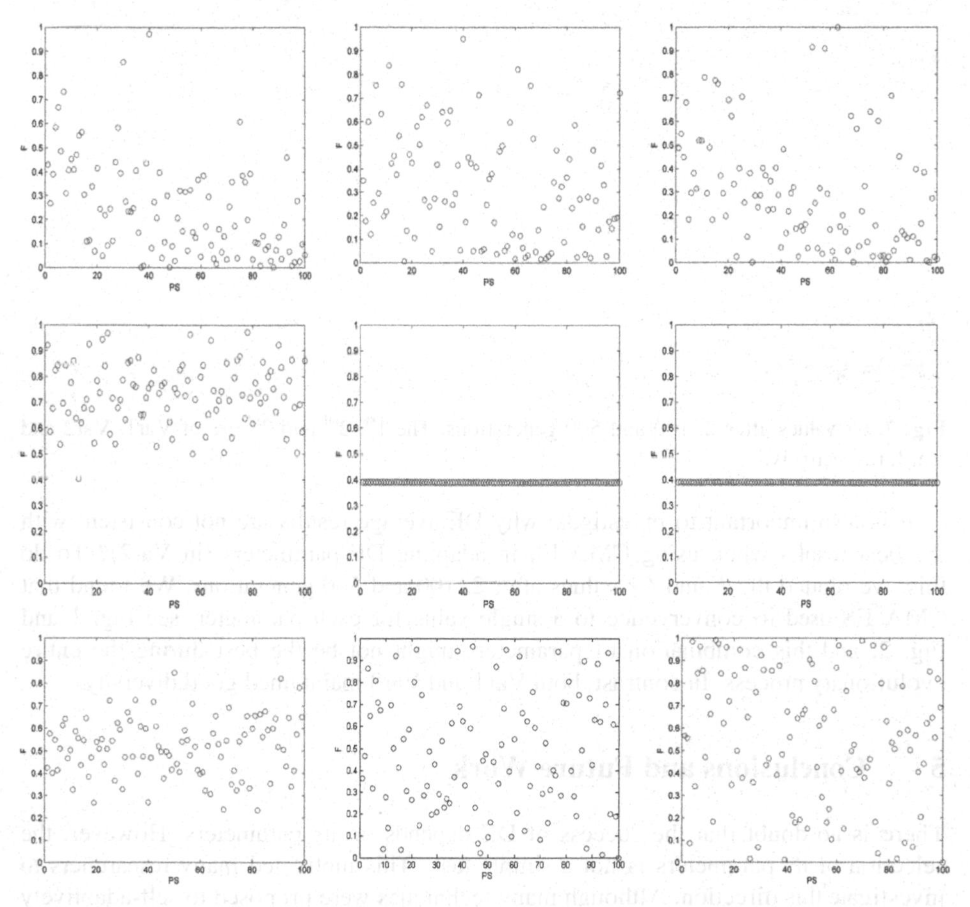

Fig. 1. F values after 2, 100 and 500 generations. The 1st, 2nd and 3rd are of Var1, Var2 and Var3, respectively.

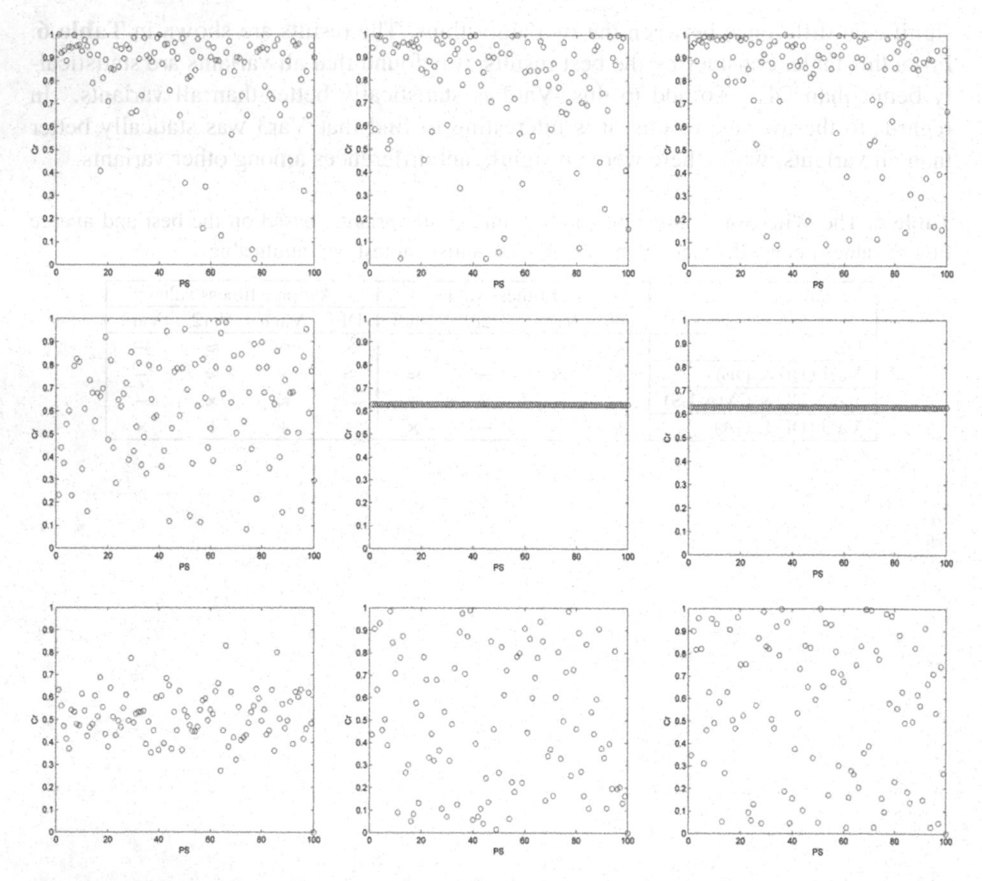

Fig. 2. Cr values after 2, 100 and 500 generations. The 1st, 2nd and 3rd are of Var1, Var2 and Var3, respectively.

It is also important to investigate why DE average results are not consistent with the best results when using CMA-ES in adapting DE parameters (in Var2)? To do this, we plotted the *F* and *Cr* values after 2, 100 and 500 generations. We found that CMA ES used to convergence to a single value for each parameter, see Fig. 1 and Fig. 2, and this combination of parameters might not be the best during the entire evolutionary process. In contrast, both Var1 and Var3 maintained good diversity.

5 Conclusions and Future Work

There is no doubt that the success of DE depends on its parameters. However, the selection of its parameters is not a simple task. This motivated many researchers to investigate this direction. Although many techniques were proposed to self-adaptively adapt its parameters, using EAs to do this process have not been fully explored. Consequently, in this paper, we compared the performance of DE when adapting its parameters using three different EAs (DE, CMA-ES and GA). From the results, we found that adapting DE's parameters using CMA-ES had the ability to obtain the best results, in terms of the best solutions obtained, in many occasion. However, its average results were inferior to the variant that considered GA.

Although, DE, with a single combination of parameters, was good in several occasions, there is a question how can we determine it? To add to this, it was interesting to find that using EAs to adapt DE's parameters might save computational time; this is because it had the ability to quickly find the optimal solutions in many occasions.

There are many open directions can be done in this direction. For instance, recent years have shown much interest in developing multi-operator DE algorithms. Doing the same trend and use multi-operator algorithms to evolve DE's parameters may be a possible future work. To add to this, providing a detailed comparison to other adaptation techniques is also important.

References

1. Goldberg, D.: Genetic Algorithms in Search, Optimization, and Machine Learning. Addison-Wesley, MA (1989)
2. Storn, R., Price, K.: Differential Evolution - A simple and efficient adaptive scheme for global optimization over continuous spaces, in Technical Report, International Computer Science Institute (1995)
3. Rechenberg, I.: Evolutions strategie: Optimierung Technischer Systeme nach Prinzipien der biologischen Evolution. Fromman-Holzboog, Stuttgart (1973)
4. Wolpert, D.H., Macready, W.G.: No free lunch theorems for optimization. IEEE Transactions on Evolutionary Computation 1(1), 67–82 (1997)
5. Abbass, H.A.: The self-adaptive Pareto differential evolution algorithm. In: IEEE Congress on Evolutionary Computation (2002)
6. Elsayed, S.M., Sarker, R.A., Essam, D.L.: Multi-operator based evolutionary algorithms for solving constrained optimization Problems. Computers and Operations Research 38(12), 1877–1896 (2011)
7. Elsayed, S.M., Sarker, R.A., Essam, D.L.: Differential evolution with multiple strategies for solving CEC2011 real-world numerical optimization problems. In: IEEE Congress on Evolutionary Computation (2011)
8. Elsayed, S.M., Sarker, R.A., Essam, D.L.: Improved genetic algorithm for constrained optimization. In: 2011 International Conference on Computer Engineering & Systems, ICCES (2011)
9. Elsayed, S.M., Sarker, R.A., Essam, D.L.: On an evolutionary approach for constrained optimization problem solving. Applied Soft Computing 12(10), 3208–3227 (2012)
10. Qin, A.K., Huang, V.L., Suganthan, P.N.: Differential Evolution Algorithm With Strategy Adaptation for Global Numerical Optimization. IEEE Transactions on Evolutionary Computation 13(2), 398–417 (2009)
11. Liu, J., Lampinen, J.: A Fuzzy Adaptive Differential Evolution Algorithm. Soft Computing - A Fusion of Foundations. Methodologies and Applications 9(6), 448–462 (2005)
12. Brest, J., Greiner, S., Boskovic, B., Mernik, M., Zumer, V.: Self-Adapting Control Parameters in Differential Evolution: A Comparative Study on Numerical Benchmark Problems. IEEE Transactions on Evolutionary Computation 10(6), 646–657 (2006)
13. Zhang, J., Sanderson, A.C.: JADE: Adaptive Differential Evolution With Optional External Archive. IEEE Transactions on Evolutionary Computation 13(5), 945–958 (2009)

14. Das, S., Konar, A., Chakraborty, U.K.: Two improved differential evolution schemes for faster global search. In: The 2005 Conference on Genetic and Evolutionary Computation, pp. 991–998. ACM, Washington, DC (2005)
15. Hansen, N., Ostermeier, A.: Completely Derandomized Self-Adaptation in Evolution Strategies. Evolutionary Computation 9(2), 159–195 (2001)
16. Storn, R., Price, K.: Differential Evolution – A Simple and Efficient Heuristic for global Optimization over Continuous Spaces. Journal of Global Optimization 11(4), 341–359 (1997)
17. Mallipeddi, R., Suganthan, P.N.: Problem definitions and evaluation criteria for the CEC, competition and special session on single objective constrained real-parameter optimization. 2010: Technical Report, Nangyang Technological University, Singapore (2010)
18. Lampinen, J., Zelinka, I.: On stagnation of the differential evolution algorithm. In: 6th Int. Mendel Conference on Soft Computing, Brno, Czech Republic (2000)
19. Vesterstrom, J., Thomsen, R.: A comparative study of differential evolution, particle swarm optimization, and evolutionary algorithms on numerical benchmark problems. In: IEEE Congress on Evolutionary Computation (2004)
20. Storn, R.: On the usage of differential evolution for function optimization. In: Biennial Conference of the North American Fuzzy Information Processing Society, NAFIPS (1996)
21. Price, K.V., Storn, R.M., Lampinen, J.A.: Differential evolution: A practical approach to global optimization. Natural Computing Series. Springer, Berlin (2005)
22. Das, S., Suganthan, P.N.: Differential Evolution: A Survey of the State-of-the-Art. IEEE Transactions on Evolutionary Computation 15(1), 4–31 (2011)
23. Elsayed, S.M., Sarker, R.A., Essam, D.L.: GA with a new multi-parent crossover for solving IEEE-CEC2011 competition problems. In: IEEE Congress on Evolutionary Computation (2011)
24. Liang, J.J., Qu, B.-Y., Suganthan, P.N.: Problem Definitions and Evaluation Criteria for the CEC, Special Session and Competition on Single Objective Real-Parameter Numerical Optimization. Computational Intelligence Laboratory and Nanyang. Technological University, China and Singapore (2014)
25. Corder, G.W., Foreman, D.I.: Nonparametric Statistics for Non-Statisticians: A Step-by-Step Approach. John Wiley, Hoboken (2009)

Individual-Based Cooperative Coevolution Local Search for Large Scale Optimization

Can Liu[1] and Bin Li[2]

[1] Nature Inspire Computation and Applications Lab (NICAL), University of Science and Technology of China, Hefei, Anhui, China
liucan@mail.ustc.edu.cn

[2] USTC-Birmingham Research Institute of Intelligent Computation and Applications (UBRI), University of Science and Technology of China, Hefei, Anhui, China
binli@ustc.edu.cn

Abstract. Decomposition methodology has been well studied and widely applied to Large Scale Global Optimization (LSGO). Cooperative Coevolution (CC) is an effective decomposition strategy and has made remarkable achievements on tackling LSGO problems. In recent studies, the role of Individual-based Local Search (ILS) has arose more and more attention, especially under the framework of Memetic Algorithms (MAs). In this paper, we investigate the validity and performance of incorporating Cooperative Coevolution strategy into Individual-based Local Search. For this purpose, a Solis and Wets' algorithm with Cooperative Coevolution (SWCC) is presented, and a comparison is made between SWCC and SW via experiments on the LSGO test suite issued in CEC'2013. Then, SWCC is embedded into Simulated Annealing algorithm (SA) and Memetic framework to investigate its effectiveness as local search operator. Experiment results show the effectiveness of SWCC on fully-separable LSGO problems and poor performance on fully non-separable problems.

Keywords: Cooperative Coevolution, Individual-based Local Search, Solis and Wets' algorithm, Simulated Annealing algorithm, Memetic Algorithms.

1 Introduction

Many real-life problems in different fields can be formulated as continuous optimization problems, and have been successfully solved with Evolutionary Algorithms (EAs) [1-4]. However, the performance of most of the available EAs deteriorates rapidly with the grown of dimensionality [5-6]. Various ideas have sprung up for solving high-dimensional problems, such as Parallel Computing, Decomposition, Screening, Mapping, Space Reduction and Visualization. Among which, decomposition methodology is most widely recognized as a promising one.

Decomposition methodology is to divide an original problem into a set of independent or coordinated sub-problems of smaller scale and has been well studied and widely applied to complex and large scale engineering problems [7-12]. Cooperative Coevolution (CC) is an implementation of divide-and-conquer idea, which has been

H. Handa et al. (eds.), *Proc. of the 18th Asia Pacific Symp. on Intell. & Evol. Systems – Vol. 1*,
Proceedings in Adaptation, Learning and Optimization 1, DOI: 10.1007/978-3-319-13359-1_41

proven an effective decomposition framework for Large Scale Global Optimization (LSGO) problems [13-14]. The procedure of Cooperative Coevolution EAs (CCEAs) can be briefly divided into three steps: First, the high-dimensional search space is decomposed into several sub-spaces with lower dimensionality. Second, each subspace is searched with a certain EA separately. Third, the obtained results of each sub-space are integrated together [16]. Most of the effective CCEAs adopted the population-based EAs as the base algorithms [16-20] due to their prominent achievement in history.

While the past experience on LSGO research imply us that the efficiency of getting better solutions, especially at the early stage of evolutionary optimization, is very important for the whole performance of the algorithm, the population-based strategy might play a non-positive role in improving the efficiency of the algorithm. In recent studies, Individual-based Local Search (ILS) methods, especially intense ILS, have shown promising ability on solving high-dimensional problems [21-23]. But, it is also shown that too quick convergence of ILS on LSGO problems, especially those with complex fitness landscapes, causes the algorithm fail to get high quality solutions.

To the best of our knowledge, the idea of incorporating CC into ILS to improve its performance on LSGO problems has not been studied carefully. In this paper, the study of the influence of CC on ILS is the main topic, rather than proposing a new algorithm. For this purpose, Cooperative Coevolution strategy is introduced into Solis and Wets' algorithm (SW), an effective ILS algorithm, to study the performance of the idea. The algorithm is named CC-based SW (SWCC). In SWCC, the decomposition strategy of multilevel CC (MLCC) [20] is adopted. The Average Fitness Increment (AFI) is selected as the performance evaluation method [15] and the decomposer is selected greedily. To investigate the performance of SWCC, it is compared with SW on test suite for algorithm competition on LSGO in CEC'2013 [31]. From the experiment results it is observed that, for most test functions, Cooperative Coevolution strategy has positive effect on the performance of ILS, especially for fully-separable functions. To further study the effectiveness of SWCC, it is embedded into a Simulated Annealing algorithm (SA) and a Memetic Algorithm (MA) as local search operator. Experiment results show the effectiveness of SWCC on fully-separable LSGO problems and poor performance on fully non-separable problems.

The rest of this paper is organized as follows: In section 2, the Individual-based Cooperative Coevolution Local Search algorithm is described in details. Experimental results and analyses are shown in section 3. Finally, the paper is concluded in section 4.

2 Individual-Based Cooperative Coevolution Local Search Algorithm

In the beginning, we review some preliminaries of CC framework. Among the whole framework of CC, the step of problem decomposition is crucial significant [17]. In the initial research, two simple decomposition methods are proposed: the one-dimensional based and splitting-in-half strategies [18-19]. The one-dimensional based strategy decomposes the high-dimensional problem into sub-components with single dimension. Without considering the dependence of each dimensionality, it cannot solve non-separable problems effectively. The splitting-in-half strategy decomposes the high-dimensional problem into two equal components. If the dimensionality of the problem is very large, this strategy cannot achieve the goal of decomposition. The strategy of random grouping for problem decomposition is proposed in [16]. This method divides the high-dimensional problem into several groups randomly according to predefined group sizes. However, it is hard to select the optimal group size for different problems and search stages. As to this consideration, a multilevel CC framework (MLCC) is presented in [20]. In MLCC, several problem decomposers based on different group sizes are designed to form a decomposer pool and a decomposer is selected from the pool according to their performance records. In this paper, the decomposition strategy of MLCC is adopted.

2.1 Cooperative Coevolution-Based Solis and Wets' Algorithm

Solis and Wets' algorithm (SW) [24] is selected as the individual-based Local search method, whose performance has been verified in many large scale optimization algorithm [22-23]. The classic SW is a randomized hill-climber with an adaptive and scalable step size. SW starts with a single point x and a deviation d is generated from a normal distribution, whose standard deviation is given by a parameter σ and mathematical expectation is 0. If either $x+d$ or $x-d$ is better, x is replaced by the better point and a record of success is made. Otherwise x remains the same and a record of failure is made. When the record of success exceeds the threshold (MaxSuccesses), σ is increased to obtain a larger step size and reset the record of success. Similarly, when the record of failure exceeds the threshold (MaxFailures), σ is decreased to obtain a smaller step size and reset the record of failure. The pseudo of SW [24] is presented in Fig. 1.

In this paper, CC strategy is incorporated into SW to form Cooperative Coevolution Solis and Wets' algorithm (SWCC). In SWCC, a decomposer pool $S=\{s_1,\cdots,s_t\}$ and the corresponding local search intensity $L=\{l_1,\cdots,l_t\}$ is designed first. s_i means that the high-dimensional problem will be divided into several groups and each group has s_i dimensions. l_i is the number of fitness evolution for each group when the group size is s_i. Then, performance record $R=\{r_1,\cdots,r_t\}$ is initialized by calculating the Average Fitness Increment (AFI) [15] of each decomposer. AFI is the performance evaluation method and its definition is as follows [15]:

$$\mathrm{AFI}=\frac{\Delta Fitness}{\Delta FEs} \tag{1}$$

In (1), $\Delta Fitness$ is the fitness increment of the selected decomposer, and ΔFEs is the consumption of the number of fitness evaluations. At the end of each search cycle, r_i is updated according to (1). In the next search cycle, the decomposer is selected greedily according to the performance record. The selected decomposer has the maximum AFI value. Thus, SWCC is able to select the optimal decomposer for different problems and search stages. The details of SWCC are presented in Fig. 2.

2.2 Simulated Annealing Embedded with SWCC

Simulated Annealing algorithm (SA) is a kind of heuristic random search algorithm based on Monte Carlo Simulation, which derived from the thermodynamics of the simulated annealing process [25]. SA not only accepts better solutions, but also accepts worse solutions stochastically according to Metropolis sample rule, which is the fundamental difference between SA and hill climbing algorithm. In this paper, SWCC, the method of generating new solutions, is embedded into SA (SA-SWCC), whose pseudo code is shown in Fig. 3.

SW

Initialization:

```
Set FEs=0, Max_FEs = 3e6, D, runs=25, Success=0, Failures=0
Set maxSuccess=3, maxFailures=5, adjustSuccess=4, adjustFailures=0.75, deviation=5
Initialize individual x and Calculate val(x)
```

Optimization Procedure:

```
while FEs< Max_FEs
   d=deviation.*randn(1,D);
   x1=x+d; x2=x-d;
   Calculate val(x1) and val(x2); FEs=FEs+2;
   [val(x3), x3]=min(val(x1), val(x2));
   if val(x3)<val(x)
      x=x3; val(x)=val(x3); Success++; Failures=0;
   else
      Failures++; Success=0;
   end if
   if Success>maxSuccess
      deviation=deviation*adjustSuccess; Success=0;
   end if
   if Failures>maxFailures
      deviation=deviation*adjustFailures; Failures=0;
   end if
end while
Return the best solution;
```

Fig. 1. Pseudo code of SW [24]

```
SWCC
Initialization:
  Set Max_FEs = 3e6, D, decomposer pool S =[1 10 20 50], local search intensity L=[10 20 50 100]
  Set maxSuccess=3, maxFailures=5, adjustSuccess=4, adjustFailures=0.75, deviation=5
  Initialize average fitness increment AFI, individual x and Calculate val(x)
Optimization Procedure:
  while FEs< Max_FEs
    Select the decomposer with maximal AFI, group size sub_dim and sub_Max_FEs;
    sub_num=D/sun_dim; index=randperm(D);
    for i=1:sub_num
      Set sub_FEs=0, Success=0 and Failures=0;
      while sub_FEs< sub_Max_FEs
        v1=(i-1)*sub_dim+1; v2=i*sub_dim;
        Conduct the optimization procedure of SW in dimensions of index(v1:v2);
      end while
    end for
    Update the value of AFI ;
  end while
  Return the best solution;
```

Fig. 2. Pseudo code of SWCC

```
SA-SW/SWCC
Initialization:
  Set Max_FEs = 3e6, D, decomposer pool S =[1 10 20 50], local search intensity L=[10 20 50 100]
  Set maxSuccess=3, maxFailures=5, adjustSuccess=4, adjustFailures=0.75, deviation=5
  Set initial temperature T=10000, inner loop lk, cooling ratio alpha=0.95
  Initialize average fitness increment AFI, individual x and Calculate val(x)
Optimization Procedure:
  while FEs< Max_FEs
    for k=1:lk
      Conduct SW/SWCC to generate new solution x_new and calculate val(x_new);
      Δval=val(x_new)-val(x);
      if Δval<0
        x=x_new; val(x)=val(x_new);
      else
        p=exp(-Δval/T);
        if rand<p
          x=x_new; val(x)=val(x_new);
        end if
      end if
    end for
    T=T*alpha;
  end while
  Return the best solution;
```

Fig. 3. Pseudo code of SA-SW/SWCC

2.3 Memetic Algorithm Embedded with SWCC

Memetic Algorithms (MAs) are a kind of population-based stochastic heuristics composed of an evolutionary framework accompanied with a set of specific Local Search (LS) operators [26]. The earliest formal definition has been presented in [27]. In recent decades, MAs have been used to solve various optimization problems and shown good performances, therefore have been becoming a popular idea for various engineering optimization tasks [28-30]. In this paper, SWCC as the local search method, is embedded into MA (MA-SWCC). In MA-SWCC, the Differential Search Algorithm (DSA) [32-33] is adopted as the global search operator and the best individual of its population is selected as the initial individual of SWCC. In each search cycle, SWCC is applied and the local search depth is a predetermined and empirical value. The pseudo of MA-SWCC is shown in Fig. 4.

```
MA-SW/SWCC
Initialization:
  Set Max_FEs = 3e6, D, decomposer pool S = [1 10 20 50], local search intensity L= [10 20 50 100],
  Set NP=30, Max_FEs_SW=500
  Initialize average fitness increment AFI, population pop and Calculate val(pop)
Optimization Procedure:
  while FEs< Max_FEs
     Apply DSA as the global search:
     Generate the matrix of map and step size R= 1/gamrnd(1,0.5);
     Conduct mutation operator to generate new population stopover;
     Select the better individuals greedily from pop and stopover to form the new population pop_new;
     Select the best individual gbest as the initial individual of SWCC;
     while FEs_SW <= Max_FEs_SW
         Conduct the optimization procedure of SW/SWCC to update gbest;
     end while
  end while
  Return the best solution;
```

Fig. 4. Pseudo code of MA-SW/SWCC

3 Experimental Studies

For investigating the effectiveness of SWCC, the specific test-suite proposed in the special session on Large Scale Continuous Global Optimization in CEC'2013 is selected. The test-suit has fifteen LSGO benchmark functions, which can be classified into four categories: Fully-separable Functions, Partially Additively Separable Functions, Overlapping Functions and Fully Non-separable Functions. Detailed information on the test-suit can be found in [31]. The dimension of each function is 1000 except F13, F14, whose dimensions are 905, the maximal Fitness Evaluation Size (FES) is set to 3×10^6. Each algorithm runs 25 times on each test function independently, and the best, worst, median, mean and standard derivation are computed.

3.1 Effectiveness of Incorporating CC into SW

The parameters setting of SW is as follow: deviation d=5, the threshold of success record (MaxSuccesses) is 3 and the corresponding adjustment coefficient is 4, the threshold of failure record (MaxFailures) is 5 and the corresponding adjustment coefficient is 0.75. As to MLCC, the pool of group sizes is set to {1, 10, 20, 50} and the corresponding Fitness Evaluation numbers (FEs) are set to {10, 20, 50, 100}. The above parameters are selected by experience [20], [23].

Table 1 contains the results of SW and SWCC on the consider test suit. From these results, we can make the following observation: For functions F1, F2, F3, F4, F7, F11, F13 and F14, SWCC performs better than SW. However, for functions F8 and F15, SWCC obtains the worse results. For the rest of the functions F5, F6, F9, F10 and F12, SWCC and SW achieve almost the same results. SWCC achieves the significant advantage on fully-separable Functions F1, F2 and F3, especially for F1, which indicates that Cooperative Coevolution strategy is suitable to solve fully-separable problems. SWCC performs quite poor on Function F15, which is a fully non-separable function. It is common recognized that decomposition strategy would not work well on such class of problems. It can be observed that along with the decrease of the separable degree of problems, that is fully-separable functions, functions with a separable subcomponent and fully non-separable function, the performance of both SW and SWCC deteriorates more and more. However, SWCC owns the larger extent.

The convergence graphs of SW and SWCC are also presented. For each functions, a single convergence curve has been plotted using the average results of 25 independent executions. From the results, we can obviously observe that these graphs can be divided into four categories. To study the characteristics, the graphs of eight typical functions (F1, F2, F5, F7, F9, F10, F14 and F15) are provided in Fig. 5. For fully-separable functions F1 and F2, the convergence curves of SWCC outperform than SW in the whole optimization procedure. However, for functions F10 and F15, the convergence curves of SW and SWCC have the exactly opposite results. Therefore, it is not a good choice to improve the performance of local search with Cooperative Coevolution strategy for these kind of functions. For functions F7 and F14, the curves of SW achieve a bigger convergence ratio in the early stage and flatten out then. In contrast, SWCC continues to improve the result in the whole stage until reaching the maximum number of fitness evaluations and obtains the better solution finally. For functions F5 and F9, the curves of SWCC and SW are almost overlapping. Therefore, Cooperative Coevolution strategy almost has no influence on this category of functions.

Table 1. Results of SW and SWCC

		F1	F2	F3	F4	F5	F6	F7	F8
SW	Best	1.08e+07	3.88e+04	2.11e+01	4.17e+09	1.73e+07	9.96e+05	2.79e+06	2.01e+13
	Median	1.23e+07	4.60e+04	2.13e+01	4.92e+09	**2.24e+07**	**9.98e+05**	2.92e+06	**3.01e+13**
	Worst	1.43e+07	5.16e+04	2.14e+01	5.98e+09	4.33e+07	1.03e+06	3.16e+06	4.59e+13
	Mean	1.22e+07	4.58e+04	2.13e+01	4.90e+09	**2.37e+07**	**1.00e+06**	2.95e+06	**3.14e+13**
	Std	7.73e+05	3.21e+03	7.26e-02	4.29e+08	5.87e+06	7.15e+03	1.08e+05	6.48e+12
SWCC	Best	1.24e-22	4.03e+02	2.00e+01	1.07e+09	1.46e+07	9.96e+05	4.93e+04	5.44e+13
	Median	**1.01e-21**	**9.63e+02**	**2.00e+01**	**3.11e+09**	**2.42e+07**	**1.01e+06**	**7.00e+04**	1.04e+14
	Worst	6.94e-21	1.37e+03	2.00e+01	7.90e+09	3.81e+07	1.06e+06	1.43e+05	2.35e+14
	Mean	**1.53e-21**	**9.50e+02**	**2.00e+01**	**3.32e+09**	**2.48e+07**	**1.02e+06**	**7.96e+04**	1.13e+14
	Std	1.54e-21	2.68e+02	0.00e+00	1.51e+09	5.23e+06	1.92e+04	3.28e+04	4.54e+13
		F9	F10	F11	F12	F13	F14	F15	
SW	Best	1.42e+09	9.06e+07	5.90e+07	1.18e+03	7.53e+06	7.73e+07	1.88e+06	
	Median	**1.75e+09**	**9.17e+07**	7.56e+07	**1.45e+03**	9.97e+06	9.48e+07	**2.34e+06**	
	Worst	2.31e+09	9.48e+07	1.19e+08	1.95e+03	1.26e+07	1.38e+08	2.96e+06	
	Mean	**1.81e+09**	**9.20e+07**	7.79e+07	**1.46e+03**	1.01e+07	9.79e+07	**2.34e+06**	
	Std	2.48e+08	1.33e+06	1.39e+07	2.00e+02	1.23e+06	1.30e+07	2.60e+05	
SWCC	Best	8.86e+08	9.20e+07	2.40e+06	7.22e+02	2.38e+06	9.14e+06	1.35e+10	
	Median	**1.86e+09**	**9.38e+07**	**7.24e+06**	**1.12e+03**	**4.28e+06**	**1.30e+07**	6.08e+10	
	Worst	2.61e+09	9.54e+07	1.67e+07	1.94e+03	6.58e+06	2.21e+07	1.11e+12	
	Mean	**1.84e+09**	**9.37e+07**	**8.07e+06**	**1.16e+03**	**4.58e+06**	**1.35e+07**	2.47e+11	
	Std	4.00e+08	1.00e+06	3.16e+06	2.42e+02	1.15e+06	3.30e+06	3.20e+11	

Table 2. Results of SA-SW and SA-SWCC

		F1	F2	F3	F4	F5	F6	F7	F8
SA-SW	Best	1.05e+07	7.61e+03	2.16e+01	4.01e+09	1.66e+07	9.96e+05	2.68e+06	2.28e+13
	Median	1.22e+07	8.26e+03	2.16e+01	4.67e+09	**2.43e+07**	**9.97e+05**	2.88e+06	**3.30e+13**
	Worst	1.37e+07	9.05e+03	2.16e+01	5.73e+09	3.68e+07	1.02e+06	3.06e+06	6.09e+13
	Mean	1.22e+07	8.35e+03	2.16e+01	4.74e+09	**2.44e+07**	**1.00e+06**	2.87e+06	**3.38e+13**
	Std	8.62e+05	4.00e+02	5.60e-03	4.85e+08	4.55e+06	6.45e+03	8.03e+04	8.54e+12
SA-SWCC	Best	2.00e+03	7.00e+01	2.03e+01	1.44e+09	1.48e+07	1.05e+06	3.88e+04	1.00e+14
	Median	**1.37e+04**	**8.94e+01**	**2.04e+01**	**2.28e+09**	**2.49e+07**	**1.06e+06**	**8.14e+04**	1.84e+14
	Worst	7.48e+04	4.25e+03	2.10e+01	7.38e+09	3.71e+07	1.06e+06	1.65e+05	3.47e+14
	Mean	**2.02e+04**	**2.55e+02**	**2.07e+01**	**2.67e+09**	**2.60e+07**	**1.06e+06**	**8.64e+04**	2.02e+14
	Std	1.77e+04	8.32e+02	3.37e-01	1.54e+09	7.13e+06	3.59e+03	3.15e+04	7.78e+13
		F9	F10	F11	F12	F13	F14	F15	
SA-SW	Best	1.23e+09	9.06e+07	6.13e+07	1.04e+03	8.85e+06	7.91e+07	1.81e+06	
	Median	**1.84e+09**	**9.10e+07**	8.38e+07	**1.04e+03**	1.01e+07	9.79e+07	**2.34e+06**	
	Worst	2.14e+09	9.44e+07	9.55e+07	1.56e+03	1.26e+07	1.19e+08	2.85e+06	
	Mean	**1.78e+09**	**9.17e+07**	7.74e+07	**1.09e+03**	1.03e+07	9.81e+07	**2.33e+06**	
	Std	3.00e+08	1.29e+06	1.37e+07	1.16e+02	1.09e+06	9.24e+06	2.59e+05	
SA-SWCC	Best	1.35e+09	9.07e+07	2.81e+06	1.01e+03	2.30e+06	9.10e+06	2.28e+06	
	Median	**1.92e+09**	**9.41e+07**	**7.89e+06**	**1.81e+03**	**4.59e+06**	**1.18e+07**	4.24e+10	
	Worst	2.35e+09	9.50e+07	3.46e+07	3.03e+03	8.31e+06	1.80e+07	7.32e+11	
	Mean	**1.88e+09**	**9.36e+07**	**9.49e+06**	**2.02e+03**	**4.47e+06**	**1.24e+07**	1.49e+11	
	Std	2.83e+08	1.22e+06	6.75e+06	5.40e+02	1.40e+06	2.06e+06	1.91e+11	

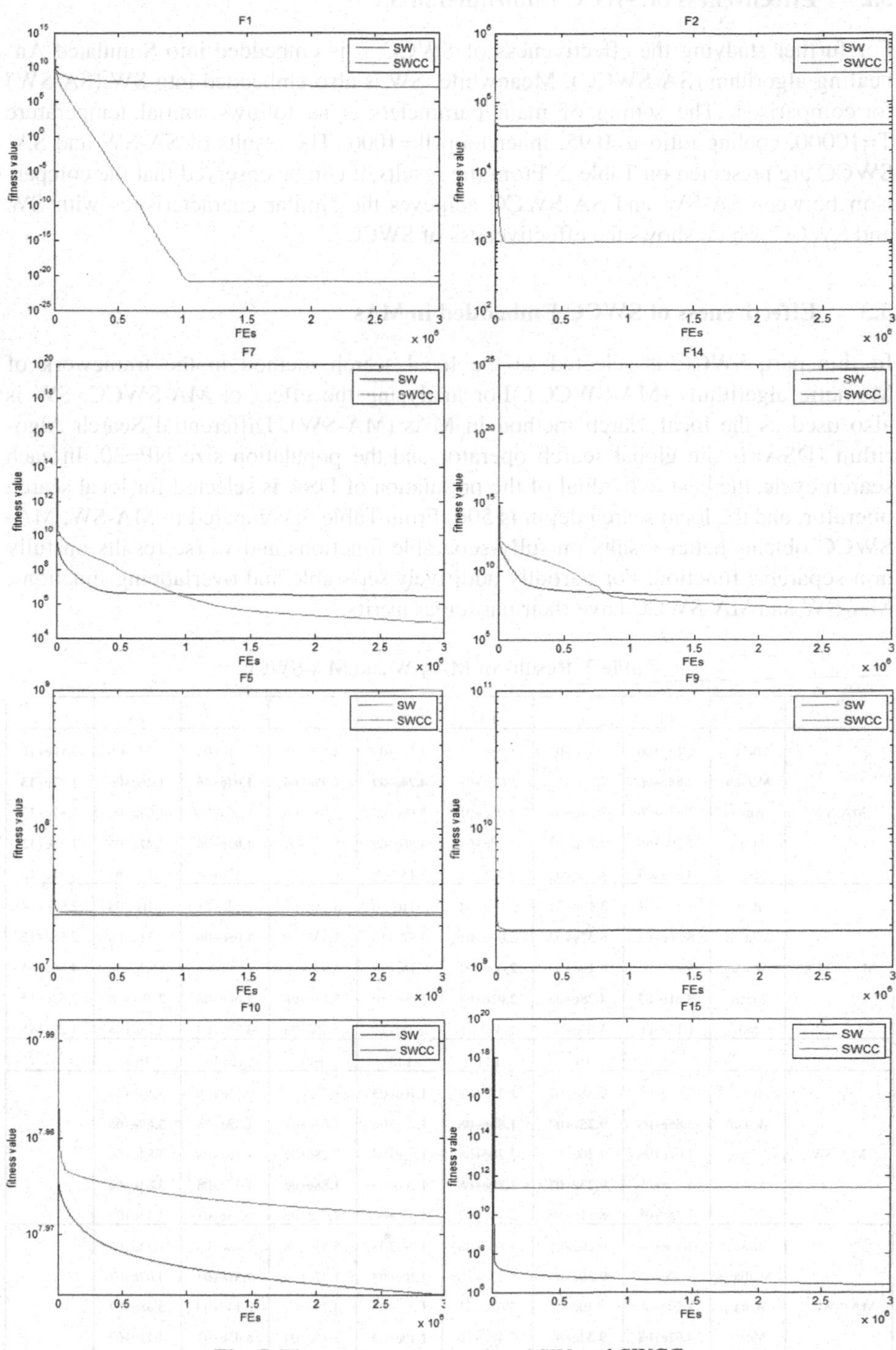

Fig. 5. The convergence graphs of SW and SWCC

3.2 Effectiveness of SWCC Embedded in SA

For further studying the effectiveness of SWCC, it is embedded into Simulated Annealing algorithm (SA-SWCC). Meanwhile, SW is also embedded into SW (SA-SW) for comparison. The setting of main parameters is as follows: initial temperature T=10000, cooling ratio α=0.95, inner loop lk=1000. The results of SA-SW and SA-SWCC are presented on Table 2. From the results, it can be observed that the comparison between SA-SW and SA-SWCC achieves the similar characteristics with SW and SWCC, which shows the effectiveness of SWCC.

3.3 Effectiveness of SWCC Embedded in MAs

In this part, SWCC is selected as the local search method in the framework of Memetic algorithms (MA-SWCC). For analyzing the effect of MA-SWCC, SW is also used as the local search method in MAs (MA-SW). Differential Search Algorithm (DSA) is the global search operator and the population size NP=30. In each search cycle, the best individual of the population of DSA is selected for local search operator, and the local search depth is 500. From Table 3, compared to MA-SW, MA-SWCC obtains better results on fully-separable functions and worse results on fully non-separable function. For partially additively separable and overlapping functions, MA-SW and MA-SWCC have their particular merits.

Table 3. Results of MA-SW and MA-SWCC

		F1	F2	F3	F4	F5	F6	F7	F8
MA-SW	Best	2.17e+06	6.47e+03	2.06e+01	4.12e+09	3.59e+06	1.01e+06	1.27e+06	7.16e+12
	Median	2.88e+06	7.26e+03	2.07e+01	**4.74e+09**	**4.99e+06**	**1.04e+06**	**1.46e+06**	**1.77e+13**
	Worst	3.63e+06	8.24e+03	2.08e+01	5.89e+09	9.19e+06	1.06e+06	1.64e+06	2.87e+13
	Mean	2.79e+06	7.27e+03	2.07e+01	**4.86e+09**	5.96e+06	**1.03e+06**	**1.47e+06**	**1.78e+13**
	Std	3.64e+05	4.37e+02	3.42e-02	5.15e+08	1.96e+06	1.52e+04	1.13e+05	5.37e+12
MA-SWCC	Best	3.97e+02	3.10e+01	2.00e+01	2.11e+10	3.18e+06	1.05e+06	4.10e+05	2.67e+14
	Median	**5.36e+02**	**6.37e+03**	**2.00e+01**	3.97e+10	5.43e+06	**1.05e+06**	1.31e+08	2.55e+15
	Worst	5.77e+06	8.19e+03	2.06e+01	9.16e+10	8.93e+06	1.06e+06	8.37e+08	4.86e+15
	Mean	**3.31e+05**	**4.28e+03**	**2.02e+01**	5.00e+10	**5.61e+06**	**1.06e+06**	2.39e+08	2.62e+15
	Std	1.17e+06	3.60e+03	2.35e-01	2.27e+10	1.46e+06	3.36e+03	2.71e+08	1.29e+15
		F9	F10	F11	F12	F13	F14	F15	
MA-SW	Best	2.67e+08	9.08e+07	1.32e+08	1.10e+03	1.55e+07	1.05e+08	3.18e+06	
	Median	**3.85e+08**	**9.25e+07**	**1.80e+08**	1.22e+03	**1.87e+07**	1.33e+08	**3.69e+06**	
	Worst	7.07e+08	9.40e+07	2.20e+08	1.63e+03	2.28e+07	1.76e+08	4.84e+06	
	Mean	4.25e+08	**9.23e+07**	**1.80e+08**	1.26e+03	**1.88e+07**	1.37e+08	**3.81e+06**	
	Std	1.25e+08	8.11e+05	2.46e+07	1.45e+02	2.00e+06	1.73e+07	4.44e+05	
MA-SWCC	Best	2.53e+08	9.20e+07	1.86e+08	1.04e+03	5.53e+08	9.48e+06	1.35e+07	
	Median	4.00e+08	**9.30e+07**	5.25e+10	**1.10e+03**	1.77e+09	**7.31e+07**	4.07e+07	
	Worst	6.20e+08	9.40e+07	3.62e+11	1.20e+03	4.79e+09	2.14e+08	5.48e+07	
	Mean	**4.07e+08**	**9.32e+07**	7.43e+10	**1.10e+03**	2.13e+09	**8.05e+07**	4.11e+07	
	Std	1.08e+08	5.39e+05	7.59e+10	4.93e+01	1.21e+09	6.93e+06	1.08e+07	

4 Conclusions

In this paper, Individual-based Cooperative Coevolution Local Search (ILS-CC) is investigated to study the effect of Cooperative Coevolution strategy on Individual-based Local Search for solving LSGO problems. A CC-based Solis and Wets' algorithm (SWCC) is presented, and is embedded into a Simulated Annealing (SA) and a Memetic Algorithm (MA) to study the effect of ILS-CC comprehensively. The experiment results show the effectiveness of SWCC on fully-separable LSGO problems and poor performance on fully non-separable problems.

Acknowledgment. The work was supported by the National Natural Science Foundation of China under Grant No. 61473271 and 61331015.

References

1. Vesterstrom, J., Thomsen, R.: A comparative study of differentialevolution, particle swarm optimization, and evolutionary algorithms onnumerical benchmark problems. In: Proc. Congr. Evol. Comput., vol. 2, pp. 1980–1987 (2004)
2. Storn, R.M., Price, K.V.: Differential Evolution – A Simple and Efficient Adaptive Scheme for GlobalOptimization over Continuous Spaces. International Computer Science Institute, Berkely, CA, USA, Tech. Rep.TR-95-012 (1995)
3. Posik, P.: Real-parameter optimization using the mutation stepco-evolution. In: Proceedings of 2005 IEEE Congress on Evol. Comput., pp. 872–879 (2005)
4. Price, K.V., Storn, R.M., Lampinen, J.A.: Differential Evolution – A Practical Approach to Global Optimization. Natural Computing Series. Springer, New York (2005)
5. Bellman, R.E.: Dynamic Programming. Dover Books on Mathematics. Princeton University Press (1957)
6. Liu, Y., Yao, X., Zhao, Q., Higuchi, T.: Scaling up fast evolutionary programming with cooperative coevolution. In: Proc. of IEEE Congress on Evolutionary Computation, pp. 1101–1108 (2001)
7. Altus, S.S., Kroo, I.M., Gage, P.J.: A genetic algorithm for schedulingand decomposition of multidisciplinary design problems. ASME J. Mech. Des. 118, 486–489 (1996)
8. Chen, L., Li, S.: Analysis of decomposability and complexity fordesign problems in the context of decomposition. ASME J. Mech. Des. 127, 545–557 (2005)
9. Kusiak, A., Wang, J.: Decomposition of the design process. ASME J. Mech. Des. 115, 687–693 (1993)
10. Michelena, N.F., Yapalambros, P.: A network reliabilityapproach to optimal decomposition of design problems. ASME J. Mech. Des. 117, 433–440 (1995)
11. Wang, Y., Li, B.: Two-stage based Ensemble Optimization for Large-Scale Global Optimization. In: Proc. the 2010 IEEE Congress on Evolutionary Computation (CEC 2010), Barcelona, pp. 4488–4495 (2010)
12. Zhang, K.B., Li, B.: Cooperative Coevolution with Global Search for Large Scale Global Optimization. In: WCCI 2012 IEEE World Congress on Computational Intelligence, Brisbane, Australia, pp. 10–15 (June 2012)

13. Potter, M.: The Design and Analysis of a Computational Model of CooperativeCoevolution. Ph.D. dissertation, George Mason University (1997)
14. Zhao, S.Z., Liang, J.J., Suganthan, P.N., Tasgetiren, M.F.: Dynamic Multi-Swarm Particle Swarm Optimizer with Local Searchfor Large Scale Global Optimization. In: Proceedings of the 10th IEEE Congresson Evolutionary Computation, pp. 3845–3852. IEEE Press (June 2008)
15. Vanneschi, L., Tomassini, M., Collard, P., Vérel, S.: Negative slope coefficient: A measure to characterize genetic programming fitness landscapes. In: Collet, P., Tomassini, M., Ebner, M., Gustafson, S., Ekárt, A. (eds.) EuroGP 2006. LNCS, vol. 3905, pp. 178–189. Springer, Heidelberg (2006)
16. Yang, Z., Tang, K., Yao, X.: Large Scale Evolutionary Optimization Using Cooperative Coevolution. Information Sciences 178(15), 2985–2999 (2008)
17. Potter, M., De Jong, K.: Cooperative Coevolution: An Architecturefor Evolving Coadapted Subcomponents. Evolutionary Computation 8(1), 1–29 (2000)
18. Liu, Y., Yao, X., Zhao, Q., Higuchi, T.: Scaling up Fast Evolutionary Programming with Cooperative Coevolution. In: Proceedings of the 2001 Congress on Evolutionary Computation, pp. 1101–1108 (2001)
19. Shi, Y.-j., Teng, H.-f., Li, Z.-q.: Cooperative co-evolutionary differential evolution for function optimization. In: Wang, L., Chen, K., S. Ong, Y. (eds.) ICNC 2005. LNCS, vol. 3611, pp. 1080–1088. Springer, Heidelberg (2005)
20. Yang, Z., Tang, K., Yao, X.: Multilevel cooperative coevolution for large scale optimization. In: 2008 IEEE Congress on Evolutionary Computation, pp. 1663–1670 (2008)
21. Tseng, L.Y., Chen, C.: Multiple Trajectory Search for LargeScale Global Optimization. In: Proceedings of the 10th IEEE Congress on Evolutionary Computation, CEC 2008, pp. 3052–3059. IEEE Press (June 2008)
22. Molina, D., Lozano, M., Herrera, F.: MA-SW-Chains: MemeticAlgorithm Based on Local Search Chains for Large Scale Continuous Global Optimization. In: Proceedings of the 2010 IEEE Congress on Evolutionary Computation, CEC 2010, pp. 1–8 (2010)
23. LaTorre, A., Muelas, S., Pefia, J.-M.: Multiple Offspring Sampling In Large Scale Global Optimization. In: WCCI 2012 IEEE World Congress on Computational Intelligence, Brisbane, Australia, pp. 10–15 (June 2012)
24. Solis, F.J., Wets, R.J.: Minimization by random search techniques. Mathematical Operations Research 6, 19–30 (1981)
25. Kirkpatrick, S.: Optimization by Simulated Annealing. Science 220, 671–680 (1983)
26. Hart, W., Krasnogor, N., Smith, J.E.: MemeticEvolutionary Algorithms. Studies in Fuzziness and Soft Computing 166, 3–27 (2005)
27. Moscato, P.: On Evolution, Search, Optimization, Genetic Algorithms and Martial Arts. Toward memetic algorithms. Tech. Rep. 826, California Institute of Technology (1989)
28. Mei, Y., Tang, K., Yao, X.: Decomposition-Based Memetic Algorithm for Multiobjective Capacitated Arc Routing Problem. IEEE Trans. Evol. Comput. 15(2), 151–165 (2011)
29. Ahn, Y., Park, J., Lee, C.-G., Kim, J.-W.: Novel Memetic Algorithm implemented With GA (Genetic Algorithm) and MADS (Mesh Adaptive Direct Search) for Optimal Design of Electromagnetic System. IEEE Trans Magnetics 46(6), 1982–1985 (2010)
30. Li, B., Zhou, Z., Zou, W., Li, D.: Quantum Memetic Evolutionary Algorithm-Based Low-Complexity Signal Detection for Underwater Acoustic Sensor Networks. IEEE Trans. Systems, Man, and Cybernetics, Part C: Applications and Reviews 42(5), 626–640 (2012)

31. Li, X., Tang, K., Omidvar, M., Yang, Z., Qin, K.: Benchmark Functions for the CEC'2013 Special Session and Competition on LargeScale Global Optimization, Technical Report, Evolutionary Computationand Machine Learning Group, RMIT University, Australia (2013)
32. Civicioglu, P.: Transforming Geocentric Cartesian Coordinates to Geodetic Coordinates by Using Differential Search Algorithm. Computers and Geosciences 46, 229–247 (2012)
33. Civicioglu, P.: http://www.pinarcivicioglu.com/ds.html (accessed October 02, 2011)

Action Recognition Using Hierarchical Independent Subspace Analysis with Trajectory

Vinh D. Luong, Lipo Wang, and Gaoxi Xiao

School of Electrical and Electronic Engineering
Nanyang Technological University
Singapore
ducvinh001@e.ntu.edu.sg, {ELPWang,EGXXiao}@ntu.edu.sg

Abstract. Action recognition in videos is an important and challenging problem in computer vision. One of the most crucial aspects of a successful action recognition system is its feature extraction component. Stacked, convolutional Independent Subspace Analysis (SC-ISA), has the best result among unsupervised learning algorithms for action recognition in Hollywood 2 (53.3%) and Youtube (75.8%). However, its performance still lags behind the current state-of-the-art, which uses computer vision-based feature engineering extraction techniques, by about 10%. In this paper, we improve SC-ISA's results by incorporating motion information into SC-ISA. By extracting blocks following motion trajectories in videos, we are able to reduce noise and increase the number of training samples without degrading the network's performance when training and testing SC-ISA. We increase SC-ISA's result by about 1%.

1 Introduction

Researchers in the field of action recognition in videos have made remarkable progress recently. As observed from the dataset aspect of the problem, the field has advanced rapidly from the limited, constrained datasets like the KTH dataset [1] to the more realistic and more challenging ones, e.g. Hollywood 2 [2], to large-scale, "in the wild" datasets such as HMDB 51 [3], UCF 101 [4], Sports-1M dataset [5]. Conventional computer vision-based techniques are currently the best methods [6], [7] to extract local, low level features for action recognition systems.

Deep learning has been a great success in object detection, localization and classification in images. In the supervised learning front, convolutional neural networks (CNN) [8] are currently the state-of-the-art in these tasks [9], [10]. As for unsupervised learning, large-scale networks have also made remarkable results such as the automatic emergence of human face and cat face detectors in the famous Google network [11]. However, in the problem of action recognition in videos, deep networks have not enjoyed such stunning progress. Unsupervised deep networks [12], [13] currently lag behind the current state-of-the-art [7] in

H. Handa et al. (eds.), *Proc. of the 18th Asia Pacific Symp. on Intell. & Evol. Systems – Vol. 1*,
Proceedings in Adaptation, Learning and Optimization 1, DOI: 10.1007/978-3-319-13359-1_42

relatively small but challenging datasets such as Hollywood 2. Furthermore, unsupervised networks have not been scaled up to tackle bigger datasets such HMDB 51 and UCF 101. Until recently, supervised networks [5] were also far behind the state-of-the-art [7]. This is really puzzling because deep networks with all the sophisticated learning algorithms have not only succeeded in the image domain but also they have made big improvements in speech recognition, a temporal domain. One possible answer is that, deep networks have not been able to incorporate motion information effectively into their network. In this paper, we will explicitly include motion information in training and testing SC-ISA network [12], which is also the current state-of-the-art for unsupervised learning algorithms in action recognition.

2 Review of Related Works

2.1 A Common Action Recognition Framework

In this paper, we limit our scope to action recognition systems that deal with local features because the local methods are the most dominant and the most accurate algorithms in the field at the moment. For global features and more comprehensive surveys of action recognition, readers should refer to [14], [15] and [16].

A common framework in action recognition for local features is as followed. First, features are extracted from training videos. Then, these features are quantized in some dictionaries using clustering and feature encoding methods such as k-means, Fisher vector [17] and VLAD [18]. After that, each video is encoded into vectors using the resulted dictionaries. Finally, a classifier, e.g. SVM, is used to train and test the videos.

2.2 Improved Trajectories

Wang et al. [6], [7], [19] explicitly reduce camera motion and use a dense optical flow algorithm [20] to track densely sampled points. They then compute Histograms of Oriented Gradients (HOG), Histograms of Optical Flow (HOF) [21], HOG/HOF [22] and Motion Boundary Histogram (MBH) [23] descriptors following the computed trajectories. They also introduce trajectory descriptor as normalized trajectory displacement. Using these newly computed descriptors, they achieve the best results in almost all datasets they test with, including challenging ones such as Hollywood 2 [2] (64.3%), Youtube [24] (85.4%) UCF-101 [4] (85.9%), HMDB51 [3] (57.2%).

2.3 Independent Subspace Analysis

Independent Subspace Analysis: Independent Subspace Analysis (ISA) is an unsupervised learning algorithm that models complex cells in V1 [25], [26]. A complex cell fires almost the same response to a grating regardless of the grating's phase. By putting linear features into groups (subspaces), features learned by ISA are able to display limited phase invariance as well as selectivity to frequency and orientation. The combination of these features is what distinguishes ISA from linear methods such as Independent Component Analysis (ICA) [27], [26] in modeling cells in V1.

ISA originates from ICA. In ICA, an input vector $\mathbf{z}^1$, which, in our case, is resulted from the whitening preprocessing of a 3-D video block as linear combination of basis vectors (features) $\mathbf{b_i}$:

$$\mathbf{z} = \sum_{i=1}^{n} s_i \mathbf{b_i} \tag{1}$$

or:

$$\mathbf{z} = \mathbf{Bs} \tag{2}$$

where $\mathbf{B}$ is the matrix consisting of vectors $\mathbf{b_i}$ as columns and s_i are the coefficients, which are random variables. ICA learns $\mathbf{b_i}$ and s_i such that s_i are nongaussian and independent.

An interesting extension of ICA is multidimensional ICA [2] [28], [25], where s_i are not all mutually independent. In fact, the model assumes s_i are uncorrelated and have unit variance. The coefficients s_i are put into groups or subspaces as followed:

$$\mathbf{z} = \sum_{k=1}^{m} \sum_{i \in S(k)} s_i \mathbf{b_i} \tag{3}$$

Here, input $\mathbf{z}$ is decomposed into the sum of m subspaces $S(k)$, each of which contains a number of the components $\mathbf{b_i}$. We assume that the total number of features $\mathbf{b_i}$ are equal to the dimension of the input vector $\mathbf{z}$ and the matrix $\mathbf{B}$ is invertible. Therefore, given an input $\mathbf{z}$, s_i can be computed as:

$$s_i = \mathbf{v_i^T z} \tag{4}$$

where $\mathbf{v_i}$ are the column vectors of the inverse matrix $\mathbf{V}$of matrix $\mathbf{B}$. Vectors $\mathbf{v_i}$ are also called feature detectors.

Note that multidimensional ICA is still a linear model, thus it can not learn invariance feature. In order to transform multidimensional ICA into a nonlinear model that learns invariance feature, Hyvärinen et al. [25] uses the principle of invariant feature subspace [29], which utilizes a linear subspace as an invariant feature within the feature space. Given an input, the value of the invariant feature is the norm projection of that input to the corresponding linear subspace:

$$e_k = \sqrt{\sum_{i \in S(k)} s_i^2} = \sqrt{\sum_{i \in S(k)} (\mathbf{v_i^T z})^2} \tag{5}$$

[1] In this paper, we assume inputs are preprocessed by the same whitening preprocessing step as in Le et al. [12]

[2] Apparently, multidimensional ICA is called ISA or general ISA in Signal Processing community nowadays

Note that, the right hand side of the equation 5 is called L2-pooling. Here, e_k is the value of the invariant feature when the input is projected into subspace S_k. e_k is also called energy detector.

Given an input $\mathbf{z}$, the sparseness of the square energy detectors e_k^2 is:

$$\sum_{k=1}^{m} h(e_k^2) = \sum_{k=1}^{m} h(\sum_{i \in S(k)} s_i^2) = \sum_{k=1}^{m} h(\sum_{i \in S(k)} (\mathbf{v_i^T z})^2) \tag{6}$$

where h is a nonlinear function, which is suitable to measure the sparseness of the distribution of e_k^2 and m is the number of subspaces. In [26], h is chosen as $h(x) = -\sqrt{x}$.

For T inputs $\mathbf{z_j}$ ($j = 1, ..., T$), the sparseness is measured as:

$$S_{sparse} = \sum_{j=1}^{T} \sum_{k=1}^{m} h(e_k^2) = \sum_{j=1}^{T} \sum_{k=1}^{m} h(\sum_{i \in S(k)} (\mathbf{v_i^T z_j})^2) \tag{7}$$

Here, feature detectors $\mathbf{v_i}$ can be learned by maximizing the sparseness S_{sparse} with regards to $\mathbf{v_i}$ subject to:

$$\mathbf{VV^T} = \mathbf{I} \tag{8}$$

where $\mathbf{V}$ is the matrix with columns as $\mathbf{v_i}$. The reason for $\mathbf{V}$ to be an orthogonal matrix is as followed. As the result of whitening, $E\{\mathbf{zz^T}\} = \mathbf{I}$. Therefore,

$$E\{\mathbf{zz^T}\} = E\{\mathbf{Bss^TB^T}\} = \mathbf{B}E\{\mathbf{ss^T}\}\mathbf{B^T} = \mathbf{I} \tag{9}$$

As we assume s_i are uncorrelated and have unit variance, so $\mathbf{BB^T} = \mathbf{I}$. Because $\mathbf{V}$ is the inverse of $\mathbf{B}$, thus $\mathbf{V}$ is also an orthogonal matrix.

Stacked, Convolutional ISA (SC-ISA): Le et al. [12] uses convolution and stacked layer idea [30] to scale up ISA into a hierarchical ISA network. In this network, each layer implements the ISA algorithm. In the first layer, they train ISA with inputs that have small spatial size. As for training of subsequent layers, in order to compute input for the next layer, the learned weights of the previous layer are copied over and convolved with input of larger spatial size. The outputs of these convolution operations will be combined and then reduced in dimension using PCA before they become the input for the next layer. The authors train the first layer to converge first before training the second layer and so on. After that, they concatenate the features learned from all layers, as previously done in [30], to create local features for further processing stages.

With this hierarchical network, they obtained the state-of-the-art results of unsupervised learning algorithms for challenging datasets such as Hollywood 2 [2] (53.5%), Youtube [24] (75.8%).

3 Proposed Improvements to Increase Performance of ISA

3.1 Trajectories

Le et al. [12] use dense sampling approach to extract blocks from videos for training and testing. We note that the way Le et al. [12] extracts blocks from input videos for training and testing might not take advantage of the dynamics of motion in videos. They randomly extract sequences of patches that have the same spatial coordinates in each frame. Such straight blocks usually do not capture many types of motion correctly. Thus, the training inputs might contain a lot of noise, which reduces the accuracy of the network learned by ISA. Inspired by the trajectory approach [6], [7], we extract blocks following motion trajectories, instead of following straight paths as in Le et al. [12].

3.2 Increase the Number of Training Inputs

We note that Le et al. [12] uses only 200 blocks per video to train, which we think is probably too small to be representative of each training video. We hypothesize that, because of too much noise from the way they extract training inputs, more training samples will only degrade the performance of the system. Blocks following motion trajectories might have less noise than straight blocks. Thus, we might be able to use more trajectory blocks per video to train the system.

4 Experiments

4.1 Baseline Code and Dataset

Le et al.'s [12] release two versions of their source code. One version can be used with low end systems where users do not have powerful NVIDIA graphics cards. This version only makes use of one resolution version of video datasets. The other version utilizes multiple resolution versions of video datasets to extract more features, thus has better classification results. Due to the limited computational power available to us, we use the former version as the baseline code in our experiments. Because of the same reason, we have to make some further reduction on the video dataset that we use by assuming that running on a smaller but challenging subset of the Hollywood 2 dataset would reflect the performance of our algorithm when running with the full dataset. The Hollywood 2 dataset has 12 actions in total and we select a challenging subset of 5 actions, in which even the baseline code has difficulties with. Working with a smaller, challenging subset would help us to save time for running more experiments and at the same time it is not likely to compromise the performance of our proposed changes.

In training and testing the baseline code and our modifications, we use the same procedure used by Le et al.'s [12], which is listed in the subsection 2.1. In vector quantization step, we run k-means 8 times and select the best results.

Table 1. Baseline code's result with the chosen subset of Hollywood 2 dataset using half-resolution version

Action name	Average Precision (AP)
HugPerson	43.0024%
Kiss	68.6265%
SitDown	72.2845%
SitUp	29.8177%
StandUp	76.3940%
Mean AP	**58.0250%**

The result for our chosen subset when running the baseline code is shown in table 1. In comparison with the mean Average Precision (mAP) of about 50.5%[3] when running the baseline code with the full Hollywood 2 dataset, it is obvious that our chosen subset is quite challenging. Even though, the number of actions is reduced by more than half, the mAP of the subset is only about 8% higher than that of the full dataset.

4.2 Trajectories

Currently, we use the dense trajectory extraction approach in [6] to extract dense trajectories from input videos. We modify the original source code to relax the constraints of a valid trajectory so that we can have enough trajectories for each input video and the trajectories are able to capture the motion dynamics of some difficult actions. For each video, once we extract dense trajectories, we extract blocks following the trajectories for training and testing. Figure 1 shows the trajectories in two videos in the subset that we experiment with. Each trajectory shown here, as a small green line or curve ended with a red point, is an optical flow of one pixel tracked across a number of frames[4].

Fig. 1. Motion trajectory

Figure 2 shows the performance of the baseline code and our trajectory modification running with various training samples (blocks) per video (BpV) using

[3] As posted in the authors' website: http://ai.stanford.edu/~wzou/. We also obtain a similar result (about 50.4%) when running the baseline code with the full Hollywood 2 dataset (half-resolution version).

[4] (The default is 15 frames

the half resolution version of the dataset. Note that, the trajectory modification's results outperforms the baseline code's best results in all settings. The baseline code's mAP increases as the number of blocks per video increases up to 600 BpV, then falls off rapidly as we adds more training blocks up to 1000 BpV. After that, the baseline code's performance goes up slightly as the number of blocks increases. In contrast, our trajectory modification's mAP decreases as we use more training blocks up to 600 BpV. After that, the performance of our trajectory modification increases as more training blocks per video are employed.

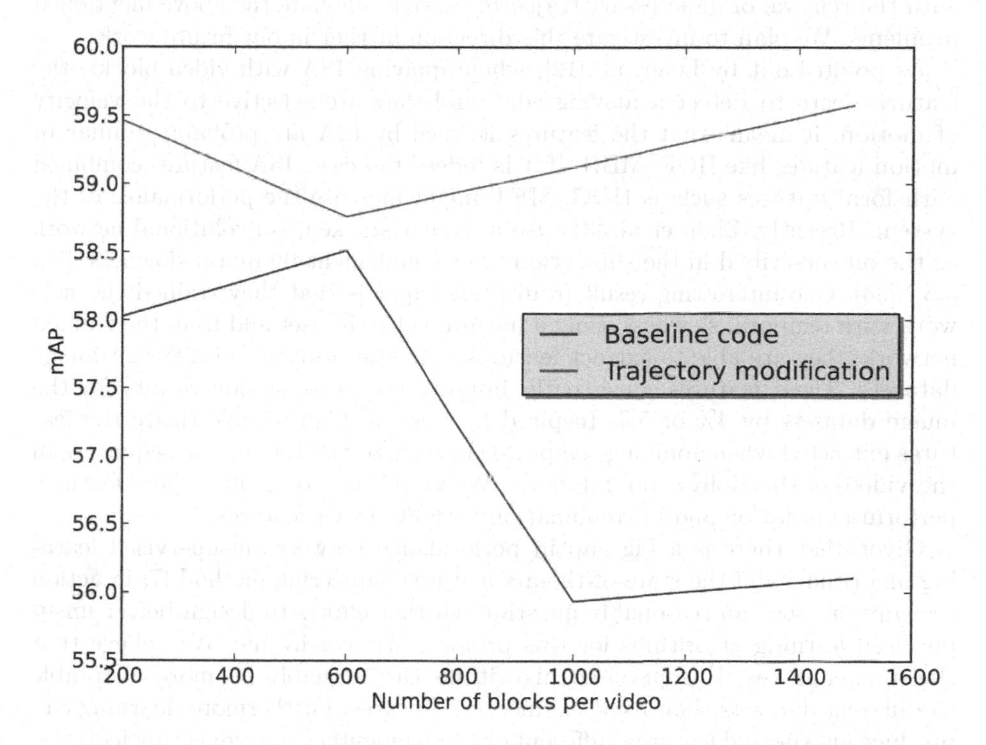

Fig. 2. Comparison of performance of the baseline code and our trajectory modification

The best result of our modification, which is about 1% (mAP) better than the best result of the baseline code, is obtained when training with the most samples per video (1500 BpV). On the other hand, too many training blocks per video reduces the performance of the baseline code significantly. As we only experiment with one version of a fixed resolution of the dataset each time, we expect the performance gain will increase when we train and test with many versions of multiple resolutions at the same time.

5 Discussion

Dense trajectory is a very powerful tool to capture motion inside a video. However, the strength of its coverage is also its weakness when applied to unsupervised learning algorithms. While dense coverage ensures that the resulting trajectories are unlikely to miss any motion in videos, it creates many irrelevant trajectories, which are not only a great computational burden but also a source of noise to train networks. We think that, for human activity datasets like Hollywood 2, a good algorithm for human detection and tracking will definitely help with the removal of unnecessary trajectories, thus alleviate the above mentioned problems. We plan to investigate this direction further in our future work.

As pointed out by Le et. al. [12], when applying ISA with video blocks, the features learn to detect a moving edge and they are selective to the velocity of motion. It means that the features learned by ISA are probably similar to motion features like HOF, MBH. If it is indeed the case, ISA features combined with form features such as HOG, SIFT might increase the performance of the system. Recently, Zhou et al. [31] use a similar stacked, convolutional network as the one described in the subsection 2.3 to implement temporal slowness [32], [33], [34]. One interesting result from their paper is that they trained the network with temporal slowness using a natural video dataset and from the trained network, they are able to extract features from static images of different image datasets. These features consistently improve the classification results for the image datasets by 4% or 5%. Inspired by this, we plan to investigate the features extracted when applying temporal slowness to tracked human sequences in the videos of the Hollywood 2 dataset. We would like to see how these features perform in isolation and in combination with SC-ISA's features.

Given that there is a big gap in performance between unsupervised learning algorithms and the state-of-the-art feature engineering method [7] in action recognition, one can reasonably question whether efforts to design better unsupervised learning algorithms for this problem are worthwhile. We believe that the answer is yes. Unsupervised algorithms can arguably be more adaptable to different datasets than fixed engineering features. Furthermore, learning can produce unexpected features, different or complementary to engineering features, thus maintaining a healthy competition between these two approaches could be much more beneficial for the progress of the field than focusing only on a single approach. In addition, we would like to point out that even the state-of-the-art combines different engineering features together. If unsupervised algorithms like SC-ISA also combine with different features either from other unsupervised algorithms or from feature engineering methods, the performance of the resulting systems will more likely increase and be comparable to the state-of-the-art. Finally, brain-inspired algorithms can make use of the proven principles in the human visual cortex, which is still the best general vision system at the moment.

6 Conclusion

We incorporate motion information into SC-ISA by training and testing SC-ISA using blocks following motion trajectories. We also show that, empirically, SC-ISA's performance degrades significantly when we train it with many more straight blocks that that of [12]. Interestingly, when SC-ISA is trained with more trajectory blocks, the performance decreases at first and then increases as more training blocks are added. Even though we only experiment SC-ISA with trajectory using one resolution version of a subset of the Hollywood 2 dataset, we expect the performance gain (1% better than the baseline code) will increase as we run with multiple resolution versions of the dataset. Unsupervised learning algorithms like SC-ISA are very interesting because potentially they can make use of the huge number of unlabeled videos available. However, we think that good human detection, or object detection in general, and tracking algorithms are needed in order to enable unsupervised learning algorithms to work with such large-scale video datasets.

References

1. Schuldt, C., Laptev, I., Caputo, B.: Recognizing human actions: A local svm approach. In: Proceedings of the 17th International Conference on Pattern Recognition, ICPR 2004, vol. 3, pp. 32–36. IEEE (2004)
2. Marszalek, M., Laptev, I., Schmid, C.: Actions in context. In: CVPR, pp. 2929–2936. IEEE (2009)
3. Kuehne, H., Jhuang, H., Garrote, E., Poggio, T., Serre, T.: Hmdb: A large video database for human motion recognition. In: 2011 IEEE International Conference on Computer Vision (ICCV), pp. 2556–2563. IEEE (2011)
4. Soomro, K., Zamir, A.R., Shah, M.: Ucf101: A dataset of 101 human actions classes from videos in the wild. arXiv preprint arXiv:1212.0402 (2012)
5. Karpathy, A., Toderici, G., Shetty, S., Leung, T., Sukthankar, R., Fei-Fei, L.: Large-scale video classification with convolutional neural networks. In: IEEE Conference on Computer Vision and Pattern Recognition, CVPR (2014)
6. Wang, H., Kläser, A., Schmid, C., Liu, C.L.: Action recognition by dense trajectories. In: CVPR, pp. 3169–3176. IEEE (2011)
7. Wang, H., Schmid, C.: Action Recognition with Improved Trajectories. In: 2013 IEEE International Conference on Computer Vision (ICCV), pp. 3551–3558 (2013)
8. LeCun, Y., Bottou, L., Bengio, Y., Haffner, P.: Gradient-based learning applied to document recognition. Proceedings of the IEEE 86, 2278–2324 (1998)
9. Girshick, R.B., Donahue, J., Darrell, T., Malik, J.: Rich feature hierarchies for accurate object detection and semantic segmentation. CoRR abs/1311.2524 (2013)
10. Sermanet, P., Eigen, D., Zhang, X., Mathieu, M., Fergus, R., LeCun, Y.: OverFeat: Integrated Recognition, Localization and Detection using Convolutional Networks. CoRR abs/1312.6229 (2013)

11. Le, Q.V., Ranzato, M., Monga, R., Devin, M., Corrado, G., Chen, K., Dean, J., Ng, A.Y.: Building high-level features using large scale unsupervised learning. In: ICML, icml.cc. Omnipress (2012)
12. Le, Q.V., Zou, W.Y., Yeung, S.Y., Ng, A.Y.: Learning hierarchical invariant spatio-temporal features for action recognition with independent subspace analysis. In: 2011 IEEE Conference on Computer Vision and Pattern Recognition (CVPR), pp. 3361–3368. IEEE (2011)
13. Taylor, G.W., Fergus, R., LeCun, Y., Bregler, C.: Convolutional learning of spatio-temporal features. In: Daniilidis, K., Maragos, P., Paragios, N. (eds.) ECCV 2010, Part VI. LNCS, vol. 6316, pp. 140–153. Springer, Heidelberg (2010)
14. Weinland, D., Ronfard, R., Boyer, E.: A survey of vision-based methods for action representation, segmentation and recognition. Computer Vision and Image Understanding 115, 224–241 (2011)
15. Poppe, R.: A survey on vision-based human action recognition. Image Vision Comput. 28, 976–990 (2010)
16. Jiang, Y.G., Bhattacharya, S., Chang, S.F., Shah, M.: High-level event recognition in unconstrained videos. IJMIR 2, 73–101 (2013)
17. Perronnin, F., Sánchez, J., Mensink, T.: Improving the Fisher Kernel for Large-Scale Image Classification. In: Daniilidis, K., Maragos, P., Paragios, N. (eds.) ECCV 2010, Part IV. LNCS, vol. 6314, pp. 143–156. Springer, Heidelberg (2010)
18. Jegou, H., Douze, M., Schmid, C., Pérez, P.: Aggregating local descriptors into a compact image representation. In: CVPR, pp. 3304–3311. IEEE (2010)
19. Wang, H., Schmid, C.: Lear-inria submission for the thumos workshop. In: ICCV Workshop on Action Recognition with a Large Number of Classes (2013)
20. Farnebäck, G.: Two-frame motion estimation based on polynomial expansion. In: Bigun, J., Gustavsson, T. (eds.) SCIA 2003. LNCS, vol. 2749, pp. 363–370. Springer, Heidelberg (2003)
21. Dalal, N., Triggs, B.: Histograms of oriented gradients for human detection. In: IEEE Computer Society Conference on Computer Vision and Pattern Recognition, CVPR 2005, vol. 1, pp. 886–893. IEEE (2005)
22. Laptev, I., Marszalek, M., Schmid, C., Rozenfeld, B.: Learning realistic human actions from movies. In: CVPR. IEEE Computer Society (2008)
23. Dalal, N., Triggs, B., Schmid, C.: Human detection using oriented histograms of flow and appearance. In: Leonardis, A., Bischof, H., Pinz, A. (eds.) ECCV 2006. LNCS, vol. 3952, pp. 428–441. Springer, Heidelberg (2006)
24. Liu, J., Luo, J., Shah, M.: Recognizing realistic actions from videos "in the wild". In: IEEE Conference on Computer Vision and Pattern Recognition, CVPR 2009, pp. 1996–2003. IEEE (2009)
25. Hyvärinen, A., Hoyer, P.: Emergence of phase-and shift-invariant features by decomposition of natural images into independent feature subspaces. Neural Computation 12, 1705–1720 (2000)
26. Hyvärinen, A., Hurri, J., Hoyer, P.O.: Natural Image Statistics: A Probabilistic Approach to Early Computational Vision, vol. 39. Springer (2009)
27. Comon, P.: Independent component analysis, a new concept? Signal Processing 36, 287–314 (1994)
28. Cardoso, J.: Multidimensional independent component analysis. In: Proceedings of the 1998 IEEE International Conference on Acoustics, Speech and Signal Processing, vol. 4, pp. 1941–1944. IEEE (1998)
29. Kohonen, T.: Emergence of invariant-feature detectors in the adaptive-subspace self-organizing map. Biological Cybernetics 75, 281–291 (1996)

30. Lee, H., Grosse, R., Ranganath, R., Ng, A.Y.: Convolutional deep belief networks for scalable unsupervised learning of hierarchical representations. In: Proceedings of the 26th Annual International Conference on Machine Learning, pp. 609–616. ACM (2009)
31. Zou, W.Y., Ng, A.Y., Zhu, S., Yu, K.: Deep Learning of Invariant Features via Simulated Fixations in Video. In: NIPS, pp. 3212–3220 (2012)
32. Hinton, G.E.: Connectionist learning procedures. Artificial Intelligence 40, 185–234 (1989)
33. Mitchison, G.: Removing Time Variation with the Anti-Hebbian Differential Synapse, Neural Computation (1991)
34. Földiák, P.: Learning Invariance from Transformation Sequences. Neural Computation (1991)

A Spiking Neural Network Model for Associative Memory Using Temporal Codes

Jun Hu[1], Huajin Tang[2], Kay Chen Tan[1], and Sen Bong Gee[1]

[1] Department of Electrical and Computing Engineering
National University of Singapore
4 Engineering Drive 3, Singapore 117576
[2] Institute for Infocomm Research
Agency for Science, Technology and Research (A*STAR)
Singapore 138632
{junhu,eletankc,a0039834}@nus.edu.sg
htang@i2r.a-star.edu.sg

Abstract. Associative memory is defined as the ability to map input patterns to output patterns. Understanding how human brain performs association between unrelated patterns and stores this knowledge is one of the most important goals in computational intelligence. Although this problem has been widely studied using conventional neural networks, increasing biological findings suggest that spiking neural network can be an alternative. The proposed model encodes different memories using different subsets of encoding neurons with temporal codes. A spike-timing based learning algorithm and spike-timing-dependent plasticity (STDP) are used to form associative memory. Simulation results show that hetero-associative memory and auto-associative memory are achievable by the synaptic modification of connections between input layer and hidden layers, and recurrent connections of hidden layers, respectively.

Keywords: Spiking Neural Networks (SNNs), associative memory, Spike-Timing-Dependent Plasticity (STDP), temporal codes.

1 Introduction

To understand how the human brain works is a fascinating challenge that requires comprehensive scientific research in collaboration with multidisciplinary fields such as biology, chemistry, computer science, engineering, mathematics, psychology, etc. In order to explore the memory function in biological systems, different parts of biological nervous system have been studied.

A few computational models focus on specific memory types or areas of the brain. Working memory models are either based on specific network structures (e.g. [1]) or neural dynamics at synaptic level (e.g. [2]). In short term memory networks, feedforward networks and feedback-based attractor networks are commonly adopted to maintain patterns of neural activity simulating different parts of the hippocampus (CA1 and CA3) [3, 4]. In addition, the theta/gamma

H. Handa et al. (eds.), *Proc. of the 18th Asia Pacific Symp. on Intell. & Evol. Systems – Vol. 1*,
Proceedings in Adaptation, Learning and Optimization 1, DOI: 10.1007/978-3-319-13359-1_43

oscillations and phase locked spikes are believed to play important roles in the hippocampus, and have been applied to memory models [5, 6].

Another trend of research is to consider the brain as an inseparable system. Inspired by the anatomical model of the brain and physiology studies on the functional roles of hippocampus and cortex, hippocampal-cortical networks have drawn great attention recently [7–9]. The hippocampal-cortical model consists of sensory system (encoding), hippocampus (short-term memory), and cortex (long-term memory). While various encoding schemes may be adopted by the sensory system, different mechanisms are used to sustain the short-term memory, and long-term memory is usually stored in the form of synaptic plasticity through a neocortical association process.

However, the organization principal of memory, which is intensively related to the learning process happening all the time throughout the brain, is still unclear. Hebbian theory, which is also called assembly theory, attempted to explain the associative learning process at the synaptic level. With the development of large-scale ensemble recording techniques, network-level functional coding units, named neural cliques, have been successfully identified in the hippocampus [10]. Since spike-timing dependent plasticity (STDP) can be understood as a special form of Hebbian learning, a possibility is that STDP process and other spike-timing based learning mechanisms are involved in the formation of neural cliques and associative learning.

By simulating artificial neuron models, neural networks are inherently close to the nature of biological nervous system and possible to mimic functions of biological neural systems. However, there is a continuing debate over which characteristics of a spike train carry useful information for processing. The major bone of contention in the debate is that whether to choose 'rate codes' or 'temporal codes' for spiking neural network. It has been shown that visual pattern classification can be carried out in 100 milliseconds [13, 14]. In addition, it has been proved that the mean fire rate fails to correctly describe the temporally varying sensory information [15]. Unlike the rate code, the temporal code assumes that the precise placement of the spikes in time carries significant information. The spiking neural network, which is inherently compatible with temporal codes, is believed to be more biologically plausible and computationally efficient than the preview two generations of artificial neural networks (ANNs).

The aforementioned studies suggest that timing of spikes might be an essential factor in neural representation of information, and spike timing may play an important role during learning. However, insufficient attention has been paid to temporal information carried by precisely timed spikes in most existing memory models using spiking neural networks. In addition, increasing biological findings support the proposal that memories are represented by invariant neural responses in the brain. All these findings and results provide substantial motivations for investigations of spiking neural networks in memory systems.

In this paper, we focus on developing associative memory models with temporal codes. Section 2 describes the computational model in detail. Section 3 presents simulation results. Discussion and conclusion are provided in section 4.

2 The Computational Model

2.1 Neuron Model and Network Architecture

Considering the computational efficiency and analyzability, all computational units in this paper are modeled as spike response model (SRM). In addition, all neurons in the proposed model are limited to fire only once. The membrane potential of the spiking neuron is given by

$$v(t) = \eta(t - \hat{t}) + \sum_i \sum w_i \epsilon(t - \hat{t}, t - t_i) \tag{1}$$

where $\hat{t}$ is the firing time of the last spike of the neuron, η defines the form of the action potential and its after-potential, w_i is the synaptic weight from the ith input neuron to the post-synaptic neuron, ϵ describes the post-synaptic potential (PSP) of an input pulse, and t_i is the arrival time of the f th input spike from ith input neuron.

As shown in Figure 1, the memory model is composed of three layers: Input layer, Layer I and Layer II. In this paper, we focus on the first two layers (Input layer and Layer I in the dashed box) and associative memory.

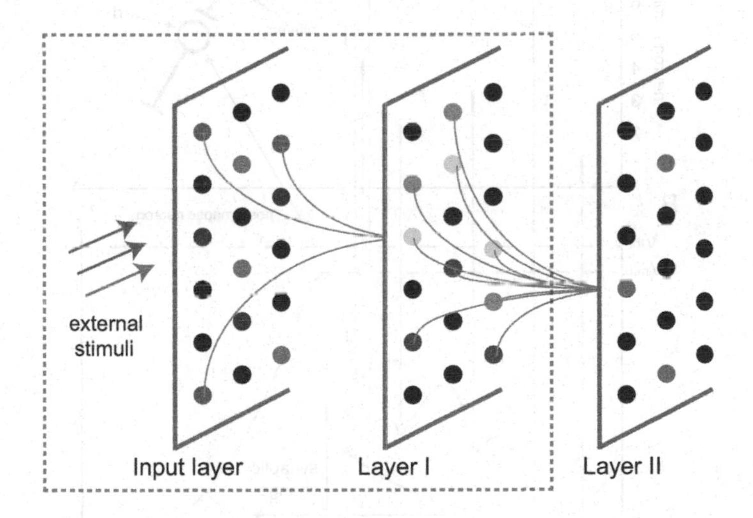

Fig. 1. Model structure

2.2 The Spike-Timing Based Learning and NMDA Channels

Since we consider time as one coding dimension for neural representation, tempotron learning is employed in training neurons to recognize upwards stimulation in the model [16]. When presented a pattern, each neuron needs to make a decision that whether the input stimulation contains certain features that have been learned. The postsynaptic neuron integrates postsynaptic potentials (PSPs) from afferent presynaptic neurons. The connections from neurons that contribute to the integrated postsynaptic membrane potential during the presentation will be enhanced according to the tempotron learning rule as described by the following equation.

$$\Delta w_i = \lambda d \sum_{s_i<0} exp(s_i) \tag{2}$$

where w_i is the synaptic weight from afferent i to the postsynaptic neuron, λ is the learning rate, d is the desired output label (either 0 or 1), and $s_i = t_i - t_{max}$ is the delay between presynaptic firing ($S_i(t) = \delta(t - t_i)$, t_i is the firing time of ith afferent neuron) and the time when postsynaptic membrane potential $v(t)$ reaches its maximal value v_{max}. The tempotron learning rule is illustrated in Figure 2.

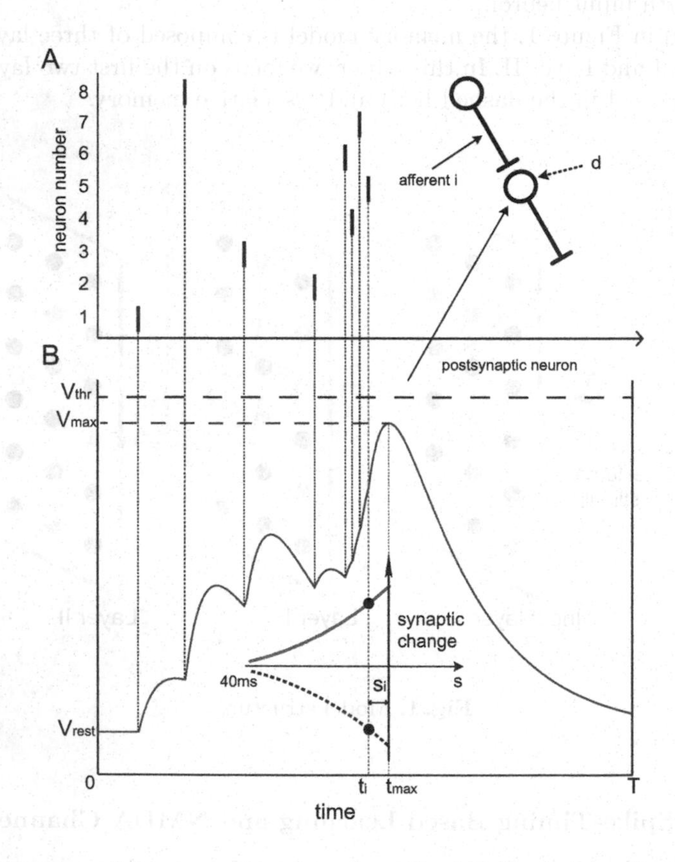

Fig. 2. Illustration of the tempotron learning rule. (A) Typical input pattern using temporal population codes. (B) Membrane potential of the postsynaptic neuron. (inset) The synaptic weight w_i changes accordingly to the time difference between s and the desired signal d.

As shown in Figure 2, the maximum value of the membrane potential is reached at t_{max}. The synaptic weight w_i changes accordingly to the time difference between s and the desired signal d as shown in the inset. If $d = 1$, $\Delta w_i \geq 0$ (solid line), or if $d = -1$, $\Delta w_i < 0$ (dashed line).

STDP is a well studied and widely accepted learning mechanism [17]. It is believed that both long-term potentiation (LTP) and long-term depression (LTD) depend on the intracellular calcium transients [18]. In addition, the calcium dependence of plasticity is related to the activation of postsynaptic NMDA receptors (NMDARs). Therefore, synaptic plasticity (LTP and LTD) relies on the activation of postsynaptic NMDARs [19, 20]. Although the biophysical and biochemical mechanisms that underpin STDP still need further investigations, these suggest that STDP could be a NMDAR-dependent mechanism. It is assumed that fast and slow NMDA channels dominate the synaptic transmission as well as plasticity in Layer I and Layer II, respectively.

Since memory items are represented by firings of neurons, the repetitive firings contribute to the enhancement of connections between activated neurons via STDP learning. As feedback inhibition from interneurons temporally separate spike volleys into individual gamma cycles, fast NMDA channel could contributes to the formation of intra-clique connections (auto-associative memory) in Layer I, and slow NMDA channel spanning over several gamma cycles enhances inter-clique connections (episodic memory) in Layer II. Based on biophysical and biochemical findings, the time course of the activation of NMDARs crucially affects long term modification, which indicates that STDP learning may perform differently with NMDA channel in different states. In order to incorporate the effect of slow NMDA, STDP learning curve with a width of $\pm 50ms$ as illustrated in Figure 3.

3 Experiment Results

Each input pattern is represented by 100 spikes using temporal population codes, and they are introduced to the network during troughs of the theta oscillation. Fast NMDA channel maintains activated state around $10ms$ after the binding of glutamate to postsynaptic cells, while slow NMDA with a slow deactivation time constant dominates the STDP process in Layer II. The inter-layer synaptic weights are updated by the tempotron learning rule, while intra-layer synaptic plasticity are modified by STDP.

Driven by input synaptic currents from afferents, increasing number of pyramidal neurons in Layer I start to fire and form different neural cliques iteration by iteration. When enough stimulation are generated by neural activities in Layer I, similar phenomena emerge in Layer II. After dozens of iterations, neural cliques response to specific patterns and repeat firing during the subsequent theta cycles (Figure 4).

We can identify three groups of neurons firing as neural cliques by their volley activities in Layer I and II, respectively (Figure 4C and 4D). Within each theta cycle, neural cliques respond selectively and repetitively to the stimulation as same order that input patterns are introduced to the network.

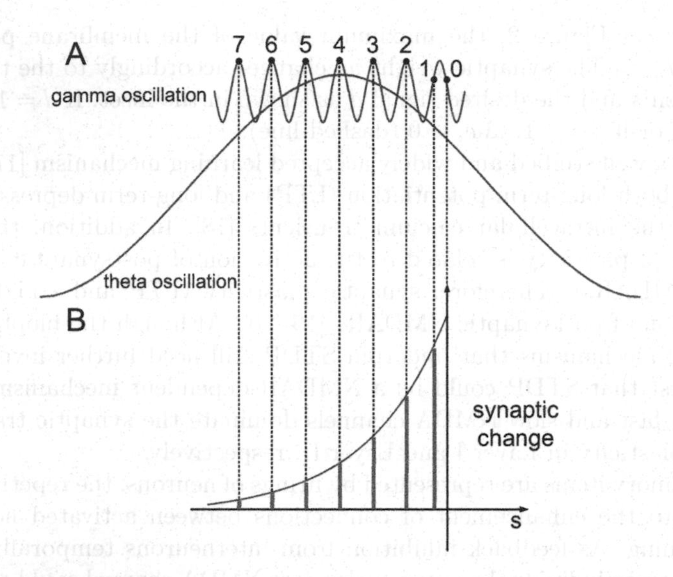

Fig. 3. LTP induced by STDP learning for different memory items. (A) Firings within each gamma cycle represent memory items 0-7. (B) Synaptic changes depend on the relative time between firings.

Figure 5 reveals the mechanism underlying repetitive firing of pyramidal neurons. The periodical firing, which functions as working memory in the model is a result of interplay between ADP and theta oscillation. After generation of the first spike by a particular neuron, its ADP starts to build up. When the slowly ramping up ADP meets near-peak theta current, the pyramidal neuron fires again in the following theta cycle. Meanwhile, inhibitory from interneurons prevents neurons coding for other patterns from firing right after the volley spikes generated by previous neural clique. As a result, spike volleys are temporally separated into individual gamma cycles (Figure 4C and 4D). When neurons initially start to fire, the spike times are randomly distributed in gamma cycles. With repetitive firings, synaptic weights between individual members of the same neural clique are strengthened with STDP learning, resulting in their synchronized firing as shown in Figure 4C.

To examine lateral connectivity caused by fast NMDA mediated STDP, interlayer connections (Input layer to Layer I) and intra-layer connections (Layer I) are examined as shown in Figure 6.

After several iterations, neurons in Layer I start to fire due to the enhancement of connections from input layer to them (Figure 6, W_{21}). Once more than one neuron starts to respond to a certain input pattern, they wire together to form neural cliques as shown in the second row (Figure 6, W_{22}).

To further study the resulted neural cliques and their connectivities, we focus on weight matrices representing synaptic connections of Layer I as shown in Figure 7. Since fast NMDA channel stays activated for several milliseconds,

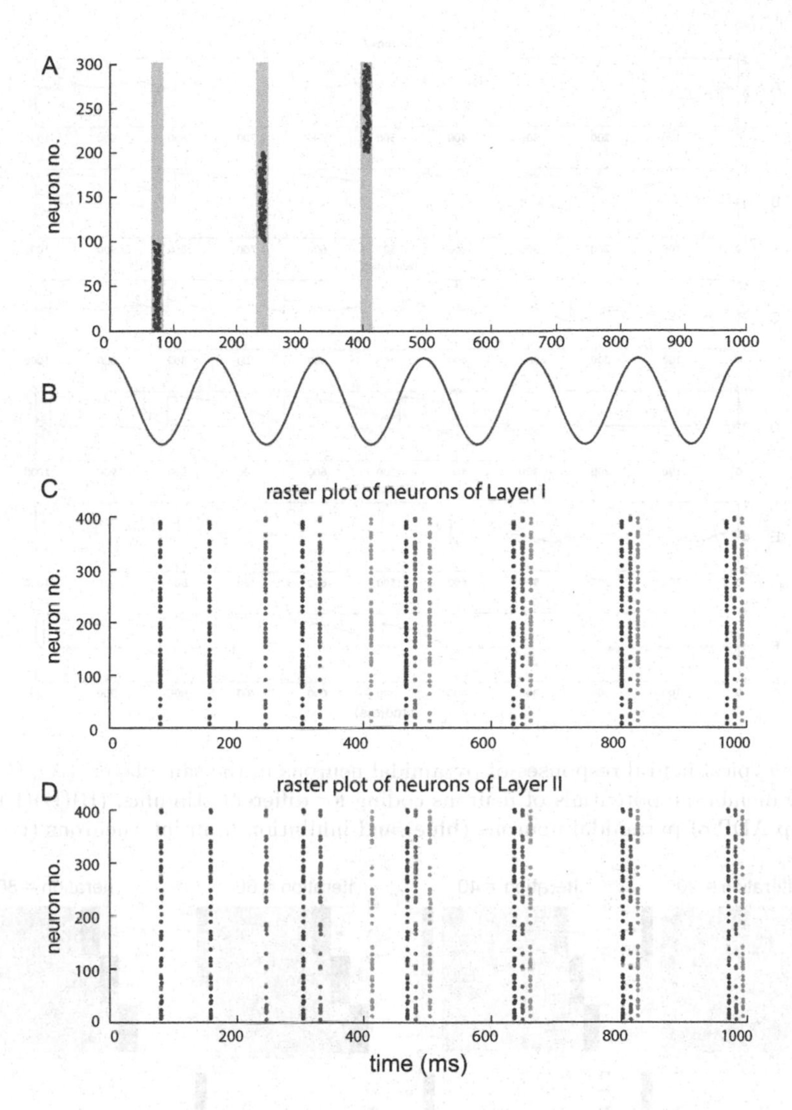

Fig. 4. Neural activity propagates through the system. (A) Input patterns. Each pattern consists of 100 neurons firing within an encoding window (gray strips). (B) Theta rhythm provides a sub-threshold current injecting to pyramidal neurons. (C) and (D) are the raster plots of the neural activities in Layer I and II.

only firings of neurons forming the same clique fall within this narrow time window. Consequently, intra-clique connections are enhanced via STDP process and salient as shown in Figure 7. Salient matrix elements at 1-50 on the x- and y-axis represent the enhanced intra-clique connections (the weights have been rearranged according to neural cliques for clear illustration). The highlighted weight matrices show that each neural assembly forms a recurrent subnetwork with auto-associative memory coded in the enhanced lateral connections.

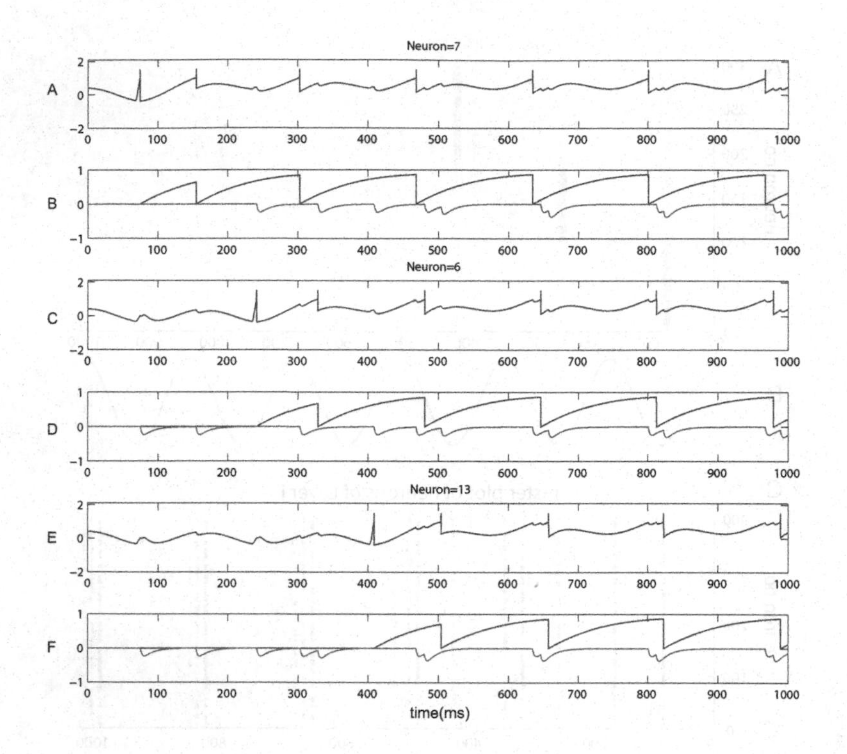

Fig. 5. Typical neural responses of pyramidal neurons in the same layer. (A), (C) and (E) are membrane potentials of neurons coding for different stimulus. (B)(D)(F) Slow built-up ADP of pyramidal neurons (blue) and inhibition from interneurons (red).

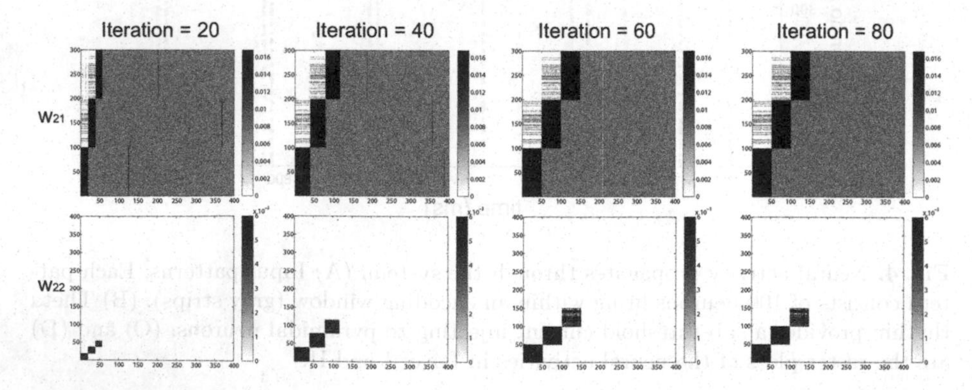

Fig. 6. Evolution of the neural connectivity of the network. Synaptic change of inter-layer connections between Input layer to Layer I (W_{21}) is shown in the first row. The generation of neural cliques in Layer I (W_{22}) is shown in the second row. The activated neurons are picked out and rearranged for clear illustration.

Intra-Clique Connections and Auto-Associative Memory. Our brain has a remarkable ability of association, despite constant changes in real-world circumstance. During perception along sensory pathways, information about

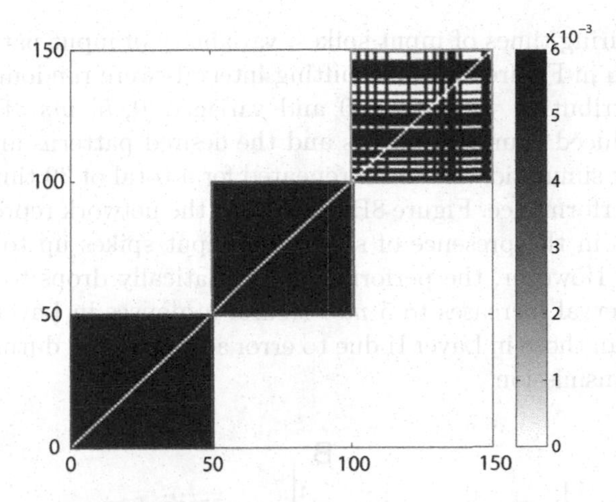

Fig. 7. Weight matrices of lateral connections in Layer I. The amplitudes of synaptic weights are normalized to the maximal value and proportional to gray-scale values. Intra-neural clique connections are highlighted by colored boxes.

external stimulation is abstracted and encoded into reliable neural activities. After training, both hetero-associative memory and auto-associative memory are stored in the connections between neurons. Input patterns are hetero-associated with neural responses in Layer I via synaptic weights between input layer and Layer I, while auto-associative memory is represented by intra-clique connections.

As neural activities can be observed as an explicit expression of stored memory, pattern completion may refer to the ability that a subset of neurons from a particular neural clique are able to arouse the rest of that clique. The trained network is expected to be competent for recalling stored neural activities upon presentation of input patterns and retaining invariant responses in presence of noises and even corruption of information. As information is distributed over neurons with population coding scheme, the information loss caused by shifting or removing spikes can be complemented with the aid of other contributing neurons. In order to investigate this capability of reproducing neural activities, time jitter and missing of spikes are considered in the following experiments. Hence, a correlation-based measure of spike timing [21] is used to calculate the distance between output pattern and target pattern to quantitatively evaluate the performance.

$$C = \frac{\overrightarrow{s_1} \cdot \overrightarrow{s_2}}{|\overrightarrow{s_1}||\overrightarrow{s_2}|} \tag{3}$$

where C is the correlation denoting closeness between two temporal coded patterns (s_1 and s_2). They are convolved with a low pass Gaussian filter of a width $\sigma = 2ms$.

By shifting firing times of input spikes, variability of input patterns was simulated as shown in Figure 8A. The shifting intervals were randomly drawn from a gaussian distribution with mean 0 and variance [0, 5] ms. The correlation between reproduced neural responses and the desired patterns are presented in Figure 8. Every simulation has been repeated for a total of 30 times to generate the averaged performance. Figure 8B shows that the network reproduces reliable neural patterns in the presence of shifting of input spikes up to 3 ms in both Layer I and II. However, the performance dramatically drops to around 0.3 as the shifting interval increases to 5 ms. Neural responses in Layer I are slightly more robust than those in Layer II due to error accumulation during the upwards information transmission.

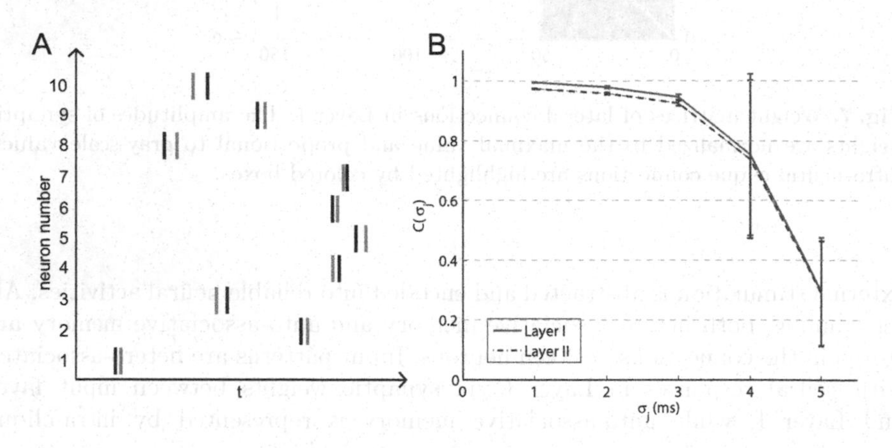

Fig. 8. (A) Illustration of shifted spatiotemporal patterns. Firing times of original input spikes (black bars) are randomly shifted with random jitters (gray bars). (B) Reliability of retrieved neural responses under different noise levels.

An additional experiment is conducted to further investigate the link between intra-clique connections and auto-associative memory. Since neurons in the same clique may provide supplementary stimulation to sustain an united activity, corruption of input patterns may not be a catastrophic error. All settings are the same as the previous experiment, whereas one out of ten spikes is removed from each training pattern. The experiment has been repeated for 20 times and the mean value of the correlation between actual output and desired pattern is calculated for each trial.

As shown in Figure 9, the result is consistent with our analysis that connections within neural cliques are responsible for the completion of patterns in Layer I. In addition, the ability of robustly reproducing patterns in presence of noise are demonstrated by testing with shifted patterns as shown in Figure 8.

Knowledge stored in the synaptic weights from the input layer to Layer I provides the capability of hetero-association by recognizing input patterns via their specific features. Meanwhile, intra-clique connections contribute to the pattern completion which is one of the most important features of auto-associative

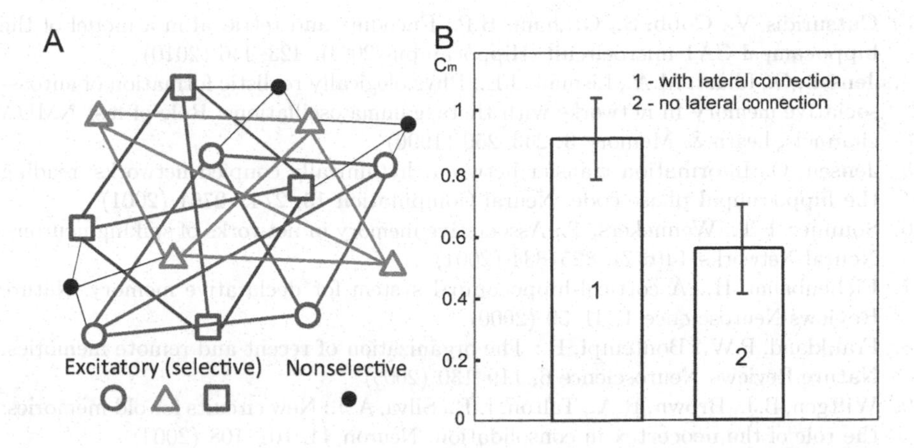

Fig. 9. (A) Illustration of neural cliques in Layer I coding for different input patterns after learning. Colored circles, triangles and squares are excited neurons coding for different patterns, while black dots are non-selective neurons. (B) Test results of the associative memory based on the correlation between retrieved and corresponding desired patterns in response to corrupted input patterns.

memory, in Layer I. Since corrupted patterns provide insufficient stimulation to neurons in the next layer, some of the trained neurons that should have been activated may not be triggered. Fortunately, lateral inputs from excited neurons can provide supplementary information to recall desired neural responses. Therefore, associative memory, which can be understood as the ability to retrieve invariant responses with partial information, relies on both the distributed storage of knowledge in synaptic connectivity between layers and within Layer I.

4 Conclusion and Future Works

In this paper, a spiking neural network model was proposed to investigate the formation of artificial cognitive memory in artificial intelligent systems. The simulation results show that the proposed model can store patterns in forms of both hetero-associative memory and auto-associative memory with temporal population codes. Since there is one more layer in this model (Layer II), investigation on the formation of episodic memory is expected for future works.

Acknowledgment. This work was supported by the Singapore Ministry of Education Academic Research Fund Tier 1 under the project R-263-000-A12-112.

References

1. Durstewitz, D., Seamans, J.K., Sejnowski, T.J.: Neurocomputational models of working memory. Nature Neuroscience 3, 1184–1191 (2000)
2. Mongillo, G., Barak, O., Tsodyks, M.: Synaptic theory of working memory. Science 319(5869), 1543–1546 (2008)

3. Cutsuridis, V., Cobb, S., Graham, B.P.: Encoding and retrieval in a model of the hippocampal CA1 microcircuit. Hippocampus 20(3), 423–446 (2010)
4. Jensen, O., Idiart, M.A., Lisman, J.E.: Physiologically realistic formation of autoassociative memory in networks with theta/gamma oscillations: Role of fast NMDA channels. Learn & Memory 3, 243–256 (1996)
5. Jensen, O.: Information transfer between rhythmically coupled networks: reading the hippocampal phase code. Neural Computation 13, 2743–2761 (2001)
6. Sommer, F.T., Wennekers, T.: Associative memory in networks of spiking neurons. Neural Networks 14(6-7), 825–834 (2001)
7. Eichenbaum, H.: A cortical-hippocampal system for declarative memory. Nature Reviews Neuroscience 1, 41–50 (2000)
8. Frankland, P.W., Bontempi, B.: The organization of recent and remote memories. Nature Reviews Neuroscience 6, 119–130 (2005)
9. Wiltgen, B.J., Brown, R.A., Talton, L.E., Silva, A.J.: New circuits for old memories: the role of the neocortex in consolidation. Neuron 44, 101–108 (2004)
10. Lin, L., Osan, R., Shoham, S., Jin, W., Zuo, W., Tsien, J.Z.: Identification of network-level coding units for real-time representation of episodic experiences in the hippocampus. PNAS 102(17), 6125–6130 (2005)
11. Shadlen, M.N., Newsome, W.T.: Noise, neural codes and cortical organization. Current Opinion in Neurobiology 4, 569–579 (1994)
12. Litvak, V., Sompolinsky, H., Segev, I., Abeles, M.: On the transmission of rate code in long feed-forward networks with excitatory-inhibitory balance. Journal of Neuroscience 23, 3006–3015 (2003)
13. Perrett, D.I., Rolls, E.T., Caan, W.: Visual neurones responsive to faces in the monkey temporal cortex. Experimental Brain Research 47(3), 329–342 (1982)
14. Thorpe, S.J., Imbert, M.: Biological constraints on connectionist modelling. In: Connectionism in Perspective, pp.63–92. Elsevier (1989)
15. Carr, C.E.: Processing of temporal information in the brain. Annual Review of Neuroscience 16, 223–243 (1993)
16. Gütig, R., Sompolinsky, H.: The tempotron: A neuron that learns spike timing-based decisions. Nature Neuroscience 9(3), 420–428 (2006)
17. Bi, G.Q., Poo, M.M.: Synaptic modifications in cultured hippocampal neurons: dependence on spike timing, synaptic strength, & postsynaptic cell type. Journal of Neuroscience 18(24), 10464–10472 (1998)
18. Sjöström, P.G., Nelson, S.B.: Spike timing, calcium signals and synaptic plasticity. Current Opinion in Neurobiology 12(3), 305–314 (2002)
19. Bear, M.F., Malenka, R.C.: Synaptic plasticity: LTP and LTD. Current Opinion in Neurobiology 4(3), 389–399 (1994)
20. Malenka, R.C., Bear, M.F.: LTP and LTD: An embarrassment of riches. Neuron 44(1), 5–21 (2004)
21. Schreiber, S., Fellous, J., Whitmer, D., Tiesinga, P., Sejnowski, T.: A new correlation-based measure of spike timing reliability. Neurocomputing 52-54, 925–931 (2003)

A Novel Adaptive ε-Constrained Method for Constrained Problem

Chunjiang Zhang, Qun Lin*, and Liang Gao

State Key Lab of Digital Manufacturing Equipment & Technology,
Huazhong University of Science and Technology, Wuhan, Hubei, China, 430074
zhchj1989@gmail.com, linqunhust@hotmail.com,
gaoliang@mail.hust.edu.cn

Abstract. Constrained optimization problems have been a research focus in the field of intelligent optimization, and massive constraint handling methods have been put forward. ε-constrained method is a very effective method proposed in recent years. The key of ε-constrained method is controlling ε level. In previous studies, the ε level is controlled by an exponential decline control method. However, considering the method is not stable and efficient for some problem, a novel adaptive ε level control method is introduced in this paper. The proposed adaptive ε level control method is combined with the basic differential evolutionary (DE) algorithm. The comparison between this new method and other ε level control methods demonstrates that the proposed novel adaptive ε level control method can avoid getting into local optimal in infeasible region at initial phase, improve the stability, and enhance the efficiency of algorithm.

Keywords: ε-constrained method, constrained optimization problem, differential evolutionary algorithm.

1 Introduction

Constrained optimization problems widely exist in engineering application and scientific research. Besides, constrained optimization has always been a research hot spot in intelligent optimization field. The difficulty of solving constrained optimization problems is how to handle the constraints. So for, many kinds of effective constraint handling methods have been proposed. According to Mezura-Montes and Coello's review article [1], some classical constrain handling methods include penalty function (PF), stochastic ranking (SR), feasibility and dominance (FAD) rules, multi-objectives concept, and ε-constrained method.

PF method is one of the first constraint handling methods, in which the violation value of constraints is added to the objective function so that the constrained optimization problems are transformed to unconstrained optimization problems. This method is simple and intuitive, but setting the coefficient value of constraint penalty term is arduous. Different constrained problems need different specific penalty coefficient

* Corresponding author.

H. Handa et al. (eds.), *Proc. of the 18th Asia Pacific Symp. on Intell. & Evol. Systems – Vol. 1*,
Proceedings in Adaptation, Learning and Optimization 1, DOI: 10.1007/978-3-319-13359-1_44

values, thus it is prone to bring out excessive penalty or insufficient penalty problems. Recently years, some adaptive control methods have been adopted to overcome this shortcoming [2, 3].

SR, as a constraint handling method, combined with evolutionary strategie and was first proposed by Runarsson and Yao [4, 5]. SR adopts bubble sort method to rank the individuals in the population, and takes use of a user-defined probability to decide whether the sorting is conducted according to the objective function value or the constraint violation value. Zhang et al [6] combined SR with differential evolution (DE) algorithm. Several mutation solutions were generated in the mutation operation process of DE, and SR was adopted to sort these mutation solutions. The main shortcoming of SR is that not all optimization algorithms need sorting. The method proposed by Zhang et al [6] that forcibly constructed population would definitely increase solving cost of the constrained problems and decrease efficiency of the algorithm.

FAD rules were first proposed by Deb in the genetic algorithm (GA) [7]. Three simple rules were utilized to compare two solutions: 1) when comparing two feasible solutions, the one with the better objective function value is chosen. 2) when comparing a feasible and an infeasible solution, the feasible one is chosen. 3) when comparing two infeasible solutions, the one with better constraints violation value is chosen. The advantage of this method is that it is easily combined with other algorithms and its results were proven to be effective in most cases. Thus this method was commonly used by researchers in recent years [8, 9]. However, there still exists a big shortcoming that it easily gets into local optimum, because this method prefers to selecting feasible solutions, and at the same time the feasible region is very small for constrained problems, so the selecting pressure is very heavy and the final results are easily converged to local optimal [10].

Multi-objective concept is one of the most outstanding methods that obtains prominent results. It considers the constraint violations as objectives, and solves the constrained problem by means of the technologies in multi-objective optimization area, pareto solutions for instance. Many researchers have done some outstanding work from this aspect [11, 12].

ε-constrained method was first proposed by Takahama and Sakai [13]. As a same with FAD rules, ε-constrained method is also used to compare two solutions. The difference is that ε level is set as threshold level when comparing two solutions. When the constraint violation values of both solutions are smaller than ε level, the one with better objective function value is selected. Otherwise, the one with smaller constraint violation value is selected. Takahama and Sakai combined this method with modified DE and local search methods [14, 15], and won the first place in the competition on evolutionary constrained real parameter single objective optimization in 2006 and 2010 IEEE Congress on Evolutionary Computation (CEC).

However, controlling ε level is a big problem. The effect of ε is to make sure that the search will be conducted in the infeasible region. If ε value is set too large, the efficiency of the algorithm will decrease, and it may fall into local optimal of infeasible region in some cases. Though ε will decrease to 0 in last phase, it still can hardly jump out of local optimum. If ε value is set too small, the exploration will be incomplete and can hardly find the global optimal. The common method to set ε level is to define the initial ε value as the violation value of a certain individual in initial population, and use an exponential decline method to change ε level with the increase of iterations [14, 16, 17]. Besides, Takahama adopted some other improved methods to control ε level, such as dynamic ε control method [18] that accelerates the decline of ε level by accelerating the iterations in exponential decline formula, truncated ε level method [19] that widens the solving scope of the ε-constrained method etc.

Our research about ε-constrained method in this paper is based on the following two theses. The first one is that the majority of ε-constrained method are combined with local search method or other improved optimization methods, whose results show the comprehensive performance of all methods. So we consider several kinds of ε-constrained methods combined with the simple DE algorithm to show the performance of each ε-constrained method by objective comparison. The second one is that almost all ε-constrained methods at present take use of exponential decline method to control ε level. On one hand, the initial ε value is set as the violation value of a certain individual in the initial population. In most cases, the initial ε value is large and the ε value changes very little at the initial phase, which will induce low search efficiency and easily getting into local optimal in the infeasible region. On the other hand, ε value decreases severely at the last phase, which will lost the infeasible solutions that near the optimal solutions. Therefore, we will abandon this kind of exponential decline ε level control method and introduce a novel adaptive ε level control method, in which the ε value will be determined according to the current population information.

The rest of this paper is organized as follows: Section 2 introduces ε-constrained method and ε level control methods, including the adaptive ε level control method proposed by us. Section 3 presents the combination of ε-constrained method and DE algorithm. Section 4 shows the experimental results and their analyses. Section 5 summarizes the conclusions of this paper.

2 ε-Constrained Method

This section will introduce ε-constrained method, including constrained optimization problems, ε level comparisons and 4 kinds of ε level control methods.

2.1 Constrained Optimization Problems

$$Min \quad f(X)$$

$$s.t. \begin{cases} g_i(X) \leq 0 & i = 1, 2, ..., m \\ h_j(X) = 0 & j = 1, 2, ..., p \\ L_k \leq X_k \leq U_k & k = 1, 2, ..., n \end{cases} \tag{1}$$

$$|h_j(X)| - \delta \leq 0 \tag{2}$$

Eq. (1) is the general form of constrained optimization problems. X is an n-dimensional decision variable. $f(X)$ is the objective function, and $g(X)$ and $h(X)$ are the inequality constraints and equality constraints, separately. The equality constraints are always transformed into Eq. (2) where δ is a very small positive value (usually 10^{-4}), called the allowance tolerance.

In dealing with constrained problems, the constrained violation value is another important indicator. The value of constraint violation can be calculated by the following Eq. (3). When the solution is feasible, the constrained violation value $\varphi(X)$ is equal to 0. Otherwise, the constrained violation value $\varphi(X)$ is larger than 0.

$$\varphi(X)=\sum_{i=1}^{m}\max(0,g_i(X))+\sum_{j=1}^{p}(\left|h_j(X)\right|-\delta,0) \tag{3}$$

2.2 ε Level Comparisons

The key of constraint handling technologies is to compare two solutions X_1 and X_2. In ε-constrained method, the comparisons are done according to the rules shown in Eq. (4).

$$(f(X_1),\ \varphi(X_1))<\varepsilon(f(X_2),\ \varphi(X_2))\Leftrightarrow\begin{cases}f(X_1)<f(X_2), & \text{if } \varphi(X_1),\varphi(X_2)\le\varepsilon\\ f(X_1)<f(X_2), & \text{if } \varphi(X_1)=\varphi(X_2)\\ \varphi(X_1)\le\varphi(X_2), & \text{otherwise}\end{cases} \tag{4}$$

If ε is equal to infinity, the comparison between the two solutions is decided by their function values $f(X)$. If ε is equal to 0, the feasible solution is better than infeasible solution, and the solution with smaller constraint violation value is better than the one with larger constraint violation value. In this case, the ε-constrained method is the same with FAD rules.

2.3 ε Level Control Methods

The final effectiveness of ε-constrained method strongly depends on the control method of ε level. In this section, four kinds of ε level control methods are introduced. The first three kinds of ε level control methods are proposed by other researchers, while the last one is proposed by us.

- Static control method

$$\varepsilon(0)=\varphi(X_\theta)$$
$$\varepsilon(t)=\begin{cases}\varepsilon(0)(1-t/T_c)^{cp}, 0<t<T_c\\ 0, \qquad t\ge T_c\end{cases} \tag{5}$$

Static control method is the simplest method to control ε level [14, 16, 17], and is also the foundation of the following two control method -- dynamic control method and truncated control method. In Eq. (5), the initial value of ε is set as the constraint violation value of top θ-th individual in initial population. When the iteration is less than T_c, ε value is updated by an exponential decline way. When the iteration is more than T_c, ε value is equal to 0. The decline speed of ε value is decided by the parameter *cp*.

- Dynamic control method

$$\varepsilon(0) = \varphi(X_\theta)$$

$$\varepsilon(t) = \begin{cases} \varepsilon(0)(1 - t'/T_c)^{cp}, 0 < t' < T_c \\ 0, \quad t' \geq T_c \end{cases} \tag{6}$$

$$t' = \begin{cases} 0, & t = 0 \\ t'+1, & \varphi(X_\eta) \geq \varepsilon(t) \\ t'+2, & \varphi(X_\eta) < \varepsilon(t)\ and\ t'+2 \geq \varepsilon^{-1}(\varphi(X_\eta)) \\ 0.5*(t'+2) + 0.5*\varepsilon^{-1}(\varphi(X_\eta)), & \text{otherwise} \end{cases} \tag{7}$$

$$\varepsilon^{-1}(\varepsilon) = (1 - \sqrt[cp]{\frac{\varepsilon}{\varepsilon(0)}})T_c \tag{8}$$

To increase the decline speed of ε value, Takahama and Sakai [18] utilized an improved iteration *t'* (*t'* > *t*) to replace *t*, as shown in Eq. (6). During the searching process, if the decrease of constraint violation values fast enough, then the increase of *t'* will accelerate. The update method of *t'* is shown In Eq. (7). Where X_η is the worst η-th individual, $\varepsilon^{-1}(\varepsilon)$ is the inverse function of $\varepsilon(t)$, as defined in Eq. (8).

- Truncated control method

On the basis of static control, Takahama and Sakai proposed a truncation mechanism to improve the usability of ε-constraint method [19]. In each iteration, after calculating the ε(*t*) in Eq. (5), the value of $\varphi_{\min}$, $\varphi_{\max}$, φ_0 are also calculated and recorded. Where $\varphi_{\min}$ and φ_{max} are the minimum and maximum of constraint violation values in current population, separately, φ_0 is the number of individuals whose constraint violation values are 0. If $\varphi_0 > ap*N$ (*N* is the total number of individuals), ε is set as 0. Otherwise, ε(*t*) is truncated to $[ap*\varphi_{\min}, ap*\varphi_{max}]$. The recommended *ap* value is 0.9.

- Adaptive control method

$$\varepsilon(t) = \begin{cases} 0, & if\ \varphi_0 > ap_1 * N \,\|\, \varphi_{max} > Th \,\|\, t > Tc \\ ap_2 * \varphi_{\max}, & else \end{cases} \tag{9}$$

Before determining ε value, we first calculate φ_0 and φ_{max}, just the same definitions as in truncated control method. If $\varphi_0 > ap_1*N$, ε is set as 0. However, the recommended ap_1 is set as 0.1, not 0.9. In fact, it is more reasonable to set a small value of ε. Suppose ap_1 is set very small, when it comes to constrained problems that only contain inequality constraints with large feasible region, generating some feasible solution is more likely, thus ε value set as 0 can avoid searching the infeasible region. When it comes to constrained problems that contain equality constraints with small feasible region, few solutions that don't violate the constraints will generate, and at the same time ε value can hardly fall to 0. Suppose ap_1 is set too large, though in the case of large feasible region, it is still hardly to generate most of the feasible solutions, thus ε value can hardly be set as 0.

When $\varphi_0 < ap_1*N$, then φ_{max} should be taken into consideration. If φ_{max} is very large, it shows that the current individuals are far away from feasible region. So it is very urgent to make the search approach the feasible region as soon as possible. To set ε value as 0 is the best choice to achieve this goal. However, using the exponential decline method to set ε value will make ε value keep at a large value for a long time, which will reduce the efficiency of algorithm, even make the algorithm converge at the local optimal in infeasible region. Therefore, we set ε value as 0 when φ_{max} is larger than a given threshold *Th*, where *Th* is recommended as 10.

When $\varphi_{max} < Th$ and $\varphi_0 < ap_1*N$, it shows that all individuals are close to feasible region, so the ε value should be not to set as 0. We refers to the truncated control method and sets $\varepsilon(t) = ap_2*\varphi_{max}$. In this way, not only is ε value set not so large, but also make ε value decline slowly so that searching is conducted completely near the feasible region. ap_2 is recommended as 0.9 as well.

3 ε-Constrained Method Based DE Algorithm

DE was proposed by Storn and Price [20]. Due to its characters of simple, efficient and easily programming, it has been widely used in continuous optimization area. The basic steps of DE include initialization, mutation, crossover and selection. According to different mutation strategies, and crossover strategies, DE has many different variants, such as DE/best/1/bin and DE/rand/1/exp. The most important two parameters in DE are mutation factor *F* and crossover probability factor *CR*.

In the previous literature about the combination of DE algorithm and ε-constrained method, the local search method or improved DE are always adopted, whose final results are much more perfect, but the nature of ε-constrained method itself can't be reflected. So in this paper, we combine ε-constrained method with the simplest DE algorithm DE/rand/1/exp. Four kinds of ε-constrained methods explained in section 2.3, namely static ε, dynamic ε, truncated ε and adaptive ε, are combined with DE. These combinations are denoted by static εDE, dynamic εDE, truncated εDE and adaptive εDE separately and will be compared with each other. The pseudo code is shown in figure 1 where $u(0,1)$ is a uniform random number generator in [0, 1]. The pseudo code is almost the same as the one in [18] except that there is a boundary treatment in our algorithm. Obviously, the boundary treatment is necessary. Otherwise, the search will be out of bounds for some problems.

εDE/rand/1/exp()
{
$P(0)$=Generate N individuals $\{x^i\}$ randomly;
$\varepsilon=\varepsilon(0)$;
for (t=1; $t \leq T_{max}$; t++) {
for (i=1; $i \leq N$; i++) {
(p_1, p_2, p_3)=select randomly from $[1, N]$
s.t. $p_1 \neq p_2 \neq p_3 \neq i$;
$x^{new} = x^i \in P(t\text{-}1)$;
j=select randomly from $[1, n]$;
k=1;
do {
$x_j^{new} = x_j^{p_1} + F(x_j^{p_2} - x_j^{p_3})$;
if ($x_j^{new} < L_k$)
$x_j^{new} = L_k + u(0,1)(x_j^{p_1} - L_k)$
else if ($x_j^{new} > U_k$)
$x_j^{new} = x_j^{p_1} + u(0,1)(U_k - x_j^{p_1})$
$j = (j+1)$ %n;
k++;
} while ($k \leq n$ && $u(0,1) < CR$);
if (($f(x^{new}), \phi(x^{new})) <_{\varepsilon} (f(x^i), \phi(x^i))$)
$z^i = x^{new}$;
else
$z^i = x^i$;
}
$P(t)=\{z^i, i = 1, 2, \ldots, N\}$
$\varepsilon=\varepsilon(t)$;
}
}

Fig. 1. The pseudo code of εDE

4 Numerical Experiment

The benchmark function set proposed in the CEC2006 competition on evolutionary constrained real parameter single objective optimization [21] is used in this paper to test these four algorithms. Each algorithm runs 25 times independently. The parameters of the four algorithms are listed in table 1 where "Common" means these parameters are commonly used in the four algorithms.

Table 1. Parameter setting of the four algorithms

Common	N=40, T_{max}=5000, F=0.7, CR=0.9, T_c=0.2T_{max}
static εDE	θ=0.2N, cp=5
dynamic εDE	θ=0.2N, cp=5, η=3
truncated εDE	θ=0.2N, cp=5, ap=0.9
adaptive εDE	Th=10, ap_1=0.1, ap_2=0.9

Table 2 and Table 3 show the results of these four algorithms on 24 standard constraint test functions. The best value, average value, worst value and standard deviation (Std) value are recorded, and the number of constraints violation on minority solutions are also recorded in the brackets after best value or worst value. These four algorithms perform very well for the majority functions, and worse results are in bold.

From the results, we can see that static εDE is the worst one. It obtains bad results on 11 test functions, and performs unstable on G01, G08, G10, G15 and G16 functions, while the other three algorithms performs very well. For G01, G08, G18 and G23 functions, some results fall into local optimal in feasible region. For G10, G15, G16 and G17 functions, some results get into local optimal in infeasible region. For G20, G21 and G22 functions, all results fall into local optimal in infeasible region.

The results of dynamic εDE on G03 and G13 functions are not so good, while the other three algorithms obtain the optimal value. It performs unstable on G17, G18 and G23 functions, and some results fall into local optimal. What should be noticed is that it performs better on G18 than other algorithms. Dynamic εDE is very difficult to solve G20, G21 and G22 functions. However, it finds few local optimal in feasible region on G21, which is better than static εDE and truncated εDE. Some results of truncated εDE on G17, G18 and G23 functions fall into local optimal in feasible region, and it also fall into local optimal in infeasible on G20, G21, and G22 functions without exception. However, truncated εDE performs better on stability than static εDE and dynamic εDE.

Different from the above three algorithms, adaptive εDE proposed in this paper can obtain optimal value of G17 function in each run. For G21 function, adaptive εDE can avoid falling into local optimal in infeasible region, and obtain optimal value in most runs. For G23 function, adaptive εDE can avoid falling into local optimal in feasible region, and will obtain optimal value if given enough iterations. For G22 that none algorithm can obtain optimal value, the number of constraints violation of adaptive εDE is less than that of other algorithms. From the results, it can be seen that adaptive εDE will not fall into local optimal in infeasible region due to the initial ε value. Only on G18 and G20, adaptive εDE has no obvious improvements.

Table 2. Results of four algorithms for G01 to G12

function/ optimal	criteria	Algorithms			
		static εDE	dynamic εDE	truncated εDE	Adaptive εDE
G01 -15	best	**-15**	-15	-15	-15
	mean	**-14.92**	-15	-15	-15
	worst	**-13**	-15	-15	-15
	std	**3.92E-01**	0.0E+00	0.0E+00	0.0E+00
G02 -0.80361	best	-0.80361	-0.80361	-0.80361	-0.80361
	mean	-0.80112	-0.80229	-0.80229	-0.80237
	worst	-0.78526	-0.79260	-0.79260	-0.79260
	std	4.92E-03	3.58E-03	3.58E-03	3.37E-03
G03 -1.0005	best	-1.0005	**-1.0005**	-1.0005	-1.0005
	mean	-1.0005	**-0.95729**	-1.0005	-1.0005
	worst	-1.0005	**-0.01342**	-1.0005	-1.0005
	std	2.77E-16	**1.93E-01**	3.88E-14	4.59E-16

Table 2. (*continued*)

G04 -30665.5	best	-30665.5	-30665.5	-30665.5	-30665.5
	mean	-30665.5	-30665.5	-30665.5	-30665.5
	worst	-30665.5	-30665.5	-30665.5	-30665.5
	std	0.0E+00	0.0E+00	0.0E+00	0.0E+00
G05 5126.496	best	5126.496	5126.496	5126.496	5126.496
	mean	5126.496	5126.496	5126.496	5126.496
	worst	5126.496	5126.496	5126.496	5126.496
	std	2.77E-12	2.77E-12	2.77E-12	1.95E-12
G06 -6961.81	best	-6961.81	-6961.81	-6961.81	-6961.81
	mean	-6961.81	-6961.81	-6961.81	-6961.81
	worst	-6961.81	-6961.81	-6961.81	-6961.81
	std	0. 0E+00	0.0E+00	0.0E+00	0.0E+00
G07 24.30620	best	24.30620	24.30620	24.30620	24.30620
	mean	24.30620	24.30620	24.30620	24.30620
	worst	24.30620	24.30620	24.30620	24.30620
	std	4.59E-08	9.10E-09	2.41E-08	2.84E-09
G08 -0.09582	best	**-0.09582**	-0.09582	-0.09582	-0.09582
	mean	**-0.08485**	-0.09582	-0.09582	-0.09582
	worst	**-0.02726**	-0.09582	-0.09582	-0.09582
	std	**2.51E-02**	8.78E-18	1.07E-17	9.21E-18
G09 680.6300	best	680.6300	680.6300	680.6300	680.6300
	mean	680.6300	680.6300	680.6300	680.6300
	worst	680.6300	680.6300	680.6300	680.6300
	std	3.23E-13	3.37E-13	3.27E-13	3.39E-13
G10 7049.248	best	**7049.248**	7049.248	7049.248	7049.248
	mean	**10597.42**	7049.248	7049.248	7049.248
	worst	**21000(2)**	7049.248	7049.248	7049.248
	std	**5.92E+03**	3.73E-09	3.63E-10	1.09E-10
G11 0.7499	best	0.7499	0.7499	0.7499	0.7499
	mean	0.7499	0.7499	0.7499	0.7499
	worst	0.7499	0.7499	0.7499	0.7499
	std	0.0E+00	0.0E+00	0.0E+00	0.0E+00
G12 -1.0000	best	-1	-1	-1	-1
	mean	-1	-1	-1	-1
	worst	-1	-1	-1	-1
	std	0.0E+00	0.00E+00	0.0E+00	0.0E+00

Table 3. Results of four algorithms for G13 to G24

function/ optimal	criteria	Algorithms			
		static εDE	dynamic εDE	truncated εDE	Adaptive εDE
G13 0.053941514	best	0.053941514	**0.053941514**	0.053941514	0.053941514
	mean	0.053941514	**0.115519289**	0.053941514	0.053941514
	worst	0.053941514	**0.438802608**	0.053941514	0.053941514
	std	1.34E-17	**1.41E-01**	1.76E-17	2.06E-17
G14 -47.76488846	best	-47.7648884	-47.7648884	-47.7648884	-47.7648884
	mean	-47.7648884	-47.7648884	-47.7648884	-47.7648884
	worst	-47.7648884	-47.7648884	-47.7648884	-47.7648884
	std	2.47E-09	1.42E-10	3.36E-09	3.98E-11
G15 961.7150223	best	**961.7150223**	961.7150223	961.7150223	961.7150223
	mean	**964.2289677**	961.7150223	961.7150223	961.7150223
	worst	**967.9998857(2)**	961.7150223	961.7150223	961.7150223
	std	**3.08E+00**	0.00E+00	0.00E+00	0.00E+00
G16 -1.905155259	best	**-1.90515525**	-1.90515525	-1.90515525	-1.90515525
	mean	**-1.91080590**	-1.90515525	-1.90515525	-1.90515525
	worst	**-2.0898279(1)**	-1.90515525	-1.90515525	-1.90515525
	std	**3.66E-02**	4.37E-16	0.00E+00	0.00E+00
G17 8853.533875	best	**8919.889862**	**8853.533875**	**8853.533875**	8853.533875
	mean	**2555.787777**	**8860.28848**	**8859.95178**	8853.533875
	worst	**1.04E-17 (2)**	**8939.240999**	**8934.636506**	8853.533875
	std	**4.10E+03**	**2.29E+01**	**2.18E+01**	0.00E+00
G18 -0.866025404	best	**-0.86602540**	**-0.86602540**	**-0.86602540**	**-0.86602540**
	Mean	**-0.79679049**	**-0.843099781**	**-0.81381773**	**-0.822102356**
	Worst	**-0.50000000**	**-0.674972761**	**-0.50000000**	**-0.5**
	Std	**1.06E-01**	**6.21E-02**	**1.11E-01**	**1.19E-01**
G19 32.65559295	Best	32.65559314	32.65559356	32.65559379	32.65559343
	Mean	32.65559553	32.65559576	32.6556117	32.65559625
	Worst	32.65560041	32.65560882	32.65574281	32.65561155
	Std	1.89E-06	3.05647E-06	2.88E-05	3.64E-06
G20 0.204979400	Best	**0.205101(17)**	**0.20031(15)**	**0.2023(18)**	**0.2002(15)**
	Mean	**0.029963046**	**0.194864563**	**0.19499096**	**0.1946139**
	Worst	**8.69E-06(8)**	**0.18042(20)**	**0.18833(17)**	**0.189 (16)**
	Std	**8.42E-02**	**6.53E-03**	**4.78E-03**	**4.09E-03**

Table 3. (*continued*)

G21 193.7245101	Best	**3.03E-136(4)**	324.7028419	**6.97E-39(4)**	**193.7245101**
	Mean	**1.91E-120**	**12.98811367**	**3.70E-26**	**246.1158428**
	Worst	**1.38E-151(4)**	**2.31E-09(2)**	**2.48E-49(4)**	**324.7028419**
	Std	**8.73E-120**	**6.36E+01**	**1.81E-25**	**6.42E+01**
G22 236.4309755	Best	**1.54E-109(20)**	**1.23E-06(20)**	**1.18E-38(20)**	**15014.93(19)**
	Mean	**8.66E-94**	**547.5489231**	**2.35E-30**	**12589.95184**
	Worst	**3.32E-107(20)**	**2.23E-14(20)**	**1.72E-33(20)**	**18931.39(19)**
	Std	**3.83E-93**	**2.68E+03**	**8.00E-30**	**5.88E+03**
G23 -400.0551000	Best	**-400.0550997**	**-400.0551**	**-400.0550999**	**-400.0550999**
	Mean	**-329.2623667**	**-381.2655762**	**-348.988144**	**-398.9685932**
	Worst	**-100.0461811**	**-100.0465997**	**-100.0459819**	**-374.7636378**
	Std	**1.16E+02**	**5.93E+01**	**1.09E+02**	**4.95E+00**
G24 -5.508013272	Best	-5.508013272	-5.508013272	-5.508013272	-5.508013272
	Mean	-5.508013272	-5.508013272	-5.508013272	-5.508013272
	Worst	-5.508013272	-5.508013272	-5.508013272	-5.508013272
	Std	0.0E+00	0.0E+00	0.0E+00	0.0E+00

To illustrate the efficiency of proposed adaptive ε level control method in detail, we draw the convergence curves of each algorithm on G21 function as a delegate, as shown in Fig.2. The static εDE, dynamic εDE, and truncated εDE cannot obtain the optimal value of function G21, while adaptive εDE obtains the optimal value 193.7245101. From the cure of ε value, it can been seen that the ε value in the first three pictures keeps at a large value (beyond 100)at the beginning phase for a long time, so as to the search is done in the infeasible region for a long time. What's worse, even though ε value declines to 0 at last, the search still stay in the infeasible region and the final solution cannot jump out of the local optimal solution in infeasible region. The cure in the fourth picture shows that ε value setting as 0 at initial phase (cannot be shown directly in the cure) still search in the infeasible region. However, with the slowly declining of ε value, the searching drops out of infeasible region and enters into feasible region. Finally, the optimal solution is found. Therefore, our improvements about ε value controlling are successful and effective.

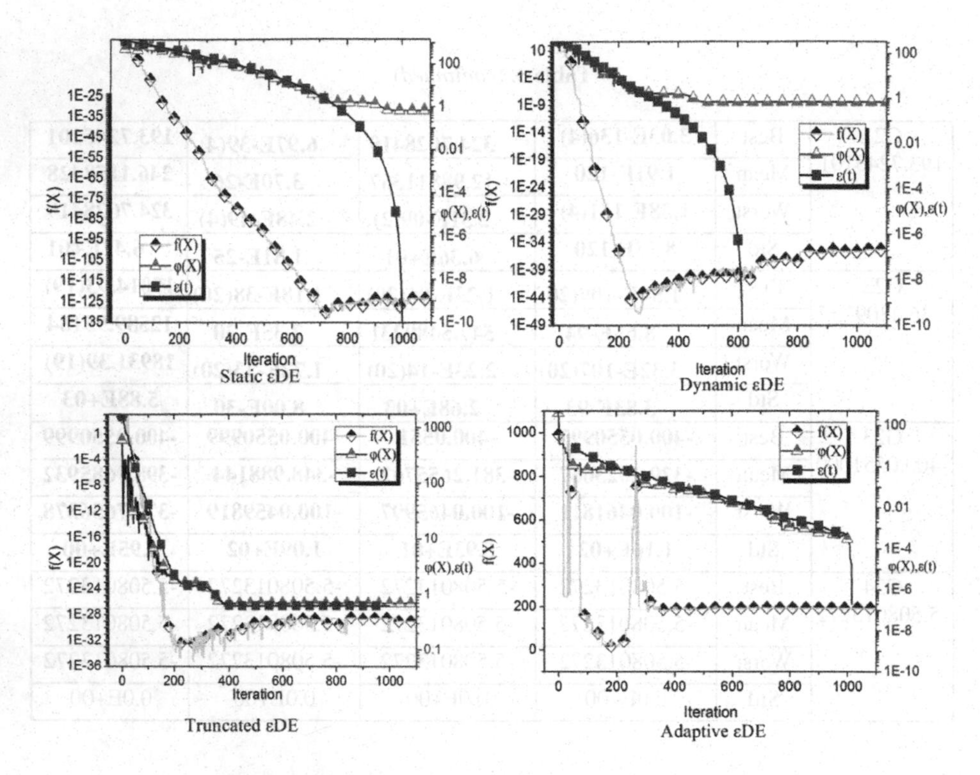

Fig. 2. Convergence curves for four algorithms on G21

5 Conclusions and Future Work

This paper proposes a novel adaptive ε constrained method, and combines it with DE/rand/1/exp. This adaptive ε constrained method no longer uses classical exponential decline method to control ε value. Instead, the current population state is used to control ε level. Compared to the other three control methods, adaptive ε method can avoid falling into local optimal in infeasible region due to the large value of initial ε. The method improves the stability, accelerate the convergence speed, and enhance the solving efficiency of algorithm. Even combined with the simplest DE algorithm, a suitable ε level control method will make the constrained optimization algorithm powerful.

The parameter *Th* in our proposed adaptive ε constrained method is set artificially. It is obvious that this parameter had better be set solely according to different problems, which is the main shortcoming of this method. How to overcome this shortcoming? Putting forward another powerful ε constrained method is a worthy direction. This ε constrained method can also be combined with other algorithms to solve other constrained problems, including some complex engineering problems.

Acknowledgement. This research work is supported by the National Basic Research Program of China (973 Program) under Grant no. 2011CB706804, and the Natural Science Foundation of China (NSFC) under Grant no. 51005088 and 51121002.

References

1. Mezura-Montes, E., Coello Coello, C.A.: Constraint-handling in nature-inspired numerical optimization: past, present and future. Swarm and Evolutionary Computation 1(4), 173–194 (2011)
2. Tessema, B., Yen, G.G.: An Adaptive Penalty Formulation for Constrained Evolutionary Optimization. IEEE Transactions on Systems Man and Cybernetics Part a-Systems and Humans 39(3), 565–578 (2009)
3. Mani, A., Patvardhan, C.: A Novel Hybrid Constraint Handling Technique for Evolutionary Optimization. In: 2009 IEEE Congress on Evolutionary Computation, vols. 1-5, pp. 2577–2583 (2009)
4. Runarsson, T.P., Yao, X.: Stochastic ranking for constrained evolutionary optimization. IEEE Transactions on Evolutionary Computation 4(3), 284–294 (2000)
5. Runarsson, T.P., Yao, X.: Search biases in constrained evolutionary optimization. IEEE Transactions on Systems Man and Cybernetics Part C-Applications and Reviews 35(2), 233–243 (2005)
6. Zhang, M., Luo, W., Wang, X.F.: Differential evolution with dynamic stochastic selection for constrained optimization. Information Sciences 178(15), 3043–3074 (2008)
7. Deb, K.: An efficient constraint handling method for genetic algorithms. Computer Methods in Applied Mechanics and Engineering 186(2-4), 311–338 (2000)
8. Mezura-Montes, E., Velazquez-Reyes, J., Coello, C.A.C.: Modified differential evolution for constrained optimization. In: 2006 IEEE Congress on Evolutionary Computation, vols. 1-6, pp. 25–32 (2006)
9. Karaboga, D., Akay, B.: A modified Artificial Bee Colony (ABC) algorithm for constrained optimization problems. Applied Soft Computing 11(3), 3021–3031 (2011)
10. Mezura-Montes, E., Coello, C.A.C.: A simple multimembered evolution strategy to solve constrained optimization problems. IEEE Transactions on Evolutionary Computation 9(1), 1–17 (2005)
11. Wang, Y., Cai, Z., Zhou, Y., et al.: An adaptive tradeoff model for constrained evolutionary optimization. IEEE Transactions on Evolutionary Computation 12(1), 80–92 (2008)
12. Venter, G., Haftka, R.T.: Constrained particle swarm optimization using a bi-objective formulation. Structural and Multidisciplinary Optimization 40(1-6), 65–76 (2010)
13. Takahama, T., Sakai, S., Iwane, N.: Constrained optimization by the epsilon constrained hybrid algorithm of particle swarm optimization and genetic algorithm. In: Zhang, S., Jarvis, R.A. (eds.) AI 2005. LNCS (LNAI), vol. 3809, pp. 389–400. Springer, Heidelberg (2005)
14. Takahama, T., Sakai, S.: Constrained optimization by the epsilon constrained differential evolution with gradient-based mutation and feasible elites. In: 2006 IEEE Congress on Evolutionary Computation, vols. 1-6, pp. 1–8 (2006)
15. Takahama, T., Sakai, S.: Constrained Optimization by the e Constrained Differential Evolution with an Archive and Gradient-Based Mutation. In: 2010 IEEE Congress on Evolutionary Computation, Cec (2010)

16. Takahama, T., Sakai, S.: Efficient Constrained Optimization by the epsilon Constrained Rank-Based Differential Evolution. In: 2012 IEEE Congress on Evolutionary Computation, Cec (2012)
17. Takahama, T., Sakai, S.: Efficient Constrained Optimization by the epsilon Constrained Differential Evolution with Rough Approximation Using Kernel Regression. In: 2013 IEEE Congress on Evolutionary Computation (Cec), pp. 1334–1341 (2013)
18. Takahama, T., Sakai, S.: Constrained optimization by ε constrained differential evolution with dynamic ε-level control. In: Chakraborty, U.K. (ed.) Advances in Differential Evolution. SCI, vol. 143, pp. 139–154. Springer, Heidelberg (2008)
19. Takahama, T., Sakai, S.: Efficient Constrained Optimization by the epsilon Constrained Adaptive Differential Evolution. In: 2010 IEEE Congress on Evolutionary Computation, Cec (2010)
20. Storn, R., Price, K.: Differential evolution - A simple and efficient heuristic for global optimization over continuous spaces. Journal of Global Optimization 11(4), 341–359 (1997)
21. Liang, J., Runarsson, T., Mezura-Montes, E., et al.: Problem Definitions and Evaluation Criteria for the CEC 2006, Special Session on Constrained Real-Parameter Optimization. Report# 2006005, Nanyang Technological University, Singapore (2005)

Sensitivity Analysis of Fuzzy Signatures Using Minkowski's Inequality

István Á. Harmati[1], Ádám Bukovics[2], and László T. Kóczy[3,4]

[1] Széchenyi István University,
Department of Mathematics and Computational Sciences
Egyetem tér 1, 9026 Győr, Hungary
harmati@sze.hu
[2] Széchenyi István University,
Department of Structural and Geotechnical Engineering
Egyetem tér 1, 9026 Győr, Hungary
[3] Széchenyi István University,
Department of Automation
Egyetem tér 1, 9026 Győr, Hungary
[4] Budapest University of Technology and Economics, Budapest
Department of Telecommunications and Media Informatics

Abstract. In this paper we give upper bounds on the change of the weighted generalized mean aggregation operator via Minkowski's inequality. Then we apply these results to characterize the sensitivity of fuzzy signatures which equipped with such aggregation operators in their nodes. A sensitivity of a real-life fuzzy signature used for ranking buildings is also examined.

Keywords: weighted mean, sensitivity analysis, fuzzy signature, buildings ranking.

1 Introduction

Fuzzy signatures are hierarchical representations of data structuring into vectors of fuzzy values [1]. A fuzzy signature is defined as a special multidimensional fuzzy data structure, which is a generalization of vector valued fuzzy sets [2], [3], [4]. Vector valued fuzzy sets are special cases of L-fuzzy sets which were introduced in [5]. A fuzzy signature is denoted by

$$A\colon X \to S^{(n)},$$

where $1 \leq n$ and

$$S^{(n)} = \times_{i=1}^{n} S_i \qquad S_i = \begin{cases} [0,1] \\ S^{(m)} \end{cases}$$

We can represent a fuzzy signature by nested vector value fuzzy sets and also by a tree graph (see Figure 1), which is much more understandable [6].

H. Handa et al. (eds.), *Proc. of the 18th Asia Pacific Symp. on Intell. & Evol. Systems – Vol. 1*,
Proceedings in Adaptation, Learning and Optimization 1, DOI: 10.1007/978-3-319-13359-1_45

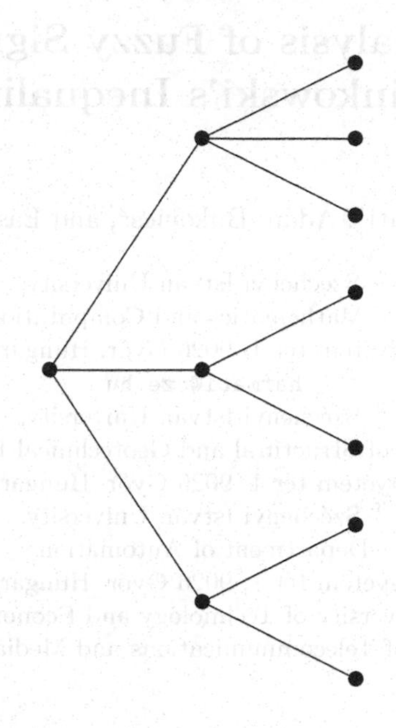

Fig. 1. A fuzzy signature graph

The goal of this article is to discuss how the membership value of the whole fuzzy set changes if the membership values in the nested vectors change. In other words, if we think of the tree graph representation, how the membership value of the root changes if the membership values of leaves change. To answer this question we have to know how to compute a membership value of a subgraph from the leaves. In this article we assume that all the operators applied on membership values in the signature are from the class of weighted generalized mean aggregation operators.

The paper organized as follows: in Section 2 we recall some mathematical tools, in Section 3 the sensitivity of WGM is discussed, in Section 4 we examine the sensitivity of fuzzy signatures.

2 Mathematical Background

The generalized mean and its generalization, the weighted generalized mean form a very large class of aggregation operators. Their various special cases often arise also in theoretical and practical problems.

Definition 1 (Generalized mean). *Let $x_1, \ldots, x_n$ be nonnegative real numbers and $p \in \mathbb{R}$ $(p \neq 0)$. Then their generalized mean with parameter p:*

$$M_p(x_1, \ldots, x_n) = \left[\frac{1}{n}\sum_{k=1}^{n} x_k^p\right]^{\frac{1}{p}} \tag{1}$$

Some special cases in p:

- $p = 1$ arithmetic mean
- $p = 2$ quadratic mean
- $p = -1$ harmonic mean

Definition 2 (Weighted generalized mean; WGM). *Let $x_1, \ldots, x_n$ and $w_1, \ldots, w_n$ be nonnegative real numbers, $w_i \geq 0, \sum_{i=1}^{n} w_i = 1$ and $p \in \mathbb{R}$ $(p \neq 0)$. Then the weighted generalized mean of $x_1, \ldots, x_n$ with weights $w_1, \ldots, w_n$ and with parameter p:*

$$M_p^w(x_1, \ldots, x_n) = \left[\sum_{k=1}^{n} w_k x_k^p\right]^{\frac{1}{p}} \tag{2}$$

The generalized mean is a special case of the weighted generalized mean with weights $w_k = \frac{1}{n}$. The limits at $\pm\infty$ regardless to the weights:

$$\lim_{p \to \infty} \left[\sum_{k=1}^{n} w_k x_k^p\right]^{\frac{1}{p}} = \max(x_i) \tag{3}$$

$$\lim_{p \to -\infty} \left[\sum_{k=1}^{n} w_k x_k^p\right]^{\frac{1}{p}} = \min(x_i) \tag{4}$$

The limit if $p \to 0$ is the weighted geometric mean:

$$\lim_{p \to 0} \left[\sum_{k=1}^{n} w_k x_k^p\right]^{\frac{1}{p}} = \prod_{i=1}^{n} x_i^{w_i} \tag{5}$$

Our goal is to give an upper bound on the changing of M if we know the changing of the input values $x_1, \ldots, x_n$. We search for such a bound for $|\Delta M|$ which depends on $\Delta \underline{x}$ or on a kind of vector norm of $\Delta \underline{x}$. First we recall the definition of the p-norm.

Definition 3 (p-norm). *Let $p \geq 1$ a real number and $\underline{x} = (x_1, \ldots, x_n) \in \mathbb{R}^n$. Then the p-norm of $\underline{x}$*

$$\|\underline{x}\|_p = \left(\sum_{k=1}^{n} |x_k|^p\right)^{\frac{1}{p}} \tag{6}$$

Some widely used p-norms:

- $p=1$ (taxicab norm) $\|\underline{x}\|_1 = |x_1| + \ldots + |x_n|$
- $p=2$ (euclidean norm) $\|\underline{x}\|_2 = \sqrt{x_1^2 + \ldots + x_n^2}$
- $p=\infty$ (maximum norm) $\|\underline{x}\|_\infty = \max(|x_1|, \ldots, |x_n|)$

Two important properties of the p-norm:

- If $1 \le p \le q \le \infty$ then $\|\underline{x}\|_q \le \|\underline{x}\|_p$.
- If $1 \le p \le q \le \infty$ then $\|\underline{x}\|_p \le \|\underline{x}\|_q \cdot n^{1/p-1/q}$.

We will use the generalization of the triangular inequality, the so called Minkowski's inequality.

Theorem 1 (Minkowski's inequality). *(see for example [7]) Let* $\underline{a}$, $\underline{b} \in \mathbb{R}^n$, $p \ge 1$, *then the following inequality holds:*

$$\|\underline{a} + \underline{b}\|_p \le \|\underline{a}\|_p + \|\underline{b}\|_p \tag{7}$$

The generalization of the reverse triangular inequalty also holds:

Corollary 1. *If* $\underline{a}$, $\underline{b} \in \mathbb{R}^n$, $p \ge 1$, *then*

$$\big|\|\underline{a}\|_p - \|\underline{b}\|_p\big| \le \|\underline{a} - \underline{b}\|_p \tag{8}$$

3 Sensitivity of the WGM for $p \ge 1$

In this section we analyse the change of the WGM under the change of its input vector. Note that we examine the case $p \ge 1$. Let we use the following notations:

$$\underline{w}^{1/p} = \left(w_1^{1/p}, \ldots, w_n^{1/p}\right) \tag{9}$$

$$\underline{w}^{1/p} \cdot \underline{x} = \left(w_1^{1/p} \cdot x_1, \ldots, w_n^{1/p} \cdot x_n\right) \tag{10}$$

If the input vector is $\underline{x} = (x_1, \ldots, x_n)$, the vector if the weights is $\underline{w} = (w_1, \ldots, w_n)$, then the weighted generalized mean with parameter p is

$$M = \left[\sum_{i=1}^n w_i x_i^p\right]^{\frac{1}{p}} = \left[\sum_{i=1}^n \left(w_i^{1/p} x_i\right)^p\right]^{\frac{1}{p}} = \left\|\underline{w}^{1/p} \cdot \underline{x}\right\|_p \tag{11}$$

If the new (maybe perturbed) input vector is $\underline{x}^* = (x_1^*, \ldots, x_n^*)$, then the new output is $M^* = \|\underline{w}^{1/p} \cdot \underline{x}^*\|_p$. So the change of the input is $\Delta\underline{x} = \underline{x}^* - \underline{x}$, the change of the output is $\Delta M = M^* - M$. In the following we give upper estimations for $|\Delta M|$.

$$|\Delta M| = \left|\left\|\underline{w}^{1/p} \cdot \underline{x}^*\right\|_p - \left\|\underline{w}^{1/p} \cdot \underline{x}\right\|_p\right| \le \left\|\underline{w}^{1/p} \cdot \underline{x}^* - \underline{w}^{1/p} \cdot \underline{x}\right\|_p \tag{12}$$

$$= \left\|\underline{w}^{1/p} \cdot (\underline{x}^* - \underline{x})\right\|_p = \left\|\underline{w}^{1/p} \cdot \Delta\underline{x}\right\|_p = \left[\sum_{i=1}^n \left(w_i^{1/p} |\Delta x_i|\right)^p\right]^{\frac{1}{p}} \tag{13}$$

We can use this formula when the precision of the inputs are known. For example if we know that the absolute value of the change is less than ε for all i ($|\Delta x_i| < \varepsilon$) then we have

$$|\Delta M| \leq \left[\sum_{i=1}^{n}\left(w_i^{1/p}\varepsilon\right)^p\right]^{\frac{1}{p}} = \left[\sum_{i=1}^{n} w_i\varepsilon^p\right]^{\frac{1}{p}} = \varepsilon\cdot\left[\sum_{i=1}^{n} w_i\right]^{\frac{1}{p}} = \varepsilon \tag{14}$$

so in this case the output value is less than ε also. An other way when we give upper bounds with the norm of the change of the input vector. Based on the previous upper estimation we get that

$$|\Delta M| \leq \left[\sum_{i=1}^{n}\left(w_i^{1/p}|\Delta x_i|\right)^p\right]^{\frac{1}{p}} \leq \left[\sum_{i=1}^{n}\left(w_i^{1/p}\right)^p\right]^{\frac{1}{p}}\cdot\left[\sum_{i=1}^{n}|\Delta x_i|^p\right]^{\frac{1}{p}} \tag{15}$$

$$= \left[\sum_{i=1}^{n} w_i\right]^{\frac{1}{p}}\cdot\|\Delta\underline{x}\|_p = \|\Delta\underline{x}\|_p \tag{16}$$

We note that in this case some information lost because only the norm of the change is used, but not the whole vector. As in the previous example if we know that the absolute value of the change is less then ε for all i ($|\Delta x_i| < \varepsilon$) then now we get weaker estimation:

$$|\Delta M| \leq \left[\sum_{i=1}^{n}\varepsilon^p\right]^{\frac{1}{p}} = [n\cdot\varepsilon^p]^{\frac{1}{p}} = n^{1/p}\cdot\varepsilon \tag{17}$$

If the parameter of the aggregation operator is p, but we would like to measure the change of the input vector in q norm, then we have to switch form p to q using the properties of p-norm. We handle the two kind of upper estimations on $|\Delta M|$ as different cases.

If the starting point is that $|\Delta M| \leq \left\|\underline{w}^{1/p}\cdot\Delta\underline{x}\right\|_p$ then

– if $p \leq q$ then

$$|\Delta M| \leq \left\|\underline{w}^{1/p}\cdot\Delta\underline{x}\right\|_p \leq n^{1/p-1/q}\cdot\left\|\underline{w}^{1/p}\cdot\Delta\underline{x}\right\|_q \leq n^{1/p-1/q}\cdot\left\|\underline{w}^{1/p}\right\|_\infty\cdot\|\Delta\underline{x}\|_q \tag{18}$$

– if $p > q$ then

$$|\Delta M| \leq \left\|\underline{w}^{1/p}\cdot\Delta\underline{x}\right\|_p \leq \left\|\underline{w}^{1/p}\cdot\Delta\underline{x}\right\|_q \leq \left\|\underline{w}^{1/p}\right\|_q\cdot\|\Delta\underline{x}\|_q \tag{19}$$

If use the estimation $|\Delta M| \leq \|\Delta\underline{x}\|_p$ then

– if $p \leq q$ then

$$|\Delta M| \leq \|\Delta\underline{x}\|_p \leq n^{1/p-1/q}\cdot\|\Delta\underline{x}\|_q \tag{20}$$

– if $p > q$ then

$$|\Delta M| \leq \|\Delta\underline{x}\|_p \leq \|\Delta\underline{x}\|_q \tag{21}$$

Easy to check that the bounds from the second estimation are weaker.

3.1 Special Case: Equal Weights

A special case worth mentioning when $w_i = 1/n$ for all i. The computations and the final formulas are much more simpler then in general case.

$$|\Delta M| = \left| \left[\sum_{i=1}^{n} \frac{1}{n} x_i^{*p} \right]^{\frac{1}{p}} - \left[\sum_{i=1}^{n} \frac{1}{n} x_i^{p} \right]^{\frac{1}{p}} \right| \tag{22}$$

$$= \left(\frac{1}{n}\right)^{1/p} \cdot \left| \left[\sum_{i=1}^{n} x_i^{*p} \right]^{\frac{1}{p}} - \left[\sum_{i=1}^{n} x_i^{p} \right]^{\frac{1}{p}} \right| \tag{23}$$

$$= \left(\frac{1}{n}\right)^{1/p} \cdot \left| \|\underline{x}^*\|_p - \|\underline{x}\|_p \right| \le \left(\frac{1}{n}\right)^{1/p} \cdot \|\Delta \underline{x}\|_p \tag{24}$$

If the change of the input vector is measured in other norm (q) then

– if $p \le q$ then

$$|\Delta M| \le \left(\frac{1}{n}\right)^{1/p} \cdot \|\Delta \underline{x}\|_p \le \left(\frac{1}{n}\right)^{1/p} \cdot n^{1/p-1/q} \cdot \|\Delta \underline{x}\|_q = n^{-1/q} \cdot \|\Delta \underline{x}\|_q \tag{25}$$

– if $p > q$ then

$$|\Delta M| \le \left(\frac{1}{n}\right)^{1/p} \cdot \|\Delta \underline{x}\|_p \le n^{-1/p} \cdot \|\Delta \underline{x}\|_q \tag{26}$$

So the general form is

$$|\Delta M| \le n^{-\min(1/p,1/q)} \cdot \|\Delta \underline{x}\|_q \tag{27}$$

4 Sensitivity of a Fuzzy Signature

Applying the results of the previous section we can analyse the sensitivity of fuzzy signatures in which the values are determined by a WGM operator in every nodes. The sensitivity bound of the whole fuzzy signature can be derived from the bounds of the WGM-s, according to the graph structure of the signature. The whole computation can be carried out from the root of the signature up to the leaves.

The sensitivity analysis of a fuzzy signature becomes much more simple if the value of the parameter p is the same for all of the WGM operators applied in the nodes. If this condition holds, the output vale of the signature is the weighted generalized mean of the input values with parameter p, where the weights are the product of the weights form the root to the leaves.

Definition 4. *A fuzzy signature is called homogeneous if all of the aggregation operators in the nodes are weighted generalized mean operators with the same value of p.*

Lemma 1. *The WGM of* $y_1, \ldots, y_k$ *with weights* $v_1, \ldots, v_k$ *and with parameter* p *where all of the* y_i*-s are WGM's of* x_{ji}*-s with weights* $w_{1j}, \ldots, w_{n_i j}$ *and with the same parameter of* p*, is the WGM of the* x*-s with weights* $v_i \cdot w_{ji}$

Proof.

$$\left[\sum_{i=1}^{k} v_i \cdot y_i^p\right]^{\frac{1}{p}} = \left[\sum_{i=1}^{k} v_i \cdot \left[\left[\sum_{j=1}^{n_i} w_{ji} \cdot x_{ji}^p\right]^{\frac{1}{p}}\right]^p\right]^{\frac{1}{p}} = \left[\sum_{i=1}^{k}\sum_{j=1}^{n_i} v_i \cdot w_{ji} \cdot x_{ji}^p\right]^{\frac{1}{p}} \tag{28}$$

So the sensitivity analysis of a homogeneous fuzzy signature is nothing else but the simple sensitivity analysis of only one weighted generalized mean aggregation operator.

4.1 Real-Life Example from Civil Engineering

In this section we give the sensitivity analysis of a fuzzy signature which was applied for status-determining and ranking buildings of similar age and structural arrangement. In Budapest city a lot of old residential buildings are available of similar age and structural arrangement. At the end of the 19th and at the beginning of the 20th centuries the number of inhabitants increased from 280000 to 730000. In this time period new city districts were constructed with the application of the technological methods, which were known at that time. A significant part of these residential buildings still constitutes the dominant element of the current townscape. It is one of the most pressing issues of the Hungarian capital that a considerable part of these buildings are in degraded condition. The modernization and renovation of these buildings and their ranking from the aspect of the urgency of their renovation are significant task due to the limited financial possibilities.

A decision-supporting model was created by applying the fuzzy signatures ([8], [9] and [10]). This model is suitable for the ranking and qualification of residential buildings. The model was used for the first time on a database, which is based on expert opinions. After that a tree-structure, necessary for the examination of the load-bearing structures of buildings, were prepared. Primary structures (main load-bearing structures) and secondary structures (so not main load-bearing structures which play an important role in the protection of the main load bearing structures) were differentiate during the research, in this article we deal only with the branch of the primary structures. With the help of this branch it is possible to make a ranking of the load bearing structures of the examined buildings based on their arrangement, materials and conditions. The examined load bearing structures used in the model are the follows: foundation structures, wall structures, floor structures, side corridor structures, step structures and roof structures. The database was prepared on the basis of the research of more hundred buildings, typical in Budapest, so the results achieved, well reflect the actual conditions of this type of residential buildings. The structure of the signature is shown in Figure 2.

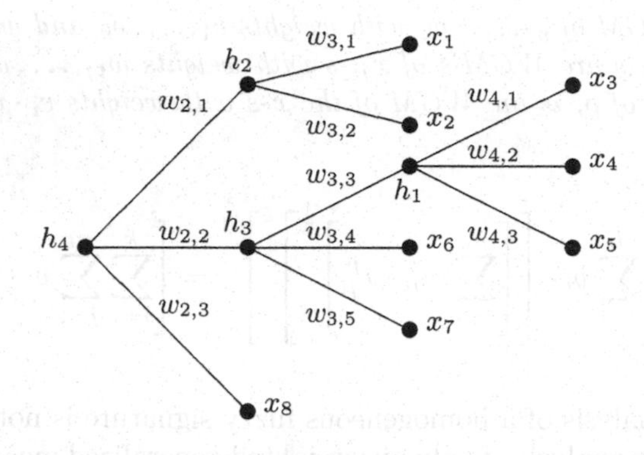

Fig. 2. A fuzzy signature for status-determining and ranking buildings

The input variables:

x_1 : foundation structures — x_2 : wall structures
x_3 : cellar floor — x_4 : intermediate floor
x_5 : cover floor — x_6 : side corridor structures
x_7 : step structures — x_8 : roof structures

The internal variables:

h_1 : floor structures
h_2 : vertical load-bearing structures
h_3 : horizontal load-bearing structures

The output variable is h_4.

This is a homogeneous fuzzy signatures with parameter $p = 1$ and with the following weights:

$$w_{2,1} = 0.4 \qquad w_{2,2} = \frac{0.6 \cdot n}{n+1} \qquad w_{2,3} = \frac{0.6}{n+1}$$

$$w_{3,1} = 0.55 - 0.05 \cdot n \qquad w_{3,2} = 0.45 + 0.05 \cdot n \qquad w_{3,3} = \frac{0.65}{0.8 + 0.2 \cdot f}$$

$$w_{3,4} = \frac{0.2 \cdot f}{0.8 + 0.2 \cdot f} \qquad w_{3,5} = \frac{0.15}{0.8 + 0.2 \cdot f}$$

$$w_{4,1} = \frac{0.35 \cdot m}{0.2 + 0.45 \cdot (n-1) + 0.35 \cdot m}$$
$$w_{4,2} = \frac{0.45 \cdot (n-1)}{0.2 + 0.45 \cdot (n-1) + 0.35 \cdot m}$$
$$w_{4,3} = \frac{0.2}{0.2 + 0.45 \cdot (n-1) + 0.35 \cdot m}$$

The possible values of the parameters:

- $n = 2, 3, 4, 5$ (number of the storeys of the building)
- $0 \leq m \leq 1$ (extend of the cellar built)
- $f = 0$ or 1 (building with or without side corridor)

The input values (x_i-s) are real numbers between 0 and 1 according to the opinion of a human expert about the status of the i-th partial structure. The final output is the membership value of h_4. If a building is surveyed by different experts then their opinion about the status of partial structures may result different values of h_4. The following question arises: if there are small differences between the ratings given by the experts to the partial structures, then how large can be the deviation between the final scores of a building? In other words, how sensitive is this fuzzy signature to small perturbations?

This is a homogeneous fuzzy signature, so we can analyse it as a simple WGM. From the results of Section 3 it follows that if the absolute values of the differences of the ratings given by the human experts less than ε for all of the variables, then the difference between the final conclusions less than ε, too.

We can conclude that this signature is not too sensitive, namely a small change in the partial opinions do not yields a large difference between the final conclusions.

5 Conclusions

The sensitivity of the weighted generalized mean aggregation operator for parameter value $p \geq 1$ was discussed via Minkowski's inequality. We also described the sensitivity of such WGMs using various vector norms. Based on these results the sensitivity of fuzzy signatures equipped with WGM-s can be done, especially in a simple case, when all of the WGM-s have the same parameter.

The sensitivity of a method for status-determining and ranking buildings was analysed, and we established that the applied fuzzy signature is not too sensitive, if the partial opinions of the experts are relatively close to each other, then their final conclusions will not differ too much.

Acknowledgement This work was supported by TÁMOP-4.2.2.A-11/1/KONV-2012-0012, and by the Hungarian Scientific Research Fund (OTKA) K105529 and K108405. This research was supported by the European Union and the State of Hungary, co-financed by the European Social Fund in the framework of TÁMOP 4.2.4.A/2-11-1-2012-0001 'National Excellence Program'.

References

1. Pozna, C., Minculete, N., Precup, R.E., Kóczy, L.T., Ballagi, Á.: Signatures: Definitions, operators and applications to fuzzy modelling. Fuzzy Sets and Systems 201, 86–104 (2012)
2. Vámos, T., Kóczy, L.T., Biró, G.: Fuzzy signatures in datamining. In: Proceedings of the Joint 9th IFSA World Congress and 20th NAFIPS International Conference, Vancouver, BC, Canada, vol. (5), pp. 2842–2846 (2001)
3. Wong, K.W., Gedeon, T.D., Kóczy, L.T.: Construction of fuzzy signature from data: an example of SARS pre-clinical diagnosis system. In: Proceedings of the IEEE International Conference on Fuzzy Systems (FUZZ-IEEE2004), Budapest, Hungary, pp. 1649–1654 (2004)
4. Ballagi, Á., Kóczy, L.T., Gedeon, T.D.: Robot cooperation without explicit communication by fuzzy signatures and decision trees. In: Proceedings of theJoint 2009 International Fuzzy Systems Association World Congress and 2009 European Society of Fuzzy Logic and Technology Conference (IFSA-EUSFLAT 2009), Lisbon, Portugal, pp. 1468–1473 (2009)
5. Goguen, J.A.: L-fuzzysets. J. Math. Anal. Appl 18(1), 145–174 (1967)
6. Kóczy, L.T., Vámos, T., Biró, G.: Fuzzy signatures. In: Proceedings of the 4th Meeting of the Euro Working Group on Fuzzy Sets and the 2nd International Conference on Soft and Intelligent Computing (EUROPUSE-SIC 1999), Budapest, Hungary, pp. 210–217 (1999)
7. Hardy, G.H., Littlewood, J.E., Pólya, G.: Inequalities. Cambridge University Press (1952)
8. Bukovics, Á., Kóczy, L.T.: Fuzzy Signature-based Model for Qualification and Ranking of Residential Buildings, XXXVIII, pp. 290–297. IAHS World Congress on Housing, Istanbul (2012)
9. Bukovics, Á.: Building Diagnostic and Pathological Analysis of Residantial Buildings of Budapest, PATORREB 2009, 3 Encontro Sobre Patologia E Reabilitacao de Edificios, Porto, pp. 1025–1030 (2009)
10. Bukovics, Á.: Pathological analysis of suspension corridor and floor structures of residential buildings, CIB 2010 World Congress. University of Salford, CIB, Manchester, UK (2010)

Large Strategy Adaptation Neighborhood Bolsters Network Reciprocity in Prisoner's Dilemma Games

Takashi Ogasawara*, Jun Tanimoto, Eriko Fukuda,
Aya Hagishima, and Naoki Ikegaya

Interdisciplinary Graduate School of Engineering Sciences, Kyushu University, Kasuga-koen, Kasuga-shi, Fukuoka 816-8580, Japan
{2es13155m,3es13023e}@s.kyushu-u.ac.jp,
{tanimoto,aya,ikegaya}@cm.kyushu-u.ac.jp

Abstract. In 2 × 2 prisoner's dilemma games, network reciprocity is one mechanism for adding social viscosity, which leads to cooperative equilibrium. Here we elucidate how strategy adaptation neighborhood affects on network reciprocity in spatial prisoner's dilemma games. Presuming an appropriate range of strategy adaptation neighborhood, we can observe the evolution of cooperation than usual network reciprocity previously reported. In the discussion based on our simulation result, we explore why those enhancements are brought, which can be summarized that varying the neighborhood ranges influences on how cooperative clusters are successfully formed and expanded in evolutionary process.

Keywords: Evolutionary game on network, Prisoner's dilemma game, Network reciprocity, Strategy adaptation neighborhood.

1 Introduction

Evolutionary games such as prisoner's dilemma (PD) have been regarded as good metaphors to model for solving the mysterious puzzle of why human beings, as well as other animal species, successfully evolve cooperation instead of egocentric defection within their societies. Many papers (for comprehensive reviews, refer to [1, 2, 3]) have discussed network reciprocity, which is one of the five fundamental mechanisms that Nowak classified [4] for resolving the dilemma; it attempts to do so by adding 'social viscosity.' Network reciprocity continues to receive huge attention because, although the central assumption of the model, i.e., 'playing with neighbors on an underlying network and copying a strategy from them,' is simple, it still seems very plausible for explaining why cooperation survives in any real context.

As the commonly-shared assumption in a spatial prisoner's dilemma (SPD) game, a gaming neighborhood, or interaction neighborhood (IN) in other words, meaning a range of neighbors to play games is presumed same as a strategy adaptation neighborhood, learning neighborhood (LN) in other words, meaning a range of neighbors from

* Corresponding author.

H. Handa et al. (eds.), *Proc. of the 18th Asia Pacific Symp. on Intell. & Evol. Systems – Vol. 1*,
Proceedings in Adaptation, Learning and Optimization 1, DOI: 10.1007/978-3-319-13359-1_46

which the focal agent copies its strategy. Also, in most of the previous works, the neighbors mean the first neighborhood from a focal agent in a certain underlying network. Thus, it has been presumed that a neighbor in the second or third neighborhood does not directly affect on the dynamics of the focal agent. But, obviously those central twofold assumptions seem too stringent to consider more realistic situations. On this point, there have been several precursors who investigated (e.g., [5, 6, 7]). Ohtsuki et al. [5, 6] concluded by a series of excellent deductions as well as simulations that consistent topologies of IN and LN brings better network reciprocity than that of inconsistent cases as a whole. Meanwhile, generally speaking, wide range of neighborhood implies agents having large degree, which means inevitably less network reciprocity, because it will approach an infinite and well-mixed population expecting no social viscosity at all. Those two things may suggest that the conventional assumption for SPD, where both IN and LN are consistently defined to be the first neighborhood on an assumed underlying network, would be most appropriate to bolster network reciprocity.

However, very recently, Xia et al. [8] delivered a rebuttal in which they explored simulations to see what happens if expanding IN and LN independently. And they found that appropriately selecting IN and LN, which are respectively larger than a first neighborhood, enhances network reciprocity. But they do not closely explain causes why appropriate IN and LN bring the results to bolster network reciprocity. Also, we guess their result was deeply affected by what strategy updating rule was presumed as well as what underlying network was assumed. The paper seems to give no mention on this. In their study, Pairwise Fermi (Fermi-PW), where a focal player i adopts a randomly chosen player j's strategy with probability calculated by a Fermi function, which is one of the most heavily used update rules in the previous studies, and is the most representative stochastic updating rule vis-à-vis deterministic ones like Imitation Max (IM).

In this paper, we explore again how changing only LN topology influences on network reciprocity by dint of deliberate simulation setting, being different from Xia's one, and give a transparent explanation on the mechanism working behind the phenomenon by using a quantitative evaluation index such as cluster characteristics (explained below).

In our recent publications, we have provided a sort of holistic discussion to answer 'what is a central mechanism to bring network reciprocity' [9, 10, 11]. The key idea in the discussion is that we should divide an evolutionary course starting from an initial random state to a final equilibrium into two periods as below, and should carefully observe what happens in those twofold periods respectively. Thus, according to Shigaki et al. [9], we use, in the present study, the term enduring period (END) to refer to the initial period in which the global cooperation fraction (P_c) decreases from an initial state in its dynamics. Perhaps the initial state has an equal number of randomly assigned cooperators and defectors on an underlying network. And we use the term expanding period (EXP) to refer to the period following END in which P_c increases since the survived cooperator's clusters for END start to expand by letting neighboring defectors become cooperators.

In this study, we try to give a straightforward exposition on why a certain appropriate range of LN fosters network reciprocity by referring to what happens in respective END and EXP periods, which persuades readers to understand what actual mechanisms to enhance network reciprocity working behind the phenomenon.

This paper is organized as follows. Section 2 describes our new model and the simulation procedure, Section 3 presents and discusses the results, and Section 4 draws conclusions.

2 Model Setup

At every time step, an agent occupying a vertex on a network plays prisoner's dilemma (PD) games with neighbors and obtains payoffs from all games. As the underlying topology, we use a two-dimensional (2D) lattice graph. The total number of agents is set to $N = 10^4$, which has been confirmed to be sufficiently large to yield simulation results that are insensitive to system size. After gaming, each agent synchronously updates his or her strategy refereeing to neighbors, defined by the LN neighborhood explained in the next section.

2.1 Underlying Neighborhood

Again, we adopt a 2D lattice as underlying topology. Relying on the so-called Moore neighborhood, respective degree of the first, second and third neighborhood correspond to k = 8, 24 and 48. We fix interaction neighborhood (IN) to be first Moore neighborhood and vary learning neighborhood (LN) among those three Moore neighborhoods; LN=1, 2 and 3 (k = 8, 24 and 48). One thing to be worthwhile to note is that we should limit Moore neighborhoods to compare each other in this study. Assuming Moore neighborhoods $k = 8$ and 24 as well as $k = 4$ that is the so-called von Newman neighborhood is an improper idea, because the latter neighborhood has potentially different network features. For example, the cluster coefficient of von Newman neighborhood $k = 4$ is 0, which obviously differs from those of Moor neighborhood graphs.

2.2 Game Description

In a PD game, a player receives a reward (R) for each mutual cooperation (C) and a punishment (P) for each mutual defection (D). If one player chooses C and the other chooses D, the latter obtains a temptation payoff (T), and the former is labeled a sucker (S). Without losing mathematical generality, we can define a PD game space by presuming $R = 1$ and $P = 0$ as follows:

$$\begin{pmatrix} R & S \\ T & P \end{pmatrix} = \begin{pmatrix} 1 & -D_r \\ 1+D_g & 0 \end{pmatrix}, \tag{1}$$

where $D_g = T - R$ and $D_r = P - S$ imply a chicken-type dilemma and stag-hunt-type dilemma, respectively [12]. A PD game is denoted with $0 \leq D_g \leq 1$ and $0 \leq D_r \leq 1$. In this study, we are primarily concerned on the Donor & Recipient (D&R) games that is denoted by $D_g = D_r$, which is one of the sub-classes belonging to PD, commonly used among rather biologists than statistical physicists who usually prefer to base the boundary games of PD with chicken game where $D_g \geq 0$ and $D_r = 0$. Obviously, the boundary game is not appropriate to discuss PDs from general point of view, because there is none of stag-hunt-type dilemma at all, which means it is just a specific PD even if one single dilemma parameter is convenient for discussions. It might be the biggest reason of why many physicists have agreed to refer to the boundary game. But also, in D&R game, there is only single parameter; $r = D_g = D_r$ to control dilemma strength. In the present study, we limit $0 \leq D_g, D_r \leq 1$.

2.3 Strategy Update

The strategy of an agent; C or D is refreshed every time-step according to Imitation Max (IM), in which the focal player i imitates the strategy with the maximum payoff among all the strategies taken by the focal player and his or her neighbors, defined by the LN.

We assume a 2D lattice as underlying network and IM for updating. We do not want to apply Fermi-PW with a lattice as the primary setting, or IM with a heterogeneous network like Scale-Free network either, for example. This is because a stochastic update rule like Fermi-PW somehow adds a masking cover over what happens in the case of assuming a homogeneous network, which makes difficult to correctly observe what exactly happens in the process, even though the case of Fermi-PW coupled with a Scale-Free network has been reported bolstered network reciprocity especially in the boundary games. Also, when we rely on a deterministic update rule like IM, presuming a heterogeneous network is an improper idea, since it has been already confirmed that the particular combination meagerly works in terms of thrusting cooperation (e.g., [11]).

Although heterogeneous topology may realize further network reciprocity by adding noise effect with some cooperative enhancement over what happening on a homogeneous network, it seems a preferable idea to assume none of stochastic effect for stating a discussion to explore what happening in the process of network reciprocity. Namely, IM as updating rule and a lattice as underlying network should be, for example.

2.4 Cluster Characteristics

For the following discussion to give an insight view on what happens in both END and EXP periods, we should define several characteristic indices to feature how emerging spatial patterns qualitatively affect the evolution of cooperation cluster (C-cluster). Three cluster characteristics: cluster size S_c, cluster number N_c and cluster shape SH_c of cooperator aggregations are employed [10]. For readers' convenience, we reprint the detailed definition of the third one. For each cluster i, we can derive SH_{ci} based on the number of C-C links, l_{CC}, within cluster i and the number of C-D links, O_{CD}, that connect cluster i with the surrounding defectors:

$$SH_{Ci} = \frac{2l_{CC} - O_{CD}}{2l_{CC} + O_{CD}}. \tag{2}$$

The value of SH_{ci} is constrained to the interval [−1, 1]. Obviously, a compact C-cluster has more links within the cluster rather than to the surrounding defectors. The value of SH_{ci} is positive, which indicates positive assortment of cooperators. While for the sparse cluster there are fewer links within the cluster but more links connecting to surrounding defectors. Thus, $SH_{ci} < 0$ and negative assortment among cooperators takes place (or positive assortment between cooperators and defectors). Moreover, in order to eliminate the influence of isolated cooperators, the cluster shape SH_c is weighed such that the weight of each cluster corresponds to its size.

2.5 Simulation Procedure

Each simulation is performed as follows. Initially, equal percentages of C and D are randomly distributed to the agents allocated on different vertices of the network. Several simulation time steps, or generations, are run until the frequency of cooperation reached quasi-equilibrium. If the cooperation fraction continued to fluctuate, we use the average fraction of cooperation over the last 250 generations of a 10,000 generation run. The results shown below are drawn from 100 runs; that is, each ensemble average is formed from 100 independent simulations.

3 Results and Discussion

Fig. 1 shows the cooperation fraction P_c averaging over 100 independent runs when LN is varied from 1 to 2 and 3. Here, we call a case of LN = 1 (k = 8) as a traditional case and cases of LN = 2 and 3 (k = 24 and 48) as LN expanded cases. For the LN expanded cases, we can find that cooperation is largely enhanced as compared with the traditional case. We summarize P_c for respective three dilemma areas as below: At first, as far as the small dilemma strength; $0 \leq r \leq 0.33$, P_c reaches high cooperation level ($P_c \approx 0.98$)

for $0.01 \leq r \leq 0.14$, or cooperation completely dominates the population ($P_c = 1$) for $0.15 \leq r \leq 0.33$. For the middle dilemma strength; $0.34 \leq r \leq 0.6$, cooperation is moderately enhanced in the LN expanded cases; where P_c increases from 0.39 to 0.53 when assuming LN = 2 and reaches from 0.11 to 0.37 when assuming LN = 3, whereas the traditional case is dominated by defectors. When it comes to the case of large dilemma strength; $0.61 \leq r \leq 1$, cooperation is not promoted. However, it can be seen some plots in which P_c tenuously increases ($P_c \approx 0.01$) when presuming LN = 2. In sum, we would say that the optimal learning neighborhood size is LN = 2.

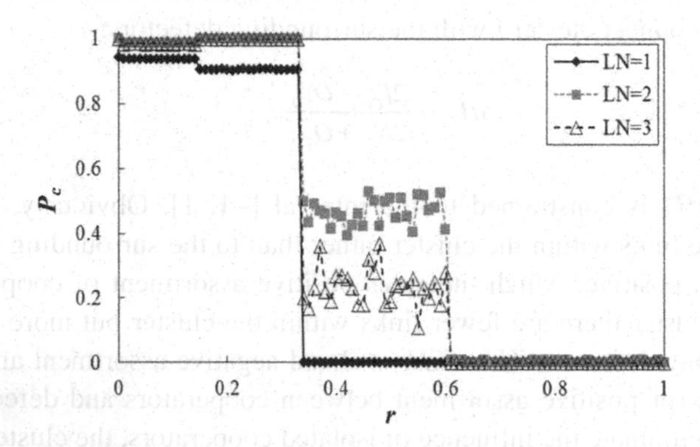

Fig. 1. The cooperation fraction P_c at the stationary state averaging over 100 independent runs as a function of dilemma strength r for different learning neighborhood LN sizes

First off, we note, again, that the LN expanded cases bolster cooperation than the traditional case does for the small dilemma strength; $0 \leq r \leq 0.33$. Fig. 2 and Fig. 3 respectively show time evolution and snapshots of representative episodes for both the traditional and the LN expanded cases for the very small dilemma strength; $r = 0.1$. We can observe from Fig. 2 and Fig. 3 that cooperators drastically decline immediately after the initial state where cooperators and defectors are equally and randomly distributed on the network. This is because cooperators copy defection from neighboring defectors who obtaining larger payoff than cooperators. This tendency is observed in not only the traditional but also LN expended cases. Fig. 3 shows that, at 2 time-step, survived cooperators successfully form C-clusters to endure invasions from defectors. This is exactly what happens in END. Following this, those survived C-clusters start to expand, of which tendency is significant with increasing LN. And this is exactly what happens in EXP. As consequence, the population in the LN expanded cases is dominated by more cooperators as compared with the traditional case at 10 time-step, despite few D-clusters remaining. On the other hand, in the population of the traditional case, more defectors than the LN expanded cases exist by forming many strips at 13 time-step. This happens by following mechanism. At the end of END, relatively larger number of C-clusters than the LN expanded cases can survive in the traditional case. This allows that many defectors who obtaining payoff Ts from neighboring C-clusters successfully survive in many gaps formed in spaces between C-clusters that start to expand in EXP. We will discuss this by referring to its detailed mechanism of emerging cooperation in case of the small dilemma strength shown in Fig. 4.

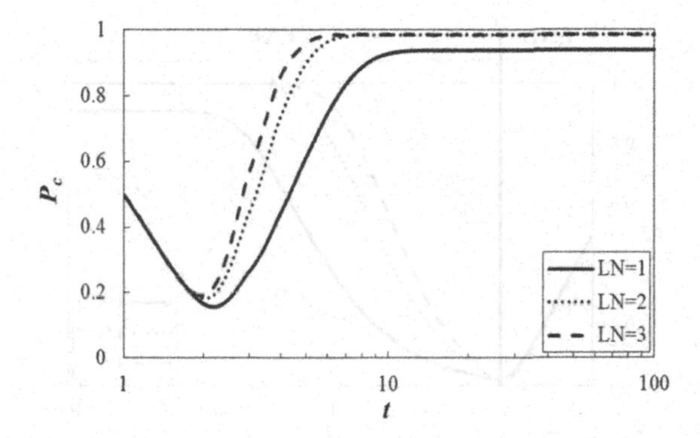

Fig. 2. Time evolution of representative episodes for different LN sizes in case of $r = 0.1$

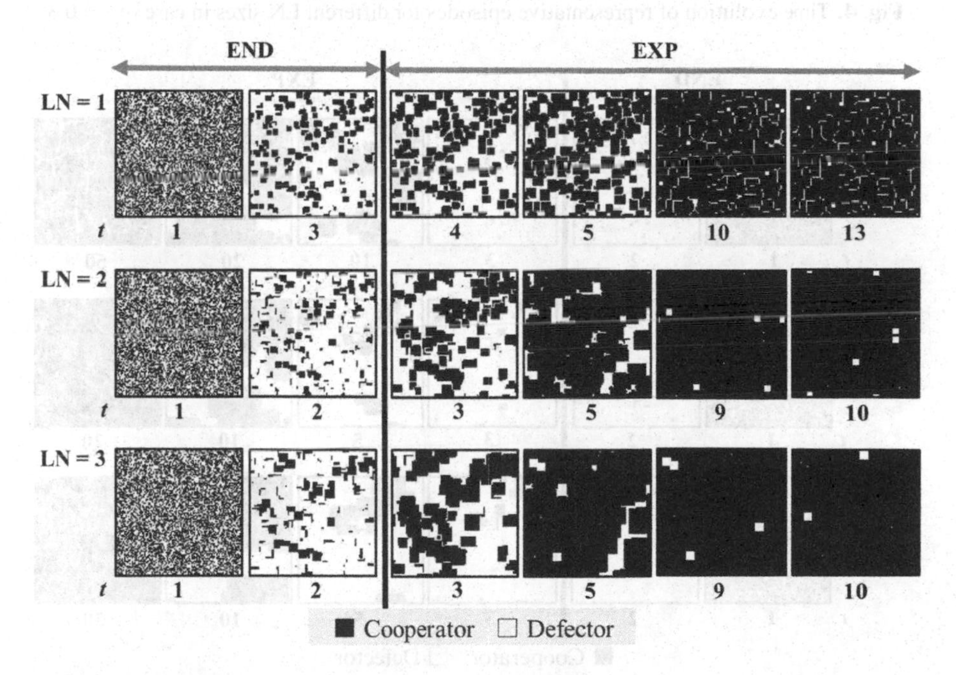

Fig. 3. Snapshots of representative episodes for different LN sizes in case of $r = 0.1$

Next, Fig. 4 and Fig. 5 show time evolution and snapshots of representative episodes for the small dilemma strength of $r = 0.3$. Comparing snapshots at the beginning of EXP shown in Fig. 3 and Fig. 5, we should note that only less number of C-clusters can survive in case of $r = 0.3$ than $r = 0.1$. In fact, P_c in case of $r = 0.3$

sharply declines from 0.5 to 0.05 as shown in Fig. 4. Subsequently, in EXP, each of C-clusters survived for END starts to expand. Particularly, whole population is dominated by cooperators in the LN expanded cases after 20 time-step. Whereas, in the traditional case, defectors somehow exist among expanded C-clusters by forming strips at 50 time-step when P_c arrives at the stationary state.

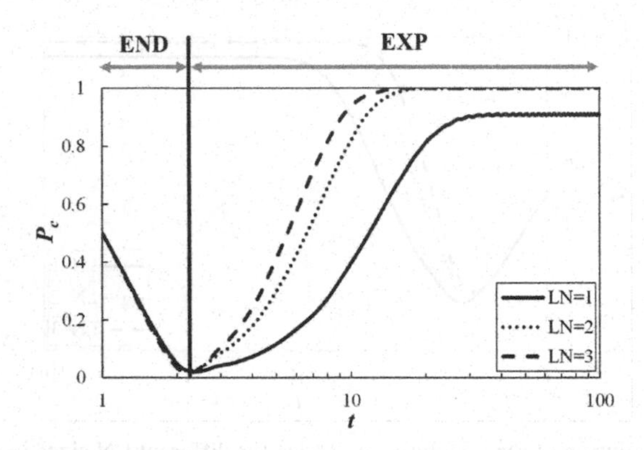

Fig. 4. Time evolution of representative episodes for different LN sizes in case of $r = 0.3$

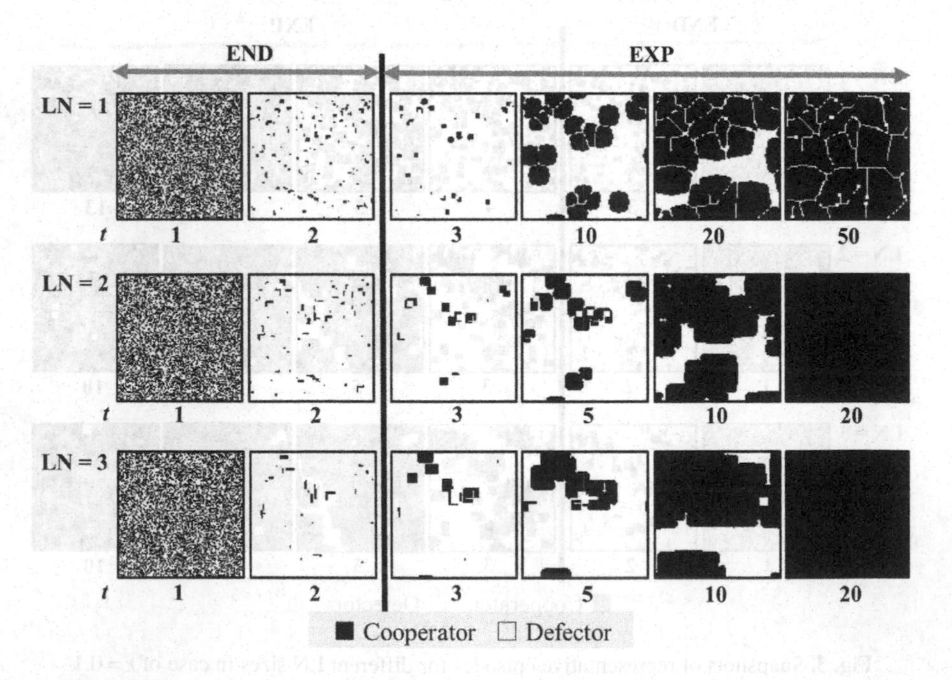

Fig. 5. Snapshots of representative episodes for different LN sizes in case of $r = 0.3$

Let us discuss a crucially interesting point. In general, P_c decreases as the dilemma strength increases. Nevertheless, in the LN expanded cases, the higher P_c is realized for the larger dilemma strength of $0.15 \leq r \leq 0.33$ than that for $0.01 \leq r \leq 0.14$.

Defectors still remain at its equilibrium for $0.01 \leq r \leq 0.14$ (Fig. 3), while the population is completely dominated by cooperators for $0.15 \leq r \leq 0.33$ (Fig. 5). This fact can be understood by follows. Less number of C-clusters at the end of END promotes the survived C-clusters in case of $r = 0.3$ smoothly expand to the whole domain occupied with lots of defectors in EXP following END. Contrariwise, the survived C-clusters for $r = 0.1$, that is relatively large number, cannot be like that, because defectors surviving between expanded C-clusters never perish.

Next, we will discuss the mechanism of why the LN expanded cases can realize moderate P_c for $0.34 \leq r \leq 0.6$ (Fig. 1). Table 1 shows survival probability of cooperators for both the traditional and the LN expanded cases when presuming $r = 0.5$. Cooperators in the traditional case survive with the high probability, whereas, for the LN expanded cases, cooperators perish in more than 50 % runs. Fig. 6 shows each snapshot of representative episodes for the traditional and the LN expanded cases in which cooperators could survive at the equilibrium when presuming $r = 0.5$. Obviously, only just a single C-cluster with small number of cooperators can survive in the traditional case at the equilibrium, while the whole domain is buried with cooperators in the LN expended cases. All those proofs lead to the following thing. Namely, the survived cooperators in the traditional case can hardly expand in EXP even though the cooperators, anyway, survive for END with high possibility. Contrariwise, in the LN expanded cases, despite low possibility of cooperators surviving for END as compared with the traditional case, those survived cooperators successfully expand to the domain in EXP, leading to relatively high average P_c as observed in Fig. 1. Thus, the LN expanded cases bring moderate P_c because there are two types of episodes among simulation runs: one is P_c reaching high level since survived C-clusters can successfully expand in EXP following END; and P_c converging to zero since the population is absorbed by all-defectors phase in END is another. Table 1 delivers us that the narrower the size of LN becomes, it is the more likely cooperators alive. This seems plausible because more wide range of LN would devastate the network reciprocity, or say, social viscosity, thus it would be natural a certain episode might be absorbed by all-defectors phase in the end. However, the case of LN = 3 somehow seems interesting because a certain episode successfully surviving for END shows all-cooperators state at its equilibrium as we confirmed in Fig. 6. A likely scenario is that each episode, despite it may be trapped with all-defectors phase in END, could produce all-cooperators phase at the end of EXP once it surviving for END. Fig. 7 shows C-cluster characteristics: S_c, N_c and SH_c of representative episodes for two types of episodes in LN = 3 for $r = 0.5$: one is an enhanced episode in which cooperators surviving for END and significantly expanded in EXP, and another is an episode failed to be enhanced, namely trapped with all-defectors phase in END. At the end of END, we easily notice that S_c and SH_c for the enhanced episode are larger than those for the episode failed to be enhanced, although no significant difference in N_c between these two types of episodes is found. The key points to bifurcate into the two scenarios is whether cooperators can survive for END, and if they once successfully surviving for END, whether C-clusters, the surviving cooperators forming, can be with expandable size and appropriate shape at the end of END.

Table 1. Survival probability of cooperators at the stationary state for different LN sizses in case of $r = 0.5$

LN size	Survival probability of cooperators
LN = 1	0.74
LN = 2	0.43
LN = 3	0.21

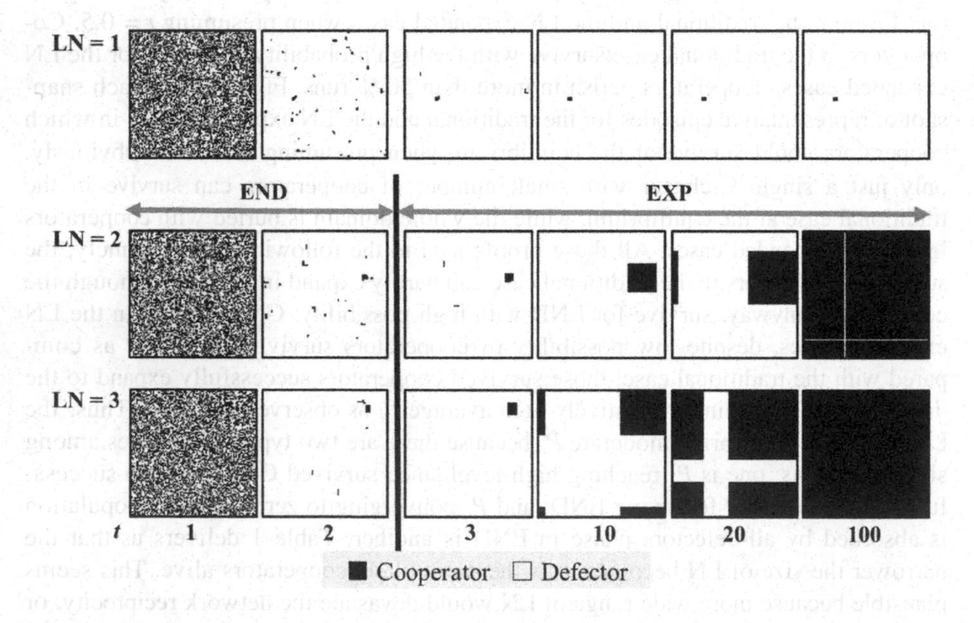

Fig. 6. Snapshots of representative episodes for different LN sizes where cooperators could survive at the stationary state in case of $r = 0.5$

Next, let us be concerned on an interesting result when P_c assuming LN = 2, which slightly increases for the large dilemma strength; $0.61 \leq r \leq 1$. Fig. 8 shows snapshot for a certain episode assuming $r = 0.99$. Although the dilemma is very large, cooperation surges very much. Interestingly, in this particular episode, only a single C-cluster survives for END, which starts to expand to surrounding area of defectors. This proves that if a surviving C-cluster has ideally proper shape and sufficient size, its episode could realize high cooperation despite large dilemma. But this is very rare from stochastic point of view. In fact, the episode shown in Fig. 8 is only one episode observed amid 100 realizations. Anyway, this fact attributes to why we could observe slight germs of weak cooperation in the large dilemma strength; $0.61 \leq r \leq 1$ when assuming LN = 2 as we confirmed in Fig. 1.

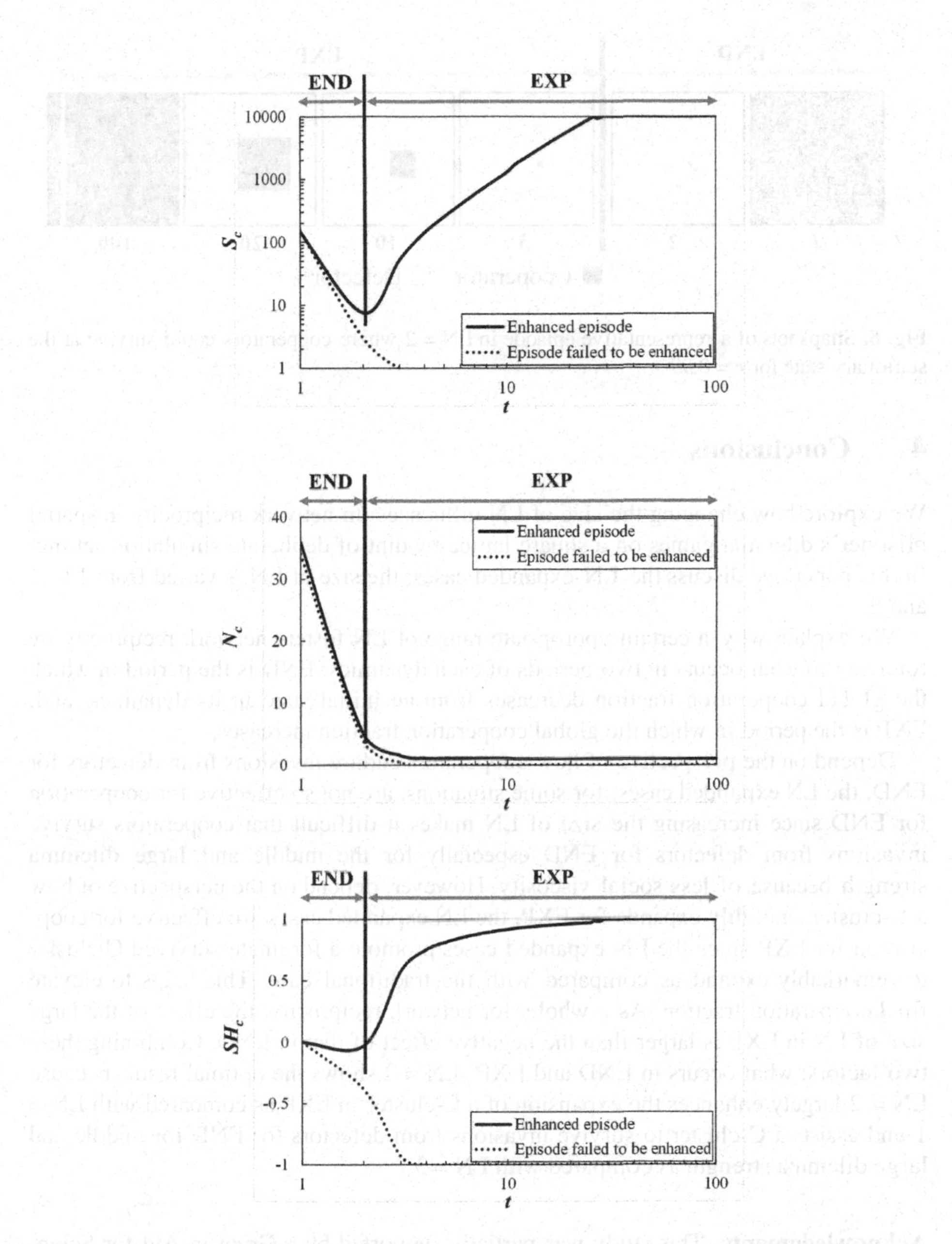

Fig. 7. C-cluster characteristics of an enhanced episode and an episode failed to be enhanced in LN = 3 for $r = 0.5$

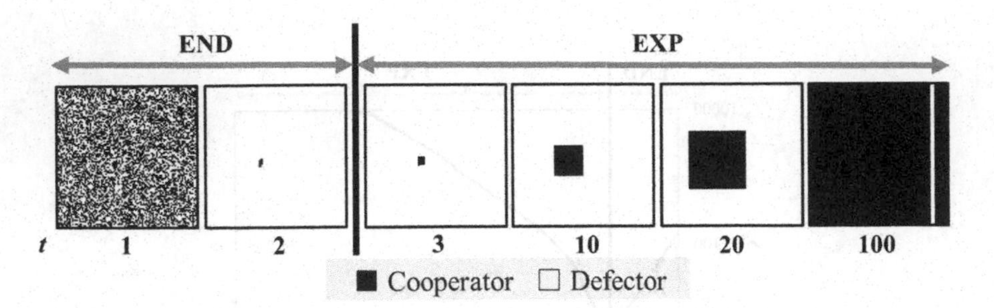

Fig. 8. Snapshots of a representative episode in LN = 2 where cooperators could survive at the stationary state for $r = 0.99$

4 Conclusions

We explore how changing the size of LN influences on network reciprocity in spatial prisoner's dilemma games on a square lattice by dint of deliberate simulation setting. In this paper, we discuss the LN expanded cases; the size of LN is varied from 1 to 2 and 3.

We explain why a certain appropriate range of LN fosters network reciprocity by referring to what occurs in two periods of each dynamics: END is the period in which the global cooperation fraction decreases from an initial state in its dynamics, and, EXP is the period in which the global cooperation fraction increases.

Depend on the perspective of how cooperators endure invasions from defectors for END, the LN expanded cases, for some situations, are not so effective for cooperation for END since increasing the size of LN makes it difficult that cooperators survive invasions from defectors for END especially for the middle and large dilemma strength because of less social viscosity. However, depend on the perspective of how a C-cluster smoothly expands for EXP, the LN expanded cases are effective for cooperation for EXP since the LN expanded cases promote a fortunate survived C-cluster to remarkably expand as compared with the traditional case. This helps to elevate final cooperation fraction. As a whole, for network reciprocity, the effect of the large size of LN in EXP is larger than the negative effect of that in END. Combining these two factors; what occurs in END and EXP, LN = 2 shows the optimal result because LN = 2 largely enhances the expansion of a C-cluster in EXP as compared with LN = 1 and assists a C-cluster to survive invasions from defectors for END for middle and large dilemma strength as compared with LN = 3.

Acknowledgments. This study was partially supported by a Grant-in-Aid for Scientific Research by JSPS, awarded to Prof. Tanimoto (#25560165), Tateishi Science & Technology Foundation. We would like to express our gratitude to these funding sources.

References

1. Szabo, G., Fath, G.: Evolutionary games on graphs. Phys. Rep. 446, 97–216 (2007)
2. Perc, M., Gomez-Gardenes, J., Szolnoki, A., Floria, L.M., Moreno, Y.: Evolutionary dynamics of group interactions on structured populations; A review. J. R. Soc. Interface 10, 20120997 (2013)
3. Perc, M., Szolnoki, A.: Coevolutionary games—a mini review. BioSystems 99, 109–125 (2010)
4. Nowak, M.A.: Five rules for the evolution of cooperation. Science 314, 1560–1563 (2006)
5. Ohtsuki, H., Nowak, M.A., Pacheco, J.M.: Breaking the symmetry between interaction and replacement in evolutionary dynamics on graphs. Phys. Rev. Lett. 98, 108106 (2007)
6. Ohtsuki, H., Pacheco, J.M., Nowak, M.A.: Evolutionary graph theory: Breaking the symmetry between interaction and replacement. J. Theor. Biol. 246, 681–694 (2007)
7. Ifti, M., Killingback, T., Doebeli, M.: Effects of neighbourhood size and connectivity on the spatial continuous prisoner's dilemma. J. Theor. Biol. 231, 97–106 (2004)
8. Xia, C., Miao, Q., Zhang, J.: Impact of neighborhood separation on the spatial reciprocity in the prisoner's dilemma game. Chaos, Solitons Fractals 51, 22–30 (2013)
9. Shigaki, K., Tanimoto, J., Wang, Z., Hagishima, A., Ikegaya, N.: Referring to the social performance promotes cooperation in spatial prisoner's dilemma games. Phys. Rev. E 86, 031141 (2012)
10. Shigaki, K., Tanimoto, J., Wang, Z., Fukuda, E.: Effect of initial fraction of cooperators on cooperative behavior in evolutionary prisoner's dilemma. PLoS One 8(11), e76942 (2013)
11. Wang, Z., Kokubo, S., Tanimoto, J., Fukuda, E., Shigaki, K.: Insight into the so-called spatial reciprocity. Phys. Rev. E 88, 042145 (2013)
12. Tanimoto, J., Sagara, H.: Relationship between dilemma occurrence and the existence of a weakly dominant strategy in a two-player symmetric game. BioSystems 90(1), 105–114 (2007)

Evolutionary Robust Optimization with Multiple Solutions

Peng Yang[1], Ke Tang[1], Lingxi Li[1], and A.K. Qin[2]

[1] USTC-Birmingham Joint Research Institute in Intelligent Computation and Its Applications,
School of Computer Science and Technology,
University of Science and Technology of China,
Hefei, Anhui, China
{trevor,lilingxi}@mail.ustc.edu.cn, ketang@ustc.edu.cn
[2] School of Computer Science and Information Technology
Royal Melbourne Institute of Technology
Melbourne, Australia
kai.qin@rmit.edu.au

Abstract. When optimizing for multiple environments, one usually needs to sacrifice performance in one environment in order to gain better performance in another. Ultimately, there may not be a single solution that meets the performance requirements for all environments. In this paper, we propose to find multiple solutions that each serve a certain group of environments. We call this formulation Robust Optimization with Multiple Solutions (ROMS). Two evolutionary approaches to ROMS are proposed, namely direct evolution and two-phase evolution. A benchmark problem generator is also suggested to produce uniform-random ROMS problems. The two approaches are then experimentally studied on a variety of synthetic problems.

Keywords: robust optimization, multiple, clustering.

1 Introduction

Sometimes, one needs to consider multiple environments in optimization, and a robust optimization formulation should be adopted [1,2,3,4]. There are several such formulations in the literature that are suitable for this purpose [1]. First of all, it is important to note that the objective function will most likely evaluate to different values under different environments for a given solution. It is possible to define an overall objective function in terms of these values, and obtain a scalar performance measure. As to designing such an overall objective function, taking the mean and worst-case objective value over[1] the different environments are the two most widely seen methods in the literature. It is also possible to consider the objective values under different[2] environments directly and obtain a multi-objective formulation [5].

* Corresponding author.

H. Handa et al. (eds.), *Proc. of the 18th Asia Pacific Symp. on Intell. & Evol. Systems – Vol. 1*,
Proceedings in Adaptation, Learning and Optimization 1, DOI: 10.1007/978-3-319-13359-1_47

All these formulations search for a single robust solution, which is too demanding in practice sometimes. More often than not, inconsistent requirements are posed on the solution by different environments. In such cases, in order to gain some performance boost in one environment, the performance in another has to be compromised [6]. Consequently, there may not be a single solution that meets the requirements for all environments.

In this paper, we propose to find multiple solutions for the environments such that each of them serves a subset of environments. By limiting the number of different environments a single solution needs to serve and by grouping similar environments together as a subset, better performance in each environment may be achieved. The number of solutions to find marks the tradeoff between robustness and performance. At one end of the spectrum, and for maximum robustness, one may aim to find a single solution for all environments, as in a conventional robust optimization formulation. At the other end of the spectrum, and for maximum performance, one may find each environment a highly specialized solution. The number of solutions is assumed to be specified by the user. We call this generalized formulation Robust Optimization with Multiple Solutions (ROMS).

The rest of this paper is organized as follows. In the next section, a formal definition of ROMS is given. A discussion of its properties follows in Section 3. Two evolutionary approaches to ROMS, namely direct evolution and two-phase evolution are then presented in Sections 4 and 5, respectively. An experimental study of the two approaches is reported in Section 6 with a suggested benchmark problem generator. Finally, conclusions are drawn in Section 7.

2 Formal Definition

The objective function is defined as $f(\boldsymbol{x}, \boldsymbol{\alpha}), \boldsymbol{x} \in \mathbb{R}^D, \boldsymbol{\alpha} \in \mathcal{C}$, where $\boldsymbol{x}$ denotes the solution and $\boldsymbol{\alpha}$ denotes the environment parameters of the objective function [7]. Given m environments to consider, $\mathcal{C} = \{\boldsymbol{\alpha}_1, \boldsymbol{\alpha}_2, \cdots, \boldsymbol{\alpha}_m\}$. For convenience, denote $f(\cdot, \boldsymbol{\alpha}_i)$ by $f_i(\cdot)$. A set of m objective functions $F = \{f_1, f_2, \cdots, f_m\}$ is then obtained. To serve these objective functions, a set of n solutions $S = \{s_1, s_2, \cdots, s_n\}$ are to be sought. Each s_i serves a subset of F, and each f_i is served by exactly one s_j. This service relationship is given by M which is a mapping from F to S such that $M(f_i) = s_j$ if and only if f_i is served by s_j. A complete solution to ROMS is thus a pair $\langle S, M \rangle$. Given some overall objective function $\bar{f}$, ROMS is formally defined as

$$\text{optimize } f(S, M) = \bar{f}\left(f_1(M(f_1)), f_2(M(f_2)), \cdots, f_m(M(f_m))\right). \quad (1)$$

As mentioned in Section 1, the overall objective function can be either the mean function or the one that returns the worst objective value as in worst-case analysis. Owing to the limitations of paper length, the following discussions and experimental studies are given assuming that the mean function is used as the overall objective function though with some minor modifications they also apply to the case when the function that returns the worst objective value is used instead.

3 Basic Observations

From the formal definition, it can be seen that a complete solution to ROMS is made up of two parts — a solution set S and a mapping M. The first observation says that given an S, the optimal accompanying M can be computed in polynomial time.

Observation 1: Given a solution set S, the optimal accompanying M is one that (assuming a minimization problem)

$$\forall f_i: [\forall s_j: f_i(s_j) \geq f_i(M(f_i))]. \tag{2}$$

In words, the optimal M is one that maps each objective function to the solution in S that performs best on it. The algorithm to compute such an M is trivial and costs $O(m \cdot n)$ objective function evaluations.

To see the second observation, first note that a partition P of F can be derived from M by putting the objective functions that are mapped to the same solution in a single part. Conversely, M can be derived from P, but with the solutions unspecified. That is, certain objective functions are known to be mapped to the same solution, but the solution itself is not determined yet. Therefore, we say P implies an M pattern. The second observation is about the optimal solution set S, given a partition P.

Observation 2: Given a partition $P = \{p_1, p_2, \cdots, p_k\}$ of F (effectively, an M pattern), and requiring that objective functions in the same part be mapped to a single solution, the optimal solution set S is one that

$$\forall p_i: (\exists s_j: s_j \text{ optimizes } p_i) \tag{3}$$

where

$$s_j \text{ optimizes } p_i \Leftrightarrow s_j \text{ minimizes} \frac{1}{|p_i|} \sum_{f_q \in p_i} f_q. \tag{4}$$

Computing such an S amounts to solving k independent conventional robust optimization problems. This second observation can be summarized as deriving an optimal S from a given M pattern, and is in some sense a reversal of the first observation.

4 Direct Evolution

A straightforward approach to ROMS is to fit it into the general framework of Evolutionary Algorithms (EAs) and solve it directly with evolutionary optimization. On the surface, the individual encoding has to account for both S and M that are of different types, leaving us a mixed-integer optimization problem to solve. But thanks to observation 1, it does not have to be so. Since given a solution set S, the optimal accompanying M can be computed directly with an exact polynomial-time algorithm, there is no need to let the EA figure it out. The idea is then to encode S only. At the time when an individual is to be evaluated, the optimal accompanying M is computed first, and fitness evaluation (FE) is then performed using the resulting pair $\langle S, M \rangle$.

This reduces the original mixed-integer problem to a real-value one. Specifically, each individual is encoded as a vector containing $n \times D$ real values, where each consecutive D values encode a single solution s_i in the solution set that contains n solutions in total. As an example, DE/rand/1/bin is employed to perform the evolutionary optimization. We denote this particular algorithm by ROMS-DE in this paper. For intuition, the pseudo code is given in Fig.1 (F and CR are parameters of differential evolution).

Algorithm 1: ROMS-DE

```
1   initialize population pop
2   evaluate pop
3   for each new generation
4       for i ← 1 to population size ps = |pop|
5           randomly select a, b, c ∈ [1..ps] that are mutually different and also different from i
6           diff ← pop[a] + F × (pop[b] − pop[c])
7           randomly select r ∈ [1..n]
8           for j ← 1 to n
9               generate a random number rnd ∈ [0, 1]
10              if rnd < CR or j = r then trial[i][j] ← diff[j]
11              else trial[i][j] ← pop[i][j]        trial[i][j] is the j-th solution of trial[i]
12      evaluate population trial
13      for i ← 1 to ps
14      pop[i] ← the better of pop[i] and trial[i]
```

Fig. 1. The pseudo code of Direct Evolution

The FE cost for ROMS-DE to evolve G generations, denoted by $Cost_{DE}(G)$, is derived as follows. With a straightforward implementation, it is easy to see that $Cost_{DE}(G) = M \cdot N \cdot PS \cdot G$, where G is the population size. A more efficient implementation can be obtained with the observation that in a newly generated solution set S, some solutions s_i are directly copied from its parent. For these solutions, there is no need to reevaluate their fitness values on each objective function in F, saving FEs thereof. This observation does not affect evaluation of the initial generation which still costs $N \cdot M \cdot PS$ FEs. But for each generation that follows, we have the following hold. For each newly generated solution set S, the expected number of solutions s_i that are not copied from its parent is $[1 + (N-1) \cdot CR]$ where CR is the crossover probability of DE/rand/1/bin. With this improved implementation, the expected FE cost for ROMS-DE to evolve G generations becomes

$$Cost_{DE}(G) = N \cdot M \cdot PS + [1 + (N-1) \cdot CR] \cdot M \cdot PS \cdot (G-1). \quad (5)$$

5 Two-Phase Evolution

The second approach is based on the interaction between the two observations made in Section 3. The first observation essentially tells a way to derive an optimal M from a given S. We denote it by $S \Rightarrow M$. Likewise, the second observation essentially tells a way to derive an optimal S from a given P or an M pattern. In the same manner, we denote it by $M \Rightarrow S$. By connecting the two, a loop that goes from S to M and back can be formed. Since the output of both procedures are optimal, by going along this loop, one is actually undergoing an optimization progress. The second approach is based exactly on this idea. We call it two-phase evolution for the optimization loop consists of two phases, namely $S \Rightarrow M$ and $M \Rightarrow S$.

A straightforward implementation of this approach goes as follows. Staring from a randomly generated S, repeat the two procedures $S \Rightarrow M$ and $M \Rightarrow S$ until some termination criterion is met. Again, as an example, DE/rand/1/bin is employed to perform the optimization involved in $M \Rightarrow S$. In accordance with direct evolution, the two-phase evolution approach can be seen as maintaining and optimizing a solution set S along the loop. But different from direct evolution, only one S is maintained by two-phase evolution. Intuitively, in each tour around the loop, the maintained S jumps from on point to another in the search space of S. The $S \Rightarrow M$ procedure is exact. Assuming no randomness in the $M \Rightarrow S$ procedure also[3], the next point to sample (jump to) is determined completely by the current one. Consequently, the whole optimization process is determined once the initial point is determined. This can become quite risky when the initial point is randomly chosen.

In an attempt to reduce the risk, we increase the initial search coverage by generating more than N solutions. Originally, a single S is randomly generated, which contains N solutions. Now, a solution pool PL of a size greater than N is randomly generated. Combinatorial optimization is then performed on the generated PL to select N solutions that together make up an optimal S with respect to the given PL [8]. Denote this procedure by $PL \Rightarrow S$. Another modification concerns the procedure $M \Rightarrow S$ that involves solving K real-value optimization problems. Since a population-based evolutionary algorithm like DE/rand/1/bin [9] is used to perform the optimization, a set of solutions is obtained for each problem. By the original design, only the best solution is selected into S. With the introduction of PL, solutions other than the best one may also be used. When the goal is to generate an updated PL that has a larger size than S, more solutions are selected. With these modifications, the original optimization loop $S \Rightarrow M \Rightarrow S$ becomes $PL \Rightarrow S \Rightarrow M \Rightarrow PL$ with increased search power to reduce risk.

Another improvement over the original design concerns incorporating an explicit mechanism to prevent the optimization process from stagnation. In order for the optimization process to continually make progress, a different M shall be reached each time around the loop. As an example, the Genetic Algorithm (GA) [6] is employed to perform the procedure $PL \Rightarrow S$. Since GA is population-based, a population of S is

[3] This does not hold strictly when the stochastic DE/rand/1/bin is employed to perform the optimization involved in this procedure.

obtained in the end, each have an accompanying optimal M. By the original design, the best S is selected with its M derived. Whereas making sense in most cases, this is not the best choice when the derived M is identical to the one derived in the previous tour around the loop. When it happens, the algorithm is simply re-producing the pervious iteration, making no progress. To fix this, instead of always selecting the absolutely best S, we first try to choose the best S among those that produce an M different from the one in the previous iteration.

With the aforementioned improvements and with the use of soft-computing optimization algorithms, quality of the maintained S is not guaranteed to be non-decreasing along the optimization loop. As a result, an external record is set up that maintains the best S ever encountered during the optimization process. This fix finally makes the two-phase evolution approach complete. We denote it by ROMS-TP. The pseudo-code is given in Fig.2.

Algorithm 2: ROMS-TP

```
1   initialize solution pool PL
2   archive ← null
3   M_last ← null
4   for each new iteration
5       pops_S ← evolve on PL          [pops_S is a population of solution sets]
6       archive ← better(archive, best individual in pops_S)
7       pops_S' ← {S ∈ pops_S | S has a different corresponding M from M_last}
8       if pops_S' ≠ ∅ then S ← best individual in pops_S'
9       else S ← best individual in pops_S
10      M_last ← the optimal M corresponding to S
11      P ← the partition defined by M_last
12      pops ← ∅
13      S ← ∅
14      for each p ∈ P
15          pop ← evolve on p          [pop is a population of single solutions]
16          pops ← pops ∪ {pop}
17          S ← S ∪ {best individual in pop}
18      archive ← better(archive, S)
19      PL ← build on pops
```

Fig. 2. The pseudo code of Two-Phase Evolution

We now describe the particular implementation used in this work and derive its computational complexity $cost_{TP}(G)$ (here G refers to the number of iterations around the loop rather than generations). Some implementation issues will also be discussed along the way. To make things simple, the size of PL is fixed to a predefined number. Line 5 is done in two steps. In the first step, each single solution in PL is evaluated in each condition, which costs us $|PL| \times m$ fitness evaluations. In the second step, a combinatorial optimization is done on PL using a generational genetic algorithm. Thanks to the pre-processing in the first step, the evolution process in the second step costs no fitness evaluation at all. Details of the genetic algorithm used are listed in Table 1. Lines 6~13 are cost-free. We use a standard DE/rand/1/bin to perform the real-value optimization at Line 15. In total, Lines 14~17 cost $|pop| \times m \times g$ fitness evaluations where g is the number of generations to evolve at Line 15 for the differential evolution. Lines 18-19 are cost-free. So, for this implementation,

$$cost_{TP}(G) = (|PL| \times m + |pop| \times m \times g) \times G \quad (6)$$

Care should be taken when implementing Line 19. There are a total of $|P| \times |pop|$ single solutions contained in pops, out of which $|PL|$ are to be selected to make up the new PL. Recall that the size of PL is fixed. Depending on the parameter settings, $|PP| \times |pop|$ may be smaller than, equal to, or larger than $|PL|$. In the first two cases, all solutions in pops go to PL with the remaining space, if any, being filled up by randomly generated solutions. In the last case, only part of $pops$ goes to PL. A proper selection of solutions should be made. We hope the selection could be drawn from each population in $pops$ as evenly as possible, so that each population is well represented with minimum information loss. To do so, we randomly select (without replacement) one solution from each population in turn repeatedly until the desired number $|PL|$ is reached.

Table 1. Details of the genetic algorithm used

Elitism	Replaces the worst individual in the next generation by the best in the current.
Encoding	An n-dimensional integer vector with each entry indexing a solution in PL.
Selection	Binary tournament selection.
Crossover	Single-point crossover.
Mutation	Vector-wise random mutation that randomly generates the whole vector.

6 Empirical Study

6.1 Experimental Setup

To get an actual feeling and experience with the ROMS problem and the two approaches proposed, experiments are done on synthetic functions. In this subsection, we first describe the problem settings, and then the algorithmic settings.

We investigate a ROMS problem from three perspectives, the number of conditions m, the desired number of solutions n, and the general function form of the conditions. Our experiments should cover all three perspectives by testing with different m, n values, and different function forms. In this work, m goes from 4 to 20 with a step-size 4, and n goes from 2 to 10 with a step-size 2. The first two columns in Table 2 list the m, n combinations we use. With respect to the function forms, we choose 5 functions, i.e., F1-F3, F6, F10, from the standard benchmark functions for the CEC-05 special session on real-parameter optimization [10]. These functions cover unimodal and multimodal functions with the latter further divided into basic functions, expanded functions, hybrid composition functions and pseudo-real problems. One of our guidelines is to choose the functions without special properties. To introduce the notion of environmental parameters, we additionally parameterize each function with a translation vector and an orthogonal matrix (matrices with orthonormal column vectors). The translation vector shifts the function while the orthogonal matrix transforms the function with a combination of rotation (around the global optimum) and reflection (via some hyperplane through the global optimum) [11]. The conditions in F are thus generated from the same function with different translation vectors and orthogonal matrices built in. We generate the environmental parameters as uniform random variables in their corresponding spaces. The generation of uniform random translation vectors is simple. Just generate each entry uniform-randomly. To generate uniform random orthogonal matrices, we follow the classical two-step algorithm [12]. First, generate each matrix entry randomly following a standard normal distribution, and then apply the Gram-Schmidt orthogonalization. All problems are minimization problems, and are tested in both $D = 2$ and $D = 30$ dimensions. The parameter settings for ROMS-TP are listed in Table 3. For differential evolution, we follow the parameter settings suggested by the original authors, i.e., with $F = 0.5$, $CR = 0.1$ and a population size of 5 times the problem dimension [9]. Notice that while for the differential evolution used in ROMS-TP, the problem dimension is D, for ROMS-DE it should be $n \times D$, which equals the length of the encoding vector of a solution set. For all experiments, a fitness evaluation budget of 1e7 is imposed. The number of iterations to run is computed accordingly to $cost_{DE}(G)$ and $cost_{TP}(G)$ respectively, as shown in Table 2.

Table 2. The first two columns list tested m, n combinations. The ratio m/n shows on average how many conditions are served by each solution, which reflects the level of requirement for robustness. The larger the ratio, the higher the requirement, and usually the lower the performance in each condition will be. The remaining four columns list the number of iterations to run for each algorithm on each m, n combination in each dimension, given a fitness evaluation budget of 1e7.

m	n	m/n	$D = 2$		$D = 30$	
			ROMS-DE	**ROMS-TP**	**ROMS-DE**	**ROMS-TP**
4	2	2	113636	3572	7575	22
8	2	4	56818	1786	3788	11
12	2	6	37878	1191	2525	8
16	2	8	28409	893	1894	6
20	2	10	22727	715	1515	5
20	4	5	9614	715	639	5
20	6	3.3	5553	715	368	5
20	8	2.5	3673	715	242	5
20	10	2	2628	715	172	5

6.2 Experimental Results

In this subsection, we present the experimental results. All experiments are repeated for 25 times with the average results reported. A 95% confidence interval is also reported for each result, assuming a normal distribution. The mean objective value over the different environments is used as the robustness measure. For convenience, we label each sub-experiment in the form Fx_m_n_D so that F1_4_2_2 refers to the experiment with function F1, $m = 4, n = 2, D = 2$. We first present the results obtained with $D = 2$ coming before $D = 30$. We evaluate the performance of each algorithm from three perspectives. The first is optimization dynamics which is demonstrated in the 2-dimensional optimization curve (also known as convergence curve in evolutionary computation) with fitness evaluation count as x-axis and the robustness measure as y-axis[4]. The second and third are scalability with respect to m and n respectively.1 For minimization problems, the robustness measure is to be minimized, which is the case in this work and somewhat counterintuitive.

Table 3. Parameter settings for ROMS-TP

	Solution pool size $\|PL\|$	$100 \times D$
GA	Population size $\|pop_s\|$	$5 \times n$
	Number of generations	$50 \times n$
	Crossover rate	1.0
	Mutation rate	0.01
DE	Population size $\|pop\|$	$5 \times D$
	Number of generations	$25 \times D$
	F	0.5
	CR	0.1

Optimization dynamics, D = 2: When n is fixed at 2, thc big picture looks similar for different m values. For brevity, we only show the results for $m = 12$. The results on different functions can be clustered into two groups. For the first group (F1~F3, F6), we take the results on F1 for example (Figure 3). Pay attention that for each algorithm, there are three lines. The middle line shows the average result, while the upper and lower ones give the bounds for the 95% confidence interval. To obtain the results, for ROMS-DE, we draw a sample, which is the best-of-generation objective value, at the end of each generation. Since differential evolution implements elitism, the best-of-generation value is also the best value seen so far. For ROMS-TP, we sample the external archive at the end of each step, resulting in two samples for each iteration. ROMS-TP's curve starts from the end of the first step, which is the start of the second step, of the first iteration. So the first drop shown by the curve, if any, is due to the second step, not the first.

[4] For minimization problems, the robustness measure is to be minimized, which is the case in this work and somewhat counterintuitive.

Now, we focus on Figure 3. The curve of ROMS-DE is typical to an evolutionary algorithm. This is expected, since ROMS-DE is essentially an application of differential evolution. The curve of ROMS-TP is more interesting. A r decrease happens at the first iteration. After that, the curve descends slowly. Besides, a careful observation will reveal that decrease happens mostly in the second step of each iteration. This is seen in the alternating "drop, level-off, drop" pattern along sample points on the curve. Comparing the two curves, we see premature convergence in ROMS-DE. A general way to counteract premature convergence is to use a larger population, but it also slows done the convergence. Figure 4 shows the corresponding result on F10 (the second group). Being contrary to the previous result, starting from the first sample point, an alternating "level-off, drop, level-off" rather than "drop, level-off, drop" pattern is seen along the curve of ROMS-TP. In other words, decrease is mainly due to the first step in each iteration. The high multimodality, which is F10's most distinct feature, may be the reason behind this phenomenon. It makes the real-value optimization involved in the second step more difficult, while having no direct impact on the first step. When the optimization effect is shifted from the second step to the first, the decrease is also more gradual. Without a dramatic decrease in the second step at the first iteration, ROMS-TP's performance is greatly weakened with a curve running always above that of ROMS-DE.

Next, we discuss the results with varying n while m is fixed at 20. Again, the results can be clustered into two groups. On functions in the second group (F3, F6, F10), the picture doesn't change much for different n values. On F3 and F6, the results are similar to Figure 3, and on F10 the results are similar to Figure 4. Taking F1 for example, Figure 5 shows typical results on the first group (F1, F2) as n increases. First, we see that the optimization process of ROMS-DE is less well developed (or less well converged) with a larger n. Recall that the population size is set to $5 \times n \times D$ which increases with n. So, a natural explanation for this observation is that with a larger population, an evolutionary algorithm just needs more fitness evaluations to converge. For ROMS-TP, the following trend can be observed. With an increasing n, the optimization effect gradually shifts from the second step of each iteration to the first. What follows is consistent with our previous observation made on the comparison between Figures 3 and 4. As the first step plays a more and more important role, descent of the optimization curve becomes more gradual, and without the remarkable decrease in the second step at the first iteration, ROMS-DE is catching up with ROMS-TP.

Scalability, $D = 2$: Figure 6 gives the results of scalability test with respect to m. The objective values are the final results obtained by the algorithms at a fitness evaluation count of approximately 1e7. As we can see, ROMS-TP has better scalability on all test functions expect F10 where the opposite seems to hold somehow. Figure 7 gives the results of scalability test with respect to n. Overall, ROMS-DE has better scalability on all test functions with the exception of F10 where, again, the opposite seems to hold. We also notice that at the given fitness evaluation count, ROMS-TP generally has got a final result no worse than that of ROMS-DE for each m, n combination on all test functions expect F10 where ROMS-DE has got the upper hand. The results on F10 are somehow abnormal when compared to the results on other test functions. Since high multimodality is the most distinct feature of F10, we suspect that the degree of multimodality has an impact on the relative performance of the two algorithms. Specifically, ROMS-DE seems to be more suitable for handling highly multimodal functions.

Optimization dynamics, D = 30: At a higher dimension D = 30, when n is fixed, the big picture looks similar with different m values and on different test functions. Figure 8 and 9 give the results corresponding respectively to Figure 3 and 4. Pay attention that, for the ROMS-TP curve in the two figures, the fitness evaluation cost and objective value decrease of the first step are so low compared to those of the second step that the sample point for the second step is visually undistinguishable from the sample point for the first step in the next iteration; they just overlapped. Compared to results in 2 dimensions, the difference is mainly seen on F10. Here, F10 doesn't distinguish itself from other test functions, and the results are just like those on other functions without any particularity. The success of the second step may due to the increase in the population size, which according to our parameter settings (see Table 3) is $5 \times D$. When m is fixed, the dynamics of ROMS-TP basically remains the same for different n values while the optimization process of ROMS-DE is becoming more and more less well developed with increasing n due to an increasing population size (just like the case in 2 dimensions).

Scalability, D = 30: Figure 8 gives the results of scalability test with respect to m in 30 dimensions, which clearly show that ROMS-TP has better scalability. Figure 11, on the other hand, gives the results in terms of n, and no clear scalability difference is observed. In 30 dimensions, ROMS-TP has always got a better final result on all experiments including the ones on F10, which is an exception in 2 dimensions.

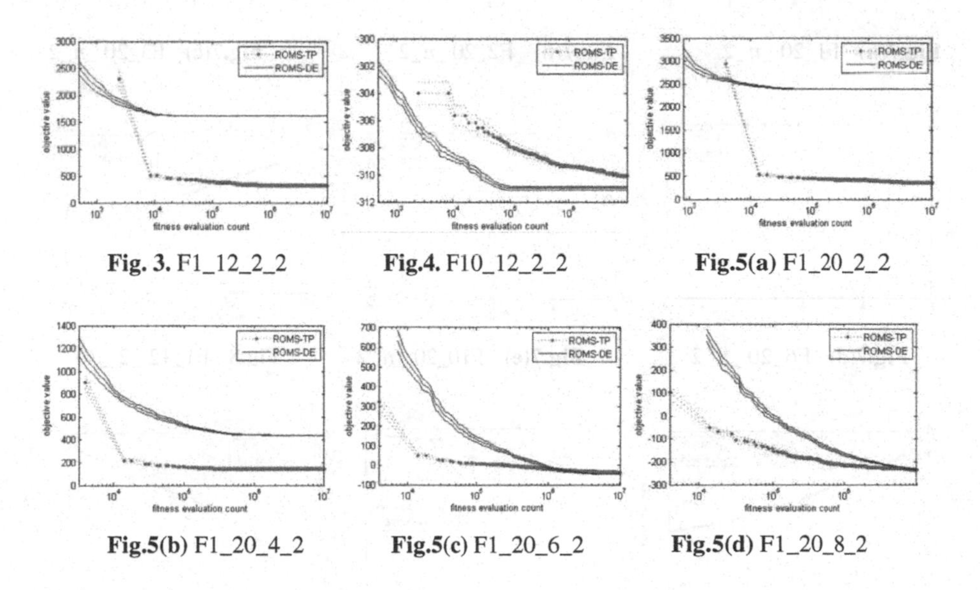

Fig. 3. F1_12_2_2 **Fig.4.** F10_12_2_2 **Fig.5(a)** F1_20_2_2

Fig.5(b) F1_20_4_2 **Fig.5(c)** F1_20_6_2 **Fig.5(d)** F1_20_8_2

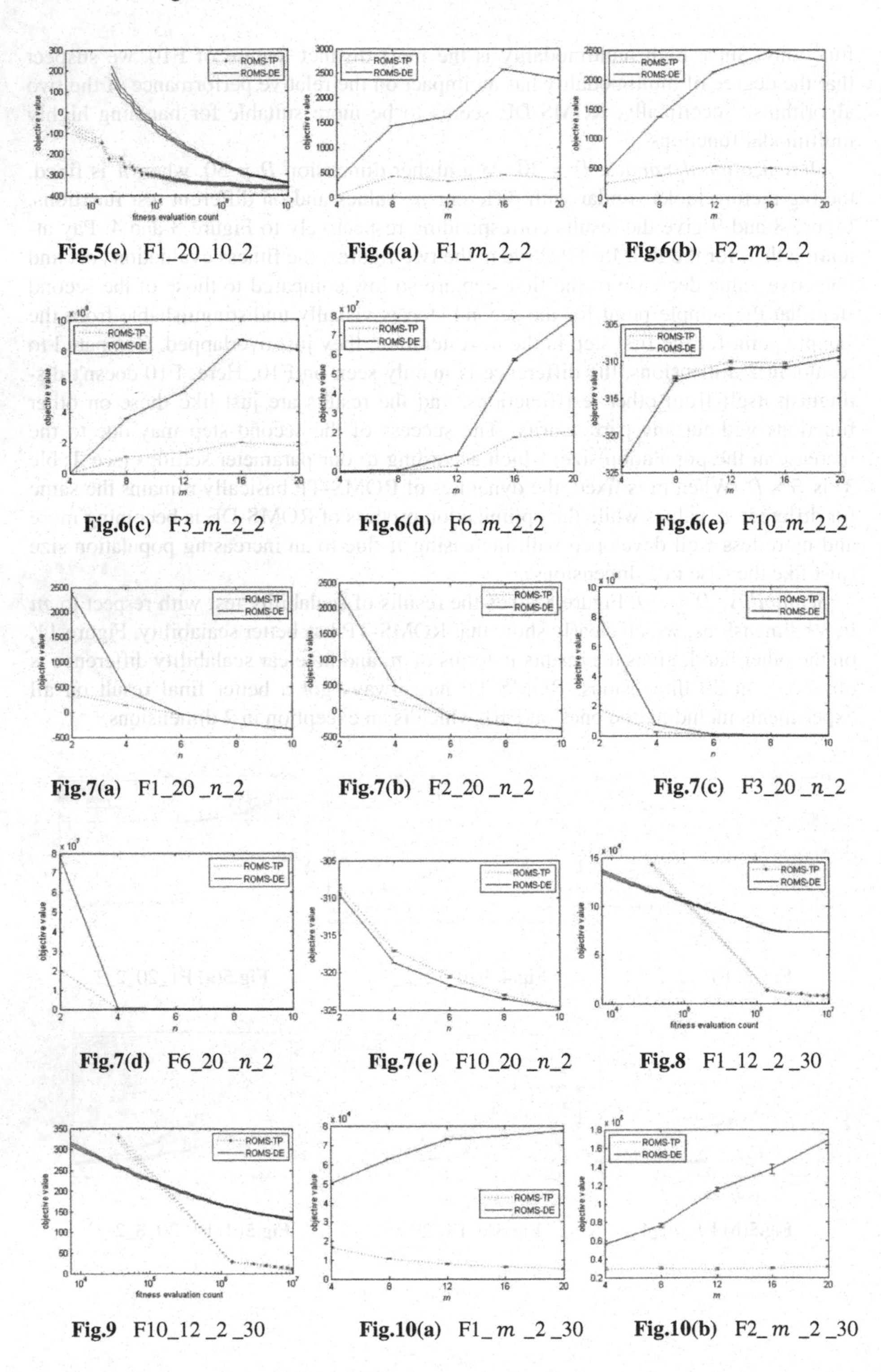

Fig.5(e) F1_20_10_2 **Fig.6(a)** F1_*m*_2_2 **Fig.6(b)** F2_*m*_2_2

Fig.6(c) F3_*m*_2_2 **Fig.6(d)** F6_*m*_2_2 **Fig.6(e)** F10_*m*_2_2

Fig.7(a) F1_20 _*n*_2 **Fig.7(b)** F2_20 _*n*_2 **Fig.7(c)** F3_20 _*n*_2

Fig.7(d) F6_20 _*n*_2 **Fig.7(e)** F10_20 _*n*_2 **Fig.8** F1_12 _2 _30

Fig.9 F10_12 _2 _30 **Fig.10(a)** F1_ *m* _2 _30 **Fig.10(b)** F2_ *m* _2 _30

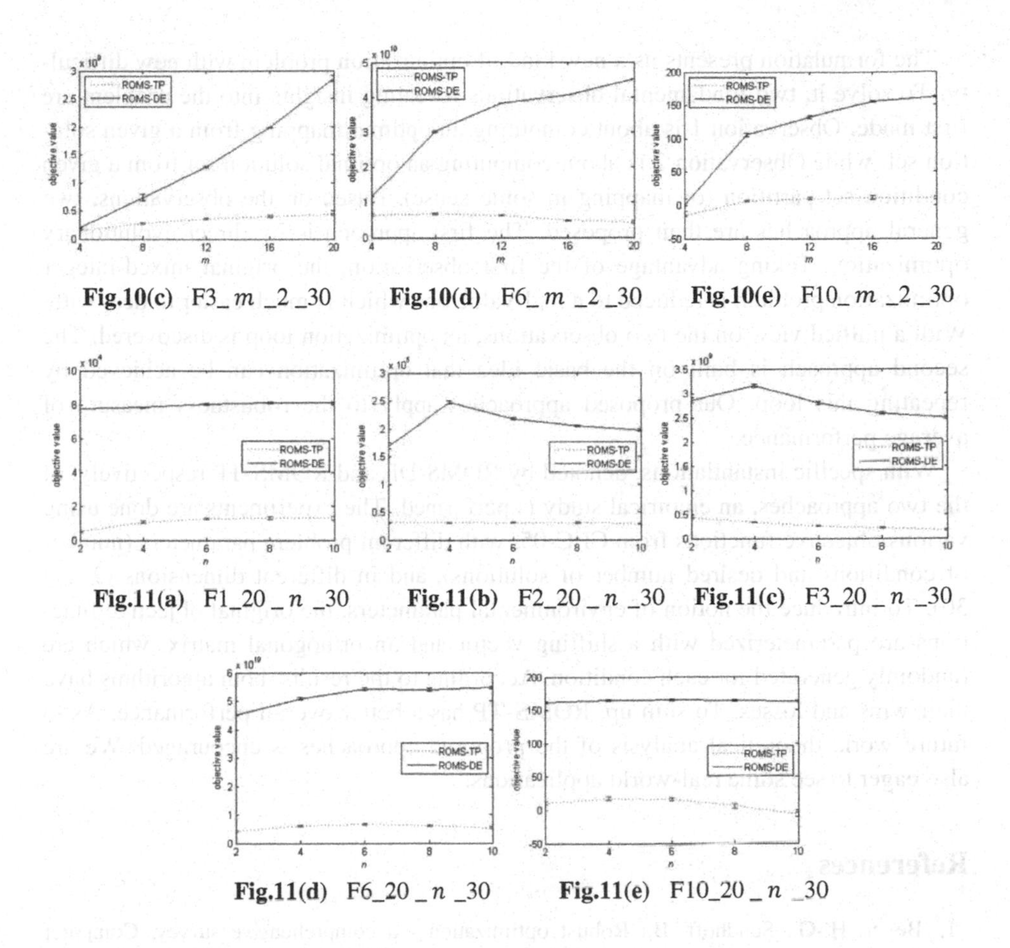

Fig.10(c) F3_ m _2 _30 Fig.10(d) F6_ m _2 _30 Fig.10(e) F10_ m _2 _30

Fig.11(a) F1_20 _ n _30 Fig.11(b) F2_20 _ n _30 Fig.11(c) F3_20 _ n _30

Fig.11(d) F6_20 _ n _30 Fig.11(e) F10_20 _ n _30

7 Conclusion

To conclude, in this work, a new formulation for robust optimization is proposed. The goal is to optimize for a variety of environmental conditions which are modeled by a set of functions. Following a traditional robust optimization formulation, a single solution is sought to cope with all different conditions, which characterizes an extremely high robustness requirement. Since robustness and performance are generally two conflicting goals, this traditional formulation may fail with way too compromised performance is some applications. To address this issue, we propose a generalized formulation, namely robust optimization with multiple solutions, in which multiple solutions, are sought to cope with the conditions. Each solution serves some subset of the conditions and the whole condition set is covered by the solution set. By tuning the desired number of solutions, the balance between robustness and performance can be flexibly adjusted.

The formulation presents us a new kind of optimization problem with new difficulty. To solve it, two fundamental observations revealing insights into the problem are first made. Observation 1 is about computing an optimal mapping from a given solution set, while Observation 2 is about computing an optimal solution set from a given condition set partition (or mapping in some sense). Based on the observations, two general approaches are then proposed. The first approach is by direct evolutionary optimization. Taking advantage of the first observation, the original mixed-integer optimization problem is reduced to a real-value one which is much easier to deal with. With a unified view on the two observations, an optimization loop is discovered. The second approach is built on the basic idea that optimization can be achieved by repeating this loop. Our proposed approaches apply to the robustness measure of average performance.

With specific instantiations, denoted by ROMS-DE and ROMS-TP respectively, of the two approaches, an empirical study is performed. The experiments are done using various objective functions from CEC-05, with different problem parameters (number of conditions and desired number of solutions), and in different dimensions (2 and 30). To introduce the notion of environmental parameters, the original objective functions are parameterized with a shifting vector and an orthogonal matrix, which are randomly generated for each condition. According to the results, both algorithms have their wins and losses. To sum up, ROMS-TP has a better overall performance. As to future work, theoretical analysis of the proposed approaches is encouraged. We are also eager to see some real-world applications.

References

1. Beyer, H.-G., Sendhoff, B.: Robust optimization - a comprehensive survey. Computer Methods in Applied Mechanics and Engineering 196(33), 3190–3218 (2007)
2. Ben-Tal, A., Ghaoui, L., Nemirovski, A.: Robust Optimization. Princeton University Press (2009)
3. Roy, B.: Robustness in operational research and decision-aiding: A multi-faceted issue. European Journal of Operational Research 200(3), 629–638 (2010)
4. Pita, J., Jain, M., Tambe, M., Ordóñez, F., Kraus, S.: Robust solutions to Stackelberg games: addressing bounded rationality and limited observations in human cognition. Artificial Intelligence 174(15), 1142–1171 (2010)
5. Jin, Y., Branke, J.: Evolutionary optimization in uncertain environments - a survey. IEEE Transactions on Evolutionary Computation 9(3), 303–317 (2005)
6. Deb, K.: Recent developments in evolutionary multi-objective optimization. Trends in Multiple Criteria Decision Analysis, 339–368 (2010)
7. Bertsimas, D., Sim The, M.: price of robustness. Operations Research 52(1), 35–53 (2004)
8. Goldberg, D.: Genetic Algorithms in Search, Optimization and Machine Learning. Addison-Wesley Professional (1989)

9. Storn, R., Price, K.: Differential evolution–a simple and efficient heuristic for global optimization over continuous spaces. Journal of Global Optimization 11(4), 341–359 (1997)
10. Suganthan, P., Hansen, N., Liang, J., Deb, K., Chen, Y.-P., Auger, A., Tiwari, S.: Problem definitions and evaluation criteria for the CEC 2005 special session on real-parameter optimization, technical report, Nanyang Technological University (2005)
11. Strang, G.: Introduction to Linear Algebra, 4th edn. Wellesley Cambridge Press (2009)
12. Diaconis, P., Shahshahani, M.: The subgroup algorithm for generating uniform random variables. Probability in the Engineering and Informational Sciences 1(1), 15–32 (1987)

Artificial Bee Colony Algorithm Based on Local Information Sharing in Dynamic Environment

Ryo Takano, Tomohiro Harada, Hiroyuki Sato, and Keiki Takadama

The University of Electro-Communications
Building W-6, 308, 1-5-1 Chofugaoka, Chofu, Tokyo
{takano,harada}@cas.hc.uec.ac.jp, {sato,keiki}@hc.uec.ac.jp

Abstract. This paper focuses on Artificial Bee Colony (ABC) algorithm which can utilize global information in the static environment and extends it to ABC algorithm based on local information sharing (ABC-lis) in dynamic environment. In detail, ABC-lis algorithm shares only local information of solutions unlike the conventional ABC algorithm. To investigates the search ability and adaptability of ABC-lis algorithm to environmental change, we compare it with the conventional two ABC algorithms by applying them to a multimodal problem with dynamic environmental change. The experimental results have revealed that the proposed ABC-lis algorithm can maintain the search performance in the multimodal problem with the dynamic environmental change, meaning that ABC-lis algorithm shows its search ability and adaptability to environmental change.

Keywords: Swarm intelligence, ABC algorithm, dynamic environment, local information sharing.

1 Introduction

Artificial bee colony (ABC) [1] algorithm designed by introducing the idea of intelligent behaviors of honey bee swarm which is one of swarm intelligence and its algorithm is effective algorithm for finding solutions in the context of to optimization. [1]. ABC algorithm is based on the intelligent behavior of honey bee swarm. The Significant feature of this algorithm is a high search capability performance for multimodal problems by using simple common control parameters [3][4] (e.g., search performance of ABC algorithm is better than PSO in Rastrigin function. Moreover, there are improvements to improve the performance of variety optimization problems. [5][6][7] This feature Advantage of high search performance in multimodal problems is expected to be effective exploring victims in disaster areas because many victims (corresponding to the peak of multimodal functions) are distributed in different places. Some researches show its effective by swarm in disaster relief. We revealed that the method using ABC algorithm for search and rescue is effectively in RoboCup Rescue simulation [8][9]. However, the original ABC algorithm was designed for focused on only static environments, meaning that it is hard to be applied to dynamic environments. To ward tackle dynamic environments, the modified ABC algorithm was

H. Handa et al. (eds.), *Proc. of the 18th Asia Pacific Symp. on Intell. & Evol. Systems – Vol. 1*,
Proceedings in Adaptation, Learning and Optimization 1, DOI: 10.1007/978-3-319-13359-1_48

proposed in [2], but it cannot follow dynamic change in multimodal problems in the case of with a large number of local minima and its search performance decreases. This is because many bees converge to some of a few local minima and cannot explore track other local minima.

To tackle this problems, this paper proposes the distributed ABC algorithm based on local information sharing (ABC-lis) for dynamic environment even when the number of local minima increases. Concretely, ABC-lis restricts communication among bees to only their close ones. This improvement enables bees to explore search all local minima and to follow dynamic change in multimodal function with a large number of local minima. To investigate the effectiveness the proposed method, we develop a minimization problem of the multimodal function which has that is time-depend N peaks. In the problem, we compare three methods, (i.e., the original ABC algorithm, the modified ABC, and ABC-lis). Besides, we verify that investigate whether it can always explore the global optimum which changes in time series can be always explored by the proposed method which aims at capturing all local minima.

This paper is organized as follows. Section 2 describes the original ABC algorithm and then Section 3 describes the modified ABC algorithm for dynamic environment, and Section 4 proposes introduces ABC-lis as the proposed method., and Section 5 describes shows the multimodal problem for of a dynamic environment, and for experiment. Section 6 conducts the experiment and discusses their results. The conclusions and future works are given described in Section 7.

2 ABC Algorithm

In ABC algorithm, the colony of bees consists of three groups of bees: *employed bees*, *onlooker bees* and *scout bees*. ABC algorithm regards food sources of bees as solutions and discovers the most important food source as the optimal solution. The importance of food sources is evaluated by objective function. In ABC algorithm, employed bees search food source where one employed bee corresponds to one food source, and each bee sends its evaluation to onlooker bees. Onlooker bees choose important food source according to evaluations sent from employed bees. If an employed bee judges the current food source have no prospect, the employed bee becomes a scout bee. A scout bee randomly chooses a new food source in search area.

In ABC algorithm, the size of population is represented as N_s. Each food source is represented as $\mathbf{x}_i$ $(i = 1,2,\cdots,N_s)$, which is D-dimensional vector, and the fitness of each food source is represented as $fit(\mathbf{x}_i)$. D is the number of optimization parameters. The main flow of the ABC algorithm is described below:

– Step0: Initialization

At the beginning, the number of iterations C is set as zero, and employed bees are randomly placed in the search area. Each bee evaluates each food source (solution) and the highest evaluation value among those solutions is memorized as $\mathbf{x}_{\mathbf{best}}$. In addition, the number of trials for each food source is $trial_i = 0$.

– Step1: Employed bees phase

In employed bees phase, the ith employed bee selects a candidate new solution $\mathbf{v_i}$ from the periphery of the current solution $\mathbf{x_i}$. A candidate new solution is produced from the current one by using the following expression (1):

$$v_{ij} = x_{ij} + \Phi(x_{ij} - x_{kj}) \tag{1}$$

where $k \in \{1,2,\cdots,N_s\}$ and $j \in \{1,2,\cdots,D\}$ are randomly chosen indexes. And Φ is a random number between $[-1,1]$. If an evaluation of the candidate new solution is better than the current one (i.e., $fit(v_i) > fit(x_i)$), the current solution is replaced with the new candidate (i.e., $x_i = v_i$) and the number of trials is initialized (i.e., $trial_i = 0$). On the other hand, if an evaluation of the candidate new solution is worse than or equal to the current one (i.e., $fit(v_i) \leq fit(x_i)$), bee stay in the current place and the number of trials is increased by 1 (i.e., $trial_i = trial_i + 1$). This flow is summarized as follows:

$$x_i = \begin{cases} v_i & if\ fit(v_i) > fit(x_i) \\ x_i & otherwise \end{cases} \tag{2}$$

$$trial_i = \begin{cases} 0 & if\ fit(v_i) > fit(x_i) \\ trial_i + 1 & otherwise \end{cases} \tag{3}$$

– Step2: Onlooker bees phase

Next is onlooker bees phase. Onlooker bees search solutions to the potential ones that are obtained by employed bee phase. Concretely, i[th] onlooker bee p_i chooses a solution using roulette wheel selection with probability calculated by the following expression (2):

$$p_i = \frac{fit(x_i)}{\sum_{n=1}^{N_s} fit(x_n)} \tag{4}$$

Solutions with a high evaluation value are frequently chosen by onlooker bees. A candidate new solution is produced by onlooker bees with the same way as the employed bees phase.

– Step3: Memorizing the best food source

When the employed bees phase and the onlooker bees phase complete, the best food source is memorized. Concretely, if the highest evaluation value in the current iteration, $fit(\mathbf{x_{ib}})$, is better than the best evaluation in the past iterations, $fit(\mathbf{x_{best}})$, $\mathbf{x_{ib}}$ is memorized as $\mathbf{x_{best}}$. Where $\mathbf{x_{best}}$ is the global optimal solution found by ABC algorithm

– Step4: Scout bees phase

If the number of trials of a solution reaches the maximum limit value, it becomes scout bee. Scout bees are randomly relocated to the search area.

– Step5: Repeating Step1 to Step4

If the above steps complete, the number of iterations is updated, $C = C + 1$. And the ABC algorithm repeats Step1 to Step4 N_{mc} times.

Motion of bees in each step of ABC algorithm is shown in Fig.1. In Fig.1, frameless red circles represent employed bees, blue triangles represent onlooker bees, and yellow circles with frame represent scout bee. In addition, star shapes represent food sources, which size expresses the evaluation value. In Fig.1, there are three employed bees and food sources. Two onlooker bees select same employed bee in the middle for search food source with high evaluation value. A scout bee is changed employed bee in the upper left and move to right in Fig.1. Besides, the flow of ABC algorithm is shown in Fig.2.

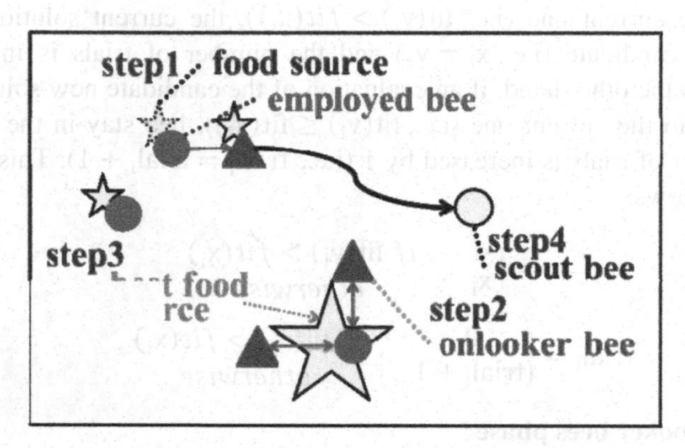

Fig. 1. Schematic diagram of ABC algorithm

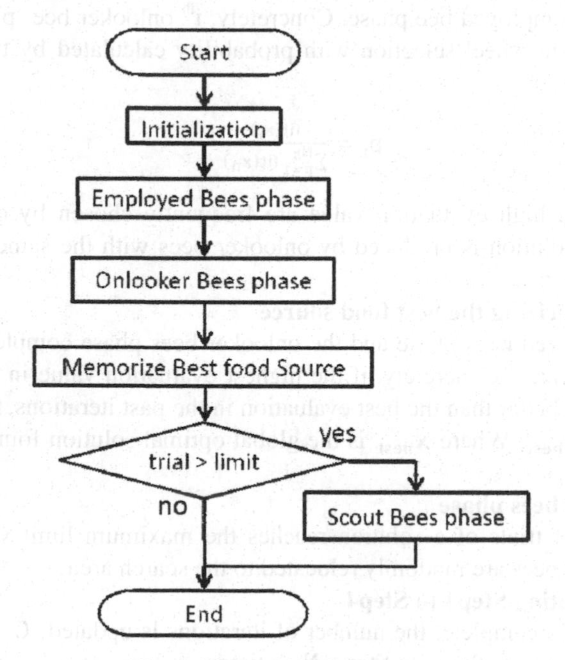

Fig. 2. Flow of ABC algorithm

3 ABC Algorithm for Dynamic Environment

3.1 Modification of ABC Algorithm for Dynamic Environment

The original ABC algorithm does not assume a dynamic environment that changes the value of the function. Because of this, it is not possible to follow the change in the evaluation value. To tackle this problem, Nishida proposed modified ABC algorithm that is adapted to dynamic environments without decreasing its search performance in a static environments [2]. Afterword, ABC algorithm with these modifications referred to as modified ABC. In the modified ABC algorithm, step1 and step3 of the original ABC algorithm are changed as follows:

– **Step1': Modified employed bees phase**
In original ABC algorithm, although the evaluation value of the candidate new solution fit($\mathbf{v_i}$) is calculated, the evaluation value of the current solution fit($\mathbf{x_i}$) is not recalculated in equation (2) and the value previously calculated is used. In static environment, since the evaluation value do not change during searching process, it is not necessary to be recalculated. However, in dynamic environment, it is a possibility to change the evaluation value during searching process. For this reason, modified ABC algorithm always recalculates the evaluation value of the previous solution, which makes it possible to follow the dynamic changes.

– **Step3': Modified memorizing the best food source**
In dynamic environment, it is not necessary to hold the best solution through all time, because the best solution continuously changes during searching process. For this reason, modified ABC algorithm chooses the best solution from only current time at step3. In particular, the best solutions at current time $\mathbf{x_{ib}}$ is always stored as the best solution without comparing it with $\mathbf{x_{best}}$.

By the modification of the above two points, ABC algorithm can be adapted to the dynamic environment without decreasing the performance in a static environment.

4 ABC Algorithm Based on Local Information Sharing

The problem of modified ABC described in Section 3 is that all bees easily converge to a local optimum and cannot follow dynamic environmental change in multimodal problems with a large number of local optima. The cause of this problem is that all bees can communicate with each other. All bees move to one local minimum within to continue the search. As a result, modified ABC loses the global search ability in multimodal problem of a dynamic environment. In order to solve this problem, we propose a novel ABC algorithm that can follow dynamical environment change without converge to a few local optima by restricting communication range among bees. Specifically, the limitation of communication range of bees is pre-defined by the Euclidean distance, and bees can communicate bees within limited range. As a result, bees share information with only nearby bees and can search wide search space even in a dynamic environment.

4.1 The Proposed ABC Algorithm

For the limitation of communication range among bees, we modify step1 and step2 of the modified ABC algorithm as follows:

– Step1": Employed bees phase with limited communication
In proposed employed bees phase, each employed bee shares information with bee within a pre-defined range d. Each bees has a circle of radius d centered at itself, and they can only share information with bees in the inside of its circle. Concretely, if any bees are within the range of the distance d from bee x_i, x_k in equation (1) is selected from them and a candidate new solution is calculated. On the other hand, if no bees are within the range of the distance d from bee x_i, a candidate new solution is randomly calculated within the range of the distance d from the current solution. A candidate new solution in these two cases is calculated as shown in the equation (5):

$$v_{ij} = \begin{cases} x_{ij} + \Phi(x_{ij} - x_{nj}) & if\ \exists x_n \in X_k, \|x_i - x_k\| < d \\ x_{ij} + \Phi d & otherwise \end{cases} \tag{5}$$

where X_k is a set of solutions within the distance d around bee xi, . As a result, a candidate new solution v_{ij} are always produced within the range of the distance d from bee x_i.

– Step2": Onlooker bees with limited communication
In conventional ABC algorithms, all solutions can be selected by onlooker bees with a probability pi calculated with equation (4). However, since the proposed ABC algorithm employs the limitation of communication range, a bee that has no bees within the range of the distance d from itself randomly searches around itself in step 1". For this reason, onlooker bees allocated to such food source are not selected. Concretely in step 2", each onlooker bee is also allocated each food source, and the selection probability pi of onlooker bee x_i is calculate as equation (6).

$$p_i = \begin{cases} p_i & if\ \exists x_n \in X_k, \|x_i - x_k\| < d \\ 0 & otherwise \end{cases} \tag{6}$$

where x_k is bee within distance d to bee x_i and X_k is a set of x_k, $X_k \in x_k$, while p_i is calculated with equation (4). In equation (6), the selection probability of bees that are out of range of the distance d is set as zero. Then, an onlooker bee that is chosen by the roulette wheel selection according to the selection probability searches a food source. In other words, an onlooker bee shares information with only employed bees that are within its neighbors and explores a food source with high evaluation value.

By these improvements, a global search performance in a static environment is limited, however, it is possible to search each local minimum in multimodal problem and to follow dynamic environmental changes. Afterword, our proposed ABC algorithm with the modification described above is named as *ABC algorithm based on Local Information Sharing* (ABC-lis for short). Motion of bees in each step of ABC-lis is shown in Fig.3, Fig.3 have almost the same meaning as Fig.1. The difference from Fig.1 is that bees only share information with other bee within distance d represented by large circles with dotted lines. Besides, the flow of ABC-lis algorithm is shown in Fig.4.

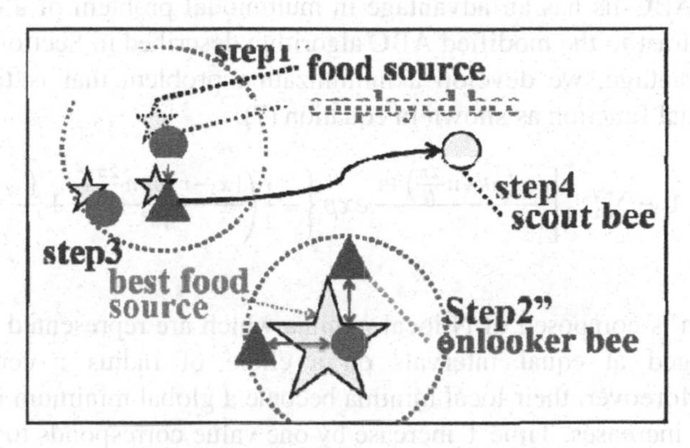

Fig. 3. Schematic diagram of ABC-lis

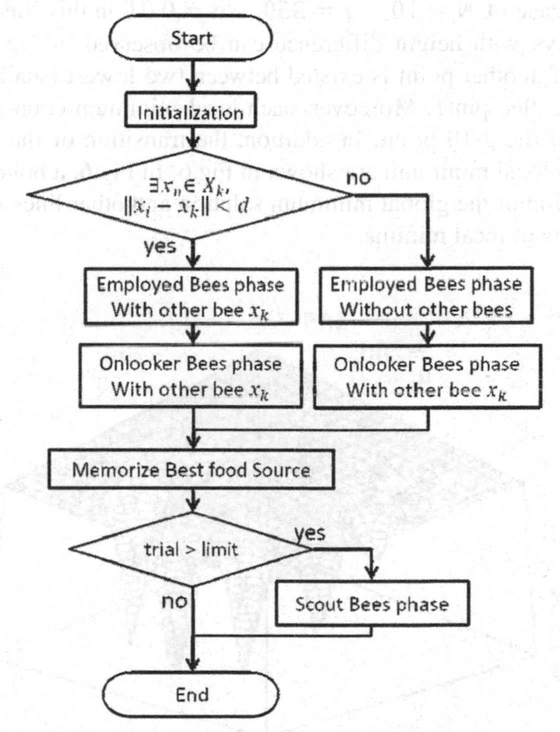

Fig. 4. Flow of ABC-lis algorithm

5 Problem Description

The proposed ABC-lis has an advantage in multimodal problem of a dynamic environment in contrast to the modified ABC algorithm described in Section 3. In order to verify this advantage, we develop a minimization problem that is time-depend N peaks multimodal function as shown in equation (7).

$$g(x_1, x_2, k) = 1 - \sum_{n=0}^{N} \left[\frac{\cos\left(\alpha k + n \cdot \frac{2\pi}{N}\right) + 1}{2} \exp\left\{ -\frac{1}{2} \left(\frac{\left(x_1 - r\cos n \cdot \frac{2\pi}{N}\right)^2}{40^2} + \frac{\left(x_2 - r\sin n \cdot \frac{2\pi}{N}\right)^2}{40^2} \right) \right\} \right] \quad (7)$$

This function is composed of N local minima which are represented by a Gaussian function arranged at equal intervals on a circle of radius r centered on the origin $(\mathbf{0}, \mathbf{0})$. Moreover, their local minima become a global minimum in the order by discrete time t increases. Time t increase by one value corresponds to the number of iterations $(C = C + 1)$ increases. Magnitude of temporal changes is controlled by $\boldsymbol{\alpha}$. The outline of a case of $N = 10$, $r = 350$, $\alpha = 0.01$ in this function is shown in Fig.5. Nine valleys with height difference can be observed in Fig.5. In addition, a local minimum of another point is existed between two lowest (shallow) valleys. The shallow point is called pint1. Moreover, each local minimum counterclockwise from point 1 are called the 2-10 point. In addition, the transition of the global minimum solution and each local minimum are shown in Fig.6. In Fig.6, a bold line at bottom is shown the transition of the global minimum solution and other lines to draw a thin sin curve is transitions of local minima.

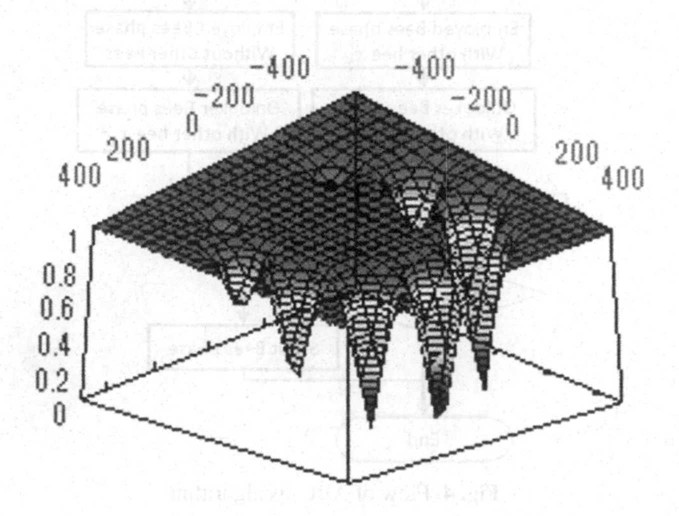

Fig. 5. Outline of the function

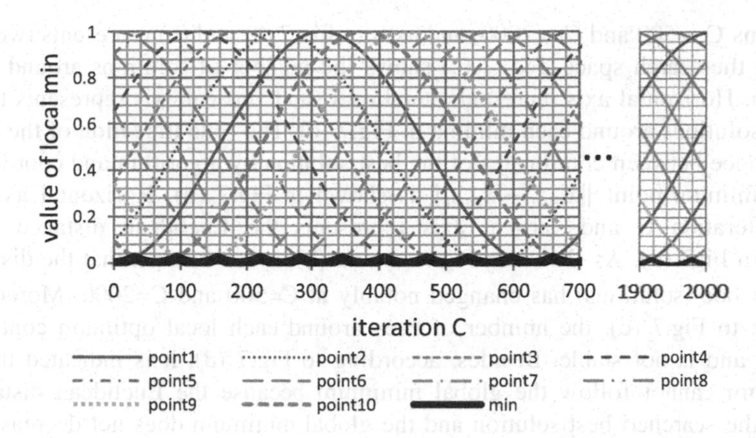

Fig. 6. Transition of each local solution

6 Experiment

6.1 Contents of Experiment

To verify the effectiveness of ABC-lis, we compare three methods, original ABC algorithm, modified ABC, and ABC-lis, in dynamic environment represented as the function $g(x_1, x_2, k)$ described in Chapter 5.

6.2 Evaluation Criteria and Parameter Setting

To evaluate three ABC algorithms, we use two evaluation criteria. One is the distribution of the number of bees for each local minimum of the function $g(x_1, x_2, k)$ to validate the global search performance for temporal change. Another is the time transition of the Euclidean distance ($\|\boldsymbol{x}_{min} - \boldsymbol{x}_{best}\|$) between coordinates of best solution by algorithm ($\boldsymbol{x}_{best}$) and coordinates of true minimum point at the time ($\boldsymbol{x}_{min}$) to evaluate the search performance of three algorithms.

In all algorithms, the common parameters are employed, in detail, $\mathrm{N_S} = 100, \mathrm{limit} = 40, \mathrm{and}\ \mathrm{N_{mc}} = 2000$. The shared area d that is specific parameter of ABC-lis is set as d = 50. In function $g(x_1, x_2, k)$, parameters are set as follows: $\mathrm{N} = 10,\quad \mathrm{r} = 350,\quad \alpha = 0.01$. Transition of the values of the local minimum and outline at that time is as shown in Fig.5and 6 of chapter 5.

6.3 Experimental Results

First, results of the original ABC algorithm is shown in Fig.7 (a) ~ (d). Fig. 7 (a) and (b) respectively show the distribution of solutions in the search space at the number of iterations C = 500 and C = 2000, and axes in Fig.7 (a) and (b) represents two vectors axes in the search space. Fig. 7 (c) shows the number of solutions around 10 local minima. Horizontal axis represents iteration C and vertical axis represents the number of solutions around each minima in Fig.7 (c). The time transition of the Euclidean distance between coordinates of the best solution by algorithm and coordinates of true minimum point $||x_{min} - x_{best}||$ is shown in Fig.7 (d). Horizontal axis represents iteration C and vertical axis represents the Euclidean distance $||x_{min} - x_{best}||$ in Fig7 (d). As shown in Fig.7 (a) and (b), it can be seen that the distribution of each bee (solutions) has changed notably at C=500 and C=2000. Moreover, according to Fig.7 (c), the number of bees around each local optimum continuously change and is not stable. Besides, according to Fig.7 (d), it is indicated that ABC algorithm cannot follow the global minimum because the Euclidean distance between the searched best solution and the global minimum does not decrease. These results indicate that x_{best} is stayed at one local optimum without searching other global solutions.

Next, results of modified ABC algorithm is shown in Fig.8 (a) ~ (d), which have the same meaning as Fig.7 (a) ~ (d). As shown in Fig.8 (a) and (b), bees are evenly converged to each local minimum at C=500, however, at C=2000, they are converging to only one local minimum. Moreover, according to Fig.8 (c), all bees begin to converge to one local optimum after 1300 iterations, and all bees completely converge on the same local optimum (point10 in Fig.8 (c)) after 1500 iterations. Thus, as show in Fig.8 (d), modified ABC algorithm loses global search capability after about 1400 iterations because all bees converge on one local minimum. These results indicate that the modified ABC algorithm decreases its search performance in a dynamic environment because of the convergence on single local optimum.

Finally, the results of the proposed ABC-lis algorithms are shown in Fig.9 (a) ~ (d). These figures have the same meaning as Fig. 7 (a) ~ (d) and Fig. 8 (a) ~ (d). As shown in Fig.9 (a) and (b), bees are quickly gathered to each local minimum after 500 iterations. In addition, bees converge on all local minimum at 2000 iterations. As show in Fig.9 (c), bees is gathered in each local minimum at about 300 iterations and it is maintained up to 2000 iterations. Moreover, the appropriate solution is always memorized as the optimal solution x_{best}, because bees are present in all local minimums. Therefore, the high search performance can be always maintained as shown in Fig.9 (d).

Thus, ABC algorithm as first of the three methods cannot follow the global minimum. In addition, modified ABC as second of the three methods loses global search capability after about 1400 iterations. Other hand, in ABC-lis as third of the three method, it is possible to maintain the global search capability in multi-modal problem of a dynamic environment.

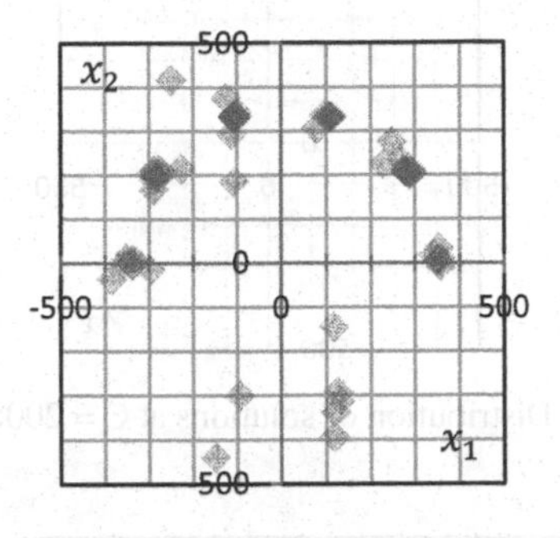

(a) Distribution of solutions at C = 500

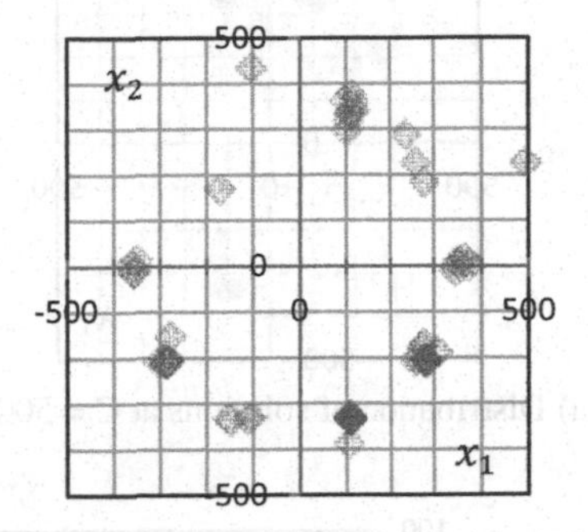

(b) Distribution of solutions at C = 2000

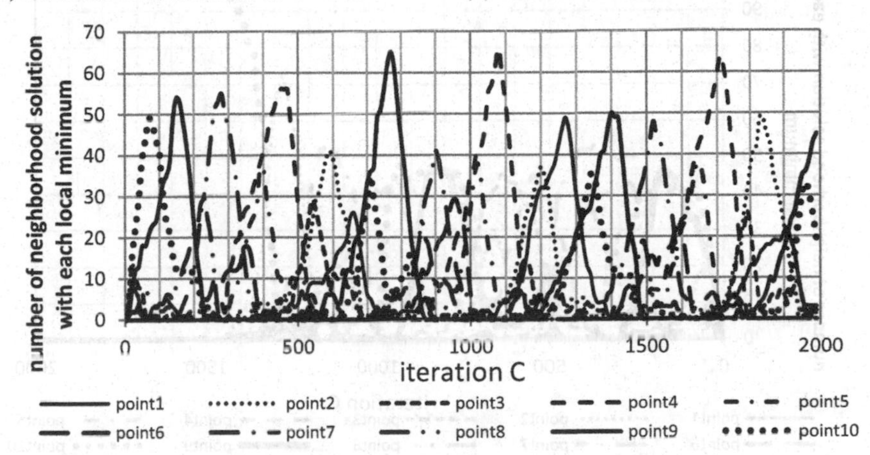

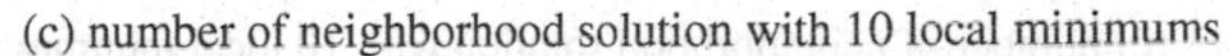

(c) number of neighborhood solution with 10 local minimums

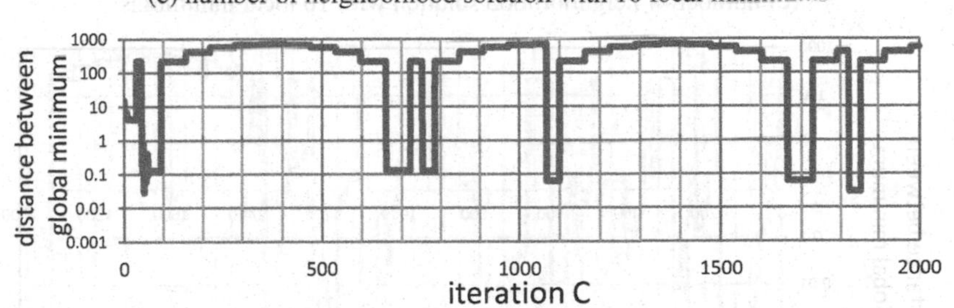

(d) transition of Euclidean distance of x_{best} and minimum solution

Fig. 7. ABC algorithm in $g(x_1, x_2, k)$

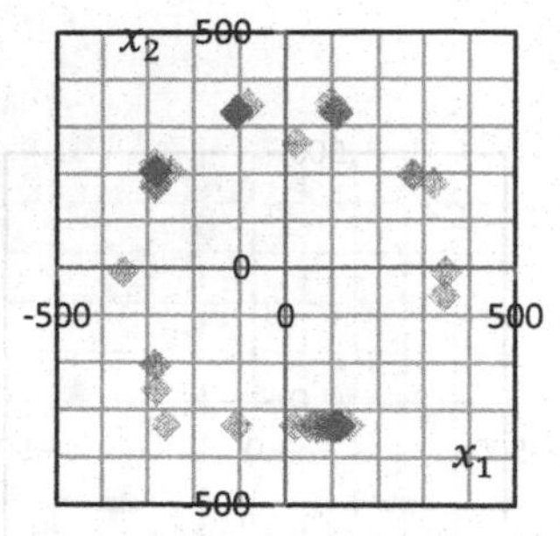

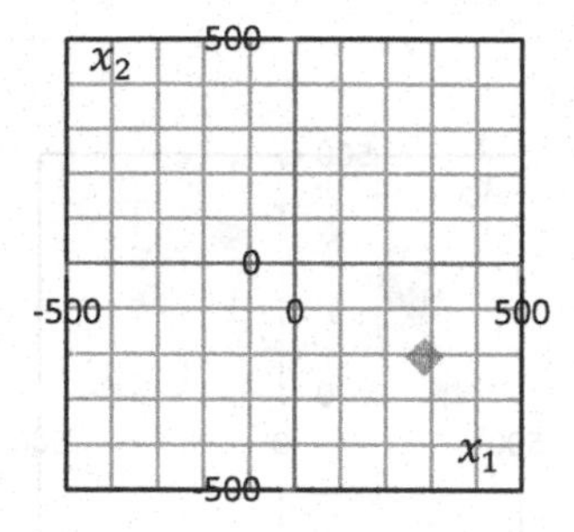

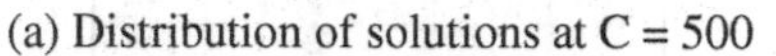

(a) Distribution of solutions at C = 500 (**b**) Distribution of solutions at C = 2000

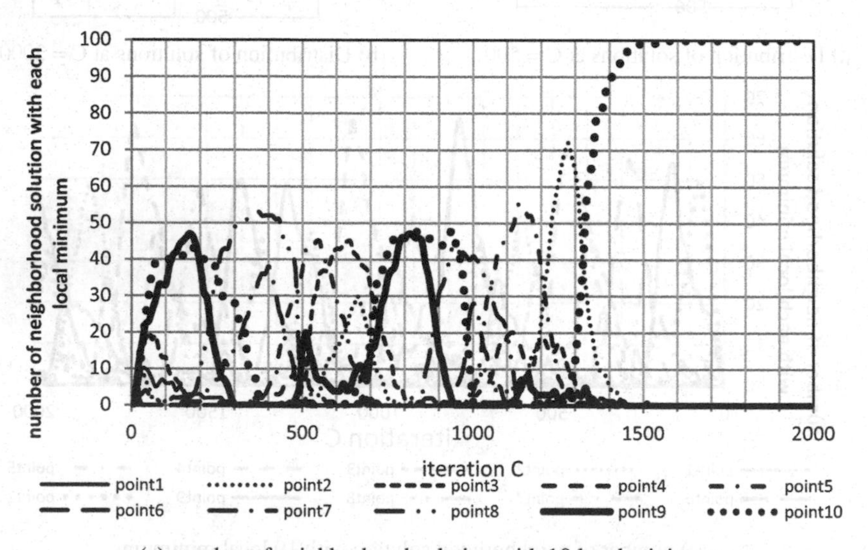

(c) number of neighborhood solution with 10 local minimums

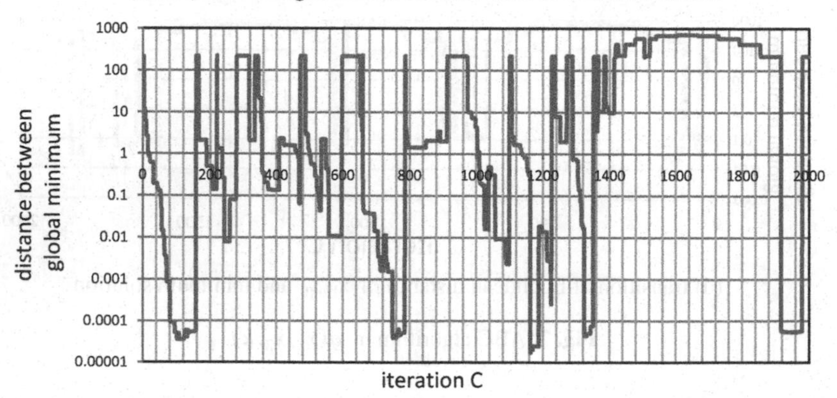

(d) transition of Euclidean distance of x_{best} and minimum solution

Fig. 8. Modified ABC in g(x_1,x_2,k)

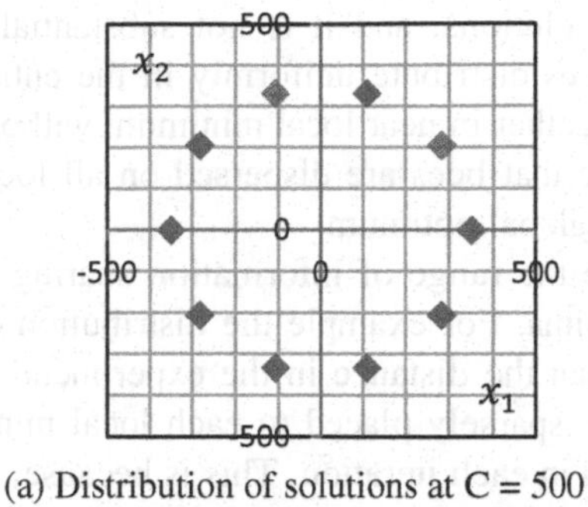

(a) Distribution of solutions at C = 500

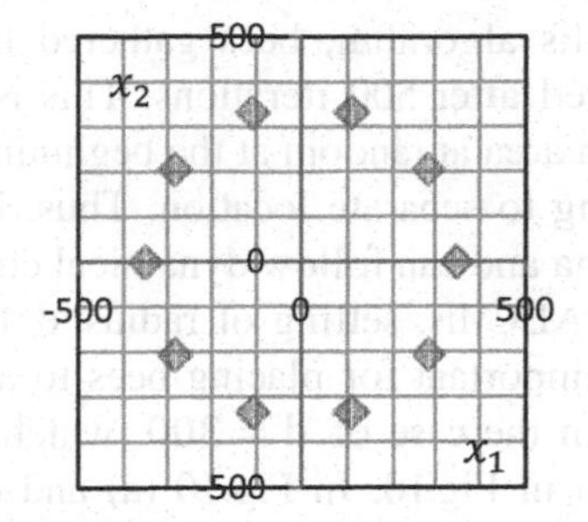

(b) Distribution of solutions at C = 2000

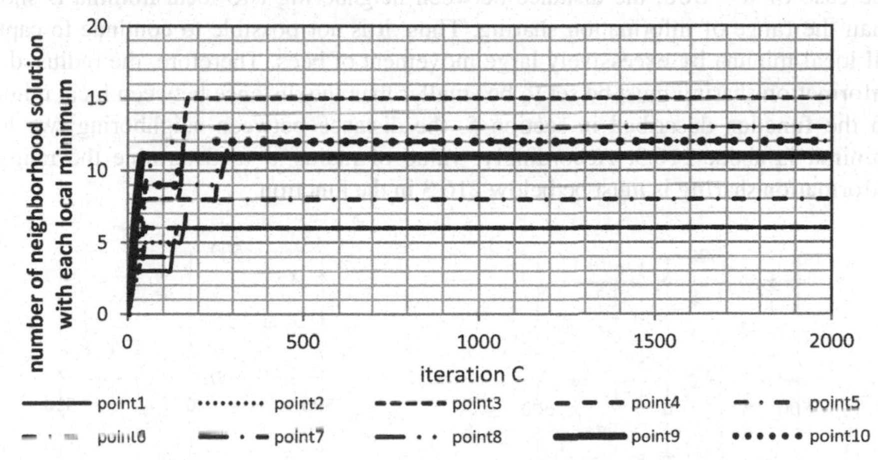

(c) number of neighborhood solution with 10 local minimums

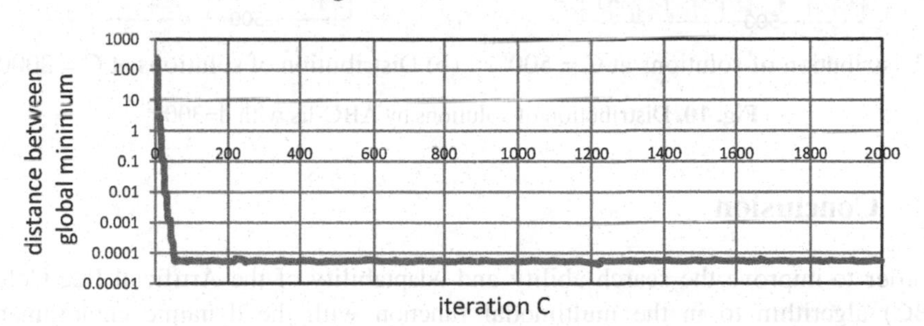

(d) transition of Euclidean distance of x_{best} and minimum solution

Fig. 9. ABC-lis in g(x_1,x_2,k)

6.4 Discussion

According to results of Fig.7 to Fig.9 in section 6.3, the proposed ABC-lis algorithm outperforms the original ABC algorithm and the modified ABC algorithm in dynamic environment. This result shows that the improvement in ABC-lis algorithm contributes to increase the search performance in dynamic environment. Especially, in

ABC-lis algorithm, bees gathered in all local solutions, and it is not substantially changed after 500 iterations. This is because bees distribute uniformly in the entire search area at random at the beginning and they gather in near local minimum without moving to separate location. Thus, it is possible that bees are dispersed on all local minima and can follow dynamical change of the global optimum.

In ABC-lis, setting of radius d to determine the range of information sharing is very important for placing bees to all local minima. For example the distribution of bees in the case of d = 300, which is larger than the distance in the experiment, is shown in Fig.10. In Fig.10 (a) and (b), bees are sparsely placed to each local minimum and some local minima cannot be searched in each iteration. This is because, in the case of d = 300, the distance between neighboring two local minima is shorter than the range of information sharing. Thus, it is not possible to continue to capture all local minima by excessively large movement of bees. Therefore, the radius d for information sharing must be set to be smaller than the distance between local minima. In the function described in section 5, the distance between neighboring two local minima is about 216.3. Accordingly, value of radius d to determine the range of information sharing is must be below 216.3 in the function.

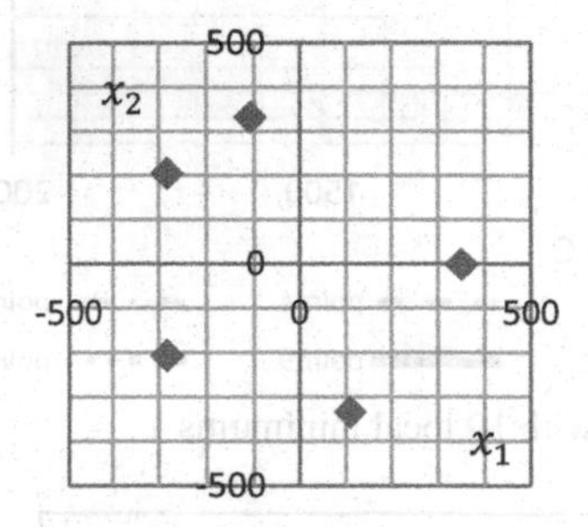

(a) Distribution of solutions at C = 500

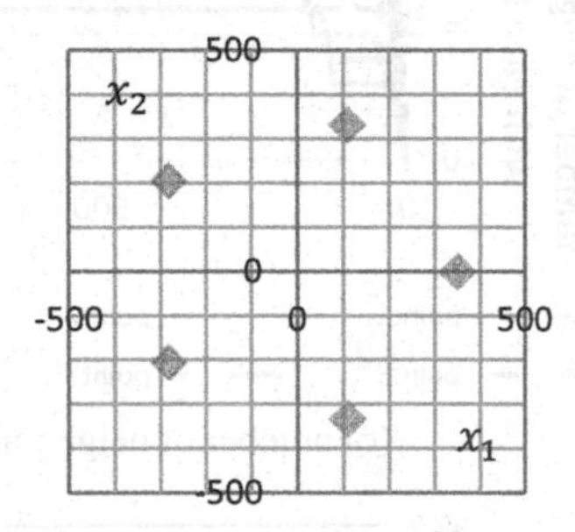

(b) Distribution of solutions at C = 2000

Fig. 10. Distribution of solutions by ABC-lis with d=300

7 Conclusion

In order to improve the search ability and adaptability of the Artificial Bee Colony (ABC) algorithm to in the multimodal function with the dynamic environmental change, this paper extends the ABC algorithm based on global information for static environment to proposed a novel the ABC algorithm based on local information sharing (ABC-lis) for dynamic environment. Unlike the conventional ABC algorithm, the ABC-lis algorithm shares only local information of solutions by that restricting communication among bees to only their close ones. To investigates the search ability and adaptability of ABC-lis algorithm to environmental change, we conducted the experiment to that compares the proposed ABC-lis algorithm with the conventional two ABC algorithms, i.e., the original one, and the modified one proposed by [2] in multimodal functions. The experimental result revealed that the following implications: (1) the proposed ABC-lis algorithm can not only maintains the search capability in the multimodal problem with dynamic environmental change, but also can increases the adaptability to dynamic environmental change global search performance, in comparison with unlike the original and modified ABC algorithms; and lost global search ability in such environment. (2) The radius d for shared among bees information sharing must be set to be smaller than the distance between local minima in the multimodal function. Because, if the radius d bigger than he distance between local minima, it is not impossible for bees to continue to capture all local minima by excessively large movement of bees.

What should be noticed here is that these results have only been obtained from one multimodal function. Therefore, further careful qualifications and justifications, such as an analysis of results using other but types of functions, are needed to generalize our results. Such important directions must be pursued in the near future in addition to the following research: We will address the following works in near future: (1) Verification of the ABC-lis algorithm in other functions; (2) Improvement of the ABC-lis algorithm to find a new optimal solution that occurs after all bees converged to all local minima; and (3) Application of the ABC-lis algorithm for rescue agents to explore victims in disaster areas.

References

1. Karaboga, D., Bastur, B.: A powerful and Efficient Algorithm for Numerical Function Optimization: Artificial Bee Colony (ABC) Algorithm. Journal of Global Optimization 39, 459–471
2. Nishida, T.: Modification of ABC Algorithm for Adaptation to Time-Varying Functions. Electronics and Communications in Japan (2012)
3. Iimura, I., Nakayama, S.: Search Performance Evaluation of Artificial Bee Colony Algorithm on High-Dimensional Function Optimization. ISCIE 24(4), 97–99 (2011)
4. Karadoga, D., Basturk, B.: On the performance of artificial bee colony (ABC) algorithm. Applied Soft Computing 8, 687–697 (2007)
5. Kang, F., Li, J., Ma, Z.: Rosenbrock artificial bee colony algorithm for accurate global optimization of numerical function. Information Sciences 181, 3508–3511 (2011)
6. Gao, W., Liu, S.: Improved artificial bee colony algorithm for global optimization. Information Processing Letters 111, 871–882 (2011)
7. Gao, W., Liu, S.: A modified artificial bee colony algorithm. Computers & Operations Research, Letters 39, 687–882 (2012)
8. Tadokoro, S., Kitano, H.: RoboCup-Rescue: Challenge to Rescue in Large-Scale Disasters (2000)
9. Takano, R., Yamazaki, D., Ichikawa, Y., Hattori, K., Takadama, K.: Multiagent-based ABC algorithm for dynamical environment: Toward cooperation among autonomous rescue agents. Computer Software - JSSST Journal 31(3), 187–199 (2014)

Regularized Boost for Semi-supervised Ranking

Zhigao Miao[1], Juan Wang[1], Aimin Zhou[2], and Ke Tang[1,*]

[1] USTC-Birmingham Joint Research Institute in Intelligent Computation and Its Applications (UBRI), School of Computer Science and Technology University of Science and Technology of China, Hefei 230027, China

[2] Shanghai Key Laboratory of Multidimensional Information Processing and Department of Computer Science and Technology, East China Normal University, Shanghai 200241, China

{mzg123,jingze}@mail.ustc.edu.cn, amzhou@cs.ecnu.edu.cn, ketang@ustc.edu.cn

Abstract. Semi-supervised ranking is a relatively new and important learning problem inspired by many applications. Motivated by the prior work on regularization in semi-supervised learning, we introduce a local smooth regularizer that can exploit the manifold structure of the data to leverage the unlabeled data. The regularizer is general and can be applied to any paired comparison ranking model. By minimizing the pairwise loss subject to this regularization penalty based on the sequential ensemble learning framework, a semi-supervised regularized boosting algorithm for ranking is derived. Each stage of boosting is fast and efficient. The proposed algorithm shares the same theoretical justification and implementation effectiveness as in RankBoost. Experimental results on benchmark datasets demonstrate that the proposed algorithm is effective and comparable to some other state-of-the-art algorithms.

Keywords: Learning to rank, Semi-supervised learning, Regularization, Boosting.

1 Introduction

Learning to rank [1], a task that seeks to induce an ordering or preference relations over a set of objects, has been drawing increasing interest in the machine learning community. Though much progress has been made in developing different algorithms for the ranking problem, the performance of ranking model is strongly affected by the number of labeled training instances. The labeling process is in general a time-consuming and a expensive task. To alleviate this problem, it is preferable to integrate unlabeled data in the training base. We naturally come to the semi-supervised paradigm which tries to construct reliable models from a small amount of labeled data and a large amount of unlabeled set.

The key component of semi-supervised learning [2] is a principle to connect the structure of the unlabeled data with the function to be learned. In this paper,

* Corresponding author.

H. Handa et al. (eds.), *Proc. of the 18th Asia Pacific Symp. on Intell. & Evol. Systems – Vol. 1*,
Proceedings in Adaptation, Learning and Optimization 1, DOI: 10.1007/978-3-319-13359-1_49

we formulate such a principle for ranking: similar items should have similar rank scores. The principle is embodied in a local smooth regularizer that tries to capture manifold structure in the data. The regularizer is general and is closely related to the graph Laplacian manifold [3] regularizer. To minimizing the combined loss, an efficient regularized boosting procedure is derived. We detail the theoretical justification and conduct experiments to validate the proposed algorithm.

The rest of the paper is organized as follows: Section 2 provides a brief literature review to the related work. In Section 3 we describe our regularized boosting algorithm for semi-supervised ranking in details and we present the results of our evaluation in Section 4. Finally, in Section 5 we conclude the paper and give directions for future work.

2 Related Work

2.1 Semi-supervised Learning to Rank

In the semi-supervised ranking problem, a set of queries $Q = \{q_1, q_2, \ldots, q_m\}$ is given. Each query q_i is associated with a list of documents $\boldsymbol{d_i} = \{d_{i1}, d_{i2}, \ldots, d_{i,n(q_i)}\}$ and a list of labels $\boldsymbol{y_i} = \{y_{i1}, y_{i2}, \ldots, y_{i,n(q_i)}\}$ where $y_{ij} \in \{r_1, r_2, \ldots, r_k\}$ is the relevance level of d_{ij} and y_{ij} is only available for the labeled items. Our task is to learn a function $f(x; w)$ with parameters w that ranks the documents for each query in the testing set.

Several methods have been developed for the task of semi-supervised ranking. For example, Amini et al. [5] proposed an adaption of supervised RankBoost [4] algorithm which builds the ranking function on the basis of two training sets: one labeled and one unlabeled. First, the algorithm uses an unsupervised method to initially assign relevance judgments to the unlabeled examples. Then an extended version of RankBoost algorithm is developed to produce a scoring function by optimizing the pair-wise loss function. Another line of work is the extension of RankSVM [6] within the semi-supervised regularization framework. Pan et al. [7] proposed Semi-RankSVM which is a combination of RankSVM and graph Laplacian regularization trick that can exploit the neighborhood manifold of the training data. The method adopts a large margin optimization approach like the traditional RankSVM, and minimizes the pair-wise loss together with a graph-based regularization term. Besides, M.szummer [10] proposed a semi-supervised version of RankNet [9] by applying a preference regularizer favoring that similar items are similar in preference to each other.

2.2 Boosting and RankBoost

Boosting is a general technique for improving the accuracies of machine learning algorithms. As a generic ensemble learning framework, boosting works by sequentially constructing a linear combination of base learners that concentrate on difficult examples, which results in a great success in supervised learning.

Inspired by boosting algorithms for classification, Freund proposed an algorithm for ranking called RankBoost, which works by combining weak rankings into a single high accurate ranking. RankBoost [4] optimize an exponential loss which is the upper bound of the ranking error, and improves the accuracy of the ranking function iteratively. The notations introduced are explained as below. $f(x)$ is the ranking function to be learned which is assumed to be a weighted sum of the weak rankers as $f = \sum_{t=1}^{T} \alpha_t h_t$. D_t is the distribution over the training data constructed by a set of preferences. The pseudo code of RankBoost algorithm is shown in Algorithm 1.

Algorithm 1: **RankBoost**

Input: training data χ : a set of preferences $(x_0 \preceq x_1)$.
- Initialize D_1.

For $t = 1, \ldots, T$

- Create weak ranker h_t using distribution D_t .
- Choose $\alpha_t \in R$.
- Update D_{t+1}

$$D_{t+1}(x_0, x_1) = \frac{P_t(x_0,x_1)\exp\left(\alpha_t(h_t(x_0)-h_t(x_1))\right)}{Z_t}$$

where $Z_t = \sum_{x_0,x_1} D_t(x_0, x_1) \exp\left(\alpha_t(h_t(x_0) - h_t(x_1))\right)$

Output : $f(x) = \sum_{i=1}^{T} \alpha_t h_t(x)$.

---

3 Our Method

As a generic framework, regularization has been used in semi-supervised learning to exploit unlabeled data by working on well-known semi-supervised learning assumptions. we introduce a regularizer that exploits the local smoothness constraints among data. Furthermore, we derived an efficient boosting algorithm to minimized the augmented cost.

3.1 Semi-supervised Regularization

Given two sample x_0, x_1, we define a loss which forces an agreement if the two samples are similar.

$$\mathcal{R}(x_0, x_1) := S(x_0, x_1) \cosh\left(f(x_0) - f(x_1)\right) \quad (1)$$

where $S(x_0, x_1) = \exp(-\frac{||x_0-x_1||^2}{\sigma^2})$ σ is the scale parameter which will be determinated in the experiment section), $\cosh(x) = \frac{1}{2}(e^x + e^{-x})$. Since $\cosh(x)$ is a symmetric function which has its minimum at $x = 0$, it measures the inconsistency between the responses and the similarity measurement. Summing up over all instance pairs, we get the penalty term $\mathcal{R}(f, D)$, which will be used as regularization term for unlabeled data. As we will see below, such a exponential penalty facilitates the derivation of boosting based algorithms.

$$\begin{aligned}\mathcal{R}(f, D) &= \sum_{x_0, x_1} \mathcal{R}(x_0, x_1) \\ &= \frac{1}{2} \sum_{x_0, x_1} S(x_0, x_1) \exp(f(x_0) - f(x_1)) + \frac{1}{2} \sum_{x_0, x_1} S(x_0, x_1) \exp(f(x_1) - f(x_0)) \\ &= \sum_{x_0, x_1} S(x_0, x_1) \exp(f(x_0) - f(x_1)) \end{aligned} \tag{2}$$

The last step of the derivation holds since $S(.)$ is a symmetric function,so $\sum_{x_0, x_1} S(x_0, x_1) \exp(f(x_0) - f(x_1)) = \sum_{x_0, x_1} S(x_1, x_0) \exp(f(x_0) - f(x_1))$. Then from the commutative law of addition, it is easy to check the following holds when x_0 and x_1 come form the same set. $\sum_{x_0, x_1} S(x_1, x_0) \exp(f(x_0) - f(x_1)) = \sum_{x_0, x_1} S(x_0, x_1) \exp(f(x_1) - f(x_0))$.

3.2 Regularized Boost Algorithm

Let us denote$I_1 = \bigcup_q \{(i, j) | x_i, x_j$ belong to the same query q, and $r_i < r_j\}$and $I_2 = \bigcup_q \{(i, j) | x_i, x_j$ belong to the same query q $\}$. The combined loss function for ranking problem is shown as below. The parameter λ is a trade-off factor to control the importance of the two parts which can be selected by cross-validation routine.

$$\begin{aligned}\mathcal{L}(f, D) &= \frac{1}{|I_1|} \sum_{(x_0, x_1) \in I_1} \exp(f(x_0) - f(x_1)) \\ &\quad + \frac{\lambda}{|I_2|} \sum_{(x_0, x_1) \in I_2} S(x_0, x_1) \exp(f(x_0) - f(x_1)) \\ &= \sum_{I_1} P(x_0, x_1) \exp(f(x_0) - f(x_1)) \\ &\quad + \lambda \sum_{I_2} Q(x_0, x_1) \exp(f(x_0) - f(x_1)) \end{aligned} \tag{3}$$

To solve the related minimization problem, we consider the case when the final ranking f is a weighted sum of the weak rankers as $f = \sum_{t=1}^{T} \alpha_t h_t$. In the following theorem, the ranking loss of f is reformulated as products of loss Z_t of every single weaker ranker h_t. This theorem also provides guidance in choosing α_t and in designing the weak learner as discussed below. The algorithm minimizing

Z_t at each step of boosting.

$$\begin{aligned} Z_t &= Z_P + \lambda Z_Q \\ &= \sum_{x_0,x_1} P(x_0,x_1)\exp\left((f(x_0) - f(x_1)\right)) \\ &+\lambda \sum_{x_0,x_1} Q(x_0,x_1)\exp\left((f(x_0) - f(x_1)\right)) \end{aligned} \tag{4}$$

Theorem 1. *For any set of α_t and h_t,the ranking loss of $f = \sum_{t=1}^{T} \alpha_t h_t$ can be rewrite as below.*

$$\mathcal{L}(f,D) = \prod_{t=1}^{T} Z_t \tag{5}$$

Proof. Unraveling the update rule, we have that

$$P_{T+1}(x_0,x_1) = \frac{P(x_0,x_1)\exp(f(x_0) - f(x_1))}{\prod_t Z_t} \tag{6}$$

$$Q_{T+1}(x_0,x_1) = \frac{Q(x_0,x_1)\exp(f(x_0) - f(x_1))}{\prod_t Z_t} \tag{7}$$

$$\begin{aligned} \mathcal{L}(f,D) &= \sum_{x_0,x_1} P(x_0,x_1)\exp\left(f(x_0) - f(x_1)\right) \\ &+\lambda \sum_{x_0,x_1} Q(x_0,x_1)\exp\left(f(x_0) - f(x_1)\right) \\ &= \sum_{x_0,x_1} P_{T+1}(x_0,x_1)\prod_t Z_t + \lambda \sum_{x_0,x_1} Q_{T+1}(x_0,x_1)\prod_t Z_t \\ &= (\sum_{x_0,x_1} P_{T+1}(x_0,x_1) + \lambda \sum_{x_0,x_1} Q_{T+1}(x_0,x_1))\prod_t Z_t = \prod_t Z_t \end{aligned} \tag{8}$$

Algorithm 2: Our derived algorithm

Input: training data χ
$I_1 = \bigcup_q \{(i,j)|x_i, x_j$ belong to the same query q, and $l_i > l_j\}$
$I_2 = \bigcup_q \{(i,j)|x_i, x_j$ belong to the same query q $\}$.

- Initialize $P_1(x_0,x_1) = \frac{1}{|I_1|}$ over I_1 and $Q_1(x_0,x_1) = \frac{S(x_0,x_1)}{|I_2|}$ over I_2.

For $t = 1,\ldots,T$

- Create weak ranker h_t with distribution P_t and $Q_t$on training data.
- Choose α_t $\qquad \alpha_t = \frac{1}{2}\ln\frac{1+r_t+\lambda(1-r_t')}{1-r_t+\lambda(1+r_t')}$
- Update P_{t+1} and Q_{t+1}

$P_{t+1}(x_0,x_1) = \frac{P_t(x_0,x_1)\exp(\alpha_t(h_t(x_0)-h_t(x_1)))}{Z_t}$

$Q_{t+1}(x_0,x_1) = \frac{Q_t(x_0,x_1)\exp(\alpha_t(h_t(x_0)-h_t(x_1)))}{Z_t}$

where $Z_t = Z_P + \lambda Z_Q$

$Z_P = \sum_{x,x_1} D_t(x_0,x_1)\exp(\alpha_t(h_t(x_0)-h_t(x_1)))$

$Z_Q = \sum_{x_0,x_1} W_t(x_0,x_1)\exp(\alpha_t(h_t(x_0)-h_t(x_1)))$

Output ranking model : $f(x) = \sum_{i=1}^{T} \alpha_t h_t(x)$.

3.3 Choosing α_t and Criteria for Weak Learners

Considering that weak ranker range from 0 to 1, we can utilize the following Equation to determine α_t.

$$e^{\alpha x} \le (\frac{1+x}{2})e^{\alpha} + (\frac{1-x}{2})e^{-\alpha} \tag{9}$$

Equation (9) holds for all real α and $x \in [-1,1]$. Thus we can approximate Z_P and Z_Q by

$$\begin{aligned} Z_P &\le \sum_{x_0,x_1} P(x_0,x_1) \times [(\frac{(1+h(x_0)-h(x_1))}{2})e^{\alpha} + (\frac{1+h(x_1)-h(x_0)}{2})e^{-\alpha}] \\ &= (\frac{1-r}{2})e^{\alpha} + (\frac{1+r}{2})e^{-\alpha} \end{aligned} \tag{10}$$

$$\begin{aligned} Z_Q &\le \sum_{x_0,x_1} Q(x_0,x_1) \times [(\frac{(1+h(x_0)-h(x_1))}{2})e^{\alpha} + (\frac{1+h(x_1)-h(x_0)}{2})e^{-\alpha}] \\ &= (\frac{1-r'}{2})e^{\alpha} + (\frac{1+r'}{2})e^{-\alpha} \end{aligned} \tag{11}$$

where

$$r = \sum_{x_0,x_1} P(x_0,x_1)((h(x_1)-h(x_0))) \tag{12}$$

$$r' = \sum_{x_0,x_1} Q(x_0,x_1)((h(x_0)-h(x_1))) \tag{13}$$

We get the following equation by combining Equation (10) with Equation (11)

$$\begin{aligned} Z &\le (\frac{1-r}{2})e^{\alpha} + (\frac{1+r}{2})e^{-\alpha} + \lambda[(\frac{1+r'}{2})e^{\alpha} + (\frac{1-r'}{2})e^{-\alpha}] \\ &= \frac{1-r+\lambda(1+r')}{2}e^{\alpha} + \frac{1+r+\lambda(1-r')}{2}e^{-\alpha} \end{aligned} \tag{14}$$

where $\alpha_t = \frac{1}{2}\ln\frac{1+r+\lambda(1-r')}{1-r+\lambda(1+r')}, r, r' \in [-1,1]$, the right hand side of Equation can be minimized. Equation (6) can be transferred to the following inequality:

$$Z \le \sqrt{(1+\lambda)^2 - (\lambda r' - r)^2} \tag{15}$$

In order to minimize Z , all we need to do is to maximize $|\lambda r' - r|$, which can be achieved by training a proper weak ranker h similarly as in RankBoost.

4 Experiment

4.1 Experiment Setup

In this section, we present an empirical evaluation of our semi-supervised ranking algorithm. The proposed algorithm Reg-RankBoost was compared with two representative semi-supervised ranking algorithm Semi-RankBoost [5], Semi-RankSVM [7] to validate our approach.

The experiments are conducted on MQ2008-semi(Million Query track) and OHSUMED datasets. The MQ2008-semi dataset [13] is public available benchmark data collections for research on semi-supervised learning to rank. There are about 2000 queries in this dataset. The OHSUMED dataset [14] consists of 106 queries with 25 features extracted from the on-line medical information database. For the OHSUMED dataset, we reduce the amount of available labeled data in the training set by forming a 10% random sample of all the labels for each query to imitate the case of limited labels. Both datasets were split in five folds with a 3:1:1 ratio for training, validation and testing. The results reported in this section are average results over multiple folds.

As to evaluate the performance of ranking models, we use Normalized Discounted Cumulative Gain (NDCG) as evaluation measures. Given a query q_i, the NDCG score at position n in the ranking list of documents can be calculated by the equation as follows. Here $r(j)$ is the rating of the jth document in the list and Z_n is the normalization constant.

$$NDCG@n = Z_n \sum_{j=1}^{n} \frac{2^{r(j)} - 1}{\log(1+j)} \tag{16}$$

4.2 Experiment Results

There are two learning parameters to be tuned in our algorithm: scale parameter σ and trade-off factor λ. We typically set the scale parameter to 10 to 20 percent of the range of the distance between examples, as suggested in many experiments. The trade-off factor λ in the loss function varied from 0 to 1 with a step of 0.05, and the best value on the validation set was chosen. The averages result across the 5 folds are summarized in the following table.

The results show that a great improvement was made in terms of the NDCG measure for both of the two datsets. This indicates that our proposed algorithm is comparable to the two baselines.

Table 1. NDCG@n Measures on the MQ2008-Semi Collection

Algorithm	NDCG@1	NDCG @3	NDCG @5	NDCG @ 7	NDCG @ 10
Semi-RankBoost	0.463	0.455	**0.449**	0.412	0.430
Semi-RankSVM	**0.495**	0.420	0.416	0.413	0.414
Reg-RankBoost	0.473	**0.467**	0.415	**0.427**	**0.437**

Table 2. NDCG@n Measures on the OHSUMED Collection

Algorithm	NDCG@1	NDCG @3	NDCG @5	NDCG @ 7	NDCG @ 10
Semi-RankBoost	0.501	**0.483**	0.431	0.459	0.426
Semi-RankSVM	0.482	0.471	0.402	0.453	**0.434**
Reg-RankBoost	**0.523**	0.462	**0.446**	**0.474**	0.432

5 Conclusion and Future Work

In this paper, we have proposed a semi-supervised boosting algorithm for ranking with unlabeled data. Based on the principle that similar documents should have similar scores closely given a query, we formulate a local smooth regularizer, and plug it into the supervised ranking model. Furthermore, we derived an efficient boosting procedure to training the model. Both theoretical analysis and experimental results demonstrate the effectiveness of our algorithm. For future work, we are working on understanding important aspects of the algorithm, in particular, generalization, error bounds, convergence and local minima.

Acknowledgements. This work was supported in part by the 973 Program of China under Grant 2011CB707006, the National Natural Science Foundation of China under Grant 61175065, the Program for New Century Excellent Talents in University under Grant NCET-12-0512, and the Natural Science Foundation of Anhui Province under Grant 1108085J16.

References

1. Liu, T.Y.: Learning to Rank for Information Retrieval. Journal of Foundations and Trends in Information Retrieval, 225–331 (2002)
2. Zhu, X.J.: Semi-supervised Learning Literature Survey. Technical Report 1530, Department of Computer Sciences, University of Wisconsin, Madison (2005)
3. Belkin, M., Niyogi, P., Sindhwani, V.: Manifold Regularization: A Geometric Framework for Learning from Labeled and Unlabeled Examples. Journal of Machine Learning Research, 2399–2434 (2006)
4. Freund, Y., Iyer, R., Schapire, R., Singer, Y.: An Efficient Boosting Algorithm for Combining Preferences. Journal of Machine Learning Research, 933–969 (2003)
5. Amini, M.R., Truong, T.V., Gutte, C.: A Boosting Algorithm for Learning Bipartite Ranking Functions with Partially Labeled Data. In: Proceedings of the 31st Annual International Conference on Research and Development in Information Retrieval, 99–106 (2008)
6. Cao, Y., Xu, J., Liu, T., Li, H., Huang, Y., Hon, H.: Adapting Ranking SVM to Document Retrieval. In: Proceedings of the 29th Annual International Conference on Research and Development in Information Retrieval, pp. 186–193 (2006)
7. Pan, Z.B., You, X., Chen, H., Tao, D.C., Pang, B.C.: Generalization Performance of Magnitude-preserving Semi-supervised Ranking with Graph-based Regularization. Journal of Information Sciences, 284–296 (2013)

8. Agarwal, S.: Ranking on Graph Data. In: Proceedings of the 23rd International Conference on Machine Learning, pp. 25–32 (2006)
9. Burges, C., Shaked, T., Renshaw, E.: Learning to Rank using Gradient Descent. In: Proceedings of the 22nd International Conference on Machine Learning, pp. 89–96 (2005)
10. Szummer, M., Yilmaz, E.: Semi-supervised Learning to Rank with Preference Regularization. In: Proceedings of the 20th ACM International Conference on Information and Knowledge Management, pp. 269–278 (2011)
11. Peng, Z., Tang, Y., Lin, L., Pan, Y.: Learning to Rank with a Weight Matrix. In: Proceedings of the 14th International Conference on Computer Supported Cooperative Work in Design, pp. 18–21 (2010)
12. Truong, V., Amini, M., Gallinari, P.: A Self-training Method for Learning to Rank with Unlabeled Data. In: Proceedings of the 16th European Symposium on Artificial Neural Networks, pp. 22–24 (2009)
13. Liu, T.Y., Xu, J., Qin, T., Xiong, W.Y.: LETOR: Benchmark Datasct for Research on Learning to Rank for Information Retrieval. Journal of Information Retrieval, 346–374 (2010)
14. Hersh, W., Buckley, C., Leone, T.J., Hickam, D.: OHSUMED: An Interactive Retrieval Evaluation and New Large Test Collection for Research. In: Proceedings of the 17th Annual International Conference on Research and Development in Information Retrieval, pp. 192–201 (1994)

Brain CT Image Classification with Deep Neural Networks*

Cheng Da[1], Haixian Zhang[2], and Yongsheng Sang[2]

[1] Institute of Automation, Chinese Academy of Sciences, Beijing 100190, P.R. China
dcfucheng@hotmail.com

[2] The Machine Intelligence Laboratory, College of Computer Science, Sichuan University, Chengdu 610065, P.R. China
{zhanghaixian,sangys}@scu.edu.cn

Abstract. With the development of X-ray, CT, MRI and other medical imaging techniques, doctors and researchers are provided with a large number of medical images for clinical diagnosis. It can largely improves the accuracy and reliability of disease diagnosis. In this paper, the method of brain CT image classification with Deep neural networks is proposed. Deep neural network exploits many layers of non-linear information for classification and pattern analysis. In the most recent literature, deep learning is defined as a kind of representation learning, which involves a hierarchy architecture where higher-level concepts are constructed from lower-level ones. The techniques developed from deep learning, enriched the main research aspects of machine learning and artificial intelligence, have already been impacting a wide range of signal and information processing researches. By using the normal and abnormal brain CT images, texture features are extracted as the characteristic value of each image. Then, deep neural network is used to realize the CT image classification of brain health. Experimental results indicate that the deep neural network have performed well in the CT images classification of brain health. It also shows that the stability of the network increases significantly as the depth of the network increasing.

Keywords: Deep neural network, Texture analysis, Gray level co-occurrence matrix (GLCM).

1 Introduction

With the emergence of the big digital image data, the development of image retrieval and image pattern recognition technology grows very fast. In recent years, image recognition technology is widely used in bioinformatics, medical engineering, military, meteorology, aerospace and other fields. Especially in the assisting medical diagnosis. With the sustained development of X-ray, CT, MRI and other

* This work was supported by the National Science Foundation of China under grant 61303015 and the Specialized Research Fund for the Doctoral Program of Higher Education under grant 20130181120075.

H. Handa et al. (eds.), *Proc. of the 18th Asia Pacific Symp. on Intell. & Evol. Systems – Vol. 1*,
Proceedings in Adaptation, Learning and Optimization 1, DOI: 10.1007/978-3-319-13359-1_50

medical imaging techniques, doctors and researchers are provided with a large number of medical images for clinical diagnosis. It can improve the accuracy and reliability of clinical diagnosis of diseases. Nevertheless, manual classification of these huge number of images is a challenging and time-consuming task. Therefore, achieving computer aided diagnosis can help to increase diagnostic approaches so as to reduce the rate of misdiagnosis. Thus, an automatic or semi-automatic classification method by computer has played an important role in assist clinical diagnosis.

Image features can be reflected in many aspects, such as texture, color, shape, spatial relations, etc. They can reflect the subtle difference in varying degree. Therefore, different selections of image features will result in different classification decisions [4][12]. Texture is a description of the gray spatial distribution. Most medical images have good texture features. Changes of medical image texture features can indirectly reflect the health of the organ. Thus, image texture is always used in feature extraction of medical images so as to complete image segmentation or image classification. As early in 1972, scientists successfully distinguished the normal and abnormal lung through the texture analysis method. Nowadays, texture analysis can be divided into four classes, namely statistics, structure, model and spectrum method. Among them, GLCM (Gray-level Co-occurrence Matrix) of statistics is widely used.

In order to realize an automatic or semi-automatic classification, we should construct an appropriate classifier. At present, the classifier can be divided into three classes. Firstly, the method based on statistics, such as Bayesian method and SVM (Support Vector Machine). Secondly, the method based on rule, such as decision tree and rough sets. Thirdly, ANN (Artificial Neural Network). Neural networks has been widely applied to image processing research, such as Hopfield neural networks can be used for unsupervised pattern classification of medical images [10]. In spite of various neural network models, the most widely used network is BP (Back propagation) neural network model.

Back-propagation (BP) neural networks have been a well-known deep neural network. Feed-forward neural networks or multi-layer perceptions(MLPs) with many hidden layers is a basic kind of deep neural networks (DNNs). Back propagation is a very popular neural network learning algorithm because it is conceptually simple, computationally efficient [1]. It is based on local gradient descent [3][7][9], and starts usually at some random initial points. However, it is assumed that back-propagation with small number of hidden layers alone did not work well in practice [2]. Using hidden layers with many neurons in a DNN can significantly improve the power of the DNN and create many optimal configurations. Even if the network is trapped into a local optimum, the resulting DNN can still perform quite well. It is assumed that because the small number of neurons are used in the network, the chance of having a local optimum becomes lower [13], [14]. Then, the network can have better computational power by using deep and wide neural architecture during the training process.

In this paper, we take normal and abnormal brain CT images as data sources. Then, the classification of brain CT images is realized by using deep neural network. The paper is organized as following. In Section 2, the model of DNN is given. In Section 3, the method of the classification of brain CT images is given in detail. Then, the experiments and discussion based on the proposed method are given in Section 4. Finally, the conclusion is given in Section 5.

2 Deep Learning Neural Networks

At beginning, deep learning is focused on the digit image classification problem [6]. The latest records of image classification are still held by deep networks. It is claimed that the new method for the unconstrained data works well by using a convolutional architecture [5], and the state-of-the-art for the knowledge free data also have a good results[11]. In the last few years, deep learning has made rapid progress in object recognition in natural images [8].

The network analysis in this paper is primarily presented in the classical multi-layer neural networks with gradient-based learning [15]. The simplest form of the deep network is shown as follows. Every network implements a function like $Z_n = F_n(W_n, Z_{n-1})$, where Z_{n-1} is a input vector, W_n is the connection matrix, and Z_n is a vector that represents the output of the network. Then, the partial derivatives of E^p with respect to W_n and Z_{n-1} can be computed using the backward recurrence if the partial derivative of E^p with respect to Z_n is known.

$$\frac{\partial E^p}{\partial W_n} = \frac{\partial F}{\partial W}\frac{\partial E^p}{\partial Z_n} \tag{1}$$

$$\frac{\partial E^p}{\partial Z_{n-1}} = \frac{\partial F}{\partial Z}\frac{\partial E^p}{\partial Z_n} \tag{2}$$

where $\frac{\partial F}{\partial W}$ and $\frac{\partial F}{\partial Z}$ are the Jacobian of F with respect to W and the Jacobian of F with respect to Z respectively evaluated at the point (W_n, Z_{n-1}). From layer N to layer 1, all the partial derivatives of the cost function with respect to all the parameters can be computed from the above equations.

The multi-layer neural network used in this paper is a special case of the above system where alternated layers of matrix multiplications and component-wise sigmoid function are used.

$$Y_n = W_n Z_{n-1} \tag{3}$$

$$Z_n = F(Y_n) \tag{4}$$

where W_n is a connection matrix. F is a vector function that applies a sigmoid function and Y_n is the vector of weighted sums.

The back-propagation equations are obtained by using the chain rule to the equation above:

$$\frac{\partial E^p}{\partial y_n^i} = f'(y_n^i)\frac{\partial E^p}{\partial z_n^i} \tag{5}$$

$$\frac{\partial E^p}{\partial w_n^{ij}} = z_{n-1}^j \frac{\partial E^p}{\partial y_n^i} \tag{6}$$

$$\frac{\partial E^p}{\partial z_{n-1}^k} = \sum_i w_n^{ik} \frac{\partial E^p}{\partial y_n^i} \tag{7}$$

The above equations can also be written in matrix form:

$$\frac{\partial E^p}{\partial Y_n} = F'(Y_n) \frac{\partial E^p}{\partial Z_n} \tag{8}$$

$$\frac{\partial E^p}{\partial W_n} = Z_{n-1} \frac{\partial E^p}{\partial Y_n} \tag{9}$$

$$\frac{\partial E^p}{\partial Z_{n-1}} = W_n^T \frac{\partial E^p}{\partial Y_n} \tag{10}$$

The gradient descent algorithm is as follows, where η is a scalar constant, a proper choice of it is important:

$$W(t) = W(t-1) - \eta \frac{\partial E}{\partial W} \tag{11}$$

3 The Method

3.1 Image Preprocessing

Image preprocessing is an essential step of image classification, which has a great impact on subsequent steps. In this paper, two preprocessing techniques are used. The details will as follows.

Texture features do not require color information, thus, it is necessary to change a colorful image into a grayscale image. In this paper, the gray value can be computed by the weighted average method.

$$P = \frac{0.3R + 0.59G + 0.11B}{3} \tag{12}$$

Thus, the value can represent a pixel information so as to reduce the amount of computation by using gray skills. Simultaneously, gray image is the base of GLCM. Then, in order to reduce the amount of calculation, we can lose some gray information so as to make the image compression. In this paper, the image are compressed to different levels, such as 8, 16 and 32 gray levels.

3.2 Feature Extraction by Texture Analysis (GLCM)

Feature extraction is the technique of extracting specific features, which can reflect the differences between two classes. These features should make within-class similarity is maximized and between-class similarity is minimized. Gray-level Co-occurrence Matrix (GLCM) based on the second conditional probability density function is currently the most popular statistical method. Therefore, GLCM is widely used in image texture analysis because of simple principle and rich information. GLCM can reflect the spatial relationship of the images. We can calculate these following features through GLCM : Energy, Contrast, Homogeneity, Correlation, Entropy, Variance, Maximum probability, Sum average, Sum variance, Sum entropy, Difference variance, Difference entropy, Cluster prominence, Cluster shade and so on. Out of these features, there are 4 main features are taken in this paper. Energy is the sum of squared elements in the GLCM. It can reflect the thickness and uniformity of image texture.

$$ASM = \sum_i \sum_j P(i,j)^2 \tag{13}$$

Entropy is a measure of randomness. Entropy can reflect the randomness and complexity of image texture. It is maximized when image is most random and complex.

$$ENT = -\sum_i \sum_j P(i,j) log P(i,j) \tag{14}$$

Contrast is a measure of the intensity contrast between a pixel and its neighbor over the whole image. It can reflect the definition and details of image texture.

$$CON = \sum_i \sum_j (i-j)^2 P(i,j) \tag{15}$$

Homogeneity is the measure of local homogeneity. It can reflect the regularity of image texture.

$$IDM = \sum_i \sum_j \frac{P(i,j)}{1+(i-j)^2} \tag{16}$$

In this paper, four GLCM features are calculated per image in four directions 0, 45, 95, 135, but only choose 0 direction in order to simplify the calculation.

3.3 Classification with Deep Neural Network

The data for classification is four GLCM features (Energy, Entropy, Contrast, and Homogeneity) from the images of 8, 16, 32 levels.

Before training the deep neural network, the data should be normalized. Normalization is a kind of data preprocessing, it can make the data mapped to the specific range such as [0, 1] so that the influence between the dimensions can be eliminated, and it also can make a better convergence of the network. In this paper, normalization is based on linear function.

$$y = \frac{x - Min}{Max - Min} \quad (17)$$

where, x is the original value, y is the normalized value. Max and Min are defined as the maximum value and minimum value of the data.

While training network, database is randomly divided into two parts in every times, 8 samples per group of 7 is selected as training samples, 1 sample for testing. Thus, 14 samples for training, 2 samples for testing. Each image is represented by four GLCM features (Energy, Entropy, Contrast, and Homogeneity). The classification result of the sample is taken as the output of network. 0 represents normal samples, and 1 represents abnormal samples. Thus, the number of output linguistic value is one in this paper. The network selects 1 and 2 layer as the network architecture, each hidden layer has 4, 9 or 14 hidden nodes.

About the activation function selection, hyperbolic tangent function ($y = tanh(x)$) is taken in hidden layer, and sigmoid function ($y = 1/(1 + e - x)$) is taken in out layer. The maximum convergence time is set to 50000, the global error limit is set to 0.001, the coefficient of inertia item is set to 0.5, and the step length is set to 0.1. The initial link weights are set to random numbers in the range of $[-1, 1]$.

In order to ensure the reliability and validity of training results, a new network is trained based on a new training set by a random algorithm. Through building the deep neural network repeatedly, we can eliminate the influence of special individual network to obtain the universal conclusion.

In this paper, we build 100 different networks, so as to obtain the average value of classification results. In one network, 14 samples are used to train the network. Then, the last 2 samples (one normal sample and one abnormal sample) input to the trained network just for testing.

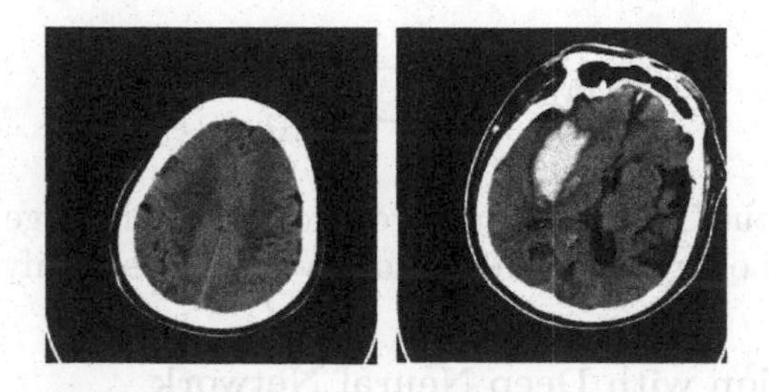

Fig. 1. Sample Data Set (left is normal, right is abnormal)

4 Experiments and Discussion

4.1 CT Image Database

The database consists of normal and abnormal brain CT images. One normal and one abnormal CT image are shown in figure 1. The CT image data is shown in figure 2, which includes normal and abnormal images.

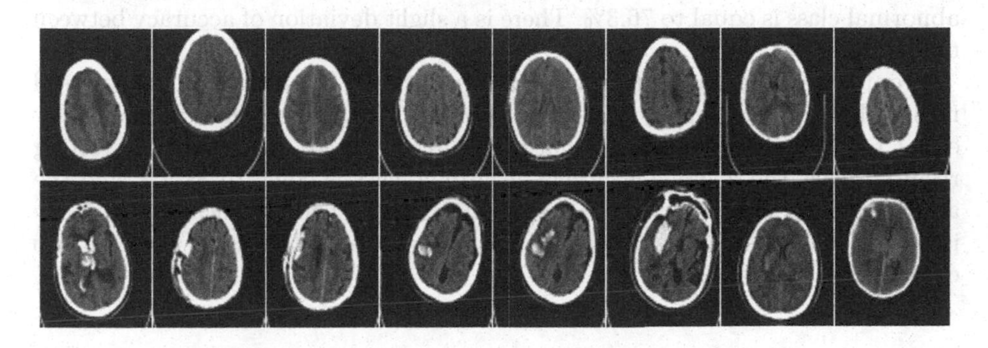

Fig. 2. Database: normal and abnormal brain CT images

4.2 Experiments and Analysis

In this paper, four GLCM features are calculated in four directions 0, 45, 95, 135, but only choose 0 direction in order to simplify calculation. Table 1 shows different features with the increase of gray level. It is shown that Contrast increased very quickly. It can reflect the image clarity very well. Thus, the Contrast of non-compressed image is highest. Based upon this value normal and abnormal brain can be differentiated.

Table 1. Features in different levels of the image shown in the left of the figure 1

Level	Energy	Entropy	Contrast	Homogeneity
8	0.4081878327	1.6661832108	2.0241771449	0.8834638442
16	0.3933533769	2.0478428541	8.9612533900	0.8324324920
32	0.3822330092	2.5023452181	37.549024593	0.77372162489
64	0.3706997971	3.0017046171	153.32908037	0.72012707619
128	0.3615409433	3.5037284921	618.92370870	0.68035517139
256	0.3563695959	3.9660440318	2486.6853658	0.65421012006

We take 8, 16, 32 levels to realize the classification. From table 2, it is easy to see that when the grayscale of CT images are compressed into 8 levels, the accuracy of the classification result on normal class is slightly lower than the accuracy of abnormal class. According to the 10 sets of testing results above, the

Table 2. Classification results with 8 levels compression

	Average Accuracy (%)
NORMAL	66.5
ABNORMAL	76.3

average accuracy of normal class is equal to 66.5%, and the average accuracy of abnormal class is equal to 76.3%. There is a slight deviation of accuracy between two classes.

From table 3, we can see that when the grayscale of CT images are compressed into 16 levels, the accuracy of the classification result on normal class is nearly equal to the accuracy of abnormal class. According to the 10 sets of testing results above, the average accuracy of normal class is equal to 81.5%, and the average accuracy of abnormal class is equal to 86.3%. The classification efficiency is better than it is shown in table 2 and the deviation of accuracy between two categories is very small.

Table 3. Classification results with 16 levels compression

	Average Accuracy (%)
NORMAL	81.5
ABNORMAL	86.3

In table 4, when the gray scale of CT image are compressed into 32 levels, the accuracy of the classification results on normal class is apparently higher than the accuracy of abnormal class. According to the 10 sets of testing results above, the average accuracy of normal class is equal to 69.2%, and the average accuracy of abnormal category is equal to 50.6%. The classification accuracy is largely decreased and there is a big deviation of accuracy between two categories.

Table 4. Classification results with 32 levels compression

	Average Accuracy (%)
NORMAL	69
ABNORMAL	50.6

Next, we will show the experimental results respectively based on different hidden nodes in 16 levels according to the method described above through 10 sets of testing data. From table 5, it can be seen that when the hidden nodes of network is variant, the deviation of the average accuracy of the classification results is small in this database. According to the 10 sets of testing results above, the average accuracy of 4 hidden nodes is equal to 82.2%, the average accuracy of 9 hidden nodes is equal to 83.9% and the average accuracy of 14 hidden nodes is equal to 83.95%. At the same time, the results of network with 14 hidden nodes are volatile, and the network training costs much more time. Therefore, the 9 hidden nodes is chosen because of the moderate time and accuracy, which is shown in table 5.

From table 6, it can be seen that there exists some differences between two kinds of the classification results when the number of hidden layers is variant. According to the 10 sets of testing results, the average accuracy of the network with 1 hidden layer is equal to 83.9%, however, the average accuracy with 2 hidden layers is only 75.5%. That is to say, as the depths of the network is increased, the classification accuracy do not be improved in such cases.

Table 5. Classification results with 16 level compression and different hidden nodes

Hidden Nodes	*Average Accuracy (%)*
4	82.2
9	83.9
14	83.95

Table 6. Classification results with 16 level compression and different hidden layers

Hidden Layers	*Average Accuracy (%)*
1	83.9
2	75.5

5 Conclusion

In this paper, we proposed a method to classify the brain CT images. Deep neural network is employed in this method, in which datasets are randomly divided into training sets and testing sets. Experimental results illustrate that the deep neural network have performed well in the CT images classification of brain health. Moreover, it also shows that the stability of the network increases significantly as the depth of the network increasing, however, the average accuracy of the classification can not be improved relatively.

Acknowledgments. This work was supported by the National Science Foundation of China under grant 61303015 and the Specialized Research Fund for the Doctoral Program of Higher Education under grant 20130181120075.

References

1. Bengio, Y.: Practical Recommendation for Gradient-based Training of Deep Architecture. In: Montavon, G., Orr, G.B., Müller, K.-R. (eds.) Neural Networks: Tricks of the Trade. LNCS, vol. 7700, pp. 437–478. Springer, Heidelberg (2012)

2. Bengio, Y., Lamblin, P., Popovici, D., Larochelle, H.: Greedy layer-wise training of deep networks. Advances in Neural Information Processing Systems 19 (NIPS), 153–160 (2007)
3. Bengio, Y.: Learning deep architectures for AI. Foundations and Trends in Machine Learning 2, 1–127 (2009)
4. Blum, M., Springenberg, J.T., Wlfing, J., Riedmiller, M.: A Learned Feature Descriptor for Object Recognition in RGB-D Data. In: Proceedings of the IEEE International Conference on Robotics and Automation (ICRA), St. Paul, Minnesota, USA (2012)
5. Ciresan, D.C., Giusti, A., Gambardella, L.M., Schmidhuber, J.: Deep neural networks segment neuronal membranes in electron microscopy images. In: Advances in Neural Information Processing Systems (NIPS), pp. 2852–2860 (2012)
6. Hinton, G., Salakhutdinov, R.: Reducing the dimensionality of data with neural networks. Science 313(5786), 504–507 (2006)
7. Hochreiter, S., Younger, A.S., Conwell, P.R.: Learning to learn using gradient descent. In: Dorffner, G., Bischof, H., Hornik, K. (eds.) ICANN 2001. LNCS, vol. 2130, pp. 87–94. Springer, Heidelberg (2001)
8. Krizhevsky, A., Sutskever, I., Hinton, G.E.: Imagenet classification with deep convolutional neural networks. In: Advances in Neural Information Processing Systems (NIPS), vol. 4 (2012)
9. Lange, S., Riedmiller, M.: Deep auto-encoder neural networks in reinforcement learning. In: International Joint Conference on Neural Networks (IJCNN 2010), Barcelona, Spain, pp. 1–8 (2010)
10. Neeraj, S., Amit, K.R., Shiru, S., Shukla, K.K., Satyajit, P., Lalit, M.: AggarwalSegmentation and classification of medical images using texture-primitive features: Application of BAM-type artificial neural network. J. Medical Physics 33(3), 119–126 (2008)
11. Rifai, S., Vincent, P., Muller, X., Glorot, X., Bengio, Y.: Contractive auto-encoders: Explicit invariance during feature extraction. In: Proceedings of the 28th International Conference on Machine Learning (ICML 2011), pp. 833–840 (2011)
12. Taleb, A.A., Doubois, P., Duquenoy, E.: Analysis Methods of CT-scan Images for the Characterization of the Bone Texture. Pattern Recognition 24, 1971–1982 (2003)
13. Williams, R.J., Peng, J.: An efficient gradient-based algorithm for on-line training of recurrent network trajectories. Neural Computation 4, 491–501 (1990)
14. Wyatte, D., Curran, T., OReilly, R.: The limits of feedforward vision: Recurrent processing promotes robust object recognition when objects are degraded. Journal of Cognitive Neuroscience 24(11), 2248–2261 (2012)
15. LeCun, Y.A., Bottou, L., Orr, G.B., Müller, K.-R.: Efficient backProp. In: Montavon, G., Orr, G.B., Müller, K.-R. (eds.) Neural Networks: Tricks of the Trade. LNCS, vol. 7700, pp. 9–48. Springer, Heidelberg (2012)

Wind Tunnel Evaluation Based Design of Lift Creating Cylinder Using Plasma Actuators

Masahiro Kanazaki, Takashi Matsuno, Kengo Maeda, and Hiromitsu Kawazoe

Tokyo Metropolitan University, Faculty of System Design,
6-6 Asahigaoka, Hino, Tokyo, Japan
kana@tmu.ac.jp, {matsuno,kawazoe}@mech.tottori-u.ac.jp,
b081054@gmail.com
http://www.sd.tmu.ac.jp/aerodesign

Abstract. A Kriging based genetic algorithm (GA) was employed to optimize the parameters of the operating conditions of plasma actuators (PAs). In this study, the lift maximization problem around a circular cylinder was considered. Two PAs were installed on the upper and the lower side of the cylinder. This problem was similar to the airfoil design, because the circular has potential to work as airfoil due to the control of flow circulation by the PAs with four design parameters. The aerodynamic performance was assessed by wind tunnel testing to overcome the disadvantages of time-consuming numerical simulations. The developed optimization system explores the optimum waveform of parameters for AC voltage by changing the waveform automatically. Based on these results, optimum designs and global design information were obtained while drastically reducing the number of experiments required compared to a full factorial experiment. An analysis of variance and a parallel coordinate plot were introduced for design knowledge discovery. According to the discovered design knowledge, it was found that duty ratios for two PAs are an important parameter to create lift.

Keywords: Plasma actuator, Genetic algorithm, Efficient global optimization, Experimental evaluation

1 Introduction

Plasma actuators (PAs, shown in Fig. 1) are flow control devices that utilize atmospheric pressure discharge [10][11][12][13]; they have gained attention in recent years, because their advantages of being fully electronically driven with no moving parts and having a simple structure and a fast response are potentially ideal for application to subsonic flow control. In [11], the drag around an airfoil can be reduced with installing PAs on the airfoil. In [16], the aerodynamic noise from the cylinder model could be reduced by PAs. Thus, PAs have remarkable benefits for the future aircraft design. Such active flow control devices have also potential to control of the circulation around arbitrary objects and produce the lift-creating object even if it is not airfoil geometry.

H. Handa et al. (eds.), *Proc. of the 18th Asia Pacific Symp. on Intell. & Evol. Systems – Vol. 1*,
Proceedings in Adaptation, Learning and Optimization 1, DOI: 10.1007/978-3-319-13359-1_51

In this study, the design problem is defined as the optimization of lift creation via flow circulation controlled by the PAs. A circular cylinder model is used as a model and two PAs are installed. A genetic algorithm (GA)-based efficient design technique was employed with wind tunnel testing to efficiently find the optimum designs. Through the design case, the applicability of the present wind tunnel testing to the multi-parameter design problem was also investigated.

Design problems are often solved by GAs based on numerical simulation, such as computational fluid dynamics (CFD) [7][15]. However, there are several difficulties with solving the flow field around PAs. First, the accuracy of existing simulation methods is still insufficient. Second, the computational cost is very high for design techniques such as GAs. Several days are needed to acquire the results for each case, whereas the actual flow physics finishes in a few seconds.

To reduce the experimental cost, Kriging surrogate model was applied to represent the input/output relationship in the experimental data. This optimization technique, which is called efficient global optimization (EGO) [1][4], enables the optimization of global parameters in a small number of experiments while simultaneously obtaining information on the design space. The EGO based on Kriging surrogate model can find efficiently near-global optimum. In [6], the three-element high lift airfoil has been designed using the time-consuming high-fidelity flow solver. This case has been efficiently solved using EGO with reducing total design cost. The EGO has been also applied to three-dimensional designs of future aircraft such as a blended-wing-body [5] and a supersonic transport [8]. The EGO is used to find the optimum installation condition of the nacelle chine which installed on the engine nacelle to improve low speed performance of an aircraft in the large scale wind-tunnel [9]. In previous studies, design tables were manually updated because model geometries and configurations were also changed manually.

In this study, Kriging surrogate model based GA performs optimization during a wind tunnel experiment in real time. The design system is automated developing the interface between the optimization and the wind tunnel testing.

2 Overview of Active Flow Control by Means of Plasma Actuator

In this research, a PA consisting of an exposed electrode and insulated electrode was used[13]. A nonconductor was placed between the two electrodes, and AC voltage was applied. Fig. 1 shows the setup; this type of PA is called a single dielectric barrier discharge (SDBA) PA. The flow around the PA can be controlled by changing the number and location of PAs and the waveform of the AC voltage. Thus, the optimal technique for solving the design problem has to handle many parameters to acquire the best flow control.

Generic home-style AC voltage has a waveform with a constant frequency. However, several studies have reported that pulse width modulation (PWM) is effective for flow control of PAs. PWM is a drive system that turns the AC

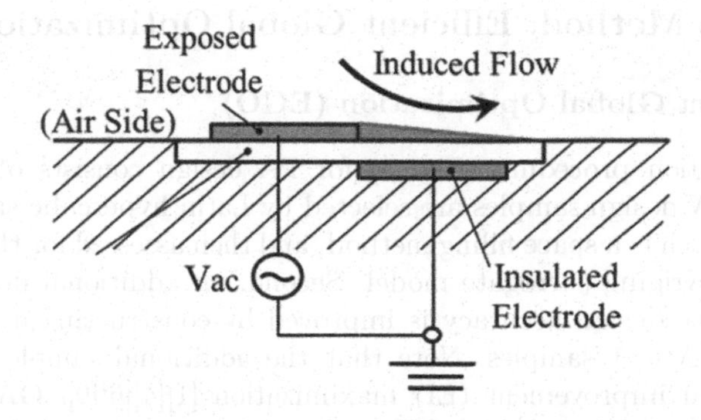

Fig. 1. Schematic of plasma actuator

voltage on or off, as shown in Fig. 2. The frequency of on/off is defined as the "modulation frequency" and is expressed by following equation:

$$f_{mod} = \frac{1}{T_1} \quad [\mathrm{Hz}] \tag{1}$$

where T_1 is the time of one cycle and T_2 is the time the AC voltage is on. The ratio of T_2 to T_1 is defined as the duty ratio, which is an important parameter for PWM. The duty ratio is expressed by the following equation:

$$D_{cycle} = 100\frac{T_2}{T_1} \quad [\%] \tag{2}$$

In this study, f_{mod} was generated from the base frequency f_p as follows:

$$f_{mod} = \frac{f_p}{20x_m} \quad [\mathrm{Hz}] \tag{3}$$

Equation 3 shows that T_1 is defined by multiples of the time of base frequency $1/f_p$. Namely, f_{mod} is determined by x_m. In this study, x_m and Dcycle were considered to be design variables.

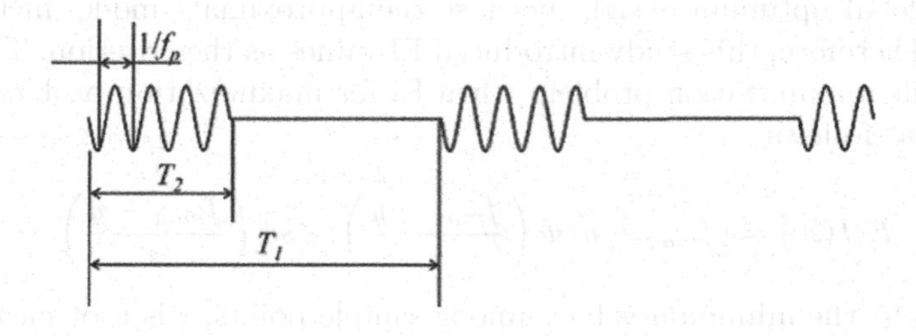

Fig. 2. Power supply by means of pulse width modulation (PWM)

3 Design Method: Efficient Global Optimization

3.1 Efficient Global Optimization (EGO)

The optimization procedure (Fig. 4) for PA design consists of the following steps. First, N design samples are selected by Latin hypercube sampling (LHS) [1][4][6][9], which is a space filling method, and then assessed for the construction of an initial Kriging surrogate model. Second, an additional design sample is added, and the design accuracy is improved by constructing a Kriging model based on all $N+1$ samples. Note that the additional sample is selected by using expected improvement (EI) maximization [1][4][6][9]. GA is applied to solve this maximization problem. This process is iterated until the improvement of the objective functions becomes negligible. Through the design procedure proposed in this paper, all samples are evaluated by the wind tunnel testing. Each technique of the optimization procedure is described in detail in the following sections.

Kriging Model. The Kriging models express the value $y(x_i)$ at the unknown design point x_i as

$$y(x_i) = \mu + \epsilon(x_i) \quad (i = 1, 2, ..., m) \tag{4}$$

where m is the number of design variables, μ is a constant of the global model, and $\epsilon(x_i)$ represents a local deviation from the global model. The correlation between $\epsilon(x_i)$ and $\epsilon(x_j)$ is strongly related to the distance between the corresponding points, x_i and x_j. In the Kriging models, the local deviation at an unknown point x is expressed using stochastic processes. Specifically, a number of design points are calculated as sample points and then interpolated using a Gaussian random function as the correlation function to estimate the trend of the stochastic process.

Expected Improvement. Once the models are constructed, the optimum point can be explored using an arbitrary optimizer. However, it is possible to miss the global optimum design, because the approximate model includes uncertainty. Therefore, this study introduced EI values as the criterion. This study solve the lift maximization problem, then EI for maximization problem can be calculated as follows:

$$E[I(x)] = (f_{max} - \hat{y})\,\Phi\left(\frac{f_{max} - \hat{y}}{s}\right) + s\phi\left(\frac{f_{max} - \hat{y}}{s}\right) \tag{5}$$

where f_{max} is the minimum values among sample points, s is root mean square error (RMSE) and $\hat{y}$ is the value predicted by Eq. 4 at an unknown point x. Φ and ϕ are the standard distribution and normal density, respectively. EI considers the predicted function value and its uncertainty, simultaneously. Therefore, by selecting the point where EI takes the maximum value, as the additional sample point, robust exploration of the global optimum and improvement of the model can be achieved simultaneously as shown in Fig. 4 because this point has a somewhat large probability to become the global optimum. In this study, the maximization of EI is carried out using GA expressed as following section.

Genetic Algorithm. GAs (Fig. 5(a)) was first proposed by Holland in the early 1970s [3] and are based on the evolution of living organisms with regard to adaptation to the environment and the passing on of genetic information to the next generation. GAs can find a global optimum because they do not use function gradients, which often lead to an exact local optimum. Thus, GA is a robust and effective method that can handle highly nonlinear optimization problems involving nondifferentiable objective functions. Owing to this advantage, GAs were applied to this experimental system. The GA used in this study [7] utilizes a real-coded representation, the blended crossover (BLX-α), and the uniform mutation. The selection probability of individuals for the crossover and mutation is expressed as follows:

$$prob = c(1 - c)rank1 - 1.0 \tag{6}$$

where $rank$ is the value of fitness ranking among the population.

In BLX-α, children are generated in a range defined by the two parents as shown in Fig. 5(b). The range is often extended equally on both sides as determined by the parameter α.

The developed system uses GA with an island model (Fig. 5(c)), which is a distributed population scheme [7][2]; it has a mechanism for global searching to avoid convergence at local optima. This model divides the population into sub-populations called islands. To retain a high degree of diversity and avoid early convergence, some individuals in each sub-population are moved to other sub-populations (migration) every k generations.

3.2 Knowledge Discovery Techniques

Analysis of Variance. To reduce the experimental cost, Kriging surrogate model was applied to represent the input/output relationship in the experimental data. This optimization technique, which is called efficient global optimization (EGO) [6][1][4][9], enables the optimization of global parameters in a small number of experiments while simultaneously obtaining information on the design space. Kriging surrogate model based GA performs optimization during a wind tunnel experiment in real time.

$$\mu_i(x_i) \equiv \int \cdots \int \hat{y}(x_1, \cdots, x_n)dx_1, \cdots, dx_{i-1}, dx_{i+1}, \cdots, dx_n - \mu \tag{7}$$

where the total mean μ is calculated as

$$\mu \equiv \int \cdots \int \hat{y}(x_1, \cdots, x_n) dx_1, \cdots, dx_n \tag{8}$$

The proportion of the variance attributed to the design variable x_i to the total variance of the model can be expressed as:

$$p \equiv \frac{\int [\mu_i(x_i)]^2 dx}{\int \cdots \int [\hat{y}(x_1 \cdots x_n) - \mu]^2 dx_1 \cdots dx_n} \tag{9}$$

The value obtained by Eq. (9) indicates the sensitivity of an objective function to the variance of a design variable.

3.3 Scatter Plot Matrix (SPM)

The solution and the design space of the multivariable design problem obtained by EGO are observed by the SPM which is one of the data mining, because the Kriging model cannot be visualized directly when the design problem has over four attribute values. SPM arranges two-dimensional scatter plots like a matrix among the objective functions and the design variables and facilitates the investigation of the design problem investigation. Each of the rows and columns is assigned attribute values such as design variables, objective functions, and constraint values. The diagonal elements show mutual same plots. Therefore, it can be said that the SPM shows scatter plots on the upper triangular part of the matrix and the correlation coefficients on the lower triangular part as additional information. Interactive Scatter Plot Matrix (iSPM) ver. 2.0 [14] developed in Japan Aerospace Exploration Agency is employed in this study.

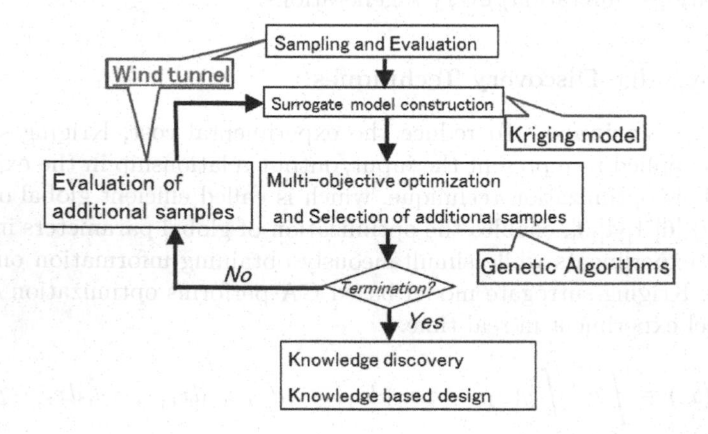

Fig. 3. Optimization procedure based on wind tunnel evaluation

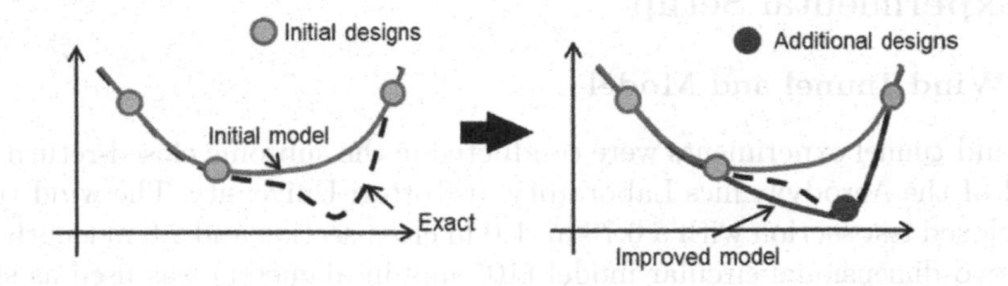

Fig. 4. Improvement of the global model by expected improvement (EI) maximization

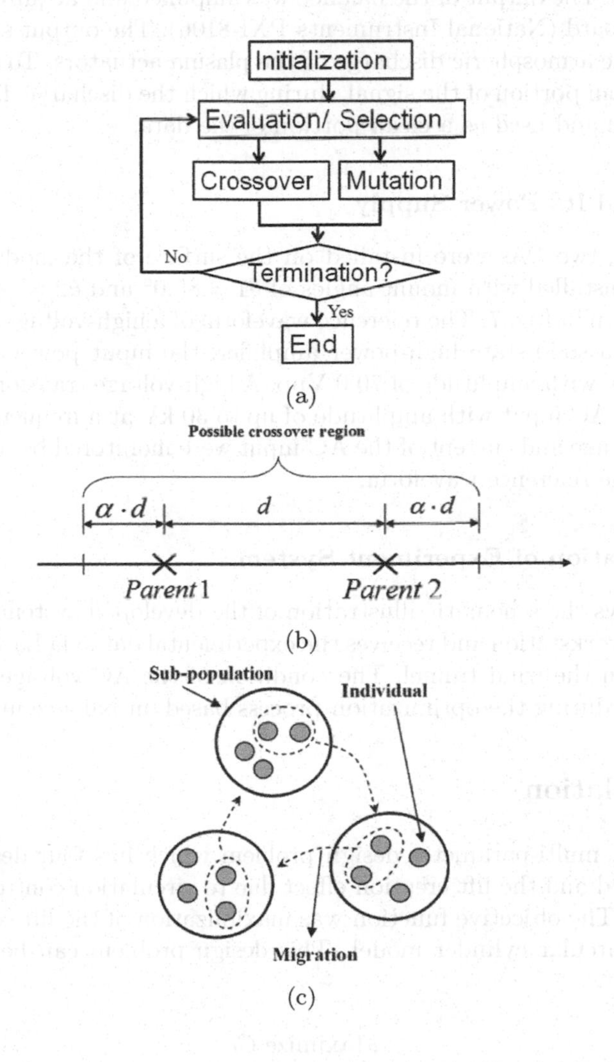

Fig. 5. Schematic illustration of genetic algorithm (GA) with the distributed scheme: (a) flowchart of GA, (b)BLX-, and (c) island model

4 Experimental Setup

4.1 Wind Tunnel and Model

The wind tunnel experiments were conducted in the subsonic closed-return wind tunnel of the Aerodynamics Laboratory at Tottori University. The wind tunnel has a closed test section with a 0.70 m 1.0 m cross-section and 2.0 m length (Fig. 6). A two-dimensional circular model (105 mm in diameter) was used as shown in Fig. 7. Model was placed on a flat plate and mounted to a support connected to a six-component external balance for measurement of the aerodynamic forces and moments. The output of the balance was amplified and acquired with a data acquisition board (National Instruments PXI-8106). The output signal contains noise from the atmospheric discharge of the plasma actuators. To eliminate this effect, the clean portion of the signal, during which the discharge did not appear, was extracted and used as a clean portion of the data.

4.2 PA and Its Power Supply

In this study, two PAs were installed on the surface of the model. PA#1 and PA#2 were installed with mount angles of $\theta1 = 85.0°$ and $\theta2 = -85.0°$, respectively, as shown in Fig. 7. The reference waveform of a high-voltage AC input was amplified by a solid-state high-power amplifier; the input power was increased up to 400.0 W with amplitude of 70.0 Vpp. A high-voltage transformer was used to achieve an AC input with amplitude of up to 30 kV at a frequency of 5.0-15.0 kHz. The voltage and current of the AC input were monitored by an oscilloscope along with the reference waveform.

4.3 Integration of Experiment System

Figure 8 shows the schematic illustration of the developed system. EGO is executed in the workstation and receives the experimental data via LabVIEW®from the balance in the wind tunnel. The condition of the AC voltage can be automatically set during the optimization process based on balance measurements.

5 Formulation

In this study, multi-parameter design problem which has four design variables was considered and the lift creation effect due to circulation control by PAs was investigated. The objective function was maximization of the lift coefficient (C_l) around the circular cylinder model. This design problem can be expressed as follows:

$$\text{Maximize } C_l \tag{10}$$

The flow velocity was set to 10.0 m/s. Eq. (6) can be written for the present design problem as follows:

Fig. 6. Test section of the wind tunnel

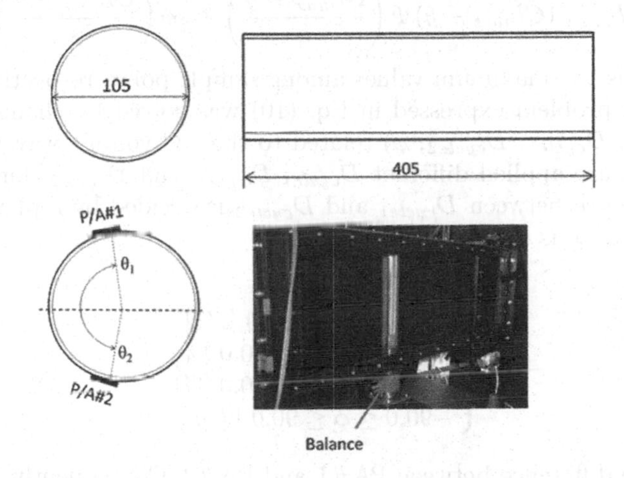

Fig. 7. Circular cylinder model and the location of plasma actuators

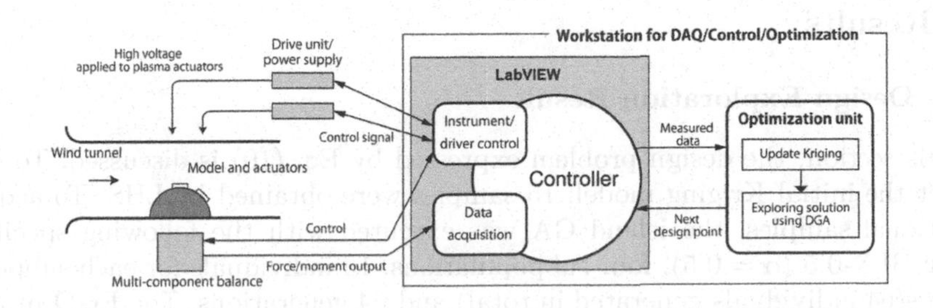

Fig. 8. Schematic diagram of the integrated optimization system

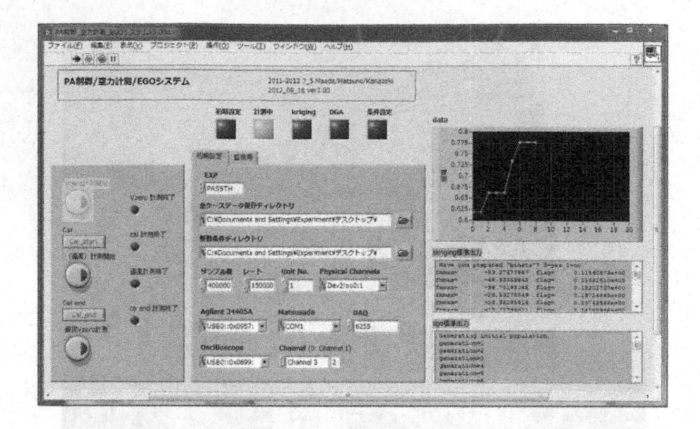

Fig. 9. Software control panel

$$EI_{Cl} = (Cl_{max} - \hat{y})\,\Phi\left(\frac{Cl_{max} - \hat{y}}{s}\right) + s\phi\left(\frac{Cl_{max} - \hat{y}}{s}\right) \tag{11}$$

where Cl_{max} is the maximum values among sample point, respectively.

The design problem expressed in Eq. (10) was solved by changing four parameters (x_m, D_{cycle1}, D_{cycle2}, ϕ) related to the AC voltage waveform. In this case, two PAs are applied different D_{cycle}; D_{cycle1} and D_{cycle2}, for each design and the difference between D_{cycle1} and D_{cycle2} is decided by a phase difference . The design space is defined as follows:

$$\begin{cases} 4.0 \leq f_{mod} \leq 25.0\ [Hz] \\ 0.0 \leq D_{cycle1} \leq 50.0\ [\%] \\ 0.0 \leq D_{cycle2} \leq 50.0\ [\%] \\ -90.0 \leq \phi \leq 90.0\ [deg.] \end{cases} \tag{12}$$

ϕ is the phase difference between PA#1 and PA#2. Consequently, the time lag can be expressed as ϕ / f_{mod}.

6 Results

6.1 Design Exploration Result

In this section, the design problem expressed by Eq. (10) is discussed. To construct the initial Kriging model, 15 samples were obtained by LHS. To acquire additional samples, the island GA was executed with the following specifications: BLX-0.5 ($\alpha = 0.5$), four subpopulations, 16 individuals for each subpopulations(64 individuals generated in total) and 64 generations. The EGO process will be stopped after five or more additional samples show better function value than that of initial samples [9][5][8].

After the objective function was converted, seven additional samples were obtained, for a total of 22 sample designs. Figure 10 shows the history of Cl values for the sampling process. According to the history, the objective function converged well with a small number of samples. Without EGO, a full factorial design of over 1000 samples would be needed to find the global optimum. The proposed system reduces the cost of the wind tunnel testing by over 99.

6.2 Design Knowledge by Analysis of Variance

Figure 11(a) shows the main effects and the two-way interaction of the design variables for objective function in this design problem. According to Fig. 11(a), D_{cycle2}, which defines the duty ratio for PA on the lower side of the cylinder, has a predominant influence on C_l. In addition, main effect of D_{cycle1} and two-way interaction of $D_{cycle1} - D_{cycle2}$ are effective to C_l. These results suggest that the circulation which creates aerodynamic lift around the model is decided by duty ratio. On the other hand,x_m and ϕ do not have influence in this analysis.

Figure 11(b) shows variances by design variables calculated by Eq. 7. (The horizontal axis is normalized design variables by minimum/maximum variables in design space, dvs_{norm}.) According to the Fig. 11(b), it is found that C_l shows higher when normalized D_{cycle1} is approximately 50% and 100%. It suggests that such duty ratio can accelerate the flow on the upper side of the model and create suction. High duty ratio requires high electronic energy, because the energizing time is higher. Thus, if the lower driving cost is required, the designer should select 50% normalized D_{cycle1}. C_l also shows the highest value when normalized D_{cycle2} is approximately 0%. The flow on the lower side should not be accelerated by the PA's volume force, because high pressure on the lower side is required

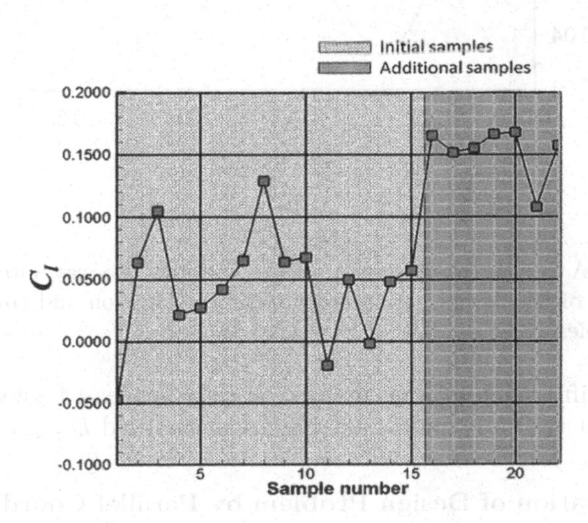

Fig. 10. Progression of objective function with sample number for the lift maximization problem

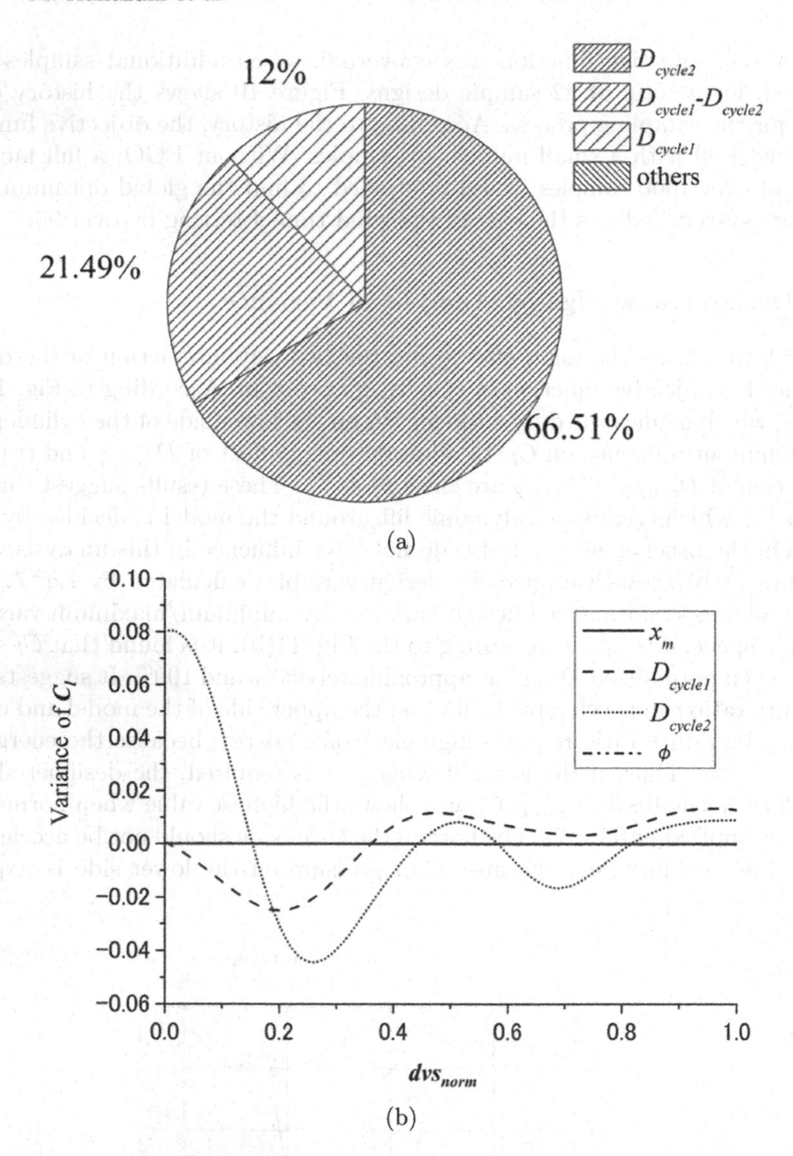

Fig. 11. ANOVA results.(a) the main effects and the two-way interactionEffect of design variables of the design variables for objective function and (b) Variance of C_l by design variables.

to create the lift. While 0% normalized D_{cycle2} achieves higher C_l, it is also remarkable that the C_l is multi-modal along normalized D_{cycle2}.

6.3 Visualization of Design Problem by Parallel Coordinate Plot

Figure 12 shows the visualization results obtained by SPM. According to Fig. 12(a), which shows the scatter plot for all parameter combinations, D_{cycle2} of

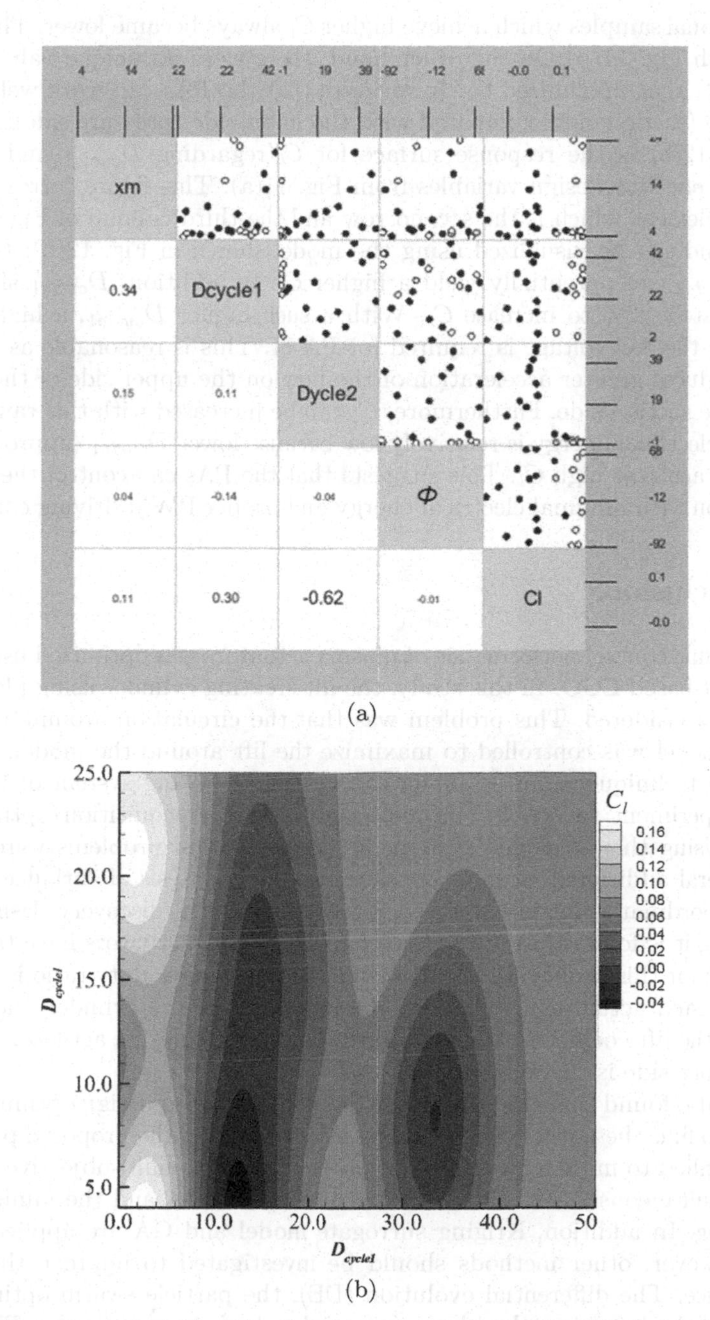

Fig. 12. Visualization of the lift maximization problem using SPM. (a) Scatter plot of all parameters, and (b) Scatter plot and response surface regarding D_{cycle1} and D_{cycle2}, colored according to C_l.(The horizontal axis is normalized design variavles by minimum/maximum variables in design space, dvs_{norm}.)

the additional samples which achieve higher C_l always became lower. This result agrees with Fig. 11(b) On the other hand, there were no unique values of x_m and D_{cycle1} that maximized C_l. It suggests that the PAs can work well for the upper side (suction side) compared with the lower side (pressure side).

Figure 12(b) is the response surface for C_l regarding D_{cycle1} and D_{cycle2}, which are effective design variables from Fig. 11(a). This figure corresponds to a matrix element which is the second row and the third column of Fig. 12. The global trend can be visualized using the model shown in Fig. 12(b); that is, a lower D_{cycle2} can potentially yield a higher C_l. In addition, D_{cycle1} should be approximately 22% to increase C_l. With a such higher D_{cycle1}, a higher total energy for the AC voltage is required for PA#1. This is reasonable as a higher D_{cycle1} induces greater acceleration of the flow on the upper side of the model, that is, the suction side. Furthermore, C_l can be increased with this range while the total electrical energy is relatively low because lower D_{cycle1} (approximately 40%) also achieves high C_l. This suggests that the PAs can control the flow for lift creation with minimal electrical energy and proper PWM driving conditions.

7 Conclusions

Aerodynamic control performance of plasma actuators was optimized using wind tunnel test-based EGO. In this study, the lift-creating cylinder using plasma actuators is considered. This problem was that the circulation around a circular cylinder model was controlled to maximize the lift around the model. The optimization technique is firstly integrated in the operating system of the wind tunnel experiment to enable automation of the data-acquisition/optimization process. Using the developed system, lift maximization problems were solved. After several additional samples are obtained, the analysis of variance and the parallel coordinate plot is employed for the knowledge discovery. Using these techniques, it is found that duty ratios for two plasma actuators have the dominant effect for this problem. It is also found that the lower duty ratio is required for the plasma actuator installed on the lower side of the cylinder and the response of the lift coefficient along the duty ratio of the plasma actuator installed on the upper side is the multi-modal.

It was also found that the wind tunnel evaluation based design optimization is effective to find the novel flow control by PAs. In future, the proposed procedure will be applied to more realistic design problem such as multi-objective problem which simultaneously solves the maximization of the lift and the minimization of the drag. In addition, Kriging surrogate model and GA are applied in this study, however, other methods should be investigated to improve the design performance. The differential evolution (DE), the particle swarm optimization (PSO) and the hybrid method have potential to be better optimizer. The radial basis function (RBF) is a candidate of the other surrogate model.

References

1. Donald, R.J., Matthias, S., William, J.W.: Efficient global optimization of expensive black-box function. Journal of Global Optimization 13, 455–492 (1998)
2. Hiroyasu, T., Miki, M., Watanabe, S.: The new model of parallel genetic algorithm in multi-objective optimization problems (divided range multi-objective genetic algorithm). In: IEEE Proc. the Congress on Evolutionary Computation, vol. 1, pp. 333–340 (2000)
3. Holland, J.H.: Adaptation in natural and artificial systems. University of Michigan Press, Ann Arbor (1975)
4. Jeong, S., Murayama, M., Yamamoto, K.: Efficient optimization design method using kriging model. Journal of Aircraft 42(2), 413–420 (2005)
5. Kanazaki, M., Hanida, R., Nara, T., Shibata, M., Nomura, T., Murayama, M., Yamamoto, K.: Challenge of design exploration for small blended wing body using unstructured flow solver. Computers & Fluids 85 (October 2013)
6. Kanazaki, M., Jeong, S.: High-lift airfoil design using kriging based moga and data mining. Korea Society for Aeronautical and Space Sciences International Journal 8(2), 28–36 (2007)
7. Kanazaki, M., Obayashi, S., Nakahashi, K.: Exhaust manifold design with tapered pipes using divided range MOGA. Engineering Optimization, Taylor&Francis 36(2), 149–164 (2004)
8. Kanazaki, M., Takagi, H., Makino, Y.: Mixed-fidelity efficient global optimization applied to design of supersonic wing. Procedia Engineering 67(1), 85–99 (2013)
9. Kanazaki, M., Yokokawa, Y., Murayama, M., Ito, T., Jeong, S., Yamamoto, K.: Nacelle chine installation based on wind tunnel test using efficient design exploration. Transaction of Japan Society and Space Science 51(173), 146–150 (2008)
10. Matsuno, T., Kawaguchi, M., Fujita, N., Yamada, G., Kawazoe, H.: Jet vectoring and enhancement of flow control performance of trielectrode plasma actuator utilizing sliding discharge (2012)
11. Matsuno, T., Kawazoe, H., Corke, T.C.: Forebody vortex control on high performance aircraft using pwm-controlled plasma actuators (2008)
12. Matsuno, T., Kawazoe, H., Nelson, R.C.: Aerodynamic control of high performance aircraft using pulsed plasma actuators (2009)
13. Matsuno, T., Ota, K.: K, Kanatani, Kawazoe, H.: Parameter design optimization of plasma actuator configuration for separation control (2010)
14. Oyama, A.: Design innovation with multiobjective design exploration (2011), `http://flab.eng.isas.jaxa.jp/monozukuri/mode/english/index.html`
15. Sato, T., Kanazaki, M., Yotsuya, Y., Matsushima, K.: Parametric airfoil representation toward efficient design knowledge discovery in various flow condition. Takamatsu Japan (2013)
16. Thomas, F.O., Kozlov, A., Corke, T.C.: Plasma actuators for cylinder flow control and noise reduction. AIAA Journal 46(8), 1921–1931 (2008)

A Preliminary Study on Quality Control of Oceanic Observation Data by Machine Learning Methods

Satoshi Ono[1], Haruki Matsuyama[1], Ken-ichi Fukui[2], and Shigeki Hosoda[3]

[1] Department of Information Science and Biomedical Engineering, Graduate School of Science and Engineering, Kagoshima University
{ono,sc109066}@ibe.kagoshima-u.ac.jp

[2] The Institute of Scientific and Industrial Research, Osaka University
fukui@ai.sanken.osaka-u.ac.jp

[3] Japan Agency for Marine-Earth Science and Technology
hosodas@jamstec.go.jp

Abstract. Argo float is a small and light-weight drifting buoy to measure oceanic temperature and salinity. More than 3,600 floats are always working for globally-covered ocean monitoring, and the accumulated big ocean observation data helps many studies such as investigation into climate change mechanism. However, the observed temperature and salinity data sometimes involves errors. Since automatic detection and correction of the errors is difficult due to ununiform observation reliability and the necessity of specifying error layers, human experts have performed manual error detection and correction. Toward the realization of high-accuracy automatic error detection method, this paper first applies Self-Organizing Map to the observation data for comprehensively understanding of the error characteristics, and then proposes a method for error detection based on Conditional Random Field. Experimental results showed that the proposed classification method based on CRF successfully detected observation errors with significantly better accuracy than the existing automatic quality control method.

Keywords: ocean observation, Argo float, self-organizing map, conditional random field.

1 Introduction

Although mechanisms of the global warming and climate changes have not been clarified sufficiently yet, the ocean is regarded as one of the most important elements in the climate system. This is because heat capacity of sea-water is 1,000 times as much as of air. To understand oceanic variability, the physical state observation of the ocean on a regular basis anywhere in the world is necessary. However, it is difficult by research vessels to conduct such ocean monitoring.

H. Handa et al. (eds.), *Proc. of the 18th Asia Pacific Symp. on Intell. & Evol. Systems – Vol. 1*,
Proceedings in Adaptation, Learning and Optimization 1, DOI: 10.1007/978-3-319-13359-1_52

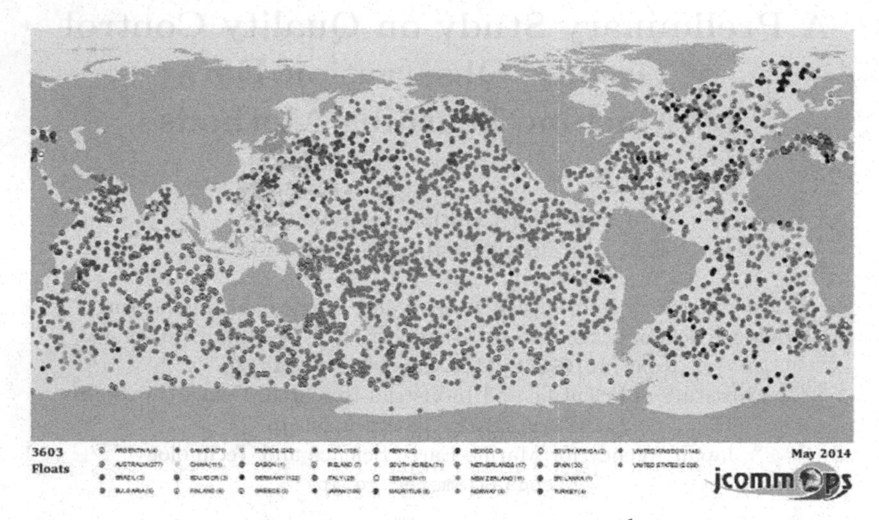

Fig. 1. Argo float distribution map[1]

Autonomous, long-term, real-time and globally-covered ocean monitoring system is essential.

Thus, the global ocean observing system Argo has started since 2000 [1–3]. This aims to construct real-time ocean monitoring system in accordance with "international Argo program"[2] , to which over 30 countries of all over the world are attending. Small and light-weight drifting buoys to measure oceanic temperature and salinity called "Argo float" have been operated in the global oceans. Observation data by Argo float is opened via World Wide Web after quality control (QC). Recently, more than 3,600 Argo floats are always working as shown in Fig. 1, resulting in accumulating big ocean observation data. This project helps to catch global variations of oceans which could not be detected by traditional observation methods, enabling many studies for investigation into climate change mechanism [4–6].

Fig. 2 shows an observation cycle of Argo float [3]. Every ten days, the float comes to the surface from 2,000 (m) water depth while measuring temperature and salinity. Observation data recorded during one flotation is called as a profile, and a profile consists of water temperature and salinity values in each depth.

Automatically observed data sometimes involves unpredicted errors. Since, in general, it is difficult to specify the reason of errors, quality labels are introduced to show the reliability of observation results [7]. Quality label assignment is performed by data management teams of various countries according to the QC method designed by international Argo program [2].

---

1 `https://plus.google.com/photos/112615107763535351524/albums/5953556558810291473?banner=pwa`

2 `http://www.argo.net/`

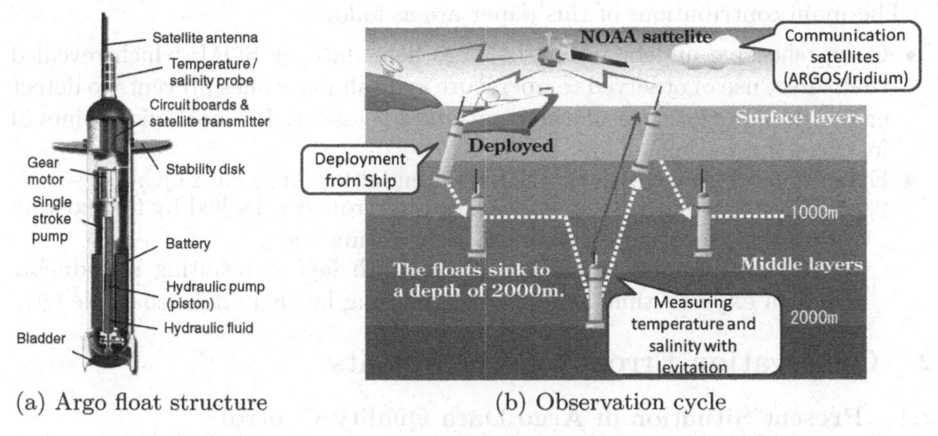

(a) Argo float structure (b) Observation cycle

Fig. 2. Structure and observation cycle of Argo float[3]

In the program, two levels of QC for Argo data are performed [7]. The first level is automatic error detection (real-time QC). The observed data and quality labels assigned by real-time QC are available to users basically within 24 hours. However, an automated QC method has not been established yet, and the existing real-time QC system sometimes causes overlook and misdetection of observation errors. This is because the measured values of water temperature and salinity occasionally involve sensor errors that are suffered from many kinds of factors, insufficiently achieving the required Argo data quality.

Therefore, human experts must visually confirm and revise real-time QC results. This process is the second level of QC (delayed-mode QC). Such manual QC by human experts cannot be performed in some countries, preventing quality regularization of the ocean observation data of all over the world. This is a serious matter of many years in the international Argo program, and so important that it may affect the reliability of globally-covered ocean monitoring.

The goal of this study is to realize an automatic method for error detection of Argo observation data, which have been performed by human experts. The subject of error detection is the sensor data of oceanic temperature and salinity obtained by Argo float (Argo data). In order to achieve this, fundamental studies are performed with two machine learning methods: Self-Organizing Map (SOM) [8] and Conditional Random Field (CRF) [10].

First, SOM is used for clustering the Argo data profiles involving errors to look down on the entire data. This helps to understand the error property and to design appropriate features in machine learning methods for Argo data QC.

Then, this paper proposes an error detection method for Argo data based on CRF since the problem of observation error detection can be regarded as a sequence labeling and SOM results indicated that cancelling large fluctuations by natural phenomena is indispensable by focusing difference between observation layers. CRF allows flexibly using various clues effective for label estimation, which are features and labels themselves represented as sequence data.

The main contributions of this paper are as follows:

- Comprehensive understanding of Argo float data by SOM, which revealed that direct use of observed temperature and salinity values prevents to detect errors. In addition, considering continuity is essential in both viewpoints of features and output quality labels.
- Experimental verification on CRF's potential for Argo data QC:
  - CRF successfully detected observation errors overlooked by the existing automatic error detection method (real-time QC).
  - CRF appropriately identified error depth layers, resulting in reducing human experts' time and effort on revising labels in delayed-mode QC.

2 Observation Errors on Argo Floats

2.1 Present Situation of Argo Data Quality Control

Measurement errors on Argo float are caused by hardware or software malfunctions, sensor stains caused by an external factor, occurrence of a communication defect, and so on. Although error patterns differ from each other, an automatic QC method has been designed for detecting some phenomena which demonstrates a tendency on the observation pattern. However, since measurement values of temperature and salinity strongly changes near the sea surface, it is difficult to determine detailed manners for automatic QC. Such technical problems make it difficult to automatically detect and correct all kind of errors. The skill inequality of technical experts over the world and limited human resources in addition to the insufficient automatic quality control prevent uniformizing the observation data quality, which is an important issue of the international Argo program.

Japan Agency for Marine-Earth Science and Technology (JAMSTEC), which manages high-accuracy QC of Argo data in Japan, have obtained more than 130,000 profiles since the Argo project's inception, and performed QC by an expert for more than 90,000 profiles – about 10,000 profiles per year. According to JAMSTEC, 5 to 10% profiles are visually confirmed for manual revision of quality labels; most of the profiles are suspected of involving an observation error by real-time QC, while other profiles do not involve any error labels but are observed by a float which has shown observation discrepancy in the past.

Human experts assign quality labels to profiles based on knowledge and experience in oceanic physics while keeping signals of variations caused by natural phenomena. Quality labels are introduced to indicate reliability of observed values in temperature and salinity. A label is assigned to each observation depth layer for each of temperature and salinity. A label value is chosen from 1 ("good"), 2 ("probably good"), 3 ("probably bad"), and 4 ("bad"). When visually confirming a profile to address the density inversion problem, an expert looks at a density graph first, and then temperature and salinity graphs. If the profile may involve an error, the expert sometimes compares its graphs with graphs of past profiles observed by the same float or graphs of its neighbor floats. In addition to detect profiles involving errors, it must be identified in which layer(s) an error is detected and what kind of situation causes the error.

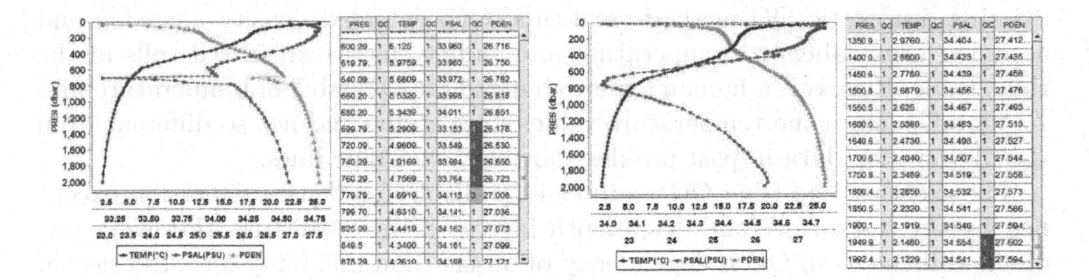

Fig. 3. Density inversion example 1 **Fig. 4.** Density inversion example 2

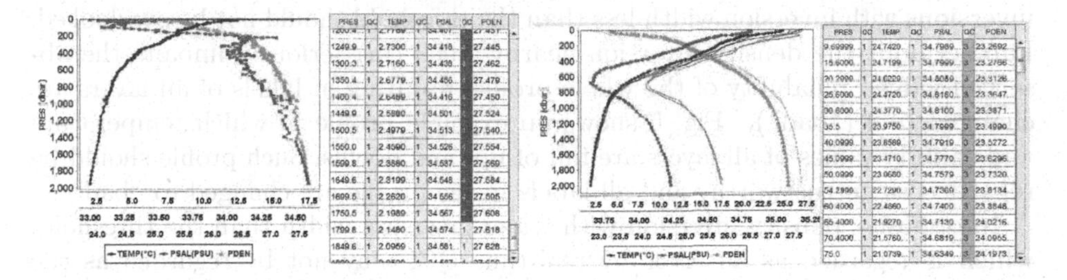

Fig. 5. Density inversion example 3. **Fig. 6.** Salinity sensor fault example

2.2 Argo Float Error Types

Since various errors occur with various patterns in Argo observation data, the existing automatic QC method (real-time QC) involves numbers of error detection rules. Some of them involve complicated conditions, which cannot be understood without comparison with past profiles of the float under consideration or other floats' profiles nearby it. Here, we focus on some major errors: *density inversion*, which occurs at the highest frequency, *salinity sensor fault*, which deeply affects the entire quality of the profile, and *same value observation*, which cannot be detected by real-time QC.

Density Inversion: Density is determined from water depth, water temperature, and salinity, and monotonically increases as depth does independent from ocean areas. If the density calculated at the greater pressure (deeper depth) is less than that calculated at the lesser pressure (shallower depth), the temperature and/or salinity values can involve be wrong [7]. Therefore, it is possible to detect an error and specify its depth by sensing reversed relationship of vertical density whose width exceeds a threshold. This irregular density change is called *density inversion.* The density inversion check is performed from more than 100 *layers*, points at which observation is made, from sea surface to 2,000 (m) depth.

Figs. 3 through 5 show example profiles with errors caused by density inversion, where blue line graphs denote water temperature (TEMP) [°C], red lines denote salinity (PSAL) [PSS-78], and green lines density (PDEN). In Fig. 3, the existing automatic QC method (real-time QC) detected density inversion and assigned error labels to temperature and salinity as shown in red cells of the right table. However, a human expert changed the QC label of temperature into 1 ("good") since the temperature values were stable and not so different from the observation data in past profiles shown by thin blue lines.

The existing real-time QC method is available for density inversion except nearby the sea surface and deep depth layers nearby 2,000 (m). However, one of the difficulties in QC is dependency of observation reliability on water depth. Example shown in Fig. 4 involved density inversion nearby 2,000 (m) depth, and the inversion could not be detected by real-time QC due to its small inversion width. Since observation at layers nearby 2,000 (m) is stable, the above inversions with inversion width less than the threshold should not be overlooked. Furthermore, the density inversion nearby 2,000 (m) seriously impairs the observed values' reliability of the whole profile, resulting in labels of all layers are changed to 4 ("bad"). Fig. 5 shows an example profile in which temperature and salinity values of all layers are full of ups and downs. Such profile should be regarded as untrustworthy and all labels in the profile are changed to "bad".

In opposite, density inversion with wider inversion width than the threshold, which is regarded as an error by real-time QC, may not be regarded as the error by an expert since observation nearby the sea surface is affected by ocean currents and weather.

In addition to the above matter, the following difficulties of label assignment lie when an inversion occurs in more than one observation layer:

- specifying the situation causes the error from various possibilities such as observation fault, natural phenomena and so on,
- specifying the exact range of layers relating to the error, and
- specifying which one or both of temperature and salinity values are wrong — most of density inversion are caused by one of them.

Salinity Sensor Fault: *Offset* is an error in which the observed data runs parallel to past or neighbor floats' profiles. Fig. 6 shows an example of the offset error. The observed salinity graph shown in the red line with rectangle markers were laterally displaced from the past observation results shown in the red lines without markers. This kind of errors can be seen on the whole layer or only on the deeper layer. Some floats show offset from the outset, and others suddenly outputs offset instead of their normal operation until then. Although correcting observation values involving an offset error, it is possible to detect such offset error by comparing to its past profiles or its neighbor floats' profiles.

Same Value Observation: When the same value is repeatedly observed in more than one layer, the observation might fail. This is due to electrical power failures of a float or other reasons, and when failed to observe temperature and/or salinity, the same values are recorded as those observed at the one layer before (deeper layer).

However, such *same value sequence* is not necessarily an error. For instance, in oceanic mixed layer, temperature and salinity tends to be made uniform. Its depth varies from ten to several hundred meters depending on ocean area and season. In opposite, the first (deepest) and its next layers occasionally have the same observation value due to the small depth difference between them. Provided that many possibilities are involved in the deepest observation layer, which are not easy even for a human expert to distinguish. Early versions of Argo float might produce the same value sequence because of the low sensor resolution.

3 Argo Data Analysis by Self-Organizing Map

3.1 Self-Organizing Map

Self-Organizing Map (SOM) [8] is a type of Neural Network (NN) for unsupervised learning, and is known as a method for clustering and analysis of multi-variate data. SOM visualizes cluster distributions in the given multi-variate data set by performing a non-linear projection to two dimenaional space. In SOM, patterns similar to each other are mapped to the same or neighbor clusters, while at the same time different patterns being mapped to distant clusters in the two-dimensional space.

This study applies SOM to Argo data to comprehensively understand the dataset property and error types, helping to select an approach based on machine leaning and to design features for realizing error detection.

3.2 Experimental Setup

Different tendency in Argo data can be seen in different areas of the ocean, and QC accuracy depends on nations and organizations. Therefore, we focus on the dataset of northern Pacific, which is qualified by experts in JAMSTEC with the world's highest standard. In detail, we select about 500 data profiles observed by the floats in the following area:

- 10N-30N, 140E-120W (240E)
- 30N-40N, 150E-130W (230E)
- 40N-50N, 155E-135W (225E)

Each profile consists of measured values of temperature and salinity ranging from 0 to 2,000 (m) at the intervals of 2 to 50 (m). Detailed depth values at which the observation is performed differ depending on profiles. Therefore, to make profile formats uniform, linear interpolation is performed for temperature and salinity data so that the observation values at intervals of 5 (m) are available. Since measured data in the 1,500 (m) depth or deeper are missing in some profiles and the data in the vicinity of the surface involves large fluctuation, this analysis targeted depth range from 400 (m) to 1,400 (m).

To see the entire data property, standard batch-type SOM was employed with standard hexagonal-cell grid, Gaussian neighbor function, decreasing strategy of the neighbor radius, and randomly initialized reference vector. Standard unified distance matrix representation [9] with 10×10 grid was adopted for visualizing SOM learning results.

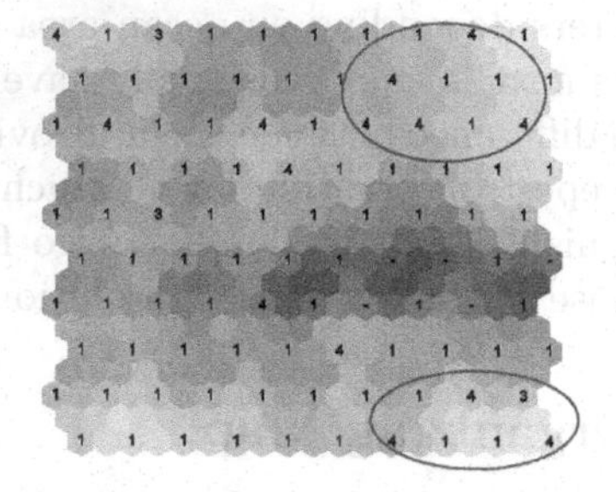

Fig. 7. SOM output with temperature labels before delayed-mode QC

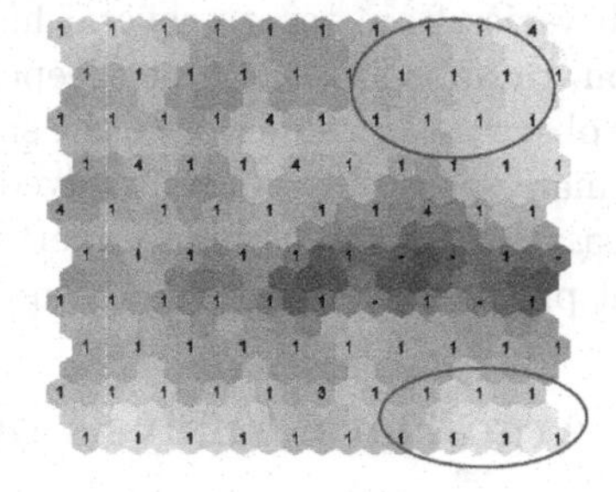

Fig. 8. SOM output with temperature labels after delayed-mode QC

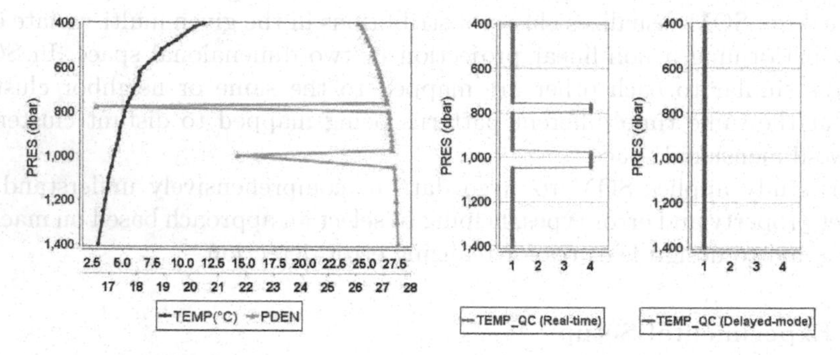

Fig. 9. Example in the cluster shown by red circle in Figs. 7 and 8

3.3 Experimental Results

Figs. 7 and 8 show SOM outputs for temperature before and after manual error detection by an expert, respectively. Figs. 10 and 11 show SOM outputs for salinity in the same manner. In the above figures, the grayscale colors denote the similarity to neighbors (brighter color indicates that the cell involves more similar reference vectors to the neighbors). Each microcluster shows the worst representative label of error types in it. Red circles indicate the area in which the expert modified the error label.

In Figs. 7 to 11, some clusters shown by red circles could be seen, in which the quality labels were revised by the expert in delayed-mode QC. That is, the clusters involve profiles with real-time QC errors. Other parts of SOM outputs were not reflective of error types due to large fluctuation components arised from natural phenomena.

Profiles belonging to two clusters, shown in red circles in Figs. 7 and 8, showed the same tendency. An example profile of the above clusters is shown in Fig. 9; orange and violet lines indicate the quality labels before and after the revision by experts, and blue, red, and green lines denote temperature, salinity, and density, respectively. At the depth range from 750 to 800 (m) and from 950 to 1,050 (m), real-time QC assigned "bad" labels to both temperature and salinity because the

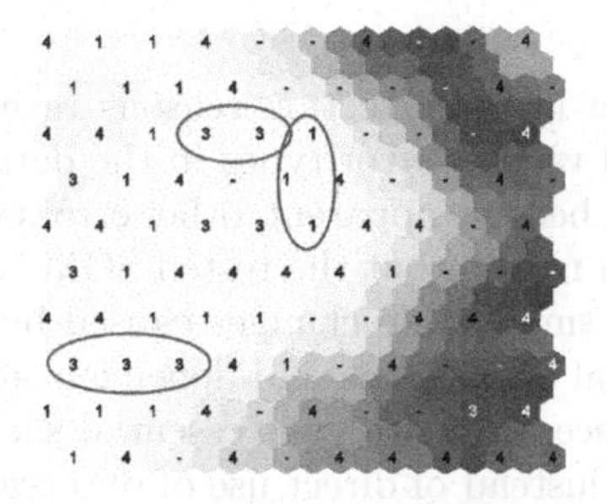

Fig. 10. SOM output with salinity labels before delayed-mode QC

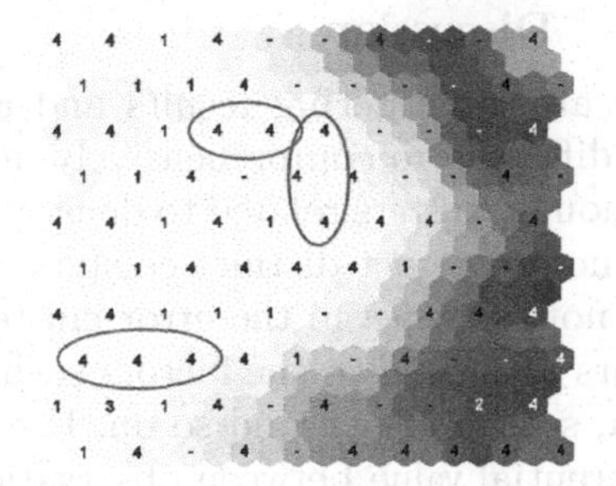

Fig. 11. SOM output with salinity labels after delayed-mode QC

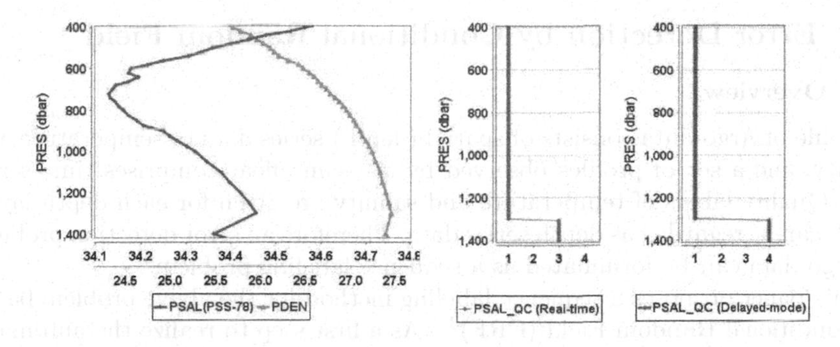

Fig. 12. Example in the cluster shown by red circle in Figs. 10 and 11

salinity values indicated abnormity to urge visual confirmation in delayed-mode QC. However, the expert changed the quality label of temperature to "good" since the temperature values were normal. The other profiles in the above two clusters showed the same type as density inversion example pattern 1 shown in Fig. 3, and those errors were caused only by inferior observation in salinity. Therefore, the temperature labels of all the above errors were turned over from 4 ("bad") to 1 ("good") by delayed-mode QC.

Profiles which are not in the above two clusters and whose labels were changed from 1 ("good") to 4 ("bad") are dotted in the temperature map. Labels of these profiles were changed according to reasons different from density inversion.

The salinity label change from 1 ("good") or 3 ("probably bad") to 4 ("bad") could be seen in three clusters shown by red circles in Figs. 10 and 11. All profiles in the above three clusters involved density inversion with near but not exceed the threshold, which is similar to density inversion example pattern 1 shown in Fig. 3. An example profile in one of the above clusters is shown in Fig. 12. In this example, since salinity observation values in depth from 1,300 to 1,400 (m) were uncommon, the QC values in salinity were revised from 3 ("probably bad") to 4 ("bad") by delayed-mode QC.

3.4 Discussion

The above clustering results and analysis of label correction reasons revealed the difficulty to comprehensively understand various error types in the dataset. Although clusters related to density inversion became apparent, other error types did not construct distinct clusters. The main reason that the tested SOM could not notably emerge the error clusters is the small value changes caused by the errors compared to the factors such as natural phenomena and dependencies on area, season, depth and so on. Use of preprocessed features is essential such as differential value between observation layers instead of direct use of observation values. In addition, it is also required to evaluate observation results of a layer under consideration with referring its neighbor layers' observation results and labels.

4 Error Detection by Conditional Random Field

4.1 Overview

A profile of Argo data consists of spatial (depth) series data of temperature and salinity, and a set of profiles observed by the same float comprises time series data. Quality labels of temperature and salinity are given for each depth layer, which can be regarded as depth series data. Therefore an error detection problem in Argo data can be formulated as a sequence labeling problem.

This paper proposes a sequence labeling method for the above problem based on Conditional Random Field (CRF)[3]. As a first step to realize the automatic error detection method for Argo data, we mainly focus on a few error types, density inversion caused by salinity observation failure and same observation value sequence, to validate the effectiveness of the proposed CRF-based method.

4.2 Conditional Random Field (CRF)

CRF is a discriminative model for sequence labeling proposed by Lafferty et al. [10]. Since sequencial observation data and dependency with precedent and/or successive labels are available in Argo data, various features effective for label estimation can be implemented in CRF. This study adopts Linear-chain model, one of the most simple CRF model.

Given an input data $\boldsymbol{x} = (x_1, x_2, \ldots, x_T)$ whose length is T, conditioned probability of $\boldsymbol{y} = (y_1, y_2, \ldots, y_T)$ is modeled as follows:

$$P(\boldsymbol{y}|\boldsymbol{x}) = \frac{1}{Z_{\boldsymbol{x}}} \exp\left(\sum_{t=1}^{T}\sum_{k} \lambda_k f_k(\boldsymbol{x}, y_t, y_{t-1})\right) \tag{1}$$

$$Z_{\boldsymbol{x}} = \sum_{\boldsymbol{y}} \exp\left(\sum_{t=1}^{T}\sum_{k} \lambda_k f_k(\boldsymbol{x}, y_t, y_{t-1})\right)$$

[3] Note that the proposed method based on CRF is independent from SOM, though the knowledge obtained by SOM outputs were referred for designing the feature functions in CRF.

Table 1. Feature functions used in the experiment

No.	Conditions	Feature value
f_1	$2 \leq PSAL_t \leq 41$	$1/T$
f_2	$PSAL_t < 2, 41 < PSAL_t$	1
f_3	$PSAL_t = PSAL_{t+1}, TEMP_t = TEMP_{t+1}, PRES_t < PRES_{t+1}, 1{,}600 < PRES_t$	1
f_4	$PSAL_t = PSAL_{t+1}, TEMP_t \neq TEMP_{t+1}, PRES_t = PRES_{t+1}$	1
f_5	$PRES_t \leq PRES_{t-1}$	1
f_6	$PRES_{t+1} \leq PRES_t$	1
f_7	$PDEN_t - PDEN_{max} < -0.01, PRES_t \leq 1{,}400$	1
f_8	$PDEN_t - PDEN_{max} < 0, 1{,}400 < PRES_t$	1
f_9	$PDEN_{t+1} - PDEN_t < -0.01, PRES_t \leq 1{,}400$	1
f_{10}	$PDEN_{t+1} - PDEN_t < 0, 1{,}400 < PRES_t$	1

where $f_k(\boldsymbol{x}, y_t, y_{t-1})$ denotes a feature function that associates input data, t-th and $(t-1)$-th output labels, λ_k is a weight parameter of feature function $f_k(\boldsymbol{x}, y_t, y_{t-1})$, and $Z_{\boldsymbol{x}}$ is a reguralization term that ensures $\sum_{\boldsymbol{y}} P(\boldsymbol{y}|\boldsymbol{x}) = 1$.

$$f(\boldsymbol{x}, y_t, y_{t-1}) = \begin{cases} \phi & \text{if } condition = \text{true} \\ 0 & \text{otherwise} \end{cases} \tag{2}$$

where ϕ is a real number called feature value.

In training phase, CRF calculates λ_k by maximum likelihood estimation from training data $D = \{(\boldsymbol{x}^{(i)}, \boldsymbol{y}^{(i)})\}_{i=1}^{N}$ in which input data $\boldsymbol{x}$ and output label $\boldsymbol{y}$ are paired. When classifying unknown input data $\boldsymbol{x}$, output labels $\boldsymbol{y}_{out}$ can be determined by solving the following maximization problem:

$$\boldsymbol{y}_{out} = \arg\max_{\boldsymbol{y}} P(\boldsymbol{y}|\boldsymbol{x}) \tag{3}$$

In this study, water depth $PRES_t$ expressed in units of sea water pressure, where t denotes depth layer index, water temperature $TEMP$, salinity $PSAL$, and density $PDEN$ were used as attributes of input data $\boldsymbol{x}$. Focusing on correcting errors caused by salinity observation failure, quality labels of salinity corresponding to $\boldsymbol{y}$ are estimated. Feature functions shown in Table 1 were manually designed with referring SOM outputs as described in Sec. 3.3. Feature functions f_1 and f_2 denote the observation values are in normal range or abnormal. f_3 and f_4 help to detect a same value sequence error. f_5 and f_6 are designed to address a depth error. f_7 through f_{10} detect density inversion, in which conditions are changed depending on depth regions, where $PDEN_{max}$ indicates the maximum value from the first to $(t-1)$-th layers except layers labeled as "bad".

4.3 Experimental Setup

This experiment aims to verify the fundamental effectiveness of CRF on Argo data from two perspectives: error detection accuracy and label assignment accuracy for error layer identification. Since label values 2 ("probably good") and

3 ("probably bad") involves some ambiguity, values 2 and 3 are unified to 1 ("good") and 4 ("bad"), respectively. That is, the problem is formulated as a binary sequence labeling problem. Labels assigned by a human expert in delayed-mode QC were used as ground truth.

Training and test datasets consist of profiles involving more than one error related to density inversion caused by salinity observation failure and same value sequence. The profiles are selected from the same ocean area as in Sec. 3. Totally, 1,642 profiles were matched the above conditions. To use a sufficient amount of training dataset, 10-fold cross validation was performed with these profiles; about 165 profiles of them were used as a test dataset, and other profiles were used for training dataset. In addition to the above dataset, an additional test dataset comprising the same number of randomly-selected profiles was used, which do not involve any errors and is not used for training. The reasons why profiles without errors were not used for training are that profiles with errors involved sufficient amount of normal observation values and that using too many profiles without errors required adjustment of value weights of feature functions. In CRF, the weight update cycle limit and the training rate were set to 100 and 0.001, respectively. The experiment is designed for evaluating the proposed method from the perspective of error detection accuracy for each profile and salinity label assignment accuracy.

4.4 Experimental Results

Error Detection: The proposed CRF-based method was compared to the existing automatic QC method (real-time QC) [7]; Table 2 shows accuracy, precision, and recall by unit of profile. "True positives" (TP) is the number of profiles correctly detected as profiles involving error, that is, at least one layer was labeled as "bad". "True negatives" (TN) is the number of profiles correctly ignored as profiles involving no error, that is, all layers were labeled as "good". "False positives" (FP) is the number of profiles which did not involve any errors and were incorrectly assigned label of "bad" to at least one layer. "False negatives" (FN) is the number of profiles which involved at least one error and were incorrectly assigned label of "good" to their all layers. The above values were averaged over all 10 folds. All of the difference on accuracy, precision, and recall were statistically significant by paired t-test at 5% significant level.

Compared to real-time QC, the proposed CRF-based method showed significantly higher accuracy and recall, indicating that the proposed method could detect errors with significantly higher sensitivity. Although precision of CRF was slightly deteriorated due to the increase of detected profiles, as a preprocess of delayed-mode QC, recall is more important than precision to prevent overlooking errors. Therefore, CRF has not merely more accurate performance but has a potential to reduce the experts workload in delayed-mode QC.

Labeling Accuracy: Table 3 shows the number of layers whose label was incorrectly assigned averaged over all 10 folds. "Depth error" indicates the number

Table 2. Experimental results on accuracy, precision, and recall by unit of profile

	TP	TN	FP	FN	Accuracy	Recall	Precision
Real-time QC	106.6	162.6	1.6	57.6	82.0%	64.9%	98.5%
CRF (proposed)	158.3	160.3	3.9	7.0	97.0%	96.4%	97.6%

Table 3. Comparison on labeling errors by unit of layer

	Profiles w/t error			Profiles w/o error
	Depth error	Misdetection	Overlooking	Misdetection
Real-time QC	69.6	216.2	83.0	283.2
CRF (proposed)	42.1	11.5	16.9	4.9

of errors by unit of layer related to an error in which value "bad" was assigned not to the appropriate layer but its neighbor layer. "Misdetection" denotes an error in which value "bad" was incorrectly assigned to a properly observed layer. As opposed to this, "Overlooking" denotes an error in which value 4 "bad" was not assigned to a layer at which observation failed.

CRF could reduce not only overlooked error to 1/5 of real-time QC but also misdetection error to 1/19 and 1/58 in profiles with and without error, respectively. This means that CRF prevents experts from overlooking errors and significantly decreases their work for label correction from "bad" to "good". Meanwhile, further improvement for CRF is necessary to specify appropriate layers to be labeled as "bad".

Table 4 shows the detailed accuracy for each error type averaged over all 10 folds; The errors are divided into density inversion, same value sequence error, and other observation errors involving outliers. The profiles of density inversion are divided into positive and negative sides (improperly high and low density values were observed, respectively). The number of profiles denotes the total number of profiles involving the categorized error type, and the number of layers does the total number of layers labeled "bad" due to the error belonging to the type.

Table 4 demonstrates that the number of errors by CRF was totally lower than that of real-time QC except the depth error of the negative side density inversion. In particular, real-time QC was quite poor at same value sequence errors and other observation errors. The reason is that real-time QC scanned the density values without directly referring salinity values, whereas the proposed CRF-based method referred the salinity values by feature functions f_1 through f_4.

4.5 Discussions

From the above results, labeling by CRF showed better performance than the existing real-time QC method. These results support CRF's promising performance for ocean observation data error detection.

Table 4. Number of errors for error types (by unit of layers)

Error type	Total number of profiles	Total number of layers	Real-time QC			CRF (proposed)		
			Depth error	Mis-detection	Over-looking	Depth error	Mis-detection	Over-looking
Density inversion (negative side)	103.5	141.5	31.9	86.4	48.6	35.2	6.8	4.4
Density inversion (positive side)	18.4	23.2	14.2	16.8	7.9	1.0	2.5	1.8
Same value sequence	29.8	61.6	15.9	94.5	26.1	5.7	2.1	8.1
observation errors	12.5	16.4	7.7	18.5	0.3	0.2	0.2	0.2

However, there is still room for improvement in the labeling accuracy. For instance, fine density inversions nearby the sea surface were overlooked, and it was difficult to exactly specify depth layers whose labels should be corrected when a density inversion error occurred over more than one layer. More flexible threshold configuration depending on water depth is necessary. Also, feature functions based on past profiles and neighbor floats' profiles should be added.

5 Conclusions

This paper focuses on error detection of ocean data observed by Argo float. First, errors are categorized and their properties are discussed. Then, Self-Organizing Map (SOM) is applied to Argo data to comprehensively understand the error distribution and obtain error pattern clusters. Although, density inversion, the most frequent error type in Argo data, was clearly seen in some clusters, there seem no additional apparent trends in SOM results. The reason is that slight error patterns are buried in large fluctuations cased by natural phenomena and that selecting the appropriate depth layers must be difficult for simple methods without considering data continuity. Therefore, it has turned out that adequate data preprocessing is necessary to generate features.

Finally, Conditional Random Field (CRF) is applied to correct quality labels to validate its fundamental effectiveness for this problem. Feature functions were designed with mainly focusing on density inversion. Experimental results have shown that CRF-based labeling method was significantly better than the existing automatic QC method (real-time QC). CRF-based labeling inclination was also discussed with error types.

In future, we plan to redesign feature functions to catch error patterns with avoiding fluctuation by natural phenomena, which has the continuity on time and space and is affected with constraints by physical phenomena. Multistage error level estimation is also an important future work.

Acknowledgement. We would like to thank Ocean Circulation group, Research and Development Center for Global Change, JAMSTEC for their help.

This study was partially supported by Kurata Grant from the Kurata Memorial Hitachi Science and Technology Foundation.

References

1. : Argo sicence team, Argo: The global array of profiling floats. In: Koblinsky, C.J., Smith, N.R. (eds.) Observing the Oceans in the 21st Century, pp. 248–258. GODAE Project Office, Bureau of Meteorology (2001)
2. Argo Data Management Team: Report of the Argo Data Management Meeting. Proc. Argo Data Management Third Meeting, Marine Environmental Data (2002)
3. Hosoda, S.: Argo Float — Innovation for Autonomous Observations of Global Oceans 9th Japanese-German Frontiers of Science Symposium, 37–38 (2012)
4. Levitus, S., Antonov, J.I., Boyer, T.P., Baranova, O.K., Garcia, H.E., Locarnini, R.A., Mishonov, A.V., Reagan, J.R., Seidov, D., Yarosh, E.S., Zweng, M.M.: World ocean heat content and thermosteric sea level change (0– 2000 m), 1955 2010, Geophys. Res. Lett. 39(10), L10603 (2012)
5. Hosoda, S., Suga, T., Shikama, N., Mizuno, K.: Global Surface Layer Salinity Change Detected by Argo and Its Implication for Hydrological Cycle Intensification. J. Oceanogr. 65(4), 579–586 (2009)
6. Kagimoto, T., Miyazawa, Y., Guo, X., Kawajiri, H.: High resolution Kuroshio forecast system — Description and its applications. In: Ohfuchi, W., Hamilton, K. (eds.) High Resolution Numerical Modeling of the Atmosphere and Ocean, pp. 209–234. Springer, New York (2008)
7. Argo Data Management Team: Argo quality control manual Version 2.7 (2012)
8. Kohonen, T.: Self-Organizing Maps. Springer-Verlag, Berlin (1995)
9. Ultsch, A.: Self-organizing neural networks for visualization and classification. In: Lausen, O.B., Klar, R. (eds.) Information and Classification- Concepts, Methods and Applications, pp. 307–313. Springer Verlag, Berlin (1993)
10. Lafferty, J., McCallum, A., Pereira, F.: Conditional random fields: Probabilistic models for segmenting and labeling sequence data. Proc. Int'l Conf. Machine Learning, 282–289 (2001)

Product Differentiation under Bounded Rationality*

B. Vermeulen[1], J.A. La Poutré[2,3], A.G. de Kok[4], and A. Pyka[1]

[1] Institute of Economics, University of Hohenheim, Stuttgart, Germany
b.vermeulen@uni-hohenheim.de
[2] CWI - Centre for Mathematics and Computer Science, Amsterdam, The Netherlands
[3] Department of Information and Computing Science, Utrecht University, The Netherlands
[4] School of Industrial Engineering, Eindhoven University of Technology, Eindhoven, The Netherlands

Abstract. We study product differentiation equilibria and dynamics on the Salop circle under bounded rationality. Due to bounded rationality, firms tend to agglomerate in pairs. Upon adding a second tier of component suppliers, downstream assemblers may escape pairwise horizontal agglomeration. Moreover, we find a persistent propensity to vertically agglomerate near suppliers as well as ubiquitous dynamic equilibria. Conclusively, to understand product differentiation equilibria and dynamics, products and their supply chains need to be disaggregated.

Keywords: product differentiation, Salop circle, discrete choice, bounded rationality, component supplier, agglomeration, dynamic equilibrium.

1 Introduction

In the classical Salop horizontal product differentiation model [9], consumers are located uniformly on a circular product-market and firms position themselves on this circle vis-a-vis competitors. Under the assumption of perfect rationality (of consumers and firms) and firm borne transportation costs, firms tend to equidistantly distribute (i.e. differentiate maximally). This equidistant distribution is a location-price equilibrium [3] (under inelastic demand with quadratic disutility), a location-quantity equilibrium [8] (under firm-borne transportation costs), and maximally entry deterring [9].

In this paper, we study product differentiation dynamics on the Salop circle under *limitations* in the location choices and thereby the *emerging* distribution. We hereby relax two assumptions on product location choice. Firstly, firms are no longer vertically fully integrated, but part of a supply chain in which they rely on an upstream supplier (or downstream retailer). In this paper, we add

* This work was realized with financial support of the Dutch science foundation NWO, grant 458-03-112, the German science foundation DFG, grant PY 70/8-1, and the Austrian science foundation FWF, grant I 886-G11.

H. Handa et al. (eds.), *Proc. of the 18th Asia Pacific Symp. on Intell. & Evol. Systems – Vol. 1*,
Proceedings in Adaptation, Learning and Optimization 1, DOI: 10.1007/978-3-319-13359-1_53

a second tier to the Salop circular landscape, where consumers' preference for product and input component strongly correlate, e.g. preference for the synchronicity of times of rental car pick-up/ drop-off and flight arrival/ departure times, the fuel efficiency of a car engine and the weight of the coachwork, the environmental friendliness of produce and biodegradability of its wrapping, or the ethical stance of a final assembler and its low-wage country component suppliers. Intuition is that suppliers of components that are sought after by consumers will draw assemblers closer, while assemblers with popular products will draw suppliers closer. So, consumer concerns cause alignment of product and input component characteristics. By comparing the two-tier results with the benchmark single tier results, we see how ignoring the vertical dimension neglects important features of industry dynamics and how supply chain alignment would affect the emerging division of the consumer market among the various competing firms. Secondly, in the traditional models, agents are perfectly rational. However, this is to be seen as a 'normative model of an idealized decision maker, not as a description of the behavior of real people' [12]. Bounded rationality hampers deciding optimally [10], thus affects managerial cognition and thereby (strategic) decisions [5], causing management to resort to heuristics and routines [7] in which firms focus primarily on immediate competitors [5]. We assume that product relocation is merely local and -given the uncertainty about competitors' moves- incrementally. Similarly, consumers often do not make the rational choice [11] and are, moreover heterogeneous because of different utility functions or individual circumstances. We model this as a *probabilistic* discrete choice among the options on offer [1].

2 Product Differentiation Model

In this section, we operationally define the product differentiation model with boundedly rational firms and consumers.

2.1 Basic Model

The basis of our model is the Salop circular location model of product differentiation, where a firm's location on the circle is associated with the specification of its product and a consumer's location is associated with its preference for these product specifications. Consumers experience disutility from a mismatch of product specifications with product preferences. Firms (re)locate on the circle to increase sales to consumers and -implicitly- evade competition, which we model as a boundedly rational two-stage game with a sales round and a relocation round.

2.2 Boundedly Rational Sales-Based Location Choices

Bounded rationality is introduced operationally in two ways. Firstly, we assume that firms have no prior ideas of what competitors will do, and thus do not anticipate competitors' moves. We assume that firms are initially located uniform randomly over the circle. In subsequent periods, firms take into account other firms only implicitly through expected future sales upon deciding on their own location. To study purely the effects of incremental myopic relocation (and, later, vertical specialization) on dynamic and emerging equilibria, we assume uniform market prices.

Secondly, firms are not able to (re)locate somewhere on the circle freely, but are forced to do so incrementally, moving in the clockwise or counterclockwise direction step by step through a myopic, ceteris paribus strategy. We assume that the steps taken by firms are of size $2\pi/N$, thereby dividing the Salop circular landscape in N discrete 'niches'[1]. Variable $\theta_i(t)$ is the location of firm i on the circle during period t.

Sales Round. Consumers are uniformly distributed over the perimeter of the Salop circle[2], fully consume and buy a single unit of product each period and are anchored to their location (i.e. have fixed preferences). Each period, each consumer buys a product based on local attractiveness of products (related to arc distance to the firm). As such, in period t, firm k sells:

$$s_k = \sum_{1 \leq n \leq N} d_k(n) \tag{1}$$

where $d_k(n)$ is the demand realization of consumers at location n purchasing product k. We assume that sales equals demand.

Relocation Round. In the relocation round at the end of the period, all of the M firms consider moving, simultaneously, based on a myopic strategy taking into account sales prospects without accounting for competitors' actual moves or relocation strategy. Each firm conducts a poll of size q per niche to make a ceteris paribus sales forecast $\widehat{s_k^-}$ of a step of one niche in counterclockwise direction and a ceteris paribus sales forecast $\widehat{s_k^+}$ of a step in clockwise direction. We do not use a sales forecast $\widehat{s}_k$ without a move, because, by a sheer sample size argument, the *actually* experienced sales $s_k(t)$ is a more accurate predictor of the future sales $s_k(t+1)$ than would be such a forecast $\widehat{s}_k$.

The firm relocates into the direction that increases the expected sales most (ceteris paribus), that is, if the predicted increase of sales $\widehat{s}' - s$ exceeds the threshold εs. The explanation is that per product investments are to be made (hence proportional to s) and that these costs are to be fully covered in the next period. When, for instance, $\widehat{s_k^+} > \widehat{s_k^-}$ and $\widehat{s_k^+} > s_k(1+\varepsilon)$, the firm move to the neighboring niche in the clockwise direction. If $\widehat{s_k^+} = \widehat{s_k^-} > (1+\varepsilon)s_k$, the firm randomly picks one of these directions.

[1] Traditionally, location models have a continuous landscape, but also discretizing is not uncommon in Hotelling [2] and Salop models [6,4].

[2] As such, firm location decisions (and thereby the emerging distribution of firms) are due to strategic positioning vis-a-vis competitors, and not due to (also) positioning vis-a-vis clusters of consumers.

2.3 Discrete Choice Demand

We assume that consumers select a product according to the discrete choice model. A consumer at niche n selects product k located at θ_k with probability:

$$p_k(n) = \frac{a(\theta_k, n)}{\sum_j a(\theta_j, n)} \tag{2}$$

where $a(\theta, n)$ is an attractiveness function. We assume that each consumer buys a product each period, so demand is inelastic and, in expectation, a fraction $d_k(n)$ proportional to $p_k(n)$ of the consumers at niche n picks product k. We assume that all demand is fulfilled, so sales equals demand.

The attractiveness function $a(\theta, n) > 0$ reflects the utility the consumer located at n derives from consuming a product located at θ. We take an attractiveness function of the following form:

$$a(\theta_k, n) = A_k\, e^{-\gamma \Delta^r(\theta_k, n)} \tag{3}$$

The distance function $\Delta(\theta, n)$ is defined as the minimum number of niches in clockwise or counterclockwise direction between θ and n, times the stepsize $(2\pi/N)$. We assume that firms are vertically undifferentiated and pick the same price $A_k = A_j = 1$ for all k and j. As the attractiveness decreases *for the consumer* in distance (and not for the firms), the consumer bears the disutility of the distance. Due to the assumption on the positivity of the attractiveness, any product has a non-zero probability of being picked by an arbitrary consumer. Parameter γ is the 'choosiness' of consumers; the higher the choosiness, the less attractive a product with a certain mismatch between product specification (firm location) and consumer preference (consumer location). If $\gamma \downarrow 0$, the probability for any product to be selected by a consumer at an arbitrary location goes to $1/M$, i.e. consumers are indifferent to what product to consume. Products are considered perfect substitutes. If $\gamma \to \infty$, consumers are utility maximizers and they are infinitely picky about which products they choose. The higher choosiness γ, the more likely it is that consumers pick a product that is nearby. As these consumers determine the actual and prospect sales, the firms nearby respond mostly to the prospect sales to these consumers. So, under high choosiness γ, firms implicitly position vis-a-vis their *immediately neighboring* competitors in their relocation to increase prospected consumer sales: competition is local. Under low choosiness γ, consumers are relatively indifferent to mismatches of own requirements and product specifications, so firms compete globally about equally fierce for an arbitrary consumer. Choosiness γ can thus be interpreted as 'competition locality'. Locality is then -as in physics- referring to the extent to which this interaction between firms is local rather than global. Alternatively, γ may be interpreted as 'specificity'; the higher γ the more specific the product is attuned to particular preferences/ niches.

Parameter r defines the nature of the attractiveness curve. We study the case $r = 2$, for which the attractiveness curve takes a bell-shape around the product location (cf. the Normal distribution).

2.4 Second Tier of Component Suppliers

We extend the circular single tier model with an additional tier. On this second tier, there are S suppliers and N niches (which is equal to the number of niches on the first tier). We associate the niche of the supplier with the technical specification of the component. Each of the M assemblers on the first tier needs to be connected to exactly one of S suppliers on the second tier.

The *composite* attractiveness for a customer at niche n of a product at location θ_k produced by firm k using a component at location $\sigma_{\pi(k)}$ produced by supplier $\pi(k)$ is:

$$a(\theta_k, \sigma_{\pi(k)}, n) := a(\theta_k, n)\, x(\sigma_{\pi(k)}, n)\, c(\theta_k, \sigma_{\pi(k)}) \tag{4}$$

This relates the assembly attractiveness a, component attractiveness x and component-product compatibility c, which are defined as:

$$a(\theta_k, n) := e^{-\gamma \Delta^r(\theta_k, n)}$$
$$x(\sigma, n) := e^{-\eta \Delta^r(n, \sigma)}$$
$$c(\theta, \sigma) := e^{-\lambda \Delta^r(\theta, \sigma)}$$

Δ is the minimum number of steps from θ_k to n in either direction over the circle. Parameter $\gamma \geq 0$ is the product design variable *specificity*. The higher specificity, the more the assembly is tailored to specific consumer niches. Parameter $\eta \geq 0$ is the component *exposure*, the more exposed consumers are to the component or how more important the component is to them (e.g. by marketing a personal computer with 'Intel inside', the consumer is exposed to the processor as an explicit component). Parameter $\lambda \geq 0$ is the product *complexity*. The higher complexity, the greater the negative effect of mismatches between input component and final assembly specifications on overall performance. Illustrations

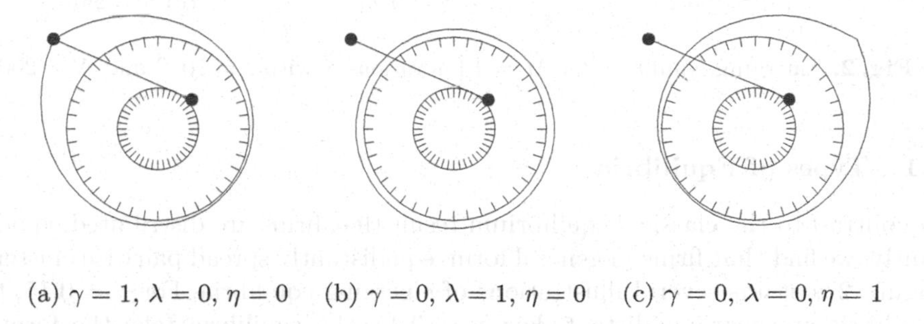

(a) $\gamma = 1$, $\lambda = 0$, $\eta = 0$ (b) $\gamma = 0$, $\lambda = 1$, $\eta = 0$ (c) $\gamma = 0$, $\lambda = 0$, $\eta = 1$

Fig. 1. Composite attractiveness curves for different values of assembly specificity γ, technical compatibility λ and component exposure η.

of the composite attractiveness curve over the circle for different parameter settings are given in Figure 1. In line with the relocation criteria in the single tier case, a firm (now assembler *or* component supplier) at location θ picks a new location in $\theta - 1, \theta, \theta + 1$ that maximizes its sales (ceteris paribus). However, the assembler now executes polls to assess $\widehat{s}^{+}$, $\widehat{s}$ and $\widehat{s}^{-}$ *whereby it also considers all possible switches to the suppliers available.* In this case, an increase in sales might be caused by moving closer to certain consumers, but also by increasing the technological compatibility, exploiting the component exposure, or both. Not only the assemblers relocate, so do the suppliers. We assume that suppliers simply move to where the income is coming from. Each period, each supplier computes the sales generated on both (strict) halves of the circle (clockwise and counterclockwise) and moves one step into the direction that generates the highest total sales. Suppliers do not relocate to/ from assemblers currently not its buyer. We ignore assembly and component redesign costs.

3 Results for the Single Tier Model

In this section, we study the emerging location distribution of firms in case of the single tier model.

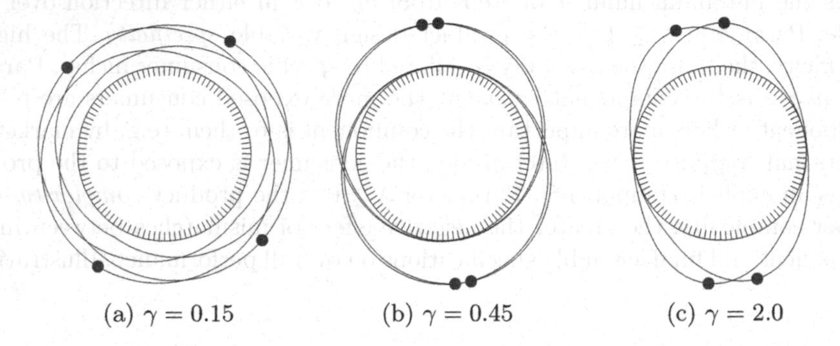

(a) $\gamma = 0.15$ (b) $\gamma = 0.45$ (c) $\gamma = 2.0$

Fig. 2. Emerging equilibria for $M = 4$ for various γ with $\varepsilon = 10^{-6}$ and $N = 200$

3.1 Types of Equilibria

In contrast to the classical equilibrium result that firms are distributed equidistantly, we find that firms in general form 'equidistantly spread pairwise clusters'. Figure 2 contains several illustrations of emerging equilibria. For $\gamma = 0.15$, the equilibria are near-equidistant. For $\gamma = 0.45$, the equilibria take the form of a two-by-two pairwise clustering, where the pairs are located at diametrically opposite sides of the circle. We use the metric D to quantify the extent to which the distribution deviates from equidistance:

$$D^2 := \frac{M}{N^2(M-1)} \sum_{1\leq i\leq M} \left(\overrightarrow{\Delta}(\theta_i, \theta_{i+1}) - \frac{N}{M} \right)^2 \quad (5)$$

For notational convenience, we take indices to reflect the location ordering ($\theta_{i+1} \geq \theta_i$), and $\theta_{M+1} = \theta_1$. We take M to be even. $\overrightarrow{\Delta}$ is the distance measured in clockwise direction. The summands are the squared differences of the distance between consecutive firms from the distance under equidistance (N/M).

Per combination in a fine-grained γ (choosiness) - ε (switching threshold) parameter landscape, we determine the average D value for 500 seed values for the emerging distribution of firms over the circle. This is plotted in density plots in Figure 3. We distinguish a number of parameter regions of interest depicted in Figure 3b.

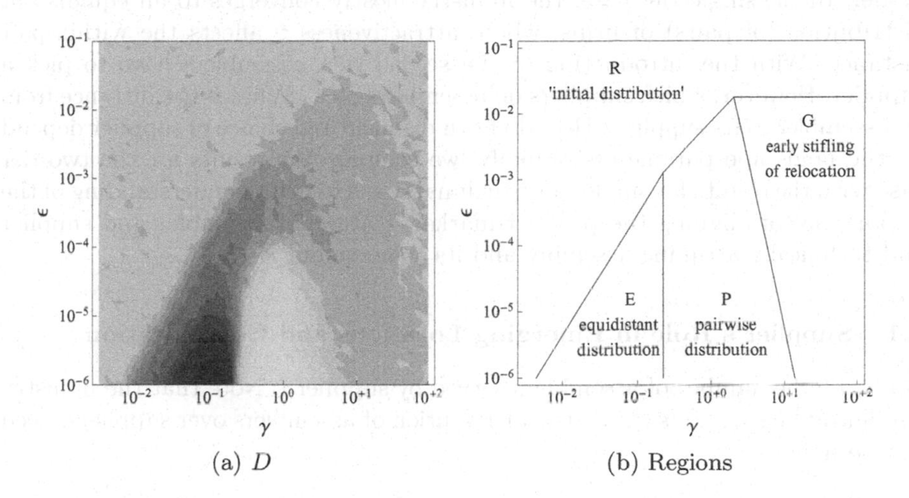

Fig. 3. Density plots (when dark (white), value is 0 (1)) of the D metric for $M = 4$ for a range of γ and ε values for $M = 4$ and $N = 500$ and 500 cases per parameter combination, including an overview of interesting regions.

In region R, the γ is so low that an increase in sales from relocating does not exceed ε, so there is little to no relocation, such that D reflects the noise of the initial random distribution. In region G, choosiness γ is so high that attractiveness increments are very limited and further relocation stops already when the within-pair distances are large (or very asymmetric). Furthermore, the higher γ, the sooner the sales increments drop below ε. In region E, D is low (the region in Figure 3a is dark), the firms in both pairs are far away from each other, so the emerging equilibrium is near-equidistant. Distances from one to the next firm are close to N/M. Competition is so global (γ is so low) that firms also capture much sales beyond the immediate neighboring firms. With higher ε, the sum of attractiveness increments needs to be larger to further diverge to equidistance, which occurs at higher γ, hence the slight tilt to the right of left-hand side boundary. In region P, D is high (the region in Figure 3a is brightly colored), the distance from firms in a pair is close to 0. The emerging equilibrium is strongly pairwise. With higher ε, the total of sales increments needs to be larger to justify relocating and thus drive convergence to tighter pairwise clustering, so already at lower choosiness γ values further relocation stops, hence the tilting to the left of the right-hand side region boundary. The boundary between region E and P is related to the location of inflection point γ^* of the attractiveness curve.

4 Results for the Two-Tier Model

In this section, we study the dynamics and emerging location distribution in case of horizontal differentiation of system assemblers and their suppliers in a two-tier model. In the single tier case, the industry mostly converges to an equidistant distribution (of pairs) of firms, where attractiveness γ affects the within-pair distance. With the introduction of the second tier, assemblers have to pick a supplier. However, which suppliers do assemblers pick? What is the distance from an assembler to its supplier? How do both distance and choice of supplier depend on the landscape parameters? Finally, we compare the results for the two-tier case with the results for single tier benchmark case to get an understanding of the importance of aligning the product-market strategy of assembler and supplier and technically attuning assembly and its components.

4.1 Supplier's Role in Emerging Locations and Configuration

Let s_i be the number of assemblers served by supplier i. Note that the industry configuration $(s_1, \ldots, s_S)$, i.e. the distribution of assemblers over suppliers, need not be uniform.

Supplier-Assembler Distance. In isolation, an assembler produces a product that uses the most compatible component available. As such, a component supplier has a natural *'part'* of the landscape stretching halfway to the next supplier on either side (clockwise and counterclockwise) of assemblers and final consumers it serves. As a general rule, an assembler placed (initially) in the part of a supplier, gradually moves closer to that initial supplier (i.e. without switching to another supplier) and eventually settles near that supplier in equilibrium. If there are other assemblers served by this supplier, each of these assemblers balances being close to the supplier and being differentiated from these immediate competitors. On top of that, firms differentiate from competitors served by other suppliers.

Supplier Switching. An assembler located in the part of one supplier may switch *to another supplier*, thereby changing the range of niches with consumers served (based on the component used). Such a switch may go at the expense of lower compatibility, but may still increase overall sales. For such '*counter-compatibility switching*' to occur, there must be asymmetry in the market shares of the various assemblers. This may be caused by an asymmetry in the configuration, i.e. an asymmetry in the number of assemblers per supplier. In case of symmetry in the configuration, it may be caused by a particular distribution of assemblers over a supplier's part. An example of the first case is that there are two suppliers ($S = 2$), one of which is busy ($s_i > M/2$) and one of which is quiet. Assemblers may then switch from their busy supplier to the quiet supplier, giving in on compatibility, but serving a different pool of consumers with the alternative component. An example of the second case (symmetry in configuration but asymmetry in market shares) is that one or multiple assemblers are at the edge of the supplier's part that they share. In this case, counter-compatibility may increase sales for the assembler at the edge of the supplier's part. This counter-compatibility switch may thereby even cause configuration asymmetry. In case of more than two suppliers, counter-compatibility switching becomes more involved. Depending on the initial locations of the suppliers, the suppliers' parts (i.e. the niches halfway to the competitors on either side) need not be of equal size. In that case, assemblers tend to switch counter to the compatibility to suppliers with a larger part. However, switching also depends on the number of assemblers currently served by the suppliers and the choices for parameters γ, η and λ.

Complexity. As soon as product technology has a certain non-trivial complexity ($\lambda > 0$) or component exposure ($\eta > 0$), assemblers take into account the location of suppliers. The higher the complexity λ, the stronger the agglomerating force of suppliers: any additional mismatch of assembly and component (e.g. due to relocating away from the supplier or switching to the other supplier) has stronger impact on overall attractiveness.

Even for low complexity levels, there is no gain in differentiating from competitors residing closer to the shared supplier and all assemblers agglomerate at their supplier's location. We find that the higher complexity, the less likely counter-compatibility switching becomes and the more likely an emerging configuration may be asymmetric.

In explaining dynamics and emerging distributions, this complexity λ is a strong force because any incompatibility between component and assembly affects demand at *all* niches across the landscape. The effect of λ is strong and rather obvious. For the remainder of this subsection, we assume components and assemblies are perfectly modular ($\lambda = 0$) and focus on component exposure η that has weaker but also less obvious effects.

Exposure. With an increase in component exposure η, components determine more of composite attractiveness and thereby expected sales on the supplier's

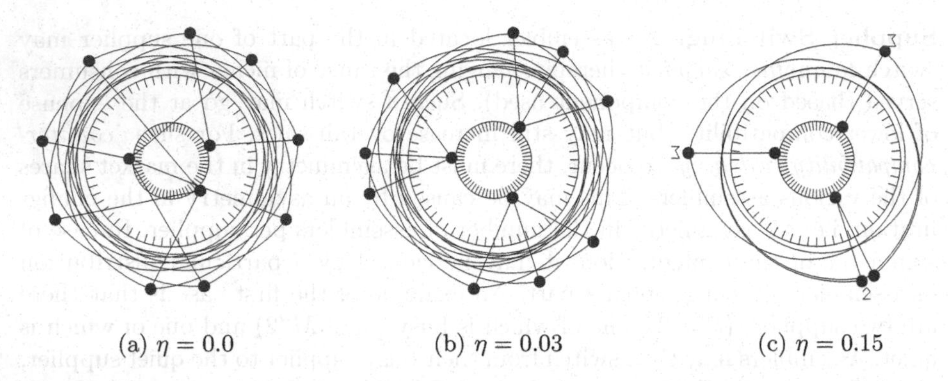

(a) $\eta = 0.0$ (b) $\eta = 0.03$ (c) $\eta = 0.15$

Fig. 4. Equilibria emerging from the same initial distribution of assemblers for different component exposure levels η under $\gamma = 0.6$. Here $M = 9$, $S = 3$, $N = 60$. The figures illustrate the sensitivity of the assembler distribution for exposure.

part. It is thereby more important for assemblers to have a high attractiveness value a of its product at this supplier's part, which is generally the case when assemblers are closer to the supplier itself. Under high exposure η relative to assembly specificity γ and complexity λ, consumers particularly value a match with *the component and not the assembly.* All the assemblers doing business with a certain supplier divide the (lion share of the) consumers in the niches in this supplier's part. In Figure 4, we see that the introduction of a component attractiveness (by picking exposure $\eta > 0$) causes assemblers to agglomerate near their supplier. In case $\eta = 0$, consumers do not care about which component is used, such that assemblers observe no change in expected sales upon switching to another supplier. In case $\eta = 0.03$, assemblers relocate towards their suppliers, but still differentiate from their immediate competitors being served by the same supplier. In case $\eta = 0.15$, consumers care strongly about the component, and firms locate near to their suppliers regardless of being close to their immediate competitors. However, having explicitly exposed components (or suppliers) does not automatically cause more agglomeration. In Figure 5, we see that an assembler may also switch to an alternative supplier (possibly counter to compatibility at first) to differentiate from immediate competitors.

4.2 Dynamic Equilibria

With the introduction of the second tier (with $\lambda > 0$ or $\eta > 0$), firms may end up in cyclic relocation patterns, i.e. emerging equilibria may be *dynamic*. In fact, extensive simulations revealed that, in certain parameter regions, the industry ends up in such dynamic equilibria in more than eighty percent of the runs.

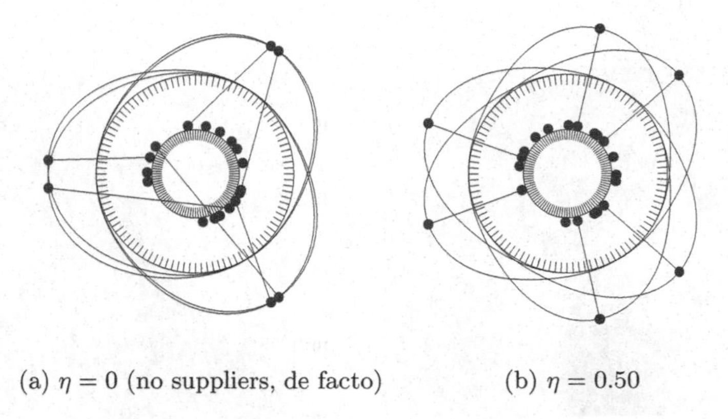

(a) $\eta = 0$ (no suppliers, de facto) (b) $\eta = 0.50$

Fig. 5. Equilibria emerging from the same initial distribution for $\gamma = 1.2$ for different component exposure levels η. Here $M = 6$, $S = 20$, $N = 100$.

We observed dynamic equilibria in the form of pulsations[3] (see the example in Figure 8.), oscillations[4] (see the example in Fig. 7), rotations[5] and combinations thereof. Apart from the choices for γ, η and λ, the types of equilibria depend on the number of suppliers S and whether suppliers are mobile or fixed, as well as the number of assemblers M. Extensive studies for $S = 1$ and $S = 2$ revealed that certain regions in the (γ, η, λ) parameter landscape robustly develop such dynamic equilibria. These equilibria generally occur if λ is low (little agglomeration) and η and γ are high such that demand is sensitive to small changes in the location distribution of assemblers and suppliers.

If demand is inelastic (i.e. does not decrease with a decrease in composite attractiveness), dynamics and emerging equilibria depend on, firstly, sales in remote niches either halfway to the next supplier, and, secondly, sales beyond the immediate competitors on either side. Whenever demand is elastic, mostly sales in local niches matters and dynamic equilibria are less likely to occur.

4.3 Emerging Distribution of Assemblers

To understand which alignment strategies should be followed by firms, we compare the emerging assembler distributions in the single and two-tier cases. To remove double dynamics effects from statistics, we *fix* supplier $i = 1, \ldots, S$ at niche $N(i-1)/i$. We study an elementary, yet non-trivial industry case that allows comparing results with the single tier industry case: $S = 2$ suppliers and $M = 4$ assemblers. The emerging industry configuration (s_1, s_2) with s_i the

[3] A two-stage process in which the distance from assemblers to their suppliers first expands and then contracts.

[4] A two-stage process in which assemblers move from the left side to the right side of their supplier and back.

[5] A perpetual process in which assemblers (and possibly suppliers) keep on relocating in the same direction.

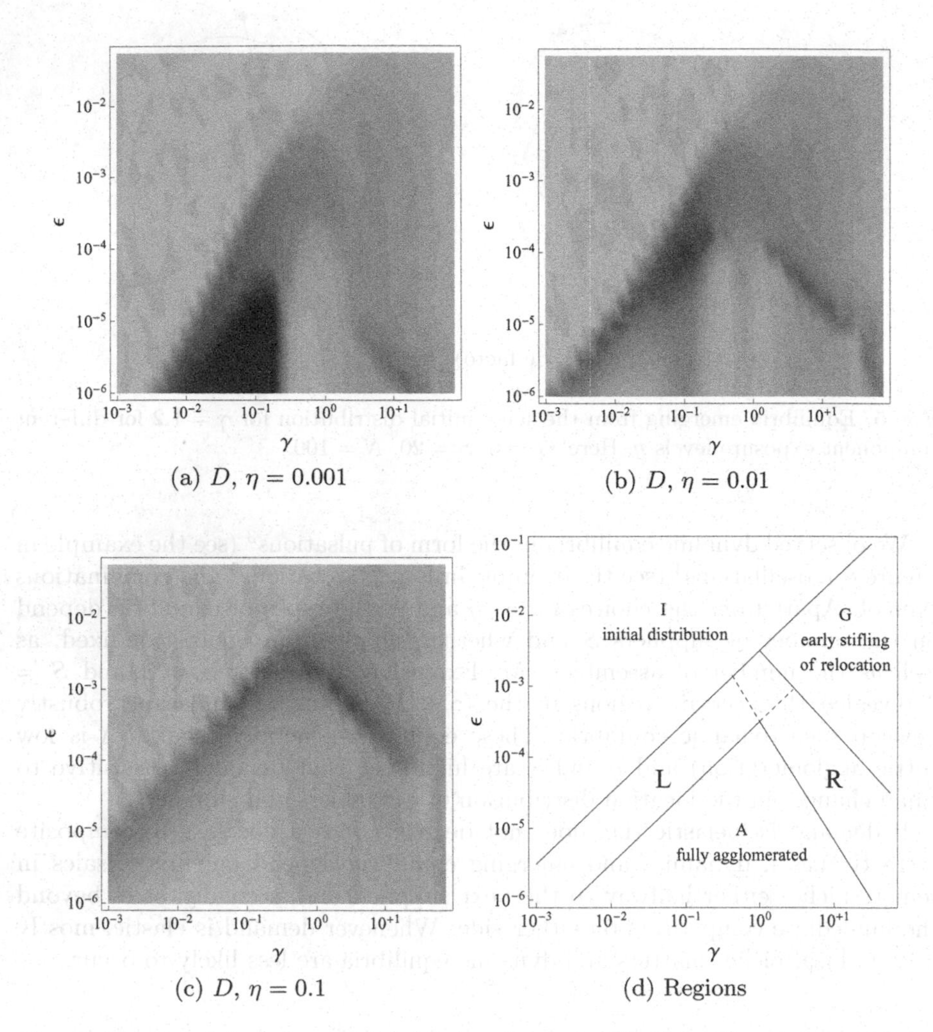

Fig. 6. Density plots (when dark (white), value is 0 (1)) of the D metric for $M = 4$ for a range of γ and ε values for $M = 4$ and $N = 500$ and 500 cases per parameter combination, including an overview of interesting regions.

number of assemblers served by supplier i may be any one of $(4, 0) \equiv (0, 4)$, $(3, 1) \equiv (1, 3)$ or $(2, 2)$.

Like in Section 3, we place all $M = 4$ assemblers randomly on the circle, have them relocate autonomously until there is an equilibrium and compute the D metric value for this equilibrium. We do this for 500 cases per γ and ϵ combination for a range of γ and ϵ values. To be able to compare the distance between firms in a pair with and without suppliers, we exclude cases with a dynamic equilibrium or a static asymmetric configuration. We then compute the average D value per combination using the 'filtered' dataset.

We repeat this for the component exposure values $\eta = 0.001, 0.01, 0.1$ for which the density plots and a schematic overview of interesting regions are contained in Figure 6.

In region I in the schematic overview in Figure 6d, the sales increments of only relocating are too small to exceed the threshold ε. In regions L and R, assemblers switch and move toward their supplier, but ε is so high compared to the sales increments realizable (related to γ) that there is early stifling of further relocation. Since we filter out the $(2, 2)$ configurations, we find many cases in which the assemblers are indeed close to equidistantly distributed. In region A, assemblers agglomerate near the supplier and the emerging equilibrium is far from equidistant. In region G, there is very little relocation, just like in region I. However, in region I, assemblers have significant sales in all niches around the circle, in expectation, while in region G, the sales is very local.

With an increase in component exposure η, the component attractiveness determines more and more of the composite attractiveness and assemblers move closer to their suppliers. Furthermore, if η is higher, the sales increments of moving closer to the supplier are higher and the ε needs to be higher to stop the assembler's relocations, so the regions L and R shift upward.

5 Conclusions

In this paper, we studied location equilibria and dynamics on a multi-tier Salop circle under bounded rationality. We started off with studying the classical single tier industry as benchmark case. In this case, the attractiveness curve in the discrete choice model mediates the emerging distribution of firms, and notably the consumer choosiness (or -alternatively- product specificity) determines whether firms disperse equidistantly or whether *pairs* of firms emerge.

In case of a two-tier supply chain in which suppliers (and their component technology) play an explicit role, the emerging equilibria are structurally different. Assemblers tend to agglomerate with head-on rivals near their supplier, particularly in case of imperfectly modular products (complexity $\lambda > 0$). However, firms also agglomerate when consumers care about the components used (components are 'exposed to consumers', $\eta > 0$), even if product technology is modular ($\lambda = 0$). In that case, relocating away from the head-on competitors and the supplier lowers sales more than sharing the sales with the head-on rivals. If consumers care much (or even mostly) about the components used, assemblers may switch to a technologically remote supplier, counter to the technological compatibility, to thereby serve an alternative market segment and gradually differentiate away from otherwise head-on competitors. The greater the exposure and the greater the upstream density (i.e. the more choice in potential suppliers), the more likely counter-compatibility switching is. Moreover, whenever the industry consists of multiple tiers, the emerging location equilibrium may be *dynamic* (i.e. a cyclic relocation pattern). The type of dynamic equilibrium depends on the number of suppliers, on whether suppliers are mobile or rather fixed, on the attractiveness curvature and parameter settings (assembly design specificity γ, component exposure η, and technological complexity λ). The existence of such dynamic equilibria hints on inefficiencies in the industry.

We have seen that pairwise clustering in the single tier case as well as dynamic equilibria in the two-tier case are due to inelasticity of demand. So, aligning a supply chain is definitely meaningful for commodity industries such as basic foodstuffs, basic clothing and gasoline, that indeed have relatively inelastic demand. However, for luxury and capital goods industries, that have relatively elastic demand, the presence of upstream suppliers frustrate the otherwise endogenous tendency to disperse maximally. So, the need for alignment is obvious under both demand regimes, albeit for different reasons.

The scientific contributions of this work are the insight that disaggregation is required to understand emerging product-market segmentation, and that product complexity and component exposure (and the role of suppliers) may cause dynamic equilibria. The implications for day-to-day management are that firms have to align their supply chain to escape lock-ins and head-on competition as well as preventing inefficient dynamic equilibria.

Appendices

1.A Illustration of Oscillation

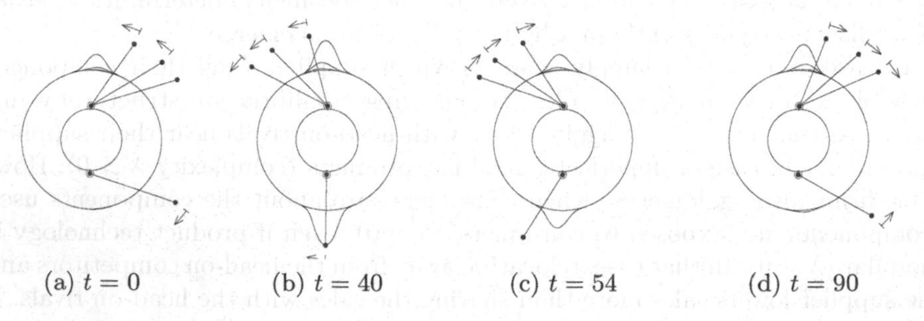

Fig. 7. Illustration of an oscillatory dynamic equilibrium, here with $\gamma = 10^{0.75}$, $\lambda = 0.0$ and $\eta = 10^{1.5}$. Each cycle takes 106 periods.

1.B Illustration of Pulsation

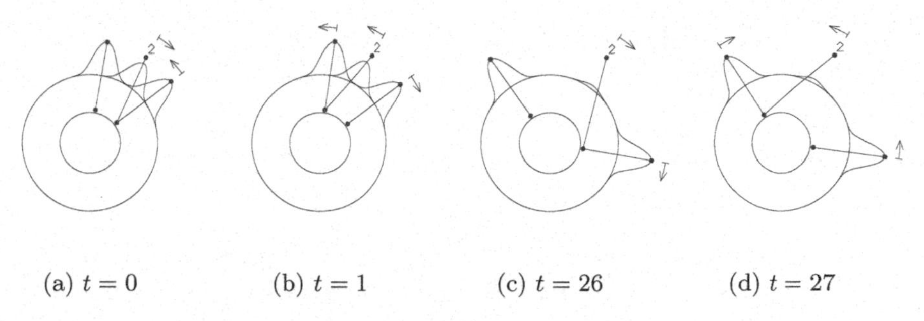

(a) $t = 0$ (b) $t = 1$ (c) $t = 26$ (d) $t = 27$

Fig. 8. Illustration of a pulsating dynamic equilibrium, here with $\gamma = 10^{1.5}$, $\lambda = 0.0$ and $\eta = 10^{0.25}$. Each cycle takes 52 periods.

References

1. Anderson, S., de Palma, A., Thisse, J.: Discrete Choice Theory of Product Differentiation. M.I.T. Press (1992)
2. Camacho-Cuena, E., Garcia-Gallego, A., Georgantzis, N., Sabater-Grande, G.: Buyer-seller interaction in experimental spatial markets. Regional Science and Urban Economics 35, 89–108 (2005)
3. Economides, N.: Symmetric equilibrium existence and optimality in differentiated product markets. Journal of Economic Theory 47, 178–194 (1989)
4. Huang, A., Levinson, D.: An agent-based retail location model on a supply chain network. Tech. rep. (2007)
5. Johnson, D., Hoopes, D.: Managerial cognition, sunk costs, and the evolution of industry structure. Strategic Management Journal 24, 1057–1068 (2003)
6. Krugman, P.: A dynamic spatial model. Working Paper 4219, NBER (1992)
7. Nelson, R.R., Winter, S.G.: An Evolutionary Theory of Economic Change. Harvard University Press (1982)
8. Pal, D.: Does Cournot competition yield spatial agglomeration? Economics Letters 60, 49–53 (1998)
9. Salop, S.: Monopolistic competition with outside goods. Bell Journal of Economics 10, 141–156 (1979)
10. Simon, H.: A behavioral model of rational choice. The Quarterly Journal of Economics 69(1), 99–118 (1955)
11. Thaler, R.: Toward a positive theory of consumer choice. Journal of Economic Behavior & Organization 1, 39–60 (1980)
12. Tversky, A., Kahneman, D.: Rational choice and the framing of decisions. The Journal of Business 59(4), S251–S278 (1986)

Author Index